A Dictionary of
Zoology

W9-BZP-435

Michael Allaby has written many books on environmental science and especially on climatology and meteorology. He is the General Editor of several Oxford Dictionaries, including the Dictionaries of *Geology and Earth Sciences, Ecology, Plant Sciences,* and *Environment and Conservation.*

( ) SEE WEB LINKS

To find recommended web links for this and many other Oxford reference titles, visit http://global.oup.com/booksites/reference/ when you see this sign.

A Dictionary of
Zoology

FOURTH EDITION

Edited by
MICHAEL ALLABY

OXFORD
UNIVERSITY PRESS

OXFORD
UNIVERSITY PRESS

Great Clarendon Street, Oxford, OX2 6DP,
United Kingdom

Oxford University Press is a department of the University of Oxford.
It furthers the University's objective of excellence in research, scholarship,
and education by publishing worldwide. Oxford is a registered trade mark of
Oxford University Press in the UK and in certain other countries

First edition published as *Concise Oxford Dictionary of Zoology* 1991
First edition published in paperback 1992
Second edition published as *A Dictionary of Zoology* 1999
Second edition reissued 2003
Third edition published 2009
Fourth edition published 2014
Impression: 1

Published in the United States of America by Oxford University Press
198 Madison Avenue, New York, NY 10016, United States of America

British Library Cataloguing in Publication Data
Data available

ISBN 978-0-19-968427-4

Printed in Great Britain by
Clays Ltd, St Ives plc

Preface to the Fourth Edition

In preparing this new edition I have added more definitions of anatomical terms. I have also incorporated recent changes in animal taxonomy and this edition contains a few more illustrations, added where I think they help a definition.

The third edition introduced links to useful web sites. Web sites come and go, and I have checked the old links, changed them where necessary, and added several new ones.

Inevitably, dictionaries grow longer with each new edition. New terms appear and although older terms may drop out of use, they do so slowly, surviving in old textbooks and articles that may still be of interest. The consequence is that new terms enter a dictionary much faster than old terms can safely be abandoned. This edition is a little longer than its predecessor.

As always, I hope the dictionary will help all those who enjoy reading books, magazines, and journal articles about the living animals with which we share the planet and the animals that are now extinct, but nevertheless fascinating, and instructive because it is from them that we reconstruct the story of animal life.

MICHAEL ALLABY
TIGHNABRUAICH, ARGYLL
www.michaelallaby.com

Contents

aardvark (ant bear; *Orycteropus afer*) *See* ORYCTEROPODIDAE.

aardwolf (*Proteles cristatus*) *See* HYAENIDAE.

abaptation The process by which an organism is fitted to its environment as a consequence of the characters it inherits, which have been filtered by *natural selection in previous environments. Because present environments seldom differ greatly from recent past environments, abaptive fitness can resemble *adaptation. In this sense, however, adaptation appears to imply advance planning, or design, which is misleading.

abdomen 1. In vertebrates, the region of the body that contains the internal organs other than the heart and lungs. In *Mammalia it is bounded anteriorly by the *diaphragm. **2.** In most *arthropods, the hind region (tagma) of the body, which contains most of the digestive tract, the gonads, and the genital openings. In *Crustacea, the abdomen bears limbs which are to a greater or lesser extent segmentally arranged and the abdomen is not homologous with that of *arachnids and *insects. The abdomen usually shows at least some trace of segmentation, though in the course of evolution this has been lost in all but one family of spiders. **3.** In insects, the segments of the body that lie posterior to the thorax. The abdominal segments carry no limbs, although there are appendages (associated with reproduction) on the terminal segments in certain exopterygote and apterygote orders (e.g. *Thysanura). Non-terminal segments may bear appendages that in some insects function as gills. In the primitive state the abdomen consists of eleven segments, but this number may be very much reduced in advanced insects.

abdominal fins 1. In fish, *pelvic fins located far back on the belly rather than in the thoracic or jugular position. **2.** Ventral fins located on the abdominal (belly) side of the body.

abduction Movement away from the midventral axis of the body. *Compare* ADDUCTION.

abductor muscle A muscle that draws a structure (e.g. a limb) away from the centre line of the body. *Compare* ADDUCTOR MUSCLE.

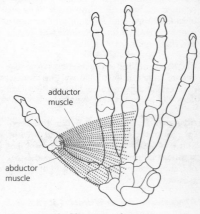

Abductor and adductor muscles

abiogenesis The development of living organisms from non-living matter, as in the origin of life on Earth, or in the concept of spontaneous generation which was once held to account for the origin of life but which modern understanding of evolutionary processes has rendered outdated.

abiotic Non-living; devoid of life. *Compare* BIOTIC.

abomasum In *Ruminantia, the fourth and final region of the specialized stomach, corresponding to the stomach in other mammals and the zone in which digestion proceeds with the usual mammalian digestive *enzymes.

aboral Away from the mouth; on the opposite side of the body from the mouth in animals that lack clear-cut *dorsal and *ventral surfaces.

Abrocomidae (rat chinchillas; order *Rodentia, suborder *Hystricomorpha) A family of medium-sized, rat-like rodents that have stiff hairs projecting over the nails of the three central digits of the hind feet. The skull is massive and narrow in the facial region, the brain case is rounded, with low occipital crests and short paroccipital processes, and the cheek teeth are *hypsodont. Rat chinchillas live in colonies in burrows or crevices, and can climb trees. They are found only in the Andes from southern Peru to northern Argentina. There are two species in a single genus, *Abrocoma*.

abyssal Applied to the deepest part of the ocean, below about 2000 m. The abyssal zone covers approximately 75% of the ocean floor. *Compare* BATHYAL; NERITIC.

abyssal fish Fish that live in the deepest part of the ocean, below about 2000 m. Many abyssal species have a prominent snout; a tapering, rat-tailed body, consisting of flabby, watery tissue; and a light-weight skeleton.

Abyssinian roller (*Coracias abyssinica*) See CORACIIDAE.

Acanthiza (thornbills) See ACANTHIZIDAE.

Acanthizidae (bristlebirds, scrub wrens, fairy warblers, thornbills, whitefaces; class *Aves, order *Passeriformes) A family of birds all of which build domed nests. They have thin, pointed bills with *basal bristles, are insectivorous, and vary from being strictly arboreal (fairy warblers) to ground-feeding (scrub wrens). The sexes are usually similar. There are 17 genera, comprising 63 species. (Acanthiza (thornbills) are sometimes placed in the *Sylviidae or *Maluridae.) They are found mainly in Australia and New Guinea.

Acanthobdellida (phylum *Annelida, class *Hirudinea) An order of parasitic worms that lack an anterior sucker. *Chaetae occur in the anterior region. There is one genus, Acanthobdella; it occurs only on salmon.

Acanthocephala (thorny-headed worms) A phylum of bilaterally symmetrical, *pseudocoelomate, worm-like organisms most of which do not have an excretory system; all lack a gut. A retractable proboscis is present, covered with the recurved spines that give the animals their common name. All acanthocephalans are endoparasitic, living in two or three marine or terrestrial vertebrates during their development. There are about 600 species.

Acanthochitonina See NEOLORICATA.

Acanthoclinidae (subclass *Actinopterygii, order *Perciformes) Family of small (about 8 cm long), agile fish that inhabit tropical waters of the Indo-Pacific region. They have *pelvic fins reduced to one spine and two soft rays.

Acanthodii Class of primitive, fossil fish that had a true bony skeleton, a *heterocercal tail fin, a persistent *notochord, *ganoid scales, and stout spines in front of the fins. The acanthodians lived from the *Silurian Period to the *Permian Period and may be related to ancestors of the more modern bony fish.

Acanthopterygii Superorder of bony fish that includes all the spiny-finned fish (e.g. *Perciformes and *Beryciformes).

acanthosoma See POSTLARVA.

Acanthostega See ICHTHYOSTEGA.

Acanthuridae (surgeonfish, doctorfish; subclass *Actinopterygii, order

*Perciformes) Large family of slab-sided fish that have a sharp spine on each side of the tail-base (hence the name 'surgeonfish'). The spine is retractable in some species. Many species exhibit striking colour patterns. There are about 80 species, found in all tropical seas; many feed on organisms that cover the surface or rocks, or on sea-grasses.

acari mites *See* ACARINA.

Acarina (acari mites, ticks; class *Arachnida) Order of small or very small arachnids most of which have a short, unsegmented *abdomen. The body comprises the *capitulum and a *prosoma. In most forms the prosoma is covered by a single carapace (*see* CEPHALOTHORAX). The respiratory organs are *tracheae. Many species are parasites of medical and veterinary importance, and many are agricultural pests. The Acarina are regarded by many zoologists as an unnatural, polyphyletic group comprising elements derived independently from different arachnid stocks. There are 20 000 known species, 2800 of which are aquatic, including some that are marine.

acarinum A small pouch in the abdomen of Old World carpenter bees of the genus *Xylocopa* (subgenus *Koptortosoma*), which provide protection for symbiotic mites of the genus *Dinogamasus*.

acceleration, evolutionary *Evolution that occurs by increasing the rate of ontogenetic (*see* ONTOGENY) development, so that further stages can be added before growth is completed. This form of *heterochrony was proposed by E. H. Haeckel as one of the principal modes of evolution.

accentors *See* PRUNELLIDAE.

accessory genitalia (secondary genitalia) Organ of intromission, present only in the males of the order *Odonata. It is situated on the *sternites of the second and third abdominal segments. It has no homologues in the animal kingdom.

accessory respiratory organ A system of air chambers formed by outgrowths from the mouth or gill region of those fish that occasionally leave the water. The uptake of oxygen from the air is facilitated by a dense network of tiny blood vessels in the skin lining these air chambers, and their possession enables such fish as labyrinth fish (*Anabantidae), snakeheads (*Channidae), or air-breathing catfish (*Clariidae) to survive outside water for some considerable time. The *swim-bladder also may serve as an accessory respiratory organ.

***Accipiter* (sparrowhawks, goshawks)** *See* ACCIPITRIDAE.

Accipitridae (hawks, eagles, buzzards, Old World vultures; class *Aves, order *Falconiformes) The largest family of birds of prey, containing many cosmopolitan genera, especially *Accipiter* (sparrowhawks and goshawks), *Buteo* (buzzards or hawks), *Circus* (harriers), *Elanus* (kites), *Haliaeetus* (fish eagles), and *Pernis* (honey buzzards), which make up nearly half the family. Other genera are more restricted in range, and more than half are monotypic. *Aquila chrysaetos* (golden eagle) has a *Holarctic distribution (there are 10 species of *Aquila*, found world-wide except for S. America and Malaysia). Hawk eagles (10 species of *Spizaetus*) occur widely in low latitude forests. Accipiters are generally carnivorous, preying on snails, insects, fish, mammals, reptiles, and birds; but one species, *Gypohierax angolensis* (palmnut vulture), feeds on oil-palm husks. Vultures (seven species of *Gyps*) have reduced talons which aid walking, inhabit open plains, mountains, and forests, feed on carrion, and nest in trees and on crags. There are 67 genera in the family, with 233 species.

acclimation A response by an animal that enables it to tolerate a change in a single factor (e.g. temperature) in its environment. The term is applied most commonly to animals used in laboratory experiments and implies a change in only one factor. *Compare* ACCLIMATIZATION.

acclimatization A reversible, adaptive response that enables animals to tolerate environmental change (e.g. seasonal climatic change) involving several factors (e.g. temperature and availability of food). The response is physiological, but may affect behaviour (e.g. when an animal responds physiologically to falling temperature in ways that make *hibernation possible, and behaviourally by seeking a nesting site, nesting materials, and food). *Compare* ACCLIMATION.

accommodation (fatigue, synaptic accommodation) The exhaustion of *neurotransmitter at the *synapse when a stimulus is repeated frequently. This may result in a decrease in behavioural responsiveness.

acentric Applied to a fragment of a *chromosome, formed during cell division, that lacks a *centromere. The fragment will be unable to follow the rest of the chromosome in migration towards one or other pole as it has lost its point of attachment to the *spindle.

acephalous Lacking a distinct head.

acetabulum The socket in the *pelvis into which the head of the *femur fits.

acetylcholine (ACh) An acetyl ester of *choline that is involved in synaptic (*see* SYNAPSE) transmission between nerve cells. It is released from *vesicles by the presynaptic *neuron and diffuses across the *synaptic cleft where it interacts with specific receptors to produce a local depolarization of the postsynaptic membrane, thus enabling the transmission of nerve impulses.

(⊕) SEE WEB LINKS
• A description of acetylcholine and other neurotransmitters.

acetylcholine esterase An enzyme present within the *synaptic cleft that hydrolyses *acetylcholine to *choline and acetic acid, thus preparing the *synapse for the passage of a new impulse.

acetyl coenzyme A (acetyl Co A) An important intermediate in the *citric-acid cycle, and in *fatty-acid and *amino-acid metabolism.

ACh *See* ACETYLCHOLINE.

acicula *See* ACICULUM.

acicular Pointed or needle-shaped.

aciculum (*pl.* acicula) In *Polychaeta, one of the chitinous (*see* CHITIN) support rods in the *parapodia.

acid According to the Brønsted–Lowry theory, a substance that in solution liberates hydrogen ions (protons). The Lewis theory states that it is a substance that acts as an electron-pair acceptor. An acid reacts with a base to give a salt and water (neutralization), and has a *pH of less than 7.

acidophilic 1. Refers to the propensity of a cell, its components, or its products to become stained by an acidic dye. 2. Applied to an organism that inhabits acid environments.

acidopore In formicine ants, a flexible, hair-fringed nozzle.

***Acinonyx jubatus* (cheetah)** *See* FELIDAE.

acinus (*pl.* acini) A cluster of cells resembling a blackberry. Acini occur in *exocrine glands and the alveoli (*see* ALVEOLUS) of the *lungs.

Acipenseridae (sturgeons; superorder *Chondrostei, order *Acipenseriformes) A family of large, fairly sluggish, bottom-feeding fish that have a low-slung mouth, toothless jaws, four *barbels in front of the upper jaw, and rows of large plates along the body. They are found in Europe, Asia, and N. America. Most of the 26 species are marine but ascend rivers to spawn. A few species, e.g. the sterlet (*Acipenser ruthenus*), live wholly in fresh water. Sturgeons are very fertile. The processed roe of ripe female common sturgeon (*Acipenser sturio*) and beluga (*Huso huso*) is renowned as caviar (the most valuable caviar is from the beluga).

Acipenseriformes (subclass *Actinopterygii, superorder *Chondrostei) An order of rather primitive fish characterized by a cartilaginous skeleton, a *heterocercal tail fin, and a head with a pointed, protruding *rostrum and ventrally located mouth. Extant members of the order either have a naked skin or possess five rows of large, rhomboid, bony scales along the body; a spiral valve is found in the intestinal tract. Apart from the sturgeon and paddlefish the order includes a number of fossil representatives dating back to the *Carboniferous Period.

Acochlidiacea (class *Gastropoda, subclass Opisthobranchia) An order of molluscs (*Mollusca) in which individuals are very small and *benthic, living between sand grains. The visceral sac is thin and spiculate, and is often much longer than the foot and markedly separate from it. There is no shell. These gastropods are deposit feeders. According to different authorities the order contains three families or only one.

Acoela 1. (class *Turbellaria, subclass *Archoophora) An order of platyhelminth worms (*Platyhelminthes) that have no gut cavity or pharynx. **2.** (class *Gastropoda, subclass *Opisthobranchia) An order of molluscs that possess no shell, *mantle cavity, or gills. Respiration is carried out by *branchiae. Dorsal outgrowths are quite common, and all have undergone complete *detorsion. *Benthic and planktonic forms occur. There is only one fossil family recorded, which appeared in the *Eocene. According to some authorities, the taxonomic use of 'Acoela' as a molluscan order should be discouraged. It represents an attempt to unite the *Notaspidea and *Nudibranchia.

acoelomate Lacking a *coelom.

Acoelomorpha A group, possibly a phylum, of about 350 species of very early bilaterian animals (*see* BILATERAL SYMMETRY) that comprises a sister clade to other bilaterians near the base of the *Deuterostomia. Acoelomorphs are flatworms related to *Xenoturbella* (*Xenoturbellida). They are flattened dorsoventrally, have a ventral mouth, are ciliated, lack a central nervous system, and possess a *statocyst at the anterior end.

(🌐) **SEE WEB LINKS**
• Description of the phylum Acoelomorpha.

acontium In *Anthozoa, an extension of a *nematocyst-bearing *mesentery, containing a nematocyst, that lies in the gastric cavity but can be protruded through the mouth in order to capture prey and possibly also for defence.

acorn barnacle *See* BALANIDAE.

acorn worms *See* ENTEROPNEUSTA.

acoustico-lateralis system The inner-ear region and *lateral-line organs located in the skin, forming a sensory system that conveys environmental information to the brain of a fish: the lateral-line organs respond to changes in water pressure and displacement, the inner ear responds to sound and gravity. Generally such a system is found in aquatic lower vertebrates (e.g. lampreys, sharks, bony fish, and one or two amphibians).

acquired characteristics Characteristics that are acquired by an organism during its lifetime. According to early evolutionary theorists (e.g. *Lamarck), *traits acquired in one generation in response to environmental stimuli may be inherited by the next generation. Thus over several generations a particular type of organism would become better adapted to its environment. The heritability of such characteristics is now discredited.

Acrania *See* BRANCHIOSTOMIDAE; CEPHALOCHORDATA.

Acrididae (short-horned grasshoppers, locusts; order *Orthoptera, suborder *Caelifera) Cosmopolitan family of small to large, short-horned grasshoppers among which *stridulation is common, the males and some females rubbing a row of pegs on the hind femora against a toughened vein of the forewing (tegmen). They have three-segmented tarsi and antennae which are shorter than the fore femora, and hind legs modified for jumping (saltatorial). There are tympanal organs (ears) at the base of the abdomen. The female has a short *ovipositor and lays eggs in the soil, or sometimes in decaying wood. All species are plant feeders and a number are important agricultural pests, most notably the locusts. Lubber, spur-throated, slant-faced, and band-winged grasshoppers are members of the family, which is the largest of the Caelifera and contains some 10 000 known species.

Acridotheres (mynas) *See* STURNIDAE.

Acrobatidae (pygmy possums, pygmy gliders; order *Diprotodontia (or *Marsupialia), superfamily *Phalangeroidea) A family containing three genera of mouse-sized marsupials, some of which have gliding membranes.

acrocentric Applied to a *chromosome in which the *centromere is located nearer to one end than to the other. During the *anaphase stage of cell division, movement of an acrocentric chromosome towards one pole results in the chromosome being shaped like a 'J', as opposed to the normal 'V' shape of a metacentric chromosome (in which the centromere is in the middle).

Acrochordidae (wart snakes and file snakes; order *Squamata, suborder *Serpentes) A family of primitive snakes with ridged scales that do not overlap. There are nasal openings on top of the snout. Usually they are found in brackish water, but occasionally offshore. They rarely come on to land. There are two species, found in India, Sri Lanka, some Indo-Australian islands, and northern Australia.

Acrocodia (Asian tapir) *See* TAPIRIDAE.

acrodont Applied to the condition in which the teeth are fused to the bones. *Compare* THECODONT; PLEURODONT.

Acroechinoidea (subphylum *Echinozoa, class *Echinoidea) A taxonomic rank, between the subclass and superorder levels, that comprises a *monophyletic group consisting of all *Euechinoidea except for the *Echinothurioida.

acromion In *Mammalia, a ventral extension of the spine of the *scapula. It articulates with the *clavicle if a clavicle is present.

Acropomatidae (subclass *Actinopterygii, order *Perciformes) A very small family of marine tropical (Indo-Pacific) fish, comprising only three species. A characteristic of this family is the anterior location, close to the *pelvic-fin base, of the vent (anus).

acrosome A thin-walled *vesicle that forms a cap on the head of a *spermatozoon. On contact with an *ovum it bursts, releasing powerful hydrolytic *enzymes which cause a localized softening of the *vitelline membrane, thus facilitating fertilization.

Acrotretida (phylum *Brachiopoda, class *Inarticulata) An order that comprises two suborders of brachiopods which are usually circular or semicircular in outline. The shell is either phosphatic or punctate calcareous. The *pedicle opening is restricted to the pedicle valve. The *shell beak is marginal to subcentral in position.

Acrotretidina (class *Inarticulata, order *Acrotretida) A suborder of inarticulate brachiopods that have phosphatic shells. Throughout life they are attached to the seabed by a *pedicle. They first appeared in the Early *Cambrian. There are three superfamilies, seven families, and about 50 genera.

ACTH *See* ADRENOCORTICOTROPHIC HORMONE.

actin A *globular protein of relative molecular mass 60 000 which is a major component of *microfilaments generally, and which is especially important in the *myofibrils of striated muscle cell.

Actiniaria (sea anemones; class *Anthozoa, subclass *Zoantharia) An order of solitary polyps that lack skeletons. They have numerous *tentacles and paired *mesenteries. There are about 200 living genera, comprising more than 800 species.

Actinistia 1. See CROSSOPTERYGII. **2.** See COELACANTHIMORPHA.

Actinoceratida (class *Cephalopoda, subclass *Nautiloidea) A large order of predominantly *orthoconic cephalopods that have large *siphuncles. Large amounts of siphuncular and *cameral deposits are present. Most are assumed to have been *nektonic. The actinoceratids ranged in age from Early *Ordovician to lower *Carboniferous.

Actinopodea (subphylum *Sarcomastigophora, superclass *Sarcodina) A class of *Protozoa that are typically spherical *amoebae with radiating *pseudopodia. They are free-living in freshwater and marine environments.

Actinopodidae See TRAPDOOR SPIDERS.

Actinopterygii (ray-finned fish; class *Osteichthyes) A subclass of ray-finned fish, that includes the majority of living bony fish of sea and fresh water. The fins are composed of a membranous web of skin supported by a varying number of spines and soft rays. This subclass includes a diversity of fish types, ranging from the sturgeon and paddlefish to the eel and tuna. They first appeared during the *Devonian Period.

Actinozoa See ANTHOZOA.

actinula larva In some members of the *Hydrozoa, a free-swimming larval form with a mouth surrounded by tentacles. In the *Hydroida the larva settles and becomes a hydroid; in other species it metamorphoses (see METAMORPHOSIS) into a *medusa.

action potential See ALL-OR-NOTHING LAW.

activational effects of hormones Changes in reproductive and social behaviours that occur in young adult animals due to the influence of sex *hormones. Such changes include *lordosis in females and mounting in males. In roosters, deprivation of *testosterone causes the comb (a secondary sexual feature) to shrink and fade in colour. In adolescent humans there are also changes in mood (e.g. irritability, mood swings, mood intensity) triggered by *adrenal *androgens, *gonadotropins, and sex *steroids. Activational effects occur only when the hormone is present. *Compare* ORGANIZATIONAL EFFECTS OF HORMONES.

activation energy (energy of activation) The energy that must be delivered to a system in order to increase the incidence within it of reactive molecules, thus initiating a reaction. It is an important feature of *enzymes that they greatly lower the activation energy of many metabolic reactions.

activator A metal ion that functions in conjunction with either an *enzyme or its *substrate in order to bring about a reaction.

active dispersal See DISPERSAL.

active evasion Fleeing a predator in order to avoid being eaten.

active immunity Resistance to a disease that is acquired by an animal as the result of the production of *antibodies in response to *antigens produced by the disease organism whilst inside the host animal.

active site Part of an *enzyme molecule, the conformation of which is such that it binds to the *substrate or substrates to form an enzyme–*substrate complex. The conformation is not absolute and may alter according to reaction conditions.

active transport The transport of substances across a membrane against a concentration gradient. Such processes require energy, the source often being the hydrolysis of adenosine triphosphate (ATP). *Protein or *lipoprotein carrier molecules are believed to be involved in the process.

actomyosin A complex of the proteins *actin and *myosin in an approximately 3:1 ratio. It is formed *in vivo* in muscle cells, or *in vitro* from purified extracts.

Aculeata (wasps, ants, bees; order *Hymenoptera, suborder *Apocrita) Division of Hymenoptera in which the *ovipositor has lost its egg-laying function and is modified as a *sting. The sting has been lost in some *aculeates (e.g. formicine ants) and is reduced or absent in some bees (e.g. the Andrenidae and Meliponini). *Compare* PARASITICA.

aculeate Prickly, pointed. The term is applied to organisms that are armed with a *sting (e.g. members of the hymenopteran division *Aculeata, which have stings). The word is derived from the Latin *aculeatus*, meaning stinging, from *acus*, needle.

acuminate Tapering to a point.

acute (of disease) Applied to a disease that develops rapidly and is of short duration; symptoms tend to be severe.

acute stress response *See* FIGHT OR FLIGHT REACTION.

adambulacral In *Echinodermata, applied to the ossicles and spines that occur at the outer ends of the ambulacral plates (*see* AMBULACRUM), with which they alternate.

Adapidae (suborder *Strepsirhini (or *Prosimii), infra-order *Lemuriformes) An extinct family of lemur-like animals; most were small, but a few had heads 10 cm or more long. The brain case was small with temporal crests and smooth cerebral hemispheres, and the tympanic ring was included in the *bulla. Dentition was full and the *incisors were not *procumbent, but in Adapis species the *canines were incisiform, the *molars resembling those of some modern lemurs. The skeleton was adapted for grasping, leaping, and perching. Old World adapids (e.g. *Adapis* and *Pronycticebus*) probably resembled lemurs and lorises, to which they may be ancestral. They were distributed in Europe and N. America in the *Eocene.

adaptation 1. Generally, the processes by which animals adjust to their environments. The adjustments may occur by *natural selection, as individuals bearing genetic traits that allow them to flourish under prevailing conditions breed more prolifically than those lacking such traits (genotypic adaptation). Alternatively, adaptation may involve non-genetic changes in individuals, such as physiological modification (e.g. *acclimatization) or behavioural changes (phenotypic adaptation). *Compare* ABAPTATION. **2. (evol.)** A characteristic or suite of characteristics that fits an organism both generally and specifically to exploit a given environmental zone (e.g. wings allow birds to fly, whereas the hooked beak and sharp talons of birds of prey are more specialized adaptations well suited to a predatory way of life). The word also implies that the feature has survived because it assists its possessor in its existing *niche. *Compare* EXAPTATION. **3.** Sensory adaptation involves a decrease over time of the frequency of the impulses leaving a sensory receptor when a stimulus is repeated frequently. *See* ACCOMMODATION; HABITUATION.

adaptive breakthrough Evolutionary change by the acquisition of a distinctive *adaptation that permits a population or *taxon to move from one *adaptive zone to another. At the most extreme such moves might be from water to land, or from land to air.

adaptive immunity (specific immunity) *Immunity conferred by specialized lymphocytes (*see* LEUCOCYTE) that produce receptors to a large number of *antigens. The immunity is adaptive because every *pathogen is remembered by a specific *antibody, allowing subsequent invasions to be dealt with swiftly.

adaptive pathway A series of small adaptive steps, rather than a single large one, which leads from one *adaptive zone across an environmental and adaptive threshold into another adaptive zone. In effect, small changes accumulate so that the organism is virtually preadapted (*see* PRE-ADAPTATION) to enter the new zone.

adaptive peaks and valleys Features on a symbolic contour map that shows the *adaptive value of genotypic combinations.

Such a map will usually display adaptive peaks and valleys occurring at points where the adaptive value is relatively strong or weak. The population of a given *species will therefore be distributed more densely at the adaptive peaks and more sparsely at the valleys.

adaptive radiation 1. A burst of evolution, with rapid divergence from a single ancestral form, that results from the exploitation of an array of *habitats. The term is applied at many taxonomic levels (e.g. the radiation of the mammals at the base of the *Cenozoic is of ordinal status, whereas the radiation of Darwin's finches in the *Galápagos Islands resulted in a proliferation of species). **2.** Term used synonymously with *cladogenesis by some authors.

adaptive type A population or *taxon that has distinctive adaptive attributes, expressed as a particular morphological theme, characteristic of a particular *habitat or mode of life. In evolutionary terms, the appearance of a new adaptive type is frequently followed by radiations that yield variants; these partition the environment and exploit it more effectively.

adaptive value (Darwinian fitness, fitness, selective value) The balance of genetic advantages and disadvantages that determines the ability of an individual organism (or *genotype) to survive and reproduce in a given environment. The 'fittest' is the individual (or genotype) that produces the largest number of offspring that survive to maturity and reproduce.

adaptive zone A *taxon that is considered together with its associated environmental regime(s), *habitat, or *niche. The adaptive specialization that fits the taxon to its environment, and hence the adaptive zone, may be narrow (as with the giant panda, which eats only certain types of bamboo shoots) or broad (as with the brown bear, which is omnivorous).

adder (*Vipera berus*) See VIPERIDAE.

additive genetic variance See HERITABILITY.

adduction Movement towards the midventral axis of the body. *Compare* ABDUCTION.

adductor muscle A muscle that draws a shell or limb of an animal towards the median axis of the body. An adductor muscle closes the shell valves in *Bivalvia and the carapace valves in *Cirripedia. *Compare* ABDUCTOR MUSCLE.

adecticous In arthropods, having non-articulated, often reduced mandibles that in most species are not used for escape from the pupal cocoon.

adelphoparasite A parasite (*see* PARASITISM) that has as its host a species closely related to itself, often within the same family or genus.

adenine A *purine base which occurs in both *DNA and *RNA.

adenohypophysis In vertebrates, part of the *pituitary gland that is derived from the *hypophysial sac during the development of the *embryo. It has two parts: the pars distalis, which forms the anterior lobe of the pituitary; and the pars intermedia (absent in some mammals and in birds), which forms part of the posterior lobe. *Compare* NEUROHYPOPHYSIS.

adenosine A *nucleoside formed when *adenine is linked to *ribose sugar.

adenosine diphosphate (ADP) A high-energy phosphoric ester, or *nucleotide, of the *nucleoside *adenosine. It can undergo *hydrolysis to adenosine monophosphate and inorganic phosphate, the reaction releasing 34 kJ/mol of energy at *pH 7.

adenosine triphosphate (ATP) A high-energy phosphoric ester, or *nucleotide, of the *nucleoside *adenosine which functions as the principal energy-carrying compound in the cells of all living organisms. Its *hydrolysis to *ADP and inorganic phosphate is accompanied by the release of a relatively large amount of free energy (34 kJ/mol at *pH 7) which is used to drive many metabolic functions.

ADH *See* VASOPRESSIN.

adherens junction A cell junction that is commonly observed in epithelial (*see* EPITHELIUM) cells (e.g. those lining the intestine and those in cardiac muscle cells). At these junctions the *cell membranes of the neighbouring cells are separated by a space of 15–25 nm which is filled with a filamentous material. Beneath this, anchoring the junction to the cell *cytoplasm, are loosely structured mats of fibres, 7 nm in diameter, thought to be *actin filaments. It is thought that these junctions may provide mobility in the regions of the cells in which they are located.

adipose Pertaining to fat.

adipose eyelid The thickened yet transparent skin that overlies the eyes of some animals. Some fish have adipose eyelids (e.g. certain herring and mullet species); apart from a small central aperture, it covers most of the eye, giving the fish a 'bespectacled' appearance.

adipose fin In some fish (e.g. members of the salmon, catfish, and a few other families), a type of second dorsal fin in the form of a small flap of fatty tissue covered with skin and lacking supporting rays.

adipose tissue *Connective tissue that contains large cells in which fat is stored.

admirals *See* NYMPHALIDAE.

adoption Investing parental care on a juvenile to which the adopter is unrelated. In some cases, observed in several species, the adoptee may belong to a different species (e.g. companion animals cared for by humans).

adoption gland An abdominal gland possessed by larvae of the rove beetle (*Staphylinidae) *Atemeles pubicollis* that releases an attractant *pheromone which stimulates brood-keeping behaviour in *Formica polyctena* ants. The adult beetle lays its eggs in the nest of the ants; when the larvae hatch the pheromone induces the ants to raise them as though they were their own.

adoral On the same side of the body as the mouth.

ADP *See* ADENOSINE DIPHOSPHATE.

adrenal gland In vertebrates, an organ that secretes certain *hormones. Many vertebrates possess multiple adrenal glands, but in *mammals there is one gland close to each *kidney. In *tetrapods, each gland consists of a central *medulla and an outer *cortex. The medulla secretes *adrenalin and *noradrenalin, hormones needed when the animal is in an excited state and must engage in strenuous activity (e.g. fighting or fleeing; *see* FIGHT OR FLIGHT REACTION). The cortex secretes sex hormones and other hormones concerned with regulating the water and salt balances of the body.

adrenalin (**adrenaline, epinephrine**) A *hormone secreted by the *adrenal *medulla and largely responsible for the 'fight or flight' response in mammals (*see* FIGHT OR FLIGHT REACTION). It stimulates the breakdown of *glycogen, thus raising the blood-sugar level, it mobilizes free *fatty acids, and it has a variety of effects on the cardiovascular and muscular systems.

adrenergic Of nerve endings, secretion of the *neurotransmitters adrenalin or noradrenaline into the *synapse on the arrival of a nerve impulse. Adrenergic nerve endings are characteristic of the *sympathetic nervous system.

adrenergic system *See* NEUROTRANSMITTER.

adrenocorticotrophic hormone (**ACTH**) A *polypeptide *hormone, secreted by the anterior lobe of the *pituitary gland, which stimulates the synthesis and secretion of hormones by the adrenal cortex.

advertisement A form of *display in which an individual makes itself as conspicuous as possible. It is used most commonly by male animals holding a *territory, in order to ward off rivals and to attract females.

aedeagus Intromittent organ or penis of males of most insect groups, which is often of great diagnostic value. Its inner wall is a continuation of the ejaculatory duct (*see* ENDOPHALLUS). It may be inflatable and everted during copulation. It has a *gonopore at its base and opens at its apex through a *phallotreme, the whole structure being surrounded by a sclerotized *cuticle. By permitting sperm to be transferred without exposing it to the air, the aedeagus frees those species that possess it from the need to mate in water.

Aëdes aegypti (yellow-fever mosquito) *See* CULICIDAE.

Aegeriidae *See* SESIIDAE.

Aegithalidae (long-tailed tits, bushtits; class *Aves, order *Passeriformes) A family of small titmice that have medium to long tails. The nest is intricate and domed, with a small entrance, and built in trees and bushes. There are three genera, comprising seven species: *Aegithalos* (long-tailed tits) is Eurasian, *Psaltriparus* (bushtit) is American, and *Psaltria* (pygmy tit) is confined to Java.

Aegithalos (long-tailed tits) *See* AEGITHALIDAE.

Aegotheles (owlet-nightjars) *See* AEGOTHELIDAE.

Aegothelidae (owlet-nightjars; class *Aves, order *Caprimulgiformes) A family of brown-grey, secretive, nocturnal birds, similar to *Podargidae (frogmouths) and *Caprimulgidae (nightjars), that are insectivorous, with small, broad bills. They possess *preen glands. They nest in holes in trees or the ground. There is one genus, *Aegotheles*, with eight species, found in Australasia (but not in New Zealand).

Aegyptopithecus zeuxis A genus and species of early *catarrhine *Primates, known from abundant remains, including several nearly complete skulls, from the early *Oligocene of the Jebel al-Qatrani Formation, Fayum, Egypt. The size of a small, living monkey, it had a long tail and could jump from branch to branch. It possessed the dental and some of the cranial characteristics of living catarrhines, but lacked many of the other cranial and most of the postcranial diagnostic features, and so represents a time when catarrhines had separated from other primates, but remained more primitive than living hominoids (*Hominoidea) or cercopithecoids (*Cercopithecoidea) and it could have been ancestral to living catarrhines.

Aepyceros melampus (impala) *See* BOVIDAE.

Aepyornithidae (elephant birds; class *Aves, order Aepyornithiformes) An extinct family of large, flightless, running birds, that stood up to 3 m tall and laid eggs more than 30 cm long. There were about eight species; their remains have been found in Madagascar.

aerobe *See* AEROBIC (2).

aerobic **1.** Of an environment: one in which oxygen is present. **2.** Of an organism: one requiring the presence of oxygen for its existence, i.e. an aerobe. **3.** Of a process: one that occurs only in the presence of oxygen.

aerotaxis A change in direction of locomotion, in a *motile organism or cell, made in response to a change in the concentration of oxygen in its immediate environment.

Aesculapian snake (*Elaphe longissima*) *See* COLUBRIDAE.

Aeshnidae (dragonflies; order *Odonata, suborder *Anisoptera) A cosmopolitan family of large, swift-flying, hawker dragonflies which are usually strikingly marked with blue or green, although several genera are normally a uniform dark brown: *Anax imperator* (emperor dragonfly) is one of the largest dragonflies in Eurasia. Aeshnids have similar triangles in the fore and hind wings, a well-defined but small anal loop in the hind wings, the eyes broadly contiguous across the top of the head, the *ocelli in a triangle around a raised tubercle, and the *ovipositor complete and not projecting beyond the long anal appendages. The *larvae have large eyes and elongate bodies. More than 430 extant species have been described.

aestival In the early summer. The term is used with reference to the six-part division of the year used by ecologists, especially in relation to the study of terrestrial and freshwater communities. *Compare* AUTUMNAL; HIBERNAL; PREVERNAL; SEROTINAL; VERNAL.

aestivation (estivation) Dormancy or sluggishness that occurs in some animals (e.g. snails and hagfish) during a period when conditions are hot and dry. Aestivation is analogous to hibernation in cold environments and normally lasts the length of the dry period or season.

Aëtosauria A mainly *Triassic group of primitive *Thecodontia ('tooth-in-socket') reptiles that resembled heavily armoured crocodiles and appear to have been specialized herbivores or possibly omnivores. They grew up to 3 m long, and their armour plating comprised rows of bony plates.

afference The reception by the brain of signals originating in sensory organs.

afferent fibres In the *autonomic nervous system, fibres that convey information from peripheral organs to the *central nervous system.

affinity index A measure of the relative similarity in composition of two samples. For example, $A = c/\sqrt{(a+b)}$, where A is the affinity index, a and b are the numbers of species in one sample but not in the other and vice versa respectively, and c is the number of species common to both. The reciprocal, $\sqrt{(a+b)}/c$, indicates the ecological distance (D) between samples.

African chevrotains (*Hyemoschus*) *See* TRAGULIDAE.

African elephant (*Loxodonta africana*) *See* ELEPHANTIDAE.

African forest elephant (*Loxodonta cyclotis*) *See* ELEPHANTIDAE.

African lungfish *See* PROTOPTERIDAE.

Afro-Tethyan mammal region A region, proposed by the biogeographer Charles H. Smith, which appears to reflect mammal distributions more satisfactorily than the traditional zoogeographic regions. It includes the *Ethiopian and *Oriental regions, as well as the Mediterranean subregion.

Afrotropical faunal region A name that is commonly used for the *Ethiopian faunal region.

afterbirth The *placenta after it has been expelled following the birth of young.

after-feather *See* AFTERSHAFT.

aftershaft (after-feather, hypoptile) A *feather that arises as a branch from the base of a contour feather. In many birds aftershafts provide important thermal insulation.

agamete In *Mesozoa, a reproductive body which divides to form a daughter within the body of the parent.

agamic generation (asexual generation) In the life cycle of some species (e.g. many members of the *Cynipidae) a generation that comprises parthenogenetically reproducing females that are genetically *diploid. The unfertilized eggs laid may be male (having undergone *meiosis) or female (not having undergone meiosis). In Cynipidae, agamic females usually emerge from stout galls designed to overwinter or to survive a period when resources are scarce.

Agamidae (agamids; order *Squamata, suborder *Sauria) A family of lizards that closely resemble iguanas (*Iguanidae), but that have *acrodont teeth.

Scales are keeled and often spiny. Typically they have a large head, long limbs, and a long tail with no *autotomy. Dorsal crests, throat sacs, and colour change are common. *Moloch horridus* (thorny devil) of the Australian desert, which feeds on ants and termites, has protective horns on the forehead formed from enlarged spines; and *Uromastyx* species (spiny-tailed lizards or dab lizards) of the Near and Middle Eastern deserts, in which the adults are herbivorous, defend themselves by lashing violently with their tails, which bear scales modified into rings of spines. Males of *Calotes versicolor* (bloodsucker or Indian variable lizard), a species of slender, agile, long-tailed, tree lizards found from Iran to southern China and Sumatra, change colour from brown to yellow and then to blood-red in rapid succession during courtship. The family includes bipedal and gliding forms (e.g. *Draco volans* (flying dragon or flying lizard) in which five or six posterior ribs are enlarged to support a gliding membrane allowing the lizard to 'parachute' from tree to tree). There are 300 species, all found in the Old World.

Agaonidae (fig wasps; suborder *Apocrita, superfamily *Chalcidoidea) Family of small or minute, sexually dimorphic, black wasps which live within the flowers on fig trees and help to pollinate them, the males rarely emerging from the fruits. The head of the winged female is long and somewhat oblong with a deep, median, longitudinal groove; males are nearly always wingless. The front and middle legs are stout, and the front tibiae have no spurs. The hind legs are slender. The males have thick antennae, with 3–9 segments. Mating occurs within flowers and there is often fighting between males.

age-and-area hypothesis The idea that, all other things being equal, the area occupied by a *taxon is directly proportional to the age of that taxon. Thus in a polytypic genus, the species with the smallest area of distribution would be the youngest in the genus. However, other things rarely are equal, and the idea has never gained acceptance as a law or rule.

Agelenidae (sheet-web spiders; order *Araneae, suborder Araneomorphae) Family of spiders that construct sheet webs with lateral, tubular retreats, and that have a somewhat elongate appearance, with a flattened thorax. The *sternum is wide, heart-shaped, and may project between the fourth *coxae. The *chelicerae are usually very convex, and nearly vertical, the outer margin having three teeth, the inner two to eight teeth. The carapace (*see* CEPHALOTHORAX) is oval, narrowed, and anteriad; and has a longitudinal depression (sometimes replaced by a dark line), with patterns of dots and bars. The head is raised and narrow, and points forward, and the eyes, usually uniform in size, are arranged in two short rows. The legs are thin and long, are covered thickly with long hairs, and are spiny, especially on the third and fourth pairs. The anterior *spinnerets are cylindrical, and more or less separated; the median spinnerets are similar or smaller; and the posterior spinnerets are usually two-segmented and long. The tracheal *spiracle is close to the spinnerets. All species in this family, including *Tegenaria domestica* (house spider), catch prey on the upper surface of the sheet web. Many species spend a long time together as couples. Distribution is world-wide. *Argyroneta aquatica* (water spider), the only species of spider to live permanently below water, is usually placed in this family, but sometimes in a family of its own.

Ageneiosidae (barbel-less catfish; subclass *Actinopterygii, order *Siluriformes) A family of freshwater fish found in tropical S. America. About 25 species are known.

age pigment *See* LIPOFUSCIN.

age polyethism *See* POLYETHISM.

agglomerative method System of hierarchical classification that proceeds by grouping together the most similar individuals, and subsequently groups, into progressively larger and more heterogeneous units. At each stage the groups or individuals linked are those giving the least increase in group heterogeneity.

agglutin *See* BLOOD GROUP.

agglutination The clumping of cells that is caused by the reaction between *antigens on their surfaces and *antibodies in their external environment. *See* RHESUS FACTOR.

agglutinogen *See* BLOOD GROUP.

aggregation The group of animals that forms when individuals are attracted to an environmental resource to which each responds independently. The term does not imply any social organization.

aggregation pheromone *See* PHEROMONES.

aggregative response The preference for consumers to spend most of their feeding time in patches containing the highest density of prey. *See also* PARTIAL REFUGE.

aggressin A toxic substance that is produced by certain micro-organisms which are pathogenic in animals or humans. Aggressins inhibit the defence mechanisms of the host organism.

aggression Behaviour in an animal that serves to intimidate or injure another animal, but that is not connected with predation.

Agnatha (Marsipobranchii; phylum *Chordata, subphylum *Vertebrata) A superclass of jawless, fish-like vertebrates that have sucker-like mouths and lack paired fins. The Agnatha includes some of the earliest primitive vertebrates as well as the extant lampreys, slime-eels, and hagfish. The superclass first appeared during the *Ordovician Period, and in the Upper *Palaeozoic they developed heavily armoured forms (e.g. *Cephalaspidomorphi, *Pteraspidomorphi).

Agonidae (poachers; subclass *Actinopterygii, order *Scorpaeniformes) A family of small, marine, cold-water fish that have an elongated body and narrow, tapering tail. There are about 20 genera.

agonistic Applied to behaviour between two rival individuals of the same species that may involve *aggression, *threat, *appeasement, or *avoidance, or that may be ritualized (e.g. as bird *song, or as certain forms of *display). Agonistic behaviour often arises from a conflict between aggression and fear.

agouti A type of hair pigmentation in which there are alternate bands of the two forms of *melanin. This kind of banded hair is well exemplified by the agouti (*see* DASYPROCTIDAE), hence the name.

Agriidae *See* CALOPTERYGIDAE.

Agriochoeridae (suborder *Ruminantia, infra-order *Tylopoda) An extinct family of N. American ruminants closely related to the *oreodonts but more slightly built and with more primitive features (e.g. a long tail and clawed feet). They lived from the upper *Eocene into the *Miocene and, judging by their claws, inhabited trees or dug for roots and tubers.

Agriotherium (superfamily *Canoidea, family *Ursidae) A genus of primitive bears, probably related most closely to the giant panda (*Ailuropoda). Its remains have been found in Europe, N. America (*Pliocene and *Pleistocene), and at an early Pliocene site in S. Africa, making this Africa's only known bear.

Agromyzidae (leaf-mining flies; order *Diptera, suborder *Cyclorrapha) Relatively large family of small, *phytophagous flies, whose larvae are leaf-miners, stem-borers, or gall-causers. The third antennal segment is short and rounded, and rarely has a sharp point. The *arista is hairy or bare. The lower fronto-orbital bristles are incurved; *vibrissae are present; and the *tibia is without pre-apical bristles. The *costa is incomplete, being broken some distance from its junction with the upper margin of vein 1, and continued as far as the apex of vein 3. The *ovipositor has a non-retractile sheath comprising the fused seventh abdominal *tergite and *sternite. They have a world-wide distribution, and 2300 species have been described.

ahermatypic Applied to corals that do not form reefs and most of which lack *zooxanthellae. Modern scleractinian corals (*see* MADREPORARIA) are less restricted environmentally than those that form reefs and consequently have a world-wide distribution. *Compare* HERMATYPIC.

AHG (antihaemophilic globulin) *See* BLOOD CLOTTING.

Ailuridae (order *Carnivora, suborder *Caniformia) A family that contains only the species *Ailurus fulgens* (red panda), otherwise included in the *Procyonidae.

Ailuropoda A genus containing the giant panda (*A. melanoleuca*) as well as several fossil species. Formerly classified in the family *Procyonidae, it is now placed in the *Ursidae (bears).

***Ailurus fulgens* (red panda)** *See* AILURIDAE; PROCYONIDAE.

air bladder *See* SWIM-BLADDER.

'air-cushion' fight A ritualized fight between two animals in which they make no physical contact, as though a cushion separated them.

air sac In birds, a thin-walled extension of the lung which penetrates into the body cavity, entering the bones and forming air spaces. The lightweight bones that result are particularly important to birds. *Anhimidae (screamers) have air spaces that penetrate to the *tibia.

airsac catfish *See* HETEROPNEUSTIDAE.

Akidolestes cifelli An extinct mammal, with no modern relatives, that lived in the early *Cretaceous (about 124.6 Ma ago). A nearly complete fossil skeleton, found in the Yixian formation in Liaoning, China, was reported in 2006 (Li, G. and Luo, Z.-X. 'A Cretaceous symmetrodont therian with some monotreme-like postcranial features,' *Nature*, 439, 195–200). Akidolestes was most closely related to *Theria, but displayed some characteristics of *Monotremata, including cervical ribs, and the structure of its *pubis and hind limbs.

ala (*pl.* **alae)** A wing or *auricle-like projection that occurs on the dorsal margin of some *Bivalvia. Alae may occur above the hinge margin on the anterior or posterior regions, or on both.

alae cordis In *Arthropoda, elastic ligaments from which the heart is suspended in a pericardial *sinus.

alamiqui (*Solenodon***)** *See* SOLENODONTIDAE.

alanine An aliphatic, non-polar *amino acid, classed as non-essential in the diet of animals because it can be synthesized in sufficient amounts within cells.

alarm pheromone A *pheromone released by a social insect that detects danger. Other members of the colony up to several centimetres away detect the pheromone and move towards its source up the concentration gradient. As they approach, and the signal becomes stronger, the insects grow increasingly agitated as they seek to remove the danger. Alarm pheromones dissipate rapidly, so unless they are refreshed they soon disappear, allowing the insects to return to their ordinary activities. Many mammals also release alarm pheromones that induce other members of their species to respond quickly by fleeing or preparing to fight.

alarm response Signals emitted by an animal that serve to warn others of a danger the individual has perceived. The signals may be visual (e.g. the white tail of a rabbit, displayed when running), aural (e.g. the call of a member of a flock of ground-feeding birds), or olfactory (substances emitted by some fish and invertebrates; *see* ALARM PHEROMONE).

alarm substance A substance released from the skin of injured fish that causes an immediate alarm reaction among fish nearby, leading to the dispersal of a school formed by members of the same species.

Originally discovered in minnows and other *Cyprinidae, the alarm substance is probably present in several fish families.

ala spuria A flat, wing-like structure.

alate Winged, or having appendages resembling wings.

Alaudidae (larks; class *Aves, order *Passeriformes) A family of small, grey-brown and buff birds, whose colours are adapted to resemble their environments. The five species of *Galerida* have conspicuous upright crests. Larks have long, sharp, hind claws that help them to walk and run rather than hop. Bush larks (26 species of *Mirafra*) are fast runners and clap their wings in their courtship flight. Larks are renowned for their song, delivered from a perch or in flight. They are seed and insect eaters. There are 15 genera, and 77 species, found in most of the Old World, *Eremophila alpestris* (shore lark or horned lark) being found also in N. America and Mexico.

albacore (*Thunnus alalunga*) See SCOMBRIDAE.

albatrosses See DIOMEDEIDAE; PROCELLARIIFORMES.

Albian A stage (99.6–112 Ma) in the *Cretaceous, underlain by the *Aptian and overlain by the *Cenomanian. It is known to contain a great variety of *Mollusca, the *Gastropoda in particular being useful zonal indicators between continents. The *Gault Clays of England are Albian.

albinism In animals, the heritable condition observed as the inability to form *melanin in the hair, skin, or vascular coat of the eyes. It is due to a deficiency in the enzyme tyrosinase, and is usually inherited as an autosomal *recessive gene (i.e. a recessive *gene on a *chromosome other than a *sex chromosome in the cell *nucleus).

Albulidae (bonefish; subclass *Actinopterygii, order *Elopiformes) A family that includes only two extant species of marine tropical fish, both of which reach lengths of up to 1 m.

albumen See ALBUMIN.

albumin (albumen) A water-soluble, *globular, simple *protein that occurs in a variety of tissue fluids, including *blood plasma, *synovial fluid, tears, egg white, *lymph, and *cerebrospinal fluid. Its functions appear to be primarily those of *osmoregulation and the transport of materials.

Alcedinidae (kingfishers; class *Aves, order *Coraciiformes) A family of brightly coloured birds that have large heads, short necks, compact bodies, short, rounded wings, and a short tail. The bill is long, straight, and massive; the toes *syndactylous. The sexes are usually alike. They are found in riverine and terrestrial habitats (the 40 species of *Halcyon* (the white-collared kingfisher, *H. chloris*, has nearly 50 subspecies), found in Africa, Asia, Australia, and the Pacific islands, inhabit dry woodland and forest areas) feeding on fish, insects, and small vertebrates. The five *Ceryle* species, of America, Africa, and Asia, are blue-grey or black and white. The 11 *Ceyx* species are blue or red with red bills and are found in southern Asia, the Philippines, Indonesia, Papua New Guinea, the Solomon Islands, and Australia. There are 14 genera in the family, and 84 species, of which the best known are *Alcedo atthis* (common kingfisher) and *Dacelo novaeguineae* (kookaburra). They are found world-wide.

Alcedo atthis (common kingfisher) See ALCEDINIDAE.

Alces alces (American elk, European moose) See CERVIDAE.

Alcidae (auks; class *Aves, order *Charadriiformes) A family of mainly black and white, small-winged, diving seabirds in which the legs are set well back, the feet are webbed, and bills vary from long and pointed to laterally compressed and high. Auks are mainly *pelagic and gregarious, breeding in burrows or crevices, or on open cliff ledges, usually colonially. *Brachyramphus marmoratus* (marbled murrelet) breeds on forest branches. The two species of *Fratercula* (puffins) have laterally compressed bills that are yellow and red during the breeding season; the horny bill plates and a horn-like structure around the eye are both shed in winter. Their feet are red with claws modified for digging burrows; they inhabit grassy island slopes and cliffs, breeding in rock crevices and burrows, and spending the rest of the year at sea. Auks feed on fish and invertebrates. There are 12–14 genera, and 22 species, found in northern regions of the Pacific and Atlantic, and in the Arctic.

alcohol A *hydrocarbon in which a hydrogen atom is substituted by a hydroxyl (OH) group. An alcohol is designated as primary, secondary, or tertiary, according to whether the carbon to which the hydroxyl group is attached is bound to one, two, or three other carbons.

Alcyonacea (soft corals; class *Anthozoa, subclass *Octocorallia) An order of corals that first appeared in the Early *Jurassic and which contains the modern genus *Alcyonium* (dead man's fingers). Members of the order are characterized by having retractable polyps with eight, branching *tentacles. The polyps are embedded in the body mass which has a skeleton of numerous, free, calcareous *spicules. Usually these corals are attached to rocks. The order contains six families, and 36 genera.

Alcyonaria *See* OCTOCORALLIA.

aldehyde An organic compound that contains the group —CHO.

alderflies *See* MEGALOPTERA; SIALIDAE.

aldose A monosaccharide or its derivative that contains an *aldehyde group.

aldosterone A *steroid *hormone that is synthesized by the adrenal cortex (*adrenal gland). It influences electrolyte balance by promoting the retention by the body of sodium ions and the excretion of those of

potassium; it also affects the rate of carbohydrate metabolism.

aldotriose A three-carbon monosaccharide that contains an *aldehyde group.

***Alectis indicus* (plumed trevally)** *See* CARANGIDAE.

alevin The yolk-bearing larva of salmon and trout species.

Aleyrodidae (whitefly; order *Hemiptera, suborder *Homoptera) Family of small, sap-sucking insects in which all generations and both sexes are similar. Pre-adult stages are immobile, and the last nymphal *instar is very like the *pupa of higher insects. There are about 1200 species, distributed world-wide.

alfalfa leaf-cutter bee (*Megachile rotundata*) *See* MEGACHILIDAE.

alfonsinos *See* BERYCIDAE.

alimentary canal In animals, a tube along which food passes and through parts of whose walls nutrients are absorbed into the body. In some animals (e.g. coelenterates) the canal has a single opening. In most animals it has two: a mouth through which food enters; and an *anus through which unabsorbed material leaves the body.

alinotum In insects, the dorsal part of the thoracic segment that bears the wings.

alisphenoid A bone that forms part of the wall of the skull.

alisphenoid canal In some *Carnivora, a channel running through the *alisphenoid bone which carries a branch of the external *carotid artery.

alivincular Applied to one of the types of bivalve (*Bivalvia) ligament which is positioned between the *cardinal areas of the two valves and is not elongated.

alkaloid One of a group of basic, nitrogenous, normally heterocyclic compounds of a complex nature. Alkaloids are derived from plants, and have powerful pharmacological effects.

allantoic bladder In vertebrates, a sac formed from the posterior region of the *alimentary canal. It functions as a urinary bladder in amphibians, receives metabolic wastes in embryonic reptiles and birds, and forms part of the *placenta in *Eutheria.

allantoic placenta In *Eutheria, a *placenta formed from the *allantois and interpenetrating foetal and maternal tissues.

allantois In tetrapod *embryos, a sac formed by the outgrowth of the posterior ventral part of the gut, as a precocious urinary bladder. In reptiles and birds it grows to surround the embryo, lying between the *yolk sac and shell; the blood vessels by which it is linked to the embryo provide the means of respiration, and the allantoic cavity receives metabolic wastes. Most of the allantois is left with the shell at hatching. In *Eutheria the allantois forms part of the *placenta, supplying it with blood for respiration, nutrition, and excretion, and the allantoic cavity may be small. Most of the allantois is detached from the embryo at birth.

allele Common shortening of the term 'allelomorph'. One of two or more forms of a *gene that arise by *mutation and occupy the same *locus on *homologous chromosomes. When in the same cell, alleles may undergo pairing during *meiosis. They may be distinguished by their differing effects on the *phenotype. The existence of two forms of a gene may be termed 'diallelism', and of many forms, 'multiple allelism'. The commonness of an allele in a population is termed the 'allele frequency'.

allele frequency See ALLELE.

allelochemicals Chemicals deposited by plants in otherwise edible tissues, which are distasteful to *herbivores.

allelomone A substance released by a member of one species that influences the behaviour of members of another species.

allelomorph See ALLELE.

Allen's rule A corollary to *Bergmann's rule and *Gloger's rule, holding that a race of warm-blooded species in a cold climate typically has shorter protruding body parts (nose, ears, tail, and legs) relative to body size than another race of the same species in a warm climate. This is because long protruding parts emit more body heat, and so are disadvantageous in a cool environment, but advantageous in a warm environment. The idea is disputed, critics pointing to many other adaptations for heat conservation which probably are more important, notably fat layers, feathers, fur, and behavioural adaptations to avoid extreme temperatures.

allergy A disorder that results when the immune system (see IMMUNITY) reacts to a substance that is ordinarily harmless.

alligator gar (*Lepisosteus spatula*) See LEPISOSTEIDAE.

Alligatoridae (alligators, caimans; class *Reptilia, order *Crocodylia) A family of crocodilians that have a broad, flat snout in which the fourth tooth of the lower jaw fits into a pit in the upper jaw and cannot be seen when the mouth is closed. There are seven freshwater species, all occurring in the New World except *Alligator sinensis* (Chinese alligator). The best known is *Alligator mississippiensis* (American alligator) of the southern USA, sluggish alligators, up to 5.8 m long, that move slowly on land, hissing if surprised. Their nest mound is guarded by the female and is opened as the young begin to 'peep' at the time of hatching. The other five species are known as caimans (caymans). The common name, spectacled caiman, of *Caiman crocodilus* is derived from the large, bony ridge between its eyes; its belly has large, overlapping, bony *scutes.

Alligator mississippiensis (American alligator) See ALLIGATORIDAE.

alligators See ALLIGATORIDAE.

Alligator sinensis (Chinese alligator) See ALLIGATORIDAE.

allochthonous Pertaining to an individual or object which originates somewhere other than the place where it is found. *Compare* AUTOCHTHONOUS.

alloenzyme *See* ALLOZYME.

allograft A *graft of tissue from a donor of one *genotype to a host of a different genotype but of the same species. If the graft takes place from one part to another part of the same individual it is called an autograft.

allogrooming *Grooming performed by one animal upon another animal of the same species. *Compare* AUTOGROOMING.

allometry A differential rate of growth, such that the size of one part (or more) of the body changes in proportion to another part, or to the whole body, but at a constant exponential rate. Strictly speaking, 'allometry' is an umbrella term describing three distinct processes. Ontogenetic allometry refers to the differential growth rates of different body parts; i.e. juveniles are not merely diminutive adults. (For example, the extinct Irish elk (*Megaceros giganteus*) was the largest of all cervids (*Cervidae), but its antlers were 2.5 times larger than would be predicted from its body size, and reached an adult span of up to 3.5 m in the largest individuals.) Static allometry refers to shifts in proportion among a series of related taxa of different size. Evolutionary allometry refers to gradual shifts in proportions as size changes in an evolutionary line (e.g. in the evolution of the horse the face became relatively longer and longer). In other cases allometry may be negative, leading to comparatively smaller parts.

alloparental behaviour Parental behaviour exhibited by an individual to young to which it is not parentally related. *See also* ADOPTION.

allopatric Applied to species or populations that occupy habitats which are geographically separated, and that do not occur together in nature. *Compare* SYMPATRIC.

allopatric speciation The formation of new *species from the ancestral species

as a result of the geographical separation or fragmentation of the breeding population. Separation may be due to climatic change, causing the gradual fragmentation of the population in a few surviving favourable areas (e.g. during glaciation or developing aridity), or may arise from the chance migration of individuals across a major dispersal barrier. Genetic divergence in the newly isolated daughter populations ultimately leads to new species; divergence may be gradual or, according to punctuationist models (*see* PUNCTUATED EQUILIBRIUM), very rapid.

allopatry The occurrence of *species or other taxa in different geographical regions. *Subspecies are allopatric by definition. *Compare* SYMPATRY.

allopolyploid A polyploid (*see* POLYPLOIDY) that is formed from the union of genetically distinct chromosome sets (usually from different species).

all-or-nothing law In the functioning of the nervous system, the rule which states that the response of an *axon is independent of the intensity of the stimulus, provided the stimulus exceeds the threshold intensity required to depolarize the nerve membrane. Depolarization reverses the electrical polarity from the resting potential (in which the inside of the axon is negatively charged and the outer membrane positively charged) to the action potential (positive charge inside, negative charge outside).

allosteric Applied to a *protein molecule that has two or more sites at which it can bind (combine) with other molecules, such that binding with one influences its binding with another. Some (allosteric) *enzymes, for example, bind in the usual way at one site while at another site they combine with a substance that inhibits their activity.

allothetic Applied to information concerning an animal's orientation in an environment that is obtained by the animal from external spatial clues. *Compare* IDIOTHETIC.

Allotriognathi Alternative name for fish now placed in the order *Lampridiformes.

allozyme (alloenzyme) A form of *protein, detectable by *electrophoresis, that is produced by a particular *allele at a single *gene *locus.

Alopex (fox) *See* CANIDAE.

Alopex lagopus (arctic fox) *See* CANIDAE; RETE MIRABILE.

Alouatta (howler monkeys) *See* CEBIDAE.

alpha amino acid An *amino acid in which the amino group is attached to the number two, or 'alpha' carbon, adjacent to the carboxyl group. Compounds of this type represent the basic building blocks of *peptides and *proteins.

alpha diversity Diversity among members of a species within a single population.

alpha helix The right-handed, or less commonly left-handed, coil-like configuration of a *polypeptide chain that represents the secondary structure of some protein molecules, particularly the *globular variety. The configuration is maintained through intrachain hydrogen bonding between)CO and)NH groups of peptide bonds.

alpine salamander (*Salamandra atra***)** *See* SALAMANDRIDAE.

alternation of generations The alternate development of two types of individual in the life cycle of an organism, one *diploid and asexual, the other *haploid and sexual, both types being capable of reproduction, implying alternating phases of *meiosis and *mitosis. No instances of such alternation are known among *animals; where two types of individual occur, both capable of reproduction, they are invariably diploid and the process is better described as *metagenesis. In many parasitic *Protozoa (e.g. the malaria parasite *Plasmodium*) the sexual phase is in one host and the asexual phase in another.

altricial Applied to young mammals (e.g. rats, mice, cats, dogs) that are helpless at birth. Their eyes and ears are sealed, and they cannot walk, maintain their body temperature, or excrete without assistance. *Compare* PRECOCIAL.

altruism Sacrifice, if necessary of life itself, so that others, commonly offspring or otherwise genetically close younger relatives, may survive or otherwise benefit. On the face of it this seems to reduce the *adaptive value of the altruist. However, by saving offspring at the cost of its own life, the altruist may 'save' more of its own genes than if the situation were reversed, particularly if the reproductive potential of the altruist is exhausted, or nearly so.

Alucitidae (many-plumed moths; order *Lepidoptera) Small family of fairly small moths in which the fore wing is divided into six plumes and the hind wing into six or seven. They are not closely related to the *Pterophoridae, which are also plumed. The larvae burrow into shoots, buds, and flowers, and may induce gall-formation. They have a world-wide distribution.

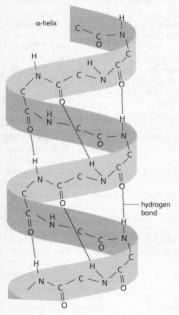

α-helix

hydrogen bond

Alpha helix

alula (bastard wing) A small projection on the leading edge of a bird's wing formed by the first digit and usually covered with *feathers. The bird ordinarily holds the alula flush with the wing, but manipulates it when landing or flying slowly.

Aluteridae *See* BALISTIDAE.

alveolus 1. In the lung of an air-breathing vertebrate, a thin-walled sac surrounded by blood vessels through whose surfaces gas exchange occurs. **2.** A sac forming an internal termination of a glandular duct. **3.** The socket in a jaw-bone into which a tooth fits.

Alytes obstetricans **(midwife toad)** *See* DISCOGLOSSIDAE.

Amandava **(waxbills, munias)** *See* ESTRILDIDAE.

Amatheusiinae Subfamily of the *Nymphalidae.

amb *See* AMBULACRUM.

Ambassidae *See* CENTROPOMIDAE.

amberjack (*Seriola dumerili*) *See* CARANGIDAE.

ambivalent Applied to apparently confused behaviour by an animal, arising most commonly from a conflict produced by impulses to behave in contradictory ways, for example to fight or to flee (*see* FIGHT OR FLIGHT REACTION).

Amblycera **(class** *Insecta,* **order** *Phthiraptera)* Suborder of chewing lice, characterized by the pedunculate first flagellomere of the *antenna, the groove of the head which conceals the antenna when at rest, the horizontal articulation of the *mandibles, and the presence of *maxillary palps. Three of the constituent families parasitize only mammals, three only birds, and one parasitizes members of both classes. The lice eat feathers, skin, sebaceous exudates, or blood. Species in one genus, *Trochiloecetes*, have the mouth-parts modified as piercing stylets (analogous to those of *Anoplura*) and

feed on the blood of their humming-bird hosts. There are 75 genera, and about 850 species.

Amblycipitidae **(torrent catfish; subclass** *Actinopterygii,* **order** *Siluriformes)* Small family of Asiatic freshwater fish that inhabit fast-flowing streams. There are three genera with 25 species.

Amblydorus hancocki **(grunting thorny catfish)** *See* DORADIDAE.

Amblyopsidae **(cavefish; subclass** *Actinopterygii,* **order** *Percopsiformes)* Family of small freshwater fish, of interest mainly because most species are typical cave dwellers, lacking functional eyes and body pigment. There are four genera and six species.

Amblypoda **(cohort** *Ferungulata,* **superorder** *Paenungulata)* An order of large, archaic, and predominantly American ungulates comprising the four distinct suborders *Pantodonta, *Dinocerata, *Xenungulata, and *Pyrotheria. The name Amblypoda is sometimes used synonymously with Pantodonta.

Amblypygi **(tail-less whipscorpions, whipspiders; class** *Arachnida)* Order of dark-coloured arachnids, in which the *abdomen is flat, oval, without a *telson, and lacking a terminal flagellum, spray glands, and *spinnerets. The first pair of legs is long, whip-like, and with many false segments; they are highly sensitive to touch and are used in water location and mating. The raptorial *pedipalps are held parallel to the ground and their last two segments may be almost *chelate. The *chelicerae are subchelate. Arthropod prey, e.g. cockroaches, crickets, beetles, and spiders, are impaled on the pedipalpal spines, pre-oral liquefaction presumably taking place. Amblypygids live in humid conditions under stones, logs, bark, and litter, and some are cavernicolous. The 60 or so species are divided between two families, and are tropical and subtropical in distribution.

Amblyrhynchus cristatus **(marine iguana)** *See* IGUANIDAE.

ambrosia beetle *See* SCOLYTIDAE.

ambulacral Applied to those areas of the body of an echinoderm (*Echinodermata) that bear tube feet.

ambulacral groove *See* AMBULACRUM.

ambulacrum (amb) In *Echinodermata, an area of the body surface, in most classes covered by calcitic plates, that lies over one of the radial canals of the internal water-vascular system and bears the *tube feet. In some echinoderms (e.g. *Asteroidea, Blastoidea, and *Crinoidea) the ambulacrum is marked by a deep linear depression, the ambulacral groove. Typically, echinoderms have five, or a multiple of five, ambulacral areas.

Ambulocetus natans The most completely known early cetacean (*Cetacea), described in 1994 by J. G. M. Thewissen, S. T. Hussain, and M. Arif, from lower to middle *Eocene beds in Pakistan. It is known by parts from most of the skeleton, showing that it had a long neck, relatively long hind limbs, and five separate (hoofed) digits on each limb. It was the size of a sea lion.

Ambystoma mexicanum (axolotl) *See* AMBYSTOMATIDAE.

Ambystomatidae (mole salamanders, axolotls; class *Amphibia, order *Urodela) A family of amphibians which have a broad head, the tongue free only at the sides, *palatal teeth, costal grooves down the body, short, strong limbs, and a laterally compressed tail. Most burrow in earth for much of the year. *Neoteny is common in some species, the best-known example being *Ambystoma mexicanum* (axolotl), confined to certain lakes around Mexico City, which normally breeds as an aquatic larva, but is capable of *metamorphosis into a terrestrial form; it is sexually dimorphic, the female being the larger, attaining up to 29 cm in length, and normally black although albinos (*see* ALBINISM) are not uncommon. There are more than 30 species, found in N. and Central America.

ameba *See* AMOEBA.

amebocyte *See* AMOEBOCYTE.

ameboid movement *See* AMOEBOID MOVEMENT.

Amebozoa *See* AMOEBOZOA.

amensalism An interaction of *species *populations, in which one population is inhibited while the other (the amensal) is unaffected. It is the opposite of *commensalism. *Compare* COMPETITION; MUTUALISM; NEUTRALISM; PARASITISM; PREDATION; PROTOCOOPERATION; *See also* SYMBIOSIS.

American alligator (*Alligator mississippiensis***)** *See* ALLIGATORIDAE.

American cockroach (*Periplaneta americana***)** *See* BLATTIDAE.

American tapir (*Tapirus***)** *See* TAPIRIDAE.

Ameridelphia One of the two cohorts into which the *Metatheria (marsupials) are generally divided in modern classifications. There are three orders, all occurring in Central and S. America: *Didelphimorphia (also represented in N. America); *Paucituberculata; and the extinct *Sparassodontia.

Ametabola *See* APTERYGOTA.

Amia calva (bowfin) *See* AMIIDAE.

Amiidae (bowfin; infraclass *Holostei, order *Amiiformes) A family with only one living representative, *Amia calva*, a fairly large (up to 6 kg) carnivorous fish that has a long *dorsal fin with only soft rays, a *gular plate, a *heterocercal tail, and a lung-like *swim-bladder. The male guards the nest and larvae.

Amiiformes (class *Osteichthyes, subclass *Actinopterygii) Order of fairly primitive bony fish consisting of several families dating back to the *Triassic Period, and considered by some to be intermediate between the *Chondrostei and the *Teleostei. Until recently they were grouped with the order *Semionotiformes in the infraclass *Holostei.

amine An organic base that is derived from ammonia (NH_3) by the replacement of one or more of the hydrogens by organic radical groups. The resultant amine is designated primary (NH_2R), secondary (NHR_2), or tertiary (NR_3) according to the number of hydrogens replaced.

aminoacetic acid *See* GLYCINE.

amino acid An organic compound that contains an acidic carboxyl (COOH) group and a basic amino (NH_2) group. They constitute the basic building blocks of *peptides and *proteins and are classified either as: (*a*) neutral, basic, or acidic; or (*b*) non-polar, polar, or charged.

amino-acid sequence The sequence of *amino acid residues in a *polypeptide chain that represents the primary structure of a *protein. This sequence is unique to each protein and influences the protein's secondary, tertiary, and quaternary structures.

aminoethanoic acid *See* GLYCINE.

amino group The radical group NH_2.

aminopeptidase An *enzyme that catalyses the *hydrolysis of *amino acids in a *polypeptide chain by acting on the *peptide bond adjacent to the essential free amino group.

amino sugar A monosaccharide (*see* SUGAR) in which an amino group has been substituted for one or more hydroxyl groups.

aminotransferase *See* TRANSAMINASE.

amitosis Cell division that occurs directly, through simple cleavage of the nucleus, without the involvement of a *mitotic spindle. *Compare* MITOSIS.

ammocoete Larva of the lamprey (family *Petromyzonidae). After hatching the ammocoete lives buried in mud for several years. It then undergoes *metamorphosis into an adult lamprey.

Ammodytidae (sand-lance, sand-eel; subclass *Actinopterygii, order *Perciformes) A family of fish that have elongated bodies and long *dorsal fins. The *pelvic fins are often absent. Feeding on animal plankton, the sand-eels themselves form an important part of the diet of cod, salmon, etc. There are about 12 species, found in marine coastal waters.

ammonites *See* AMMONOIDEA.

Ammonoidea (phylum *Mollusca, class *Cephalopoda) A subclass of cephalopods which generally have planispirally, tightly coiled, septate shells (although some are coiled loosely or spirally). The *protoconch is globular; the shells may be either *involute or *evolute. Some forms have marked ventral keels; ribs and nodes may also be present. The *siphuncle is variable but mainly ventral in position. Sutures are often very complex. *Cameral deposits are absent. The Ammonoidea were probably tetrabranchiate cephalopods. They constitute the largest cephalopod subclass, with 163 families, including the ammonites, in which the suture lines form very complex patterns; the ceratites, in which part of the suture line is frilled; and the goniatites, which have relatively simple suture lines. They range in age from *Devonian to Late *Cretaceous. All members are now extinct.

ammonoids *See* AMMONOIDEA.

ammonotelic Applied to organisms (e.g. many aquatic invertebrates and teleost fish) that excrete nitrogenous waste derived from amino-acid catabolism in the form of ammonia. *See also* UREOTELIC; URICOTELIC.

amnion An inner cellular layer (*epithelium) surrounding the fluid-filled *embryo sac of reptiles, birds, and mammals. The wall of this sac has two layers of epithelium, with *mesoderm and coelomic space between. The outer layer is usually called the *chorion. The amnion and chorion develop as folds in the *ectoderm and mesoderm which grow up around the sides of the embryo, eventually covering it. The amniotic fluid within the sac provides a liquid environment for the embryo, necessary for animals that reproduce on land. It may also cushion the mammalian embryo against distortion by maternal organs pressing on it. Sometimes the term 'amnion' is applied to the entire embryo sac.

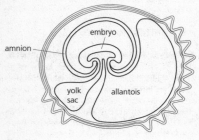

Amnion

amniote Applied to a type of development, typical of higher vertebrates (reptiles, birds, and mammals), in which the *amnion surrounds the *embryo in a bag of (amniotic) fluid. The amnion is primitively associated with a shell, and is capable of gaseous exchange; its development thus enabled eggs to be laid on dry land for the first time in vertebrate evolution.

amniotic fluid *See* AMNION.

amoeba (ameba) Any single-celled *eukaryote that is naked and changes shape due to the irregular extension and retraction of *pseudopodia.

amoebiasis Any disease in which the causal agent is an *amoeba. In humans, the most common form of amoebiasis is amoebic dysentery, caused by *Entamoeba histolytica*. This disease varies in severity from mild diarrhoea to severe or fatal dysentery. *See* AMOEBOZOA.

amoebic dysentery *See* AMOEBIASIS.

Amoebida (superclass *Sarcodina, class *Rhizopodea) An order of *Protozoa, lacking a *test, in which each cell contains a single nucleus and is of indeterminate shape. The cells typically form *lobopodia (rarely *filopodia or anastomosing) which serve for locomotion and feeding. Most species are free-living in freshwater, moist-soil, or marine environments; a few are parasitic. There are several families, and many genera, one of which is the genus *Amoeba*. *A. proteus*, a popular subject for study in elementary biology courses, measures up to about 0.6 mm across, and is found in slow-moving or still, freshwater habitats.

amoebocyte (amebocyte) A cell capable of *amoeboid movement found in the bodies of some invertebrates, e.g. *Echinodermata, *Mollusca, and *Porifera. They perform various functions, including digestion, the transport of nutrients and metabolic wastes, and defence against *pathogens.

amoeboid movement (ameboid movement) Movement in the fashion of an *amoeba, i.e. by pseudopodia (*see* PSEUDOPODIUM).

Amoebozoa (Amebazoa) A phylum of amoeboid *protozoa. They are *eukaryotes which move by means of pseudopodia (*see* PSEUDOPODIUM) and feed by *phagocytosis. They vary in size and shape. Most are free-living in shallow water, but some are parasites; *Entamoeba histolytica* causes amoebic dysentery in humans.

Amorphochilus *See* FURIPTERIDAE.

Amorphoscelidae (subclass *Pterygota, order *Mantodea) The smaller of the two families of mantids, comprising insects which are small and mottled in colour, and which have raptorial fore legs devoid of spines on their external, lower sides. Many species have wingless females. Most live on rough-barked tree trunks against which they are well camouflaged. The *pronotum tends to be shorter than that of members of the *Mantidae. Most of the 50 or so species are found in Australia.

amphetamine 1-phenyl-2-aminopropane, a drug that stimulates the central nervous system and prevents sleep.

Amphibia (amphibians; phylum *Chordata, subphylum *Vertebrata) A class represented today by just three groups, the *Apoda (caecilians), *Urodela (salamanders), and *Anura (frogs and toads). The class was much more varied formerly and in the *Triassic some forms, e.g. *Mastodontosaurus*, grew up to 6 m long. Amphibians first appeared in the *Devonian, having evolved from the *Rhipidistia (lobe-finned fish). They flourished in the *Carboniferous and *Permian but declined thereafter. The first modern types were established during the Triassic. They are *poikilothermic vertebrates. The majority are terrestrial but develop by a larval phase (tadpole) in water. The skin is soft, naked (non-scaly), rich in mucous and poison glands, and important in *cutaneous respiration. There are about 7000 extant species. Most are found in damp environments and they occur on all continents except Antarctica.

amphibians See AMPHIBIA.

amphibiotic 1. Applied to an organism that can live with a host organism either parasitically or mutualistically. See MUTUALISM; PARASITISM. **2.** Living in water (larval form) and later on land (adult form).

amphiblastula A poriferan (sponge) larva in which the area of flagellate cells is equal to the area of non-flagellate cells. At *metamorphosis the flagellated cells move to the interior and become *choanocytes.

amphicoelous vertebrae Condition in which the central part of each vertebra (the centrum) is concave on both anterior and posterior. See LEIOPELMATIDAE. *Compare* HETEROCOELUS VERTEBRA.

Amphicyonidae A family of very large, heavily built dogs, which appeared during the *Oligocene, spread to all the northern continents, and became extinct in the *Pliocene. They have been called 'dog-bears', but apart from their size they had few truly bear-like features.

amphid In *Nematoda, an anterior sense organ that is believed to be a *chemoreceptor.

amphidetic In *Bivalvia, applied to a ligament located in the anterior and posterior areas, around the *shell beaks.

amphidisc One of the small, flesh *spicules of poriferans (sponges), if the spicule has recurved rays at the end of the rod.

Amphidiscophora (phylum *Porifera, class *Hexactinellida) A subclass of sponges in which the small, flesh *spicules occur as *amphidiscs.

amphidromous Applied to the migratory behaviour of fish moving from fresh water to the sea, and vice versa. Such migration is not for breeding purposes, but occurs regularly at some stage of the life cycle (feeding, overwintering, etc.). *Compare* DIADROMOUS. *See also* ANADROMOUS; CATADROMOUS; POTAMODROMOUS.

Amphineura (phylum *Mollusca) A class of elongate, bilaterally symmetrical, marine molluscs in which the shell, if present, consists of seven or eight overlapping, calcareous plates on the dorsal surface. The head is poorly differentiated, if at all. Some possess an effective *radula. The posterior *mantle cavity contains the *anus and *gills. A broad, flat foot occurs ventrally and is used for creeping. The anatomy is typically molluscan, with a primitive, ganglionic nervous system. All are blind and lack *tentacles. Sexes are generally separate. The class first appeared in the Late *Cambrian. There are approximately 940 and 430 fossil species.

Amphioxiformes *See* BRANCHIOSTOMIDAE.

amphioxus (lancelet; *Branchiostoma lanceolatum*) *See* BRANCHIOSTOMIDAE.

amphipathic Applied to a molecule that has a negatively charged phosphate group in the polar head, which is *hydrophilic, and non-polar *fatty acid chains in the tail, which are *hydrophobic.

amphiplatyan Applied to *vertebrae in which the centrum is flattened on both the anterior and posterior sides.

amphipneustic Applied to insect larvae in which the spiracular openings of the respiratory system are restricted to a pair each on the *prothorax and the posterior of the *abdomen. The condition is common in larval *Diptera.

Amphipnoidae (cuchia; subclass *Actinopterygii, order *Synbranchiformes) A family of fish that includes only one species, *Amphipnous* (*Monopterus*) *cuchia*, of southern Asia. It is of interest because it has two lung-like air sacs connected with its gill chambers.

Amphipoda (sandhoppers, beach hoppers, beach fleas, water lice; phylum *Arthropoda, subphylum *Crustacea, superorder *Peracarida) Order of mainly laterally compressed crustaceans, with the *cephalothorax devoid of a *carapace and the whole head fused to one thoracic segment, or occasionally to two. The first three *pleopods are usually many-segmented *swimmerets, the posterior ones modified for jumping and comprising a few segments. The gills are borne on the thoracopods. There are 4000 species.

Amphipterygidae (damselflies; order *Odonata, suborder *Zygoptera) Family of damselflies that resemble the *Anisoptera (true dragonflies) in their habit of resting with their wings open. They are distinguishable from all other zygopteran families by having five or more *antenodals, of which the first two are thickened and the rest are incomplete. The robust larvae are adapted for clinging to rocks in fast-flowing streams, and have large, saccoid, tracheal gills. The family is mainly tropical, and 19 extant species have been described.

Amphisbaenidae (worm lizards; order *Squamata, suborder *Sauria) A family of subterranean lizards that have a small, wedge-shaped head, adapted for digging. The body is cylindrical, with loose skin and rings of scales. The tail is short. Worm lizards are generally limbless, but well-developed fore limbs are present in *Bipes*, found in Mexico, and are used to scratch a cavity prior to tunnelling, while the wedge-shaped head compacts the soil as the lizard burrows. Eyes and ears lie beneath the skin. There are about 130 species, found in America, the Mediterranean region, and Africa.

amphistyly A type of jaw suspension in which the upper jaw is attached by ligaments to both the *chondrocranium and to the second *visceral (hyoid) arch. Amphistyly results in a large gape with the jaw extending back behind the *cranium and is present in primitive members of the *Elasmobranchii (e.g. *Cladoselache*). *Compare* HYOSTYLY.

amphitropical species Species that have disjunct distribution patterns, one part of the range being to the north of the Equator, the other to the south, the different parts being geographically quite separate. These disjunctions probably arose in the *Pleistocene when the climatic belts were telescoped, and migration across the Equator would have been easier.

Amphiumidae (Congo eels, lamper eels; class *Amphibia, order *Urodela) A family of eel-like, semi-larval salamanders that have a cylindrical body up to 1 m long, rudimentary limbs and lungs, small, lidless eyes, no tongue, and external gills that are lost in adults. They are nocturnal feeders on small invertebrates. They may leave the water in wet weather. Eggs are laid in a nest. Fertilization is external, with no *spermatophore. There is a single genus, *Amphiuma*, and three species, occurring in freshwater swamps in the south-eastern USA.

Amphymerycidae A family of primitive ruminants from the late *Eocene and early *Oligocene of the Old World. Some authorities believe these small animals were in the direct line of descent to the advanced ruminants.

amplexus A type of *pseudocopulation, e.g. in frogs and toads, in which the male grasps the female with his front legs while releasing sperm that fertilizes eggs released by the female.

Ampulicidae (cockroach wasps, digger wasps; order *Hymenoptera, superfamily *Sphecoidea) The most primitive family of sphecid *hunting wasps, in which the antennae are inserted low on the face, with the sockets touching or close to the fronto-clypeal suture (*see* CLYPEUS). The *pronotum has a high collar, and is long to very long and often tuberculate. The *propodeum is long to very long, with the enclosed area entirely dorsal, and U-shaped to nearly triangular. The mid- *tibiae have two apical spurs, and the claws have an inner tooth. The marginal cell of the fore wing is *acuminate apically, and apendiculate. The fore wing has two or three submarginal cells, and two recurrent veins (the first recurrent received by submarginal 1 or 2, the second received by 2 or 3). The *abdomen is sessile, or with a petiole (*see* GASTER) composed of the *sternum and *tergum. There are about 160 species, mainly tropical.

ampulla An approximately spherical enlargement of a canal or duct in which material is stored. The rectal ampulla stores *faeces in the *rectum prior to defecation; the ampulla of Vater occurs where ducts from the *liver and *pancreas enter the small *intestine.

ampullae of Lorenzini Jelly-filled tubes located in the heads of sharks. Open at the surface, the deeper part of the ampullae contain sensory cells that respond to electrical gradients such as those produced by potential prey hidden in a sandy sea-bottom.

ampullary organ An organ comprising *electroreceptors, which are cells found in weakly electric teleosts (*Teleostei) that detect tonic (steady) electrical discharges. *Compare* TUBEROUS ORGAN.

ampulliform Flask-shaped or bottle-shaped.

amygdaliform Almond-shaped.

amylase A member of a group of *enzymes that hydrolyse starch or *glycogen by the splitting of glycosidic bonds, so giving rise to the sugars *glucose, dextrin, or *maltose. Amylases are widely distributed in plants and animals, occurring in microorganisms and, for example, in pancreatic juices and *salivary glands.

amylolytic Capable of digesting starch.

amylopectin A branched-chain *polysaccharide that is found in native starches composed of *glucose units joined by α-1,4 glycosidic bonds and at points of branching by α-1,6 bonds.

amylose A long, unbranched-chain polysaccharide component of native starch that is composed of *glucose units joined by α-1,4 glycosidic bonds.

Amynodontidae (amynodonts; suborder *Ceratomorpha, superfamily *Rhinocerotoidea) An extinct family of rhinoceros-like animals that lived during the *Eocene and *Oligocene in America and Eurasia and persisted into the *Miocene in Asia. They were analogous in size and shape to the hippopotamus and, like them, probably lived in rivers.

amynodonts *See* AMYNODONTIDAE.

Anabantidae (climbing gouramis, labyrinth fish; subclass *Actinopterygii, order *Perciformes) A family of fish capable of aerial respiration because of the presence of an *accessory respiratory organ. There are about 40 species, the best known of which is *Anabas testudineus* (climbing gourami or walking fish) of southern Asia. It lives in freshwater ponds and swamps, but can crawl over land in search of a new source of water during the dry season.

***Anabas testudineus* (climbing gourami)** *See* ANABANTIDAE.

Anablepidae (four-eyed fish; subclass *Actinopterygii, order *Atheriniformes) A small family of fish that includes the species *Anableps anableps*, unusual because its bulbous eyes are actually divided into upper and lower lobes so that the two eyes actually function as four eyes. It is a predominantly freshwater fish from Central America.

anabolism *See* METABOLISM.

Anacanthini Older name for the cod-like fish species now included in the order *Gadiformes.

anachoresis The habit of living in holes or crevices as a means of avoiding predators.

anaconda (*Eunectes murinus*) *See* BOIDAE.

anadromous Applied to the migratory behaviour of fish that spend most of their lives in sea, but then migrate to fresh water to spawn (e.g. salmon and lamprey). *Compare* CATADROMOUS; POTAMODROMOUS. *See also* AMPHIDROMOUS; DIADROMOUS.

anaerobe *See* ANAEROBIC (2).

anaerobic 1. Of an environment: one in which oxygen is absent. **2.** Of an organism: one able to exist only in the absence of oxygen, i.e. an anaerobe. Organisms may be facultative anaerobes (e.g. yeasts) or obligatory anaerobes (e.g. sulphur bacteria). **3.** Of a process: one that can occur only in the absence of oxygen.

anagenesis In the original sense, evolutionary advance; the term is now often applied more widely, to virtually all sorts of evolutionary change, along a single, un-branching lineage.

anal fin The unpaired fin located on the ventral side of the body of a fish, posterior to the anus. It plays an important role in the swimming movements of sharks and bony fish.

anal glands (anal sacs) In all members of the Carnivora other than bears (*see* UR-SIDAE), a pair of sacs located on either side of the *anus between the interior and exterior muscles controlling the *sphincter. *Glands in the lining of the sacs secrete a strong-smelling liquid used for identifying individuals within the species.

analogous variation Features with similar functions which have developed independently in unrelated taxonomic groups, in response to a similar way of life, or similar method of locomotion, or similar food source, etc. Thus the wings of birds and insects are analogous.

anal sacs *See* ANAL GLANDS.

analysis of variance A statistical procedure that is used to compare more than two means.

(●) SEE WEB LINKS
• A description of analysis of variance.

anamniote Applied to a type of development typical of lower vertebrates (fish and amphibians). The egg lacks a shell and protective embryonic membranes: consequently it must be laid in water or in a suitably damp environment.

Anamorpha *See* CHILOPODA.

anamorphic growth *See* ANAMORPHOSIS.

anamorphosis (anamorphic growth) A developmental phenomenon in some *Protura and *Chilopoda in which the young do not possess the full, adult complement of body segments and legs on hatching; these are developed successively in later stages, after moulting. Centipedes that hatch with only part of the adult complement of segments take several years to reach sexual maturity. *Compare* EPIMORPHOSIS.

anaphase The stage of *mitosis or *meiosis at which the *centromere becomes functionally double, and daughter *chromatids separate from the equator, moving towards the opposite poles of the *spindle. The spindle then elongates and pushes the two groups of chromosomes further apart. In the second stage (anaphase II) of meiosis, centromeres do not divide and *homologous chromosomes are separated.

anapophysis In many *Mammalia including humans, a small process arising on the thoracic and lumbar *vertebrae.

anapsid Applied to a *cranium that has no openings near the temples. The condition occurs in *Amphibia and *Anapsida. *Compare* DIAPSID, SYNAPSIDA.

Anapsida (class *Reptilia) A subclass of reptiles characterized by a skull that lacks apertures in the temple regions behind the eyes. The only living representatives are the turtles (*Chelonia) but three other groups flourished in the late *Palaeozoic, two of them surviving into the *Triassic.

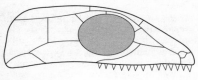

Anapsida

Anaptomorphidae (suborder *Haplorhini (or *Prosimii), infra-order *Tarsiiformes) An extinct family of tarsier-like animals that showed signs of tarsioid specialization. The brain was similar to that of a modern tarsier but the olfactory regions were more highly developed, and the temporal and occipital lobes enlarged. The eyes were large, the face short. There were at least 20 genera, distributed in Europe and America from the *Palaeocene to *Oligocene.

Anarhichadidae (wolf-fish; subclass *Actinopterygii, order *Perciformes) A family of marine fish which are eel-like in shape, have a very long *dorsal fin, lack *pelvic fins, and have powerful jaws with strong teeth used to crush hard-shelled molluscs and crustaceans. There are about six species, living in cold to Arctic waters.

***Anas* (dabbling ducks)** *See* ANATIDAE.

Anasca (class *Gymnolaemata, order *Cheilostomata) Suborder of relatively primitive bryozoans in which the tentacles are extruded by a simple hydrostatic system without a compensation sac (*see* AScoPHORA). They occur from the *Jurassic to the present.

Anaspidea (Aplysiacea, Aplysiomorpha; class *Gastropoda, subclass *Opisthobranchia) An order of gastropods which have an internal and much reduced shell. Only a small *mantle cavity is present. The order contains the sea hares.

Anaspidiformes (subphylum *Craniata, class *Cephalaspidomorphi) An extinct order of fossil, fish-like vertebrates that are considered to be related to the lampreys (*Petromyzoniformes). They had a *fusiform body with a *hypocercal tail, and up to 15 pairs of external gill openings. Extant during the *Silurian and *Devonian Periods, these small (up to 15 cm in length) animals lived predominantly in fresh water.

***Anas platyrhynchos* (mallard)** *See* ANATIDAE.

anastomose To join together structures, such as blood vessels, separated by the earlier branching of a single structure, forming loops or a network.

Anatidae (ducks, geese, mergansers, pochards, sawbills, swans; class *Aves, order *Anseriformes) A family of mainly aquatic birds that have flat, *lamellate bills, except for the six species of *Mergus*, the mergansers or sawbills, which feed on fish and have long, narrow, serrated bills, and shelducks (seven species of *Tadorna*), which are fairly large birds that resemble geese, and have short, slightly upturned bills. The front toes are webbed (the more terrestrial and non-migratory *Branta sandvicensis*, the Hawaiian goose or ne-ne, has reduced webbing and short wings). Many Anatidae show *sexual dimorphism. They have thick feathers with insulating down. (Stifftails (six species of *Oxyura*) have long, stiff tail feathers and the males have long, blue bills.) The flight feathers are moulted simultaneously after breeding. They feed on vegetable and animal foods and nest on the ground or in holes in trees, among rocks, or in the earth, and the nest is usually lined with down. Eiders (three species of *Somateria*) are sea ducks, found in estuaries and coastal areas, as are scoters (three species of *Melanitta*) although these breed inland. Whistling ducks (eight species of *Dendrocygna*) are partially nocturnal. The largest genus, with 36 species, is *Anas* (dabbling ducks); *A. platyrhynchos* (mallard) is the ancestor of most domestic ducks. There are nine or ten species of *Anser* (geese); *A. anser* (greylag goose), *A. cygnoides* (swan goose), and *Cygnus olor* (mute swan) are also extensively domesticated. *C. atratus* (black swan) has been introduced to New Zealand. *Branta canadensis* (Canada goose) has been introduced into Europe. There are 12 species of *Aythya* (pochards), some of which feed in sea water. *Anseranas semipalmatis* (magpie goose) occurs on the floodplains of northern Australia. There are 43 species, with cosmopolitan distribution. There are 40 genera of Anatidae, with about 146 species.

Anatolepis heintzi Possibly one of the earliest jawless fish, known from a number of associated scales and fragments discovered in Early *Ordovician (Arenig) strata of Spitzbergen.

Anax imperator (emperor dragonfly) See AESHNIDAE.

anchovy (*Engraulis encrasicolus*) See ENGRAULIDAE.

Ancylopoda (chalicotheres; superorder *Mesaxonia, order *Perissodactyla) An extinct suborder (sometimes called Chalicotheriidae) of horse-like animals, that were distributed widely in N. America and Europe in the *Eocene and survived in Africa until the *Pleistocene. *Moropus*, usually taken as representative of the group, resembled a horse with front legs longer than the back legs. It had three toes with terminal phalanges that were cleft and undoubtedly bore claws rather than hoofs. Probably chalicotheres fed on roots and tubers and the claws and long fore limbs were used in digging, or possibly to pull down leafy branches.

Andean condor (*Vultur gryphus*) See CATHARTIDAE.

Andrenidae (solitary mining bees; subclass *Pterygota, order *Hymenoptera) Large family of short-tongued, solitary bees. Each antennal socket is connected to the *clypeus by two sutures. The tongue is pointed. The segments of the *labial palps are similar, or sometimes the first or second are elongate and flattened. The *coxae of the mid-leg are externally shorter than the distance from their apices to the posterior wing bases. The pollen *scopa is a dense fringe of hind tibial, branched hairs, often augmented by a coxal floccus of long, curved hairs. The brood cells are lined with a wax-like secretion of the *Dufour's gland.

Andrias japonicus (Japanese giant salamander) See CRYPTOBRANCHIDAE.

androchronial scale See ANDROCONIAL SCALE.

androconial scale (androchronial scale) In male *Lepidoptera, a wing or body scale that is modified for the dispersal of a sexual scent (*pheromone). Glandular cells in the wing membrane or integument of the body connect with the base of the scale, but just how the scent is dispersed is not fully understood. The scent may diffuse via the hollow body of the scale to the fimbriae (minute hairs) at the scale tip, or may simply pass on to the surface of the scale. Androconia may be scattered singly over the surface of the wing, but are sometimes grouped together in tufts.

androgens A generic term of the *hormones, secreted principally by the testis, that regulate the development of male *secondary sexual characteristics.

anemonefish See POMACENTRIDAE.

aneucentric Applied to an aberrant *chromosome that possesses more than one *centromere.

aneuploid Applied to a cell or organism whose nuclei possess a *chromosome number that is greater by a small number than the normal chromosome number for that species. Instead of having an exact multiple of the *haploid number of chromosomes, one or more chromosomes are represented more times than the rest.

angelfish *See* CHAETODONTIDAE; CICHLIDAE.

angel shark *See* SQUATINIDAE.

anglerfish 1. *See* LOPHIIDAE; MELANO-CETIDAE. **2. (frogfish)** *See* ANTENNARIIDAE.

Anguidae (lateral fold lizards, slow-worms, glass snakes; order *Squamata, suborder *Sauria) A family of reptiles whose members vary from four-legged lizards to legless snake-like forms (e.g. *Anguis fragilis*, the slow-worm or blindworm, and *Ophisaurus apodus*, the European glass lizard or glass snake). All have a forked tongue (unlike skinks of similar form), mobile eyelids, and tail *autotomy. Some (e.g. *A. fragilis*) are *ovoviviparous, others (e.g. *O. apodus*) egg-layers. There are about 40 species, found in America, Europe, Asia, and N. Africa.

Anguillidae (eels; superorder *Elopo-morpha, order *Anguilliformes) A family of snake-like fish with long, cylindrical bodies, narrow, vertical gill slits, and a long *dorsal fin continuing through the *tail fin into the *anal fin. The skin bears minute scales. Widely distributed in tropical and temperate regions, the adults tend to live in fresh water for a number of years, finally migrating to the sea to spawn. Being a valuable food fish, eels are marketed in large numbers in many countries. The common eel, *Anguilla anguilla*, is known to spawn in the Sargasso Sea (western Atlantic), the juveniles (elvers) taking up to three years to reach the coast of Europe. There is one genus, *Anguilla*, with about 15 species.

Anguilliformes (subclass *Actinop-terygii, superorder *Elopomorpha) An order of eel-like fish with elongate, cylindrical bodies, narrow gill apertures, a *tail fin confluent with the low *dorsal and *anal

fins, no ventral fins, and the skin either naked or with minute scales. The order comprises about 22 families, including the *Anguillidae (eels), *Muraenidae (moray eels), and *Congridae (conger eels).

anguilliform swimming A type of swimming practised by highly flexible fish such as eels, in which most of the body length undulates such that over a half a sinusoidal wave is formed. *Compare* BALLISTIFORM SWIMMING; CARANGIFORM SWIMMING; LABRIFORM SWIMMING; OSTRACIIFORM SWIMMING; RAJIFORM SWIMMING.

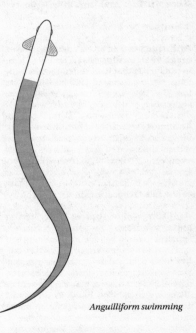

Anguilliform swimming

Anguis fragilis (blindworm, slow-worm) *See* ANGUIDAE.

angular In most *Gnathostomata, a bone which forms the angle of the lower jaw. In mammals it forms the *tympanic bone.

angwantibo (*Aretocabus calabaren-sis*) *See* LORIDAE.

Anhimidae (screamers; class *Aves, order *Anseriformes) A family of black or grey, short-necked, fairly long-legged, ungainly birds in which the bill is chicken-like, the toes are long and semipalmate, the wings broad with a spur on the leading edge. There are subcutaneous *air sacs. The ribs lack *uncinate processes. These birds are found in marshy habitats where they wade or swim. They are vegetarian and nest on the ground near water. There are two genera, with three species, found in tropical and subtropical S. America.

***Anhinga* (darters)** *See* ANHINGIDAE.

anhingas *See* PELECANIFORMES.

Anhingidae (darters; class *Aves, order *Pelecaniformes) A family of long-necked birds that have an elongated body, long, narrow, pointed bill, and long, stiff tail. They are *totipalmate. The food consists mainly of fish, caught under water, and darters often swim with only the head and neck above water. These birds are excellent fliers, and are migratory in some areas. They inhabit the wooded shores of lakes and rivers, breeding colonially in trees. There is one genus, *Anhinga*, with four species, often considered to be one, found in all the major continents except Europe.

Aniliidae (pipe snakes; order *Squamata, suborder *Serpentes) A family of subterranean snakes that are similar in general form to the blind snakes (*Typhlopidae) but have enlarged belly scales. The small eyes are not covered by scales, and remnants of hind limbs remain as claws near the anus. There are 10 species, found in S. America and south-east Asia. *See also* CORAL SNAKES.

animal A multicellular, heterotrophic organism that develops from an *embryo derived from *gametes produced in specialized organs or surrounded by somatic cells. Typically, animals are *motile, at least during some stage of the life cycle, and have sensory apparatus with which to detect changes in their immediate environment. *Protozoa are unicellular but otherwise resemble animals in many ways (although

there are plant-like protozoons) and were formerly classified as an animal phylum; they are now more usually classified in the kingdom *Protoctista. *See also* ANIMALIA.

animal behaviour *See* ETHOLOGY.

animalcule An old name for a microscopic, animal-like organism, particularly a protozoon.

Animalia (Metazoa) Multicellular organisms that develop from *embryos; one of the three kingdoms of multicellular organisms (the other two being Fungi and Plantae, the plants). The *gametes form within multicellular sex organs and never within unicellular structures. The kingdom includes all animals other than protozoons (some of which are colonial); *Porifera (sponges) are sometimes excluded because their structure differs markedly from that of other animals. Animals first appeared in the *Precambrian, the sponges (Porifera) from one kind of *protist forebear, and all other animals from another (or possibly more than one other) protist. The oldest fossils are burrows of a *coelomate in rocks rather less than 700 Ma old.

animal pole The end (or pole) of an egg that contains the most *cytoplasm and the least *yolk. It is also the point on the surface of the egg that is nearest to its *nucleus. The opposite end is called the vegetal pole. These two poles form the anchor points from which *spindle fibres extend to attach to the *centromere of each of the two *homologous chromosomes during cell division.

animal starch *See* GLYCOGEN.

anis (*Crotophaga*) *See* CUCULIDAE.

aniso- Prefix meaning unequal, from the Greek *an* and *isos*, 'not' and 'equal'.

anisodactylous In birds, the condition in which three toes are free and face forward, and the hind toe is in the same plane and opposable to them. This foot morphology is common among *passerines. *Compare* ZYGODACTYLOUS.

anisogamy (heterogamy) The union of two *gametes that differ in size or form, e.g. *ovum and *spermatozoon.

anisomyarian Applied to the condition in *Bivalvia in which the two *adductor muscles are not clearly equal in size.

Anisoptera (dragonflies; class *Insecta, order *Odonata) One of the three suborders of dragonflies, comprising insects which are generally more robust, and fly more strongly, than those in the other suborders. The anisopteran hind wing has a broader base than the fore wing, both parts being held outspread when the insect is at rest. The larvae are aquatic and robust, with an elaborate system of *tracheal gills arranged in longitudinal rows within the rectum. Adults are often brightly coloured, and fly near water. This cosmopolitan suborder has nine families (*Aeshnidae, *Cordulegasteridae, *Corduliidae, *Gomphidae, *Libellulidae, Macromiidae, Neopetaliidae, *Petaluridae, and *Synthemidae), containing more than 2500 species.

Anisozygoptera (dragonflies; class *Insecta, order *Odonata) Smallest of the three suborders of dragonflies, comprising insects which are superficially similar to the *Anisoptera, but whose wings have narrow bases (a feature of *Zygoptera). The larvae also show features possessed by both the other suborders. The Anisozygoptera flourished during the *Mesozoic, when they replaced the ancestral dragonfly order Protanisoptera. Today there are only two, poorly known, living species, found in India and Japan. *See also* EPIOPHLEBIIDAE.

Ankylosauria (ankylosaurs; subclass *Archosauria, order *Ornithischia) A suborder of *Cretaceous dinosaurs that were heavily armoured with bony plates, rather in the manner of armadillos. They had small heads with insignificant teeth, and some forms were toothless.

ankylosaurs *See* ANKYLOSAURIA.

Annelida A phylum of coelomate (*see* COELOM) worms that have a definite head and good *metameric segmentation. Anatomically they are more complex than the *Platyhelminthes. They have vascular, respiratory, and nervous systems which are well developed. The body is elongate, and each segment has *chaetae. They occur in marine, freshwater, and terrestrial environments. Some are parasitic. They are represented today by the earthworms, sandworms, and leeches. Their fossils are found in rocks dating from the *Cambrian, and possible fossil annelid worms are known from *Precambrian sediments in southern Australia. There are about 12 000 species.

Anniellidae (shovel-snouted legless lizards; order *Squamata, suborder *Sauria) A family of limbless, burrowing lizards that inhabit sandy ground. The head is wedge-shaped with a countersunk lower jaw, small eyes with lids, and no *tympanic membrane. The scales are smooth. The diet consists mainly of insect larvae. There are two species, both found only in California.

annual fish Certain small fish (*Cyprinodontidae) whose adult life is confined to a single year. Spawning takes place before the onset of the dry season; the adults then die, but the hardy eggs survive, hatching with the arrival of the rains.

annular In the shape of a ring or series of rings.

annulate Having ring-shaped markings.

annulate lamellae Flat, membranous *cisternae, bearing regularly spaced pores, that are derived from the *nuclear envelope and apparently represent an intermediate stage between this and the *endoplasmic reticulum.

annulus One of a series of concentric rings or bands of varying width and opacity which are formed in the scales of bony fish. Winter rings are often narrower and denser than summer rings. The number of annuli is indicative of the age of a fish.

Anobiidae (furniture beetles; class *Insecta, order *Coleoptera) Family of small,

a

*pubescent, brown beetles, 2–7 mm long, in which the head is hooded by the *pronotum. The antennae have the last three segments elongate. Larvae are white, fleshy, and C-shaped; they have tiny legs. They bore into the wood and bark of dead trees. Larvae of *Anobium punctatum* (woodworm) attack furniture and structural timbers; distinctive holes are caused by the emergence of adults. *Xestobium rufovillosum* (death-watch beetle) bores into structural timbers. Adults make a 'ticking' noise, by tapping the head against wood as a signal to the mate; this is sometimes heard by those keeping vigil with the dying and so is thought to presage death. There are 1100 species.

Anobium punctatum (woodworm) See ANOBIIDAE.

anoestrus In female *Mammalia, the period between *oestrus cycles, during which the sexual organs are quiescent and breeding does not occur.

anogenital Applied to the region of the body containing the *anus and genitalia.

anointing A behaviour in which an animal covers itself in a chemical substance it acquires from its environment. The behaviour is best known in hedgehogs (*Erinaceoidea), which chew toxic substances to make a froth with which they anoint their spines, but it also occurs in other species, e.g. California ground squirrels (*Spermophilus beecheyi*) and rock squirrels (*S. variegates*), which chew shed rattlesnake skins, then lick their fur. Predator deterrence is the likely purpose of the behaviour.

anoles (*Anolis*) See IGUANIDAE.

Anolis (anoles) See IGUANIDAE.

Anomalodesmata (phylum *Mollusca, class *Bivalvia) A subclass of equivalve, or nearly equivalve, bivalves most of which are relatively elongate. They have an aragonitic shell and many genera have a *nacreous *endostracum. If present, the ligament is *opisthodetic, and the musculature is nearly *isomyarian. There is a thickened hinge margin, and dentition is *desmodont or *edentulous. Living members have *eulamellibranchiate, or occasionally septibranchiate, gills. The subclass first appeared in the Middle *Ordovician. There is only one order, the Pholadomyoida.

Anomalopidae (lantern-eye fish; subclass *Actinopterygii, order *Beryciformes) A family of primarily warm-water, Indo-Pacific fish that have a cream-coloured light organ beneath each eye. The fish can turn their 'lanterns' off at will by masking the light organ with a piece of black skin. *Anomalops katopteron* can 'flick' its light at the rate of about four to five times a minute.

Anomaluridae (scaly-tailed squirrels; order *Rodentia, suborder *Sciuromorpha) A family of arboreal, squirrel-like rodents, usually brightly coloured, and in all genera but *Zenkerella* possessing a *patagium supported laterally by an *olecranon cartilage, used in gliding. The tail is bushy, with a double row of raised scales on the under side, and in some species it is long. The limbs are *pentadactyl, the claws strong. The skull is high, the palate narrow, the eyes and ears large. The cheek teeth are rooted, *brachydont with four or five cusps. There are about 12 species in four genera: *Anomalurus, Anomalurops, Idiurus*, and *Zenkerella*. They are found in W. and Central Africa.

Anopheles See CULICIDAE.

Anopla (phylum *Nemertini) A class of nemertinid worms in which the mouth is posterior to the brain. A simple proboscis is present. The class contains two orders, *Palaeonemertini and *Heteronemertini.

Anoplopomatidae (sable fish; subclass *Actinopterygii, order *Scorpaeniformes) A family, comprising only two species, of marine fish with a streamlined body and two *dorsal fins. They are found in the N. Pacific region.

Anoplura (sucking lice; class *Insecta, order *Phthiraptera) Suborder of lice characterized by piercing mouth-parts (made up of three *stylets), and by the fusion of all three parts of the thorax. All species are obligate parasites of placental mammals and feed on blood, this being digested with the aid of symbiotic bacteria. Some species are vectors of disease in humans and other animals. There are 15 families, 43 genera, and about 500 species known. Approximately 840 species of mammal are known to harbour Anoplura, and more are expected to be discovered.

Anoptichthys jordani (blind cavefish) The former name for *Astyanax fasciatus* (or *A. mexicanus*). See CHARACIDAE.

Anostomidae (head stander; subclass *Actinopterygii, order *Cypriniformes) Small family of freshwater fish very similar to the *Chilodontidae. They often swim head-down. They occur in S. America. They first appeared in the *Palaeogene and there are more than 140 species.

Anostraca (fairy shrimps, brine shrimps; subphylum *Crustacea, class *Branchiopoda) Order of mainly freshwater branchiopods lacking a carapace, and with an elongate body comprising 20 or more segments, with limbs borne by segments 11–19. The brine shrimps (*Artemia* species) are found in salt lakes throughout the world and tolerate very high salinities. There are 175 species.

Anser (geese) See ANATIDAE.

Anseranas semipalmatis (magpie goose) See ANATIDAE.

Anseriformes (screamers, waterfowl; class *Aves) An order of birds in which the palate is *desmognathous. They lay unspotted eggs which hatch to produce *nidifugous, downy young. There are two families, *Anhimidae and *Anatidae.

antagonistic resources Two or more resources that can substitute for one another, but when taken together some partially offset the effects of others. The consumer requires more of the resources when they are taken together than when they are taken separately. *Compare* COMPLEMENTARY RESOURCES.

ant bear (aardvark; *Orycteropus afer*) See ORYCTEROPODIDAE.

antbirds See FORMICARIIDAE.

ant-eaters See MYRMECOPHAGOIDEA.

antebrachium The lower section of the vertebrate forelimb; in humans the forearm.

Antechinus (antechinus; infraclass *Metatheria, order *Marsupialia) A genus of mouse-size marsupials, comprising about eight species, typical of the eucalypt woodlands and rain forests of eastern Australia. They are voracious predators. They and a very few other related genera exhibit *semelparity.

antelope See BOVIDAE.

antenna One of a pair of sensory structures which grow from the head of an invertebrate animal. Antennae may be long and filiform as in cockroaches, feathery as in male moths, or club-shaped as in some *Diptera. The antenna is richly supplied with nerves and is covered with a battery of sense organs, including various types of mechanoreceptors and chemoreceptors. The antenna is a dead-end space, but haemolymph is circulated through it by means of a pump or heart.

antennal gland A sac opening into the base of an *antenna, found in *Crustacea.

Antennariidae (anglerfish, frogfish; subclass *Actinopterygii, order *Lophiiformes) A family of balloon-shaped, relatively short-bodied fish, with the naked and loose skin often possessing wart-like processes. The first of the three separate anterior dorsal spines is usually modified into a movable rod, the 'fishing pole' often provided with a fleshy lure used to attract prospective prey. The family comprises about 60 species of marine frogfish found in all tropical and subtropical seas. *See also* BRACHIONICHTHYIDAE.

antennate In insects, to communicate information by touching *antennae. *See* DANCE LANGUAGE.

antennule A small *antenna; in *Crustacea, one of the first *uniramous pair of antennae.

antenodal In *Odonata, a cross-vein between the wing base and nodus, which runs between vein C (costal vein) and vein R1 (median vein). The number of antenodals is an important diagnostic character.

anteriad Pointing forward.

anterior chamber *See* AQUEOUS HUMOUR.

Anthidium *See* MEGACHILIDAE.

anthocodium In colonial coelenterates, the exposed part of the feeding *polyp.

Anthocopa *See* MASON BEE.

Anthocyathea (Irregulares; phylum *Archaeocyatha) A class of solitary, rarely colonial organisms found in Early, Middle, and Late *Cambrian rocks. The conical cup is from cylindrical to discoid in outer form, often with an irregular outline, having one, or more usually two, porous walls. The *intervallum contains rods and bars or *septa, always with *dissepiments and commonly with *tabulae.

Anthomedusae *See* GYMNOBLASTINA.

Anthomyiidae (order *Diptera, suborder *Cyclorrapha) Moderately sized family of small to tiny flies with slender bodies, whose larvae are phytophagous, feeding in plant shoots or inside galls. Some genera, e.g. *Erioischia* and *Hylemyia*, are economically important as pests of cabbage and wheat respectively. More than 1000 species have been described.

Anthonomus grandis **(cotton boll weevil)** *See* CURCULIONIDAE.

Anthophoridae (order *Hymenoptera, suborder *Apocrita) Large, diverse, cosmopolitan family of mainly solitary bees, with a few species which show some degree of sociality. Most species are long-tongued and have a rapid, darting flight. A pygidial plate (*see* PYGIDIUM) is present in the females of almost all species, and in most males. The *clypeus is usually protuberant, and the anterior *coxa is only slightly broader than long. The pollen *scopa consists of hairs and is restricted to the hind *tibiae and *basitarsi. The family includes some of the largest bees. With the exception of the carpenter bees (*Xylocopa*, *Ceratina*, and related genera), which bore into solid wood or plant pith, all anthophorids are ground nesters. They line their brood cells with a water-proofing secretion of the *Dufour's gland. One subfamily, the five genera comprising the Nomadinae, consists entirely of parasitic or cuckoo bees, the females of which have lost the pollen scopa and lay their eggs in the nests of other bee species.

Anthozoa (Actinozoa; sea anemones, corals, sea pens; phylum *Cnidaria) A class of exclusively *polypoid, marine cnidarians that are solitary or colonial, and usually sedentary. The oral end is expanded as an oral disc with a central mouth, with one or more rings of hollow *tentacles. A well-developed *stomodaeum leads from the mouth to a gastrovascular cavity, generally with one or two vertical grooves. The gastrovascular cavity is partitioned into compartments by complete or incomplete *mesenteries, some bearing endodermal gonads. Anthozoans first appeared in the *Ordovician.

Anthracosauria A group of early (*Carboniferous and Early *Permian) reptiles that were transitional from the *Labyrinthodontia.

Anthracotheriidae (superorder *Paraxonia, order *Artiodactyla) A family of large, primitive artiodactyls that were especially common in the middle *Tertiary in the Old World. They were pig-like in general appearance. From the nature of the sediments in which they are fossilized, it seems that they were amphibious. This supports the view based on anatomical evidence that they were the ancestors of the hippopotami. They arose in the *Eocene and became extinct in the *Pleistocene.

anthrax A disease of herbivorous mammals that occasionally infects other mammals and some birds. Humans can contract it through contact with infected animals or exposure to infected or contaminated animal products. In humans anthrax takes three forms: cutaneous, gastrointestinal-tract, and pulmonary, acquired by infection through a cut or abrasion, ingestion, and inhalation respectively. The disease is caused by the bacterium *Bacillus anthracis*.

Anthrenus (carpet beetle, museum beetle, woolly bear) *See* DERMESTIDAE.

Anthreptes (sunbirds) *See* NECTARINIIDAE.

Anthribidae (class *Insecta, order *Coleoptera) Family of robust beetles, 2–15 mm long, with a short, flattened *rostrum. The antennae are short and clubbed, or filiform and very elongate, but without *scape. Larvae are curved and fleshy, with legs and antennae absent or reduced, and feed inside fungi, dead wood, and plant galls. *Brachytarsus* larvae feed parasitically inside scale insects. There are 2400 species, mostly tropical.

Anthropogene *See* QUATERNARY.

anthropoid A member of the *Simiiformes.

Anthropoidea *See* SIMIIFORMES.

anthropomorphism The attribution of human characteristics to non-human animals, most commonly by supposing non-human behaviour to be motivated by a human emotion that might motivate superficially similar human behaviour.

Antiarchi (class *Placodermi) Group or order of fossil, *Devonian Period, fish, possibly related to sharks. The trunk shield, which was larger than the head shield, possessed a pair of movable, spine-like *pectoral fins.

The fossils may reach a length of about 30 cm, and show a subterminal mouth as well as a *heterocercal tail.

antibody A complex *protein that is produced in response to the introduction of a specific *antigen (which is normally foreign to it) into an animal. Antibodies are usually highly specific, combining only with antigens of a particular kind. Antibodies belong to a class of proteins called immunoglobins, which are formed by plasma cells in the blood as a defence mechanism against invasion by parasites, notably bacteria and viruses, either by killing them or rendering them harmless. The specificity of their binding reaction with a particular antigen is due to the configuration of a particular small area, known as the active site, on the surface of the antigens. Thus when a parasite (or its poisonous products) enters the tissues of its host, the antigens deriving from the parasite each produce a particular response according to the specific antibody that binds to that antigen. This 'recognition' by the host of the species or strain of parasite which has entered is sometimes applicable to other parasites should they share the same antigen. For example, the vaccinia and smallpox viruses share the same antigen so that immunity to one confers immunity to the other. Antibodies may persist in the body long after the disappearance of the antigen, so conferring immunity to any new infection by the same strain or species of parasite. Vaccination or inoculation provides immunity to an organism by the injection of particular foreign proteins (not necessarily from a parasite) which then stimulate the production of antibodies.

anticoagulant A substance that inhibits the clotting of blood.

anticodon A triplet sequence of *nucleotides in *transfer-RNA that during protein synthesis binds by base-pairing to a complementary sequence, the *codon, in *messenger-RNA attached to a *ribosome.

anticuckoldry tactics Behaviours by which a male seeks to ensure he is the father of offspring in whose care he must invest resources. For example, in langurs (*Cercopithecidae) and lions, a male that achieves dominance over a troop or pride will kill infants fathered by his predecessor. This removes the risk that he will protect and provide for young that are not his own, and induces sexual receptivity in the mothers of the infants he has killed. In general, the amount of parental care males provide is proportional to the certainty they have that they are the fathers.

antidiuretic hormone *See* VASOPRESSIN.

antienzyme A substance that retards or inhibits the action of an *enzyme.

antigen A molecule, normally a *protein, *glycoprotein, or *polysaccharide, usually found on the surface of a cell, which can interact with receptors found on lymphocytes or antibodies. *See also* IMMUNOGEN.

antihaemophilic globulin (AHG) *See* BLOOD CLOTTING.

Antilocapridae (infra-order *Pecora, superfamily *Bovoidea) A family of 'antelope-goats', which first appeared in the middle of the *Miocene. In general appearance, the earliest representatives probably resembled roe deer. The horns have two branches in bucks, and are usually simple in does but sometimes have a small prong. The horns have a hard bony core, and a soft, keratinous covering which is shed. The feet possess two digits. Antilocaprids graze or browse in herds. There is one surviving species, in one genus (*Antilocapra*, the pronghorn), found in western N. America.

antilocaprids *See* ANTILOCAPRIDAE.

anti-oxidant A compound, usually organic in nature, that prevents or retards the oxidation by molecular oxygen (autoxidation) of materials such as food, rubber, and plastics. It acts by scavenging the free radicals generated in autoxidation chain reactions, and thus provides an alternative oxidation pathway. It does not act indefinitely as it is destroyed in the process.

Antipatharia (black corals; phylum *Cnidaria, class *Anthozoa) An order of colonial, branching corals in which the *polyps are small, with few *tentacles (six simple, or eight branched) and few unpaired *mesenteries (6, 10, or 12). The skeleton is a branched, thorny, chitinous (*see* CHITIN), axial structure, developed from the *ectoderm, extending throughout the colony, and surrounded by a fleshy *coenosarc containing the polyps. Black corals are non-reef dwellers in depths of 2000–35 000 m. About 20 extant genera are known, one of which is also known from the *Miocene.

antiphilandery tactics In species that invest heavily in the care of young, and in which mothers depend on the fathers for support, behaviours by which females aim to discourage males from abandoning them. Among wolves, for example, females may drive potential female rivals from the pack or kill pups that may have been fathered by the fathers of their own pups. In some bird species, females spend time with their male partner prior to mating, which discourages the male from leaving.

antiport A *membrane transport system that carries different substances in opposite directions. *See also* COTRANSPORT.

antiserum An immune serum containing specific *antibodies that is prepared from the blood of a human or of another animal that has been immunized against *antigens from bacteria, viruses, or other parasites.

antitoxin An *antibody that neutralizes or inactivates a specific *toxin (forming an *antigen).

antitragus In *Mammalia, the lower posterior part of the outer ear, which lies opposite the *tragus.

antlers Bony growths borne on the heads of deer (*Cervidae). Antlers are shed each year at the end of the *rut and regrown the following year, producing more branches with each year. The antlers are covered with a highly vascular skin called velvet that is shed when their growth is complete, shortly before the rut. Only male deer grow antlers, except for reindeer (*Rangifer tarandus*), in which both sexes grow them.

ant lions *See* MYRMELIONTIDAE; NEUROPTERA.

ant-loving crickets *See* MYRMECOPHI-LIDAE.

antpittas *See* FORMICARIIDAE.

Antrodiaetidae *See* TRAPDOOR SPIDERS.

ants *See* APOCRITA; FORMICIDAE; HYMENOPTERA.

antshrikes *See* FORMICARIIDAE.

anucleate Lacking a nucleus. For example, erythrocytes (red blood cells) in mammals lack nuclei, although they are nucleated in birds.

Anura (Salientia; anurans, tail-less amphibians, frogs and toads; subphylum *Vertebrata, class *Amphibia) An order of amphibians in which the head joins directly into the trunk. The vertebral column is short and there is no tail in the adult. The fore limbs are stout and the long, webbed, hind limbs are used for jumping and swimming. The eyes are large and high on the head. There are superficial ear membranes posterior to the eyes. All anurans lay eggs, have external fertilization, and develop from tailed tadpoles. There is no *neoteny. *Sexual dimorphism is common, the males often being smaller and having vocal pouches. There are 2600 species, widespread in tropical and temperate zones.

anurans *See* ANURA.

anus In most *animals, but not in coelenterates or *Platyhelminthes, the exterior opening of the *alimentary canal through which are expelled the unabsorbed remainder of food items, solid excretion products, and bacteria (*see* BACTERIA) associated with both. In some aquatic invertebrates the posterior end of the alimentary canal is also used in respiration.

anvil *See* INCUS.

aorta In *Mammalia, the main artery conveying blood to the whole body.

aortic arches In fish, the arteries supplying the gills, passing from the ventral aorta and then uniting to form the dorsal aorta. In tetrapods they are modified and reduced in number.

Aotus (night monkey) *See* CEBIDAE.

Apachyidae (order *Dermaptera, family *Labiduridae) Subfamily (or in some classifications family) of large, winged, extremely flattened, and brightly coloured earwigs. Fewer than 12 species have been recorded, from Australia and south-east Asia, where they are usually found in rain forest.

Apatosaurus (order *Saurischia, suborder *Sauropoda) One of several gigantic dinosaurs recorded from the Late *Jurassic, *Apatosaurus* was a quadruped with a long neck whose total weight reached 30 tonnes. The name *Apatosaurus*, meaning 'deceptive lizard' (derived from the Greek *apate* meaning 'deceit' and *saura* meaning 'lizard'), is a senior synonym of *Brontosaurus*. Animals of 22 m in length have been recorded from the Morrison Formation of N. America.

ape A name originally (in medieval times) applied to the Barbary macaque (*Macaca sylvanus*) of N. Africa (as were the Latin *simia* and Greek *pithecus*) and, by extension, applied to other primates as these were made known in Europe. As long-tailed monkeys ('tailed apes', or cercopitheci) became better known, 'ape' came to mean primarily 'tail-less ape', and today commonly denotes a member of the *Hominoidea, comprising lesser apes (gibbons) and great apes (orangutan, gorilla, chimpanzee, and, in some usages, human).

apertural Applied to the end or area of a mollusc shell that contains the aperture out of which soft, internal body parts may emerge.

Aphaniptera An old name for the *Siphonaptera (fleas).

aphicide Natural or synthetic chemical substance that is toxic to aphids (*Aphididae).

Aphididae (aphids, greenfly, blackfly; order *Hemiptera, suborder *Homoptera) Family of soft-bodied insects that feed on plant sap using the *rostrum that arises between the front pair of legs. Typically, one generation a year reproduces sexually, laying overwintering eggs, usually on woody plants, and several generations reproduce asexually and viviparously, either on the woody host or on herbs. Many species are tended by ants which feed on the *honeydew. Some are serious pests. There are more than 4000 species, occurring mainly in the temperate northern hemisphere.

aphids See APHIDIDAE.

aphotic Of an environment, without light.

apical system At the centre of the *aboral surface of some *Echinodermata, a double ring of plates surrounding a hole (*periproct) that contains the *anus.

Apidae (bumble-bees, honey-bees, orchid bees, stingless bees; order *Hymenoptera, suborder *Apocrita) Family of bees, in which the pollen-transporting apparatus is on the outer face of the hind *tibia and comprises a *corbiculum. The hind tibial spurs are absent, except in the primitively *eusocial bumble-bees (genus *Bombus*). The stingless bees (genera *Melipona* and *Trigona*), and the four species of honeybees (*Apis cerana, A. dorsata, A. indica,* and *A. mellifera*) have achieved the highest levels of insect sociality, apart from those found among the termites and ants, with a morphologically distinct *worker caste.

Apionidae (weevils; class *Insecta, order *Coleoptera) Insects very closely related to *Curculionidae, but distinguished by having non-elbowed, clubbed antennae.

They have highly convex, punctured *elytra, 1.5–2.0 mm long; and the head is produced into a long, curved or stout *rostrum. The larvae are legless and grub-like, and feed inside seed pods, stems, or plant roots. There are 1060 species.

Apis See APIDAE; HIVE; HONEY-BEE; HONEYCOMB.

Apistogramma ramirezi (butterfly cichlid) See CICHLIDAE.

Aplacophora (Solenogastres; phylum *Mollusca, class *Amphineura) A subclass of *benthic, worm-like molluscs which lack a dorsal covering of valves. Calcareous *spicules may occur in the *mantle that covers the body. The anatomy and nervous system are very similar to those of the *Polyplacophora. The ventral foot is represented as a groove in the order Neomeniida, but is absent in the second order, Chaetodermatida. They are entirely marine, and have a world-wide distribution. This group does not occur in the fossil record.

aplanetism Condition in which no motile stage is formed.

aplanogamete A non-motile *gamete.

aplasia The failure of an organ to develop.

Aploactinidae (Haploactinidae; velvetfish; subclass *Actinopterygii, order *Scorpaeniformes) A family of marine, coastal fish of the western Pacific. The sharply tapering body has a long *dorsal fin originating on the head, and the *pelvic fins are reduced to a weak spine and two or three rays. The scaleless skin bears papillae, giving it a velvety appearance. There are 17 genera and about 40 species.

Aplochitonidae (Haplochitonidae; southern smelts; subclass *Actinopterygii, order *Salmoniformes) A family of freshwater and brackish-water fish that have an elongate body and a single *dorsal as well as an *adipose dorsal fin. There are two genera, *Aplochiton* occurring in S. America and *Lovettia* in Tasmania, and three species, *A. zebra*, *A. taeniatus*, and *L. sealii*. Some authorities rank the aplochitonids as a subfamily (Aplochitoninae) of the *Galaxiidae. *Prototroctes maraena* (Australian grayling) is sometimes included in the Aplochitonidae, but otherwise in the *Retropinnidae.

Aplodactylidae (sea carps; subclass *Actinopterygii, order *Perciformes) A small family of marine, herbivorous fish, somewhat blunt-nosed, with a single but deeply notched *dorsal fin. Sea carps occur in Australia, New Zealand, and S. America. There is one genus, *Aplodactylus*, with five species.

Aplodontia (mountain beaver) *See* APLODONTIDAE.

Aplodontidae (mountain beaver; order *Rodentia, suborder *Sciuromorpha) A family of burrowing rodents in which the body is thickset and heavy, the limbs and tail short, and the claws large. The eyes and ears are small. The skull is flat, wide posteriorly, and the palate broad. The animal is found only in the N. American Rockies. There is one genus (*Aplodontia*), and one species.

Aplonis (glossy starlings) *See* STURNIDAE.

Aplousobranchiata *See* ASCIDIACEA.

Aplysiacea *See* ANASPIDEA.

Aplysiomorpha *See* ANASPIDEA.

apnoea Nonventilation of the lungs.

apocrine gland A type of gland in which the apical part, from which the secretion is released, breaks down during the secretion process. The gland opens at the surface of the skin near, but not within, hair follicles. From the opening a long tubular *invagination extends into the dermis (*see* INTEGUMENT). Odourless secretions produced by the gland are converted to odorous products by surface bacteria (*see* BACTERIA). The scent and musk glands of *Mammalia are

thought to be modified apocrine glands. *Mammary glands and *eccrine glands also resemble apocrine glands in structure, development and mode of secretion. *Compare* HOLOCRINE GLAND.

Apocrita (ants, bees, wasps; subclass *Pterygota, order *Hymenoptera) The larger of the two suborders of Hymenoptera, whose members have a constriction between the *thorax and *abdomen, occurring between the narrowed first abdominal segment (which is attached to the thorax) and the rest of the abdomen. Wing *venation in some minute forms is greatly reduced, and the *ovipositor can be of two types: a non-stinging type normally kept outside the abdomen, and a stinging type which is retractable. The Apocrita comprises 11 superfamilies, with about 105 000 species, many of which are of benefit to humans. Bees, gall wasps, and some *Chalcidoidea have herbivorous larvae, but the majority of Apocrita are parasites of other insects and spiders.

Apoda (Gymnophiona, caecilians; class *Amphibia) An extant order of amphibians, now confined to the tropics and subtropics, that have become adapted to a burrowing life and look like large earthworms. They are the only living amphibians some species of which have retained scales, although the scales are vestigial and hidden in folds of the skin. They have no limbs or limb girdles, eyes that are rudimentary, vestigial hearing apparatus, a sensory feeler beneath the eye, no gills or gill slits in the adults, and a tail that is small or absent. Fertilization is internal, as a cloacal copulatory organ is developed in the male. There are three families: *Caecilidae; *Ichthyophiidae; and *Typhlonectidae.

Apodacea (subphylum *Echinozoa, class *Holothuroidea) A subclass of sea cucumbers that have simple or digitate *tentacles. The *tube feet are reduced or absent, there are no pharyngeal retractor muscles, and the *test is vestigial. Apodaceans first appeared in the *Carboniferous.

apodeme Piece of inflected *cuticle that forms the attachment for muscles in the body of an insect. All muscles attach to

a

elements of the *exoskeleton, which are frequently inflected, either to act as tendons, or, in effect, as an *endoskeleton.

Apodemus sylvaticus (field mouse, wood mouse) *See* MURIDAE; SPERM TRAIN.

Apodes Alternative name for the group or order of eels and related fish (*Anguilliformes).

Apodida (class *Holothuroidea, subclass *Apodacea) An order of tropical reef-dwelling holothuroids which lack *respiratory trees, gas exchange occurring through the general body surface. *Tube feet and *radial canals are also absent.

Apodidae (swifts, swiftlets; class *Aves, order *Apodiformes) A family of medium to small birds that have small, short bills and a wide gape. They have long, curved, narrow wings, a short tail, often square-ended or spine-tipped, some with white rumps or bellies, and short, weak legs with reversible *hallux. They are highly aerial, fast-flying, and insectivorous (the 21 species of *Collocalia* (swiftlets), of south-east Asia, Australia, and the islands of the Indian and Pacific Oceans, use *echo-location to fly in darkness). Their nests are built on trees, buildings, rocks, and in caves, often using saliva. *Apus horus* (horus swift) nests in sand burrows. Many authorities do not recognize the genera *Hirundapus*, *Raphidura*, *Telacanthura*, and *Zoonavena*, including them within the genus *Chaetura*, eight species of small, bluish-black or brown swifts, with paler throats or rumps, and spine-tipped tails, found in America. There are 12 or more genera in the family, containing about 80 species. Nearly all are tropical, northern-breeding birds migrating to the tropics in winter, but they are found world-wide. The 15 *Apus* species are found mainly in Africa, but also in Europe, Asia, and Australia.

Apodiformes (swifts, tree swifts, humming-birds; class *Aves) An order of birds all of which have a short, thick *humerus, a short *ulna, and short secondary feathers. They are high-speed fliers. The order comprises three families. *Trochilidae

(humming-birds) are placed in a different suborder from *Apodidae (swifts) and *Hemiprocnidae (tree swifts).

apodous Without legs.

apoenzyme The portion of a conjugated enzyme that is a *protein. *See* CONJUGATED PROTEIN.

Apogonidae (cardinal fish; subclass *Actinopterygii, order *Perciformes) A large family of mainly marine fish, none more than 20 cm long, that have two separate *dorsal fins and relatively large eyes and mouth. Most cardinal fish display the unusual habit of *buccal incubation. In some species the males are the mouth-brooders, in others the female performs this duty. There are about 170 species. They are shallow-water fish of all temperate and tropical seas.

Apollo butterflies *See* PAPILIONIDAE.

apomixis A form of reproduction that superficially appears to be sexual but actually occurs without the fusion of a male and female *gamete.

apomorphic Applied to an evolutionarily advanced character state. Apomorphic is the opposite of *plesiomorphic. The long neck of the giraffe is apomorphic; the short neck of its ancestor is plesiomorphic. Apomorphic features are those possessed by a group of biological organisms that distinguish those organisms from others descended from the same ancestor. The term is taken from the Greek *apo*, 'from' or 'away', and *morphe*, 'form', and means 'new-featured'. It refers to 'derived' characters which have appeared during the course of evolution. Apomorphic features may be *autapomorphic (uniquely derived) or *synapomorphic (shared-derived).

apophysis 1. In vertebrates, a projection from a bone, usually for muscle attachment. Vertebrae have pairs of apophyses. **2.** In echinoids, an internal projection from one of the interambulacral plates (*see* INTERAMBULACRUM) around the *peristome; apophyses serve for the attachment of the muscles supporting the *Aristotle's lantern.

apoptosis Programmed cell death, which occurs as a necessary component of growth and development. The process results in cell shrinkage, followed by the uptake of dying cells by *macrophages, without the possibly harmful release of cellular contents. Apoptosis allows control of cell growth and removal of possibly harmful cells. Breakdown of the system can result in uncontrolled cell growth, giving rise to tumours.

apopyle In asconoid (*see* ASCON) *Porifera, a pore through which water passes from a *choanocyte into an adjacent canal.

aposematic coloration (**aposematism**) Warning coloration in which conspicuous markings on an animal serve to discourage potential predators. Usually the animal is poisonous (e.g. has a venomous bite or sting) or unpalatable. Examples include the conspicuous bands or stripes on the back and sides of venomous snakes (e.g. coral snakes), and contrasting markings on the wings of distasteful monarch butterflies. *See also* PAPER WASPS.

aposematism *See* APOSEMATIC COLORATION.

apostatic selection A type of selection which operates on a polymorphic species. Classically, the term is used in relation to prey species that have several different morphological forms. It occurs when, in proportion to their frequency in the population, rare forms of a species are preyed on less than common forms, thus conferring a selective advantage on them. Such selection may produce a stable genetic *polymorphism.

Apostomatida (**class *Ciliatea, subclass *Holotrichia**) An order of ciliate *Protozoa in which mature forms have spirally arranged body *cilia. The organisms live as parasites or commensals and often have two hosts, one of which is commonly a marine crustacean (e.g. *Foettinga* lives inside the gastrovascular cavity of sea anemones; mo-

tile young leave the sea anemone and encyst on the outside of small crustaceans; when the crustaceans are eaten by sea anemones the young emerge from their cysts).

appeasement Behaviour by an animal that serves to reduce *aggression shown towards it by another member of the same species, but that does not involve *avoidance or escape.

appeasement gland An abdominal gland possessed by adult *Atemeles pubicollis* rove beetles (*Staphylinidae). When the beetle enters the nest of *Formica polyctema* ants and the ants approach the invader, the beetle offers its abdomen, which exudes a *pheromone that inhibits hostile behaviour.

appendicular skeleton The part of the vertebrate *endoskeleton that is composed of paired *fin or limb bones, and the pelvis. Fins of vertebrate fish have a basic morphology composed of three elements: a small number of fan-like basal elements support a greater number of cylindrical radials, which in turn support a large number of *fin rays. These may be soft and flexible (soft rays) or hard and inflexible (fin spines), supporting the fin web. The limbs of tetrapods articulate with a pectoral (*scapula) or pelvic (*pelvis) girdle and consist of five segments: the propodium (*humerus or *femur); epipodium (*ulna/*radius or *fibula/tibula); mesopodium (*carpus or *tarsus); metapodium (*metacarpus or *metatarsus); and *phalanges. Although pectoral and pelvic fin and limb appendages have an apparently homologous structure, they are derived serially from different body segments. *Compare* AXIAL SKELETON. *See also* PENTADACTYL.

appendix In some *Mammalia, a vermiform, cul-de-sac termination to the *caecum, located close to the junction of the large and small intestines. It contains a concentration of lymphoid tissue. In herbivorous animals (e.g. *Lagomorpha) whose large intestine is involved in the digestion of cellulose, the caecum and appendix are large. In most *Primates it is absent and the caecum ends bluntly. In *Hominoidea, the caecum is very small, but the appendix is large.

appendix dorsalis *See* ARCHAEO-GNATHA.

appetite A complex phenomenon, not to be confused with hunger, that in humans is the comparatively pleasant, though at times compelling, anticipation of certain foods. In other animals, appetite is studied by observing feeding behaviour and its relation to specific feeding stimuli, with no assumption of subjective experiences.

appetitive A general and rather imprecise adjective that is applied to the behaviour exhibited by an animal that is exploring its environment or seeking a goal. *See* CONSUMMATORY.

aptation A character that suits its possessor to its *environment; it may be an *abaptation, *adaptation, or *exaptation.

apterium An area of bare skin between feathered areas on a bird.

Apteronotidae (subclass *Actinopterygii, order *Cypriniformes) A family of freshwater fish that have a body tapering down to a narrow tail with a tiny tail fin. Related to the electric eels, the apteronotids can also produce weak electric charges. They occur in S. America.

apterous Literally, without wings (from the Greek *a*, not, and *pteron*, wing), and applied to insect species that are primitively wingless, or secondarily so due to a parasitic lifestyle.

Apterygidae (kiwis; class *Aves, order *Apterygiformes) A family of fowl-sized, brown-grey *ratites. They have short, muscular legs, and a long flexible bill with nostrils at the tip. Their feathers are loose and hair-like. They have vestigial wings and no tail. They are small-eyed and mainly nocturnal, inhabiting thick forests and nesting in burrows. They feed on grubs and worms. The family comprises one genus (*Apteryx*)

containing three species, all of them endemic to New Zealand.

Apterygiformes (kiwis; class *Aves) An order that comprises one family, *Apterygidae.

Apterygota (Ametabola; subphylum *Uniramia, class *Insecta) The smaller subclass of insects, containing two orders (*Archaeognatha and *Thysanura), whose members are primitively wingless (in contrast to secondarily wingless insects, e.g. fleas, lice, etc.), terrestrial insects with *ectognathous, biting mouth-parts. The subclass formerly included the *Protura, *Collembola, and *Diplura, which are now regarded as distinct subclasses. Most Apterygota are free-living; elongate or oval; 3–15 mm long; and with direct larval development, moulting occurring throughout the life of the animal. They feed on fungi, lichens, pollen, etc., domestic species subsisting on a diet of *carbohydrates, dextrins, and a small amount of protein from glue and sizes. Approximately 580 species occur world-wide, and are found almost everywhere, even at high altitudes and in polar regions.

Apteryx (kiwis) *See* APTERYGIDAE.

Aptian One of the stages of the Early *Cretaceous Period in Europe, followed by the *Albian, and dated at 125–112 Ma. It is named after Apt, in France.

Apulicida *See* SIPHONAPTERA.

Apus (swifts) *See* APODIDAE.

Apus horus (horus swift) *See* APODIDAE.

aqueous humour A transparent liquid that fills the anterior and posterior chambers of the vertebrate *eye and the anterior chamber of the eye in *Cephalopoda. The anterior chamber lies between the *iris and *cornea; the posterior chamber between the iris and lens. Aqueous humour is produced constantly and drains via the posterior and anterior chambers to the eye surface. *Compare* VITREOUS HUMOUR.

***Aquila* (eagles)** *See* ACCIPITRIDAE.

Aquitanian Defined first from the Aquitanian Basin, France, the Aquitanian Age marks the beginning of the *Miocene, 23.03 Ma ago; it ended 20.43 Ma ago. It also marks the start of the upper *Cenozoic (*Neogene) Period of time. The Aquitanian is itself characterized by the appearance of the planktonic foraminiferid *Globigerinoides primordia*. *See* FORAMINIFERIDA; GLOBIGERINA; GLOBIGERINA OOZE.

***Ara* (macaws)** *See* PSITTACIDAE.

arabinose An *aldose that contains five carbon atoms.

Arachnida (Amblypygi, harvestmen, mites, Palpigradi, pseudoscorpions, scorpions, spiders, Uropygi, whipscorpions; phylum *Arthropoda, subphylum *Chelicerata) Class of chelicerates which have *book lungs or *tracheae derived from gills, indicating their aquatic derivation. They have invaded most terrestrial habitats and have secondarily invaded aquatic habitats, although to a very much smaller extent (there is only one species of aquatic spider, and only one species of mite in ten is aquatic). Except for many plant and animal parasites found among the mites, and some scavenging harvestmen (order Opiliones or Phalangida), most arachnids are predatory. Scorpions have been recorded from the *Silurian Period and a Silurian scorpion, *Palaeophonus nuncius*, was perhaps the first terrestrial animal. The first fossil spiders are known from the *Devonian. The class is extremely diverse, but apart from the mites the body is in two portions, the prosoma (anterior portion) bearing the four pairs of legs, the eyes, the *pedipalps, and the *chelicerae; and the opisthosoma (posterior portion) containing most of the internal organs and glands. The two portions may be broadly jointed, or connected by a *pedicel. The prosoma has a dorsal shield (carapace), and the opisthosoma is segmented in most orders, but not in spiders and mites and only very weakly in harvestmen. The number of eyes varies up to 12 in some scorpions, but generally vision is poor, many species being nocturnal and equipped with sensory hairs to detect prey. Pedipalps function as hands, and the chelicerae as jaws or teeth. In all arachnids the mouth is small, and food is generally predigested by enzymes from the mid-gut. Reproductive organs are placed on the ventral surface of the abdomen, and courtship may be complex and prolonged, with parental care of the young common to all. All arachnids are *dioecious. The production of silk and poison is characteristic of some orders, but the methods of production and their origins are varied. Silk is produced from abdominal glands in spiders, from the mouth region in mites, and from the chelicerae in pseudoscorpions. Poison is produced from the chelicerae of spiders, the tails of scorpions, and the pedipalps of pseudoscorpions. There are 11 orders, with more than 42 000 species described and an estimated 60 000–170 000 awaiting description. Members of five orders occur in northern Europe, the remainder being tropical in distribution.

arachnoid Resembling a spider's web.

arachnoid layer *See* MENINGES.

Aradidae (barkbugs; order *Hemiptera, suborder *Heteroptera) Family of bugs most of which live under the bark of dead trees and logs and feed on fungi that attack both broad-leaved and coniferous trees, although a few suck the sap of living trees. There are several hundred species, occurring from the lowland tropics to the tree-line.

aragonite A form of calcium carbonate ($CaCO_3$) from which the shells of many invertebrates (e.g. *Mollusca) are formed. It is less stable than calcite and in many fossil shells aragonite has been converted to calcite or been replaced by other minerals. The name is taken from the Aragon province of Spain.

Aramidae (limpkin; class *Aves, order *Gruiformes) A monotypic family (*Aramus guarauna*) which is a brown-olive bird, streaked with white, that is structurally crane-like and rail-like in appearance. The bill is long, laterally compressed, and *decurved. The bird has a long neck, broad wings, a short tail, and long legs with long, unwebbed toes. It feeds on snails, nests on the ground or in bushes, and is found in wooded swamps in the southern USA, and in Central and S. America.

Aramus guarauna (limpkin) *See* ARAMIDAE.

Araneae (spiders; subphylum *Chelicerata, class *Arachnida) Very large and diverse order of predatory arachnids most of which feed on arthropods, although the large *tarantulas (mygalomorphs) will catch nestling birds and other vertebrates, and some will catch tadpoles and small fish. The prosoma has a uniform dorsal carapace bearing two to eight eyes. The *chelicerae are large but have a fang (rather than a pincer) which bears the opening of a poison gland lying entirely within the chelicera in primitive species, or extending into the prosoma. The *pedipalps are leg-like and the complex structures of the male pedipalp are crucial in sperm transfer, the pedipalp introducing the sperm into the female epigyne with a pumping action. The *abdomen (opisthosoma) is usually unsegmented and attached to the prosoma by a narrow, stalk-like *pedicel. The abdomen is extremely manoeuvrable, bears the *spinnerets, and contains two pairs of respiratory organs (*book lungs and posterior *tracheae), the reproductive organs, and internal glands. Almost all species use their eight walking legs for running rapidly, and some jump by means of rapid rises in internal pressure. The legs have seven segments, the distal segment (pretarsus) having two comb-like claws and often a hook-like median claw for silk manipulation. Spider sensory apparatus varies from sensory *setae, trichobothria, and slit sense organs to chemical and tactile receptors on the *tarsi and pedipalpi. Vision is important and acute in the *Salticidae (jumping spiders). In many species prey items are caught by special webs, sheets, or snares of silk which may be stationary or thrown, and may have drops of sticky adhesive; while other species hunt on the ground, pouncing on their prey. Poison and digestive juices are injected and the digested contents are sucked out. All spiders are *dioecious, usually with smaller males, and exhibit elaborate courtship displays involving sexual signalling, nuptial gifts, and some risk to the males in some species. Most species enclose their eggs in a silken sac and guard the eggs and young. As well as using silk for prey capture and egg sacs, spiders may 'balloon', using long threads of silk to carry them as high as 1000 m or more, and may travel 100 km. Spiders are found in all terrestrial niches and there is one aquatic species, some shore dwellers, and others which hunt on the surface film of water, their tarsi being specially adapted with hydrophobic hairs. The more than 50 000 or so known species are found on all continents except Antarctica, and in all climates.

Araneidae (orb-web spiders; order *Araneae, suborder *Araneomorphae) Family of spiders which spin typical orb webs and whose legs are armed with many long spines. The *chelicerae possess many teeth, and many species of orb spinners have dwarf males. There are more than 2500 species, of which *Araneus* is the largest genus. *Araneus diadematus* is the garden-cross spider or garden spider.

Araneus diadematus *See* ARANEIDAE.

Arapaima gigas *See* OSTEOGLOSSIDAE.

Arbacioida (class *Echinoidea, superorder *Echinacea) An order of *regular echinoids in which the imperforate, non-*crenulate primary *tubercles are poorly developed. Adventitious skeletal material on the *test surface, simulating tubercles, is commonly developed, but does not bear spines. Arbacioids first appeared in the Middle *Jurassic.

arborization The formation of a branched or tree-like form in a *dendrite.

Arborophila (tree partridges) *See* PHASIANIDAE.

arbovirus A (non-taxonomic) term applied to a virus that can replicate both in vertebrates and in arthropod vectors.

Arcellinida (Testacida; superclass *Sarcodina, class *Rhizopodea) An order of shelled (testate) *amoebae. The *test has an aperture through which the *pseudopodium emerges. The organisms are found primarily in freshwater habitats.

Archaea (Archaebacteria) 1. In the widely used five-*kingdom system of classification, one of the two subkingdoms within the kingdom *Bacteria (*see also* EUBACTERIA). **2.** In the three-*domain system of classification, one of the domains, comprising organisms formerly known as archaebacteria and placed in a kingdom of that name. The former kingdom Archaebacteria has been split into two kingdoms: Crenarchaeota and Euryarchaeota.

(((⊕))) SEE WEB LINKS
• An introduction to Archaea.

Archaean The eon, formerly known as the Archaeozoic or Azoic, during which life first appeared. It lasted from approximately 3800 to 2500 Ma ago and comprises part of the former *Precambrian.

Archaebacteria *See* ARCHAEA.

Archaeoceti (ancient whales; cohort *Mutica, order *Cetacea) An extinct suborder comprising the oldest and most primitive cetaceans, which flourished in the *Eocene and may have originated in Africa. Most were comparable in size with modern porpoises, had an elongated snout, and nostrils on top of the skull. The brain case was long and low. The front teeth were peg-like, the cheek teeth *heterodont and characteristic of primitive carnivores. There were 44 teeth in all. The hind legs in most were reduced to vestiges, but in some early genera (*Ambulocetus, *Basilosaurus) still protruded from the body wall. They were fish-eating carnivores that had adopted an aquatic life to which they were more highly adapted than e.g. modern seals. The term archaeocete really means any primitive

cetacean and probably does not designate a natural monophyletic (*see* MONOPHYLY) group.

(((⊕))) SEE WEB LINKS
• Description of Cetacean palaeobiology

Archaeocyatha An extinct phylum of reef-forming organisms known only from the *Cambrian. They were cup-like, usually 10–30 mm in diameter and up to 50 mm high. The cylindrical, conical, or discoid cup often had an irregular outline or outgrowths. The outer wall had simple pores or was partially displaced by *dissepiments. The inner wall usually had one longitudinal row of pores between intercepts. The *intervallum contains dissepiments, rods, and bars or *septa. Some were solitary, others colonial. In some respects they were similar to both sponges and corals, and may represent a true advance in evolutionary grade over the former. It is possible that they lived in a symbiotic relationship with some trilobites. The cause of their extinction is not known.

Archaeogastropoda (class *Gastropoda, subclass *Prosobranchia) An order of gastropods that first appeared in the Early *Cambrian and that includes the extant *Patella vulgata* (limpet). Gastropods may be subdivided according to their respiratory structures; archaeogastropods, the most primitive members of the class, have just two gills, one of which may be reduced or absent, and some forms have a marginal slit near the aperture through which water and wastes are removed.

Archaeognatha (class *Insecta, subclass *Apterygota) One of the two apterygote orders, comprising fusiform, subcylindrical insects that can run fast and jump by flexion of the *abdomen. The body is covered with pigmented scales, and the *thorax is strongly arched. *Ocelli are present, and the *compound eyes are large and contiguous. The *mandibles are *monocondylar, and the *maxillary palps are long and seven-segmented. The abdomen has ventral *styles on segments 2–9, used for running on sloping ground, and the posterior portion of the abdomen bears a pair of *cerci and a single, much longer, dorsal appendage (the appendix dorsalis). These insects are free-living and nocturnal, feeding on lichen, algae, and vegetable debris. Postembryonic development is slow (six months to two years), and moulting occurs throughout the entire life-span. The 250 or so species are of little economic importance.

Archaeopteryx lithographica The first known bird, of which only six specimens and one feather imprint have been found, all of them from the Lithographic Limestone of the Solnhofen region of Bavaria, Germany. The bird was first described by H. von Meyer in 1861 and is of Middle *Kimmeridgian (Late *Jurassic) age. Work by several palaeontologists tends to support the theory that the birds, through *Archaeopteryx*, evolved from coelurosaur dinosaurs (*Coelurosauria) similar to *Compsognathus*. The species *A. lithographica* possesses several primitive characters such as teeth, as well as specialized features such as feathers and hollow bones. It is a good example of a connecting species which exhibits a mosaic of evolutionary features.

Archaeozoic *See* ARCHAEAN.

archenteron A primitive digestive cavity of the *embryo at the *gastrula stage of development in *animals. It is formed by *invagination of *mesoderm and *endoderm cells, opens to the outside by a *blastopore, and finally develops into the gut cavity.

archerfish *See* TOXOTIDAE.

Archiannelida (phylum *Annelida) A class of small, simple, annelid worms which are metamerically segmented (*see* META-MERIC SEGMENTATION) and most of which have narrow, tube-like bodies. In some, the nervous system is primitive. *Parapodia and *tentacles may be present. Separate male and female forms generally occur. The group was formerly regarded as a distinct phylum; most authorities now suggest it represents an assortment of largely unrelated annelids, most of which are marine and members of the *interstitial fauna of sand grains. There are three families, all found in aquatic habitats.

Archilochus colubris **(ruby-throated humming-bird)** *See* TROCHILIDAE.

archipterygium In some fish, a leaf-shaped, narrow-based, paired fin.

Archonta A group of placental mammals (*Eutheria) that is sometimes recognized. It contains the orders *Chiroptera, *Dermoptera, *Primates, and *Scandentia.

Archoophora (phylum *Platyhelminthes, class *Turbellaria) A subclass of entirely marine worms, which comprises the two orders *Acoela and *Polycladida.

Archosauria (class *Reptilia) A subclass of *diapsid reptiles that includes the orders *Crocodilia, *Ornithischia, *Saurischia, *Pterosauria, and *Thecodontia. Thecodontians were ancestral to the other groups and appeared at the base of the *Triassic. Archosaur is derived from the Greek *arkho*, to rule, and *saura*, lizard.

Arcoida (phylum *Mollusca, class *Bivalvia) An order of sedentary, *epifaunal bivalves in which the adults are *byssate or free-living. Typically they are *isomyarian, *taxodont, and equivalved. The shells are normally trapezoid and composed of crossed-lamellar calcium carbonate. They have *filibranchiate gills and flat *cardinal areas on the dorsal margin. They first appeared in the Early *Ordovician.

arctic fox (*Alopex lagopus***)** *See* CANIDAE, RETE MIRABILE.

Arctiidae (tiger moths, ermines, footmen, tussocks, woolly bears; subclass *Pterygota, order *Lepidoptera) Family of small to medium-sized moths, many of which are brightly coloured, and which are often distasteful to predators. The larvae are often hairy (and so called 'woolly bears'), and hairs are used to construct the cocoon. There are about 10 000 species with a worldwide distribution.

Arctocyonidae (superorder *Protoungulata, order *Condylarthra) An extinct family of unspecialized primitive mammals (*Eutheria), with more than 20 genera, which lived from the *Cretaceous to the *Eocene but were most abundant in the *Palaeocene. Some were the size of modern bears and were probably omnivorous; others were smaller and herbivorous. The body was slim, the limbs slender, the feet clawed. The brain case was small, and the teeth primitive, with moderately sharp cusps. *Protoungulatum* of the N. American Late Cretaceous is the oldest known ungulate.

Arctogea A traditional *zoogeographical region that comprises the northern continents, Africa, and Indo-China. It is usually subdivided into the *Palaearctic and *Nearctic (together Holarctica), *Ethiopian, and *Oriental regions. *Compare* NEOGEA; NOTOGEA. *See* FAUNAL REGION.

arcualia In the development of jawed vertebrates (*Gnathostomata), the paired *cartilages that fuse to form the centrum of a *vertebra.

arcuate Curved or arched.

***Ardea* (herons)** *See* ARDEIDAE.

Ardeidae (bitterns, egrets, herons; class *Aves, order *Ciconiiformes) A family of long-legged, long-necked, wading birds that have spear-like bills and long toes, the middle toe being *pectinate. Many have long, plumed feathers and all have powder-down patches (many egrets (13 species of *Egretta*) occur in colour phases varying from white to black). They take a variety of animal food, especially fish. Night herons (three species of *Nycticorax*) are nocturnal. Ardeids nest in trees or reed-beds. There are 17 genera in the family, with 61 species, and they are found in marshy areas and on shores, with cosmopolitan distribution. The 11 species of typical herons comprise the genus *Ardea*. The four species of bitterns, comprising the genus *Botaurus*, nest on platforms of reeds and have a loud, booming call. *Cochlearius cochlearius* (boat-billed heron) is often placed in its own family, *Cochleariidae.

Ardipithecus ramidus The earliest known member of the human lineage, discovered in 1993 by Tim White, Gen Suwa, and Berhane Asfaw at Aramis, Ethiopia, and dated to 4.4 million years bp. The *canine teeth are somewhat reduced from the primitive ape-like condition, but not so much as in *Australopithecus* (*australopithecines); the *enamel on the teeth is thin; the *deciduous *molars are intermediate between those of a human and a chimpanzee. The postcranial skeleton indicates that it was, at least to some degree, bipedal.

area-effect speciation *Speciation that is associated with an increased differentiation of two *subspecies that have incompatible gene complexes, so that *hybrids are strongly selected against.

area-restricted search A foraging pattern in which a consumer responds to an intake of food by slowing down its movement and remaining longer in the vicinity of the most recently located food item. This behaviour causes consumers to remain longer in areas where the density of food items is high than in areas where it is low.

arenavirus A genus of *RNA viruses (family Arenaviridae), most of which are maintained in a particular rodent species (*Rodentia). Fifteen species infect non-humans and five cause a range of diseases in humans, including haemorrhagic fevers.

areolar tissue Loose, flexile *connective tissue, found throughout the vertebrate body that cushions and supports the organs and blood vessels.

areolate Divided into small areas (areolae) by cracks or lines (e.g. the wing of an insect).

***Aretosabus calabarensis* (angwantibo)** *See* LORIDAE.

argentines (herring smelts) *See* ARGENTINIDAE.

Argentinidae (herring smelts, argentines; subclass *Actinopterygii, order *Salmoniformes) A family of small, marine, large-eyed, silvery fish that are found in coastal areas of the oceans. There are about 12 species.

arginine An aliphatic, basic, polar *amino acid that contains the *guanido group.

Argiopidae The now suppressed familial name for orb-web spiders (*Araneidae).

Argyroneta aquatica See AGELENIDAE.

Argyropelecus lychnus (silvery hatchetfish) See STERNOPTYCHIDAE.

argyrophil fibres See RETICULAR FIBRES.

Ariidae (Tachysuridae; sea catfish; subclass *Actinopterygii, order *Siluriformes) A family of mainly marine catfish with a forked tail fin, *adipose *dorsal fin, naked skin, and several pairs of *barbels around the mouth. They are found in tropical and subtropical seas around the world. Many species carry out *buccal incubation, the males incubating fertilized eggs in the mouth for up to six weeks. There are many species.

arista Bristle-like extension of the third antennal segment in the dipteran suborders *Brachycera and *Cyclorrapha. The 'bristle' is borne dorsally and may be visibly segmented, with or without *pubescence or cilia (see CILIUM).

aristogenesis An outmoded theory holding that evolution proceeds along a determined path. (The modern view is that *natural selection does not direct evolution towards any particular kind of organism or physiological attribute, nor is there any mysterious inner guiding force.) See also ENTELECHY; NOMOGENESIS; ORTHOGENESIS.

Aristotle's lantern In *regular echinoids, the jaw apparatus, which consists of five, strong jaws, each with one tooth, that form a structure shaped like a lantern inside the mouth. The teeth are used for scraping algae and other material from the surface on which the animal feeds.

Arixeniidae (class *Insecta, order *Dermaptera) Family of earwigs that are hairy, wingless, and robust, with short, rodlike *forceps. Only two species are known, and they are associated with two species of bats found in Malaysia, Indonesia, and the Philippines.

Armadillidae (pill bugs; class *Malacostraca, order *Isopoda) Family of terrestrial pill bugs which have five pairs of *tracheal gills and lack a water-transport system. All members of the family can roll up completely in response to disturbance or in the presence of predators. There are 695 species, widely distributed in tropical and subtropical regions.

armadillos See DASYPODOIDEA.

armoured catfish See CALLICHTHYIDAE; LORICARIIDAE.

armpit effect The recognition by an animal of its own body odour and its use of that knowledge to distinguish between close relatives and strangers.

army ant See DORYLINAE.

army worms See NOCTUIDAE; SCIARIDAE.

arolium Median pad found between the *tarsal claws of an insect.

arousal 1. The transition from the sleeping to the waking state. **2.** An increase in the responsiveness of an animal to sensory stimuli.

arrhenotoky Production of males from unfertilized eggs, a phenomenon characteristic of all *Hymenoptera.

Arripidae (Australian salmon; subclass *Actinopterygii, order *Perciformes) A small family, confined to the temperate waters of Australia and New Zealand, that comprises two species of fish which are often abundant in the coastal areas, and of considerable economic importance. The Australian salmon, *Arripis trutta* (not related to the true salmon), is a large carnivorous fish, often exceeding 5 kg in weight.

arrow-poison frogs (*Dendrobates*) See RANIDAE.

arrowtooth eels See DYSOMMIDAE.

arrow worms See CHAETOGNATHA.

Artamidae (wood-swallows; class *Aves, order *Passeriformes) A family of highly social, smallish, brown or grey birds that have black and white markings and long, pointed wings. They are the only *passerines with powder-down feathers. They are insectivorous, mainly aerial feeders, and inhabit clearings and open country near water, nesting on ledges or in trees. There is one genus (*Artamus*) containing 10 species, found in India, southern China, south-east Asia, Australasia, and some S. Pacific islands.

Artamus (wood-swallows) See ARTAMIDAE.

artefact (artifact) **1.** Man-made object. **2.** Something observed that is not naturally present but that has arisen as a result of the process of observation or investigation.

arteriole A small branch of an *artery.

artery A blood vessel that conveys blood away from the heart.

Arthrochirotida (subphylum *Echinozoa, class *Holothuroidea) An extinct order of sea cucumbers, known with certainty only from the *Devonian, and characterized by an articulated axial skeleton of stout *sclerites in the *tentacles.

arthrodial membrane See EXOSKELETON.

Arthrodira (Arthrodiriformes; class *Placodermi) Group or order of fossil (*Devonian) fish in which the body was covered with bony plates, including a heavily armoured head shield; the gills opened between the head and body armour.

Arthrodiriformes See ARTHRODIRA.

arthrophragm In *Decapoda, a *septum between articulations.

Arthropleona See COLLEMBOLA.

Arthropoda A highly diverse phylum of jointed-limbed animals, which includes the crustaceans, arachnids, and insects as the major components, as well as the classes *Symphyla, *Pauropoda, *Chilopoda (centipedes), *Diplopoda (millipedes), and the extinct *trilobites and eurypterids (*Merostomata). They first appeared in the *Cambrian, already well diversified with such forms as the trilobites, trilobitoids, *ostracodes, and crabs present, implying an earlier, hidden history, reaching back into the Precambrian. The Arthropoda comprise more than 75% of all animal species that have been described. Embryological evidence shows that they are derived either from primitive *polychaete worms, or from ancestors common to both. Arthropods share with *annelid worms a metamerically segmented body (*see* METAMERIC SEGMENTATION), at least in the embryo, a dorsal heart, a dorsal anterior brain, and a ventral nerve cord that has segmental, ganglionic swellings. The limbs of all arthropods are paired, jointed, and segmental, and the body has a chitinous *exoskeleton. Primitively, the limbs and cuticular plates correspond to the metameric segmentation of the body, but in many groups there is considerable loss and/or fusion of segments.

arthropodin A water-soluble *protein that forms part of the *exoskeleton of arthropods (*Arthropoda).

articulamentum The lower of the two layers (the upper is the tegmentum) of calcareous shell material possessed by the *Polyplacophora. It is generally a hard layer that extends forward to form insertion plates between the valves.

articular bone In all *Gnathostomata except *Mammalia, the bone of the lower jaw that articulates with the upper jaw. In

Mammalia the articular has become the *malleus of the *middle ear.

articular capsule (joint capsule) An envelope that surrounds a *synovial joint.

Articulata 1. (phylum *Brachiopoda) A class of brachiopods in which the calcareous valves are impunctate, *punctate, or pseudo-punctate. The fibrous or prismatic secondary layer and non-fibrous primary layer are well differentiated. Hinge teeth and dental sockets are developed. *Lophophore support is formed from modifications of socket ridges to give *crura, loops, or spires. The alimentary canal ends blindly. Articulate brachiopods first appeared in the Early *Cambrian. There are six orders, with 37 superfamilies. **2.** (subphylum *Crinozoa, class *Crinoidea) A subclass containing all living, stalked crinoids, plus the feather stars, which are identified by the arrangement of the skeletal plates forming the *calyx. Articulation between the radial and brachial plates, and in the majority of brachials, is muscular, with a well-developed fulcral ridge. The arms are always uniserial. All post-Palaeozoic crinoids belong to the Articulata.

articulation A joint between bones, the limbs of an arthropod (*Arthropoda), or the *valves of a bivalved mollusc (*Mollusca) or brachiopod (*Brachiopoda).

artifact See ARTEFACT.

artificial selection Selection by humans of individual plants or animals from which to breed the next generation, because these individuals exhibit the most marked development of the required attributes. Typically, the process is repeated in successive generations until those attributes are fixed in the descendent offspring.

Artiodactyla (cohort *Ferungulata, superorder *Paraxonia) The even-toed ungulates, an order of mammals that includes the camels, pigs, and ruminants, together with numerous extinct varieties. They are the most successful of the hoofed animals. They are descended from the *Condylarthra, and underwent a spectacular burst of adaptive radiation in *Eocene and early *Oligocene times. They are terrestrial or amphibious herbivorous mammals. Except for the camels the gait is *unguligrade. The axis of the foot is *paraxonic; the first digit is absent, the second and fifth often reduced or lost. The digits terminate in hoofs, the third and fourth of equal size, usually flattened on the inner and ventral surfaces. The dentition is specialized; the upper *incisors are lost in some species, the lower incisors biting against the hardened gum of the *premaxilla; the *canines may form tusks; the *premolars are not molarized (i.e. adapted for grinding, as in many *Perissodactyla), elongated *hypsodont *molars providing grinding surfaces, the four cusps often being developed into longitudinal ridges. The tongue is large, mobile, protrusible, pointed, and the papillae are often horny. The stomach is elaborate. The brain is moderately well developed in later forms, although the cerebral hemispheres only partly cover the *cerebellum, and in early and some modern forms (e.g. hippopotamus and pig) the brain is small. The olfactory sense is important. Visual, auditory, olfactory, and tactile signals are important in communication.

Ascaphidae (tailed frogs; class *Amphibia, order *Anura) A monotypic family (*Ascaphus truei*) of amphibians, found in mountain streams in western N. America, that, like *Leiopelmatidae, have nine *amphicoelous vertebrae and free ribs. A muscular extension of the *cloaca ('tail') is used in copulation. Tadpoles cling to rocks with a sucking mouth.

ascaris Nematode worms (*Nematoda), all of which are parasites, some of mammals, others of birds, and some of invertebrates. The group includes *Toxocara canis* and *Ascaris lumbricoides* (the dog and human ascarids respectively). In both these species, eggs are voided and begin to develop in *faeces. If ingested by another host, the eggs hatch in the small intestine, releasing small worms that penetrate the gut wall, enter the blood stream, and are carried to the lungs. There they break into alveoli (*see* ALVEOLUS), move along air passageways to the *oesophagus, and are swallowed, returning as adults to the intestine, where they lay eggs.

Aschelminthes The name sometimes given to a phylum of worm-like animals that are typified by possessing a *pseudocoelom. True segmentation is not present. The intestine is straight with a highly developed *pharynx and *anus. Cell constancy is characteristic of the group: after the young has hatched (or been born) the number of cells in the body never increases although the size of the body usually does. There are six classes, some free-living and some parasites. The group contains several miscellaneous groups whose classification has always been controversial; the classes are now generally thought to be unrelated and are usually regarded as distinct phyla. *See* GASTROTRICHA; KINORHYNCHIA; NEMATODA; NEMATOMORPHA; PRIAPULIDA; ROTIFERA.

Ascidiacea (sea squirts; subphylum *Urochordata) A class of sea squirts, most of which inhabit shallow waters where they attach themselves to such structures as rocks, pilings, the bottoms of ships, and coral reefs. Deep-water species inhabit soft substrates and are known to occur at depths exceeding 2000 m. Most ascidians are inconspicuous and are often covered by smaller sessile animals. *Halocynthia pyriformes*, of the Atlantic coast of N. America, is known as the 'sea peach' because it resembles the fruit in size, shape, and colour. The class contains four orders: Aplousobranchiata; Aspiraculata; Phlebobranchiata; and Stolidobranchiata.

Ascoglossa *See* SACOGLOSSA.

ascon The body structure of a poriferan (sponge) if it consists of one, rather large, chamber.

Ascophora (class *Gymnolaemata, order *Cheilostomata) Suborder of cheilostome bryozoans in which the tentacles are protruded and withdrawn by means of a hydrostatic organ (the compensation sac). The *operculum is hinged in such a way that it controls the flow of water into the interior. The suborder occurs from the *Cretaceous to the present.

ascorbic acid The water-soluble vitamin C that occurs in large quantities in fruit and vegetables, a deficiency of which causes scurvy in humans.

aseptate Lacking partitions (septa).

aseptic meningitis A form of *meningitis in which there is an increase in the number of lymphocytes in the cerebro-spinal fluid. Usually it is caused by one of several types of virus, although bacterial causal agents are also known.

asexual generation *See* AGAMIC GENERATION.

asexual reproduction Reproduction without the sexual processes of *gamete formation. It occurs by fission in many protozoan animals but also in some multicellular ones (e.g. corals); by budding (gemmation) in many coelenterates and sea squirts (*Urochordata); and by *parthenogenesis in some vertebrates (e.g. lizards).

Asian chevrotains (*Tragulus*) *See* TRAGULIDAE.

Asian elephant (*Elephas maximus*) *See* ELEPHANTIDAE.

Asian land salamanders *See* HYNOBIIDAE.

Asian tapir (*Acrocodia*) *See* TAPIRIDAE.

Asilidae (assassin flies, robber flies; order *Diptera, suborder *Brachycera) Family of medium to large, elongate, bristly flies, which have a horny proboscis adapted for piercing. The legs are powerful and flexible, with large *pulvilli, and a bristle-like *empodium. A prominent tuft of hairs, or 'beard', and protruding eyes are also characteristic of the asilids. Virtually all adults are predators of other insects, using their long legs to grasp and carry off prey. The proboscis, used to pierce the prey and suck it dry, can easily enter the hardest of cuticles, and in certain cases may inject venom. Asilid larvae are either predators or scavengers in soil, leaf-litter, or other ground substrates. Larvae are *amphipneustic; their mouthparts comprise a hook-shaped *labrum, sharp *mandibles, and broad *maxillae with two-segmented *palps. The Asilidae is the largest family of Brachycera, and contains at least 4000 described species.

Asio (owls) *See* STRIGIDAE.

asities (*Philepitta*) *See* PHILEPITTIDAE.

asp (*Vipera aspis*) *See* VIPERIDAE.

aspartic acid An aliphatic, acidic, polar *alpha amino acid.

aspect ratio Of a fin or wing, the ratio of length to width. A high-aspect-ratio fin or wing tends to be long and thin, producing a high lift- or thrust-to-drag ratio.

aspic viper (*Vipera aspis*) *See* VIPERIDAE.

Aspidochirotacea (subphylum *Echinozoa, class *Holothuroidea) A subclass of sea cucumbers that have 10–30 shieldshaped *tentacles which spoon organic debris from the sea floor into the mouth. There are no pharyngeal retractor muscles. The *test is vestigial, consisting of scattered *spicules in the dermis. The subclass first appeared in the *Carboniferous.

Aspidochirotida (class *Holothuroidea, subclass *Aspidochirotacea) An order of sea cucumbers that have large, branched *respiratory trees.

Aspidogastraea *See* ASPIDOGASTREA.

Aspidogastrea (Aspidogastraea; phylum *Platyhelminthes) A class, or in some classifications an order of *Trematoda, of parasitic worms in which the gut, pharynx, intestine, and ventral sucker are present and the anterior sucker absent. Mature individuals also lack *cilia. All are endoparasitic, and development may involve one or two hosts.

Aspiraculata *See* ASCIDIACEA.

Aspredinidae (banjo catfish; subclass *Actinopterygii, order *Siluriformes) A family of mainly small freshwater catfish whose common name is derived from the peculiar shape of the flat, triangular head region and the long tail. In some species the female takes care of the fertilized eggs by carrying them along in a modified region of the skin of the belly. There are about 50 species, found in S. America.

asp viper (*Vipera aspis*) *See* VIPERIDAE.

ass (*Equus*) *See* EQUIDAE.

assassin bug General name for bugs of the family *Reduviidae, but in Britain usually restricted to *Reduvius personatus,* which lives in unheated buildings, and preys on small arthropods. The larval stages mask themselves with a covering of debris.

assassin flies *See* ASILIDAE.

assemblage A collection of plants and/or animals characteristically associated with a particular environment that can be used as an indicator of that environment.

assemblage zone (coenozone, faunizone) A stratigraphic unit or level of strata that is characterized by an assemblage of animals and/or plants.

assimilate The portion of the food energy consumed by an organism that is metabolized by that organism. (Some food may pass through the organism without being used.)

assimilation The synthesis of *protoplasm and other complex substances from simpler molecules that are derived from food by digestion and absorbed into the living cells of an organism.

assisted migration The gathering of a sufficiently large sample of a *population and relocation of the animals in a *habitat where they may have a better chance of surviving.

association 1. A group of animals that share a home range and spend a significant proportion of their time in each other's company. The association may be interspecific, comprising members of two species, or polyspecific, involving several species. Both species in an interspecific association benefit from improved exploitation of their shared resources. Polyspecific associations are common among callitrichid primates (*Callitrichidae). For example, in Bolivia saddle-backed tamarins (*Sanguinus fuscicollis*) and red-bellied tamarins (*S. labiatus*) were observed to spend approximately half of their time together and were sometimes joined by a group of Goeldi's monkeys (*Callimico goeldii*). The association allows each species to benefit from the experience of the others in locating and exploiting food sources. **2.** A learned connection between a type of event and a neutral stimulus with which it is paired (e.g. the sensation of hunger that may follow the chiming of a clock, denoting the time at which a meal is customarily presented). *See also* CONDITIONING.

association analysis A hierarchical method of classification (*see* HIERARCHICAL and NON-HIERARCHICAL CLASSIFICATION METHODS), which is divisive and *monothetic. The method uses χ^2 as a measure of association between pairs of species (attributes) found at a range of sample sites (individuals). The species or attribute with the highest overall sum of χ^2 values (i.e. the strongest links) with all other species is selected as the basis for subdivision into two groups of sites or individuals. The process is then repeated for each new group until no further subdivision is required.

assortative mating Sexual reproduction in which the pairing of male and female is not random, but involves a tendency for males of a particular kind to breed with females of a particular kind, or the converse. If the two partners in each pair tend to be more alike than is expected by chance, then it is referred to as positive assortative mating. Assortative mating for some traits is common: examples include positive assortative mating for skin colour and height in hu-

mans, and for time of development to sexual maturity in many insects with one generation per year; and negative assortative mating for different plumage colour-phases in arctic skuas. *Compare* PANMIXIS.

aster A group of *microtubules that radiates out from each *centriole pair during cell division in animal cells.

Asteroidea (starfish; subphylum *Asterozoa, class *Stelleroidea) A subclass which includes all extant starfish, characterized by the possession of poorly demarcated, broad arms containing a large coelomic cavity. The open *ambulacral grooves bear rows of large *tube feet which are used especially in feeding and locomotion. Respiration takes place through external gills (papulae) which are restricted to the *aboral surface. Asteroids occur from the *Ordovician to the present.

Asterozoa (phylum *Echinodermata) A subphylum whose members have radial symmetry of projecting rays and a star-shaped body. It contains the one class Stelleroidea.

Astian The land-mammal stage that spans the early *Pliocene (3.6–2.58 Ma ago) in Europe.

astogenetic heterochrony Among colonial animals, *heterochrony that affects the colony as a whole. *Compare* MOSAIC HETEROCHRONY; ONTOGENETIC HETEROCHRONY.

Astomatida (class *Ciliatea, subclass *Holotrichia) An order of ciliate *Protozoa which have uniformly arranged body *cilia and which appear to lack a *cytostome. Most are parasitic in invertebrates, especially in *Oligochaeta.

astomatous Lacking a mouth or *cytostome.

astragalus The ankle-bone.

astral rays During cell division, the arrangement of *microtubules that form the *aster.

Astrapotheria (cohort *Ferungulata, superorder *Protoungulata) An extinct order of S. American ungulates that appeared during the *Eocene and flourished in the absence of herbivore competition or carnivore predation during the *Oligocene and *Miocene. The skull was short, the upper *incisors were lost and the *premaxillae much reduced, the lower incisors were present, the upper and lower *canines were large and persistently growing, and the rearmost *molars were large. A slit-like nasal opening on top of the skull suggests that astrapotherids (e.g. the Miocene *Astrapotherium*, which was the size of a rhinoceros) had elephant-like proboses. In contrast to the heavily built front legs, head, neck, and anterior part of the body, the spine, hindquarters, and hind legs were feeble. The feet were small, perhaps resting on pads, but unlike those of any other group. Possibly, therefore, the animals were amphibious.

Astroblepidae (subclass *Actinopterygii, order *Siluriformes) A family of freshwater fish that have naked skin, small eyes, and a ventrally positioned mouth. There are about 35 species, occurring in S. America.

Astrocoeniina (subclass *Zoantharia, order *Madreporaria) A suborder of colonial corals composed of small *corallites, which first appeared in the Middle *Triassic. They are characterized by having *septa formed from relatively few (not more than eight) *trabeculae and usually they have no more than two cycles of *tentacles arranged in a single ring around the mouth, which leads to a smooth *stomodaeum. The suborder contains five families, with 46 genera.

Astronesthidae (snaggle tooth; subclass *Actinopterygii, order *Salmoniformes) A family of small, black, deep-sea fish found in tropical and temperate waters. They have well-developed fangs and a comb-toothed edge on the upper jaw. A distinct *barbel is attached to the chin. There are about 27 species.

atavism The reappearance of a character after several generations, the character being the expression of a recessive gene or of complementary genes. The character, or individual possessing this character, is sometimes referred to as a 'throw-back'.

Ateles (spider monkey) *See* Cebidae.

Atelocerata (phylum *Arthropoda) A former subphylum that contained those classes formerly assigned to the subphylum *Uniramia and now placed in *Hexapoda.

Atelocynus microtis (small-eared dog) *See* Canidae.

Atelopodidae (class *Amphibia, order *Anura) A family of small, brightly coloured, often poisonous, tree frogs. The *Bidder's organ found in males indicates a bufonid relationship. Like toads, most walk rather than hop. There are two genera, with 30 species, found in Central and S. America.

Atelostomata (subphylum *Echinozoa, class *Echinoidea) A superorder of irregular echinoids characterized by the lack of compound *ambulacra, *Aristotle's lantern, girdle, or gill slits. The apical system and *peristome are rarely opposite, and the interambulacra are wider than the ambulacra on the lower surface. The superorder includes the orders Disasteroida, *Holasteroida, and *Spatangoida.

Athecata *See* Gymnoblastina.

Atherinidae (silversides, sand smelts; subclass *Actinopterygii, order *Atheriniformes) A family of rather small, slender fish that have relatively large eyes, a terminal mouth, and two widely separated *dorsal fins. Most species occur in schools in inshore seas, with some found in estuaries or freshwater systems. There are more than 150 species.

Atheriniformes (class *Osteichthyes, subclass *Actinopterygii) An order of bony fish that includes fairly large marine individuals (e.g. the flying fish and the sauries) as well as many small freshwater species (e.g. killifish, guppy, and mosquitofish). Some authors have classified members of this order in other orders, including the Beloniformes, and the Cyprinodontiformes.

Atlantic mackerel (Scomber scombrus) See SCOMBRIDAE.

Atlantic pomfret See BRAMIDAE.

Atlantic saury (Scomberesox saurus) See SCOMBERESOCIDAE.

Atlantic scad (Trachurus trachurus) See CARANGIDAE.

Atlantic torpedo (Torpedo nobiliana) See TORPEDINIDAE.

Atlantic tuna (Thunnis thynnus) See SCOMBRIDAE.

Atlantic velvet belly (Etmopterus spinax) See SQUALIDAE.

Atlantic winter flounder (Pseudopleuronectes americanus) See PLEURONECTIDAE.

atlas moths See SATURNIIDAE.

atlas vertebra See VERTEBRA.

Atopogale See SOLENODONTIDAE.

ATP See ADENOSINE TRIPHOSPHATE.

Atrichornis clamosus (noisy scrub-bird) See ATRICHORNITHIDAE.

Atrichornithidae (scrub-birds; class *Aves, order *Passeriformes) A family of small, brown birds that have short, round wings and long, graduated tails. They have powerful voices, especially *Atrichornis clamosus* (noisy scrub-bird). They are insectivorous, strongly territorial, and are found only in the dense undergrowth of two isolated areas of coastal Australia. There is one genus, with two species.

atriopore See ATRIUM (1).

atrium 1. In *Urochordata and *Cephalochordata, a chamber surrounding the pharyngeal region which receives water from the pharyngeal clefts and opens to the exterior through an atrial opening (sometimes called an atriopore). **2.** See AURICLE.

atrophy Of a structure, limb, organ, tissue, etc., to diminish in size.

atropine A substance that, by competing with *acetylcholine for post-synaptic membrane receptor sites, inhibits the passage of nerve impulses across a *synapse.

Attila (attilas) See TYRANNIDAE.

Attini (family *Formicidae, subfamily *Myrmicinae) Tribe of ants in which foliage cut by foraging workers provides a substrate for the growth of a fungus, localized in fungus gardens within the soil nest. Fungal gongylidia (hyphae with swollen tips) are harvested by workers and fed to larvae. There are 200 species, restricted to the New World where most occur in the neotropics.

Atypidae (purse-web spiders; order *Araneae, suborder *Mygalomorphae) Family of spiders, all less than 3 cm long, which have three *tarsal claws, six *spinnerets, massive *chelicerae, and two pairs of *book lungs. New World species construct open silk tubes on the sides of trees; European species construct the tubes on the ground. In all cases, prey walking over the tube is impaled on the fangs and dragged through the purse web. The few species in this family may live for up to nine years.

aubemouth See AULORHYNCHIDAE.

Auchenipteridae (subclass *Actinopterygii, order *Siluriformes) A family of freshwater catfish, distinguished by having a scaleless body, a zigzag *lateral line along the body, a tiny *adipose fin, and pairs of *barbels around the mouth. There are about 25 species, found in tropical S. America.

auditory bulla In *Mammalia, a hollow, bony structure that encloses the inner ear. In cats (*Felidae), it includes an ossified segment that is absent in all other members of the *Carnivora, thus distinguishing felids from other carnivores.

auditory meatus A canal that leads from the external ear to the *tympanic membrane.

aufwuchs *See* PERIPHYTON.

auks *See* ALCIDAE; CHARADRIIFORMES.

aulodont Applied to echinoids in which the *Aristotle's lantern has teeth which are longitudinally grooved and broadly U-shaped in cross-section.

Aulopodidae (subclass *Actinopterygii, order *Myctophiformes) A small family of marine, warm- to temperate-water fish. All species possess a rather large head, big jaws, a high *dorsal fin, and a slender body. There are about eight species.

Aulorhynchidae (aubemouth, tube-snout; subclass *Actinopterygii, order *Gasterosteiformes) A family of marine coastal fish that have a very elongate body and tube-like snout. The first *dorsal fin consists of about 25 short, isolated spines. There is one species, *Aulorhynchus flavidus*, occurring in the northern Pacific.

Aulorhynchus flavidus (tube-snout) *See* AULORHYNCHIDAE.

Aulostomidae (flutemouth, trumpet-fish; subclass *Actinopterygii, order *Syngnathiformes) A family of marine tropical fish characterized by a very elongate body, a long compressed snout, a series of isolated dorsal spines, and a rounded tail fin. Found mainly in reefs or inshore rocky areas, they can be encountered swimming in an almost vertical position, hidden by sea-grass leaves or coral branches. There are about four species.

auricle 1. (atrium) The chamber of a heart into which blood flows from veins and from which blood flows to a ventricle. **2.** External ear. **3.** An internal projection from

one of the ambulacral plates around the *peristome of echinoids which serves for the attachment of muscles supporting the *Aristotle's lantern.

auricularia larva In *Asteroidea and *Holothuroidea, a larval form in which a sinuous, ciliated band outlines the body.

Australian chats *See* EPHTHIANURIDAE.

Australian faunal subregion A region that is distinguished by a unique marsupial (*Marsupialia) fauna, including herbivores, carnivores, and insectivores. These evolved in isolation from the placental mammals (*Eutheria) which now dominate the other continental faunas. In addition to marsupials there are also very primitive mammals (*Monotremata), the spiny ant-eater and the platypus; and small rodents which are relatively recent (probably *Miocene) immigrants.

Australian grayling (*Prototroctes maraena*) *See* APLOCHITONIDAE, RETROPINNIDAE.

Australian lungfish *See* CERATODIDAE.

Australian magpie (*Gymnorhina tibicen*) *See* CRACTICIDAE.

Australian salmon (*Arripis trutta*) *See* ARRIPIDAE.

Australian tree-creepers *See* CLIMACTERIDAE.

Australian wrens *See* MALURIDAE.

Australidelphia (Austridelphia) One of the two cohorts into which the *Metatheria (marsupials) are often divided. There are five orders: *Dasyuromorphia; *Diprotodontia; *Dromiciopsia; *Peramelemorphia; and *Syndactyliformes. All except the Dromiciopsia occur in Australia and New Guinea. Because representatives of both cohorts are found in S. America, the hypothesis of a neotropical origin for marsupials (which spread only later into the Australian–New Guinea region and Antarctica) is supported.

australopithecines Literally, 'southern apes', early members of the human lineage that lived from about 4 to about 1 million years ago in Africa. The so-called 'robust australopithecines' are nowadays placed in a separate genus, *Paranthropus*. The other ('gracile') australopithecines are also a very diverse group of species, some very primitive and perhaps ancestral to all later hominids, others probably specialized sidelines. The species usually recognized are *Australopithecus anamensis* (4.1–3.9 million years bp) and *A. afarensis* (3.75–3.0 million years bp) from E. Africa, *A. bahrelghazali* (about 3.4–3.0 million years bp) from W. Africa, and *A. africanus* (about 3.0–2.4 million years bp) from S. Africa. Probably some of these species should be placed in different genera.

Australopithecus *See* AUSTRALOPITHECINES.

Austridelphia *See* AUSTRALIDELPHIA.

autapomorphic Applied to an *apomorphic character state that is unique to a particular *species or lineage in the group under consideration.

autecology The *ecology of individual organisms or species, including physiological ecology, animal behaviour, and population dynamics. Usually only one species is studied. *Compare* SYNECOLOGY; *See also* POPULATION ECOLOGY.

autocatalysis The process whereby symbiotic bacteria in an animal's gut degrade food particles (especially the fermentation of cellulose and hemicellulose in plant cell walls) and so assist in its nutrition.

autochthonous Applied to an individual or object that originates in the place where it is found. *Compare* ALLOCHTHONOUS.

autoecious (monoxenous) Applied to a parasitic organism that can complete its life cycle in a single host species.

autogamy The process of self-fertilization found in some *Protozoa (e.g. *Paramecium*). The nucleus of an individual divides

into two parts which then reunite, resulting in homozygosis (the production of identical *alleles at any *gene locus on *homologous chromosomes).

autograft *See* ALLOGRAFT; GRAFT.

autogrooming Grooming of an animal by itself. *Compare* ALLOGROOMING.

auto-immunity An immune reaction in an organism in which the same individual acts as the source of both *antigens and *antibodies. While it may arise as a result of a pathological and therefore undesirable condition, it is also an integral part of the mechanism in the body for eliminating unwanted cells (e.g. old erythrocytes).

autolysis The destruction of a cell or some of its components through the action of its own hydrolytic enzymes (*See* HYDROLYSIS). It is a process that is particularly marked in organisms undergoing *metamorphosis and involves *lysosomes.

automimicry Mimicry that exploits a *polymorphism for palatability to predators. In the monarch butterfly, for example, unpalatability arises from the food plants which are chosen by the ovipositing female. After hatching, the feeding larvae take in substances that are toxic to birds but not to the insects themselves. This renders them unpalatable. The situation is used to advantage by palatable members of the species, which mimic the coloration of the unpalatable ones and so obtain protection. The polymorphism is frequency-dependent, and the unpalatable insects must be more abundant (and hence more frequently encountered by the bird predators) than the palatable insects.

autonomic Applied to behavioural responses (e.g. pallor and sweating in humans in response to fear) that are produced by the *autonomic nervous system. Certain ritualized forms of behaviour used in communication may have originated as autonomic behaviour.

autonomic nervous system The part of the nervous system which controls involuntary activities such as *homoeostasis. The autonomic nervous system is regulated by the *hypothalamus and comprises two antagonistic parts, the *parasympathetic and *sympathetic nervous systems. *See also* CENTRAL NERVOUS SYSTEM; PERIPHERAL NERVOUS SYSTEM.

autophagic vacuole A membrane-bound *vacuole, derived from within the cell, that contains intracellular material (e.g. redundant *organelles) that is to be digested. *See also* LYSOSOME.

autophagy The digestion within a cell of material produced by the cell itself but which it no longer requires.

autopoiesis The processes by which a discrete unit such as a cell or *taxon organizes itself without external assistance. This may involve compensation for environmental and genetic changes, as a result of which a taxon undergoes minimal morphological alteration (e.g. *Plethodon*, a salamander genus that has changed little morphologically over 60 million years, despite frequent speciation and major changes to the environments its species inhabit, because of its behavioural flexibility and generalized tongue and teeth). *See also* CANALIZATION.

autopolyploid A *polyploid organism that originates by the multiplication of a single *genome, such that all the *chromosomes come from sets within one single species.

autoradiograph A photographic image of the radioactive areas in a specimen (e.g. in *DNA fingerprinting).

autosome Any *chromosome in the cell nucleus other than a *sex chromosome.

autostylic The condition of having the upper jaw connected directly to the *cranium or articulating directly with it, found in lungfishes (*Dipneusti) and the ancestors of tetrapods (*Tetrapoda).

autotheca One of the three types of graptolite (*Graptolithina) *thecae, possibly containing the female *zooid. *Compare* BITHECA; STOLOTHECA.

autotomy The voluntary severance by an animal of a part of its body (commonly one of its own limbs), usually to escape capture by a predator that has seized that part. The part then regrows. Autotomy of the claws occurs in some *Crustacea and of the tail in some *Lacertilia (lizards). It is a defensive mechanism: in some lizards the detached tail continues to wriggle, distracting the predator while the lizard escapes.

autotroph An organism that uses carbon dioxide as its main or sole source of carbon. *Compare* HETEROTROPH.

autoxidation *See* ANTI-OXIDANT.

autumnal In the autumn (fall). The term is used with reference to the six-part division of the year used by ecologists, especially in relation to studies of terrestrial and freshwater communities. *Compare* AESTIVAL; HIBERNAL; PREVERNAL; SEROTINAL; VERNAL.

auxiliary spiral *See* CATCHING SPIRAL.

Avahi (woolly indri) *See* INDRIDAE.

aversive pheromone A *pheromone in the urine of males of some mammal species (e.g. rodents) that discourages other males, especially subordinate ones, from closely investigating the scent-marked area and inhibits aggressive behaviour in them. The effect is confined to males of the same strain.

Aves (birds; subphylum *Vertebrata, superclass *Gnathostomata) The class that comprises all the birds. The late *Jurassic *Archaeopteryx lithographica* is still the best-known *Mesozoic bird, but others have been described since the 1980s: *N**oguerornis*, from the lowermost *Cretaceous of Spain; the slightly later and more advanced *Concornis* and *Iberomesornis*, known by complete skeletons, also from Spain; and, also early Cretaceous, *Sinornis* and *Cathayornis* from China. There were also some curious, specialized, Late Cretaceous birds, such as the flightless *Mononykus* from

Central Asia, in which the forelimbs were reduced to stubby claws, and *Hesperornis*, a diving form. All these early birds had teeth and long, bony tails. Birds arose from within the theropod dinosaurs (*Theropoda) and so should properly be classified as a subgroup of them; those closest were the Dromaeosauridae (the family which includes the famous *Velociraptor* of *Jurassic Park*).

avicularium In some polymorphic (*See* POLYMORPHISM) colonial *Bryozoa, a *zooid shaped like the head of a bird.

Avimimus An intriguing fossil vertebrate (*Cretaceous) that is usually classified as a theropod dinosaur (*Theropoda) but probably would be more correctly classed as a primitive bird. It is incompletely known but many of its anatomical features are more advanced, in a bird-like sense, than those of *Archaeopteryx*.

avirulent Not virulent.

avocet *See* RECURVIROSTRIDAE.

avoidance Behaviour that tends to protect an animal by reducing its exposure to hazard. Avoidance behaviour may be learned (e.g. when an animal does not enter an area in which a predator has been encountered, or when it does not eat an item that previously made it ill), or innate (e.g. when the young of some bird species utter distress calls and seek to hide when they see the shadow of a hawk-like object, although they have no experience of predatory birds).

axenic Applied to a culture that consists of only one species.

axial skeleton (somatic skeleton) The part of the vertebrate *endoskeleton that comprises the *cranium, *visceral skeleton, *notochord, *vertebrae, and *ribs. *Compare* APPENDICULAR SKELETON.

Axinellida (class *Demospongiae, subclass *Tetractinomorpha) An order of sponges which possess large amounts of organic fibres that envelop the *spicules. An axial skeleton normally occurs. All species are oviparous.

axis vertebra *See* VERTEBRA.

axolotl (*Ambystoma mexicanum*) *See* AMBYSTOMATIDAE.

axon A single, long, straight, relatively unbranched process arising from the cell body of a *neuron, which may or may not be *myelinated. It is responsible for conducting the nerve impulse away from the cell body.

axoneme The '9+2' formation of *microtubules that forms the principal structural feature of *cilia and *flagella.

axoplasm The *cytoplasm of an *axon.

axopodium A type of *pseudopodium which is slender and possesses a central filament or core. Axopodia extend radially from the cell bodies of some *Sarcodina.

aye-aye (*Daubentonia madagascariensis*) *See* DAUBENTONIIDAE.

Aysheaia pedunculata An extinct invertebrate, known from fossils found in the Middle *Cambrian *Burgess Shale, that, although marine, is considered a likely ancestor for the extant members of the phylum *Onychophora that exist in humid tropical forests. It possessed an elongate, worm-like body with a terminal mouth and frontal pupillae. The body was segmented and bore 10–43 pairs of walking legs.

Aythya **(pochards)** *See* ANATIDAE.

ayu *See* PLECOGLOSSIDAE.

Azoic *See* ARCHAEAN.

B

babblers *See* TIMALIIDAE.

baboon (*Papio*) *See* CERCOPITHECIDAE.

babysitting Taking care of a young animal in the temporary absence of its mother by another adult. Babysitters are probably closely related to the young they tend, but babysitting requires them to forfeit opportunities for foraging or other activity.

Bacillus A genus of *Bacteria in which the cells are rod-shaped, often *motile, and typically Gram-positive (*see* GRAM REACTION). Some species can grow only in the presence of air; others can grow in either the presence or absence of air. There are many species, found in a wide range of *habitats. Some species can cause disease in vertebrate animals (e.g. *anthrax), or in insects. *Bacillus thuringiensis* is used in the control of insect pests, and certain crops are genetically modified to express *B. thuringiensis* toxins.

backcross A cross of an *F_1 *hybrid or *heterozygote with an individual whose *genotype is identical to that of one or other of the two parental individuals. Matings involving a hybrid genotype and a recessive parental genotype are used in genetic analyses to determine *linkage and cross-over values.

back mutation A *reverse mutation in which a mutant gene (called the non-wildtype form) reverts to the original standard form (the wild-type form).

backswimmer *See* NOTONECTIDAE.

bacon beetle (*Dermestes lardarius*) *See* DERMESTIDAE.

bacteraemia The condition in which bacteria are present in the blood.

Bacteria A taxonomic *kingdom comprising the subkingdoms *Archaea and *Eubacteria or, in the three-*domain classification, the only kingdom in the domain Eubacteria. Bacteria are *prokaryotes, most of which are single-celled and most having a rigid cell wall. They are almost universal in distribution and play many important roles as agents of decay and mineralization and in the recycling of elements such as nitrogen. Some species cause illness in animals. There are eleven principal groups: Gram-positive bacteria (*See* GRAM REACTION); purple bacteria; *cyanobacteria; green non-sulphur bacteria; spirochaetes; flavobacteria; green sulphur bacteria; Planctomyces; Chlamydiales; Deinococci; and Thermatogales.

Bactrian camel (*Camelus bactrianus*) *See* CAMELIDAE.

baculum The name sometimes given to the penis bone of some mammals.

badger (*Meles, Taxidea*) *See* MUSTELIDAE.

Baermann funnel A device used to extract nematodes (*Nematoda) from a soil sample or plant material. A muslin bag containing the sample is submerged in water in a funnel sealed at the lower end by a rubber tube and clip. Being heavier than water, the nematodes pass through the muslin and sink to the bottom. After about 12 hours they can be collected by drawing off the bottom centimetre of water. The efficiency of the device is increased by gentle warming which immobilizes the nematodes.

Bagridae (bagrid catfish; subclass *Actinopterygii, order *Siluriformes) A large family of freshwater catfish that have a large *adipose *dorsal fin, small eyes, and four pairs of *barbels around the mouth. There are 30 genera and 210 species distributed widely in tropical Africa and Asia.

bagrid catfish *See* BAGRIDAE.

bagworms *See* PSYCHIDAE.

baiji (*Lipotes vexillifer*) *See* PLATANISTOIDEA.

Baikal oilfish *See* COMEPHORIDAE.

Baikal sculpins *See* COTTOCOMEPHORIDAE.

Bailey's triple catch A *mark–recapture technique, devised in the 1950s by N. T. J. Bailey, for estimating the size of an animal population in which a sample of animals is caught, marked, and released. This procedure is followed on two subsequent occasions. Provided each sample contains more than 20 individuals, the population estimate (P) is given by $(a_2n_2r_4)/r_1r_3$, where a_2 is the number of marked individuals released from the second catch, n_2 is the number caught in the second sample, r_4 is the number of individuals in the third catch that were marked in the first catch, r_1 is the number of individuals marked in the first catch that were caught in the second, and r_3 is the number of individuals caught in the third sample that were marked in the second. If the sample numbers fewer than 20 individuals, an adjustment is made, giving: $P=(a_2(n_2+1)r_4)/(r_1+1)(r_3+1)$.

Balaena glacialis (right whale) *See* BALAENIDAE.

Balaena mysticetus (bowhead whale) *See* BALAENIDAE.

Balaeniceps rex (shoe-bill) *See* BALAENICIPITIDAE.

Balaenicipitidae (shoe-bill; class *Aves, order *Ciconiiformes) A monospecific family comprising a grey, stork-like bird (*Balaeniceps rex*) that has a huge, broad, flattened bill. It is found in swamps feeding on fish, and it nests on the ground. It occurs in tropical E. Africa.

Balaenidae (order *Cetacea, suborder *Mysticeti) A family that comprises the right and bowhead whales. The body is robust, the snout blunt, the throat ungrooved. A dorsal fin is present in some species. The flippers are broad and short. The posterior margin of the tail is notched at the mid-line. The cervical vertebrae are fused. The *baleen plates are long and narrow, and left and right rows are separated anteriorly. The animals are dark grey or black and feed on *plankton. *Balaena mysticetus* (bowhead whale) is confined to northern waters. There are three species of *Eubalaena*: *E. glacialis* (North Atlantic right whale), *E. japonica* (Pacific right whale), and *E. australis* (southern right whale). Bowhead whales grow to 15 m, right whales up to 20 m.

Balaenoptera acutorostrata (minke, lesser rorqual) *See* BALAENOPTERIDAE.

Balaenoptera borealis (sei whale) *See* BALAENOPTERIDAE.

Balaenopteridae (order *Cetacea, suborder *Mysticeti) A family that includes the fin, humpback, and blue whales. The body is streamlined, the snout pointed, the throat and chest grooved. The dorsal fin is hooked and set far back on the body. The tail is slightly notched at the mid-line, and scalloped along the posterior edge. The cervical vertebrae are not fused. The *baleen plates are broad and short, the left and right rows being joined anteriorly. The animals are typically grey on the dorsal surface, with some white on the ventral surface; but *Megaptera novaengliae* (humpback) may be black above and below, with white flippers. Members of the family feed on krill and small fish, and occur in both hemispheres. They grow to 7–30 m depending on species. There are six species, in two genera: *Balaenoptera* (*B. acutorostrata*, minke or lesser rorqual; *B. borealis*, sei; *B. physalus*, fin; *B. musculus*, blue; *B. edeni*, Bryde's); and *Megaptera novaengliae* (humpback).

balanced polymorphism A genetic *polymorphism that is stable, and is maintained in a population by *natural selection, because the *heterozygotes for particular *alleles have a higher *adaptive value (fitness) than either *homozygote. This condition is also referred to as *overdominance, as opposed to under-dominance where the heterozygote has a lower fitness, giving rise to unstable equilibrium.

Balanidae (acorn barnacles; order *Thoracica, suborder Balanomorpha) Family of radially symmetrical, balanomorph barnacles in which the third pair of *cerci resemble the second rather than the fourth, and in which there are no caudal appendages. The family includes the familiar barnacles exposed at low tide, of the genus *Balanus*, which has been divided into 10 subgenera.

Balanus (acorn barnacle) *See* BALANIDAE.

Balbiani ring *See* CHROMOSOMAL PUFF.

bald crows *See* PICATHARTIDAE.

Baldwin effect The idea that *phenotypic plasticity is a crucial element for the operation of *natural selection. In the 1950s, some blue tits in Britain learned to peck away the caps of milk bottles and drink some of the milk. Those capable of learning this were strongly advantaged; in each generation, those which were more and more focused on milk-bottle-opening were selected for, so that a mere 'capacity to learn' became a 'drive to open milk bottles'. The effect was proposed in 1896 by the American psychologist James Mark Baldwin (1861–1934).

(⊕) **SEE WEB LINKS**
• 'Social Learning and the Baldwin Effect' by David Papineau.

baleen In *Mysticeti, sheets of *keratin which hang transversely from the roof of the mouth, their lower edges fringed with hairs which form a comb-like structure used to filter *plankton.

baleen whales *See* MYSTICETI.

Balistidae (triggerfish, filefish; subclass *Actinopterygii, order *Tetraodontiformes) A large family of marine fish, all of which have rather compressed, deep bodies with a large, stout, first dorsal spine that can be locked into a vertical position when erected. Some species show *crypsis, others have brilliant colour patterns. There are at least 120 species, many of which are sometimes considered to belong to an alternative family (Aluteridae or *Monacanthidae). They occur in tropical and subtropical waters.

ballistiform swimming A type of swimming practised by fish in which movement is effected by undulation of medium length fins only. *Compare* ANGUILLIFORM SWIMMING; CARANGIFORM SWIMMING; LABRIFORM SWIMMING; OSTRACIIFORM SWIMMING; RAJIFORM SWIMMING.

Ballistiform swimming

ballooning *See* ARANEAE; LINYPHIIDAE.

Baluchitherium *See* RHINOCEROTIDAE.

bamboo rats *See* RHIZOMYIDAE.

bamboo worms *See* SEDENTARIA.

band A vertical stripe on a *polytene chromosome that results from the specific association of a large number of *chromomeres. The most commonly examined polytene chromosomes are in the salivary gland nuclei of insects, and recent studies of *Drosophila* suggest that each band contains the genetic material of a single gene. However, the significance of the bands in human and other chromosomes is not known.

banded ant-eater (*Myrmecobius fasciatus*) *See* MYRMECOBIIDAE.

bandfish *See* CEPOLIDAE.

bandicoot *See* PERAMELOIDEA.

band-winged grasshopper *See* ACRIDIDAE.

banjo catfish *See* ASPREDINIDAE.

Banjosidae (subclass *Actinopterygii, order *Perciformes) A monospecific family of marine fish which have a deep, compressed body and fairly large eyes, the head showing a steep profile. The first *dorsal fin, with 10 spiny rays, is separated by a deep notch from the second dorsal fin, consisting of 12 soft rays. The second of the three anal spines is much larger than the other two. Greenish along the back and silvery along the sides, *Banjos banjos* (30 cm) is of some commercial importance to local fisheries along the coasts of China and Japan.

Baptornis advenus (class *Aves) A grebe-like, fish-eating bird from the Late *Cretaceous Period of N. America. Like modern divers, it had rather poorly developed wings. It is known only from disarticulated bones.

barb One of the branches arising from the central shaft (rachis) of a *feather.

barbel A fleshy protuberance near the mouth of some fish, especially suctorial feeders (e.g. sturgeons and catfish). Basically a sensory structure, it is provided with tactile and chemical *receptor cells, and is therefore useful in the location of food.

barbel-less catfish *See* AGENEIOSIDAE.

barbets *See* CAPITONIDAE.

Barbus tor (mahseer) *See* CYPRINIDAE.

barbicel One of the hooks by which *barbules are attached to each other.

barbule One of the branches arising from a *feather *barb.

bark beetle *See* SCOLYTIDAE.

barkbug *See* ARADIDAE.

barklice *See* PSOCOPTERA.

barnacles 1. *See* CIRRIPEDIA; CRUSTACEA; THORACICA. **2. (acorn barnacles)** *See* BALANIDAE. **3. (goose barnacles)** *See* LEPADIDAE. **4. (parasitic barnacles)** *See* RHIZOCEPHALA.

barn owls (*Tyto*) *See* TYTONIDAE.

barn swallow (*Hirundo rustica*) *See* HIRUNDINIDAE.

barophilic Living in or preferring environments subject to high pressures (e.g. deep-sea environments).

barracuda *See* SPHYRAENIDAE.

barracudina *See* PARALEPIDIDAE.

Barr body The condensed, single *X-chromosome, appearing as a densely staining mass, that is found in the nuclei of *somatic cells of female mammals. It is named after its discoverer, Murray Barr (1908–95), and is derived from one of the two X-chromosomes which becomes inactivated. The number of Barr bodies is thus one fewer than the number of X-chromosomes. Barr bodies are commonly referred to as sex chromatin. The human abnormalities called Kleinefelter's syndrome and Turner's syndrome both result from an unnatural presence or absence of a Barr body. In the case of the former, the male possesses a Barr body that it would normally not have, and in the latter case the Barr body is absent. *See* MARY LYON EFFECT.

barreleye *See* OPISTHOPROCTIDAE.

basal body A structure located in the *cytoplasm of cells that is essential for the formation of *cilia and *flagella and from which they project. Normally it is composed of nine sets of *microtubules, each set arranged in triplets, embedded in a dense, granular matrix.

basal bristle In birds, a small modified feather with little or no vane found at the base of the bill. *See also* RICTAL BRISTLE.

b

basal ganglia (basal nuclei) Highly interconnected clusters of nerve cells that form structures inside the *cerebrum. The principal ganglia are the caudate nucleus, putamen, and globus pallidus. They receive and process signals from various regions of the cerebral *cortex, then return the information to the *thalamus, from where it passes to the motor cortex. Together with another loop in the *cerebellum, this process initiates and regulates coordinated voluntary movements.

basal lamina One of two layers (the other is the reticular lamina) that form the basement membrane. It is an amorphous, or filamentous, sheet of material that underlies epithelial cells (*see* EPITHELIUM), composed of a special kind of *collagen and a *carbohydrate. In addition to support, it is thought to provide a barrier controlling the exchange of molecules between the epithelial layer and underlying tissue.

basal metabolic rate (BMR) The minimum *metabolic rate needed to sustain the life of an organism that is in an environment at a temperature equal to its own. The BMR is measured in organisms in a post-absorbtive, resting state.

base analogue A *purine or *pyrimidine base that differs slightly in structure from the normal base, but that because of its similarity to that base may act as a mutagen when incorporated into DNA. Once in place, these bases, which have pairing properties unlike those of the bases they replace, can cause mutations by causing insertions of incorrect *nucleotides opposite them during replication. Though the original base analogue exists only in a single strand, it can cause a nucleotide-pair substitution that is replicated in all DNA copies descended from that original strand. An example is 5-bromo-uracil (5BU), an analogue of *thymine that has bromine at the C-5 position in place of the CH_3 group found in thymine.

basement membrane A thin sheet of fibre that underlies *epithelium and *endothelium. *See* BASAL LAMINA.

base pair 1. Two *nucleotides on separate DNA strands that are connected through hydrogen bonds. **2.** A unit of measurement of a length of double-stranded DNA.

base pairing Complementary binding by means of hydrogen bonds of a *purine to a *pyrimidine base in *nucleic acids. In DNA, such binding occurs between cytosine and guanine, and between adenine and thymine; while in RNA, uracil substitutes for thymine.

basic dye A dye that consists of an organic cation which combines with and stains negatively charged macromolecules, e.g. *nucleic acids. It is used particularly for staining cell nuclei which contain nucleic acids.

basic rank In an animal society where rank is determined by alliances, the rank an individual possesses in a one-to-one encounter with no allies present. Basic rank was first described in Japanese macaques (*Macaca fuscata*). *Compare* DEPENDENT RANK.

***Basileuterus* (wood-warblers)** *See* PARULIDAE.

***Basiliscus* (basilisks)** *See* IGUANIDAE.

basilisks *See* IGUANIDAE.

***Basilosaurus* (order *Cetacea, suborder *Archaeoceti)** One of the best-known archaeocetes; it grew to approximately 20 m in length and lived during the upper *Eocene. It was discovered recently that *Basilosaurus* retained small hind limbs, useless for locomotion, but perhaps usable as *claspers during copulation.

basipodite In the segmented walking limbs of *Crustacea, the segment closest to the body.

basitarsus First tarsal segment in the leg of an insect, which articulates with the *tibia proximally, and with the other tarsal segments, or tarsomeres. *See* TARSUS.

basitibial At the base of the *tibia.

basking shark (*Cetorhinus maximus*) *See* LAMNIDAE.

Basommatophora (class *Gastropoda, **subclass *Pulmonata*)** An order of gastropods that are predominantly aquatic; certain groups have secondary gills. Eyes are present at the base of a single pair of tentacles. The order first appeared in the *Mesozoic.

basophilic Applied to a cell, its components, or products that can be stained by a basic dye.

basopodite *See* PROTOPODITE.

bass *See* SERRANIDAE.

basslets *See* GRAMMIDAE.

bastard wing *See* ALULA.

bat *See* CHIROPTERA.

Bateman gradient A linear relationship between the number of mates an animal has and the number of its offspring. The relationship, or gradient, is much larger for males than for females, therefore, male reproductive success should increase with multiple mating but female reproductive success should not. It was discovered in 1948 by the English geneticist Angus John Bateman (1919–96), experimenting with fruit flies (*Drosophila melanogaster*), who summarized the relationship as: 'Variance in number of mates is . . . the only important cause of sex difference in variance of fertility'.

Bates, Henry Walter (1825–92) An English naturalist who, in 1848, accompanied A. R. Wallace on an exploration of the Amazon, where he collected nearly 15 000 species of insects, 8000 of which were new to science. His studies of them led him to propose the way in which the mimicry named after him can arise among unrelated species. He received enthusiastic support from Darwin and from Sir Joseph Hooker and Darwin wrote an introduction to his only book, *The Naturalist on the River Amazon*, published by John Murray in 1863.

Batesian mimicry Mimicking of brightly coloured, or distinctively patterned, unpalatable species by palatable species. This helps protect the mimics from predators. It is named after H. W. Bates.

batfish *See* EPHIPPIDAE; OGCOCEPHALIDAE.

Bathonian A Middle *Jurassic stage, about 167.7–164.7 Ma ago, commonly represented by carbonate sediments in many areas of Europe. It is known to contain an abundant fauna of invertebrates.

bathyal Applied to those parts of the ocean where the depth of water is 200–2000 m. It is the region where the continental shelf slopes from the shallower *neritic zone on the landward side toward the deeper *abyssal zone.

Bathyclupeidae (subclass *Actinopterygii, order *Perciformes) A small family of physically fragile, silvery, deep-bodied, large-eyed, mid-water, deep-sea fish whose mouth gape is almost vertical. There are about four species, found at depths of about 500 m in tropical waters.

Bathydraconidae (dragonfish; subclass *Actinopterygii, order *Perciformes) A family of mainly deep-water Antarctic fish that have a flattened head, tapering body, and rounded tail fin. Some individuals have been caught at depths exceeding 2000 m. There are about 15 species.

***Bathydraco scotiae* (dragonfish)** *See* BATHYPELAGIC FISH.

Bathyergidae (mole rats; order *Rodentia, suborder *Hystricomorpha) A family of mole-like rodents that live in shallow tunnels. Most of the hair is lost, the legs are short, strong, and *pentadactyl, with strong claws. The skull is short and strong, with occipital and sagittal crests prominent, and paroccipital processes short. The cheek teeth are *hypsodont, and rooted. The teeth are reduced in number. The animals are distributed throughout Africa south of the Sahara. There are about 22 species, in five genera. At least one species, *Heterocephalus glaber* (naked mole rat) is known to be *eusocial like some insects, with a giant queen who is the sole breeder and a horde of small but adult individuals who perform specialized, non-breeding roles.

Bathylagidae (deep-sea smelts, black smelts; subclass *Actinopterygii, order *Salmoniformes) A family of small (15 cm) open-ocean fish with large eyes, a small mouth, and varying body shape. Probably they undertake vertical migrations between different ocean depths. There are about 35 species.

Bathymasteridae (searcher, ronquil; subclass *Actinopterygii, order *Perciformes) A small family of marine coastal fish of the north Pacific. The seven known species have a somewhat slender body with long *dorsal and *anal fins, and do not exceed 30 cm in length.

bathypelagic fish Deep-sea fish that live at depths of 1000–3000 m, where the environment is uniformly cold and dark. A swim-bladder is often absent, and many species (e.g. *Bathydraco scotiae*, dragonfish, and *Chiasmodon niger*, swallower) possess *photophores.

Bathypteroidae (subclass *Actinopterygii, order *Myctophiformes) A small family of mainly deep-water, bottom-living fish that have a flattened head, minute eyes, slender body, and very long first *pectoral and *pelvic fins, the latter being used occasionally to 'walk' over the sea-bottom. There are about eight species.

Batoidea Older name for the group or order of fish comprising the rays and skates (*Rajiformes).

Batoidimorpha (class *Chondrichthyes, subclass *Elasmobranchii) A superorder of living and fossil cartilaginous fish, including all rays and skates.

Batrachoididae (toadfish; superorder *Paracanthopterygii, order *Batrachoidiformes) A fairly large family of bottom-dwelling, inshore marine fish of tropical and subtropical waters. They have a wide (toad-like) mouth, a large head with dorsally positioned eyes, a short, stout body, and small gill openings. Some species can produce underwater sounds by means of special muscles acting on the wall of the swim-bladder. There are about 55 species.

Batrachoidiformes (subclass *Actinopterygii, superorder *Paracanthopterygii) An order of marine bony fish that have large heads, dorsally positioned eyes, a wide mouth, and *pelvic fins in the *jugular position. The order includes the various types of toadfish.

***Batrachostomus* (frogmouths)** *See* PODARGIDAE.

batrachotoxin The most powerful poison produced by any vertebrate animal. It is secreted by the skin of S. American arrow-poison frogs of the genus *Dendrobates*.

Bauplan (*pl.* Baupläne) The generalized body plan of an archetypal member of a major taxon.

bay owls *See* TYTONIDAE.

bay-winged cowbird (*Molothrus badius*) *See* ICTERIDAE.

B cell A type of lymphocyte (*leucocyte) manufactured in the *bone marrow (hence the name) which secretes *antibodies and is involved in the *humoral response. *Compare* T CELL.

Bdelloida (phylum *Aschelminthes, class *Rotifera) An order of freshwater rotifers which have slender, jointed bodies, and a free-swimming mode of life. The cuticle may bear longitudinal depressions. There are no males present in this order.

Bdellonemertini (phylum *Nemertini, class *Anopla) An order of worms which possess a relatively simple *proboscis. All have a contorted intestine.

beach flea *See* AMPHIPODA.

beach hopper *See* AMPHIPODA.

beach salmon *See* LEPTOBRAMIDAE.

beaded lizard *See* HELODERMATIDAE.

beak 1. The bill of a bird. **2.** *See* SHELL BEAK.

bean weevil (*Sitona lineatus*) *See* CURCULIONIDAE; BRUCHIDAE.

bearded reedling (*Panurus biarmicus*) *See* PARADOXORNITHIDAE.

bearded tits *See* PARADOXORNITHIDAE.

beardfish *See* POLYMIXIIDAE.

beard worms *See* POGONOPHORA.

bears *See* URSIDAE.

beaver *See* APLODONTIDAE; CASTORIDAE.

bedbug (*Cimex lectularius*) *See* CIMICIDAE.

bee bread Mixture of pollen, honey, and in some species plant oils and *Dufour's gland exudates, that comprises the food of bee larvae.

bee dance *See* DANCE LANGUAGE.

bee-eaters *See* MEROPIDAE.

beefalo *See* CATTALO.

bee-flies *See* BOMBYLIIDAE.

bee humming-bird (*Mellisuga helenae*) *See* TROCHILIDAE.

bee lice *See* BRAULIDAE.

bees 1. *See* APIDAE; APOCRITA; HYMENOPTERA; SPHECOIDEA. **2. (solitary mining bees)** *See* ANDRENIDAE.

beeswax Wax secreted from glands beneath the abdominal *terga or *sterna of bees of the family *Apidae, which is used in nest construction.

beetle-hunting wasp *See* HUNTING WASP.

beetles *See* COLEOPTERA.

behavioural assay A technique used to determine an animal's response to a stimulus and its sensory powers. Geese, gulls, and other birds roll into their nests any egg they find outside the nest. Herring gulls (*Larus argentatus*) react more strongly to large eggs than to small ones regardless of their shape and prefer stippled eggs to those without stippling. Karl von Frisch used food rewards to train minnows to distinguish colours and catfish to respond to a whistle, demonstrating that minnows are not colour-blind and catfish are not deaf.

behavioural ecology 1. The study of the behaviour of an organism in its natural habitat. **2.** The application of behavioural theories (e.g. *game theory) to particular activities (e.g. *foraging).

behavioural thermoregulation The maintenance of a constant body temperature by means of basking, sheltering, shivering, etc. *See also* ECTOTHERM.

behavioural unit (unit of behaviour) An orderly sequence of muscular contractions that results in a recognizable pattern of behaviour, such that an animal will repeat the behaviour whenever the circumstances triggering it occur, and other members of its species perform similar behaviour.

belemnites *See* BELEMNITIDA.

Belemnitida (belemnites; class *Cephalopoda, subclass *Coleoidea) An order of extinct cephalopods that had an internal shell composed of a *phragmocone, a *rostrum, and a pro-*ostracum. There are five families. They appear in upper *Carboniferous rocks and persist generally until the end of the *Cretaceous, and in one area into the *Eocene.

bell animalcules *See* PERITRICHIA.

bell moths *See* TORTRICIDAE.

bellowsfish *See* MACRORHAMPHOSIDAE.

Belonidae (garfish, long tom, needle-fish; subclass *Actinopterygii, order *Atheriniformes) A family of mainly marine, tropical and subtropical fish with very long, slender bodies, beak-like jaws, and *dorsal and *anal fins placed close to the tail fin. Often swimming in schools, they are known to leap out of the water when frightened. There are about 26 species.

Belontiidae (gourami, labyrinth fish; subclass *Actinopterygii, order *Perciformes) A family of small, freshwater fish that have a protrusible upper jaw, vestigial *lateral line, 10 or fewer soft rays in the *dorsal fin, and often an elongate pelvic ray in the *jugular *pelvic fins. Labyrinth-like *accessory respiratory organs can be found in the gill chambers. Prior to spawning, males often make a foamy bubble nest at the surface of the water. There are approximately 12 genera and 46 species, indigenous to Africa and southern Asia.

Belostomatidae (order *Hemiptera, suborder *Heteroptera) Family of large, aquatic bugs, which are ferocious predators of invertebrates, small fish, and frogs, seizing their prey with the front pair of legs. The family includes some of the largest aquatic insects. They are often attracted to artificial light. They occur mainly in the tropics.

belt transect A strip, typically 1 m wide, that is marked out across a habitat and within which species are then recorded to determine their distribution in the habitat. *See* TRANSECT. *Compare* LINE TRANSECT.

beluga 1. (*Huso huso*) *See* ACIPENSERIDAE. **2. (white whale, *Delphinapterus leucas*)** *See* MONODONTIDAE.

Bengalese finch (*Lonchura striata*) *See* ESTRILDIDAE.

Bentheuphausiidae *See* EUPHAUSIACEAE.

benthic fish Fish that live on or near the sea-bottom, irrespective of the depth of the sea. Many benthic species have modified fins, enabling them to crawl over the bottom; others have flattened bodies and can lie on the sand; others live among weed beds, rocky outcrops, and coral reefs.

benthos (*adj.* benthic) In freshwater and marine *ecosystems, the organisms attached to or resting on the bottom sediments.

Bergmann's rule The idea that the size of *homoiothermic animals in a single, closely related, evolutionary line increases along a gradient from warm to cold temperatures—i.e. that races of species from cold climates tend to be composed of individuals physically larger than those of races from warm climates. This is because the surface-area:body-weight ratio decreases as body weight increases. Thus a large body loses proportionately less heat than a small one. This is advantageous in a cold climate but disadvantageous in a warm one. *See also* ALLEN'S RULE; GLOGER'S RULE.

Beroida *See* BEROIDEA.

Beroidea (Beroida; phylum *Ctenophora, class *Nuda) An order of ctenophorids which are mainly cylindrical in shape. The body contains a very large mouth. *Tentacles do not occur at any stage in development.

Berycidae (alfonsinos; subclass *Actinopterygii, order *Beryciformes) A small family of marine, offshore fish that are deep-bodied, with strongly forked tails, large eyes, and often a reddish skin indicative of fish living at moderate depths: in temperate Australia, *Trachichthodes affinis* (nannygai) are trawled commercially and are sometimes sold as 'redfish'. There are about 10 species, with world-wide distribution.

Beryciformes (class *Osteichthyes, subclass *Actinopterygii) Order of bony fish, most of which have short, deep bodies, rough scales, sharp spines on the head, and usually more than five soft rays in the ventral fins. With more than 140 species, this order is divided into several families, e.g. *Berycidae (alfonsinos), *Holocentridae (squirrelfish), and *Anomalopidae (lantern-eye fish).

beta-galactosidase An inducible *enzyme that is responsible for the catalytic *hydrolysis of *beta-galactoside.

beta-galactoside A *glycoside of the beta-stereoisomer (a stereoisomer is an *isomer in which groups form a square-shaped plane around a single atom or group) of *galactose.

beta pleated sheet A structural configuration, normally found in fibrous proteins, in which *polypeptide chains are partially extended and held together by interchain bonds between —NH and —CO groups on all *peptide bonds.

bet-hedging The behavioural response of a *species *population to a K-selecting environment in which occasional fluctuations in conditions affect the mortality rate of juveniles to a much greater extent than that of adults. In such a situation adults release young into several different environments to maximize the chance that some will survive. *See also* K-SELECTION; r-SELECTION.

Bethylidae (suborder *Apocrita, superfamily Bethyloidea) Family of small to medium-sized wasps (usually less than 8 mm long), which are usually black, with an elongate head, and slightly elbowed, 12- or 13-segmented antennae. The wing venation is reduced, and the hind wing has a jugal (*see* JUGUM) lobe. The females of many species are *apterous and ant-like, and a few will sting. The habits of the family are very diverse, but in most cases no true nest is made; the host larva (a small lepidopteran or coleopteran) is dragged to a sheltered location where more than one egg may be laid. The larvae are external parasites, and having matured will spin a silken cocoon. These wasps are often considered intermediate between true parasitoid (*see* HYPERPARASITE) and *fossorial families.

Bibionidae (fever flies, March flies, St Mark's fly; order *Diptera, suborder *Nematocera) Family of somewhat variable flies which has been subdivided by some authors into as many as five subgroups. The adults are, typically, robust flies with shorter legs and wings than those of other families of Nematocera. Their antennae have 8–16 bead-like segments and are usually shorter than the *thorax. Wings are large, with prominent anterior veins. Males have *holoptic eyes, and *ocelli are present. The larvae live in soil, roots, or decaying vegetation. The larvae of some species are among the most primitive of all dipteran larvae, having 12 segments, a large head, well-developed mouth-parts, and no *pseudopods. More than 700 species have been described.

Bibymalagasia An order of *mammals described in 1994 by R. D. E MacPhee, based on the enigmatic genus (*see* ENIGMATIC TAXON) *Plesiorycteropus, known from sub-Recent fossil material from Madagascar. The order is distinguished by having large, perforating, transarcual canals in the neural arches of the lumbar, posterior thoracic, and anterior sacral *vertebrae; a posteromedial process on the *astralagus, with a ventral groove for flexor tendons; and large ischial expansions (*see* ISCHIUM). *Plesiorycteropus* was formerly assigned to the order *Tubulidentata.

bichir (*Polypterus*) *See* BRACHIOPTERYGII; POLYPTERIDAE.

Bidder's organ A rudimentary ovary found in the males of some frogs and toads (*Anura).

bifid Split in two.

bifurcate Forked, with two branches.

big-bang reproduction *See* SEMELPARITY.

big-eye *See* PRIACANTHIDAE.

bighead *See* MELAMPHAEIDAE.

big-headed turtles *See* PLATYSTERNIDAE.

bigmouth sleeper (*Gobiomorus dormitor*) *See* ELEOTRIDAE.

bigscale *See* MELAMPHAEIDAE.

Biharian A faunal stage of the lower *Pleistocene (1.8–1.5 Ma ago), corresponding to the early part of the *Calabrian.

bilateral symmetry The arrangement of the body components of an animal such that one plane divides the animal into two halves which are approximate mirror images of each other. Bilateral symmetry is associated with movement in which one end of the animal constantly leads. *Compare* RADIAL SYMMETRY.

bilbies *See* PERAMELOIDEA.

bile A secretion, produced in the *liver and stored in the *gall bladder, that comprises some waste products of *metabolism and a surface-active agent. It is discharged via the bile duct into the small intestine, from which it is excreted; it also serves to emulsify fats during digestion and so aids their transport across the gut membrane.

bilharzia *See* SCHISTOSOMIASIS.

bilirubin A golden-yellow *bile pigment that is derived from the degradation of *porphyrins, in particular *haem compounds, by cells of the *reticuloendothelial system, notably those of the *liver.

biliverdin A green *bile pigment that is usually formed as a result of the oxidation of *bilirubin in the *gall bladder.

billing During birds' courtship, a behaviour in which a pair touch or clasp each other's bills. This appears to reinforce the pair bond (*see* BONDING). In gannets (*Sula bassana*) the pair raise their bills and clatter them together repeatedly. A male puffin (*Fratercula arctica*) touches the female's bill with a nibbling action.

bilophodonty In some mammals, a condition in which the four cusps on the molar teeth are joined by two transverse ridges. *See also* LOPHODONT.

binary fission The division of one cell into two similar or identical cells; it is a common method of *asexual reproduction in single-celled organisms.

binocular vision Vision that results from the ability of an animal to view an object using both eyes simultaneously, and which probably enables the animal to judge distances. In those *Mammalia and *Aves which possess binocular vision, the *orbits are directed forward. In some arboreal *Serpentes, in which the orbits are directed laterally, binocular vision may result from the possession of eyes whose pupils are long, horizontal slits located eccentrically and close to the anterior margins of the orbits.

binomial classification (binominal classification) The systematic description of *species by means of two names, both in Latin. The first name, with an initial capital letter, is that of the genus into which the species is placed, the second that of the species itself (e.g. the name *Homo sapiens* (human being) comprises the species *sapiens* within the genus *Homo*). This method of classification was introduced by Carolus Linnaeus. The precise procedures and rules for naming newly discovered animals are laid down in the *International Code of Zoological Nomenclature.

binominal classification *See* BINOMIAL CLASSIFICATION.

biochemical oxygen demand (biological oxygen demand, BOD) An indicator of the pollution of water by organic matter where the decomposition of the organic matter by micro-organisms depletes the water of dissolved oxygen. It is measured as the mass (mg) of oxygen per litre of water that is taken up when a sample of water containing a known amount of oxygen is kept in darkness at 20°C, usually for five days (when the result is described as the BOD_5).

biochrome *See* PIGMENT.

biochronology The measurement of units of geological time by means of biological events. Biochronologists often derive their correlations from widespread and distinctive events in the biological history of the world based on the first and last appearances of organisms.

biocoenosis The living part of a *biogeocoenosis, comprising the *phytocoenosis, the *zoocoenosis, and the *microbiocoenosis. It is equivalent to the biotic component, or *biome, in an *ecosystem.

biodegradable Applied to substances that are easily broken down by living organisms.

biodiversity A term that came into widespread use in the 1980s, which describes all aspects of biological diversity, especially *species richness, *ecosystem complexity, and genetic variation.

bio-energetics The study of energy transformations in living organisms, in particular the formation of *ATP and its subsequent use in metabolic processes.

biogenesis The principle that a living organism can arise only from another living organism, a principle contrasting with concepts such as that of the spontaneous generation of living from non-living matter. The term is currently more often used to refer to the formation from or by living organisms of any substance (e.g. coal, chalk, or chemicals).

biogenetic law The early stages of development in animal species resemble one another, the species diverging more and more as development proceeds. The law was formulated by the embryologist E. K. von Baer (1792–1876).

biogenic Applied to the formation of rocks, traces (fossils), or structures as a result of the activities of living organisms.

biogeocoenosis Term equivalent to '*ecosystem' that is often used in Central European literature. Comprises a *biocoenosis together with its habitat (the ecotope).

biogeographical barrier A barrier that prevents the migration of species. The

various disjunctive geographical groupings of plants and animals are usually delimited by one or more such barriers which may be climatic, involving temperature and the availability of water, or physical, involving e.g. mountain ranges or expanses of sea water.

biogeographical province A biological subdivision of the Earth's surface, usually on the basis of taxonomic rather than ecological criteria, and embracing both faunal and floral characteristics. The hierarchical status of such a unit, and the total number of such units, varies from one authority to another.

biogeographical region A biological subdivision of the Earth's surface that is delineated on the same general principles as a *biogeographical province but has superior taxonomic status. The provinces are grouped into regions, of which the following are generally recognized: Antarctic, Australasian, *Ethiopian, *Nearctic, *Neotropical, Oceanian, *Oriental, and *Palaearctic. A variant to this grouping has been proposed for mammals (*see* MAMMAL REGIONS).

biogeography A diverse subject, that focused traditionally on the geographical distribution of plants and animals at different taxonomic levels, past and present. Modern biogeography also lays great stress on the ecological character of the world vegetation types, and on the evolving relationship between humans and their environment.

biological clock An endogenous, physiological mechanism, whose exact nature has not been determined, that keeps time independently of external events, enabling organisms to determine and to respond to daily, lunar, seasonal, and other periodicities. Its existence has been inferred from the observation of organisms which retain rhythmic activity under constant conditions. *See also* CIRCADIAN RHYTHM.

biological conservation Active management to ensure the survival of the maximum *diversity of *species, and the maintenance of genetic variety within species. The term also implies the maintenance of *biosphere functions, e.g. biogeochemical recycling, without which the basic resources for life would be lost. Biological conservation embraces the concept of long-term sustained resource use or sustained yield from the biosphere, which may conflict with species conservation in some circumstances. Conservation of species and biological processes is unlikely to succeed without simultaneous conservation of *abiotic resources.

biological control The control of a pest species by the introduction of other organisms (e.g. the control of rabbits in Australia by the introduction of the myxomatosis virus, and the control of citrus scale insects in California by the introduction of an Australian species of ladybird). The term is also applied to the introduction of large numbers of sterilized (usually irradiated) males of the pest species, whose matings result in the laying of infertile eggs (e.g. to control the screw worm fly in the USA).

biological oxygen demand *See* BIOCHEMICAL OXYGEN DEMAND.

biological species concept The view that the *species comprises populations (or groups of populations) that are reproductively isolated from each other, i.e. that species should always be *biospecies. This has been mistakenly interpreted as implying that either they cannot interbreed, or that *hybrids between them are sterile; but there are many *reproductive isolating mechanisms. The concept was proposed by Ernst Mayr (1904–2005) in the 1940s, but many biologists have come to regard it as too restrictive. For example, species that are not *sister groups may interbreed in nature. Further, the professional taxonomist, usually working in a museum, must make (often unwarranted) judgements about whether two taxa might or might not be capable of interbreeding in nature. Other species concepts (the *recognition, *cohesion, and, especially, the *phylogenetic species concepts) have come to be employed more and more in recent years.

bioluminescence The production by living organisms of light without heat. Bioluminescence is a property of many types of organism (e.g. certain (mostly marine) bacteria, dinoflagellates, and fireflies).

biomass The total mass of all living organisms, or of a particular set (e.g. species), present in a *habitat or at a particular *trophic level in a *food chain, and usually expressed as dry weight or, more accurately, as the carbon, nitrogen, or calorific content, per unit area.

biome A biological subdivision of the Earth's surface that reflects the character of the vegetation. Biomes are the largest geographical biotic communities that it is convenient to recognize. They broadly correspond with climatic regions, although other environmental controls are sometimes important. They are equivalent to the concept of major plant formations in plant ecology, but are defined in terms of all living organisms and of their interactions with the environment (and not only with the dominant vegetation type). Typically, distinctive biomes are recognized for all the major climatic regions of the world, emphasizing the adaptation of living organisms to their environment, e.g. tropical-rainforest biome, desert biome, tundra biome.

biomechanics The scientific study of the structure and function of biological organisms and their component parts, interpreted through the laws of mechanics.

biometrics *See* BIOMETRY.

biometry (biometrics) Quantitative biology, i.e. the application of mathematical and statistical concepts to the analysis of biological phenomena.

bionomic strategy Characteristic features of an organism or population (e.g. size, longevity, fecundity, range, and migratory habit) that give maximum fitness for the organism in its environment.

biomineralization The incorporation of inorganic compounds, such as salts, into biological structures, often to lend them hardness or rigidity. Biomineralization first occurred in, and defines, the *Cambrian period, beginning about 542 million years ago, in *brachiopods, *trilobites, *ostracods, and graptolites (*Graptolithina). In vertebrates, *hydroxyapatite usually occurs, in invertebrates inorganic minerals are more varied: calcite and aragonite (a harder, less stable form of calcite) are common, permeating chitin to form the hard exoskeletons of *Arthropoda and also forming the calcareous material of shells; in *Radiolaria and some *Porifera, the skeleton is made of opaline silica; radiolarians occasionally have a strontium sulphate instead of siliceous skeleton.

biophage *See* CONSUMER ORGANISM.

biospecies A group of interbreeding individuals that is isolated reproductively from all other groups.

biosphere The part of the Earth's environment in which living organisms are found, and with which they interact to produce a steady-state system, effectively a whole-planet *ecosystem. Sometimes it is termed 'ecosphere' to emphasize the interconnection of the living and non-living components.

biosphere reserve One of a series of conservation sites designated by the United Nations Educational, Scientific, and Cultural Organization (UNESCO) in an attempt to establish an international network of protected areas encompassing examples of all the Earth's major vegetation and physiographic types.

(((⊕))) SEE WEB LINKS

• Description of biosphere reserves at the UNESCO website.

biostratigraphy The branch of stratigraphy that involves the use of fossil plants and animals in the dating and correlation of the stratigraphic sequences of rock in which they are discovered. A zone is the fundamental division recognized by biostratigraphers.

biosynthesis The formation of compounds by living organisms.

biota Plants and animals occupying a place together, e.g. marine biota, terrestrial biota.

biotelemetry (radiotelemetry) Remote-sensing method for monitoring animal movements using small transmitters attached externally to the body of the animal or implanted within the body cavity. It enables the precise locations of animals to be followed, and may also monitor changes in heartbeat, e.g. in response to stress. It is especially useful for studying migratory animals in remote areas and is an important aid to the study of bird migrations. *See also* RADAR TRACKING; RADIO TRACKING.

biotic Applied to the living components of the *biosphere or of an *ecosystem, as distinct from the *abiotic physical and chemical components.

biotic indices *Indicator species, when used as a guide to the level of a particular *abiotic factor. For example, the presence of certain invertebrate groups in fresh water can be awarded a score that indicates the quality of the water.

biotic potential (intrinsic rate of natural increase) The maximum reproductive potential of an organism, symbolized by the letter r. The difference between this and the rate of increase that actually occurs under field or laboratory conditions reflects the environmental resistance. *See also* LOGISTIC EQUATION.

biotin Vitamin B_7 (*See* VITAMIN); it serves as a *prosthetic group in *enzymes involved in carboxylation–decarboxylation reactions (e.g. in the *citric-acid cycle).

biotope An environmental region characterized by certain conditions and populated by a characteristic *biota.

biotopographic unit 1. A small *habitat unit with distinctive topography formed by the activities of an organism (e.g. an ant hill). 2. A small topographic unit that, by its aspect, position, or other characteristics, generates a distinctive micro-environment for living organisms. Examples include solar or shade slopes, windward slopes, and similar units in sand-dunes.

biotroph *See* PARASITISM.

biotype 1. A naturally occurring group of individuals with identical *genomes. **2.** A physiological race, i.e. a group of individuals identical in structure but showing differences in physiological, biochemical, or pathogenic characters.

biozone The total range of a given species defined within specific time limits.

bipectinate Resembling a comb, in arrangement or shape, in which the 'teeth' occur on both sides of the main stem. The term is most commonly applied to insect antennae.

Bipes (two-legged worm lizards) *See* AMPHISBAENIDAE.

bipinnaria larva In *Asteroidea, a larva that is derived from the *auricularia but differs from it in the development of two ciliated lateral projections used for locomotion and feeding.

biradial symmetry The arrangement of the body components of an animal such that similar parts are located to either side of a central axis and each of the four sides of the body is identical to the opposite side but different from the adjacent side.

biramous Two-branched.

birds *See* AVES.

birds of paradise *See* PARADISAEIDAE.

birdwing butterflies *See* PAPILIONIDAE.

birth cord *See* UMBILICAL CORD.

biserial Side by side.

biseriate Arranged in rows.

bisexual Applied either to a species comprising individuals of both sexes, or to a hermaphrodite organism (in which an individual animal possesses both ovaries and testes).

bishops *See* PLOCEIDAE.

Bison *See* BOVIDAE.

Biston betularia (peppered moth; order *Lepidoptera) Moth which has been the subject of classical studies of *industrial melanism, where there is selective predation by birds upon light moths that are conspicuous against a dark, sooty background, and, conversely, upon dark moths (the *carbonaria* form) against a light, soot-free background.

bitheca One of the three types of graptolite (*See* GRAPTOLITHINA) thecae, possibly containing the male *zooid. *Compare* AUTOTHECA; STOLOTHECA.

biting lice *See* MALLOPHAGA.

biting midges *See* CERATOPOGONIDAE.

Bitis arietans (puff adder) *See* VIPERIDAE.

Bitis nasicornis (rhinoceros viper, river-jack) *See* VIPERIDAE.

Bittacidae *See* MECOPTERA.

bittern *See* ARDEIDAE.

Biuret reaction A chemical test that detects the presence of *peptide bonds. The sample is treated with a strong base (usually sodium hydroxide, NaOH) and aqueous copper sulphate ($CuSO_4$). Peptides react with the reagent to form a blue-violet complex. The total concentration of *protein is indicated by the intensity of the colour.

(((⊕))) SEE WEB LINKS
• Instructions for performing the Biuret test.

bivalent In genetics, applied to two *homologous chromosomes when they are paired during the *prophase of *meiosis.

Bivalvia (Pelecypoda, Lamellibranchia; phylum *Mollusca) A class of molluscs in which the body is laterally compressed, there is no definite head, and the soft parts are enclosed between two oval or elongated, calcareous valves. The valves are not equilateral, being united on the dorsal side by a toothed hinge. Most are *bilaterally symmetrical along the plane of junction of the two valves. The valves are closed ventrally by the contraction of one or two *adductor muscles, and opened by a horny, elastic ligament. Large, modified, ciliated gills combine with ciliated *labial palps in food collection. The stomach is extremely elaborate, with a crystalline *style. The circulatory, excretory, and reproductive systems are less complex. There is a ganglionic nervous system. Inhalant and exhalant siphons are present. *Trochophore and *veliger larvae are produced by marine forms. The sexes are generally separate; some are hermaphroditic and *protandrous. Bivalves are entirely aquatic and are adapted to various modes of life, as burrowing, boring, sessile, and free-living organisms. They first appeared in the Early *Cambrian and form the second largest molluscan class, with more than 20 000 species.

bivouac See DORYLINAE.

Blaberidae (class *Insecta, order *Blattodea) Family of medium to large, broad-bodied cockroaches, most of which have short, stout legs, and relatively short antennae. Some species burrow in the ground, or live in rotting wood. *Leucophaea maderae* (Madeira cockroach) is able to *stridulate, and produces an offensive odour when disturbed. *Gromphadorhina* species can produce sound by forcing air through the *spiracles. This mainly tropical group includes the largest cockroaches known.

blackbird 1. (*Turdus merula***)** See TURDIDAE. **2. (New World blackbirds)** See ICTERIDAE.

black box system A system about whose structure nothing is known beyond what can be deduced from its behaviour. Statistical relationships between inputs to the system and outputs from it can be deduced by manipulating the inputs.

black coral See ANTIPATHARIA.

black dragonfish 1. See IDIACANTHIDAE. **2.** See MELANOSTOMIATIDAE.

blackfish (black ruffe; *Centrolophus niger*) See CENTROLOPHIDAE.

black-flies See SIMULIIDAE.

blackfly See APHIDIDAE.

black grouse (*Lyrurus*) See TETRAONIDAE.

blackhead A disease of turkeys caused by the protozoon *Histomonas meleagridis*. The disease affects the caecum and liver of the bird, and the mortality rate may be high. The disease is transmitted by a nematode worm parasitic in the intestines of turkeys.

black kingfish See RACHYCENTRIDAE.

black rhinoceros (*Diceros bicornis*) See RHINOCEROTIDAE.

black vruffe (blackfish; *Centrolophus niger*) See CENTROLOPHIDAE.

black salamander (*Salamandra atra*) See SALAMANDRIDAE.

black smelts See BATHYLAGIDAE.

black swan (*Cygnus atratus*) See ANATIDAE.

black widow spider (*Latrodectus mactans*) See THERIDIIDAE.

bladder A bag in which metabolic products, or air, may be stored.

blastocoel (segmentation cavity) A cavity that appears during embryonic development in animals at the end of *cleavage of the egg. It is found within the mass of cells that makes up the *blastula, and is filled with fluid.

blastocyst A mammalian *embryo at the stage at which, after *cleavage, it is implanted in the wall of the *uterus. It consists of a thin-walled, hollow sphere (trophoblast) enclosing a *blastocoel. Attached to the inner surface of the trophoblast at one end there is a sheet of cells that will become the embryo proper.

blastodisc (germinal disc) A disc-shaped layer of *cytoplasm that is formed at the *animal pole by the *cleavage of a large, yolky egg (e.g. that of a bird or reptile). *Mitosis within the blastodisc gives rise to the *embryo and its membranes.

blastomere One of the cells formed by the division of the fertilized egg during *cleavage and prior to *gastrulation. The blastomeres may differ in size, the larger ones being termed macromeres and the smaller ones micromeres.

blastopore A mouth-like opening of the *archenteron on the surface of an *embryo in the *gastrula stage. In many animals the blastopore becomes the *anus, although in some it closes up at the end of *gastrulation and may appear again at or near the same site. The blastopore is formed by an inward movement of the endoderm and mesoderm cells of the archenteron during gastrulation. Sometimes this movement is incomplete, so that an open pore does not develop; this explains the *primitive streak of a bird or mammal embryo during gastrulation. In that case it is referred to as a 'virtual' blastopore. *See also* DEUTEROSTOMY; PROTOSTOMY.

Blastozoa (phylum Echinodermata) A subphylum, whose validity is not universally accepted, that includes the extinct classes Eocrinoidea, Rhomoifera, Blastoidea, and Parablastoidea.

blastozooid A member of a colony of animals which are produced by asexual budding.

blastula An early embryonic stage of an animal, around the time of *cleavage, that usually consists of a hollow sphere of cells. It is formed immediately prior to *gastrulation.

Blattellidae (class *Insecta, order *Blattodea) Large, cosmopolitan family of generally small cockroaches, which have long antennae and slender legs. Many species are fully winged in both sexes, and some species are domestic pests.

Blattidae (class *Insecta, order *Blattodea) Cosmopolitan family of medium to large cockroaches some of which are domestic pests (e.g. *Periplaneta americana*, the American cockroach).

Blattodea (cockroaches, roaches; phylum *Arthropoda, class *Insecta) Order of medium to large, *exopterygote insects in which the fore wings are modified as rather strongly sclerotized *tegmina. Wings are absent in some groups. The legs are *gressorial, and the body is dorsoventrally compressed. Auditory organs are absent. Females have a small, concealed *ovipositor, and eggs are laid in an *ootheca. Most cockroaches are nocturnal, and during the day are usually found hiding in crevices. Burrowing and cave-dwelling species are known, and some species are even amphibious. Cockroaches may be adapted to desert conditions, or live in tropical rain forest. It is thought that most cockroaches are omnivorous. Of the approximately 3500 species known, some are of economic importance as domestic pests.

blending inheritance Inheritance in which the characters of the parents appear to blend to form an intermediate state in the offspring, and in which there is no apparent segregation in later generations. The concept was proposed originally by biologists in the 19th century, including *Darwin, but later it was discredited as a model of inheritance after the results of *Mendel's experiments had been recognized.

Blenniidae (blennies; subclass *Actinopterygii, order *Perciformes) A very large family of small, mostly marine fish found in shallow waters, even in tidal pools, in large numbers. They are identifiable by their long *dorsal and *anal fins, virtually naked skin, and two pronounced rays in each of the *pelvic fins located below the gills; often they have tufts of skin (tentacles) above the head. Many species rely on skin coloration for protection. There are about 550 species, distributed world-wide in tropical and temperate waters.

blepharoplast A cylindrical *basal body of flagellates that is composed of parallel peripheral rods connected to the axial filaments of *flagella or *cilia.

blind cavefish (*Anoptichthys jordani*) See CHARACIDAE.

blind snakes See TYPHLOPIDAE.

blindworm (*Anguis fragilis*) See ANGUIDAE.

blister beetle See MELOIDAE.

blood In animals, a fluid circulated through the body by muscular activity and usually containing respiratory pigments conveying oxygen, food materials, excretory products, cells that produce *antibodies (lymphocytes), and cells that invade tissue to attack invading organisms.

blood clotting A response to the injury of a blood vessel in which the escape of liquid blood is prevented by its conversion into a solid mass, or clot, composed of fibres of *fibrin in which blood cells are enmeshed. In vertebrates, a soluble *protein, *fibrinogen, found in the *blood plasma is acted upon by the enzyme *thrombin. As a result, a negatively charged *peptide is split off from the fibrinogen molecule, leaving monomeric fibrin; this is capable of rapid polymerization to produce the tangle of insoluble threads that forms a clot. Active thrombin is formed from inactive prothrombin, also found in blood plasma, by a very complex process involving a lipoprotein factor liberated by rupturing blood platelets, a protein plasma thromboplastin component (PTC), antihaemophilic globulin (AHG), calcium ions, etc. Removal of free calcium from the blood, e.g. by adding citrate or oxalate, inhibits the formation of thrombin.

blood group A group of individuals (usually humans, but the term may also refer to other primates) classified according to the occurrence of *agglutination of the red blood cells when bloods from incompatible groups are mixed. Agglutination occurs when blood from any two groups is mixed: a reaction occurs between agglutinogens (*antigens) in red blood cells and agglutins (*antibodies) in *blood plasma. The classical blood groupings and their frequencies in the human population are: A (41%); B (9%); AB (3%); and O (47%), although very many other groups have also been identified, e.g. Duffy, Kidd, Lewis, Lutheran, MN, P, and Rh. See also RHESUS FACTOR.

blood plasma The almost colourless fluid that remains when all corpuscles have been removed from blood (present as a suspension after centrifugation of whole blood). It contains the *protein *fibrinogen, which is acted upon by the enzyme *thrombin to cause *blood clotting; and *antibodies responsible for immune reactions on combining with *antigens.

blood platelet See PLATELET.

bloodsucker (*Calotes versicolor*) See AGAMIDAE.

blow flies See CALLIPHORIDAE.

blubber In cetaceans (*Cetacea), seals, sea lions and walruses (*Pinnipedia), and sea cows (*Sirenia), a thick layer of *dermal fat that provides thermal insulation and may also serve as a reserve of food and possibly water.

bluebottles See CALLIPHORIDAE.

blue butterflies See LYCAENIDAE.

blue cod (*Parapercis colias*) See MUGILOIDIDAE.

bluefish (*Pomatomus saltator*) See POMATOMIDAE.

blue marlin (*Makaira nigricans*) See ISTIOPHORIDAE.

blue parrotfish (*Scarus coeruleus*) See SCARIDAE.

blue shark (*Prionace glauca*) *See* CARCHARHINIDAE.

bluethroats *See* TURDIDAE.

blue whale (Sibbalds's rorqual; *Balaenoptera musculus*) *See* BALAENOPTERIDAE.

BMR *See* BASAL METABOLIC RATE.

Boa constrictor (boa constrictor) *See* BOIDAE.

boarfish *See* CAPROIDAE; PENTACEROTIDAE.

boas *See* BOIDAE.

boat-billed heron (*Cochlearius cochlearius*) *See* ARDEIDAE; COCHLEARIIDAE.

BOD *See* BIOCHEMICAL OXYGEN DEMAND.

body louse (*Pediculus humanus*) *See* PEDICULIDAE.

Bohr effect A decrease in the affinity of *haemoglobin for oxygen and a shift of the *oxygen-dissociation curve to the right, brought about either by a decrease in *pH or by an increase in the partial pressure of carbon dioxide, such that a high concentration of carbon dioxide causes more oxygen to be given up at any given oxygen pressure (e.g. in the tissues). It is an important mechanism in the release of oxygen by haemoglobin in the blood in respiratory tissues, where the partial pressures of carbon dioxide are high.

Boidae (constricting snakes; order *Squamata, suborder *Serpentes) A family of large, mainly tropical snakes that kill their prey by constriction. They have a flexible skull, paired lungs, and enlarged ventral scales. Rudimentary hind limbs are present internally and continue externally as anal spurs. The family includes the pythons (all Old World in distribution) and the boas (mainly New World in distribution). *Boa constrictor* occurs in a wide range of habitats from Mexico to Argentina and on some W. Indian islands, but seldom enters water. The exclusively aquatic *Eunectes murinus*

(anaconda), of S. and Central America, is the largest extant snake, allegedly reaching 11 m in length. *Python reticulatus* (reticulated python) of the Indo-Malayan region is similar to the anaconda, the largest authenticated specimen being 8.5 m long, and is often found near water, sometimes swimming out to sea. There are about 70 species of Boidae.

Bojanus, organ of One of a pair of nephridial (*see* NEPHRIDIUM) excretory organs found in bivalve *Mollusca; they are situated on either side of the body below the *pericardium. They were discovered by the German physician Ludwig Heinrich Bojanus (1776–1827).

bombardier beetle (*Brachinus* species) *See* CARABIDAE.

Bombay duck *See* HARPADONTIDAE.

bomb calorimeter A device for measuring the energy content of material. The sample is burned in an oxygen-rich atmosphere inside a sealed chamber that is surrounded by a jacket containing a known volume of water. The rise in temperature of the water is recorded and used to calculate the amount of heat produced.

Bombina bombina (fire-bellied toad) *See* DISCOGLOSSIDAE.

Bombus (bumble-bees) *See* APIDAE.

Bombycidae (silk moths; subclass *Pterygota, order *Lepidoptera) Family of moths with broad and generally brown wings. The antennae are *bipectinate. The larvae spin very dense silk cocoons. There are few species, found mainly in tropical Africa and Asia.

Bombycilla (waxwings) *See* BOMBYCILLIDAE.

Bombycillidae (hypocolius, silky flycatchers, waxwings; class *Aves, order *Passeriformes) A family of variable, softplumaged, mostly crested birds that inhabit forests, plains, and deserts, feeding on fruit and nesting in trees. The three species of

Bombycilla (waxwings) are brown-grey birds with crests, and *secondaries that have red, wax-like tips. The tail is tipped with yellow or red. They have short, stout bills, and short legs. Waxwings are gregarious in winter. *Hypocolius ampelinus* (hypocolius) is a blue-grey bird with a black mask, white-tipped black *primaries, and a black-tipped tail. Its bill is slightly hooked, and its legs are short and strong. It inhabits open scrub with trees, feeds on berries and insects, and nests in trees and bushes. It occurs in south-western Asia and is sometimes placed in a family of its own, Hypocolidae. There are five genera in the family, with eight species, although the silky flycatchers are sometimes placed in a separate family Ptilogonatidae. They are found in N. and Central America, Europe, and Asia.

bombykol A *pheromone, 10-trans-12-cis-hexadecadien-1-ol, that a female silk moth (*Bombycidae) releases from the tip of her abdomen. Bombykol was discovered in 1959 by Adolf Butenandt (1903–95), who was awarded the 1939 Nobel Prize in chemistry for his work on pheromones, and it was the first pheromone to be characterized chemically.

Bombyliidae (bee-flies; order *Diptera, suborder *Brachycera) Family of flies, most of which are medium to large, with densely *pubescent bodies and slender legs. The proboscis is elongate, points forward, and is adapted for feeding on nectar from flowers with long corollas. The third antennal segment is simple, with the style small or absent. Their common name refers to their appearance and adult behaviour, and to their larvae which are parasitic upon bees and wasps, although some species have larvae which parasitize other insects. Eggs are laid near the entrance to the nest of the specific bee host parasitized by that species of bee-fly. The tiny larvae enter the nest and usually wait until the bee larva has pupated before metamorphosing from a small, mobile animal to a smooth, fat larva which feeds on the bee pupa. The bee-fly pupa is dual-phased. The first pupa is 'normal'; the second stage has a sharp battering ram with which to break down the nest cell wall, made

by the adult bee when closing the cell. The adult bee-fly emerges once the second stage pupa has ruptured the seal. More than 2000 species of bee-flies have been described.

bonding Social behaviour that tends to keep individuals together as cohesive groups (e.g. as herds, flocks, or schools), or as pairs (as in parent–infant or mating pairs).

bone The skeletal tissue of *vertebrates. Bone is composed of about 70% inorganic calcium salts, mostly *hydroxyapatite but also carbonate, citrate, and fluoride amines are present. The organic component is mostly made up of the structural *protein *collagen. *See also* OSSIFICATION.

bonefish *See* ALBULIDAE.

bone marrow Spongy vascular tissue found inside the long bones of vertebrates. In young animals the marrow is red and produces *erythrocytes (red blood corpuscles). In older animals yellow marrow replaces the red marrow except at the upper ends of the *femur and *humerus. Yellow marrow consists of fatty tissue. Stem cells produced in the yellow marrow develop into *T cell lymphocytes (*see* LEUCOCYTE).

bonnetfish *See* EMMELICHTHYIDAE.

bonobo (*Pan paniscus*) *See* PONGIDAE.

bony fish (*Osteichthyes, *Teleostomi) Fish in which the *endoskeleton is at least partly ossified, including the dermal bones of the upper and lower jaws. Characteristic features are the terminal mouth, *homocercal tail, *operculum covering the gills, scales on the skin, and the presence of a *swimbladder (although sometimes this has been lost in the course of later evolution).

bony labyrinth (osseous labyrinth) A structure that forms the outer wall of the inner ear of vertebrates; it contains the *membranous labyrinth from which it is separated by *perilymph.

bony tongue *See* OSTEOGLOSSIDAE.

boobies See SULIDAE.

booklice See PSOCOPTERA.

book lungs One of the two kinds of respiratory organ (*compare* TRACHEAL SPIRACLES) found in spiders. They are located in the *abdomen, just below the *pedicel, and are composed of many fine leaves. Blood is passed over a large surface area to absorb oxygen. The openings of the book lungs (branchial opercula) are situated on the ventral surface of the abdomen and may be closed to prevent water loss. Book lungs are believed to be the ancestral type of respiratory organ in arachnids.

boomslang (*Dispholidus typus*) See COLUBRIDAE.

Boopiidae (order *Phthiraptera, suborder *Amblycera) Family of chewing lice which are parasitic chiefly on Australian marsupials. Only two species are not found on marsupials: *Heterodoxus spiniger* has become established on dogs, presumably having transferred to dogs since humans introduced them to Australia, and is now found all around the world between latitudes 40°N and 40°S; and *Therodoxus oweni* is found on the cassowary. Members of the family differ from other Amblycera in possessing a *seta borne on a protuberance on each side of the *mesonotum, and in the fusion of the *metanotum to abdominal *tergum I. There are eight genera, with about 35 species.

boreal Pertaining to the north (from Boreas, the Greek god of the north wind).

Boreidae See MECOPTERA.

Borhyaenoidea (infraclass *Metatheria, order *Marsupialia) An extinct superfamily of S. American marsupial carnivores, the largest about the size of a wolf, which lived from the *Palaeocene to *Pliocene. Borhyaenoids became extinct following the colonization of S. America by placental mammals.

Bos (cattle) See BOVIDAE.

Bostrichidae (branch borer, limb borer, shothole borer; class *Insecta, order *Coleoptera) Family of black or brown beetles, 3–20 mm long, with the head tucked under a rounded, *rugose *thorax. The *elytra are sharply declined at the rear, with several projecting, horizontal spines. The antennae have a loose club. Larvae are curved, white, and fleshy, with a tiny head and enlarged thorax. They bore into felled or dying trees. There are many generations per year. There are 434 species, largely tropical.

Botaurus (bitterns) See ARDEIDAE.

bot fly See GASTEROPHILIDAE; OESTRIDAE.

Bothidae (left-eye flounder; subclass *Actinopterygii, order *Pleuronectiformes) A large family of mainly marine, shallow-water flatfish, in which both eyes lie on the left side of the body. The *dorsal fin originates in front of the eyes and spines are absent from all the fins. Many species are of small to moderate size. Others, e.g. *Scophthalmus maximus* (European turbot) and *Paralichthys californicus* (California halibut), are of considerable economic value. Flatfish are known for their remarkable colour adaptations to the prevailing background. There are about 425 species, with world-wide distribution.

Bothriocephalidea (phylum *Platyhelminthes, class *Cestoda) An order of parasitic worms which possess a relatively simple *scolex. All members of the order inhabit three hosts during their development.

bothrium In some *Cestoda, a sucking groove on the 'head'.

botryoidal tissue In *Hirudinea, *parenchyma and *connective tissues that invade the *coelom. It consists of grape-like masses of cells containing a brown pigment and may serve an excretory function.

bottle-nosed dolphin (*Tursiops truncatus*) See DELPHINIDAE.

bottle-nosed whale (*Hyperoodon ampullatus*) See ZIPHIIDAE.

boundary-zone A time line that is based on either the appearance or the disappearance of a key species or fauna. Associated faunas and sediments may transgress a zonal boundary.

Bourgueticrinida (sea lilies; class *Crinoidea, subclass *Articulata) An order of articulate sea lilies in which the compact calyx (cup) is made up of five basal and five radial plates only. The column consists of elliptical or rounded *ossicles which are united by *synarthrial articulations. Five or ten arms are present. The order first appeared in the Late *Cretaceous.

Bovichthyidae (subclass *Actinopterygii, order *Perciformes) A family of small, mainly marine fish, restricted to the cooler waters of the southern hemisphere, including oceanic islands (e.g. Tristan da Cunha). A freshwater representative, *Pseudaphritis urvilli* (congolli) of S. Australia, lives in rivers, where it reaches a length of some 35 cm. There are about six species.

Bovidae (oxen, antelopes, goat antelopes, gazelles, sheep, goats; infraorder *Pecora, superfamily *Bovoidea) The largest by far of the artiodactyl families, including cattle, bison, musk-ox, sheep, goats, antelopes, and allied groups, descended from the *Oligocene *Traguloidea. They appeared as a distinct type in the *Miocene and a major diversification followed. Most are grazers but some are browsers. Horns, present in the male and often in the female, are covered with *keratin and are not shed. The feet have two or four digits. Bovidae are of world-wide distribution. The family includes *Bison bison* (N. American bison), *Bison bonasus* (European wisent, sometimes known, incorrectly, as the 'aurochs'), *Bos taurus* (all breeds of European cattle), *Bos indicus* (Zebu cattle), *Bos grunniens* (yak), *Capra* (goats, domesticated breeds being descended from *C. aegagrus*, the pasang), *Ovis* (sheep, descended from several wild members of the genus), *Ovibos moschatus* (musk-ox), *Tragelaphus* (formerly *Taurotragus*), *Oryx* (eland), *Aepyceros melampus* (impala), and *Saiga tatarica* (the saiga antelope of Siberia) and others, generally lumped as 'antelopes', but forming a number of different subfamilies. In all there are 44 genera.

bovine spongiform encephalitis (BSE) A transmissible, neurodegenerative, fatal brain disease of cattle. It affects the brain and spinal cord, producing lesions with a spongy appearance that are visible under a light microscope. Researchers believe BSE is caused by a self-replicating *protein known as a *prion. The disease was first identified in the United Kingdom in November 1986 and subsequently appeared in many other countries. BSE is closely related to scrapie, a similar disease of sheep.

Bovoidea (order *Artiodactyla, infraorder *Pecora) A superfamily that comprises two families, *Antilocapridae and *Bovidae, sometimes grouped together as a single family, Bovidae. The pronghorn (*Antilocapra*) and numerous extinct allied forms have been distinct from the bovids since the *Miocene and in fact may have evolved from a cervid line. Bovoids are pecorans with lateral toe digits reduced in some species to a nodule of bone. The upper *canines are reduced or absent, the cheek teeth high-crowned. Horns are present, usually in both sexes though more developed in males, and covered with *keratin rather than skin (as in cervids).

bowerbirds See Ptilonorhynchidae.

bowfin (*Amia calva*) See Amiidae.

bowhead whale (*Balaena mysticetus*) See Balaenidae.

Bowman's capsule See nephron.

boxfish (*Ostracion lentiginosum*) See Ostraciontidae.

box model See compartment model.

bp Before the present.

brachial Of or pertaining to the arm or an arm-like structure.

brachiation In some arboreal *Primates, a form of locomotion in which an animal swings hand over hand from branch to branch. The only true brachiators are the gibbons (*Hylobatidae), which are said to be able to move more quickly through the trees than a human can walk on the ground below, but the spider monkey (*Ateles*, *see* CEBIDAE) and its relatives use brachiation a good deal in their locomotion. One hypothesis holds that the *Hominidae (including humans) are descended from brachiating ancestors.

brachidium In *Brachiopoda, the calcified support for the *lophophore.

Brachinus (bombardier beetle) *See* CARABIDAE.

brachiolaria In some *Asteroidea, a larval form, bearing three pre-oral adhesive processes, which develops from the *bipinnaria larva.

brachiole In certain *Echinodermata, a food-gathering arm; they occur side by side and resemble the arms of *Crinoidea.

Brachionichthyidae (warty angler; subclass *Actinopterygii, order *Lophiiformes) A small family of marine fish, differing from the related frogfish (*Antennariidae) in having a thick membrane, separate from the rest of the *dorsal fin, above the eyes. The *pectoral and *pelvic fins are arm-like in shape, making the fish more frog-like than the frogfish. There are a few species, found in the southern half of Australia.

Brachiopoda (lamp shells) A phylum of solitary, marine, bivalved, coelomate invertebrates that live attached to the sea-bed by a muscular stalk (pedicle), or are secondarily cemented, or are free-living. Usually they consist of a larger, pedicle (ventral) valve and a brachial (dorsal) valve, lined by reduplications (mantle lobes) of the body wall which enclose the large *mantle cavity. They are *bilaterally symmetrical about the posterior–anterior mid-line of the valves. The *lophophore surrounds the mouth and is covered by ciliated *tentacles. Brachiopods first appeared in the Early *Cambrian and since then have undergone numerous *adaptive radiations, the first involving the inarticulate brachiopods. The articulate groups became more important after the Cambrian. Many of the larger articulate groups are now extinct.

Brachiopterygii (class *Osteichthyes) Group (subclass) of bony fish that includes the curious bichirs (*Polypterus* species) and reedfish (*Calamoichthys* species) of central Africa. These fish possess a number of primitive and specialized features, e.g. a series of spinous dorsal *finlets, *ganoid scales, lobate *pectoral fins with special skeletal supports, a *spiracle, a spiral intestinal valve, and a pair of ventral, lung-like swim-bladders. The systematic position of the bichirs is not yet certain: the 11 living species, grouped together as the order *Polypteriformes, have also been included in the subclass *Actinopterygii.

brachium 1. The upper arm, from shoulder to elbow, or a *homologous structure, e.g. in a wing or flipper. 2. Any structure reminiscent of an arm, e.g. fibres forming the auditory pathway or linking the optic tract to the brain.

Brachycera (class *Insecta, order *Diptera) Suborder of flies whose members are recognizable from their wing *venation and *porrect *palps. With one exception (*Stratiomyidae) all Brachycera larvae have the head retractile within the thorax. The antennae usually have fewer than six segments, with exceptions where segments have fused as a secondary feature. There are 16 families.

brachydont Applied to teeth in which the crowns are low or short and the roots well developed, with narrow canals.

Brachygalba (jacamars) *See* GALBULIDAE.

Brachypteraciidae (groundrollers; class *Aves, order *Coraciiformes) A family of brightly coloured, long-legged, short-winged birds, which have the anterior toes joined at the base. They are ground-dwelling in humid forest, except for *Uratelornis chimaera* (long-tailed groundroller), which inhabits arid desert. They are insectivorous, also taking small amphibians and reptiles. There are three genera, with five species, found only in Madagascar.

brachypterous Applied to insects in which both pairs of wings are reduced.

Brachypteryx (shortwings) *See* TURDIDAE.

Brachyrhamphus marmoratus (marbled murrelet) *See* ALCIDAE.

Braconidae (suborder *Apocrita, superfamily Ichneumonoidea) Family of insects which are distinguished from the *Ichneumonidae on the basis of their wing *venation. Braconids are small (2–15 mm long) parasitic wasps, mostly brownish or black. Eggs are laid in or upon the body of the host insect, which is consumed before the parasite pupates (pupation sometimes occurring outside the host body in a silken cocoon). Braconids parasitize the larvae of *Lepidoptera, *Symphyta, *Coleoptera, *Diptera, etc., and some are important as biological-control agents.

bradycardia A condition in which the heart rate is reduced substantially.

Bradypodidae (tree sloths; infra-order *Pilosa, superfamily *Bradypodoidea) A family of toothless mammals that are fully adapted for arboreal life and unable to walk on the ground; they use their limbs to hang upside-down from branches. The head is short and rounded. They are not actually toothless, but the five teeth have no enamel and by growing continuously provide a grinding surface. The fore limbs are longer than the hind limbs, the digits bearing hooked claws used in hanging. The tail is short. The fur is long, coarse, and grooved; cyanobacteria grow in the grooves, giving the animals a greenish colour. There are six cervical vertebrae in *Choloepus*, nine in *Bradypus*. The stomach is large and chambered, the rectum very large, retaining faeces for several days. All movements are slow.

The body temperature is low and variable, sloths neither shivering nor sweating, but dying from hypothermia at temperatures below 4°C. They are distributed throughout the forests of tropical America. There are two genera, and five species.

Bradypodoidea (tree sloths; order *Edentata, infra-order *Pilosa) A superfamily comprising the family *Bradypodidae (tree sloths). These animals are unknown as fossils, probably because their humid tropical habitat fails to preserve them after death. Structurally they appear to be related to the ground sloths (*Megalonychoidea).

bradytely An exceedingly slow rate of evolution, manifest by slowly evolving lineages which survive much longer than would normally be expected. Living fossils (e.g. the *coelacanth) and groups that have remained relatively stable with time (e.g. opossums and crocodiles) represent the low end of the bradytelic range. *Compare* HOROTELY; TACHYTELY.

brain In animals, the anterior section of the *central nervous system associated with the aggregation of sense organs. In invertebrates, the brain is represented by well-developed cerebral *ganglia. In vertebrates, the brain comprises three main bodies, the *cerebrum, the *brain stem, and the *cerebellum. *See also* PROSENCEPHALON; RHOMBENCEPHALON.

brain stem The part of the *brain which still conforms with the organization of the spinal cord. The brain stem is divided into four regions anterior to posterior: *diencephalon, *mesencephalon, *metencephalon, and *myelencephalon.

Brama brama (Ray's bream) *See* BRAMIDAE.

brambling (*Fringilla montifringilla*) *See* FRINGILLIDAE.

Bramidae (subclass *Actinopterygii, order *Perciformes) A family of deep-bodied, moderate to large oceanic fish. Oval in outline, the body has long *dorsal and *anal fins, a forked tail, and large, wing-like *pectoral fins. *Brama brama* (Ray's bream or Atlantic pomfret), reaching a length of up to 70 cm, lives in depths of 100 m or below. There are about 18 species, found in all but polar seas.

branch borer *See* BOSTRICHIDAE.

branched hydrocorals *See* STYLASTE-RINA.

branchiae In some aquatic invertebrates, thin-walled extensions of the body wall which function as gills.

branchial Of the gills.

branchial arch A single gill arch which supports a single pair of gill slits. Together the gill arches form the branchial basket which is part of the *visceral skeleton.

branchial basket 1. *See* BRANCHIAL ARCH. 2. *See* RECTAL GILL.

Branchiopoda (gill-footed shrimps; phylum *Arthropoda, subphylum *Crustacea) Diverse class of small, mainly freshwater, filter-feeding crustaceans, all characterized by flattened, leaf-like trunk appendages in which the *coxa bears a flattened *epipodite which functions as a gill. The Branchiopoda includes the *Anostraca (fairy shrimps), *Notostraca (tadpole shrimps), *Conchostraca (clam shrimps), and *Cladocera (water fleas). Many are pink or red owing to the presence of *haemoglobin. They are often inhabitants of temporary bodies of water, and have drought-resistant eggs. There are 800 species.

Branchiostegidae Older name for a family of fish now known as *Malacanthidae.

Branchiostomidae (lancelets; order Amphioxiformes) A family of small, fish-like, marine, coastal animals. Lancelets are the only living representatives of the subphylum *Cephalochordata. Although definitely not true fish, lancelets (e.g. *Branchiostoma lanceolata*) are considered fairly close to the vertebrate ancestral lineage. They have an almost transparent body, no eyes or brains, a permanent *notochord extending into the head and a dorsal hollow *nerve cord, gill slits, and segmented muscle blocks. They spend most of their lives buried in the sediments beneath the world's oceans.

Branchiura (phylum *Arthropoda, subphylum *Crustacea) Small class of crustaceans, all of which are parasites, and which differ from the *Copepoda in having a pair of *sessile *compound eyes, and a large, shield-like *carapace covering the head and *thorax. Branchiurans are common, blood-sucking ectoparasites on the skin or in the gill cavities of marine and freshwater fish and a few amphibians. Attachment to the host is via a large, claw-like modification of the first antennae, and by a pair of large suckers derived from the bases of the first *maxillae, the rest of the maxillae being vestigial. A sucking mouth cone is formed from the *labrum and *labium. (In *Argulus* a hollow spine in front of the mouth cone pierces the skin of the host and injects a supposedly poisonous secretion.) There are no maxillipeds, and the second maxillae are *uniramous, well developed, and have terminal claws. There are four large, *biramous, thoracic appendages, armed with swimming *setae used when swimming between hosts. There are 75 species.

***Branta* (geese)** *See* ANATIDAE.

***Branta canadensis* (Canada goose)** *See* ANATIDAE.

***Branta sandvicensis* (Hawaiian goose, ne-ne)** *See* ANATIDAE.

Brassolinae (owl butterflies) A subfamily of the *Nymphalidae.

Braula *See* BRAULIDAE.

Braulidae (bee lice; order *Diptera, suborder *Cyclorrapha) Family of abnormal flies which lack wings and resemble members of the *Hippoboscidae. The antennae are set in lateral grooves, and the eyes are reduced in size. The *thorax appears to be joined to the *abdomen. The *tarsi are adapted to cling to the bodies of *Hymenoptera, especially bees. Bee lice eggs are laid near honey- or pollen-storage cells inside a bee nest, and the larvae feed on these substances and pupate within the nest of the host. The pupa is unique among the Cyclorrapha in being enclosed in the unmodified last (final *instar) larval *cuticle. There are eight species in two genera (*Braula* and *Megabraula*).

breakage and reunion In genetics, the established model of *crossing-over by the physical breakage and crosswise reunion of broken *chromatids during the pairing of *homologous chromosomes at the *prophase stage of *meiosis. The points at which chromosomes cross over are termed *chiasmata. The result of crossing-over is the mutual exchange between chromatids of parts which contain corresponding *loci.

breed An artificial mating group derived (by humans) from a common ancestor, usually for agriculture (e.g. domesticated animals and crop plants), or for genetic analysis, or for pleasure (e.g. cats and dogs).

breeding dispersal *See* DISPERSAL; PHILOPATRY.

breeding size The number of individuals in a *population that are involved in reproduction during a particular generation. It does not include the non-breeding element of the population.

breeding true Producing offspring whose *phenotypes are identical to those of the parents. Homozygous individuals (*see* HOMOZYGOTE) necessarily breed true (unless mutations arise), whereas *heterozygotes rarely do so.

Bregmacerotidae (codlets; subclass *Actinopterygii, order *Gadiformes) A small family of marine, oceanic, warmwater fish that have an elongate body, deeply notched *dorsal and *anal fins, the first dorsal ray reduced to one long spine on the head, and outer *pectoral-fin rays trailing beside the body. The seven species are of world-wide distribution. Although not a true deep-sea fish, the Indo-Pacific *Bregmaceros macclellandi* (up to 12 cm long) has been caught at depths exceeding 500 m.

Brenthidae (class *Insecta, order *Coleoptera) Family of narrow beetles, up to 40 mm long, often with an exceptionally lengthened head and *rostrum; the *elytra of some species comprise only one-quarter of the total body length. Many are sexually *dimorphic. The females' long rostrum is used for boring into wood in which eggs are laid; the jaws are tiny. Males have a broad rudimentary rostrum but large jaws possibly for holding the female during mating. Larvae are fleshy, with short legs, and are found in decaying wood. There are 1230 species, largely of tropical woodland.

Breviceps adspersus (rain frog, shorthead) *See* MICROHYLIDAE.

breviconic Applied to the shell of a cephalopod (*Cephalopoda) when it is short and has a fairly wide aperture.

Brevoortia tyrannus (menhaden) *See* CLUPEIDAE.

brimstones *See* PIERIDAE.

brine shrimps *See* ANOSTRACA.

bristlebirds *See* ACANTHIZIDAE.

bristlemouth *See* GONOSTOMATIDAE.

bristletail Common, collective name for members of the *Thysanura, sometimes extended to include the *Archaeognatha, and which is applied incorrectly to *Diplura.

bristleworms *See* POLYCHAETA.

brittle-stars *See* EURYHALINA; OPHIOMYXINA; OPHIURIDA; OPHIUROIDEA; PHRYNOPHIURIDA.

broadbills *See* EURYLAIMIDAE.

broken wing display A form of *deception in ground-nesting birds in which a bird staggers away from its nest, dragging a wing that appears to be broken and drawing the predator behind it. When the predator is a safe distance from the nest the bird with the 'broken wing' flies off.

bronchial tree See BRONCHUS.

bronchiole See BRONCHUS.

bronchus (*pl.* bronchi) In air-breathing vertebrates, the part of the respiratory tract that connects the *trachea (windpipe) to the lungs. The base of the trachea divides into two bronchi, carrying air into the two lungs. These divide further into bronchioles and then into many smaller passages that form a bronchial tree carrying air to the alveoli. See ALVEOLUS.

Brontosaurus See APATOSAURUS.

brontothere See BRONTOTHERIIDAE.

Brontotheriidae (brontotheres, titanotheres; class *Mammalia, order *Perissodactyla) A suborder (ranked by some authorities as a family of the suborder *Hippomorpha) of rhinoceros-like animals that flourished in the *Eocene and *Oligocene, and then became extinct. Probably they first appeared in N. America, but they spread to Asia and some lived in eastern Europe. Brontotheres were large, some reaching 2.4 m at the shoulder (as large as modern elephants). The skull was long and low, the brain small. Bony horns were present in later forms, often large and presumed larger in males than females, carried as a pair, side by side on the front of the head. Some (e.g. *Brontotherium*) had Y- or V-shaped horns on their noses. The limbs were *graviportal. The teeth were primitive, the *incisors and a *premolar absent in some forms, the *molars large, low-crowned, and *bunolophodont. The brontotheres seem to have evolved rapidly to large size, but their teeth must have restricted their diet to soft plant material, perhaps making them vulnerable to minor environmental changes.

brood 1. *noun.* All of the offspring that hatch from a single *clutch of eggs. **2.** *verb.* To incubate eggs.

brood parasite (nest parasite) An animal which lays its eggs in the nest of a member of its own or another species. The recipient (host) raises the young which, in some species (e.g. *Cuculus canorus*, cuckoo), eject from the nest or kill the natural offspring of the host.

brood pouch A sac in which some animals (e.g. *Malacostraca and some *Syngnathidae) *brood their eggs. See also MARSUPIUM.

brotula See OPHIDIIDAE.

brown lemur (*Eulemur*) See LEMURIDAE.

browns See SATYRINAE.

Bruchidae (seed weevils; class *Insecta, order *Coleoptera) Family of small, compact beetles, 1.5–5 mm long, with the head curved under the body. The *elytra are dark, with a pattern of white or yellow hairs. Bruchidae are a common pest of leguminous crops. Eggs are laid on developing seeds. The larva is fleshy, grub-like, and completes its development inside a single seed. There are 1200 species. They are also known, incorrectly, as pea weevils and bean weevils (see CURCULIONIDAE).

Brunhes A palaeomagnetic *chron, characterized by normal magnetization, in which we are living at present. It began about 780 000 years ago.

Brunner's glands In *Eutheria, *glands in the *proximal *duodenum and proximal small *intestine which secrete *mucus that lubricates the intestinal tract. They were first described by the Swiss anatomist Johann Conrad Brunner (1653–1727).

brush-footed butterflies See NYMPHALIDAE.

Bryde's whale (*Balaenoptera edeni*) See BALAENOPTERIDAE.

Bryozoa (Ectoprocta; moss-animals) A phylum of small, aquatic, colonial animals, related to the *Brachiopoda, many of which possess a well-developed, *calcite skeleton which comprises microscopic, box-like divisions, each housing an individual animal possessing ciliated *tentacles and a *coelom. Food is collected by the tentacles which surround the mouth and are borne on the *lophophore. (The lophophore does not surround the *anus, distinguishing bryozoans from the *Entoprocta.) Reproduction takes place by asexual *budding, and by the release of larvae which give rise to new colonies. Bryozoans have occurred from the *Ordovician to the present day. Fossilized branched colonies are common in some rocks. They were important reef builders in the *Phanerozoic, and underwent several great *adaptive radiations.

BSE *See* BOVINE SPONGIFORM ENCEPHALITIS.

Bubo **(eagle owls)** *See* STRIGIDAE.

buccal cavity Mouth cavity.

buccal force pump Respiratory system, typical of the *Amphibia, whereby air is forced into the lungs by raising the floor of the mouth while the valvular nostril is closed. *Compare* COSTAL RESPIRATION.

buccal incubation (mouth brooding) The incubation of eggs in the mouth of one of the parents. In a number of species of fish, the male or the female carries the fertilized eggs in the mouth until some time after hatching. During incubation the parent involved takes no food, and so may appear rather lean at the end of the incubation period.

Bucconidae (puffbirds; class *Aves, order *Piciformes) A family of small, black or brown and white birds that have thick necks and large heads. The bill is stout with a hooked, often *bifid, tip. There are *rictal bristles, and the feet are *zygodactylous. They are insectivorous and nest in holes in banks (one of the four species of *Notharchus* nests in tree-termite nests). There are 10 genera, with 32 species, found in tropical forests in Central and S. America.

buccopharynx The mouth and *pharynx.

Bucerotidae (hornbills; class *Aves, order *Coraciiformes) A family of medium to large, typically black and white birds that have huge, curved, red or yellow bills, a casque on the *culmen, short legs, and a long tail. Some of the 15 species of *Tockus*, found in Africa and southern Asia, have black bills. Hornbills are arboreal and terrestrial. They nest in holes in trees, the females being walled in with mud. They are omnivorous, feeding on fruit and insects. There are 12 genera, with 46–9 species, found in Africa, southern Asia, Malaysia, the Philippines, Indonesia, and the Solomons.

Buddenbrockia plumatellae An animal of uncertain affinities, described by O. Schröder in 1910, but not reported since. A parasite of *Bryozoa, it is 3.6 mm long, unciliated, poorly worm-like, and has no gut or nerves, but the presence of true muscle fibres indicates it does not belong to the *Mesozoa, with which it might otherwise be classified.

budding **1. (gemmation)** A form of *asexual reproduction in which a new individual develops within the body wall or *cell membrane of the parent, causing a bud-like swelling, then detaches itself to commence an independent life. In certain single-celled micro-organisms a new cell is formed by extrusion or outgrowth from an existing cell. **2.** In *Hymenoptera, the migration of a group of workers and brood to a new nest site.

budgerigar *See* PSITTACIDAE.

buffalo Common name for several species of cattle-like mammals (*Bovidae) occurring in Africa south of the Sahara and in south-eastern Asia. The N. American Bison is sometimes incorrectly called 'buffalo'.

buffalo gnats *See* SIMULIIDAE.

Bufo bufo (common toad, European toad) *See* BUFONIDAE.

Bufo calamita (natterjack toad) *See* BUFONIDAE.

Bufo marinus (marine toad, cane toad) *See* BUFONIDAE.

Bufonidae (toads; class *Amphibia, order *Anura) A family of squat-bodied amphibians which have short, powerful limbs for crawling. The skin is generally dry and warty. Teeth are absent. Vertebrae are *procoelous, the pectoral girdle bow-shaped. A *Bidder's organ is present in males. Bufonids are usually nocturnal or *crepuscular land-dwellers, only breed-ing in water. *Sexual dimorphism (females larger than males) occurs in *Bufo bufo* (common toad or European toad); their col-our is changeable, red-brown to grey, and males develop *nuptial pads on three inner fingers prior to mating. *B. calamita* (natter-jack toad) is tolerant of brackish water; its hind limbs are short and it runs rather than jumps. *B. viridis* (green toad or changeable toad) has a greyish body with green patches and red warts; it may breed in brackish water and tolerates dry surroundings. *B. marinus* (marine toad, known as cane toad in Australia), of tropical America and intro-duced into Caribbean islands, Hawaii, Aus-tralia, and New Guinea (in most of which places it has become a pest), is the largest bufonid, females reaching 25 cm in length; it is mottled yellow and brown, with red warts, and has large *parotoid glands that secrete a powerful toxin. There are some 300 species, found native in all continents except Aus-tralia and Antarctica.

Bufo viridis (green toad, changeable toad) *See* BUFONIDAE.

bugs *See* HEMIPTERA.

bulbuls *See* PYCNONOTIDAE.

bulla In most mammals, a thin-walled, bony projection of the skull, often hemi-spherical, which encases the *middle ear.

bullate Bearing blister-like swellings; blister-like in appearance.

bulldog ant *See* MYRMECIINAE.

bulldog bat (*Noctilio*) *See* NOCTILIONIDAE.

bullfinches (*Pyrrhula*) *See* FRINGILLIDAE.

bullfrog (*Rana catesbiana*) *See* RANIDAE.

bullhead shark *See* HETERODONTIDAE.

Bullomorpha *See* CEPHALASPIDEA.

bullseye *See* PEMPHERIDAE; PRIACAN-THIDAE.

bumble-bee Any member of the genus *Bombus* (family *Apidae).

bumble-bee bat (*Craseonycteris thonglongyai*) *See* CRASEONYCTERIDAE.

Bungarus (kraits) *See* ELAPIDAE.

bunodont Applied to teeth in which the cusps are rounded.

bunolophodont Applied to teeth in which the cusps are rounded and are linked by ridges (lophs).

Bunopithecus (hoolock) *See* HYLOBATI-DAE.

buntings *See* EMBERIZIDAE.

Buprestidae (jewel beetles; class *In-secta, order *Coleoptera) The most beauti-ful of beetle families, often used in jewellery. The more spectacular species are tropical. The *elytra and *thorax are elongate, with bright metallic colours; they are often of a consid-erable size. The eyes are large, the antennae short and *filiform. Adults are daytime nectar feeders; larvae are club-shaped, legless, and called 'flat-headed borers': the head is tiny, the *prothorax unusually broad, the body narrow and tapering. The larvae live in trees or roots, where they gnaw flattened tunnels. Some mine leaves or stems. There are 11 500 species.

burbot (*Lota lota*) *See* GADIDAE.

Burdigalian A stage of the early *Miocene, about 20.43–15.97 Ma ago, underlain by the *Aquitanian and overlain by the Langhian. Mammals, including elephants, are impor-tant in the stratigraphic definition of this stage.

Burgess Shale An horizon from the *Cambrian of British Columbia which has yielded an exceptionally preserved fauna. It was originally discovered by C. D. Walcott in 1909. Apparently, the fauna was deposited in deep water on or near a submarine fan. Many of the animals are *Arthropoda, but other groups are also represented, some of them bizarre. Their identification has been aided by the discovery of other Cambrian faunas, especially those in China, Greenland, and Australia. Some forms, including *Anomalocaris* and *Opabinia*, form a *metazoan group between the *Onychophora and arthropods, but the fauna as a whole represents a rapid radiation and was probably a typical Cambrian fauna, unusual only in the preservation of soft-bodied forms.

Burhindiae (thick-knees, stone-curlews; class *Aves, order *Charadriiformes) A family of sandy or grey-brown birds that have large heads and eyes, shortish, stout bills, and long legs with thickened *tibiotarsal joints. The feet are *semipalmate with three toes. They are *crepuscular and nocturnal, and live in open, stony and sandy country. They feed on invertebrates, mice, and frogs, and are ground-nesting. There are two genera, with nine species, found in Africa, Europe, Asia, Australia, Central and S. America.

burnet See ZYGAENIDAE.

Burramyidae (order *Diprotodontia (or *Marsupialia), superfamily *Phalangeroidea) A family of mouse-sized Australian possums, including the ground-dwelling *Burramys*, known first as a fossil and discovered to be living in 1967.

Burramys See BURRAMYIDAE.

burrowing goby See TRYPAUCHENIDAE.

bursa A small sac lined with *synovial membrane and filled with *synovial fluid that forms a cushion between *bones and *tendons or the *muscles surrounding a joint.

burying beetle See SILPHIDAE.

bush-baby (*Galago*) See LORISIDAE.

bush cricket See GRYLLIDAE.

bush dog (*Speothos venaticus*) See CANIDAE.

bush katydid See TETTIGONIIDAE.

bush larks See ALAUDIDAE.

bush rat See OCTODONTIDAE.

bush-shrikes (*Tchagra*) See LANIIDAE.

bushtits (*Psaltriparus*) See AEGITHALIDAE.

bush warblers See SYLVIIDAE.

bustards See OTIDIDAE.

butcherbirds See CRACTICIDAE.

Butenandt, Adolf See BOMBYKOL.

Buteo (buzzards, hawks) See ACCIPITRIDAE.

butterfish See PHOLIDIDAE; SCATOPHAGIDAE; STROMATEIDAE.

butterflies See LEPIDOPTERA.

butterfly cichlid (Apistogramma ramirezi) See CICHLIDAE.

butterfly fish See CHAETODONTIDAE; OSTEOGLOSSIFORMES; PANTODONTIDAE.

buttonquails See TURNICIDAE.

buzzards See ACCIPITRIDAE.

Byrrhidae (pill beetles; class *Insecta, order *Coleoptera) Family of dull, *pubescent, highly convex beetles, 1.5–10 mm long. The antennae are clubbed. When disturbed, the legs are withdrawn completely into grooves on the under side of the abdomen. Larvae are fat and cylindrical, with a large, broad *prothorax, and a pair of retractable false legs on the terminal segment. Larvae are found among roots and in moss. There are 270 species.

byssal gland *See* BYSSATE.

byssate Applied to the condition found in certain *Bivalvia of having strands of byssus (strong, horny threads) by which the animals are often attached to the *substrate or to objects. A byssal gland located in the foot secretes a fluid *protein which passes down a groove along the under side of the foot; the secretion hardens on exposure to sea water.

byssoid Consisting of fine threads.

byssus *See* BYSSATE.

caddis flies *See* TRICHOPTERA.

caecilians *See* APODA.

Caecilidae (caecilids; class *Amphibia, order *Apoda) A family of caecilians that are mainly terrestrial as adults. Development is *oviparous or *ovoviviparous, often with a free-swimming larval stage. *Stapes are present. The tail is absent. There are more than 100 species, in localized populations in the Old and New World tropics.

caecotrophy The passing of food through the alimentary canal twice. Rabbits and some other small mammals take soft faecal pellets directly from the *anus at night and store them in the *stomach to be mixed with food taken during the day, probably obtaining metabolites essential for digestion that are produced in the *caecum. The animals will die if they are prevented from ingesting these pellets. *Compare* COPROPHAGY.

caecum In the *alimentary canal of vertebrates, a pouch which in some animals (e.g. *Leporidae) contains bacterial populations involved in the digestion of cellulose. In humans the caecum is a vestigial organ and poorly developed.

Caelifera (class *Insecta, order *Orthoptera) Suborder of grasshoppers whose antennae have fewer than 30 segments. The auditory organs, when present, are found on the sides of the first abdominal segment, and *stridulation is usually produced by friction between the hind *femora and either modified lateral veins on the fore wing, or areas on the *abdomen. The *ovipositor, when present, is short, and consists of four separate valves. There are eight families: *Acrididae; *Cylindrachetidae; *Eumastacidae; *Pneumoridae; *Proscopiidae; *Pyrgomorphidae; *Tetrigidae; and *Tridactylidae.

Caenogastropoda (phylum *Mollusca, class *Gastropoda) A combination of the orders *Mesogastropoda and *Neogastropoda.

Caenolestidae *See* CAENOLESTOIDEA.

Caenolestoidea (opossum rats; infraclass *Metatheria, order *Marsupialia) A superfamily (or order, *Paucituberculata) comprising a single family (Caenolestidae) of shrew-like marsupials, known since the *Eocene. Opossum rats possess four upper and three or four lower *incisors, the first lower pair being large. The hind feet are *didactylous and the first hind digit has a claw. There is no *marsupium. Opossum rats live in the Andes forests. There are seven species in three genera: *Caenolestes*; *Lestoros*; and *Rhyncholestes*.

caiman *See* ALLIGATORIDAE.

***Caiman crocodilus* (spectacled caiman)** *See* ALLIGATORIDAE.

Cainotheriidae (superorder *Paraxonia, order *Artiodactyla) A family of small artiodactyls from the late *Eocene and *Oligocene of Europe that lacked gnawing front teeth but otherwise seem to have resembled hares and rabbits. Their hind legs were markedly elongated, implying a hopping gait. They became extinct in the *Miocene, apparently displaced by early members of the true *Lagomorpha.

Cainozoic *See* CENOZOIC.

Calabrian An early *Pleistocene marine stage, about 1.806–0.781 Ma ago; it is noted for major evolutionary changes in mammalian faunas, which are split between two land-mammal stages (Upper *Villafranchian and *Biharian).

calamistrum In some spiders, a row of bristles borne on the metatarsus of each of the fourth pair of legs and used to comb out silk.

Calamoichthys (reedfish) *See* BRACHIOPTERYGII.

calcaneum The heel bone.

calcar A hollow spur. In *Chiroptera, a process of the *fibulare that helps to support the *uropatagium.

Calcarea (Calcispongia; phylum *Porifera) A class of marine sponges that have a skeleton of calcareous *spicules. Forms displaying *leucon structure are common, and *sycon and *ascon forms also occur. Individuals vary in shape from tube colonies to massive forms. They first appeared in the *Cambrian.

Calcaronea (phylum *Porifera, class *Calcarea) A subclass of sponges whose larvae have an area of flagellate cells equal to the area of non-flagellate cells. The *flagellum emanates from the nucleus of the cell.

calcichordates A group of early *Palaeozoic animals with calcareous *exoskeletons, interpreted by some as *carpoid echinoderms (*Echinodermata), and by others as ancestors of the *Chordata.

calciferol A compound, designated vitamin D_2, obtained through the ultraviolet irradiation of ergosterol.

Calcinea (phylum *Porifera, class *Calcarea) A subclass of sponges whose larvae have only a very small area lacking flagellate cells. The *flagellum is independent of the cell nucleus.

Calcispongea *See* CALCAREA.

calcite A very common form of calcium carbonate ($CaCO_3$) that is the principal ingredient of many sedimentary rocks (e.g. limestones, marble, and chalk).

Calidris (sandpipers, stints) *See* SCOLOPACIDAE.

California condor (*Gymnogyps californianus*) *See* CATHARTIDAE.

California halibut (*Paralichthys californicus*) *See* BOTHIDAE.

Callaeas See CALLAEIDAE.

Callaeidae (wattlebirds; class *Aves, order *Passeriformes) A family of medium-sized, brown and black or grey birds that have paired fleshy *wattles at the corners of their mouths, revealed when at full gape. They have rounded wings, long first *primaries, and long tails, and are arboreal and tree-nesting. They feed on insects and fruit. Two monotypic genera, *Callaeas* and *Creadion*, are both found in New Zealand. A third species, *Heteralocha acutirostris* (huia), noted for extreme sexual differentiation of the bill, is now extinct.

Callanthidium See MEGACHILIDAE.

Callichthyidae (armoured catfish; subclass *Actinopterygii, order *Siluriformes) A large family of small, bottom-dwelling, freshwater fish. The body is covered with tile-like scales, the head shows a rather steep profile, and the ventrally positioned mouth is surrounded by *barbels. Several species, when exposed to oxygen-deficient water, can gulp in air and store it in the gut. There are about 130 species, occurring in the tropical rivers of S. America.

calling hare (*Ochotona*) *See* OCHOTONIDAE.

Callionymidae (dragonets; subclass *Actinopterygii, order *Gobiesociformes) A family of marine, warm-water fish. Mainly shallow-water bottom-dwellers, they have somewhat flattened bodies, a small mouth, and narrow gill openings on the upper side of the head. The males are often brightly coloured and have larger fins than the females. The male of the European dragonet (*Callionymus lyra*, up to 30 cm long) is known to initiate a 'nuptial dance' prior to spawning. There are about 40 species, some of them found at depths of several hundred metres. A number of authors consider the dragonets to belong to the order *Perciformes.

Callionymus lyra (European dragonet) *See* CALLIONYMIDAE.

Calliphoridae (bluebottles, greenbottles, blow flies; order *Diptera, suborder *Cyclorrapha) Large family of medium-sized to large flies whose larvae are *saprophages or carnivores. The family includes the subfamily Calliphorinae, insects that are often metallic blue or green, with a markedly plumose *arista; many species are important as pests, disease-carriers, or bloodsuckers. The *Sarcophagidae is sometimes regarded as a subfamily (Sarcophaginae) of the Calliphoridae. More than 1000 species are known.

Calliphorinae *See* CALLIPHORIDAE.

Callitrichidae (marmosets, tamarins; suborder *Simiiformes, superfamily *Ceboidea) A family of very small, squirrel-like, insectivorous and fruit-eating, New World monkeys that have thick fur, non-prehensile tails, claws on all but the first hind digits, and a quadrupedal gait. There are only two *molar teeth, except in the genus *Callimico*, and no cheek pouches. The nostrils are *platyrrhine. The thumb is not opposable. Litters commonly consist of two or three young, and the placental *yolk sac grows larger than in most higher primates. The Callitrichidae live in family groups of three to eight individuals, but sometimes in larger groups. Fully arboreal, they inhabit the forests of tropical S. America. There are five genera, with some 25–30 species.

Callorhynchidae (elephant fish, plownose chimaera; subclass *Holocephali, order *Chimaeriformes) A small family of marine fish related distantly to the sharks and rays. They possess a cartilaginous skeleton and the club-shaped body terminates in a flexible process above the mouth resembling the trunk of an elephant, especially in the males. Unlike sharks, the elephant fish have a movable gill cover over the gill openings, and no *spiracle. The males possess copulatory *claspers on the ventral fins and the females produce eggs that are protected by a horny case. Fertilization is internal. The few species known are found in the southern hemisphere.

Calopterygidae (formerly Agriidae; damselflies; order *Odonata, suborder *Zygoptera) A widely distributed family of damselflies which have complex courtships. They have indistinguishable primaries among their many antenodal wing veins. Many species have metallic-coloured bodies, and heavily tinted wings. The larvae are usually found in streams, and are slender. More than 180 extant species have been described.

calorific value The gross calorific value of a substance is the number of heat units that are liberated when a unit weight of that substance is burned in oxygen, and the residual materials are oxygen, carbon dioxide, sulphur dioxide, nitrogen, water, and ash. The energy content of biological materials has been expressed traditionally in calories (c) or kilocalories (C) per gram dry weight. Sometimes results are expressed more significantly in terms of ash-free dry weight, i.e. in terms of organic constituents only. Contemporary studies of *ecological energetics express results in terms of the SI energy unit, the joule (4.182 J=1 calorie).

Calotes versicolor (bloodsucker, Indian variable lizard) *See* AGAMIDAE.

Calyptoblastina (Leptomedusae, Thecata; class *Hydrozoa, order *Hydroida) A suborder of *Cnidaria in which the colony is covered by a chitinous *perisarc with cup-like extensions, the hydrothecae and gonothecae, which protects the feeding and reproductive *polyps respectively. *Medusae have gonads on the radial canals. (*Obelia* is commonly used as an examination specimen.)

calyx 1. The cup-shaped, plated body in the *aboral body wall of *Crinozoa. It consists of skeletal *ossicles by which members of the Crinozoa may be identified. **2.** A bowl-shaped depression at the top of a calcareous coral skeleton, usually formed by the upper edges of the *septa.

CAM *See* CELL-ADHESION MOLECULE.

camara *See* CAMERA.

camaral *See* CAMERA.

Cambrian The first of six periods of the *Palaeozoic Era, from about 542–488.3 Ma ago, during which sediments deposited include the first organisms with mineralized skeletons. Common fossils include *Brachiopoda, *Trilobitomorpha, *Ostracoda, and, late in the period, *Graptolithina. Trilobites are important in the stratigraphic subdivision of the period.

((⊕)) **SEE WEB LINKS**
- Detailed information about life in the Cambrian.

camel *See* CAMELIDAE.

camel cricket *See* RHAPHIDOPHORIDAE.

Camelidae (camels; order *Artiodactyla, infra-order *Tylopoda) A family that includes the dromedary (*Camelus dromedarius*), Bactrian camel (*C. bactrianus*), llama (*Lama glama*), guanaco (*L. guanicoe*), and vicuña (*L. vicugna*). The dromedary is known only as a domestic animal; the Bactrian camel and guanaco have domestic forms (the domestic guanaco is called the llama). Another domestic camelid, the alpaca of the Andes, may be descended either from the guanaco or from a guanaco×vicuña hybrid. Camelids are ruminants in which the stomach has three chambers. Horns and tusks are absent. Hoofs, present in earlier forms, have been replaced by wide pads on the two digits of each foot. The gait is swaying, as fore and hind legs of each side move together. There is a vacuity between the *lachrymal and nasal bones. Camelids appear first as fossils in N. American rocks of

upper *Eocene age. Until nearly the end of the *Tertiary they were almost exclusively N. American, but in the *Pliocene they entered S. America and Asia and then spread into Africa. The late-Tertiary American forms included a 'gazelle camel' and 'giraffe camels'. Camels became extinct in N. America toward the close of the *Pleistocene. Today *Camelus* species are native to the Old World, where they are nearly extinct in the wild, and *Lama* species to New World. There are two genera, with four wild species.

camera (camara; *adj.* **cameral, camaral,** *pl.* **camerae)** One of the chambers within a chambered mollusc, e.g. a nautiloid or ammonite.

camerae *See* CAMERA.

cameral *See* CAMERA.

camouflage An adaptation that matches the shape, design, or colour of an animal to a particular type of background, so the animal is hidden from predators when it rests against that background with its markings aligned with those of the background.

c-AMP (cyclic AMP) A cyclic form of *adenosine monophosphate, formed from ATP (*adenosine triphosphate) in a reaction mediated (catalysed) by adenyl cyclase, which has numerous functions in cells. It can act variously as a genetic regulator, as a mediator in the activity of some *hormones, as an *enzyme activator, as a secondary messenger, and as a chemical attractant.

campanulate Shaped like a bell.

Campephagidae (cuckoo-shrikes, greybirds, minivets; class *Aves, order *Passeriformes) A family of small to medium-sized, grey or brightly coloured birds that are long-winged and long-tailed. The bill is medium-sized, often hooked, and they have well-developed *rictal bristles and spiny feathers on the lower back and rump. They are arboreal, tree-nesting, and insectivorous. The 41 species of *Coracina* are found, occasionally in groups, in open country as well as forests. There are nine genera in the family, with 72 species, found in Africa, Asia, and Australasia. *Lalage sueurii* (white-winged triller) is unusual in having two annual moults.

Campodeidae *See* DIPLURA.

campodeiform Applied to a type of larval form found in *Neuroptera, *Trichoptera, *Strepsiptera, and some *Coleoptera. The campodeiform larva is predacious, and has a well sclerotized (*see* SCLERITE) body, flattened dorsoventrally, with six legs.

Campodeiform

Canada goose (*Branta canadensis*) *See* ANATIDAE.

canalization A developmental process that is held within narrow bounds despite both genetic and environmental disturbing forces. Development is such that all the different *genotypes have a standard *phenotype over the range of *environments common to that *species.

canalizing selection The elimination of *genotypes that render developing individuals sensitive to environmental fluctuations. Genetic differences may be revealed in organisms by placing them in a stressful environment, or if a severe *mutation stresses the developmental system.

canary (*Serinus canaria*) *See* FRINGILLIDAE.

cancer cell A malignant cell within which the normal controls on growth, division, and cell-surface recognition have broken down. Such cells often give rise to cell masses (*tumours) that can invade and destroy adjacent tissues.

Cancridae (crabs; class *Malacostraca, order *Decapoda) Family of broad, oval or hexagonal crabs which have teeth on the front margin of the *carapace. The first antennae are folded longitudinally or obliquely, and the outermost section is hairy. The last pair of *pereopods is modified.

cane rat (*Thryonomys*) *See* THRYONOMYIDAE.

cane toad (*Bufo marinus*) *See* BUFONIDAE.

Canidae (dogs; order *Carnivora, superfamily *Canoidea) A family of carnivores, apparently descended from ancestral forms that appeared before the end of the *Eocene, that comprises several extinct genera and all extant dogs, jackals, and foxes. They are long-legged and *digitigrade, with the *pollex and *hallux reduced. The claws are blunt, straight, and non-retractile; the teeth are unspecialized, the *canines being large, and the cheek teeth adapted for crushing, except for the last upper *premolar and first lower molar (the carnassials) which are adapted for cutting. The face is long. The *alisphenoid canal is present, the paroccipital process long. The diet is mainly carnivorous but some species eat significant amounts of plant material. There are 13 genera, with about 41 species. *Canis* species (e.g. the wolf, *C. lupus*, from which domestic dogs (*C. familiaris*) are descended, several species of jackals, and the coyote, *C. latrans*) are distributed world-wide except for Madagascar and some islands. They have been introduced to S. America and Australasia, where some are now feral (e.g. *C. dingo*, the dingo). *Lycaon pictus* (Cape hunting dog) is confined to Africa. *Chrysocyon brachyurus* (maned wolf) is restricted to parts of tropical S. America. *Atelocynus microtis* (small-eared dog, or zorro) is also S. American. *Cuon alpinus* (dhole) is distributed throughout much of Asia. The bush dog (*Speothos venaticus*) hunts in small packs in the S. American tropical forests. *Nyctereutes procyonoides* (raccoon dog) is a fox-like animal with facial markings reminiscent of those of a raccoon; it is distributed throughout much of Eurasia. Foxes of the *Vulpes* (e.g. *V. vulpes*, red fox) and *Alopex* (e.g. *A. lagopus*, arctic fox) genera are distributed throughout the *Holarctic and N. Africa. *Dusicyon* 'foxes' or culpeos are confined to S. America. The fennec (*Fennecus* or *Vulpes zerda*) is confined to Africa; the crab-eating fox (*Cerdocyon thous*) and several related species to tropical S. America.

Caniformia In some classifications, a suborder of *Carnivora that includes the *Canoidea and the *pinniped families.

canine 1. Pertaining to a dog. **2.** A conical, pointed tooth situated between the *incisors and *premolars, particularly well developed in carnivorous mammals which use their canine teeth to seize prey. In some mammals the canines are enlarged to form tusks, in others they are reduced or absent.

cannibalism The eating by an animal of members of its own species. It is known to occur in nearly 140 species, most commonly among invertebrates, and often (but not always) in response to stress or to a reduction in the availability of other food.

Cannomys *See* RHIZOMYIDAE.

Cannon, Walter Bradford *See* FIGHT OR FLIGHT REACTION.

cannon bone In *Artiodactyla, a single bone formed usually by the fusion of the third and fourth distal *tarsal bones, and in horses by the enlargement of the third distal tarsal bone.

Canoidea (superorder *Ferae, order *Carnivora) A superfamily that comprises dogs (*Canidae), bears (*Ursidae), raccoons (*Procyonidae), weasels (*Mustelidae), and probably the *pinniped families. These are relatively unspecialized carnivores whose evolution shows tendencies to move from a carnivorous diet to one containing increasing amounts of plant material, and eventually (e.g. some bears, and pandas) to a fully herbivorous diet.

Cantharidae (soldier beetles, sailor beetles, leather-winged beetles; class *Insecta, order *Coleoptera) Family of soft-bodied, *pubescent beetles, 3–18 mm long, in which the *elytra are parallel-sided, without striae, and often brightly coloured, commonly red or black. The antennae are long, and *filiform. The adults are frequently found at flowers. The larvae are rather flattened, with a well-marked head, long legs, and a two-pronged tail; usually they are dark and pubescent. They are found under bark and logs and in loose soil. They are active in winter, hence their name 'snow-worms'. Both adults and larvae are predatory on insect grubs, though some are omnivorous. There are 3500 species.

Cape jumping hare (*Pedetes*) *See* PEDETIDAE.

Cape petrel (Cape pigeon, *Daption capense*) *See* PROCELLARIIDAE.

capercaillie *See* TETRAONIDAE.

Caperea marginata (pygmy right whale) *See* NEOBALAENIDAE.

capillary In the blood-circulation system of vertebrates, the narrowest blood vessel, with walls, only one cell thick, through which molecules can pass to transfer oxygen and nutrients to cells and to remove waste products.

Capitonidae (barbets; class *Aves, order *Piciformes) A family of small, stout, brightly coloured, large-headed birds that have heavy, sharp bills, feather tufts around the nostrils, and *rictal and chin bristles. Their wings and legs are short. They are arboreal, nesting in tree holes or banks, and feeding mainly on fruit, berries, and seeds. There are 12 genera, with over 80 species, found in Central and S. America, Africa, and southern Asia.

capitulum (gnathosoma) 1. In goose barnacles (*Lepadidae), the main part of the body, including the *carapace, that is mounted on the *peduncle. **2.** In *Chelicerata, the part of the body that carries the mouth-parts.

***Capra* (goat)** See BOVIDAE.

***Capra aegagrus* (pasang)** See BOVIDAE.

***Capreolus* (roe deer)** See CERVIDAE.

Caprimulgidae (night-hawks, night-jars; class *Aves, order *Caprimulgiformes) A family of grey-brown, long-winged birds that have white patches on the wings and tail, elongated *retrices, and in some species inner *primaries. The middle toe is *pectinate. They inhabit deserts, open wooded areas, and forests, and are ground-nesting. Many are migratory. *Caprimulgus* species are typical nightjars: aerial insectivores, they are nocturnal or *crepuscular and make a variety of churring and whistling sounds. *Caprimulgus vociferus* (whip-poor-will) is named after its call. There are 19 genera, with 75–80 species, found world-wide.

Caprimulgiformes (frogmouths, night-hawks, nightjars, owlet-night-jars, oilbirds, potoos; class *Aves) An order of grey-brown, cryptically plumaged birds that have small bills with large gape and *rictal bristles, short legs, and a short hind toe. They are nocturnal, terrestrial, and mainly insectivorous. There are five families: *Aegothelidae, *Caprimulgidae, *Nyctibiidae, *Podargidae, and *Steatornithidae. They are found world-wide.

***Caprimulgus* (nightjars)** See CAPRIMULGIDAE.

***Caprimulgus vociferus* (whip-poor-will)** See CAPRIMULGIDAE.

Caprinae (goat antelopes; family *Bovidae, superfamily *Bovoidea) A subfamily of bovids that range in size from the goral (*Nemorhaedus goral*), 1.06–1.17 m from head to tail, to the musk-ox (*Ovibos moschatus*), 2.45 m long. Most are gregarious and several have adapted to extreme environments (subarctic, very rugged, or semidesert). Goat antelopes first appear in the *Miocene. There are 11 genera and 26 species, distributed world-wide.

Caproidae (boarfish; subclass *Actinopterygii, order *Zeiformes) A small family of marine fish often found at moderate depths, which is why the fish often have a reddish colour and large eyes. Generally they have deep, almost diamond-shaped bodies, a pointed snout, small jaws, and minute scales. They have a world-wide distribution.

Capromyidae (coypus, hutias; order *Rodentia, suborder *Hystricomorpha) A family of arboreal, ground-dwelling, or semi-aquatic rodents whose short limbs are *pentadactyl but with the first digit of the fore limb reduced. The claws on the other digits are strong and the digits are webbed in some semi-aquatic forms. The eyes and ears are small, the skull massive, the occipital ridges prominent. The cheek teeth are *hypsodont. There are eight species in four genera: *Capromys*, *Geocapromys*, *Plagiodontia*, and *Myocaster*. They are native to the W. Indies and temperate S. America.

capsid The *protein case or shell that encloses the *genome (RNA or DNA) in an individual *virus particle.

capsid bug See MIRIDAE.

capsomere The *protein 'building block' of a *virus *capsid.

captacula In *Scaphopoda, prehensile *tentacles that arise from the base of the *proboscis.

Captorhinomorpha (subclass *Anapsida, order *Cotylosauria) A suborder of reptiles which appeared in the *Carboniferous and became extinct after the early *Permian. The suborder includes the earliest known reptiles, retaining some amphibian features, but distinctly reptilian in the form of their skulls, and differing from amphibians in the structure of the pelvic girdle. Because they were ancestral to all later reptile groups, the cotylosaurs are known as 'stem reptiles'. *See* HYLONOMUS LYELLI.

capuchin monkey (Cebus) *See* CEBI-DAE.

capybara (Hydrochoerus hydrochaeris) *See* HYDROCHOERIDAE.

Carabidae (ground beetles; class *Insecta, order *Coleoptera) Major family of generally black, brown, or metallic beetles, whose size ranges widely, from 2 to 35 mm long. They are often flightless, with fused *elytra. The legs are long and slender, and adapted for running; in some species they are specialized for digging. The *mandibles are prominent. Most ground beetles are carnivorous, but some species feed on plant material. Many species are nocturnal, hiding under stones during the day. Tiger beetles (*Cicindela* species, which some authorities classify as a family, Cicindellidae) are thought to be the fastest-running land beetles, with speeds of almost 2.4 km per hour. Bombardier beetles (*Brachinus* species) deter attackers with puffs of caustic chemicals, dispelled audibly from the abdomen. There are 25 000 species, distributed world-wide.

caracaras *See* FALCONIDAE.

Carangidae (jacks, scad, trevally; subclass *Actinopterygii, order *Perciformes) A large family of marine *pelagic fish. All have streamlined bodies, with two separate *dorsal fins, and a forked tail; and some (but not *Seriola*) have a series of scutes or comb-like scales along the *caudal peduncle. The two anal spines are detached from the rest of the *anal fin. Distributed widely in warm and temperate waters, many of these fish, especially the younger ones, travel in vast schools, but some of the larger ones are solitary. The young often live in small shoals beneath jellyfish. Carangids range in size from the relatively small *Trachurus trachurus* (eastern Atlantic scad), measuring up to 35 cm, to the much larger *Seriola dumerili* (amberjack). Some, e.g.

Elagatis bipinnulatus (rainbow runner), have a torpedo-shaped body. Others, e.g. the Indo-Pacific *Alectis indicus* (plumed trevally), have a compressed, deep body.

carangiform swimming A type of swimming practised by fish in which undulations are limited to the caudal (tail) regions with the body bending into less than one half of a sinusoidal wave form. *Compare* ANGUILLIFORM SWIMMING; BALLISTIFORM SWIMMING; LABRIFORM SWIMMING; OSTRACIIFORM SWIMMING; RAJIFORM SWIMMING.

Carangiform swimming

carapace 1. The domed top of a chelonian shell (*Chelonia), composed of two layers: (*a*) an inner body capsule of interlocking plates (the median neurals associated with the vertebrae; the lateral costals associated with the ribs, and encircled by the marginals); and (*b*) a covering of horny *scutes, forming a similar pattern but not corresponding exactly to the underlying bones. **2.** *See* CEPHALOTHORAX.

Carapidae (pearlfish; subclass *Actino-pterygii, order *Gadiformes) A family of small, somewhat eel-like, marine fish with very long *dorsal and *anal fins, but lacking *pelvic and *caudal fins. Many species have the peculiar habit of hiding in the body cavity of sea cucumbers. A fairly common species, *Carapus bermudensis*, has a semi-transparent body and leaves its host at dusk to feed; when returning it enters the posterior of its host, tail first. There are about 28 species, distributed in tropical and temperate seas.

carbohydrate A generic term for molecules that have a basic empirical formula of $C_x(H_2O)_y$, where x and y are variable numbers. They are normally classified as mono-, oligo-, or polysaccharides, depending upon whether they are single sugars, short-chain molecules, or large polymers respectively.

carbonaria *See* BISTON BETULARIA.

carbonic anhydrase An *enzyme, found largely in *erythrocytes, which catalyses the reaction between carbon dioxide and water to form carbonic acid.

Carboniferous The penultimate period of the *Palaeozoic Era, from about 359.2–299 Ma ago, preceded by the *Devonian and followed by the *Permian. In Europe the lower part of the period is termed the Dinantian. It is divided into three stages and is characterized by marine limestones with a rich coral-brachiopod fauna. In contrast the upper part, the Silesian, which is subdivided into four stages, is noted for the deposition of terrestrial and freshwater sediments. The vast forests of the upper Carboniferous gave rise to the rich coal measures of S. Wales, England, Scotland, and many other areas worldwide. N. American geologists subdivide the Carboniferous into two subperiods. Of these the lower is named the Mississippian and is the equivalent of the Dinantian stages (the Tournaisian, Visean, and Serpukhovian). The upper period, the Pennsylvanian, is the equivalent of the Silesian.

(⊕) SEE WEB LINKS
• Detailed information about life in the Carboniferous.

carboxylase An *enzyme that catalyses reactions in which a molecule of carbon dioxide is incorporated into an organic compound.

carcajou (glutton, skunk bear, wolverine; *Gulo gulo) *See* MUSTELIDAE.

Carcharhinidae (whaler shark, blue shark, tiger shark; subclass *Elasmobranchii, order *Lamniformes) A large family of mainly *pelagic sharks, in which the last of the five gill slits are placed over the origin of the *pectoral fin; the *spiracle is narrow and slit-like, and the first and larger of the two *dorsal fins is placed well forward. These sharks have a generally drab brown to blue colour, being lighter underneath. Some species (e.g. *Prionace glauca*, blue shark, which can reach a length of 3.5 m), have a world-wide distribution, avoiding only the polar seas. *Galeocerda cuvieri* (tiger shark) also inhabits many oceans and can reach a length of 5 m. It is *live-bearing, producing up to 46 live young at a time. There are about 85 species, including species such as *Mustelus mustelus* (smooth hound, or dogfish), which sometimes is placed in another family (Triakidae).

Carchariidae *See* ODONTASPIDIDAE.

***Carcharodon carcharias* (white shark)** *See* LAMNIDAE.

carder bees *See* MEGACHILIDAE.

cardiac cycle The series of muscular contractions which squeeze blood through the *heart. In *Amphibia and higher vertebrates, blood is pumped through a double circulatory system. The right side of the heart pumps blood through the pulmonary circuit to reach the lungs, the left side pumps oxygenated blood through the systemic circuit which supplies the body. Deoxygenated blood enters the right atrium via the vena cava and is pumped into the right ventricle. The right ventricle pumps the blood into the pulmonary artery which divides and leads to both lungs. Oxygenated blood returns from the lungs via the pulmonary vein to the left atrium and in turn to the left ventricle.

From the left ventricle, the most muscular of the heart chambers, the blood is pumped around the body where it becomes deoxygenated and returns to the right atrium. *Amphibia have only one ventricle, which must pump blood around both circuits; deoxygenated blood is kept separate from oxygenated blood by a combination of timing and complex architecture. During the phase of muscle relaxation (diastole), blood flows freely into the atrium until it is full, whereupon the sino-atrial node sends out an electrical pulse causing the cardiac muscle of the atrium to contract (systole). Blood is forced into the ventricle, which is in diastole, until it is full, whereupon the atrioventricular node sends out an electrical pulse along the Purkinje fibres around the right ventricle and the bundle of His (specialized muscle fibres) around the left ventricle, causing the ventricle to enter systole and force blood out into the arteries of the circulatory system. Nerves of the *sympathetic nervous system from the spinal cord connect to the sino-atrial node and can speed up the rate of heartbeat, while the vagus nerve from the medulla also connected to the sino-atrial node slows it down.

cardiac glycoside A generic term for a group of *steroid *glycosides that includes *oubain and *digitalis, which increase cardiac output. They act by stimulating systolic contraction of the heart muscle (see SYSTOLE). They can also inhibit *in vivo* the *sodium-pump mechanism of cell membranes.

cardiac muscle The specialized muscle from which the vertebrate heart is composed. The muscle fibres are branched and interlock and the muscle undergoes spontaneous (i.e. without stimulation) rhythmic contractions.

cardinal 1. In *Bivalvia, applied to the area around the hinge. The word is derived from the Latin *cardinalis* meaning 'hinge'. **2. (*Cardinalis cardinalis*)** See CARDINA-LIDAE.

cardinal beetle See PYROCHROIDAE.

cardinal fish See APOGONIDAE.

Cardinalidae (cardinals, grosbeaks, saltators; class *Aves, order *Passeriformes) A family of medium to large, heavy-bodied finches that have large conical bills; the males are often highly coloured (e.g. the six species of *Pheucticus* are yellow and black, orange and black, or red, white, and black). They inhabit woodland, feeding on fruit, seeds, blossoms, and insects, and are tree-nesting. *Cardinalis cardinalis* (cardinal) is a common cage bird. There are 8–13 genera, with 39 species, often combined with *Fringillidae or *Emberizidae, and found in N., Central, and S. America.

***Cardinalis cardinalis* (cardinal)** See CARDINALIDAE.

cardinal tooth A large hinge tooth, present in some bivalves, which is immediately below the *umbo. More than one may be present on each valve.

***Caretta caretta* (loggerhead turtle)** See CHELONIIDAE.

Carettochelyidae (pitted-shell turtle; order *Chelonia, suborder *Cryptodira) A monospecific family (*Carettochelys insculpta*) of turtles found in the rivers of southern New Guinea. The bony shell is complete, but skin-covered, lacking horny scales. The nose has a short breathing proboscis, but lung respiration is supplemented by cutaneous surfaces (see CUTANEOUS RESPIRATION). The family is believed to be intermediate between *Emydidae and *Trionychidae.

Cariama See CARIAMIDAE.

Cariamidae (seriemas; class *Aves, order *Gruiformes) A family of long-necked, long-legged, greyish-brown birds that have rounded wings, long tails, and a crest. They have small and *semipalmate feet, and are rapid runners but poor fliers. They inhabit grassland and forests, feeding on insects, reptiles, small mammals, and vegetable matter, and nest in trees and bushes. There are two monotypic genera, *Cariama* and *Chunga*, found in Central and S. America.

caribou (reindeer; *Rangifer tarandus*) *See* CERVIDAE.

caries A condition in which bones or teeth decay and (eventually) disintegrate.

carina 1. In air-breathing vertebrates, a ridge of *cartilage in the *trachea, aligned anterior to posterior and located between the two primary bronchi. **2.** *See* KEEL.

cariogenic Applied to any factor that leads to or encourages the formation of *caries.

carnassial In many *Carnivora, a modification of *premolar or *molar teeth, commonly the lower first molar and the upper last premolar, giving them a scissor-like shearing action used for cutting flesh.

carnitine A compound found in animal cells that functions in both fatty-acid synthesis and oxidation. It also facilitates the transport of fatty-acyl groups across the inner mitochondrial membrane.

Carnivora (cohort *Ferungulata, superorder *Ferae) An order that comprises the modern carnivorous placental mammals and their immediate ancestors. It used to be divided into two suborders, the *Fissipedia (mainly land-dwelling) and *Pinnipedia (seals, sea lions, walrus), but a more modern classification is into *Caniformia (dog-like) and *Feliformia (cat-like), with the 'pinnipeds' belonging to the former. The carnivores are descended from a single stock of the probably insectivorous, placental mammals of the early *Cretaceous, the change being reflected in their dentition. Strong *incisors for biting, and *canines for piercing, were retained from the insectivorous forms, but in general carnivores acquired modified cheek teeth (carnassials) specialized for shearing. These subsequently became reduced in those carnivores which adopted a herbivorous diet. Hoofs have rarely developed, and claws are used for seizing prey, and digits are never greatly elongated (and, apart from the *pollex and *hallux, they are not reduced). The first true carnivores were the weasel-like *Miacidae of the *Palaeocene, which had diverged by the end of the *Eocene to give the *Canidae (dogs) and *Mustelidae (weasels and their allies) as one branch and the *Viverridae (Old World civets) and *Felidae (cats) as another. According to some authors, the Mustelidae later branched again to give the *Phocidae (seals); and the Canidae diversified widely to produce such forms as the *Amphicyonidae ('dog-bears'), *Otariidae (sea lions), *Procyonidae (raccoons and pandas), and, ultimately, *Ursidae (bears); but molecular studies seem to indicate that the Phocidae and Otariidae are descended from a single ancestor which was related to the mustelid–ursid–procyonid stem. Finally, the *Hyaenidae (hyenas) emerged in the late *Miocene from viverrid stock; this is the youngest of the carnivore families, but the skunks and stink badgers, formerly placed in the Mustelidae, are now placed in a family of their own, the *Mephitidae.

carnivore 1. Any heterotrophic, flesh-eating animal. *See also* FOOD CHAIN. **2.** A member of the *Carnivora.

carnosaur A saurischian (*Saurischia), carnivorous dinosaur that was bipedal, powerfully built, and possessed large, dagger-like teeth. *Tyrannosaurus rex*, of the Late *Cretaceous, represented the culmination of the carnosaur line.

Carolina box turtle (*Terrapene carolina*) *See* EMYDIDAE.

carotene One of a group of *hydrocarbon *carotenoids.

carotenoid A generic term for water-soluble, polyisoprenoid pigments which often function as accessory photosynthetic pigments in higher plants and photosynthetic *bacteria. The group includes the carotenes. In the vertebrate *liver carotene is changed into (i.e. is a precursor of) *retinol (vitamin A).

carotid artery One of a pair of arteries supplying blood to the head.

carp (*Cyprinus carpio*) *See* CYPRINIDAE.

carpal One of the bones of the wrist that articulate with the digits.

carpal spur A sharp, horn-covered bone, situated on the *carpus of birds, and used in combat. It is found in some *Anatidae and *Charadriidae, and in all *Chionididae, *Jacanidae, and *Anhimidae.

carpenter bees *See* ANTHOPHORIDAE.

carpenter moths *See* COSSIDAE.

carpet beetle (*Anthrenus*) *See* DERMESTIDAE.

carpet moths 1. *See* GEOMETRIDAE. **2.** *See* TINEIDAE.

carpet shark *See* ORECTOLOBIDAE.

Carpodacus (rosefinches) *See* FRINGILLIDAE.

carpoids An informal, collective term describing the homalozoan (Homalozoa) classes Homoiostelea, Homostelea, and Stylophora.

Carpolestidae (order *Primates, suborder *Plesiadapiformes) An extinct family of *Palaeocene and *Eocene primates that were animals about the size of shrews, probably lived on the forest floor, and may have been herbivorous. They had nails rather than claws on the opposable thumb and big toe, enlarged, chisel-like *incisors, and enlarged lower *premolars. The ear region of the skull is reminiscent of lemuroids.

carpus The wrist joint, which in a bird forms the outer joint of the wing.

carrion beetle *See* SILPHIDAE.

carrying capacity The maximum population of a given organism that a particular environment can sustain; the K (saturation) value for *species populations showing S-shaped population-growth curves. It implies a continuing yield without environmental damage.

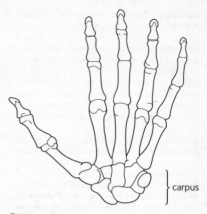

Carpus

cartilage In vertebrates, flexible skeletal tissue formed from groups of rounded cells lying in a matrix containing *collagen fibres. It forms most of the skeleton of *embryos and in adults is retained at the ends of bones, in intervertebral discs, and in the *pinna of the ear; in *Elasmobranchii calcified cartilage rather than true bone provides the entire skeleton.

cartilaginous fish Fish in which the skeleton, including the skull and jaws, consists entirely of *cartilage and never, even in the adult stage, comprises *bone. In mature sharks, part of the skeleton may calcify due to impregnation with calcium salts, but no bone is ever formed. Sharks and rays (*Chondrichthyes) have a cartilaginous skeleton, as have other lower vertebrates, e.g. the jawless lampreys and hagfish (*Agnatha).

caruncle A fleshy protuberance (e.g. the *wattles of some birds).

Carybdeida Alternative name for *Cubomedusae.

Caryophyllidea (phylum *Platyhelminthes, class *Cestoda) An order, or in some classifications family, of parasitic worms which possess one set of genitals. The *scolex may have suckers. Most inhabit two hosts during their development.

casque Literally, a helmet; in some vertebrates, a bony process on the top of the head.

Cassiduloida (subphylum *Echinozoa, class *Echinoidea) An order of irregular echinoids that have *petaloid *ambulacra on the upper surface. The *periproct is outside the *apical system. There are large pores and swellings around the *peristome, and no jaws or gill slits. The Cassiduloida first appeared in the *Jurassic.

Cassowari *See* CASUARIIDAE; CASUARI-IFORMES.

caste 1. In social insects, the existence of more than one functionally different form (polymorphism) within the same sex in the same colony, characterized by morphological features, age, or both. In bees and wasps, the female morphological castes are queens and workers. Within hive bees there are three castes, each performing a special function in the colony: non-reproductive workers (females which are usually sterile but help with provisioning of food for the queen in the hive); drones (reproductive males); and queens (fertile females). In termites, castes are not distinguished by sex: worker and soldier termites may be sterile males or females. There are several different kinds of workers among ants, all of which are sterile females, and ants also have a soldier caste. **2.** A system of social classification in humans, in which membership is determined culturally by birth and remains fixed; the group is ranked in a hierarchy of groups in the system.

caste determination In social insects, the process by which embryological development is influenced by physiological and environmental factors, giving rise to the various *castes.

caste polyethism *See* POLYETHISM.

Castniidae (subclass *Pterygota, order *Lepidoptera) Fairly small family of medium-sized, colourful, day-flying moths, often with clubbed antennae. They resemble some butterflies. Frequently the hind wing, but not the fore wing, is coloured. The adults feed on flowers; the larvae feed on plant stems and roots. There are about 200 species. The family is represented in the *Neotropical, Australasian, Indo-Malayan, and *Madagascan regions (or subregions).

Castoridae (beaver; order *Rodentia, suborder *Sciuromorpha) A family of semi-aquatic rodents that have dense fur, valvular nostrils and ears, enlarged, webbed hind feet, strongly clawed *pentadactyl limbs, horizontal, scaly, paddle-like tails, and small eyes and ears. Both sexes possess paired anal scent glands (*see* ANAL GLANDS). The skull is massive, the cheek teeth rooted and *hypsodont. There are two species in a single genus, *Castor*, which has a *Holarctic distribution.

Casuariidae (cassowaries; class *Aves, order *Casuariiformes) A family of large, heavy-bodied *ratites that have black, hair-like plumage; the wing feathers have bare quills and there are no tail feathers. The head and neck are featherless with a *casque, the bill is short and strong, and there are *wattles on the neck. The legs are stout with three toes, the inner toe having a long sharp claw. Cassowaries are rain-forest dwellers, feeding mainly on fruit and berries and nesting on the ground. There are three species in the genus *Casuarius*, found in Australia, New Guinea, and neighbouring islands.

Casuariiformes (cassowaries, emus; class *Aves) An order of large *ratites that have coarse, hair-like plumage, and three toes. There are two families, *Casuariidae and *Dromaiidae, confined to Australasia.

Casuarius (cassowaries) *See* CASUARIIDAE.

cat *See* FELIDAE.

catabolism *See* METABOLISM.

catadromous Applied to the migratory behaviour of fish that spend most of their lives in fresh water but that travel to the sea in order to breed there (e.g. *Anguilla anguilla* (common eel) which breeds in the Sargasso Sea). *Compare* ANADROMOUS;

POTAMODROMOUS. *See also* AMPHIDRO-
MOUS; DIADROMOUS.

Catagonus wagneri (Chacoan peccary)
See TAYASSUIDAE.

catalase A haemoprotein, oxidizing *enzyme, present in all aerobic organisms,
which catalyses the decomposition of hydrogen peroxide to water and oxygen. It is
the fastest-acting enzyme known, with a
*turnover number of 6×10^6. It is found in the
tissues of animals and is particularly concentrated in the *peroxisomes of cells.

catalysis A chemical term, literally meaning 'breaking down', and in digestive physiology referring specifically to the breaking
down of ingested food by an animal's own
digestive enzymes. *Compare* AUTOCATALY-
SIS.

**Catamblyrhynchidae (plush-capped
finch; class *Aves, order *Passeriformes)** A monotypic family (*Catamblyrhynchus diadema*), a short-billed finch
that has grey upper-parts and chestnut
underparts. Its fore-crown has stiff, erectile
yellow feathers. It inhabits forest edges and
clearings, and is insectivorous. It is found in
S. America.

**Catamblyrhynchus diadema (plush-
capped finch)** *See* CATAMBLYRHYNCHIDAE.

catarrhine In *Primates, applied to nostrils that are close together and open
downward.

**Catarrhini (cohort *Unguiculata, order
*Primates)** A suborder (or infra-order) of
primates that includes Old World monkeys, apes, and humans. Many authorities
place the Catarrhini with the *Platyrrhini
and *Tarsioidea in a single suborder, *Haplorhini. The earliest-confirmed catarrhines
are known from the *Oligocene of Egypt,
though possible representatives are known
from the *Eocene of Burma.

**catastrophic evolution (catastrophic
speciation)** A theory proposing that environmental stress might lead to the sudden
rearrangement of *chromosomes, which in
self-fertilizing organisms may then give rise
*sympatrically to a new *species. Probably
this explanation applies only to some special cases.

catbirds *See* PTILONORHYNCHIDAE.

catching spiral In the orb webs constructed by orb-web spiders (*Araneidae),
the sticky thread that catches prey. Having
built the basic frame and radii of her web,
the spider moves outward from the hub,
spinning an auxiliary spiral as she goes and
attaching it to the radii, until she reaches the
outer periphery. She then retraces her steps,
using the auxiliary spiral as a guide for laying
the sticky catching spiral and removing the
auxiliary spiral as she goes, until she reaches
the hub.

catecholamine A dihydroxyphenyl-
alkylamine derivative of the *amino acid
*tyrosine: *adrenalin and *noradrenalin are
two such derivatives. Many catecholamines
act as *neurotransmitters.

caterpillar Larval stage of *Lepidoptera,
although sometimes the term is also applied
to larval *Symphyta (sawflies).

caterpillar-hunting wasps *See* HUNT-
ING WASP; POMPILIDAE.

catfish 1. (air-breathing catfish) *See*
CLARIIDAE. **2. (airsac catfish)** *See* HETER-
OPNEUSTIDAE. **3. (armoured catfish)** *See*
CALLICHTHYIDAE; LORICARIIDAE. **4.** *See*
AUCHENIPTERIDAE. **5. (bagrid catfish)**
See BAGRIDAE. **6. (banjo catfish)** *See* AS-
PREDINIDAE. **7. (barbel-less catfish)** *See*
AGENEIOSIDAE. **8. (electric catfish)** *See*
MALAPTERURIDAE. **9. (long-whiskered
catfish)** *See* PIMELODIDAE. **10.** *See*
PANGASIIDAE. **11. (parasitic catfish)**
See TRICHOMYCTERIDAE. **12. (sea cat-
fish)** *See* ARIIDAE. **13.** *See* SCHILBEIDAE.
14. *See* SISORIDAE. **15. (torrent catfish)**
See AMBLYCIPITIDAE. **16. (upside-down
catfish)** *See* MOCHOKIDAE. **17. (wels)** *See*
SILURIDAE. **18. (whale catfish)** *See* CE-
TOPSIDAE.

Cathartidae (New World vultures, condors; class *Aves, order *Falconiformes) A family of large, long-winged, brownish-black vultures that have bare heads which are often coloured. Their nostrils are *perforate and many have a ruff of feathers around their neck. Their feet are weak with a small hind toe. They inhabit forests, mountains, and plains, feeding on carrion and nesting in caves or hollow trees. *Gymnogyps californianus* (California condor) is nearly extinct; *Vultur gryphus* (Andean condor) is the largest of all birds of prey. There are five genera, with eight species, found in N., Central, and S. America.

cathemeral Applied to an activity pattern in which an animal is neither prescriptively *nocturnal, nor *diurnal, nor *crepuscular, but irregularly active at any time of night or day, according to prevailing circumstances.

cathepsin One of a group of intracellular proteinases (*see* PROTEASE) that occur in most animal tissues, but especially in the *kidney, *liver, and *spleen. They are often active only in the presence of certain reducing substances, e.g. *cysteine, hydrogen cyanide, *ascorbic acid, and hydrogen sulphide. They occur in *lysosomes and are responsible for *autolysis after death.

Catostomidae (suckers; subclass *Actinopterygii, order *Cypriniformes) A fairly large family of freshwater fish: they resemble minnows, but have a protrusible sucker-mouth which is used to ingest organic material from the river bottom. Many have large scales, a single *dorsal fin, and a body that is round in cross-section. Schools of mature suckers are known to undertake spawning migrations upstream in the river. Although most of the 58 species live on the N. American continent, two species are found in eastern Siberia and China.

catshark *See* SCYLIORHINIDAE.

cattalo (beefalo) A fertile *hybrid between a domestic bull (*Bos taurus*) and an American bison cow (*Bison bison*).

cattle *See* PECORA; BOVIDAE.

cauda equina Literally 'horse's tail' (Latin). In most vertebrates, a bundle of the nerve roots at the end of the *spinal cord.

caudal Pertaining to the tail.

caudal fin The tail fin of a fish, used for steering, balancing, or locomotion.

caudal furca In some *Crustacea, a forked extension of the *telson.

caudal gill *See* TRACHEAL GILL.

caudal gland A gland located near the tail. In marine nematodes (*Nematoda) it secretes through spinnerets a substance the nematode uses to attach itself to an object. In swordtail characin fish (*Corynopoma riisei*) the caudal gland secretes a *courtship *pheromone. In mammals its secretions serve for identification.

caudal peduncle The tail region of the body of a fish, located between the *anal fin and the origin of the *caudal fin; a 'tail stalk'.

caudal vertebra *See* VERTEBRA.

Caulolatilus princeps (ocean whitefish) *See* MALACANTHIDAE.

cave cricket *See* RHAPHIDOPHORIDAE.

cavefish *See* AMBLYOPSIDAE.

cavernicolous Cave-dwelling.

cave salamander (*Proteus anguinus*) *See* PROTEIDAE.

Caviidae (order *Rodentia, suborder *Hystricomorpha) A family of ground-dwelling rodents in which the body is sturdy and there are four digits on the fore limbs and three on the hind limbs. The tail is vestigial. The cheek teeth are *hypsodont. The Caviidae are distributed widely in S. America. There are 12 species in five genera, including *Cavia* (guinea-pigs) and *Dolichotis* (Patagonian hares).

caviomorph Applied to rodents of the family *Caviidae.

cayman *See* ALLIGATORIDAE.

Cebidae (suborder *Simiiformes, superfamily *Ceboidea) A family (perhaps artificial) that comprises the New World monkeys. They lack cheek pouches, and the nostrils are separated and face laterally. The gait is quadrupedal but some use a form of *brachiation. Some have prehensile tails. The family includes *Aotus* species (night monkey, owl monkey, or douroucoulis) which are nocturnal. They have eyes with pure rod retinas, and lack prehensile tails. They may be closer than any other monkeys to the ancestral anthropoid form. Capuchin monkeys (several *Cebus* species) dwell in the forest canopy, feed on fruit and insects, and have an elaborate social structure. Howler monkeys (*Alouatta* species) feed on leaves and often use the prehensile tail in locomotion. Spider monkeys (*Ateles* species) sometimes move by a form of brachiation, with the aid of their prehensile tails, using long swings and jumps. They dwell in the canopy in small or large troops, and feed on fruit and nuts. Cebids are found throughout the tropical Central and S. American forests. There are 11 genera, with about 55 species.

Ceboidea (New World monkeys; order *Primates, suborder *Simiiformes) A superfamily that comprises the families *Callitrichidae and *Cebidae. Isolated, probably since the *Eocene, the ceboids show certain differences from the Old World monkeys. Their noses are flat, the nostrils facing laterally (platyrrhine) due to ceboids' possession of a larger nasal apparatus than Old World forms. Facial *vibrissae are present but are usually small. The *tympanic bone forms a ring fused with the *petrosal and the *bulla is large. The second *premolar tooth is retained. Social organization is elaborate. Communication is mainly vocal, in some genera (e.g. *Alouatta*) incorporating distinct sounds each of which has a precise meaning. All species are fully arboreal, some possessing prehensile tails with ridged tail pads used as tactile sense organs

and linked to a large area in the cerebral cortex, but the thumb is not fully opposable in any species.

Cebus (capuchin monkey) *See* CEBIDAE.

Cecidomyidae (gall midges, gall gnat; order *Diptera, suborder *Nematocera) Large family of flies, all of which are minute and delicate, with long, monofiliform antennae with obvious whorls of hair. *Ocelli may be present or absent. The wings have few longitudinal veins, most of which are unbranched. The antennae and simple wings are good points of recognition. There are three types of larval habit: parasitism; saprophagy; and phytophagy. Some cecidomyids form galls on plants which they attack. Some show pupal *paedogenesis. Many are considered to be serious pests, as their larvae are often phytophagous. More than 4000 species have been described.

cell 1. The fundamental autonomous unit of plant and animal bodies, consisting of, at least, a *cell membrane containing *cytoplasm and nuclear material, but often having a more complex structure. Simple organisms are unicellular, but more complex organisms consist of many co-operating cells. Cells may be *eukaryote or *prokaryote. **2.** *See* MASS PROVISIONING.

cell-adhesion molecule (CAM) A surface signalling molecule in the *cell membrane that mediates adhesion between adjacent cells. The molecule has an NH_2 end, a COOH end, and a middle region containing sialic acid. The middle section changes during the development of the animal from embryo to adult.

cell culture A mass of *cells derived either from a single cell, or from a small group of cells from the same tissue or organ, that is maintained *in vitro* using solid or liquid nutrient media.

cell cycle The sequence of events that occurs between the formation of a *cell and its division into daughter cells. There are three phases: *interphase; *mitosis; and *cytokinesis (cell division).

cell differentiation The process by which descendants of a single *cell produce structural and functional specializations and maintain these during the course of the life of that individual. Specializations arise during early development of the individual by differential DNA–RNA transcription; within a species, particular cells differentiate into particular forms, each with their own functions.

cell division See CYTOKINESIS.

cell fractionation The separation of cell *organelles into pure groups by homogenization followed by centrifugation.

cell fusion The experimental fusion of *nuclei and *cytoplasm from different *somatic cells to form a single hybrid cell. Cells used for fusion often come from tissue cultures derived from different species: fusion is facilitated by the modification of the surface of cells by adsorption of certain *viruses (e.g. Sendai virus).

cell growth An increase in the size of a cell: it occurs during the *interphase of the *cell cycle, during which both the *nucleus and the *cytoplasm enlarge.

cell line A lineage or pedigree of *cells that are related through asexual division. The cell line of an organ traces the course taken by cell division from a single-celled *zygote to the formation of that particular multicellular organ. A cell line may also be derived in the laboratory from a primary culture.

cell-mediated response A type of immunological reaction carried out by cytotoxic *T cells. A cell infected with foreign *antigens, possibly through viral infection, presents the antigen on the surface of the cell bound to a major histocompatibility complex (MHC) protein. The T cell receptor recognizes both the antigen and the MHC protein and consequently may attack the infected cell or invading extracellular particles. Compare HUMORAL RESPONSE.

cell membrane (plasmalemma, plasma membrane, protoplast) A sheet-like membrane, 7.5–10 nm thick, that forms a selectively permeable barrier enclosing and delimiting the *protoplasm of a *cell. It is a living structure consisting of *lipid molecules in a fluid bilayer, and associated *protein. Water molecules can pass freely through this structure, but the passage of most other molecules can be controlled. It may also have other functions, as in *prokaryotic cells, where it is associated with oxidative metabolism and cell division.

cell theory The basic theory (proposed by M. J. Schleiden and T. Schwann in 1838) that all animals and plants are made up of *cells, and that growth and reproduction are due to division of cells.

cellulase A highly specific endoglucosidase *enzyme that catalyses the *hydrolysis of *cellulose into *glucose by attacking β-1,4 linkages.

cellulolytic Able to break down or digest *cellulose.

cellulose A straight-chain, insoluble polysaccharide composed of *glucose molecules linked by β-1,4 glycosidic bonds. It is the principal structural material of plants, and as such is the most abundant organic compound in the world. It has also been found in certain sea squirts (*Urochordata).

cellulytic Able to break open (lyse) cells.

cement A bone-like substance which coats the surface of that part of a tooth which is embedded in the jaw and which sometimes coats the enamel of the exposed part of the tooth.

Cenomanian One of the stages of the Late *Cretaceous Period in Europe, about 99.6–93.5 Ma ago.

Cenozoic (Cainozoic, Kainozoic) An era of geological time that began about 65.5 Ma ago and that continues to the present. It includes the *Palaeogene and *Neogene Periods (formerly the *Tertiary and *Quaternary sub-eras). Molluscs and microfossils are used in the stratigraphic subdivision of the era. The Alpine orogeny (mountain-building episode) reached its climax during this period of geological time.

• Detailed information about life in the Cenozoic.

centipedes See CHILOPODA.

central limit theorem A theorem stating that the arithmetic-mean values for a series of similar-sized, fairly large samples ($n > 30$) taken from a large population will be distributed approximately normally about the true population mean (μ), irrespective of the actual distribution pattern of the individual counts.

central nervous system (CNS) The main *ganglia of the nervous system, which make up the *brain and the *spinal cord. The central nervous system is encased within sheets of tissue called *meninges. *Compare* PERIPHERAL NERVOUS SYSTEM. *See also* AUTONOMIC NERVOUS SYSTEM.

Centrarchidae (sunfish; subclass *Actinopterygii, order *Perciformes) A fairly large family of freshwater fish in which the first of the two united *dorsal fins is strongly spined, the scales feel rough because of the comb-like edges, the tail is broad (deep), and the tail fin is slightly notched. The larger species, e.g. *Micropterus salmoides* (largemouth bass) may reach a weight of 8 kg. It has been introduced into certain European rivers. There are about 30 species, all found in the eastern half of N. America.

centric fusion The whole-arm fusion of *chromosomes.

centrifugal speciation The principle that new *species are likely to arise towards the centre of the range of the present species, rather than at the periphery. In practice, it is very often observed that primitive species are located on the edges of the distribution of a species-group or genus.

centrifuge An apparatus used for the separation of substances by the application of centrifugal force: this is generated by high-speed rotation of a vessel containing tubes filled with a fluid in which the substances are suspended but not dissolved. Separation occurs because different substances have different rates of sedimentation according to their molecular size and shape.

centriole A hollow, cylindrical structure, normally one of a pair lying at right angles to one another, adjacent to the *nuclear envelope in animal cells, and composed of nine sets of *microtubules, each set arranged in triplets. Centrioles are thought to be organizers of microtubular structures in these cells, and during cell division a pair is found at each pole of the *mitotic spindle.

Centriscidae (razorfish, shrimpfish; subclass *Actinopterygii, order *Syngnathiformes) A family of small, marine, tropical fish which often swim in a vertical, head-down position. The compressed, thin body is covered by a bony casing produced from modified scales. The ventral side is produced into a hard and sharp keel ('razor blade'). They have a long, tubular snout with a small mouth at the end of it, which is used to suck up minute organisms. At least four species occur in the Indo-Pacific region, *Centriscus aristatus* (smooth razorfish) reaching a length of up to 25 cm.

Centriscus aristatus (smooth razorfish) See CENTRISCIDAE.

Centrolenidae (class *Amphibia, order *Anura) A small family of mainly arboreal frogs which are probably closely related to the *Leptodactylidae. The feet have pads and extra digital cartilaginous elements. The *tibia, *fibula, and *calcaneum are all fused. The pectoral girdle is bow-shaped. There are 40 species, occurring in Mexico and S. America.

Centrolophidae (trevalla, ruffe; subclass *Actinopterygii, order *Perciformes) A family of medium-sized to large marine fish, with long *anal and *dorsal fins, fairly small *pelvic fins, and a slightly forked tail fin. The few spines of the dorsal fin may be absent or short, or grade into the many-rayed, soft second dorsal. Centrolophids tend to be *pelagic fish, living near the edge of the continental shelf in temperate waters. *Centrolophus niger* (blackfish or black ruffe) occurs in the N. Atlantic. Among those living in the southern hemisphere are *Hyperoglyphe porosa* (deepsea trevalla) and the warehou (a name thought to be of Maori origin), *Seriolella brama*. There are about 22 species.

Centrolophus niger (blackfish, black ruffe) *See* CENTROLOPHIDAE.

centromere The point at which the two halves of a *chromosome, the *chromatids, are joined. *See* KINETOCHORE; SPINDLE ATTACHMENT.

Centropomidae (Ambassidae; perchlet, snook; subclass *Actinopterygii, order *Perciformes) A family of basically marine fish, but with representatives living in brackish or fresh water, with a wide range in size. All tend to have two separate *dorsal fins, strong spines in the first dorsal and *anal fins, and a *lateral line extending on to the tail. The smaller species have large eyes and a forked tail. There are about 30 species.

Centropus (coucals) *See* CUCULIDAE.

centrosome In a *cell, the distinct part of the *cytoplasm that contains the *centriole.

centrum *See* VERTEBRA.

Cepaea A genus of snails, including *C. hortensis* and *C. nemoralis*, that show visible *polymorphism for banding, shell colour, and shell height. They have formed the subject of extensive studies in ecological genetics, where it has been shown that different selective processes relating to various factors, including differential predation and differential response to sunlight, are responsible for the maintenance of the different morphs in the population.

Cephalaspidea (Bullomorpha; phylum *Mollusca, class *Gastropoda) An order of *opisthobranch gastropods in which the *mantle cavity is reasonably well developed. The bubble-shaped shell is internal and very small in some genera. A wide foot

and head shield are present. All are hermaphrodites.

Cephalaspidiformes (class *Cephalaspidomorphi, superorder *Cephalaspidomorpha) An order of fish-like, fossil vertebrates which lived during the late *Silurian to *Devonian Periods. They had somewhat flattened bodies, with a broad head, 10 pairs of ventral gill openings, a ventral jawless mouth, a *heterocercal tail, and a single median nostril communicating with a blind-ending *nasohypophysial sac. The dorsally located eyes and flat belly suggest a bottom-dwelling way of life. Together with the still-living lampreys (order *Petromyzoniformes) they have been placed in the class Cephalaspidomorphi, although other classifications have also been made.

Cephalaspidomorpha (superclass *Agnatha, class *Cephalaspidomorphi) A superorder of jawless vertebrates including the order *Petromyzoniformes (lampreys).

Cephalaspidomorphi (superclass *Agnatha) A class of jawless fish-like animals including the extant lampreys and their fossil relatives.

cephalic Pertaining to the head.

cephalins A group of *phospholipid compounds, which includes phosphatidyl ethanolamine and phosphatidyl serine. Both are major phosphoglycerides in animals and higher plants, in which they are important constituents of membranes; they are particularly common in brain membranes.

cephalization An evolutionary trend in animals whereby nerve centres and sensory organs become concentrated at the anterior end of the body, eventually forming a distinct head bearing the mouth, ears, and eyes.

Cephalocarida (phylum *Arthropoda, subphylum *Crustacea) The most primitive class of crustaceans, first described in 1955, cephalocarids are shrimp-like and small (all less than 4 mm long). The body comprises a semicircular head and a long trunk of 19 segments, of which only the anterior nine bear limbs. Eyes are absent and both pairs of antennae are short. All the trunk limbs are similar, and they resemble the second pair of *maxillae. They are unusual in that each bears a *pseudopipodite. Cephalocarids are hermaphroditic, and live in bottom silts which they filter for organic matter. The class comprises four genera, with seven species, and has been found in coastal waters off eastern and western N. America, the W. Indies, and New Zealand.

Cephalochordata (Acrania; phylum *Chordata) A subphylum that contains only *Branchiostoma lanceolata* (amphioxus, or lancelet), probably the most primitive of living chordates, although some soft-bodied *Cambrian Period fossils are dubiously referred to the group. *See also* BRANCHIOSTOMIDAE.

Cephalodiscida (phylum *Hemichordata, class *Pterobranchia) An order of free, unattached associations of individual *zooids. The *mesosoma bears several pairs of arms. The gonads are paired, with a pair of gonadial openings. There is one pair of *branchial pores. The skeleton (coenoecium) is very variable and usually irregular in form. The first known occurrence of cephalodiscids is in the Early *Ordovician.

cephalon In some *Arthropoda (e.g. *Trilobitomorpha and *Pycnogonida) the anterior portion of the body, or head, comprising fused segments.

Cephalopoda (phylum *Mollusca) Literally 'head-foot' (from the Greek *kephale*, head, and *pod-*, foot), a class of molluscs, exclusively marine, that are related to the *Bivalvia and *Gastropoda. Cephalopods are *bilaterally symmetrical and have a circlet of prehensile tentacles. The exhalation of a current of water through a muscular funnel provides them with a form of jet propulsion. The shell is either internal or external. Most are active predators. The class includes the *Nautiloidea (nautiloids), *Sepioidea (cuttlefish), *Teuthoidea (squids), *Octopoda (octopuses), and the extinct *Ammonoidea (goniatites, ceratites, ammonites) and *Belemnitida (belemnites). The earliest forms belonged to the Nautiloidea and date from the Late *Cambrian. There are 17 000 extinct species and 800 living species.

((())) SEE WEB LINKS
• Description of the Cephalopoda

cephalothorax In some members of the phylum *Arthropoda, the fused head and *thorax. It is found in members of the *Chelicerata (classes *Merostomata, *Arachnida, and *Pycnogonida), and in most *Crustacea. In spiders, it is the prosoma, which is covered by a sclerotized plate (carapace), and which bears the walking limbs, *pedipalps, *chelicerae, and eyes. The opisthosoma (abdomen) is joined to the cephalothorax by a stalk (pedicel).

cephalotoxin A *toxin produced by the *salivary glands of *Cephalopoda that stuns prey.

Cephidae (suborder *Symphyta, superfamily Cephoidea) Small family of stem-boring symphytans the adults of which are 9–13 mm long, usually black, and with a laterally slightly flattened abdomen. The larvae tunnel into the stems of grasses and other plants.

Cepolidae (bandfish; subclass *Actinopterygii, order *Perciformes) A small family of marine fish that have eel-like bodies, tapering towards the tail, and long *dorsal and *anal fins which are continuous with the tail fin. They occur at depths of more than 100 m, but also closer inshore in the eastern Atlantic, Indo-Pacific, and off Australia and New Zealand. There are about seven species.

Ceractinomorpha (phylum *Porifera, class *Demospongiae) A subclass of sponges which have *parenchymular, incubated larvae. Either organic fibres or *spicules may dominate the skeleton. The *spongin fibres may contain detritus and spicules.

Cerambycidae (longicorn beetles, longhorn beetles, timber beetles; class

*Insecta, order *Coleoptera) Family of elongate, *pubescent beetles, usually strikingly marked, up to 20 cm long. The antennae are *filiform, at least two-thirds of the body length, and often greater. Larvae are fleshy and elongate, with powerful jaws and reduced legs. They feed on wood of dead or dying trees, and may take four years to reach maturity. There are 20 000 species.

Cerapachyinae (suborder *Apocrita, family *Formicidae) Subfamily of ants that are sometimes classified as a tribe within the subfamily *Ponerinae. They possess spines on the *pygidium and short, thick antennae; and they lack dorsal thoracic sutures. They are predators of other ant species. There are about 200 species, found in the tropics.

cerata See NUDIBRANCHIA.

Ceratiidae (deep-sea devil; subclass *Actinopterygii, order *Lophiiformes) A family of short, thickset, deep-sea fish, with a large mouth, minute eyes, and a long, fleshy 'lure' above the eyes. The much smaller males are attached to the females, remaining in a parasitic association. There are two genera, *Ceratias* and *Cryptopsaras*, and four species.

Ceratina (carpenter bees) See ANTHOPHORIDAE.

ceratites See AMMONOIDEA.

Ceratodidae (Australian lungfish; subclass *Dipneusti, order *Ceratodiformes) A monospecific family comprising the Australian lungfish which has a fairly compressed body, with flipper-like *pectoral and *pelvic fins, and large scales. It occurs in a number of rivers in Australia, mainly south-eastern Queensland. The lung-like *swim-bladder can be used to utilize atmospheric oxygen.

Ceratodiformes (subclass *Dipneusti, superorder *Ceratodimorpha) An order comprising the one family *Ceratodidae.

Ceratodimorpha (class *Osteichthyes, subclass *Dipneusti) A superorder of bony fish that includes the orders *Ceratodiformes and *Lepidosireniformes.

Ceratomorpha (superorder *Mesaxonia, order *Perissodactyla) A suborder of mammals that comprises the superfamilies *Tapiroidea and *Rhinocerotoidea (fossil and recent Old and New World tapirs and rhinoceroses), perissodactyls in which the nasal bones are stout and projecting, the cheek teeth are low-crowned, and the limbs are short and thick, with three digits on the hind feet and three or four on the fore feet. Rhinoceroses, but never tapirs, may have horns.

Ceratopogonidae (biting midges; order *Diptera, suborder *Nematocera) Family of small to minute, gnat-like flies, in which the males have *plumose antennae, the head is not concealed by the *thorax, and *ocelli are absent. The short mouthparts are of a piercing type. All adults of this family are predatory, sucking blood or eating smaller insects. There are two types of larval habit. One group has aquatic, vermiform larvae, the other has terrestrial larvae which live under bark, or in decaying material. Adults of the genus *Culicoides* will bite mammals, and can transmit certain bloodborne diseases. They are avid biters at sunset in certain parts of the *Holarctic, and in some places are great pests. In the northern parts of the USA and Canada they are known as 'no-see-ums' on account of their small size and vicious bite. There are more than 500 described species. Their distribution is mainly Holarctic, but there are representatives throughout the world.

Ceratopsia (subclass *Archosauria, order *Ornithischia) A suborder of horned dinosaurs which had beak-like jaws, from the Late *Cretaceous. The head accounted for about one-third of the total length of the body because of the development of a large, bony frill which protected the neck and shoulders. *Triceratops* is perhaps the best-known member of the group. It was 5–6 m long and had three forward-projecting horns, one over each eye, and the third over the nose.

Ceratotherium simum (white rhinoceros) See RHINOCEROTIDAE.

cercariae *See* SCHISTOSOMIASIS.

cerci In many insects, a pair of cylindrical or conical appendages at the posterior end of the abdomen, often serving a sensory function.

Cercopidae (frog-hoppers, spittle-bugs, cuckoo-spit insects; order *He-miptera, suborder *Homoptera) Family of homopterans in most species of which the nymphs live on the shoots or roots of plants and produce a protective mass of froth around themselves. Adults of some species have several different colour forms, and some species are pests of crops or pasture plants. There are about 2500 species, distributed world-wide.

Cercopithecidae (Old World monkeys; suborder *Simiiformes, superfamily *Cercopithecoidea) A family that, in some arrangements, comprises all existing Old World monkeys; some authors restrict the family to the group that is otherwise classified as the subfamily Cercopithecinae, animals with the characteristics of the superfamily. The family, as commonly conceived, includes two subfamilies: Cercopithecinae, including *Papio* (baboons), *Mandrillus* (mandrills), *Theropithecus* (geladas), and *Cercopithecus* (guenons) of Africa, and *Macaca* (macaques) of Asia; and Colobinae (sometimes raised to the family rank as Colobidae), including *Colobus* (colobus monkeys) of Africa, and *Presbytis* and related genera (e.g. *P. entellus*, the entellus langur of India, and *Pygathrix nemaeus*, the douc langur of south-east Asia), *Rhinopithecus* (snub-nosed monkeys), and *Nasalis* (proboscis monkeys) of Asia. *Rhinopithecus* species, found in high forests of Tibet and north-western China, are langurs with short, projecting noses; they may be the 'abominable snowmen' of legend. *Nasalis larvatus* of Borneo has the cercopithecid nose developed to an extreme: up to 17 cm long in adults. There are 8–20 genera, with about 80–85 species, distributed throughout Africa and Asia.

Cercopithecinae *See* CERCOPITHECIDAE.

Cercopithecoidea (order *Primates, suborder *Simiiformes) A superfamily that comprises the extinct and extant monkeys of the Old World. Most are larger than *Ceboidea. *Ischial callosities are present, often surrounded by brightly coloured skin which becomes enlarged in females prior to ovulation. Cheek pouches are present and are used to store food. The tail is never prehensile. The olfactory *turbinal bones are reduced, the nostrils close together and pointing downward. The second *premolar tooth is absent. Some species are ground-dwelling. Thumbs may be opposable. Cercopithecoids may be herbivorous or omnivorous. The most characteristic specialization is *bilophodonty. As well as the living *Cercopithecidae, there is an early *Miocene family, Victoriapithecidae (with two genera, *Victoriapithecus* and *Prohylobates*), which includes very primitive monkeys with only partially developed bilophodonty.

***Cercopithecus* (guenon)** *See* CERCOPITHECIDAE.

***Cerdocyon thous* (crab-eating fox)** *See* CANIDAE.

cere In birds, an area of skin at the base of the upper *mandible, surrounding the nostrils, that is found in *Falconiformes, in which it is bare, in *Psittaciformes, in which it is usually feathered, and in *Columbidae, in which it forms a swollen flap above the nostrils.

cerebellum One of the three major parts of the *brain. The cerebellum forms from a dorsal outgrowth from the metencephalon region of the *brain stem and is involved in co-ordination and regulation of motor activities, such as balance or escape movements. The *grey matter of the cerebellum receives impulses from various other co-ordination centres as well as from tendon stretch receptors and acoustic areas of the *myelencephalon.

cerebral cortex A region of the brain, comprising *grey matter, that covers the surface of the *cerebrum in all mammals. It is derived from the *pallium, consists of up to six layers, and in large animals is deeply folded.

cerebriform Resembling a brain; convoluted.

cerebroside Ceramide monosaccharide, a substance that occurs most abundantly in the *myelin sheath of nerves.

cerebrospinal fluid A clear, colourless liquid, produced in the brain, that surrounds and cushions the brain and spinal column.

cerebrum A pair of hemispheres originating from the anterior end of the fore brain: in primitive vertebrates it is concerned mainly with the olfactory sense, but in mammals it forms the largest part of the brain, with an increasingly convoluted cortex (grey matter), and is concerned with the analysis of sensory information and the instigation of responses. *See also* TELENCEPHALON.

Ceriantharia (phylum *Cnidaria, class *Anthozoa) A small order of solitary, semi-burrowing sea anemones, that lack a skeleton. *Mesenteries are complete but unpaired. *Tentacles are arranged in two rings, marginal and oral.

Ceriantipatharia (phylum *Cnidaria, class *Anthozoa) A subclass which includes the orders *Antipatharia and *Ceriantharia, on the basis of the very weak or indefinite musculature of the *mesenteries, and the insertion of new mesenteries only in the dorsal intermesenterial space.

Certhia **(tree-creepers)** *See* CERTHIIDAE.

Certhiidae (tree-creepers; class *Aves, order *Passeriformes) A family of one genus (*Certhia*) of small, brownish birds, streaked above and white below, that have slightly curved bills. They are short-winged, have long claws, stiff, pointed tail feathers, and are weak fliers. They inhabit forests, nesting in crevices, and are insectivorous, foraging by climbing vertically up tree and rock surfaces. There are six species, found in N. America, Europe, and Asia. The

monotypic family *Salpornithidae (spotted creeper) is sometimes included.

cervical vertebra *See* VERTEBRA.

Cervidae (deer; infra-order *Pecora, superfamily *Cervoidea) A family of browsing or grazing animals, which appeared in Eurasia in *Miocene and early *Pliocene times and had radiated widely by the end of the Pliocene. Some of the later *Pleistocene representatives (e.g. *Megaceros giganteus*, the Irish elk) were giant types, much larger than the surviving forms. The horns (antlers) are complex in many species but simple in ancestral and some primitive species, and absent in *Hydropotes* (Chinese water deer); usually they are present only in the male (but in *Rangifer* they are present in both sexes). Usually they are shed annually. Ancestral and some modern species (e.g. *Hydropotes*) have canine tusks. The feet have four digits. Most species (but not all) are gregarious, living in herds with elaborate social organization. They have a *Holarctic distribution. There are about 16 genera, and 43 species, including *Cervus* (red deer and wapiti or American elk are included in *C. elaphus*), *Rangifer* (reindeer or caribou is *R. tarandus*), *Alces* (European elk and American moose are both *A. alces*), *Capreolus* (roe deer), *Dama* (fallow deer), and *Muntiacus* (muntjaks). *Moschus* (musk-deer), formerly placed in Cervidae, nowadays is universally referred to a separate family, *Moschidae.

Cervoidea (order *Artiodactyla, infra-order *Pecora) A superfamily that comprises the ancestral deer and their modern descendants: the families *Moschidae, *Cervidae, and *Giraffidae.

Cervus **(red deer, wapiti, American elk)** *See* CERVIDAE.

Ceryle **(kingfishers)** *See* ALCEDINIDAE.

Cestida *See* CESTIDEA.

Cestidea (Cestida; phylum *Ctenophora, class *Tentaculata) An order of ctenophorids which have greatly compressed bodies. Some forms may be up to

1.5 m long. There are two main *tentacles, and smaller tentacles surround the anterior edge.

Cestoda (tapeworms; phylum *Platyhelminthes) A class of parasitic worms all of which lack a gut. The genitalia are normally repeated. The majority of forms possess a *scolex which may have suckers. Mature individuals lack *cilia. All are endoparasitic (*see* PARASITISM), most mature individuals living within vertebrates.

Cetacea (whales; infraclass *Eutheria, cohort *Mutica) An order that comprises the one extinct (*Archaeoceti) and two extant (*Odontoceti and *Mysticeti) suborders of whales. The earliest whales (Archaeoceti) are known from *Eocene rocks in Africa and South Asia (*see* BASILOSAURUS; AMBULOCETUS), and are descended from early *Artiodactyla, probably more closely related to hippopotamus (*Hippopotamidae) than to other groups. Of the two existing groups of whales, the Odontoceti (toothed whales) can be traced back to ancestral forms in the upper Eocene, while the first definitive Mysticeti (baleen whales) occur in *Oligocene strata. Whales are streamlined, almost hairless, entirely aquatic. The fore limbs are modified to form paddles without visible digits, the hind limbs are absent, the pelvis is vestigial, except in some Archaeoceti. The tail fin is horizontal and used for propulsion. The skull is modified, with the nasal openings far back on the dorsal surface except in *Physeteridae. The diet comprises fish and molluscs (Odontoceti) or mainly plankton (Mysticeti).

Cetomimidae (flabby whalefish; subclass *Actinopterygii, order Beryciformes) A small family of marine, oceanic fish that have naked skin, rudimentary eyes, and lack *pelvic fins. There are about 10 species.

Cetopsidae (whale catfish; subclass *Actinopterygii, order *Siluriformes) A small family of freshwater catfish that have a scaleless body, three pairs of *barbels around the mouth, small eyes, a small single *dorsal and no *adipose fin, and a low *anal

fin. There are about 12 species, found in S. America.

***Cetorhinus maximus* (basking shark)** *See* LAMNIDAE.

***Cettia* (bush warblers)** *See* SYLVIIDAE.

***Ceyx* (kingfishers)** *See* ALCEDINIDAE.

chachalacas *See* CRACIDAE.

Chacoan peccary (*Catagonus wagneri*) *See* TAYASSUIDAE.

chaetae Bristles made from *chitin that are characteristic features of *Oligochaeta and *Polychaeta.

Chaetodermatida (class *Amphineura, subclass *Aplacophora) An order of aplacophorans which lack a ventral groove. The simple, bell-shaped *mantle cavity contains a pair of, or several, true gills. Individuals are unisexual. They are *Holocene in age.

***Chaetodon capistratus* (four-eye butterfly fish)** *See* CHAETODONTIDAE.

Chaetodontidae (butterfly fish, angelfish; subclass *Actinopterygii, order *Perciformes) Large family of marine, deep-bodied, and almost discus-shaped fish. They have a small, protractile mouth, minute teeth on the jaws (the name is derived from the Greek *khaite*, hair, and *odont-*, tooth), a single, continuous *dorsal fin with scales along the base, and a round to notched tail fin. The family includes numerous beautiful, brightly coloured species, which are very popular among sea-water aquarium hobbyists. Among the more widely distributed butterfly fish are *Chaetodon capistratus* (four-eye butterfly fish) of the western Atlantic, and *C. vagabundus* (vagabond butterfly fish) of the Indo-Pacific and northern Australia. Angelfish, e.g. *Pomacanthus imperator* (emperor angelfish) of the Indo-Pacific, have a characteristic spine at the angle of the preopercle (part of the gill cover), and are often placed in a separate family, the Pomacanthidae.

Chaetodon vagabundus (vagabond butterfly fish) *See* CHAETODONTIDAE.

Chaetognatha A phylum that comprises the 'arrow worms', first encountered in the fossil record in *Carboniferous rocks. They are characterized by the possession of horizontal fins that flank the trunk and tail. The body is semi-transparent, streamlined, and divided into a head, trunk, and post-anal tail region. The head bears eyes and food-catching spines. Arrow worms are common among marine *plankton.

Chaetonotoida (phylum *Aschelminthes, class *Gastrotricha) An order of microscopic worms in which the male genitalia are always lacking. Most inhabit fresh-water habitats.

Chaetorhynchus papuensis (Papuan mountain drongo) *See* DICRURIDAE.

chaetotaxy (setation, trichiation) The arrangement of hairs on the body of an insect, which is often used as a taxonomic guide.

Chaetura (swifts) *See* APODIDAE.

chafer *See* SCARABAEIDAE.

chaffinches *See* FRINGILLIDAE.

Chagas's disease A disease that can affect humans and other animals. It occurs chiefly in Central and S. America. The causal agent is a protozoon, *Trypanosoma cruzi.* The pathogen is transmitted by blood-sucking bugs. Symptoms may include anaemia and various signs of heart, gland, and nervous-system involvement. The disease is named after the Brazilian physician Carlos Chagas (1879–1934). Charles Darwin may have been infected with this disease in S. America, and some have suggested that his continual ill-health in later years was due to it.

chain response A sequence of behaviour in which each item produces a situation that evokes the next (e.g. in *courtship rituals the correct response of one partner is likely to evoke the next behaviour in the other).

Chalcidae (chalcid wasps; suborder *Apocrita, superfamily *Chalcidoidea) Family of fairly common wasps most of which are 2–7 mm long, generally uniformly dark in colour, and with greatly enlarged hind *coxae and femora (*see* FEMUR). The femora are often toothed ventrally. The wings are not folded longitudinally at rest, and the *ovipositor is short. Members of the family are larval parasites of *Lepidoptera, pupal parasites of *Diptera, and some are egg parasites of cockroaches. When disturbed, chalcids often roll up and feign death.

Chalcidoidea (chalcid wasps; order *Hymenoptera, suborder *Apocrita) Very large superfamily of mostly small to minute, parasitic (sometimes phytophagous) wasps which occur almost everywhere and have very diverse forms and habits. Adults are 0.2–3.0 mm long, the majority being 1–3 mm. Wing venation is greatly reduced, the *pronotum squarish, the antennae elbowed and usually short, and the *ovipositor short. Adults are usually black, blue-black, or greenish, and many have a metallic sheen. The wings are normally held flat over the abdomen although some are *brachypterous or *apterous. The larvae may be parasitic, hyperparasitic, or phytophagous, and many species exhibit *polyembryony. Many chalcidoid species have a modified middle leg and can jump with agility.

chalcid seed flies *See* TORYMIDAE.

chalcid wasp Common name often given to any member of the superfamily *Chalcidoidea, but applied strictly only to members of the family *Chalcidae.

Chalicodoma *See* MASON BEE.

Chalicotheriidae *See* ANCYLOPODA.

chalone In mammalian epithelial tissue (*see* EPITHELIUM), a tissue-specific *protein of moderate size, or a simple *polypeptide, that in the presence of *adrenalin inhibits its *mitosis in cells. The loss of chalones from damaged cells stimulates and directs wound healing. Chalones are thought to be important in ageing and cancer.

Chamaea fasciata (wren-tit) *See* CHAMAEIDAE.

Chamaeidae (wren-tit; class *Aves, order *Passeriformes) A monotypic family (*Chamaea fasciata*), comprising a small, brown, long-tailed bird that inhabits scrub, nests in bushes, and feeds on insects and berries. It is found along the Pacific coast of N. America and northern Mexico. It is often placed in the *Timaliidae.

Chamaeleo chamaeleon (common chameleon, European chameleon, Mediterranean chameleon) *See* CHAMAELEONTIDAE.

Chamaeleontidae (chameleons; order *Squamata, suborder *Sauria) A family of arboreal lizards, descended from the *Agamidae, in which the eyes are protruding, with the lids fused to a small central aperture, and move independently. There is no *tympanic membrane. The tongue is sticky and may equal the body in length. Horns are common. The toes are grasping and opposable. The tail is prehensile. *Chamaeleo chamaeleon* (common chameleon, European chameleon, or Mediterranean chameleon) is basically greenish but can change colour; it basks on branches of trees and shrubs and occurs throughout the Mediterranean region. There are 86 species in the family, all confined to the Old World, and found mainly in Africa and Madagascar.

Chambers, Robert (1802–71) The author of *Chambers's Encyclopaedia*, who in 1844 published anonymously a book called *Vestiges of a Natural History of Creation*, in which he revived the idea of evolution first proposed by Lamarck 30 years earlier. The book's popularity and notoriety refocused attention on this issue and so paved the way, among the general public, for DARWIN's *Origin of Species*.

changeable toad (*Bufo viridis*) *See* BUFONIDAE.

Chanidae (milkfish, salmon-herring; subclass *Actinopterygii, order *Gonorynchiformes) A monospecific family (*Chanos chanos*), comprising a fish whose streamlined body and long, deeply forked tail fin give it a graceful appearance. Because of its fairly small, toothless mouth, the single *dorsal fin, and the bright, silvery colour, it is sometimes also known as the giant herring. Milkfish often form large schools, swimming near the surface. Along the shores of many tropical countries it is an important source of protein for humans. The young are caught in large numbers, placed in fish ponds, and kept until they are of marketable size. It is found in the tropical waters of the Indo-Pacific and S. American regions, where it can grow to at least 1 m in length.

Channichthyidae (crocodile fish, icefish; subclass *Actinopterygii, order *Perciformes) A family of marine fish that have a depressed head and long snout, large mouth and eyes, and long *dorsal and *anal fins. In many species the blood contains few red blood cells. These fish are apparently rather sluggish, living in the cold waters of the South Atlantic and Antarctic. There are about 16 species.

Channidae (snakeheads; subclass *Actinopterygii, order *Perciformes) A family of tropical, freshwater fish that have an elongated body, snake-like head, very long *dorsal and *anal fins, rounded tail and *pectoral fins, no fin spines, and the lower jaw protruding beyond the upper. *Accessory respiratory organs enable the fish to live out of water for some time. There are about 11 species, found in tropical Africa and southern Asia.

Chanos chanos (milkfish) *See* CHANIDAE.

Chaoboridae (gnats; order *Diptera, suborder *Nematocera) A small family of gnats in which the adults resemble mosquitoes but have poorly developed mouth-parts. The family includes *Chaoborus crystallinus* whose carnivorous larvae (ghost, or phantom, larvae) are almost transparent, and have no *spiracles, respiring by absorbing air from two conspicuous air sacs, which also adjust the density of the larva so it matches that of the water in which it lives.

Chaoborus crystallinus (ghost larva, phantom larva) *See* CHAOBORIDAE.

Characidae (characin, tetra, piranha; subclass *Actinopterygii, order *Cypriniformes) A very large family of freshwater fish. Most are small in size and some are very colourful. All have a single *dorsal fin on the back, often a small *adipose dorsal fin, a forked tail fin, and an *anal fin which often is larger than the dorsal fin. Many species are extremely popular aquarium fish. *Astyanax fasciatus* (or *mexicanus*), formerly known as *Anoptichthys jordani* (blind cavefish) lives in cave waters; its body lacks pigment and its eyes are vestigial. The family also includes the voracious piranhas, powerfully built fish with strong jaws and teeth. *Serrasalmus rhombeus* (white piranha) can grow to a length of 30 cm and appears to be a good table fish. There are about 500 species, found in Central and S. America and in Africa.

characin *See* CHARACIDAE.

character Any detectable attribute or property of the *phenotype of an organism. Defined heritable differences in the character may exist between individuals within a species.

character displacement The principle that two species are more different where they occur together (i.e. are sympatric) than where they are separated geographically (i.e. are allopatric). *See* ALLOPATRY; SYMPATRY.

character states Particular versions of a *character. Thus, the character 'horns' may have the character states 'straight', 'curly', etc. The proper elucidation of character states and their *polarity is one of the major concerns of *cladistic analysis.

charade *See* DANCE LANGUAGE.

Charadriidae (lapwings, plovers;class *Aves, order *Charadriiformes) A family

of medium-sized brown, grey, or black and white waders, in which the head and neck are often boldly marked. Some are crested or have *wattles. They have short necks and long wings, some have spurs, and their legs are short to long with a reduced or absent *hallux. They inhabit open, bare areas, nesting on the ground and feeding on a variety of animal matter. Most are migratory. The three species of *Pluvialis* breed in upland and tundra areas in N. America, Europe, and Asia, and winter on grassland and coastal estuaries in S. America, Africa, and Australia. There are 11–13 genera in the family, with 60–65 species (of which 20–30 are placed in the genus *Charadrius* and 24 in the genus *Vanellus*), found world-wide.

Charadriiformes (waders, gulls, skuas, terns, auks, plovers; class *Aves) A large, diverse order that contains 14 families of birds that live on, or near, water, and whose young are active and covered in down at hatching: *Jacanidae, *Rostratulidae, *Haematopodidae, *Charadriidae, *Scolopacidae, *Dromadidae, *Burhinidae, *Glareolidae, *Thinocoridae, *Chionididae, *Stercorariidae, *Laridae, *Rhynchopidae, and *Alcidae. Some authorities also include the *Recurvirostridae (avocets, stilts) as a family within this order.

Charadrius (plovers) *See* CHARADRIIDAE.

chats *See* TURDIDAE.

Chattian The final stage of the *Oligocene Epoch, from about 28.4 to 23.03 Ma ago. It is dated by means of sea-floor spreading rates, small mammals, and *plankton.

Chauliodontidae (viperfish; subclass *Actinopterygii, order *Salmoniformes) A small family of deep-sea fish that have an elongated body, tapering down toward the tail, and a small *dorsal fin that is not larger than the *adipose fin but has a very long first dorsal ray that can be used as a lure. The upper and lower jaws have long, needle-like teeth, giving the head a ferocious appearance. There are about six species, some having a row of luminous organs along the belly.

Chaunacidae (sea toad; subclass *Actinopterygii, order *Lophiiformes) A small family of deep-sea angler fish that have a balloon-shaped body, an oblique mouth, and the gill opening behind the *pectoral fin. There are two genera, *Chaunacops* and *Chaunax*, and 15 species, distributed worldwide at depths up to 2500 m.

cheek teeth *Premolar and *molar teeth.

cheetah (*Acinonyx jubatus*) *See* FELIDAE.

***Cheilinus undulatus* (giant maori wrasse)** *See* LABRIDAE.

Cheilodactylidae (morwong;subclass *Actinopterygii, order *Perciformes) A family of fairly large (up to 1.2 m long) marine fish. The body is very deep just behind the head, but tapers sharply toward the tail. The continuous, spinous, and soft-rayed *dorsal fin has a very long base, but the *anal fin is short. The tips of the lower pectoral rays project well beyond the fin membrane. The fish are found in the temperate waters of the southern hemisphere, and near Japan. They are well represented in southern Australia and New Zealand. There are about 30 species.

Cheilostomata (subphylum *Ectoprocta, class *Gymnolaemata) A large order of bryozoans, whose members are highly variable in gross morphology and are characterized by the presence of a small, non-calcified *operculum which covers the aperture when the soft parts are withdrawn. The order occurs since the *Jurassic.

Cheirogaleidae (order *Primates, suborder *Strepsirhini) A family of small lemurs in which the eyes are large and the muzzle short. They are entirely nocturnal, and during the dry season some species *aestivate in nests made from leaves, sustaining themselves on reserves of body fat accumulated mainly at the base of the tail. There are four genera, *Cheirogaleus* (dwarf lemurs), *Microcebus* (mouse lemurs), *Allocebus* (the very rare hairy-eared mouse lemur), and *Phaner* (fork-crowned lemur).

Cheirolepis trailli An early representative of the primitive bony fish, the palaeoniscids, known from the Middle *Devonian.

chela Prehensile claw or pincer, e.g. in Crustacea.

chelate 1. *adj.* Pincer-like or claw-like. **2.** *noun* A ring structure formed as a result of the reaction of a metal ion with two or more groups on a *ligand. *Haemoglobin and chlorophyll are chelate compounds in which the metal ions are iron and magnesium respectively.

chelicera One of the first pair of the six pairs of appendages on the prosoma of an arachnid (*Arachnida), which has no more than three segments. In most orders the terminal segment is *chelate, while in spiders and amblypygids it is subchelate. In mites, especially parasitic species, the chelicerae become narrowed and lose the chelate finger, becoming a piercing structure. The chelicerae are generally held parallel and anterior to the body, working alternately. If they are large, as in spiders, *Solifugae, and harvestmen, they serve as prehensile organs, squeezing and killing prey, and are also used in defence and in digging. In spiders, a poison gland opens at the apex of the terminal article, and in *Pseudoscorpiones a silk gland opens in the same location.

Chelicerata (phylum *Arthropoda) A subphylum of arthropods in which the body comprises a *cephalothorax or prosoma and an *abdomen or opisthosoma, the first pair of feeding structures are the *chelicerae that give the subphylum its name, and there are no antennae. The chelicerates are known as fossils from the early *Palaeozoic, and originated as marine organisms, but today there are only five marine species: the horseshoe crabs (*Merostomata). The subphylum includes the classes *Arachnida, *Merostomata, and *Pycnogonida.

Chelidae (snake-necked turtles; order *Chelonia, suborder *Pleurodira) A family of freshwater turtles in which the head and neck may be longer than the *carapace.

The neck vertebrae fold only sideways. The turtles' diet is mixed. *Chelus fimbriatus* (matamata), which inhabits stagnant pools in tropical S. America, has a large and flattened head with nostrils at the end of a slender proboscis, a long neck, and a carapace up to 45 cm long, with three ridges and covered with small knobs; small prey are rapidly sucked into the large mouth cavity. There are 31 species in the family, occurring in S. America, Australia, and New Guinea.

chelifore *See* PYCNOGONIDA.

cheliped *See* DECAPODA.

Chelisochidae (class *Insecta, order *Dermaptera) Small family of earwigs, generally less than 3 cm long, which show no striking distinguishing features. Most of them are found in the tropics.

Chelonia (Testudines; turtles, terrapins, tortoises; class *Reptilia, subclass *Anapsida) An order of reptiles in which the body is enclosed in a shell of bony plates covered by horny scales with an upper *carapace and lower *plastron. The carapace is often fused to the vertebrae and ribs. The jaws are toothless and horny. There are some 250 species, including marine, freshwater, and terrestrial forms.

Chelonia mydas (green turtle) *See* CHELONIIDAE.

chelonians *See* CHELONIA.

Cheloniidae (marine turtles;order *Chelonia, suborder *Cryptodira) A family of turtles in which the shell consists of a *carapace which is flat and streamlined, and a *plastron which is reduced. The fore limbs are broad and flat, with one or two claws. The hind limbs are flattened and rudder-like. Neither the head nor legs retract into the shell. These turtles are mainly carnivorous, although *Chelonia mydas* (green turtle) is basically herbivorous, occasionally eating molluscs and crustaceans. There are five species, with a cosmopolitan distribution in warm seas. *Caretta caretta* (loggerhead turtle) has a chestnut-coloured carapace

up to 1.3 m long and a yellowish plastron; it occurs in all warm seas and occasionally reaches Britain. *Eretmochelys imbricata* (hawksbill turtle), the smallest member of the family (up to 90 cm long), has multicoloured, translucent carapace scales, overlapping towards the back, used in the oriental tortoiseshell industry. Marine turtles come ashore only to lay eggs.

chelophore In *Pycnogonida, one of the appendages located on either side of the proboscis: it bears a *chelicera at the tip and functions as a mouth-part.

***Chelus fimbriatus* (matamara)** *See* CHELIDAE.

Chelydridae (snapping turtles; order *Chelonia, suborder *Cryptodira) A family of freshwater turtles in which the head is large, with knobs of skin around the face. The *carapace is incompletely ossified, and the *plastron is reduced to a cross-strut. The tail is long, with lateral spines. Snapping turtles are carnivorous and obtain food on the bottom of ponds and rivers. There are two species, found in the USA.

chemiosmotic theory A theory concerning *oxidative phosphorylation in which it is proposed that the *electron-transport chain is arranged such that it generates an energy-rich proton gradient across the inner membrane of a *mitochondrion, and electrons cross the membrane by a mechanism reminiscent of that in which molecules of solvent cross a *semipermeable membrane in *osmosis. The energy is then used to drive the phosphorylation of *ADP through a membrane-bound ATP-ase.

chemoreceptor A type of *receptor cell that responds to chemical substances, as in the taste, touch, and smell senses. *See also* MECHANORECEPTOR; RADIORECEPTOR.

chemotaxis In a *motile organism or cell, a change in the direction of locomotion that is made in response to a change in the concentrations of particular chemicals in its environment.

chequered beetle See CLERIDAE.

chevrotain (*Tragulus, Hyemoschus*) See TRAGULIDAE.

chewing lice See MALLOPHAGA.

chiasma (*pl.* chiasmata) In genetics, a cross-shaped structure that forms the points of contact between non-sister *chromatids of *homologous chromosomes first seen in the *tetrads of the diplotene stage of meiotic *prophase. Chiasmata are thus the visible expression of *crossing-over of genes. There are usually one or more chiasmata per chromosome per meiosis. *See also* BREAKAGE AND REUNION.

chiasma interference The non-random frequency of more than one *chiasma in a *bivalent segment during *meiosis. If the frequency of occurrence is higher than that expected from purely chance events then it is termed negative chiasma interference; if the frequency is lower than expected, it is referred to as positive chiasma interference.

chiasmata See CHIASMA.

Chiasmodon niger (swallower) See BATHYPELAGIC FISH.

Chiasmodontidae (swallower; subclass *Actinopterygii, order *Perciformes) A small family of deep-sea fish that have a very large, toothed mouth. Swallowers are thought to be capable of swallowing prey larger than themselves. There are about 15 species, occurring in deep oceans.

chickadees See PARIDAE.

Chilodontidae (head stander; subclass *Actinopterygii, order *Cypriniformes) A small family of freshwater fish that have a single *dorsal fin and an *adipose fin, a small mouth, and a forked tail fin. These fish often swim in a head-down position. They occur in southern Australia.

Chilopoda (centipedes; phylum *Arthropoda, subphylum *Atelocerata) A class of *uniramous arthropods that have segmented bodies with one pair of legs to each segment. Some centipedes are adapted to burrowing in soil but most run on the surface. All are believed to be predators. The head bears long antennae and beneath the mouth-parts there is a large pair of claws (maxillipeds, forming part of the first body segment) that inject poison. Below the *mandibles a pair of *maxillae forms a functional lip, overlain by a second pair of maxillae. The name, Chilopoda, is derived from the Greek *cheilos*, lip, and *pod-*, foot. There are about 3000 species, found world-wide, and grouped in two subclasses (Epimorpha and Anamorpha) and four principal orders: Geophilomorpha; Scolopendromorpha; Lithobiomorpha; and Scutigeromorpha.

chimaera See CHIMERA.

***Chimaera monstrosa* (rabbitfish)** See CHIMAERIDAE.

Chimaeridae (ghostfish, ratfish; subclass *Holocephali, order *Chimaeriformes) A family of oceanic fish that have large eyes, a short and rounded snout, and a strongly tapering body which ends in a narrow, filament-like tail. The first of the two *dorsal fins has a strong spine at the leading edge; the *pectoral fins are large and wing-like. The rabbitfish of western Europe (*Chimaera monstrosa*) lives close to the bottom, at 100–500 m, and may grow to 1.2 m. It is of some commercial importance in a number of countries. There are about 15 species, with a world-wide distribution.

Chimaeriformes (class *Chondrichthyes, subclass *Holocephali) An order of marine fish that have a cartilaginous skeleton, a gill cover over the four gill openings, a naked skin, and no *spiracle opening. The males have a pair of pelvic *claspers and a peculiar clasping structure on the head. The order includes a number of fossil families as well as the living ghostfish. They are not considered to be closely related to either the sharks or the bony fish.

chimera (chimaera) Tissue containing two or more genetically distinct cell types, or an individual composed of such tissues. It arises in animals as a result of mutation or abnormal distribution of *chromosomes, affecting a particular cell during development and hence all its descendants.

chimpanzee (Pan troglodytes) *See* HOMINIDAE.

chinch bug *See* LYGAEIDAE.

Chinchillidae (chinchillas, vizcachas; order *Rodentia, suborder *Hystricomorpha) A family of rodents which shelter in crevices or burrows. The hind limbs are long, with three or four digits, the fore limbs short with four or five digits. The eyes are large, the ears rounded, and in some species large. The tail is long, and in some species bushy. They are found in southern S. America. There are six species in three genera: *Chinchilla*; *Lagidium*; and *Lagostomus*. The Chilean chinchilla (*C. laniger*), an animal about 23 cm long excluding the tail, is raised in captivity for its fur, which has been prized at least since Inca times; in the past chinchillas have also been a source of meat.

Chinese pygmy dormouse (Typhlomys) *See* PLATACANTHOMYIDAE.

Chinese water deer (Hydropotes) *See* CERVIDAE.

chin-leafed bat (Mormoops, Pteronotus) *See* MORMOOPIDAE.

Chionididae (sheathbills; class *Aves, order *Charadriiformes) A family of white, pigeon-like birds that have short, stout legs and unwebbed feet. The stout bill is black, or yellow and black, and covered by a horny sheath. Chionidids have bare and *wattled cheeks. Their wings have sharp *carpal spurs. They are gregarious and mainly terrestrial, omnivorous, and hole-nesting in crevices and burrows. There are two species, found on sub-Antarctic Atlantic and Indian Ocean islands.

chipmunk (Tamias, Eutamias) *See* SCIURIDAE.

Chirocentridae (wolfherring; subclass *Actinopterygii, order *Clupeiformes) A monospecific (*Chirocentrus dorab*) family of marine, warm- to temperate-water fish. The wolfherring has an elongate body, a large mouth with fang-like teeth, and a single *dorsal fin located close to the forked tail fin. A voracious carnivore, it is a very large, silvery fish, reaching a length of up to 3.5 m, and is widely distributed in Indo-Pacific waters.

Chirocentrus dorab (wolfherring) *See* CHIROCENTRIDAE.

Chiromyiformes (order *Primates, suborder *Strepsirhini) A group of primates that contains only the family *Daubentoniidae.

Chironemidae (kelpfish; subclass *Actinopterygii, order *Perciformes) A family of marine coastal fish. Because of the high-backed body profile, and the lower pectoral ray tips projecting free from the fin membrane, the kelpfish resemble the morwongs (family *Cheilodactylidae) but have a truncate tail fin and fewer (less than nine) soft *anal-fin rays. Their somewhat marbled colour pattern allows them to hide easily in shallow rocky areas. There are about four species, found in temperate waters around the Tasman Sea.

Chironex fleckeri (sea wasp) *See* CUBOMEDUSAE.

Chironomidae (non-biting midges; order *Diptera, suborder *Nematocera) Family of delicate, gnat-like flies, in which the males have conspicuous, *plumose antennae, and those of females are *pilose. The head is usually overhung, and often concealed, by the *thorax. Mouth-parts are poorly developed. Anterior wing veins are more prominent than posterior ones. Adults bear some resemblance to *Culicidae, but can be distinguished by their unscaled wings. Adults fly in large numbers at sunset near standing bodies of water. The swarms comprise mainly males, and mating occurs when a female flies into a swarm. Eggs are laid in a mass, or ribbon of clear, mucus-like jelly. Larvae are almost all aquatic, or mud-dwellers that live in tubes. Some species have larvae containing *haemoglobin as an adaptation to life in anoxic substrates. Some species can reproduce parthenogenetically, and *paedogenesis has been recorded in others. There are some 2000 described species.

Chiroptera (bats; infraclass *Eutheria, cohort *Unguiculata) An order that comprises the only true flying mammals, possessing features parallel to those of birds (e.g. active metabolism and economy of weight). Insectivores possibly ancestral to the bats are known from the *Paleocene. The first undoubted bats are preserved in middle *Eocene deposits in both Europe and N. America. Differentiation of the modern lineages was far advanced by the Eocene–*Oligocene transition. The wing is a *patagium supported by four, or in some species all five, elongated digits of the fore limb, and attached along the sides of the body, the legs (leaving the feet free), and in most species the tail. The pelvis is weak but adapted to enable the legs to rotate to the rear so that the bat can hang head-down. The orbit is rarely closed behind. The *incisors are often specialized or reduced. There are two suborders: *Megachiroptera and *Microchiroptera.

chiropterophily Pollination by bats (Chiroptera).

chi-squared test (χ^2) A statistical test that is used to determine whether data obtained by sampling agree with those predicted hypothetically, and thus to test the validity of the hypothesis.

chitin Linear homopolysaccharide of N-acetyl-d-glucosamine, found as the major component of the *cuticle of an arthropod, in which the molecules are laid down in chains. Depending on their orientation, these chains can be cross-linked to yield a very strong, lightweight material from which, with the addition of *protein, the *exoskeleton is constructed.

Chitinodendron franconianum A primitive *foraminifer from the Late *Cambrian of Wisconsin, USA, which had an external 'chitinoid' membrane, making it one of the first protozoans with an external skeleton.

chitinophosphatic Applied to the shells of invertebrates that are composed of: (*a*) alternating layers of *chitin and phosphate salts, the phosphatic layers being thicker where they overlie the body cavity (e.g. *Brachiopoda); or (*b*) chitin impregnated with phosphate salts.

chitons See POLYPLACOPHORA.

Chlamydoselachidae (frilled shark; subclass *Elasmobranchii, order *Hexanchiformes) A monospecific (*Chlamydoselachus anguineus*) family of marine shark, which has six pairs of gill clefts, the first one continuing below the throat to join its counterpart on the other side. The single *dorsal and the *anal fins are set far back on the body. The frilled shark has a somewhat snake-like appearance due to its elongate body, flat head, and large mouth. Probably it lives in deep water. It has a world-wide distribution.

Chlamydoselachus anguineus **(frilled shark)** See CHLAMYDOSELACHIDAE.

chloragogen tissue Tissue composed of brown or greenish cells, located in the intestinal wall or heart of *Annelida, that is an important centre of metabolism and the synthesis of *haemoglobin, and that may also have an excretory function.

chloride cells Cells located in the gills of *teleosts which pump sodium and chloride ions out into the sea against a concentration gradient. To reduce osmotic dehydration due to *hyposmotic body fluid, teleosts drink large quantities of sea water, absorbing in the gut sodium and chloride ions which serve to raise the osmotic pressure in the blood and cause water to flow in from the gut. Subsequently the excess of ions absorbed are pumped out of the body via the chloride cells.

chlorocruorin A green respiratory pigment, of molecular weight 3.5×10^6, that contains iron in its *prosthetic group. It occurs only in the *blood plasma of certain *Polychaeta.

Chlorocyphidae (damselflies; order *Odonata, suborder *Zygoptera) Family of damselflies which can be identified by the many *antenodal veins on the wings. The insects are often very colourful, and fly by streams or rivers in shady areas. In the larva, the medial *tracheal gill is reduced to a spike. The family occurs in the Old World tropics. More than 230 extant species have been described.

Chlorolestidae (Synlestidae, damselflies; order *Odonata, suborder *Zygoptera) Family of damselflies which breed in running water. They are distinguished by the arching forward of the posterior *cubitus vein as it leaves the distal end of the quadrilateral in the wing. These damselflies are often elongate, and metallic in colour. The family is mainly tropical, and 33 extant species have been described.

Chlorophthalmidae (cucumberfish, green-eye; subclass *Actinopterygii, order *Myctophiformes) A small family of marine, coastal fish that have an elongate body, large eyes (appearing green in live fish), a single *dorsal fin followed by a small *adipose fin, and a forked tail fin. In Australia, *Chlorophthalmus nigripinnis* (cucumberfish), found in the cooler southern waters, derives its name from the smell emanating from the skin. There are about 20 species, with world-wide distribution.

Chlorophthalmus nigripinnis (cucumberfish) *See* CHLOROPHTHALMIDAE.

Chloropidae (frit flies, grass flies, stem flies; class *Insecta, order *Diptera) Moderately large family of small flies, of some economic importance as pests of plants. Adults are usually black, yellow-black, or green-black, and almost without bristles. They may be identified by the large, triangular, *ocellar plate on the dorsal surface of the head. Males have lost abdominal *tergites 7 and 8. The larvae are mainly phytophagous, the exceptions being predacious. 1180 species have been described.

Chloropsis (leafbirds) *See* IRENIDAE.

choanae Internal nostrils; paired openings in the roof of the mouth which are connected to the exterior.

Choanichthyes (Sarcopterygii) The name given by some zoologists to a group (subclass) comprising the *Crossopterygii (lobe-finned fish) and *Dipneusti (lungfish). Members of this group were thought to share functional lungs, external and internal nares (choanae), and narrow-based, paired fins with fleshy lobes.

choanocyte In *Porifera (sponges), one of the flagellated cells surrounded by a collar-like sheath of *protoplasm that form a layer lining the internal chambers. Choanocytes are very similar in structure to choanoflagellate *Protozoa.

Choeropsis liberiensis (pygmy hippopotamus) *See* HIPPOPOTAMIDAE.

cholecystokinin A *hormone secreted by epithelial cells (*see* EPITHELIUM) in the *duodenum that stimulates the release of digestive *enzymes from the *pancreas and *bile from the *gall bladder, and also acts to suppress appetite.

cholesterol The most abundant sterol in animal tissues, which strengthens *cell membranes. It is derived endogenously from *acetyl coenzyme A, or exogenously from food. It is a precursor of *steroid hormones and *bile acids.

choline A basic, nitrogenous, organic compound that functions as a methyl-group donor in some *phospholipids and in *acetylcholine. Although it has no known *coenzyme function, it is normally classified as one of the B group of vitamins.

cholinergic Applied to nerve endings that secrete the *neurotransmitter acetylcholine into the synapse on the arrival of a nerve impulse. Cholinergic nerve endings are characteristic of the *parasympathetic nerve system, although preganglionic synapses of the *sympathetic nerve system also use acetylcholine.

cholinergic system *See* NEUROTRANS-MITTER.

cholinesterase A group of *enzymes, some of which catalyze the *hydrolysis of *acetylcholine to *choline and acetate, and others that catalyze the conversion of acyl-cholines to choline and a carboxylate (weak acid). The hydrolysis of acetylcholine allows an activated *cholinergic *neuron to return to its resting state.

Chondrichthyes Class of vertebrate animals characterized by a cartilaginous *endoskeleton, a skin covered by *placoid scales, the structure of their fin rays, and the absence of a bony operculum (*see* GILL COVER), lungs, and *swim-bladder. It includes the subclasses *Elasmobranchii (sharks and rays) and *Holocephali (ghost-fish). The group extends back to the Late *Devonian. It is not clear whether the cartilaginous skeleton is a primitive feature or the result of young fish reaching reproductive age prematurely, before a bony skeleton develops.

chondrification The formation of cartilage.

chondrin A gelatinous *protein that forms the matrix of *cartilage.

chondroblast A *mesenchyme stem cell that develops into a *chondrocyte.

chondocranium *See* CRANIUM.

chondrocyte One of the cells forming healthy *cartilage.

Chondrophora (Chondrophorina; phylum *Cnidaria, class *Hydrozoa) An order of Cnidaria whose members are characterized by having a single, large *polyp possessing a chitinous, gas-filled float supporting the other *zooids. They have free *medusae.

chondrophore In certain genera of *Bivalvia which possess *desmodont dentition, an internal process with a depressed surface that supports the ligament.

Chondrophorina Alternative name for the *Chondrophora; in this form it is regarded as a suborder of *Siphonophora.

Chondrostei A group of bony fish belonging to the subclass *Actinopterygii. Often ranked as a superorder, the chondrosteans have a partly cartilaginous skeleton, a *heterocercal tail, a *spiracle, and an intestinal spiral valve. Apart from a number of fossil orders, it includes the extant order *Acipenseriformes.

Chonotrichida (class *Ciliatea, subclass *Holotrichia) An order of *Protozoa in which mature individuals lack body *cilia but have cilia in the oral region. Immature forms have body cilia. Chonotrichida are found attached to crustaceans in various types of aquatic environment.

chorda dorsalis *See* NOTOCHORD.

chordamesoderm In a developing vertebrate *embryo, those cells of the *mesoderm which give rise to the *notochord.

Chordata (kingdom *Animalia) A large phylum comprising the animals that possess a rod of flexible tissue (*notochord), which is protected in higher forms by a vertebral column. The phylum includes *Urochordata (tunicates), amphioxus, fish, amphibians, reptiles, birds, and mammals. The first chordates and the earliest vertebrates are both found in *Cambrian rocks.

chorion 1. The outer cellular layer of the *embryo sac of reptiles, birds, and mammals. It consists of two epithelial layers, an ectoderm surrounding a mesoderm layer. In mammals, the outer epithelium (trophoblast) of the chorion forms the *placenta, maintaining close contact with the maternal tissues. *See also* AMNION. **2.** The structurally complex outer coat of an insect egg, which is often externally patterned, frequently reflecting the pattern of follicle cells in the ovary which laid it down.

Choristida (class *Demospongiae, subclass *Tetractinomorpha) An order of sponges which have complex body shapes.

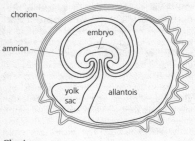

Chorion

Organic fibre and *spicules are normally present, and the spicules often have long shafts. A *cortex is developed. The group first appeared in the *Carboniferous.

choroid In the eye of a vertebrate, a layer of tissue lying immediately outside the *retina that contains pigment and blood vessels.

chromatid One of the two daughter strands of a *chromosome that has undergone division. Chromatids are joined together by a single *centromere, usually positioned in the centre of the pair as they lie beside one another. When the centromere divides at the *anaphase of *mitosis or the second stage of *meiosis (meiosis II), the sister chromatids become separate chromosomes.

chromatid interference The non-random participation of non-sister *chromatids of a *tetrad in successive *crossings-over of *meiosis. It results in a deviation from the expected 1:2:1 ratio for the frequencies of two-, three-, and four-strand double crossovers.

chromatin The substance of *chromosomes, which includes DNA, chromosomal *proteins, and chromosomal *RNA. Chromatin stains strongly with *basic dyes.

chromatography An analytical technique for separating the components of complex mixtures, based on their repetitive distribution between a mobile phase (of gas or liquid) and a stationary phase (of solids or liquid-coated solids). The distribution of the different component molecules between the two phases is dependent on the method of chromatography used (e.g. gel-filtration, or ion-exchange), and on the movement of the mobile phase (which results in the differential migration and therefore separation of the components along the stationary phase).

chromatophore In many animals, a cell containing pigment granules; by dispersing or contracting such granules certain animals are able to change their colour.

chromomere A small, bead-like structure that is visible in a *chromosome during *prophase of *meiosis and *mitosis, when it is relatively uncoiled (in particular at the leptotene and zygotene stages of meiosis). In *polytene chromosomes, the chromomeres lie in parallel, giving the chromosome its banded appearance. Chromomeres in corresponding positions on *homologous chromosomes pair during meiosis in many organisms.

chromosomal puff A localized swelling at a specific site along the length of a *polytene chromosome, where DNA–RNA transcription takes place. Large RNA puffs are called Balbiani rings.

chromosome A DNA-histone *protein thread, usually associated with RNA, occurring in the *nucleus of a *cell. Although chromosomes are found in all animals and plants, bacteria and viruses contain structures that lack protein and contain only DNA or RNA; these are not chromosomes, though they serve a similar function. The presence of pairs of *homologous chromosomes is referred to as the diploid state; in diploid organisms the chromosomes associate in a particular way during *meiosis. *Gametes are haploid, with only one member of each pair in their nuclei. Each species tends to have a characteristic number of chromosomes (46 in humans), found in most nucleated cells within most organisms. Usually chromosomes are visible only during *mitosis or meiosis when they contract to form short thick rods coiled into a spiral. Each chromosome possesses two sets of *chromomeres and a *centromere, and some contain a nucleolar organizer. Chromosomes contain a line of different genes, a spindle-attachment at some point along their length, and regions of *heterochromatin, which stains strongly with *basophilic dyes.

chromosome map A map that shows the locations (loci) of *genes on a *chromosome, deduced from genetic-recombination experiments. For example, the frequency of cross-overs between pairs of genes indicate their relative positions or linear order, the distances being given in units of cross-over frequency.

chromosome polymorphism The presence of one or more *chromosomes in two or more alternative structural forms within the same interbreeding population.

chromosome substitution The replacement of one or more *chromosomes by others (totally or partially *homologous) from another source (either a different strain of the same species or a related species that will permit hybridization) by a suitable crossing programme.

chromosome theory of heredity The unifying theory put forward by W. S. Sutton in 1902 that Mendel's laws of inheritance may be explained by assuming that *genes are located in specific sites on *chromosomes.

chron A subdivision of geological time based on the direction of magnetization, as preserved in rocks of the period. Within many chrons there are shorter periods of alternate magnetization (subchrons). The two most recent chrons in Earth history are the *Matuyama (reversed) and *Brunhes (normal).

chronospecies (evolutionary species) According to one view of evolution (*phyletic gradualism), a group of organisms that is derived from its ancestor by a process of slow, steady, evolutionary change and is not regarded as a member of the same *species as its ancestor.

chronostratigraphy The branch of stratigraphy that is linked to the concept of time. In chronostratigraphy intervals of geological time are referred to as chronomeres. These may be of unequal duration. Intervals of geological time are given formal names and grouped within a Chronomeric Standard hierarchy. The formal terms are: eon, era, period, epoch, age, and chron. Of these, the last four are the equivalent of system, series, stage, and chronozone in the Stratomeric Standard hierarchy. The terms are often written with initial capital letters when accompanied by the proper names of the intervals to which they refer. Some geologists hold that the term 'chronostratigraphy' is synonymous with 'biostratigraphy', but most agree that the two branches are separate.

chrysalis (*pl.* **chrysalises or chrysalides**) Pupal stage of a butterfly, having a hard outer covering beneath which the forming legs, wings, and antennae can sometimes be seen. Chrysalises are usually attached by stems to the underside of leaves, or on pieces of rock. They are frequently camouflaged to resemble their environment. *See also* LEPIDOPTERA; PUPA.

Chrysididae (ruby-tailed cuckoo wasps, ruby-tailed wasps; suborder *Apocrita, superfamily Bethyloidea) Family (according to some authors, within its own superfamily, the Chrysidoidea) of wasps which are 6–12 mm long (although some are up to 22 mm long), with bright, metallic blue, red, or green bodies. The *abdomen has four segments, or fewer, and is concave ventrally. The hind wings do not have closed cells and possess a basal lobe. Adults lay their eggs in the burrows of solitary wasps or bees (particularly mud-dauber wasps) and some parasitize symphytans. Most larvae are external parasites, development occurring only after the host is completely consumed, although some larvae are *inquilines and others feed on a provisioned supply of food. The adults, which do not sting, often curl up into a ball when disturbed. They are fairly common and their common names refer to the often bright-red abdomen of many species, and their parasitic and inquiline habits in the nests of solitary bees and wasps.

Chrysidoidea See CHRYSIDIDAE.

Chrysochloridae (golden mole; sub-order *Lipotyphla, superfamily *Sori-coidea) A family of insectivores in which there are no external ears and the eyes are covered with skin. The fur has a metallic appearance. The tail is rudimentary. The fore limbs and paws are modified for digging, the second and third digits bear large claws, and the fifth digit is absent. They are distributed in Africa from the Equator south. There are nine genera, with 21 species. Their mole-like adaptations have been evolved independently of those of the *Talpidae (true moles) and it has been proposed, on molecular grounds, that golden moles do not belong to Lipotyphla at all, but are part of a *clade of African mammals that includes *Proboscidea, *Sirenia, *Hyracoidea, and *Tubulidentata.

Chrysocyon brachyurus (maned wolf) See CANIDAE.

Chrysomelidae (leaf beetles;class Insecta, order Coleoptera) Family of beetles whose members occur in a great variety of shapes and sizes, from 1.5 to 22 mm, but usually they are robust, compact beetles with smooth, brightly coloured, or metallic *elytra. Larvae are grub-like, often coloured, and have short legs; all are plant feeders. Tortoise beetles have expanded sides to the body, giving a flat, shell-like appearance; the larvae conceal themselves with cast skins and faeces. Many chrysomelids are pests. Flea beetles, adults and larvae, 'graze' on upper leaf surfaces; adults, which have enlarged hind femora (see FEMUR), are able to spring. Larval juices of some tropical species (poison beetles) are used to tip arrows by Kalahari Bushmen. There are 20 000 species.

Chrysopidae (green lacewings; sub-class *Pterygota, order *Neuroptera) Family of insects which have pale-green bodies, iridescent, often greenish wings, and prominent, reddish-golden eyes. The eggs are laid on foliage at the end of tiny stalks, and the larvae are highly predacious on aphids and plant lice, which they suck out with their hollowed, sickle-shaped *mandibles. The larvae often camouflage themselves with the empty skins of their prey and small plant fragments. Pupation occurs in a small, round, silk cocoon. Green lacewings are common insects, found on grass, shrubs, and weeds. They often give off an offensive odour when handled.

Chunga See CARIAMIDAE.

chymotrypsin A peptidase *enzyme found in the small intestine that acts on the interior *peptide bonds of *proteins.

cicadas See CICADIDAE.

Cicadellidae (leaf hoppers; order *Hemiptera, suborder *Homoptera) Family of homopterans which occur on almost every kind of herbaceous and woody plant. Many of them are serious pests of crops which they damage either directly through feeding on the sap or cell contents, or indirectly through the effects of toxic saliva, or by transmitting *virus diseases. There are many thousands of species, distributed worldwide.

Cicadidae (cicadas; order *Hemiptera, suborder *Homoptera) Family of large homopterans in which the nymphs live underground, burrowing with the aid of crab-like fore legs, and feeding on the sap of roots. Adults live in trees, and the males produce a loud 'song'. The life cycle may occupy several years and one well-known species in the USA spends 17 years underground. The nymphs of this species, like some others, construct 'chimneys' of earth above their burrows in which to complete the final moult. There are about 2000 species, occurring mainly in the tropics.

Cichlidae (cichlids; subclass *Actinopterygii, order *Perciformes) A very large family of freshwater fish, many of which are rather deep-bodied. They have a single, continuous *dorsal fin which can be very high (e.g. in Pterophyllum species, angelfish) or low (e.g. in Symphysodon species, discus fish). The *lateral line is interrupted and there is a single pair of nostrils. Cichlids vary in size: Apistogramma ramirezi (butterfly cichlid) does not exceed 6 cm, but other species (e.g. *Sarotherodon aureus*) grow to 35 cm. Many cichlids are cultivated to supply aquarium hobbyists, being colourful, fairly easy to keep, and showing interesting behavioural traits. Other species of the genera *Tilapia* and *Sarotherodon*, e.g. *S. mossambicus*, are a very important source of protein in many countries. There are probably at least 700 species: most occur in Africa, but they are also found in S. America, India, and parts of the Middle East.

cichlids *See* CICHLIDAE.

Cicindellidae *See* CARABIDAE.

Ciconia *See* CICONIIDAE.

Ciconia ciconia (white stork) *See* CICONIIDAE.

Ciconiidae (storks; class *Aves, order *Ciconiiformes) A family of large wading birds that have stout-based bills and long necks, some with distensible throat pouches (e.g. two of the three species of *Leptoptilos*, found in Africa and Asia) and bare faces, heads, or necks. They have long, broad wings, and long legs with *semipalmate toes. They are found in open, often wet, country, frequently breeding colonially in trees. *Ciconia ciconia* (white stork, one of five *Ciconia* species) frequently breeds on nest platforms on buildings. The three species of Ibis (wood storks), found in Asia and Africa, are usually placed in the genus *Mycteria*. There are six genera in the family, with 17 species, found nearly world-wide.

Ciconiiformes (herons, storks, ibises, spoonbills, flamingos; class *Aves) An order of medium to large wading birds, many of which have specialized bills. Their toes are unwebbed or partially webbed and the middle claw is often *pectinate. There are six families: *Ardeidae, *Cochleariidae, *Balaenicipitidae, *Ciconiidae, *Threskiornithidae, and *Phoenicopteridae. They are found world-wide.

Cidaroida (class *Echinoidea, subclass *Perischoechinoidea) An order of regular echinoids that have a *radially symmetrical, subspherical *test. The *ambulacra are in two columns, with each plate bearing a single pore pair. The *interambulacra are conspicuously wider than the ambulacra, each plate bearing one enlarged primary *tubercle carrying a primary spine. The teeth are not keeled. There are no gill slits or *spheridia. Cidaroida first appeared in the Late *Silurian. There are six subfamilies, within two families, extant.

ciguatera A type of poisoning caused by eating fish whose flesh contains ciguatoxin, a substance that is probably obtained initially from marine *cyanobacteria eaten by small fish, in turn eaten by larger fish in which the toxin accumulates. The symptoms include nausea, vomiting, numbness of parts of the body, and even coma. Sometimes, therefore, it is probably wise not to eat certain species (e.g. mackerel, grouper, jack, and snapper).

cilia *See* CILIUM.

Ciliatea *See* CILIOPHORA.

Ciliophora (phylum *Protozoa) A subphylum of protozoa in which *cilia are present during at least one stage of the life cycle and in which two different types of nucleus (macronucleus and micronucleus) are normally present. Cells reproduce by binary fission. Sexual processes occur. Most species are free-living, although some are parasitic. There is one class, Ciliatea, containing four subclasses, many orders, and numerous genera.

cilium (*pl.* cilia) A short, hair-like appendage, normally 2–10μm long and about 0.5μm diameter, usually found in large numbers on those cells that have any at all. Cilia have a microtubular skeletal structure enclosed by an extension of the *cell membrane. The *microtubules are arranged in nine sets of doublets around the circumference, with two single tubules in the centre, the so-called '9+2' construction. In certain *Protozoa, cilia function in locomotion and/or feeding. They generate currents in the fluid surrounding the cell by beating in a coordinated manner.

Cimbicidae (suborder *Symphyta, superfamily *Tenthredinoidea) Family of moderate- to large-sized (18–25 mm long) symphytans which resemble hairless bumble-bees. The antennae possess seven or fewer segments and are slightly clubbed. The larvae feed on elm, willow, poplar, hawthorn, etc. and can be serious pests of trees. They may be recognized by their greenish-yellow colouring, with black spiracles and a dorsal stripe. The larvae of many species can eject an offensive secretion from glands just above the spiracles. Cimbicidae occur in Europe and N. America.

***Cimex lectularius* (bedbug)** See CIMICIDAE.

Cimicidae (order *Hemiptera, suborder *Heteroptera) Family of bugs comprising several dozen flightless forms which suck the blood of birds and mammals, especially bats, and a few hundred species that are predators of other small arthropods. *Cimex lectularius* is the bedbug; *Oeciacus hirundinis* (martin bug) is a nest parasite of birds (especially house martins) that may enter houses and bite humans after its principal hosts have moved to their winter quarters. A few cimicids feed on pollen. They occur in all parts of the world.

Cinclidae (dippers; class *Aves, order *Passeriformes) A family of brown, grey, or black, thrush-like birds that have white on the head or under-parts. They have short tails, long, stout legs, and dense, waterproof plumage. They are found by swift streams: they build a domed nest with a side entrance in a hollow near a stream, and feed on aquatic larvae, crustaceans, and fish, taken under water. They swim well and walk under water. There is one genus, *Cinclus*, with five species, found in Europe, N. Africa, Asia, and N., Central, and S. America.

Cinclus See CINCLIDAE.

Cingulata (order *Edentata, suborder *Xenarthra) An infra-order that comprises one extinct (*Glyptodontoidea) and one extant (*Dasypodoidea) superfamily of armadillo-like animals, distinguished by their banding from the furred ant-eaters and sloths to which they are related.

circadian rhythm The approximately 24-hourly pattern of various metabolic activities seen in most organisms. The rhythmic patterns may persist even when the organism is removed from exposure to 24-hour cycles of light and dark. In a natural habitat the rhythm is 24-hourly; in constant conditions it becomes slightly longer or shorter than 24 hours. The rhythm is controlled by an endogenous *biological clock that is reset by cues (e.g. dawn) from the environment. The word is derived from the Latin *circa*, 'about', and *dies*, 'day'. The different types of activity rhythm are classified as *cathemeral, *crepuscular, *diurnal, and *nocturnal.

circle flight In bird *courtship, a ritual in which the pair fly in circles around their territory. *See also* FLAP-FLIGHT DISPLAY.

***Circus* (harriers)** See ACCIPITRIDAE.

Cirrhitidae (hawkfish; subclass *Actinopterygii, order *Perciformes) A family of fairly small marine fish that have a robust body, a rounded tail fin, and tiny frills (cirri) at the tips of the spines of the first *dorsal fin. Living among rocks or reefs, their brown-spotted skin makes these fish hard to see. One of the larger species, *Cirrhitus alternatus* of Hawaii, grows to at least 25 cm. Although widely distributed in tropical seas, most species are found in the Indo-Pacific region. There are about 35 species.

***Cirrhitus alternatus* (hawkfish)** See CIRRHITIDAE.

cirri See CIRRUS.

Cirripedia (barnacles; phylum *Arthropoda, subphylum *Crustacea) The only class of crustaceans, apart from parasitic groups, in which the adults are wholly sessile. Cirripedes are the familiar barnacles that settle on rocks, submerged timbers, corals, shells, and the undersides of ships. The 900 or so described species are entirely marine. Cirripedes have free-swimming larvae, the last of which is called a *cypris larva. The larval carapace persists and in adult barnacles is often termed the mantle. In the common barnacles (*Thoracica) calcareous plates develop and cover the mantle externally, and there are both stalked and sessile groups. Except for the cement glands, the first antennae are vestigial, and the second antennae are present only in the larvae. Typically, there are six pairs of long, *biramous, thoracic feeding appendages (cirri) which are projected through the mantle opening to scoop up plankton. The body comprises a cephalic region and an anterior trunk (or thoracic region), with very little indication of external segmentation. Apart from the typical 'barnacle' type, there are forms (order Acrothoracica) modified for boring into shells and corals, and naked forms (orders Ascothoracica and Rhizocephala) which parasitize other marine organisms.

cirrus 1. (*pl.* **cirri)** In certain ciliate *Protozoa, an *organelle, formed by the fusion of a group of *cilia, which usually functions in locomotion. **2.** In many invertebrates, a slender bodily appendage, often resembling a tentacle. In *Polychaeta it bears *cilia. **3.** In some flatworms and trematodes, an *eversible copulatory organ. **4.** In some fishes (e.g. *Creediidae) a tuft of skin on the jaw. **5.** *See* CIRRIPEDIA.

cisternae Large, flattened, fluid-filled, membranaceous sacs that occur in a number of cell *organelles (e.g. in the *Golgi body) and in the *endoplasmic reticulum.

Cisticola **(cisticolas)** *See* SYLVIIDAE.

cisticolas (*Cisticola***)** *See* SYLVIIDAE.

Cistothorus **(marsh wrens)** *See* TROGLODYTIDAE.

cis-trans **test** *See* COMPLEMENTATION TEST.

cistron A section of the *DNA molecule that specifies the formation of one *polypeptide chain. A *cis-trans* test is performed to determine whether two mutant sites of a gene are in the same cistron or in different

ones. In the *cis* configuration, both mutants are on one *homologous chromosome and both *wild types are on the other ($a^1a^2/++$), producing a wild-type *phenotype. In the *trans* configuration each homologue has a mutant and a non-mutant (a^1+/a^2+), producing a mutant phenotype.

Citharidae (subclass *Actinopterygii, order *Pleuronectiformes) A small family of marine flatfish, with short *pelvic-fin bases, these fins carrying one spine and five soft rays. In *Citharus* and *Citharoides* the eyes are on the left side of the body; in *Brachypleura* and *Lepidoblepharon* the eyes are on the right side. Citharidae occur in the Mediterranean and Indo-Pacific regions. There are four genera and seven species.

Citharinidae (subclass *Actinopterygii, order *Cypriniformes) A small family of freshwater fish that have a very deep body, relatively small head, forked tail fin, large *dorsal fin, and small *adipose fin near the tail. There are three genera and six species, found in the rivers of central Africa.

Citharoides *See* CITHARIDAE.

Citharus *See* CITHARIDAE.

citric-acid cycle (Krebs's cycle, tricarboxylic-acid cycle) A cyclic series of reactions which take place in *mitochondria, in the presence of oxygen. The cycle represents the principal means by which most living cells provide electrons for the generation of *ATP via the electron-transport chain and *oxidative phosphorylation. The sequence is initiated by the condensation of *acetyl coenzyme A with oxaloacetic acid to form citric acid. This then passes through a series of reactions wherein oxaloacetic acid is reformed, and 2 moles of CO_2 and water and 12 moles of ATP are synthesized.

civet *See* VIVERRIDAE.

Cixiidae (order *Hemiptera, suborder *Homoptera) Family of homopterans which resemble small *cicadas in the adult state. The nymphs feed on plants at or below the ground surface, but the adults live in most exposed situations. There are about 1000 species, distributed world-wide.

clade In *cladistics, a lineage branch that results from splitting in an earlier lineage. A split produces in theory two distinct new taxa, each of which is represented as a branch in a phylogenetic diagram. The term is derived from the Greek *klados*, 'twig' or 'branch'.

Cladista (Cladistia) Older name for either the order *Polypteriformes or the subclass *Brachiopterygii.

cladistic analysis The method of analysis which aims to discover *clades and their interrelationships. For each taxon of the group being analysed, *character states are ordered by their *polarity, to find which taxa are united by most *derived states. Only the sharing by two taxa of *derived character states is evidence that they belong to the same clade.

cladistics The application of *phylogenetic systematics to produce a taxonomic system that is applied to the study of evolutionary relationships. In the *cladograms used to portray these relationships, *cladogenesis always creates two equal sister groups: the branching is dichotomous. Thus each pair of sister groups constitutes a *monophyletic group with a common stem *taxon, unique to the group. Monophyletic groups are deduced by identifying *synapomorphic character states.

Cladocera (water fleas; class *Branchiopoda, order *Diplostraca) Suborder of branchiopods in which the carapace encloses the trunk but not the head. The strongly *biramous second antennae are the sole means of a rather jerky locomotion, the trunk appendages being reduced to five or six pairs, at least some of which bear food-filtering *setae. The post-abdomen is bent ventrally and forward, and bears claws and spines modified for cleaning the carapace. Many cladocerans undergo *cyclomorphosis. Cladocera are important food for many freshwater invertebrates and fish, and one genus, Daphnia, is sold commercially as food for aquarium fish. There are 425 species.

Cladodontiformes (class *Chondrichthyes, subclass *Elasmobranchii) An order of fossil sharks that had an elongate body and one *dorsal fin without a spine. They existed from the Middle *Devonian to *Carboniferous periods.

cladogenesis In *cladistics, the derivation of new taxa that occurs through the branching of ancestral lineages, each such split forming two (possibly more) equal sister taxa that are often considered taxonomically separate from the ancestral taxon, though this is no longer considered obligatory.

cladogram (phylogenetic tree) A diagram that delineates the branching sequences in an evolutionary tree.

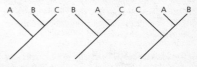

Cladogram

Cladoselache A genus of shark-like fishes, recorded from Europe and N. America, that were 0.5–1.2 m long and are noted for the presence of a large ventral fin. The fishes existed during the *Devonian and *Carboniferous periods and numerous specimens have been collected from the Cleveland Shales (Late Devonian) in N. America.

Cladoselachiformes (class *Chondrichthyes, subclass *Elasmobranchii) An order of fossil sharks (e.g. *Cladoselache*) that had an elongate body, the two *dorsal fins each bearing a spine. They lived from the *Devonian to the *Carboniferous.

***Clamator glandarius* (great spotted cuckoo)** *See* CUCULIDAE.

clam shrimp *See* CONCHOSTRACA.

Clarias batrachus (walking catfish) *See* CLARIIDAE.

Clarias mossambicus *See* CLARIIDAE.

Clariidae (air-breathing catfish; subclass *Actinopterygii, order *Siluriformes*) A large family of freshwater catfish that have elongate to eel-shaped bodies. The head carries four pairs of *barbels, the *dorsal and *anal fins are very long and spineless, and the tail fin is rounded. Typically clariids have a labyrinthic structure above the gill arches used as an *accessory respiratory organ. *Clarias batrachus* (walking catfish) can move over land for short distances. One of the larger species, *Clarias mossambicus*, grows to 1.2 m and is known to survive in poorly oxygenated African freshwater systems. There are about 100 species, found in Africa and tropical Asia.

claspers 1. The *pelvic fins of male sharks and rays, modified to serve as copulatory organs. *See also* BASILOSAURUS. 2. In the males of some insect species, structures at the tip of the abdomen that are used to hold the female during copulation.

classical conditioning *See* CONDITIONING.

classification methods Any scheme for structuring data that is used to group individuals, or sometimes attributes. In ecological and taxonomic studies especially, quite sophisticated numerical classification schemes have been devised, and the methods developed in these disciplines are being applied increasingly in other fields, notably pedology and palaeontology. Various classificatory strategies may be used, e.g. *hierarchical or non-hierarchical; and where hierarchical schemes are used, these may be divisive or *agglomerative, and *monothetic or *polythetic, with the divisive polythetic approach being considered the optimum strategy. *Compare* ORDINATION METHODS.

clathrin A family of *proteins that coat depressions, called coated pits, that appear on the cytoplasmic (*see* CYTOPLASM) side of the *cell membrane in the early stages of *endocytosis.

Clathrinida (class *Calcarea, subclass *Calcinea) An order of sponges in which the body has an *ascon structure. There is no *cortex (outer layer) or dermis (inner layer).

clavate Club-shaped; thicker toward one end.

clavicle In many vertebrates, a bone on the ventral side of the shoulder girdle. In humans it is commonly called the collar bone.

clawed frog (*Xenopus laevis*) *See* PIPIDAE.

clawed toad (*Xenopus laevis*) *See* PIPIDAE.

cleaner fish Fish (e.g. the cleaner wrasse *Labroides dimidiatus*) that remove ectoparasites (*see* PARASITISM) from other fish (which usually are much larger in size). Cleaner fish have conspicuous colour markings which make them easily recognizable to their temporary 'host', allowing them to come close with little risk of being eaten. About 45 species of cleaner fish are known.

cleaner wrasse (*Labroides dimidiatus*) *See* CLEANER FISH; LABRIDAE.

cleaning station A *territory that is defended by a member of a certain species of *cleaner fish and to which their customers go to have their parasites (*see* PARASITISM) and dead surface tissue removed. At the cleaning station, cleaner fish and their customers recognize one another by performing *displays.

clearwings *See* SESIIDAE.

cleavage (furrowing, segmentation) The process by which a dividing egg cell, following fertilization, gives rise to all the cells of the organism. In animals, this division forms a cellular mass called a *blastula. If cleavage follows a definite pattern it is said to be determinate (and hence permits the tracing of cell lineages). In some species, however, the pattern is lost after the first few cell divisions.

cleavage furrow (division furrow) During *cytokinesis and *karyokinesis, a furrow that appears in the *cell membrane and deepens until the cell separates into two *daughter cells.

clegs *See* TABANIDAE.

cleidoic egg An egg that is enclosed by a shell which effectively isolates it from the outside environment and prevents the loss of moisture (i.e. the egg of a land-dwelling animal).

Cleptidae (suborder *Apocrita, superfamily *Bethyloidea) Family of wasps which parasitize the prepupae of symphytans and the eggs of *Phasmatodea. Marked sexual *dimorphism occurs, the males being fully winged but not active; the females, which are seldom seen, often being *brachypterous, with their fore wings much reduced.

Cleridae (chequered beetles; class *Insecta, order *Coleoptera) Family of elongate, soft-bodied, *pubescent beetles, up to 40 mm long. Usually they are brightly coloured, but some are cryptic. The antennae are short, and loosely clubbed. They often occur in woodland, where they prey upon adult bark beetles (*Scolytidae). The larvae are cylindrical and coloured blue, pink, or orange, and feed on bark-beetle larvae. There are 3400 species, mostly tropical.

click-beetle *See* ELATERIDAE.

Climacteridae (Australian tree-creepers; class *Aves, order *Passeriformes) A family of small, brown birds that have slightly *decurved bills and large feet and claws. They are found in forest and scrub and are mainly insectivorous, feeding in an upward spiral on tree trunks, and occasionally on the ground. They nest in a hollow branch or trunk. There are two genera: *Climacteris* with five species and *Cormobates* with two species, found in Australia and New Guinea.

Climacteris *See* CLIMACTERIDAE.

climate The average weather conditions experienced at a particular place over a long period (usually 30 years).

Climatiiformes (spiny sharks; class *Acanthodii) One of the larger orders of fossil fish that had bony jaws and body skeleton, *ganoid scales, a *heterocercal tail, and stout spines located before the *dorsal, *anal, *pectoral, and *pelvic fins. One of the better-known forms, *Climatius*, carried five additional pairs of fins along the belly and reached a length of only 8 cm. These fish may actually be half-way between the sharks and the bony fish.

climax The final stage in a plant succession in which the vegetation attains equilibrium with the environment and, provided the environment is not perturbed, the plant community becomes more or less self-perpetuating. Subsequent changes occur much more slowly than those during earlier successional stages.

climbing gourami (*Anabas testudineus*) *See* ANABANTIDAE.

clinal speciation A type of *allopatric speciation in which a geographic barrier falls across a *cline, cutting a *species (which already shows some variation) into two segments that continue to diverge in their new isolation.

cline A gradual change in gene frequencies or *character states within a species across its geographic distribution.

clingfish *See* GOBIESOCIDAE.

Clinidae (scaled blennies; subclass *Actinopterygii, order *Perciformes) A large family of marine, mainly tropical fish which differ from the true blennies (*Blenniidae) in having fully scaled bodies and slender spines in the long *dorsal and *anal fins. They are small, bottom-dwelling fish, and occur in inshore situations. One of the larger species is *Clinus superciliosus* (klipfish), of S. Africa, which grows to 30 cm. There are about 175 species, found world-wide.

Clinus superciliosus (klipfish) *See* CLINIDAE.

clitellum A swollen, glandular, saddle-like region on the body of *Oligochaeta and *Hirudinea, which has a reproductive function. It assists in bonding together the two copulating worms and after copulation it secretes a cocoon in which the eggs are deposited.

cloaca In most vertebrates, including *Monotremata but excluding other mammals, the terminal part of the gut into which the alimentary, urinary, and reproductive systems open, leading to a single aperture in the body.

clock, internal A physiological mechanism, governed by 'clock genes', that regulates the production of *proteins that control cycles of sleeping and waking, body temperature, heart activity, *hormone secretion, blood pressure, oxygen consumption, *metabolism, and other bodily functions according to rhythmic patterns. The master clock, which regulates the 24-hour *circadian rhythm, is located in the suprachiasmatic nuclei in the *hypothalamus of the brain. Migrating birds and foraging bees navigate partly by reference to the position of the sun; their internal clocks monitor the passage of time and inform them of the anticipated position of the sun at any time of day. *See also* CLOCK SHIFTING.

clock shifting An experimental procedure in which homing pigeons are kept for some time in a light-proof room and exposed to periods of light and darkness that differ from those of natural light. The birds experience the same length of daylight as they would outdoors, but at a different time of day. When released, the birds are said to be 'clock-shifted' and fly in the direction that relates to the position of the sun predicted by their internal clocks, and not the sun's actual position.

clonal-selection theory A theory of *antibody specificity to account for the very great variety of immunoglobins that an organism may produce, and the absence of antibodies that react with the organism's

own cells. The theory suggests that during embryonic development a vast population of lymphoid cells is produced, each possessing the ability to synthesize a different antibody. There follows selection by cloning of only those cells that produce antibodies with configurations that are not complementary to any *antigens normally present in the *embryo.

clone A group of genetically identical cells or individuals, derived from a common ancestor by asexual mitotic division (*see* MITOSIS). If a section of DNA is engulfed into the *chromosome of a bacterium, phage, or plasmid vector and is replicated to form many copies, each copy is referred to as a DNA clone.

closed population A population in which there is a barrier to *gene flow. *Compare* OPEN POPULATION.

clothes moth *See* TINEIDAE.

clouded leopard (*Neofelis nebulosa*) *See* FELIDAE.

Clupea harengula (herring) *See* CLUPEIDAE.

Clupeidae (herring, shad; subclass *Actinopterygii, order *Clupeiformes) A large family of mainly marine fish which have compressed, streamlined bodies with a single, soft-rayed, *dorsal fin half-way down the back, and a forked tail fin. Nearly all clupeids have protruding scales, giving the ventral side a rough or serrated appearance. The family includes many commercially important species, e.g. *Clupea harengula* (herring), *Sardina pilchardus* (pilchard), and *Brevoortia tyrannus* (Atlantic menhaden). Some species, e.g. *Dorosoma cepedianum* (American gizzard shad), are found only in freshwater systems. Like many other clupeids, they move in large schools, feeding on *plankton. They are found world-wide, in temperate and tropical waters. There are 66 genera and 216 species.

Clupeiformes (subclass *Actinopterygii, superorder *Clupeomorpha) An order of herring-like, marine or freshwater fish, that have a streamlined, silvery body and soft-rayed fins. The ventral fins are in an abdominal position opposite the single *dorsal fin; the *pectoral fins are inserted below the body axis. The order comprises the families Denticipitidae (denticle herring), *Clupeidae (herring), *Engraulidae (anchovy), and *Chirocentridae (wolfherring).

Clupeomorpha A superorder of bony fish that includes the order Clupeiformes.

clutch All of the eggs that are laid during a single episode of laying.

Clypeasteroida (sand-dollars; subphylum *Echinozoa, class *Echinoidea) An order of irregular echinoids with *ambulacra that are wider than the interambulacra on the lower surface. Accessory *tube feet are developed outside the ambulacral petals. The order first appeared in the *Palaeocene.

clypeus A cuticular (*see* CUTICLE) area of the head of an invertebrate animal, which lies between the *frons and the *labrum, with which there is an articulation.

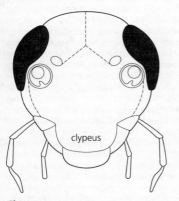

Clypeus

Cnidaria A phylum that comprises the sea anemones, jellyfish, and corals, and which is known from the late *Precambrian. Cnidar-

ians are basically *radially symmetrical and have *tentacles. The body contains a *gastrovascular cavity and the body wall is *diploblastic. Cnidarians occur as two distinct types: *polyps and *medusae, which in many taxa constitute alternating generations. All are aquatic and most are marine. There are about 9000 extant species.

cnidoblast A rounded or oval, epidermal cell with a basal nucleus, typical of *Cnidaria, that contains *nematocysts.

cnidocil *See* NEMATOCYST.

cnidosac In certain *Nudibranchia, one of the internal sacs located near the tips of the dorsal projections from the body surface (cerata), each of which contains a *nematocyst derived from coelenterate prey. In the alimentary canal, nematocysts are separated from the nutritive parts of the meal and passed to the cnidosacs; should a predator bite through the cerata, nematocysts are discharged, with the probable consequence that the predator abandons its attack.

Cnidospora (phylum *Protozoa) A subphylum of protozoa which form complex spores containing threads (polar filaments) which can be extruded, and function in attachment. There are two classes. They are parasites in a range of vertebrate and invertebrate hosts.

co-adaptation The development and maintenance of advantageous genetic *traits, so that mutual relationships can persist. Predator–prey and flower–pollinator relationships often exhibit examples of co-adaptation, which is an aspect of *coevolution.

coagulation The clumping together of colloidal particles to form a large mass; it may be caused by heating (e.g. the cooking of an egg causes the *albumin (the 'white') to solidify) or by the addition of ions that neutralize the electrical charge which stabilize the colloid. *See also* BLOOD CLOTTING.

coarctate Applied to a *pupa in which the puparium is composed of the *cuticle of the final larval *instar.

coated pits *See* CLATHRIN.

coated vesicle The structure that forms during *endocytosis when a coated pit (*see* CLATHRIN), containing both *ligands and *receptors, pinches off from the *cell membrane.

coati *See* PROCYONIDAE.

coat-of-mail shells *See* POLYPLACOPHORA.

cobalamine A form of vitamin B$_{12}$, necessary to many organisms, that contains cobalt and lack of which interferes with cell division. In humans, lack of cobalamine causes pernicious anaemia.

cobia *See* RACHYCENTRIDAE.

Cobitidae (loaches; subclass *Actinopterygii, order *Cypriniformes) A family of mainly small, slender-bodied, freshwater fish, that have tiny scales and toothless jaws. The ventrally positioned mouth bears at least three pairs of *barbels. Cobitidae occur in swift-flowing mountain streams as well as in slow-moving rivers. When the oxygen concentration of the surrounding water falls to a low level, loaches can gulp in air and extract oxygen via well-vascularized parts of the gut. Many species are sensitive to changes in atmospheric conditions. *Misgurnus fossilis* (European weatherfish), growing to at least 20 cm, has been used in the past as a barometer because it becomes more active when a weather depression approaches. There are nearly 150 species, found mainly in Europe and Asia.

cobra, king (*Ophiophagus hannah*) *See* ELAPIDAE.

cobras *See* ELAPIDAE.

cobweb spiders *See* THERIDIIDAE.

Coccidae (scale insects, mealybugs; order *Hemiptera, suborder *Homoptera) Family of homopterans which feed on plant sap, and some species of which are pests. Females are wingless. Males have a single pair of wings and do not feed; usually they are very small. The newly hatched nymphs (crawlers) are the main dispersal stage, and are blown about by the wind. Most older nymphs and females are sedentary, except in the mealybugs and their relatives, in which some mobility is retained. There are about 4000 species occurring mainly in the tropics. The Coccidae belong to a group that is usually split into about 12 families (*see also* LACCIFERIDAE).

Coccidia (subphylum *Sporozoa, class *Telosporea) A subclass of parasitic *Protozoa which grow and reproduce inside the cells of an animal host. There are two orders, and many species.

coccidiosis Any disease caused by a protozoon of the suborder Eimeriina. These organisms seldom cause disease in humans, but they may cause important economic losses among domestic animals. The disease usually affects the intestine, and symptoms often include diarrhoea with bloody stools.

Coccinellidae (ladybirds, ladybugs, lady-beetles; class *Insecta, order *Coleoptera) Family of rounded, convex, usually shiny beetles, coloured red, black, or yellow, with a pattern of spots or lines. The head is largely covered by the *pronotum. The antennae are short, and clubbed. Larvae are very active, with well-developed legs, and often grey with a pattern of spots. Some are covered with spines, or a white, waxy coat. If provoked, adults and larvae exhibit reflex bleeding, producing a sticky orange fluid from 'knees' or spines: presumably this has a deterrent effect. Economically they are very important, and are used in biological control. The majority feed on aphids and plant pests, some are fungus feeders, and others (e.g. *Epilachna varivestis*, the Mexican bean beetle) are phytophagous and attack crop plants. There are 5000 species.

Coccothraustes (hawfinches) *See* FRINGILLIDAE.

coccyx In tailless *Primates, the final section of the vertebral column, comprising 3–5 fused vertebrae below the *sacrum.

***Coccyzus* (cuckoos)** *See* CUCULIDAE.

cochineal A dye (cochinealin or carminic acid, $C_{22}H_{20}O_{13}$) plus a wax and a fat, all three of which have commercial uses, obtained from the dried bodies of female *Dactylopius coccus* (family *Coccidae), hemipteran insects which feed on cacti, are native to Mexico and Peru, and are cultivated in parts of Central America, southern Europe, and N. Africa.

cochlea Part of the inner ear of some reptiles, birds, and mammals. It is concerned with the analysis of the pitch of received sound. In mammals other than *Monotremata it is spirally coiled.

Cochleariidae (boat-billed heron; class *Aves, order *Ciconiiformes) A monotypic family (*Cochlearius cochlearius*), sometimes placed with *Ardeidae, comprising a medium-sized heron that has a grey back and a black, long-feathered crown. Its bill is very broad, flat, and slightly hooked, and its middle toe is *pectinate. It inhabits marshes, feeds on fish, crabs, and amphibians, and nests in trees. It is found in Central and S. America.

***Cochlearius cochlearius* (boat-billed heron)** *See* ARDEIDAE; COCHLEARIIDAE.

cockatoos *See* PSITTACIDAE; PSITTACIFORMES.

cockchafer (*Melolontha melolontha*) *See* SCARABAEIDAE.

cock-of-the-rock (*Rupicola*) *See* COTINGIDAE.

cockroaches *See* BLABERIDAE; BLATTELLIDAE; BLATTIDAE; BLATTODEA; CRYPTOCERCIDAE; DICTYOPTERA; POLYPHAGIDAE.

cockroach wasp *See* AMPULICIDAE.

cocoon Silken sheath which the *pupae of many insects, e.g. moths and bees, spin around themselves for protection during this vulnerable stage in their life cycle. In the case of the silk moth, this covering can be unwound and spun into thread for the making of silk cloth.

cod (*Gadus morrhua*) *See* DEMERSAL FISH; GADIDAE.

cod icefish *See* NOTOTHENIIDAE.

codlets *See* BREGMACEROTIDAE.

codominant A *heterozygote that shows fully the phenotypic effects of both *alleles at a gene *locus. For example, humans of the AB *blood group show the phenotypic effect of both IA and IB codominant genes.

codon A triplet sequence of *nucleotides in *messenger-RNA that acts as a coding unit for an *amino acid during protein synthesis. It binds by *base pairing to a complementary sequence, the anticodon, in *transfer-RNA.

coefficient of coincidence The experimental value of the observed number of double recombinants (cross-overs) divided by the expected number.

coefficient of inbreeding The probability that two allelic genes forming a *zygote are both descended from a gene found in an ancestor common to both parents. It is also used for the proportion of loci at which an individual is homozygous. The coefficient of consanguinity of an individual is the probability that two homologous genes drawn at random, one from each of the two parents, will be identical, and therefore homozygous in an offspring. The term is sometimes known as Wright's inbreeding coefficient (F), after the geneticist Sewall Wright (1889–1988) who formulated it. $F=\frac{1}{2}$ for a selfed mating; $\frac{1}{4}$ for fullsib mating; $\frac{1}{8}$ for uncle × niece and aunt × nephew, or double first cousins (i.e. cousins by both parents); and $\frac{1}{16}$ for first cousins.

coefficient of variation A measure of variability within a sample, representing a *population or a *species; it is calculated as the *standard deviation×100 divided by the mean. Experience shows that within a homogeneous population of mature individuals the coefficient of variation rarely exceeds 10.

coelacanth *See* COELACANTHIFORMES; LATIMERIIDAE.

Coelacanthiformes (subclass *Crossopterygii, superorder *Coelacanthimorpha) An order of bony fish thought to have been extinct for 65 million years (since the end of the *Mesozoic Era) until the discovery of the coelacanth *Latimeria chalumnae* off S. Africa in 1938 (*see* LATIMERIIDAE). Both fossil and living coelacanths are bulky marine fish with a *diphycercal or trilobed tail fin, the second *dorsal, *anal, *pectoral, and *pelvic fins being very unusual as they are supported by movable stalks or lobes. The much smaller fossil species lived from the *Devonian to the *Cretaceous. They were initially freshwater fish, but in the *Triassic marine representatives also evolved.

Coelacanthimorpha (Actinistia; class *Osteichthyes, subclass *Crossopterygii) A superorder of bony fish that includes the order *Coelacanthiformes.

Coelenterata The name formerly given to a phylum comprising both *Cnidaria and *Ctenophora. Today, when these two groups are universally separated in different phyla, the term 'Coelenterata' is sometimes used as a synonym for Cnidaria alone.

coelenteron (enteron) In *Cnidaria and *Ctenophora, the *gastrovascular cavity, with a single opening, the mouth. In some groups eggs and sperm are discharged into it. The coelenteron may be separated by *mesenteries or may form a canal system.

***Coelodonta* (woolly rhinoceros)** *See* RHINOCEROTIDAE.

coelom A fluid-filled (*see* COELOMIC FLUID) body cavity that originates by the splitting of the *mesoderm of *triploblastic animals. It separates the muscles of the body wall from the gut, allowing them to move independently, and provides an area for the enlargement of internal organs, thus permitting the gut to be differentiated for various functions. In many animals, the coelom plays an important part in collecting excretions and acts as a storage site for the maturation of *gametes. It is well developed in *Vertebrata, *Echinodermata, and *Annelida, but in *Arthropoda and *Mollusca it is reduced and its role replaced by the *haemocoel.

coelomate Possessing a *coelom.

coelomic fluid The fluid within the *coelom; in some animal groups it functions as a *hydrostatic skeleton and it may also serve as a circulatory medium.

coelomoduct In invertebrates, a duct formed from the lining (*mesoderm) of the *coelom, and connecting the coelom with the exterior. In some animals this is used to convey products of the gonads; in others to convey metabolic wastes.

Coelophysis *See* COELUROSAURIA.

Coelopidae (kelp flies, seaweed flies; order *Diptera, suborder *Cyclorrapha) Small family of true flies which are often hairy or bristly, and somewhat flattened. The antennae are often held in hollows, and the post-vertical bristles are convergent. The legs may be bristly, or woolly with pre-apical bristles. These flies are usually associated with the sea-shore, as their larvae feed on rotting seaweed on the strand-line. Some species make short mass migrations, and may also be attracted to aromatic chemicals. Only 48 species have been described.

Coelurosauria (order *Saurischia, suborder *Theropoda) An infra-order of carnivorous, bipedal dinosaurs that had small skulls, sharp, serrated teeth, long necks and tails, and hands with long, grasping fingers. They were the most persistent of the infra-ordinal dinosaur groups, extending from *Triassic to *Cretaceous times. The largest measured about 3 m long, although the great majority of these rather slender, agile dinosaurs were much smaller; *Coelophysis*, from the Late *Jurassic of N. America, was less than 2 m long and weighed about 23 kg.

Coenagriidae *See* COENAGRIONIDAE.

Coenagrionidae (Coenagriidae, dam-selflies; order *Odonata, suborder *Zygoptera) Cosmopolitan family of damselflies, most of which are small and brightly coloured and can be seen skimming over the surface of still water. The larvae live among aquatic vegetation in still water, and have slender *tracheal gills. More than 1290 extant species have been described.

Coendou (tree porcupine) *See* ERETHIZONTIDAE.

coenenchyma A substance, secreted by the *coenosarc, that links the colonial *polyps of *Octocorallia. It consists of a thick mass of *mesogloea perforated by tubes that are continuous with the *gastrovascular cavity of the colony. It is sometimes fleshy but often calcified. Cells of the coenenchyma secrete the skeletal material of Octocorallia, therefore the skeleton is internal.

coenoecium *See* RHABDOPLEURIDA.

coenosarc Within the stems of colonial *polyps, material that links the individuals, containing the *gastrovascular cavity and surrounded by the *perisarc. The coenosarc secretes the *coenenchyma.

Coenothecalia (class *Anthozoa, subclass *Octocorallia) An order of colonial corals which first appeared in the *Cretaceous and which includes the modern genus *Heliopora* (blue coral) common in tropical Pacific reefs. Members are characterized by a massive skeleton of fibrocrystalline *aragonite, the skeleton being perforated by wide tubes which house the *polyps, and by narrow tubes containing blind downgrowths of solenial tissue (*see* SOLENIUM). The order contains two families, and three genera.

coenzyme An organic substance that acts as a *cofactor for an *enzyme.

co-evolution The complementary evolution of closely associated *species. The interlocking adaptations of many flowering plants and their pollinating insects provide some striking examples of co-evolution. In a broader sense, predator–prey relationships also involve co-evolution (when an evolutionary advance in the predator triggers an evolutionary response in the prey). *See also* CO-ADAPTATION.

cofactor A non-protein component that is required by an *enzyme for its function, and to which it may be either tightly or loosely bound. Cofactors that are tightly bound are known as *prosthetic groups. Cofactors are required by many *enzymes: they may be metal ions (activators) or organic molecules (*coenzymes), and generally act as donors to or acceptors from the substrate (or substrates) of functional groups of atoms. Examples are *NAD, *NADP, and *ATP. These are used in the laboratory, after *electrophoresis, in histochemical staining for the specific enzymes.

cognition The mental processes that are presumed to be occurring within an animal but which cannot be observed directly. Cognition may be important in *insight learning.

cognitive map A mental model (or map) of the external environment which may be constructed following *exploratory behaviour.

cohesion species concept A proposal, made by Alan R. Templeton in 1989, to broaden the *biological species concept by including asexual organisms and downplaying interbreeding in sexually reproducing organisms. A species, in this view, is a *population, or series of populations, that has genetic or demographic cohesion.

cohort 1. A group of individuals of the same age. **2.** In animal taxonomy, a group of orders.

Coilostega (order *Cheilostomata, suborder *Anasca) A division of bryozoans in which the lateral dorsoventral muscles (moving the frontal membrane) pass through special notches or holes in the *zooid. The division occurs since the Middle *Jurassic.

Coleoidea (phylum *Mollusca, class *Cephalopoda) A subclass of cephalopods which have one pair of gills. The true shell, if it occurs at all, is internal. The head possesses eyes and a mouth that bears a pair of *mandibles and a *radula. Eight arms extend from the head. Some forms also have two additional, retractile tentacles. A *hyponome is also present. There are four extant orders and one extinct order. The subclass first appeared in the lower *Carboniferous.

Coleophoridae (subclass *Pterygota, order *Lepidoptera) Widespread family of very small to small moths. The wings are fringed with long, hair-like scales. At rest the antennae are held *porrect. Larvae are leaf-miners in the first *instar, and then construct cases. As case-bearers, the larvae feed externally or are leaf-miners. Pupation usually occurs in the larval case.

Coleoptera (beetles; subclass *Pterygota, division *Endopterygota) An order of insects which have biting mouth-parts and in which the fore wings are modified into more or less horny, rigid *elytra which meet medially when at rest and partly or wholly cover the hind wings and *abdomen. When present, the hind wings are membranous and much folded when at rest. They are the sole means of flight, although the elytra are said to provide some lift in some groups. The head and well-developed *prothorax form a distinct fore body, while the hind body comprises the meso- and metathorax and abdomen, covered by the elytra. The mesothorax is usually reduced, and the abdominal sterna (*see* STERNUM) are more strongly sclerotized than the *tergites. The larvae have a distinct head capsule, with antennae and well-developed *mandibles, and may or may not have thoracic legs. The pupae are *adecticous and *exarate. The Coleoptera is the largest order of animals and contains at least 350 000 species.

colies *See* COLIIDAE.

Coliidae (mousebirds, colies; class *Aves, order *Coliiformes) A family of brown or grey, short-winged and very long-tailed birds that are crested, with bare skin around the eye, and have short, stout, curved bills and short, strong feet, the hind toe being reversible so that all four toes point forward. They are found in scrub and bushes, feed on fruit and vegetable matter, and nest in trees and bushes. There is one genus, *Colius*, containing six species, all of which are confined to Africa.

Coliiformes (mousebirds, colies; class *Aves) An order of birds that comprises a single family, *Coliidae.

colinearity The correspondence between the order of *nucleotides in a section of DNA (*cistron) and the order of *amino acids in the *polypeptide that the cistron specifies. The evidence for this is that the positions of mutation within a particular cistron correspond to, and are in the same order as, the positions at which amino-acid substitutions are found in the polypeptide coded by the cistron.

Colius *See* COLIIDAE.

collagen A fibrous *scleroprotein that is almost inert, of high tensile strength, and relatively inelastic, which is a major constituent of *connective tissue and the organic material in *bone. It may represent up to 6% of the total body weight. It is unusual among proteins in that glycine, proline, and hydroxyproline represent over 50% of the total amino-acid residues; and these are arranged in a tight, triple left-handed helix. When boiled, collagen yields gelatin.

collared dove (*Streptopelia decaocto*) *See* COLUMBIDAE.

collared peccary (*Tayassu tajacu*) *See* TAYASSUIDAE.

Collembola (springtails; subphylum *Hexapoda, class *Insecta) Subclass and order of primitive, eyeless, wingless, small insects, exhibiting simple metamorphosis, which have *entognathous, biting mouth-parts, short antennae, six-segmented *abdomens (sometimes fused), and legs which lack tarsi (*see* TARSUS). Most species are less than 6 mm long, and can leap by means of a special, forked, springing organ (the *furcula) which is held up against the under side of the abdomen by a catch (*retinaculum, *hamula, ventral tube, or collophore): when the furcula is moved downwards suddenly, it hits the substrate and propels the animal through the air. There are two suborders, the elongate Arthropleona, and the globular Symphypleona whose members have the mouth-parts tucked under the head. Springtails occur throughout the world and live in soil, leaf-litter, under bark and decaying wood, and among fungi and vegetation, and are very common, attaining densities of 60 000/m². Some species are pests of leguminous crops, some feed on the blood of vertebrates, but most are important as scavengers and agents of nutrient recycling. Some species live on the water surface of ponds, or even on alpine snowfields where they feed on pollen and other debris. There are about 2000 species.

Colletes See COLLETIDAE.

Colletidae (order *Hymenoptera, suborder *Apocrita) Cosmopolitan family of primitive bees, which is best represented in the Australian region. The tongue is short, broad, and *emarginate apically, modified as a kind of paintbrush and used to apply the waterproof cell lining of macrocyclic lactones, which are secreted by the *Dufour's gland. Females of the subfamilies Hylaeinae and Eurglossinae have no pollen *scopa, and transport pollen in the crop. The principal ground-nesting genus, *Colletes*, is absent from Australia but is otherwise cosmopolitan. Most hylaeines nest in pithy stems or insect borings in wood.

colliculus A small raised area. In vertebrates, one of the raised areas (superior and inferior colluculi) in the *mesencephalon concerned with auditory and visual reflexes.

colloblast In *Ctenophora, an adhesive cell on the *tentacle used to capture food.

Collocalia (swiftlet) See APODIDAE.

collophore See COLLEMBOLA.

Colobidae See CERCOPITHECIDAE.

Colobinae See CERCOPITHECIDAE.

Colobus (colobus monkeys) See CERCOPITHECIDAE.

Cololabis saira (Pacific saury) See SCOMBERESOCIDAE.

colon In vertebrates, the large intestine, concerned in many animals with the absorption of water from the contents of the intestine, but in many *Mammalia it contains symbiotic (*see* SYMBIOSIS) *bacteria that are concerned with the digestion of *cellulose and with the synthesis of some *vitamins.

colostrum Milk that is secreted in the first few days after parturition. It differs in its physical and biological properties from that secreted later, containing a large number of *globulins that represent all the *antibodies in the maternal blood and therefore conferring temporary *passive immunity upon the new-born mammal.

Coluber viridiflavus (dark-green whip snake, European whip snake, western whip snake) See COLUBRIDAE.

Colubridae (colubrid snakes; order *Squamata, suborder *Serpentes) The largest family of snakes, most of which are harmless and have solid teeth, but some of which are venomous, with teeth at the back of the jaws grooved for venom. The left lung is small, or absent. They have no trace of hind limbs. Members of the genus *Natrix* are usually found near water, where they feed on amphibians and small mammals. *N. maura* (viperine snake), of the Mediterranean region, grows to almost 1 m and bears markings similar to those of the adder, from which it can be distinguished by the round pupils of its eyes (those of an adder are vertical). *N. natrix* (grass snake or ringed snake) is widely distributed in Europe. *N. tessellata* (dice snake or tessellated snake) whose range extends from central Europe to central Asia rarely exceeds 1 m in length. *Coluber viridiflavus* (dark-green whip snake, European whip snake, or western whip snake), found mainly in France, southern Switzerland, and Italy, grows to a maximum of about 1.9 m and is harmless to humans, but aggressive and agile. *Coronella austriaca* (smooth snake) of Europe and western Asia rarely exceeds 70 cm in length; it is *ovoviviparous. *Elaphe longissima* (Aesculapian snake) of central and southern Europe grows to 2 m, *E. quatuorlineata* (four-lined snake), whose range extends into southern Russia and Iran, grows to 2.5 m, and *E. situla* (leopard snake), of southern Europe and the Near East, grows to 1 m. *Dispholidus typus* (boomslang) is the only member of the family whose bite may be fatal to humans; it is an agile tree snake (its common name means 'tree snake' in Afrikaans), up to 1.8 m long, found in the African savannah. There are about 1000 colubrid species, distributed throughout the world. *See also* EGG-EATING SNAKES.

colugo (*Cynocephalus*) *See* CYNO-CEPHALIDAE.

colulus *See* SPINNERET.

***Columba* (pigeons)** *See* COLUMBIDAE.

Columbidae (pigeons, doves; class *Aves, order *Columbiformes) A family of small to large birds that have soft, dense plumage, plump bodies, and small heads. The bill is short to medium with a bare *cere, and the legs are generally short. Most are arboreal, though some are terrestrial, and feed mainly on seeds and fruit, regurgitating a milky fluid for their young. Green pigeons (23 species of *Treron*, mainly bright green birds, some with yellow, orange, and mauve markings, found in Africa, southern Asia, and Indonesia) feed on figs; their gizzards can grind up the seeds they ingest. Pigeons usually nest in a tree or on a ledge. They are gregarious, and many are migratory. Fruit doves (49 species of *Ptilinopus*, found in southern Asia, Australia, and the Pacific islands) are mainly green or sometimes orange with areas of contrasting colours. Quail doves (13 species of *Geotrygon*, found in Central and S. America) live in forests. Imperial pigeons (36 species of *Ducula*, found in southern Asia, Australasia, and the Pacific region) are arboreal and feed on fruit and leaves. There are 40 genera in the family, with 296 species, found world-wide. The 51 species of *Columba* comprise the typical pigeons (*C. livia* (rock dove) is the wild ancestor of the domestic pigeon) and there are seven species of *Columbina* (ground doves) in America. *Ectopistes migratorius* (passenger pigeon) is one of many extinct species. During the second half of the 20th century *Streptopelia decaocto* (collared dove), one of 16 *Streptopelia* species (turtle doves), has spread westward from Asia to colonize Europe.

Columbiformes (pigeons, doves, sandgrouse; class *Aves) An order that comprises two families of birds, *Columbidae and *Pteroclidae. A third family, Rhaphidae, is now extinct.

***Columbina* (ground doves)** *See* COLUMBIDAE.

columella 1. In corals, a rod-like structure that is formed from the swollen end of a *septum. It may form a central structure within the skeleton. **2.** In *Gastropoda, a spiral, rod-like structure that is formed by the fusion of their inner surfaces in shells that are coiled very tightly around an axis.

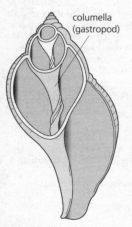

Columella

columella auris In amphibians, reptiles, and birds, a bony or cartilaginous rod connecting the *tympanic membrane with the inner ear and transmitting sound. It is homologous with the hyomandibular bone in fish.

columnar cell One of the three types of epithelial (*see* EPITHELIUM) cells (*see also* CUBOIDAL CELL; SQUAMOUS CELL) that is columnar in shape.

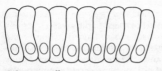

Columnar cell

Comatulida (feather stars; class *Crinoidea, subclass *Articulata) An order of feather stars that have no *pedicle and are free-living.

comber (*Serranus cabrilla*) *See* SERRANIDAE.

comb-footed spiders *See* THERIDIIDAE.

comb jellies *See* CTENOPHORA.

Comephoridae (Baikal oilfish; subclass *Actinopterygii, order *Scorpaeniformes) A small family of freshwater fish found only in Lake Baikal (Russia). They have a scaleless body, a snake-like head, and very long *pectoral fins. Almost translucent, the body has a high fat content. There is one genus and two species.

comfort behaviour Behaviour (e.g. yawning, scratching, and *grooming) that increases the physical comfort of an animal.

commensal *See* COMMENSALISM.

commensalism An interaction between *species populations in which one species (the commensal) benefits from another (sometimes called the host) that is not affected itself. For example, a hydroid (*Hydractinia echinata*) living on a whelk shell occupied by a hermit crab is carried by the crab to sites where it can feed but it does not deprive the crab because the two species have different food requirements. *Compare* MUTUALISM; PARASITISM.

commissure 1. A seam or junction. **2.** A band of nerve fibres that links the two hemispheres of the brain or the two sides of the central nervous system (e.g. in *Annelida and *Arthropoda it connects the *ganglia of the double nerve cord).

common bellowsfish (*Macrorhamphosus elevatus*) *See* MACRORHAMPHOSIDAE.

common chameleon (*Chamaeleo chamaeleon*) *See* CHAMAELEONTIDAE.

common crane (*Grus grus*) *See* GRUIDAE.

common cuckoo (*Cuculus canorus*) *See* CUCULIDAE.

common eel (*Anguilla anguilla*) *See* ANGUILLIDAE.

common frog (*Rana temporaria*) *See* RANIDAE.

common hippopotamus (*Hippopotamus amphibius*) *See* HIPPOPOTAMIDAE.

common house gekko (*Hemidactylus bibronii*) *See* GEKKONIDAE.

common iguana (*Iguana iguana*) *See* IGUANIDAE.

common kingfisher (*Alcedo atthis*) *See* ALCEDINIDAE.

common lizard (*Lacerta vivipara*) *See* LACERTIDAE.

common mullet (*Mugil cephalus*) *See* MUGILIDAE.

common musk turtle (*Sternotherus odoratus*) *See* KINOSTERNIDAE.

common myna (*Acridotheres tristis*) See STURNIDAE.

common pheasant (*Phasianus colchicus*) See PHASIANIDAE.

common pony-fish (*Leiognathus equula*) See LEIOGNATHIDAE.

common rhea (*Rhea americana*) See RHEIDAE.

common sturgeon (*Acipenser sturio*) See ACIPENSERIDAE.

common tegu (*Tupinambis teguixin*) See TEIIDAE.

common toad (*Bufo bufo*) See BUFONIDAE.

common viper (*Vipera berus*) See VIPERIDAE.

common wombat (*Vombatus ursinus*) See VOMBATIDAE.

common wryneck (*Jynx torquilla*) See PICIDAE.

communicating junction See GAP JUNCTION.

communication The transfer of information from one animal to another through the sense organs, resulting in behavioural changes that have survival value to one or both of the animals.

community Generally, any grouping of populations of different organisms that are found living together in a particular environment; essentially, the *biotic component of an *ecosystem. The organisms interact and give the community a structure.

community ecology An approach to ecological study which emphasizes the living components of an *ecosystem (the *community). Typically, it involves description and analysis of patterns within the community, employing methods of *classification and *ordination, and examines the interactions of community members (e.g. in the partitioning of resources).

compartment model (box model) A modelling approach that emphasizes the quantities and materials in different compartments of a system, and which may also express connections between compartments by some form of transfer coefficient. The approach is frequently used for studies of whole *ecosystems.

compass orientation The ability to head in a particular compass direction without reference to landmarks.

competition Interaction between individuals of the same *species (intraspecific competition), or between different species (interspecific competition) at the same *trophic level, in which the growth and survival of one or all species or individuals is affected adversely. The competitive mechanism may be direct (active), as when one organism releases a chemical substance that inhibits another, or indirect, as when a common resource is scarce. Competition leads either to the replacement of one species by another that has a competitive advantage, or to the modification of the interacting species by selective adaptation (whereby competition is minimized by small behavioural differences, e.g. in feeding patterns). Competition thus favours the separation of closely related or otherwise similar species. Separation may be achieved spatially, temporally, or ecologically (i.e. by adaptations in behaviour, morphology, etc.). The tendency of species to separate in this way is known as the *competitive-exclusion principle.

competitive-exclusion principle (exclusion principle, Gause principle) The principle that no two *species will occupy the same ecological *niche; i.e. two or more resource-limited species, having identical patterns of resource use, cannot coexist in a stable environment because one species will be better adapted and will outcompete or otherwise eliminate the others. The concept was derived mathematically from the logistic equation by *Lotka and Volterra, working independently, and was first demonstrated experimentally by G. F. *Gause in 1934 using two closely related species of *Paramecium*. When grown separately, both species populations showed normal *S-shaped growth curves; when grown together, one species was eliminated.

competitive inhibition The reversible inhibition of the activity of an *enzyme, brought about by the presence of an inhibitor molecule structurally similar to the normal substrate.

competitive release An expansion in the food preferences and foraging range of an animal that follows a reduction in the intensity of competition by other species.

complementary genes Mutant *alleles at different *loci which complement one another to give a *wild-type*phenotype. Dominant complementarity occurs where the dominant alleles of two or more genes are required for the expression of a particular trait. Recessive complementarity is the case of suppression of a particular trait by the dominant allele of either gene, so that only the homozygous double recessive displays the trait.

complementary resources Two or more resources that can substitute for one another and, when taken together, augment one another, so that the consumer requires less of them when taken together than when taken separately. *Compare* ANTAGONISTIC RESOURCES.

complementation map A gene map in which each mutation is represented by a line or 'bar' that overlaps the bars for other mutations with which it will not complement. Non-complementing mutants are represented by overlapping, continuous lines. Each bar probably represents the region of the *polypeptide that is distorted by an amino-acid substitution. Complementation maps are usually linear, and the positions of mutants on the complementation and genetic maps usually accord.

complementation test A test to determine whether two mutant sites of a gene are on the same *cistron. It is performed by introducing two mutant chromosomes into the same cell and observing whether the *wild-type *phenotype will be expressed, which it will if each chromosome complements the defect of the other. If complementation occurs, the two sites must be in different cistrons. The *genotype of the complementing *hybrid would be symbolized as a+/+b. This test is also known as the *cis-trans* test.

complete penetrance The case in which a specified *genotype always manifests itself at the phenotypic level. For this to happen, not only must a dominant gene always produce a phenotypic effect; so must a recessive gene in the homozygous state.

complex gradient A gradient of environmental factors that are linked in a complex fashion (e.g. the interrelated changes in rainfall, windspeed, and temperature, found along a transect from high to low elevation).

composite species concept An attempt, by Dina Kornet, to define *species cladistically and to formalize the concept in the fossil record. A species, in this definition, is a segment of a lineage in which a new *character state becomes fixed, from the point where it arises (by *cladogenesis) to the point where a new lineage (in which another character state becomes fixed) arises by cladogenesis.

compound eye Eye that is made up of *ommatidia, whose number may vary from a few dozen in some insects to several thousand in animals with good eyesight, e.g. bees and dragonflies. Many insects and crustaceans, some chelicerates, and a few annelids have compound eyes.

Comulatida (feather stars; class *Crinoidea, subclass *Articulata) An order of crinoids which lack a stem or column in the adult state and which swim freely. The uppermost *ossicle of the column is modified into a conical or discoidal plate (a centrodorsal). This bears numerous cirri (*see* CIRRUS), which are used to grasp substrata. The order has existed since the *Jurassic.

conchin *See* CONCHIOLIN.

conchiolin (conchin) A *protein secreted by the *mantle of *Mollusca that forms the organic component of the shell.

Conchorhagida (phylum *Aschelminthes, class *Kinorhyncha) An order of worms in which the head is covered on retraction by two plates on the third body segment.

Conchostraca (clam shrimps; class *Branchiopoda, order *Diplostraca) Suborder of branchiopods called 'clam shrimps' because the body is almost or completely enclosed in a bivalved carapace, which is folded, rather than being hinged dorsally, and closed by a transverse adductor muscle. The *compound eyes are sessile, and the second antennae are well developed, *biramous, and *setose, and augment the trunk limbs in locomotion. There are 10–32 trunk segments, each with a pair of appendages. There are 180 species.

concolorous Of the same colour (as some other specified structure).

concolours (*Nomascus*) *See* HYLOBATIDAE.

conditional stimulus *See* CONDITIONING.

conditioning (classical conditioning, Pavlovian conditioning) A form of learning in which an animal comes to associate an unconditional (significant) stimulus (e.g. the smell of food) with a conditional (neutral) stimulus (e.g. a sound), so that the previously conditional stimulus evokes a response that is rarely identical to the unconditional response but that is nevertheless appropriate to the unconditional stimulus. The method for studying this form of conditioning is derived from the work of I. P. Pavlov. *Compare* OPERANT CONDITIONING.

condors *See* CATHARTIDAE.

Condylarthra (cohort *Ferungulata, superorder *Protoungulata) An extinct order comprising seven families of primitive ungulates, including some that resemble the original eutherian stock so closely that their classification has presented many difficulties; some authors break the order into several different groups, assigned to other orders. Condylarths appear to be transitional between insectivores and true ungulates. They ranged from the late *Cretaceous to the latter part of the *Miocene and were a major part of the *Palaeocene fauna. Some forms possessed claws and may have been arboreal (*Hyopsodontidae), others became highly specialized, but the development of hoofs and ungulate dentition can be traced to families in the order.

condyle A knob of bone, round or ellipsoid in shape, that fits into a socket of an adjacent bone to form a joint.

condylobasal length The length of a skull, measured from the anterior points of the *premaxilla to the posterior surfaces of the *occipital condyles.

cone In a vertebrate eye, the less common (*compare* ROD) of the two types of light-receptor cell, which is sensitive only to relatively high levels of stimulation. Many cones are connected each to a single bipolar cell and thence to a *ganglion cell whose axon forms a nerve fibre in the optic tract. Since cones respond only to high levels of stimulation and since each stimulation produces an individual nerve impulse, cones provide sharp images.

cone-headed grasshopper *See* TETTIGONIIDAE.

coneworms *See* SEDENTARIA.

coney *See* CONY.

conflict The condition in which *motivations urge an animal to perform more than one activity at a time. Conflict may lead to one motivation becoming dominant; or to unresolved, *ambivalent behaviour; or to *irrelevant behaviour.

congenital Existing at birth; applied to inherited traits that become apparent only during growth.

conger eels *See* CONGRIDAE.

Congiopodidae (pigfish; subclass *Actinopterygii, order *Scorpaeniformes) A small family of fairly deep-bodied marine fish in which the head has a steep profile, and a produced, pig-like snout. The long and continuous *dorsal fin originates above the eyes. There are about seven species found in the colder waters of the southern hemisphere.

Congo eel *See* AMPHIUMIDAE.

congolli (*Pseudaphritis urvilli*) *See* BOVICHTHYIDAE.

Congridae (conger eels; subclass *Actinopterygii, order *Anguilliformes) A large family of marine fish which in profile resemble the common eels, but in which the *dorsal fin originates above the tip of the *pectoral fin. The skin bears no scales (that of common eels possesses minute scales). Congers live mainly in rocky coastal waters and are voracious predators. There are about 100 species, with a world-wide distribution.

Congrogadidae (eelblennies; subclass *Actinopterygii, order *Perciformes) A small family of marine fish which have elongate to eel-like bodies and long *dorsal and *anal fins; the *pelvic fins may be absent. There are about eight species, occurring in the Indo-Pacific region.

Coniophis precedens An early snake first identified in 1892 from a single vertebra by the American palaeontologist Othniel Charles Marsh (1831–99) in the upper Maastrichtian (early *Cretaceous, 65.5–70.6 Ma ago) Lance Formation of eastern Wyoming. The more recent discovery of a *Coniophis* skull, with teeth, revealed that the animal moved like a snake, but was unable to feed like a snake because it could not disarticulate its jaws to accommodate large vertebrate prey. It lived on dry land and fed on fairly large, soft-bodied prey. It was transitional between lizards and true snakes and provides evidence that snakes evolved from terrestrial burrowing lizards, rather than from marine lizards.

conjugant One of a pair of organisms that are undergoing *conjugation.

conjugated protein A *protein that contains a non-protein component, which may be a metal ion or an organic substance.

conjugation A process whereby organisms of identical species (but opposite *mating types) pair and exchange genetic material (DNA): i.e. usually each individual both gives and receives material. This process of sexual reproduction is found only in unicellular organisms such as *Bacteria and *Protozoa, although it is sometimes also applied to the union of *gametes, particularly in *isogamy. Details of the process differ greatly between different organisms. For example, in bacteria only DNA is transferred from one cell to another, while in some *eukaryotic microbes the process may involve the fusion of two entire cells. The transfer of DNA may be unidirectional (as in the case of *Escherichia coli*), in which case one cell is called the donor and the other the recipient, or bidirectional (as in *Paramecium aurelia*).

conjunctiva The *mucous membrane that lines the inside of the eyelid and the white part of the eye (sclera).

connective tissue In vertebrates, supportive or packing tissue composed of fibres, mainly of white *collagen but with some made from more elastic, yellow material, containing scattered cells and blood and lymph vessels, all lying in a matrix containing polysaccharide (starch-like *carbohydrate). *See also* ADIPOSE TISSUE.

connexin A *protein that forms hexagonal arrays lining the walls in the channel of a *gap junction, with their hydrophilic residues facing the centre of the channel.

connexivum In many *Hemiptera (true bugs), the series of *laterotergites along the sides of the abdomen.

connexon The channel of a *gap junction, lined with *connexin.

Conocyemidae *See* DICYEMIDA.

Conodontophora The category into which *conodonts were formerly placed. Conodonts are tiny, phosphatic, tooth-shaped *fossils that occur in rocks from the *Cambrian to the *Triassic.

conodonts Small, phosphatic, *fossil teeth, common in rocks from the *Cambrian to *Triassic (and formerly placed in the category *Conodontophora), that belonged to elongated, fish-like animals closely related to the vertebrates.

Conopidae (thick-headed flies; order *Diptera, suborder Cyclorrapha) Family of rather elongate, medium-sized flies, which have a long, often jointed proboscis, used to obtain nectar from flowers. The *palps are reduced and the ptilinal groove (*see* PTILINUM) is sometimes shortened. The male sixth, seventh, and eighth abdominal *tergites are lost, exposing the genitalia. The *abdomen may be down-curved posteriorly. In general, the adults resemble bees and solitary or social wasps in colour and appearance, and may be mistaken for them. Their eggs are specially adapted to cling to host wasps, on which the larvae will feed as endoparasites. Adult conopids are regular visitors to flowers. The family is distributed throughout the world, and there are more than 800 species.

Conopophagidae (gnateaters; class *Aves, order *Passeriformes) A family of small, short-tailed, long-legged birds, mainly brown above and white below, many of which have a white post-ocular stripe. They are secretive and ground-feeding insectivores, building nests at the base of bushes. There is one genus, with eight species, often placed with the *Formicariidae, found in S. America.

consanguinity (*adj.* consanguineous) A genetic relationship in which individuals share at least one ancestor in the preceding few generations. Matings between related individuals may reveal deleterious recessive *alleles. For example, first-cousin marriages amongst humans account for about 18–24% of albino children and 27–53% of children with Tay–Sachs disease, both of which are rare recessive conditions.

consensus sequence In an alignment of *homologous sequences of DNA (*see* DEOXYRIBONUCLEIC ACID) or *amino acids, that sequence which represents the most common *character state at each site.

conservation The artificial control of ecological relationships in an environment in order to maintain a particular balance among the species present.

consistency index In *cladistic analysis, a measure of *homoplasy in a phylogenetic tree (or *cladogram), calculated as the number of steps (i.e. *character state changes) in the cladogram divided by the smallest possible number of steps. The index therefore runs from 0 to 1. A low consistency index (less than 0.5) tends to indicate that much homoplasy has occurred.

conspecific Applied to individuals that belong to the same *species. *Compare* HETEROSPECIFIC.

constellation diagram A representation of *species affinities that is based on χ^2 as a measure of the association between species. The reciprocal of the χ^2 value for each species pair is used to plot the diagram, so that highly positively associated species with large χ^2 values are positioned closely together. Thus clusters of similarly distributed species may emerge, while the transitional affinities of a species with those of another main focal area or cluster will be evident. The constellation diagram exemplifies a simply calculated ordination method.

constitutive enzyme An *enzyme that is always produced whether or not a suitable *substrate is present (enzymes functioning by forming a complex with the substrate, *see* LOCK-AND-KEY THEORY). Such enzymes are sometimes produced by particular regulatory mutants which, though not affecting the structure of the enzyme, instead affect the process by which its synthesis occurs. An example is the *lac*-operon, which controls the synthesis of three enzymes (betagalactosidase, permease, and acetylase) that are involved in the lactose metabolism of the bacterium *Escherichia coli*.

constricting snakes *See* BOIDAE.

consumer In the widest sense, a *heterotroph that feeds on living or dead organic material. Two main categories are recognized: (*a*) macroconsumers (mainly animals), which wholly or partly ingest other living organisms or organic particulate matter; and (*b*) microconsumers (mainly bacteria and fungi), which feed by breaking down complex organic compounds in dead protoplasm. Sometimes the term 'consumer' is confined to macroconsumers, microconsumers being known as 'decomposers'. Consumers may then be termed 'primary' (herbivores), 'secondary' (herbivore-eating carnivores), and so on, according to their position in the *food chain.

consummatory In animals, applied to behaviour associated with the achievement of a goal (e.g. the eating of food) as distinct from *appetitive behaviour (e.g. searching for food).

contact inhibition The cessation *in vitro* of both movement and replication in a cell on making contact with other cells, such that a confluent monolayer is formed in the culture. Probably it occurs as a result of the formation of cytoplasmic bridges between cells. In many cancer cells this inhibition is absent.

contest competition *Competition for a resource that is partitioned unequally, so that some competitors obtain all they need and others less than they need (i.e. there are winners and losers). *Compare* SCRAMBLE COMPETITION.

contingency table (two-way table) A table of data for two methods of classification of the same individuals (e.g. hair colour and eye colour). This type of data can then be analysed statistically for association between these properties using a χ^2 test.

(⊕) SEE WEB LINKS
• Explanation of the contingency table method.

continuous distribution Data that yield a continuous spectrum of values. Examples are measurements of wing length of a bird or weight of a mammal.

continuous variation An assemblage of measurements of a phenotypic character which form a continuous spectrum of values. Examples are body weight, height, or shape, reproductive rate, and various behaviour traits. The continuity of *phenotype is a result of two phenomena: (*a*) each phenotype does not have a single phenotypic expression but a norm of reaction that covers a wide phenotypic range; (*b*) there may be many segregating *loci whose *alleles make a difference to the phenotype being observed.

contractile vacuole An *organelle which appears to function in the removal of excess water from the cells of *Protozoa. The contractile vacuole expands as it fills with water from the *cytoplasm, and contracts as this water is emptied to the exterior.

controlling gene A *gene that is involved in turning on or off the *transcription of *structural genes. Two types of genetic element exist in this process: a regulator and a receptor element. A receptor element is one that can be inserted into a gene, making it a mutant, and can also exit from the gene (the mutation is thus unstable). Both of these functions are under the influence of the regulator element, and are non-autonomous.

conures (*Pyrrhura*) *See* PSITTACIDAE.

convergent evolution The development of similar external morphology in organisms which are unrelated (except through distant ancestors) as each adapts to a similar way of life. Sharks (fish), dolphins (mammals), and ichthyosaurs (extinct reptiles) provide good examples of convergence in the aquatic habitat.

cony (coney) The name given in England to the rabbit in medieval, Tudor, and Stuart times; in the King James Bible it is used to mean the hyrax (*Hyracoidea). Today the name is usually reserved for the pika (*see* OCHOTONIDAE).

Cooloola monster (*Cooloola propator*) *See* COOLOOLIDAE.

Cooloola propator (Cooloola monster) *See* COOLOOLIDAE.

Cooloolidae (order *Orthoptera, suborder *Ensifera) Family of cricket-like insects, which was first described in 1980 and is known from only one genus and species, *Cooloola propator* (Cooloola monster), found in south-eastern Queensland, Australia. Its common name refers to its remarkable, robust form, and it is unusual among ensiferans in having short antennae with 10 segments. The structure of the mouth-parts is unique among Orthoptera. *C. propator* lives in sandy, moist soil. Males wander about on the surface at night, but females appear to be wholly subterranean.

co-operation In animals, mutually beneficial behaviour that involves several individuals (e.g. collaborative hunting, and care of the young). Co-operation may involve *altruism. Co-operation among members of different species is usually called '*symbiosis'.

co-operativity An interaction between binding sites within a *protein molecule whereby the binding of a ligand to one site influences the affinity of other sites on the same molecule for further ligands.

coots (*Fulica*) *See* RALLIDAE.

Copepoda (phylum *Arthropoda, subphylum *Crustacea) Large and diverse class of mostly marine crustaceans, though there are many freshwater species and a few which live in water films between mosses and soil particles. Many are ectoparasites of fish and whales. In free-living forms, the body is usually short and cylindrical, the 10-segmented trunk comprising a *thorax and an *abdomen. The head may bear a pointed *rostrum, or be rounded anteriorly. A median *nauplius eye is typical of all copepods, as is the absence of *compound eyes. The first antennae are *uniramous, and well developed; the second pair are smaller. The first pair of thoracic appendages are feeding structures and are modified to form maxillipeds. With the exception of the last one or two pairs, the remaining thoracic appendages are similar, and symmetrically *biramous. The posterior of the head is fused to the first thoracic segment and also, occasionally, to the second. The tapered thorax comprises three to five unfused segments with appendages. The narrow, cylindrical abdomen is devoid of appendages; the anal segment bears two caudal *rami. Many copepods are planktonic, occur in vast numbers, and—as the major food of many marine animals—are the main link between phytoplankton and the higher trophic levels. Most species are tiny, ranging from less than one to several millimetres in length, though the species of *Penellus*, which are ectoparasites of fish and whales, may attain 32 cm or more. There are more than 7500 species.

copepodid larva Series of larval stages, usually five, following the six *nauplius stages in *Copepoda. Typically, the first copepodid larva shows no abdominal segmentation, and there may be only three pairs of thoracic limbs. Otherwise, copepodid larvae display the general features of the adult.

Cope's rule In 1871, the American palaeontologist Edward Drinker Cope (1840–97) noted a phylogenetic trend towards increased body size in many animal groups, including mammals, reptiles, arthropods, and molluscs. This came to be known as Cope's rule. It remained unchallenged until a study of more than 1000 insect species in 1996 and was finally disproved in 1997, by a study in which David Jablonski made more than 6000 measurements on 1086 species of Late *Cretaceous fossil molluscs spanning

16 million years and found as many lines led to decreased size as increased. Evolutionary lineages show no overall tendency to greater size, but if the extant survivor happens to be larger than its immediate ancestor (e.g. the horse) this coincidence appears to validate Cope's rule.

coppers *See* LYCAENIDAE.

coprophagy Feeding by the ingestion of faecal pellets that have been enriched by microbial activity during exposure to the external environment. *Compare* CAECOTROPHY.

copulation Sexual intercourse in which the penis of the male penetrates the female and introduces sperm that may fertilize the egg.

Coracias abyssinica (Abyssinian roller) *See* CORACIIDAE.

Coraciidae (rollers;class *Aves, order *Coraciiformes) A family of brightly coloured birds that have short necks, long wings, and stout, slightly hooked bills. Their common name refers to their somersaulting display flight. Their tails are quite long, some species (e.g. *Coracias abyssinica*, Abyssinian roller) having long outer tail feathers. Their legs are short, with the second and third toes joined basally. They are arboreal and mainly insectivorous, sometimes dropping on to their prey from an exposed perch. They nest in holes in trees and banks. Some are migratory. There are two genera, with 11 species, found in Europe, Africa, and from southern Asia to Australasia.

Coraciiformes (kingfishers, todies, motmots, bee-eaters, rollers, hornbills, hoopoes; class *Aves) An order of birds in which the three anterior toes are united. Most nest in holes and lay white eggs. The order comprises the families *Alcedinidae, *Brachypteraciidae, *Bucerotidae, *Coraciidae, *Leptosomatidae, *Meropidae, *Momotidae, *Phoeniculidae, *Todidae, and *Upupidae.

Coracina (cuckoo-shrikes, greybirds) *See* CAMPEPHAGIDAE.

coracoid In the shoulder girdle of vertebrates, a bone on the ventral side extending from the *scapula to the *glenoid cavity. In mammals other than *Monotremata it is reduced to a small process of the scapula.

coral-billed nuthatch (*Hypositta corallirostris*) *See* HYPOSITTIDAE.

Corallimorpharia (class *Anthozoa, subclass *Zoantharia) An order of solitary or colonial, anemone-like *polyps. They lack a skeleton, and the radially arranged *tentacles terminate in a small knob.

corallite The skeleton formed by an individual coral *polyp, which may be either solitary or part of a colony.

corallum The skeleton of a colonial coral, made up from individual *corallites.

coral reef A massive, wave-resistant structure, built largely by coral (*see* CORALS), and consisting of skeletal and chemically precipitated material. Coral reefs extend over an area of more than $175 \times 10^6 km^2$ in tropical and subtropical seas, being best developed where the mean annual temperature is 23–5°C; they do not develop significantly at less than 18°C. Surface illumination is important and reefs do not grow in regions of high sedimentation, their skeletal formation depending on the activity of symbiotic algae and *zooxanthellae.

corals 1. (black coral) *See* ANTIPATHARIA. **2. (branched hydrocorals)** *See* STYLASTERINA; TELESTACEA. **3. (hydrocorals)** *See* HYDROCORALLINA. **4. (massive hydrocorals)** *See* MILLEPORINA; OCTOCORALLIA. **5. (soft corals)** *See* ALCYONACEA; ANTHOZOA; ASTROCOENIINA; COENOTHECALIA; CNIDARIA; HOLAXONIA. **6. (stony corals)** *See* MADREPORARIA. **7. (tetracorals)** *See* SCLERAXONIA; STOL-ONIFERA.

coral snakes Venomous snakes (*Elapidae) which are characterized by bright coloration, generally in bands running around the body. They occur in the New World, but the name is sometimes applied to African and Asian forms of similar appearance. The warning coloration of the American genus *Micrurus* is mimicked by a harmless pipe snake *Anilius scytale* (*Aniliidae), and some *Lampropeltis* species, including the milk snake (*L. triangulum sinaloae*), Californian kingsnake (*L. getulus),* and Californian mountain kingsnake (*L. zonata*).

corbiculum Area on the outer face of the hind *tibia of female bees of the family *Apidae, which is specialized for the transport of pollen and of nest-building materials, e.g. resin, mud, or dung. The corbiculum comprises a smooth, shiny depression fringed by stiff bristles, into which pollen or building materials are compacted for transport to the nest. The corbiculum is absent in apids which live as cuckoo bees.

Corcoracidae *See* GRALLINIDAE.

Cordulegasteridae (dragonflies; order *Odonata, suborder *Anisoptera) Small family of large, black and yellow dragonflies, in which the anal loop in the wings is compact but well developed, and the posterior corners of the hind wings are sharply angled in the male. Females have spade-like *ovipositors which they use to deposit eggs in stream beds. The larvae are elongate and hairy, and lurk in the silt of streams. There are 75 described extant species, found in the Palaearctic, south-east Asia, and N. America.

Corduliidae (dragonflies; order *Odonata, suborder *Anisoptera) Cosmopolitan family of medium-sized dragonflies which have slender, often metallic-green bodies. They have a slight projection on the rear margin of the *compound eye, and a foot-shaped anal loop on the wings. The larvae occur in a wide variety of aquatic habitats, and are difficult to distinguish from those of the *Libellulidae. More than 200 extant species have been described.

Cordylidae (girdle-tailed lizards, plated lizards; order *Squamata, suborder *Sauria) A family of terrestrial, insectivorous lizards, the majority of which have transverse rows of ossified, rectangular scales which may be elongated into spines. Some species are flattened and can enter rock crevices, others are snake-like with reduced limbs. *Cordylus giganteus* (sungazer, Lord Derby's zonure, or giant girdled lizard) of southern Africa is up to 40 cm long, with a triangular head, powerful limbs, sturdy body, and large, curved spines protecting the neck, back, and tail. There are about 50 species, occurring in the African savannah.

***Cordylus giganteus* (sungazer, Lord Derby's zonure, giant girdled lizard)** *See* CORDYLIDAE.

core area Part of a range in which an animal or group of animals may rest securely, in which young may be raised, and to which in some species food may be taken to be eaten. It is likely to be defended against members of the same species that do not share the range.

Coreidae (order *Hemiptera, suborder *Heteroptera) Family of bugs which feed on the shoots, buds, fruits, and unripe seeds of many plants throughout the world. The family includes the N. American squashbug, which is a pest of pumpkins and related plants. There are about 2000 species.

co-repressor A metabolite that, in conjugation with a repressor molecule, binds to the operator gene present in an *operon and prevents the synthesis of a repressible *enzyme.

corium 1. The lower skin (dermis) lying beneath the *epidermis. **2.** The thickened base of an insect wing.

Corixidae (order *Hemiptera, suborder *Heteroptera) Family of aquatic bugs which swim by means of the hair-fringed hind pair of legs, with their backs uppermost. They filter edible animal and vegetable matter from the debris at the bottom of ponds with their scoop-shaped and bristle-fringed front pair of legs. There are more than 200 species distributed world-wide.

cormidium In *Siphonophora, an assemblage of *polyps of different forms. Often the assemblage is repeated along the stem.

Cormobates See CLIMACTERIDAE.

cormorants See PHALACROCORACIDAE.

corncrake (Crex) See RALLIDAE.

cornea The layer of transparent tissue that forms the front of the eye of a vertebrate; it refracts light on to the lens.

cornetfish See FISTULARIIDAE.

Coronatida (class *Scyphozoa, subclass *Scyphomedusae) An order of free *medusae that are distinguished by a circular, horizontal, coronal groove. They are found mostly in the deeper parts of the ocean; surface forms are also known in warmer water.

Coronella austriaca (smooth snake) See COLUBRIDAE.

corpora allata See CORPUS ALLATUM.

corpora quadrigemina In the midbrain of mammals, two pairs of lobes extending from the dorsal surface, the anterior pair being optic centres and the posterior pair auditory centres.

corpora trapezioidea A band of nerve fibres connecting the right and left sides of the *cerebellum.

corpus allatum (pl. corpora allata) One of the glandular organs found on either side of the oesophagus in most insects, although they may become fused to form a single organ, as in *Diptera. The corpora allata, which release *juvenile hormone, have a nervous connection with the corpus cardiacum, which in turn connects with neurosecretory cells in the brain.

corpus callosum In *Eutheria, a band of nerve fibres connecting the left and right cerebral hemispheres (*cerebrum).

corpus luteum A mass of yellow tissue that remains after ovulation when a mature *Graafian follicle ruptures from the ovary of a mammal. It is formed by ingrowth of the follicle wall as a result of the action of *luteinizing hormone (which in turn requires the presence of *prolactin from the *pituitary). If ovulation does not result in fertilization, then the corpus luteum quickly degenerates. If fertilization occurs, then the corpus luteum persists and continues to secrete luteal tissue during part or all of pregnancy.

correlated progression The hypothesis that evolutionary change of characters occurs by *correlated response. A change in one character may influence change in another, such that the rate of change of the two characters is not independent. *Compare* MOSAIC EVOLUTION.

correlated response A change in a *character which occurs as an incidental consequence of the *selection for an apparently independent character (e.g. selection for increased bristle number in *Drosophila* may also result in reduced fertility).

correlation coefficient A statistic that is used to measure the degree of relationship between two variables.

corridor dispersal route As originally defined by the American palaeontologist G. G. *Simpson in 1940, a corridor is a *migration route that allows more or less uninhibited faunal interchange. Thus many or most of the animals of one faunal region can migrate to another one. A dispersal corridor has long existed between W. Europe and China via central Asia.

cortex 1. An outer layer. In vertebrates, the cerebral cortex is a layer of grey matter lying above each cerebral hemisphere (*cerebellum) in the brain. **2.** In *Heliozoa, an outer sphere of *ectoplasm, often greatly vacuolated. **3.** In some *sessile, aquatic invertebrates, a fibrous outer layer of the *periderm. **4.** In some *Porifera (e.g. *Calcarea), an outer surface consisting of a *syncitium formed from the interconnecting *pseudopodia of *amoebas.

corticosteroid hormones *Hormones produced by the adrenal cortex of the *adrenal gland that are involved in a range of physiological systems, including immune response, metabolism of *carbohydrates, and inflammation.

corticosterone A *corticosteroid hormone, produced by the adrenal cortex in the *adrenal gland, that in many animal species regulates the metabolic release of energy, immune reactions, and response to *stress.

Corti, organ of In mammals, the structure in the inner ear that contains sensory cells. It was first described by the Italian anatomist Alfonso Giacomo Gaspare Corti (1822–76).

cortisol A *corticosteroid hormone, produced by the adrenal cortex in the *adrenal gland, which increases blood pressure and blood sugar levels, and has an immunosuppressive function. It is involved in the response to *stress. Cortisol levels indicate social status and the amount of stress an individual is experiencing.

cortisone A *steroid hormone produced by the adrenal cortex and used medicinally to suppress allergies and other manifestations of the immune response.

Corvidae (crows, jays, magpies, nutcrackers, rooks; class *Aves, order *Passeriformes) A family of small to large black, black and white, or brightly coloured birds, that have large, heavy bills and round nostrils covered by feathers. Some are crested, some are long-tailed. All have strong legs. Corvids are typically gregarious, and inhabit forest, woodland, and open country. They are omnivorous, and their nests are usually open and bulky, in a tree or on a cliff. Many species are kept as cage birds and are good vocal mimics. The genus *Corvus* comprises 40 species of crows, ravens, and rooks (the common rook is *C. frugilegus*). There are three species of jays (*Garrulus*) found in Europe, N. Africa, and Asia; *G. glandarius* (jay) exists in five distinct forms across its range, while *G. lidthi* (Lidth's jay) is endemic to Japan. Nutcrackers (two species of *Nucifraga*) have white-spotted, brown, or grey

and black plumage, inhabit coniferous forest, feed on conifer seeds, nuts, and insects, and nest in trees. There are 26 genera in the family, with about 110 species, found worldwide, except for New Zealand and Pacific Ocean islands.

***Corvus* (crows, ravens, rooks)** *See* CORVIDAE.

Corydalidae (Dobson flies, fishflies; subclass *Pterygota, order *Megaloptera) Family of large, soft-bodied insects which have a fluttering flight pattern and are often found near streams. Adults possess *ocelli, and the larvae (called hellgramites and used as fish bait), which are found in streams and under stones, have a pair of hooked anal *prolegs, but lack a terminal filament. Male Dobson flies of the genus *Corydalus* have *mandibles three times the length of the head.

***Corydalus* (Dobson fly)** *See* CORYDALIDAE; MEGALOPTERA.

Coryphaenidae (dolphinfish, dorado; subclass *Actinopterygii, order *Perciformes) A small family of oceanic fish which have a high forehead (almost vertical in males) and a strongly tapering body ending in a forked tail fin. They are very fast swimmers, capable of reaching speeds of more than 60 km/h. Feeding mainly on flying fish and squid, the larger male dolphinfish can grow to more than 1.5 m in length. There are two species, with a world-wide distribution.

cosmine A type of *dentine that is perforated by branching canals.

cosmoid scale A type of scale found only in fossil lungfish (*Dipneusti), and in *Crossopterygii, including the living coelacanth. The thick scales are composed of layers of vitrodentine, followed by *cosmine, and finally by layers of vascular and laminated bone underneath.

cosmopolitan distribution The occurrence of an organism throughout the world or in many, widely separated places. Apart from opportunist plants (weeds), commensal animals (*see* COMMENSAL-ISM), and some non-vascular plants (such as algae, mosses, and liverworts), there are relatively few organisms that occur on all six continents.

Cossidae (goat moths, carpenter moths; subclass *Pterygota, order *Lepidoptera) Family of small to very large moths in which the mouth-parts are much reduced, the antennae are often *bipectinate, and the medial vein is forked in the discal cell of both wings. A single female may lay many eggs. The larvae generally bore into trees, and pupation occurs in the tunnel. The larvae of *Cossus cossus* (goat moth) smell like a goat. The family has a widespread distribution and many species are found in Australia.

Cossus cossus (goat moth) *See* COSSI-DAE.

costa One of the main longitudinal veins, which generally forms the leading edge of an insect wing.

costal respiration Respiration, typical of most higher vertebrates, in which movements of the muscles attached to the ribs causes lung ventilation. *Compare* BUCCAL FORCE PUMP.

costimulation An interaction that enhances signalling to an *antigen *receptor in developing an immune response, or a signal to a *B cell or *T cell that has no stimulatory effect alone, but that acts synergistically with the antigen receptor to fully activate a naïve lymphocyte.

(((⊕))) SEE WEB LINKS
• Description of activation and inhibition of lymphocytes by costimulation.

cotingas *See* COTINGIDAE.

Cotingidae (cotingas, cock-of-the-rock; class *Aves, order *Passeriformes) A family of small to large, brightly coloured birds, many of which have facial *wattles, crests, and modified wing feathers. They are arboreal, feeding on fruit and insects, which are taken mainly in flight. They usually nest in trees and bushes, building an open cup (but *Rupicola* (cock-of-the-rock) builds a mud-and-vegetable nest on rock faces). There are 25 genera, with 65 species, found in Central and S. America.

cotransport A process in which two substances are transported across a *cell membrane simultaneously by a *protein or protein complex. If both substances move in the same direction the process is known as symport; if they move in opposite directions it is antiport.

Cottidae (sculpin, father lasher; subclass *Actinopterygii, order *Scorpaeniformes) A very large family of marine and freshwater fish, all of which have a bony support beneath the eye, and the eyes placed high on the head. There is a deep notch between the two *dorsal fins, and the large *pectorals are wing-like. Most sculpins are found in inshore situations; some species (e.g. *Gymnocanthus tricuspis*, staghorn sculpin) live in cold polar waters. *Myoxocephalus scorpius* (father lasher), of the N. Atlantic, is a voracious predator but is used as a source of bait and fishmeal. There are about 300 species, widespread in non-tropical regions. *See also* PSYCHROLUTIDAE.

Cottocomephoridae (Baikal sculpins; subclass *Actinopterygii, order *Scorpaeniformes) A family of sculpin-like freshwater fish with large eyes, and very large *pectoral fins. They are found in the Lake Baikal drainage system. There are three or four genera and seven species.

cotton boll weevil (*Anthonomus grandis***)** *See* CURCULIONIDAE.

cotton stainer *See* PYRRHOCORIDAE.

cottontail Common name for several species of N. American lagomorphs of the genus *Sylvilagus*. *See also* LEPORIDAE.

Cottunculidae (subclass *Actinopterygii, order *Scorpaeniformes) A family of marine, sculpin-like fish that have a rough skin and continuous *dorsal fin. There are about six species, occurring in temperate to cold waters.

Cotylosauria (class *Reptilia, subclass *Anapsida) An order of stem reptiles which appeared in the *Carboniferous, having diverged from the *Labyrinthodontia. They flourished in the remainder of the *Palaeozoic but rapidly dwindled to extinction in the *Triassic.

couas *See* CUCULIDAE.

coucals (*Centropus*) *See* CUCULIDAE.

counter-current exchange 1. In organisms possessing two fluid systems separated by a permeable membrane across which exchange of gases or ions occurs, the occurrence of opposing fluid flow on either side of the membrane (e.g. the interface between the blood circulatory system and the gills of *teleosts). Counter-current exchange maximizes the rate of exchange. **2.** *See* RETE MIRABILE.

coural *See* LEPTOSOMATIDAE.

coursers *See* GLAREOLIDAE.

courtship The behaviour that precedes the sexual act and involves *displays and posturings by the male partner.

courtship feeding The presentation of food, or food-like objects, by a male to a female during *courtship. The gift is sometimes of nutritional significance but more usually the presentation is an act of appeasement, and it is often highly ritualized (*see* RITUALIZATION).

Couvinian *See* DEVONIAN.

covalent bond A bond in which a pair (or pairs) of electrons is shared between two atoms. The bond is often represented by drawing a single line between the symbols of the two atoms that have bonded together. Sometimes the bonding is between atoms of different elements (e.g. hydrogen chloride, H—Cl), and sometimes between atoms of the same element (e.g. fluorine, F—F). The

name 'molecule' is used to describe any uncharged particle containing covalently bonded atoms.

cowbirds *See* ICTERIDAE.

cowfish *See* OSTRACIONTIDAE.

cow killer (*Dasymutilla occidentalis*) *See* MUTILLIDAE.

Cowper's gland In mammals, a small gland that discharges a mucus, possibly serving as a lubricant, into the male urethra.

cow shark *See* HEXANCHIDAE.

coxa Uppermost joint of the leg of an insect, which articulates with the *thorax proximally, and with the *trochanter distally. The coxa is usually a stout, truncated cone of *cuticle, freely mobile with respect to the thorax; but in *Lepidoptera the middle and hind coxae are fused to the thorax. Muscles that move the coxae originate in the thorax.

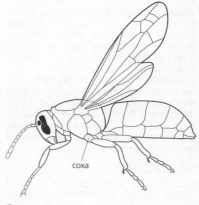

coxa

Coxa

coxal gland In some *Arthropoda, an excretory gland opening at the base of the legs.

coxal vesicle In some uniramous *Arthropoda, an *eversible sac associated with the *style or the basal segment of the legs.

coxopodite In some *Crustacea, the basal segment of the *biramous abdominal limb.

coyote (*Canis latrans*) *See* CANIDAE.

coypu (*Myocaster coypu*) *See* CAPROMYI-DAE.

crab-eating fox (*Cerdocyon thous*) *See* CANIDAE.

crab plover (*Dromas ardeola*) *See* DROM-ADIDAE.

crabs 1. *See* CRUSTACEA; DECAPODA; CANCRIDAE. **2. (hermit crabs)** *See* PAGU-RIDAE. **3. (land crabs)** *See* GECARCINIDAE. **4. (mud crabs)** *See* XANTHIDAE. **5. (spi-der crabs)** *See* MAJIDAE. **6. (stone crab, *Menippe mercenaria*)** *See* XANTHIDAE. **7. (swimming crabs)** *See* PORTUNIDAE.

crab spiders *See* THOMISIDAE.

Cracidae (curassows, guans, chacha-lacas; class *Aves, order *Galliformes) A family of black, black and white, or brown, long-tailed, pheasant-like birds that have short, rounded wings and strong legs with large feet. Their bills are variable, often hav-ing an ornate knob or *wattle, and many have crests and bare throat skin which is coloured and pendulous (e.g. the 15 spe-cies of guans (*Penelope*) found in rain for-est; *P. albipennis* (white-winged guan) has white *primaries). Chachalacas (nine *Or-talis* species) have a bare, red throat patch; curassows (12 species of *Crax*) have ornate, coloured bills or crests of *decurved feath-ers. Cracids are mainly arboreal, feeding on fruit and leaves, and nesting in trees. There are eight genera in the family, with 41 spe-cies, found in the southern USA, and in Central and S. America. Many species are endangered.

Cracticidae (butcherbirds, curraw-ongs; class *Aves, order *Passerifor-mes) A family of medium-sized, grey and white birds that have stout bills, sometimes hooked, and short, pointed wings. They are gregarious and arboreal, feeding on large insects, small vertebrates, and fruit, taken on the ground. They build an open nest in a tree. The family includes *Gymnorhina tibicen* (Australian magpie), a bird with a black-tipped, white bill that becomes tame in parks and gardens. There are three gen-era, with 10 species, found in Australia and New Guinea.

crakes *See* RALLIDAE.

craneflies *See* TIPULIDAE.

cranes *See* GRUIDAE.

Craniata (kingdom *Metazoa, phylum *Chordata) The subphylum that comprises animals that have a bony or cartilaginous skull and a dorsal vertebral column. *See also* VERTEBRATA.

(((⊕))) SEE WEB LINKS
• Description of the Craniata.

cranium (skull) In the *axial skeleton of vertebrates, the bony structure that en-cases the *brain. The cranium comprises three parts: dermatocranium; chondro-cranium; and splanchnocranium. The der-matocranium is derived from the *neural crest cells of the *integument and includes the roof of the cranium, the area around the orbits, and the jaw. The chondrocra-nium is derived mostly from cartilaginous material by endochondrial *ossification and includes the floor of the cranium. The splanchnocranium is derived from skel-etal elements derived from the gut and gives rise to the *visceral skeleton (gill arches and derivatives such as larynx and trachea).

Cranoglanididae (subclass *Actinop-terygii, order *Siluriformes) A small fam-ily comprising three species of freshwater catfish that have a scaleless body, four pairs of *barbels, a short *dorsal fin, and a forked tail fin. They are found in Asia.

Craseonycteridae (order *Chiroptera, suborder *Microchiroptera) A family of very small, tail-less bats that have long, separate ears in which the *tragus is tapered and visible, a simple nose with a *narial pad and open, non-valvular nostrils, and the second digit of the fore limb partly free in the leading edge of the wing. They were discovered in 1973, only in Thailand. There is one species, *Craseonycteris thonglongyai* (Kitti's bat or bumble-bee bat), which is the smallest living mammal.

Crax (curassows) *See* CRACIDAE.

crayfish *See* DECAPODA.

Creadion *See* CALLAEIDAE.

creatine phosphate (phosphocreatine) A high-energy, phosphorylated, nitrogenous compound that acts as an energy store in muscles and helps to maintain a relatively constant level of *ATP in the muscle during contraction. In a reversible reaction in the presence of creatine kinase: creatine phosphate +ADP<—>creatine+ATP.

creationism A modern variant of *special creation, in which it is maintained that all 'kinds' of organisms were created during one week, 6000–10 000 years ago, exactly as is stated in the book of Genesis. Creationism involves a rejection not only of the concept of evolution but also of the whole of geology and radiometric dating. *See also* CREATION 'SCIENCE'.

creation 'science' The extreme form of *creationism, in which it is maintained that science does not refute the Genesis stories of the creation and flood, but confirms them. Many analyses of creation 'science' in recent years have shown that it engages in the twisting or ignoring of scientific evidence, as well as special pleading, and that it is intellectually dishonest.

Creediidae (subclass *Actinopterygii, order *Perciformes) A small family of slender-bodied, marine coastal fish that have a protractile mouth, spineless, long, *dorsal and *anal fins, and the lower jaw lined with skin tufts (cirri). There are about three species, found in S. Africa and Australia.

Crenarchaeota *See* ARCHAEA.

crenate With a round-toothed or scalloped edge or margin.

crenulate Finely notched.

Creodonta (cohort *Ferungulata, superorder *Ferae) The more ancient of the two placental mammalian carnivorous orders, an extinct order that comprises two families (*Oxyaenidae and *Hyaenodontidae) which appeared in the late *Palaeocene and dwindled to extinction in the *Pliocene. Oxyaenids were rather mustelid, although some were large; hyaenodonts had more narrow skulls, longer legs, and well-developed *carnassials. They diversified into forms reminiscent of the dogs, cats, and hyenas of the order *Carnivora. Only the hyaenodonts survived the *Eocene and filled the role of scavengers until they were displaced by the modern hyena. Creodonts were small-brained and slow-moving, and are not related closely to modern carnivores.

creodont-like teeth Teeth resembling those of the *Creodonta, in which *carnassials were formed by the *molars, rather than *premolars and molars as in modern *Carnivora.

crepuscular Of the twilight; applied to animals that are active at dusk.

crested newt (*Triturus cristatus*) *See* SALAMANDRIDAE.

crested shelduck (*Tadorna cristata*) *See* ANATIDAE.

crestfish (*Lophotus capelei*) *See* LOPHOTIDAE.

Cretaceous The third of the three periods that are included in the *Mesozoic Era, about 145.5–65.5 Ma ago. It is noted for the deposition of the chalk of the White Cliffs of Dover, England, and for the mass extinction, at the end, of many invertebrate and vertebrate stocks. Among these were the *dinosaurs, *mosasaurs, *ichthyosaurs, and *plesiosaurs.

(⊕) SEE WEB LINKS
• Detailed information about life in the Cretaceous.

Crex (corncrake) See RALLIDAE.

cribellum A sieve-like structure which lies just anterior to the *spinnerets of some spiders. In these (cribellate spiders) there is also a row of small, curved bristles on the *calamistrum. The calamistrum combs out the very fine strands of silk from the cribellum, to form a characteristically bluish silk.

cribiform organs Specialized, dorsoventrally positioned *fascioles that occur between some or all of the marginal ossicles in some *Asteroidea (e.g. *Paxillosida).

Cribrimorpha (order *Cheilostomata, suborder *Anasca) A division of bryozoans in which the frontal (i.e. ventral) region of individuals is partially roofed by a network of fused, calcite spines, whose arrangement is important in classification. The division was abundant in the *Cretaceous, when it appeared, and it continues to the present day.

Cricetidae (order *Rodentia, suborder *Myomorpha) A large and very successful family that comprises the hamsters, voles, lemmings, gerbils, and the New World rats and mice. The family is almost certainly *paraphyletic. Most are terrestrial (and some of these are burrowing) but others are semi-aquatic. Their form varies according to their adaptation to a particular way of life. *Peromyscus leucopus* is the whitefooted deer mouse of N. America, which bears a striking resemblance to the field mouse (*Apodemus sylvaticus*) of Europe, which belongs to the *Muridae. The family also includes *Mesocricetus auratus* (golden hamster), and *Ondatra zibethicus* (musk-rat or musquash). Their distribution is worldwide except for Australasia and Malaysia. There are about 97 genera, with some 567 species.

Crick, Francis Harry Compton (1916–2004) The British geneticist who, with J. D. *Watson and M. Wilkins, won the 1962 Nobel Prize for Physiology or Medicine for their modelling of the *DNA molecule. Crick and Watson worked at the Cavendish Laboratory, Cambridge, and in 1977 Crick moved to the Salk Institute, California.

(⊕) SEE WEB LINKS
• Biography of Crick.

crickets See ENSIFERA; GRYLLIDAE; GRYLLOTALPIDAE; MYRMECOPHILIDAE; ORTHOPTERA; PROPHALANOPSIDAE; RHAPHIDOPHORIDAE; SCHIZODACTYLIDAE; STENOPELMATIDAE; TRIDACTYLIDAE.

Crinoidea (feather stars, sea lilies; phylum *Echinodermata, subphylum *Crinozoa) The most primitive living class of echinoderms, whose members have a long stalk (or, rarely, are sessile without a stalk, or free-swimming), a *calyx (lower surface) composed of regularly arranged plates, well-developed, movable arms, mouth and arms on the upper surface, radial food-grooves on the arms, leading to the mouth, and *tube feet on the arms. The more primitive types are attached to the sea floor by stalks, the more highly evolved types are free-swimming. Crinoids are known with certainty, as Eocrinoidea, from the Early *Ordovician onwards, and were fully modern by the end of the *Palaeozoic.

Crinozoa (phylum Echinodermata) A subphylum of echinoderms that have radial symmetry, a tendency to produce a cup-shaped or globoid, plated *test (*theca), and appendages (brachioles or arms) which support exothecal extensions of feeding *ambulacra. The subphylum includes eight classes, seven of which were extinct by the end of the *Palaeozoic, with only the *Crinoidea surviving to the present day.

criss-cross inheritance The transmission of a *gene from mother to son or father to daughter; e.g. in X-chromosome linkage. Classical examples are wing colour in the magpie moth (*Abraxus*), and feather pattern (barred v. non-barred) in chickens.

crista (*pl.* **cristae**) An infolding of the inner membrane of a *mitochondrion that projects into the matrix of the *organelle. Cristae bear numerous mushroom-shaped bodies, variously called respiratory assemblies, stalk bodies, or elementary particles, which contain certain of the enzymes involved in *ATP synthesis.

critical *See* RARITY.

crochets *See* PROLEGS.

crocodile fish *See* CHANNICHTHYIDAE.

crocodile lizard (*Shinisaurus crocodilurus***)** *See* XENOSAURIDAE.

crocodiles *See* CROCODILIA; CROCODYLIDAE.

Crocodilia (crocodilians; class *Reptilia, subclass *Archosauria) An order that includes the crocodiles, *Alligatoridae (alligators and caimans), and *Gavialidae (gavials), derived from thecodontian (*Thecodontia) ancestors and closely related to the dinosaurs and pterosaurs. They are first recorded from *Triassic rocks and were the only archosaurs to survive the *Mesozoic Era, although strictly speaking the birds (*Aves) are also archosaurs. Modern crocodilians are large, tropical, aquatic reptiles. The snout is long with strong conical teeth, raised valvular nostrils, and a secondary palate (which allows air-breathing with the mouth open in the water). The body has an armoured skin, short, powerful limbs, and a laterally flattened tail with a double row of raised scales merging towards the tip.

crocodilians *See* CROCODILIA.

Crocodylidae (crocodiles, dwarf crocodiles; subclass *Archosauria, order *Crocodilia) A family of crocodilians in which the fourth tooth of the lower jaw fits into a notch in the upper jaw, being visible when the mouth is closed. *Crocodylus niloti-*

cus (Nile crocodile) basks on river banks and may draw its legs beneath its body in order to move rapidly on land; it occurs throughout central and southern Africa and Madagascar, although it is now rare due to hunting. Crocodiles sometimes exceed 6 m in length. *C. porosus* (salt-water crocodile or estuarine crocodile), found from India to northern Australia, tolerates salt water and swims in the open sea; it has ridges from the eyes along the length of the snout and salt-secreting glands around the eyes. The false gharial (*Tomistoma schlegeli*) has been shunted back and forth between the Crocodylidae and the *Gavialidae and its affinities are still obscure; there is even a view that the true *gharial should be included in the Crocodylidae. There are 13 species in the family, occurring in fresh, brackish, and salt water in the Old and New World tropics.

Crocodylus niloticus (Nile crocodile) *See* CROCODYLIDAE.

Crocodylus porosus (saltwater crocodile, estuarine crocodile) *See* CROCODYLIDAE.

crop A thin-walled extension of the *oesophagus of a bird or insect, used for food storage. It is particularly well developed in grain-eating birds.

crop milk *See* PIGEON MILK.

cross-agglutination test One of a series of tests commonly used in blood typing, in which *erythrocytes from a donor of unknown type are mixed with *sera of known types.

crossbills *See* FRINGILLIDAE.

cross-breeding Usually, outbreeding or the breeding of genetically unrelated individuals. In animals it refers to the mating of individuals that are not consanguineous (*see* CONSANGUINITY).

crossing-over The exchange of genetic material between *homologous chromosomes by *breakage and reunion. This occurs during pairing of chromosomes at *meiosis, and in some organisms (e.g. certain fungi, and insects such as *Drosophila*) may also occur at *mitosis. The temporary joins between chromosomes during crossing-over are called chiasmata.

Crossopterygii (Actinistia; class *Osteichthyes) A subclass of bony fish that includes both fossil and living lobe-finned or tassel-finned fish, including the *Coelacanthimorpha and *Rhipidistia.

cross-over region The segment of a *chromosome lying between any two particular genes that are used as markers, e.g. during gene mapping (*see* GENETIC MAP) by recombination experiments.

Crotalidae (pit vipers; order *Squamata, suborder *Serpentes) A family of venomous snakes which are closely related to true vipers (*Viperidae) but which have heat-sensitive pit organs just behind the nostrils. Their diet consists mainly of birds and small mammals. Tail-rattling mechanisms have been developed in advanced forms (rattlesnakes); these are formed from loosely connected horny segments more of which are added to the 'rattle' at each moult. *Crotalus horridus* (timber rattlesnake) grows up to 1.8 m long and occurs in woodland in the eastern USA. *C. cerastes* (sidewinder) is small (up to 60 cm) and occurs in the deserts of the southwestern USA; it has a small horn above each eye and its common name is derived from its method of locomotion across soft sand: it proceeds diagonally by throwing its body forward in loops ('sidewinding' also occurs in some vipers). There are approximately 130 species in the family, occurring in the Americas and tropical Asia.

Crotalus cerastes (sidewinder) *See* CROTALIDAE.

Crotalus horridus (timber rattlesnake) *See* CROTALIDAE.

Crotophaga (anis) *See* CUCULIDAE.

crotovina *See* KROTOVINA.

crow butterflies *See* DANAINAE.

crown group In *cladistic analysis, the extant taxa descended from a common ancestor. *Compare* STEM GROUP.

crows *See* CORVIDAE.

cruciate (cruciform) Cross-shaped.

cruciform *See* CRUCIATE.

crura *See* CRUS.

crura cerebri A band of nerve fibres, one on each side of the brain, connecting the *cerebrum and the *cerebellum.

crus (*pl.*** crura)** Part of the calcified brachial support in *Brachiopoda which attaches it to the interior beak region of the brachial valve.

Crustacea (crabs, lobsters, shrimps, slaters, woodlice, barnacles; phylum *Arthropoda) Diverse class of arthropods which have two pairs of antennae, one pair of *mandibles, and two pairs of *maxillae. The limbs are *biramous, and are adapted for a wide range of functions. Closely placed *setae on the limbs function as filters in filter-feeding species. Respiratory gills are situated on the appendages, but vary greatly in location and number; they are absent only in very small species. In addition to the antennae, sense organs include a pair of *compound eyes, and a small, dorsal, median, *nauplius eye, comprising three or four closely applied *ocelli. The nauplius eye, characteristic of crustacean larvae, is absent in many adults; and some groups lack the compound eyes. Nitrogenous excretion is via a pair of *maxillary glands. Most of the 31 400 species are marine, but there are many freshwater species, and a relatively small number have invaded the land. A few marine species are parasites of other Crustacea; and one group, the Cyamidae, are ectoparasites of whales (whale lice). The first representatives of the group are known from *Cambrian rocks.

crypsis *See* CRYPTIC COLORATION.

crypt- A prefix derived from the Greek *kruptos*, 'hidden'.

Cryptacanthodes maculatus (Atlantic wrymouth) *See* CRYPTACANTHODIDAE.

Cryptacanthodidae (wrymouth; subclass *Actinopterygii, order *Perciformes) A small family comprising one genus (*Cryptacanthodes*) with four species of coastal marine fish that have an elongate body, long *dorsal and *anal fins, oblique mouth, and lack *pelvic fins. *Cryptacanthodes maculatus* (Atlantic wrymouth), 1 m long, often hides in burrows beneath the surface of the seabottom. Wrymouths are found in temperate waters of the Atlantic and Pacific.

cryptic behaviour A modification of behaviour that makes an animal more difficult for predators to detect. This may involve restricting activity to periods when predators are inactive, for example at night or, if predators are also nocturnal, on moonless nights. Alternatively, prey species may form groups in which individuals are concealed by other individuals and the group is less conspicuous than its members would be if they were isolated.

cryptic coloration (crypsis) Coloration that makes animals difficult to distinguish against their background, so tending to reduce predation. The effect of cryptic coloration may be to cause the appearance of the animal to merge into its background (e.g. the absence of all colour in some *pelagic fish larvae) or to break up the body outline (e.g. the spotted patterns of many bottom-dwelling flatfish). Both effects often occur in the same animal.

Cryptobranchidae (giant salamanders; class *Amphibia, order *Urodela) A family of what are probably neotenous (*see* NEOTENY), *hynobiid descendants. *Metamorphosis is incomplete, the gills are degenerate (hence the name, meaning 'hidden gills'), the eyelids never develop, and larval teeth are retained throughout life. The head and body are broad and flattened, the tail laterally compressed, the skin smooth and slimy, and the habit totally aquatic. The largest extant amphibian, *Andrias japonicus* (Japanese giant salamander), is more than 1.5 m long and some are longlived (*A. japonicus* individuals are known to have lived for 50 years). *Tertiary fossil species are distributed widely in the northern hemisphere, but there are only two living genera, comprising three species, in N. America and Asia.

Cryptocercidae (wood-eating cockroach; subclass *Pterygota, order *Blattodea) Monotypic family comprising the semi-social *Cryptocercus punctatus* (wood-eating cockroach), an insect which is 2–3 cm long, and is notable for feeding on rotting wood. It shows remarkable similarities to the *Isoptera, and both share a similar gut fauna of protozoans that break down cellulose. The wood-eating cockroach occurs in N. America and eastern Asia.

Cryptocercus punctatus (wood-eating cockroach) *See* CRYPTOCERCIDAE.

Cryptodira (hidden-necked turtles; subclass *Anapsida, order *Chelonia) A suborder of chelonians which withdraw the head straight back into the shell by flexing the neck vertically. There are 10 families, including marine, freshwater, and terrestrial forms, and they are cosmopolitan, except for Australia.

Cryptodonta (phylum *Mollusca, class *Bivalvia) A subclass of bivalves most of which have thin, equivalve shells composed of *aragonite. They have an *amphidetic to *opisthodetic external ligament. The hinge plate is narrow or absent, with most forms having an *edentulous hinge margin; some members are *taxodont. They have an *infaunal mode of life. They first appeared in the *Ordovician.

Cryptoprocta *See* VIVERRIDAE.

cryptozoa Invertebrate animals, large enough to be visible to the naked eye, which live between litter and the soil.

Crypturellus (tinamous) *See* TINAMI-
DAE.

ctenidial axis The junction of an inner
and outer *demibranch in molluscan
demibranchiate gills. Two V-shaped demibranchs are joined to form a W-shaped gill.
Gills of other shapes that are symmetrical
about an axis may also be said to have a ctenidial axis.

ctenidium 1. In aquatic *Mollusca, the
respiratory (or in *Bivalvia food-collecting)
surface or gill situated in the *mantle cavity. **2.** In some *Arthropoda, a fan-shaped,
eversible comb of spines. *See* NYCTER-
IBIIDAE.

ctenii *See* CTENOID SCALE.

Ctenizidae *See* TRAPDOOR SPIDERS.

**Ctenodactylidae (gundis; order *Ro-
dentia, suborder *Hystricomorpha)** A
family of small, ground-dwelling rodents
of arid, rocky environments in which the
legs are short and the digits are reduced to
four on each foot, the two inner digits of the
fore feet bearing bristles. The eyes are large,
the ears rounded. The tail is short and fully
haired. There are four genera, and eight species, occurring in N. Africa.

ctenoid scale In bony fish, a type of scale
that has many tiny, tooth-like processes
(ctenii) in the posterior, or outer, segment.
The word is derived from the Greek *kteis*,
'comb'.

**Ctenoluciidae (subclass *Actinop-
tery-gii, order *Cypriniformes)** A small
family of freshwater fish that have an elongate, pike-like body, large mouth, and the
single *dorsal and anal fins set far back
on the body. They occur in Central and S.
America.

**Ctenomyidae (tuco-tucos; order *Ro-
dentia, suborder *Hystricomorpha)** A
family of burrowing rodents in which the
body is compact, the head large, the legs
short and *pentadactyl with strong claws,

and the soles of the feet bordered with bristles. The eyes and ears are small. The skull
is strong and flattened, the occipital crests
prominent, the paroccipital processes
curved beneath the *bullae. The cheek teeth
are *hypsodont, the last *molar being vestigial. There is one genus (*Ctenomys*) and about
26 species, distributed widely throughout
southern S. America.

Ctenophora ('comb jellies') A small
phylum of carnivorous, hermaphroditic,
marine animals, in which the body is *bi-
radially symmetrical and can be divided
into two hemispheres, and into equal sections by eight ciliated bands, the 'combs'
from which the phylum derives its common
name. The *cilia provide locomotive power
in most species, although some lobate
species swim by contracting the lobes. In
the class *Tentaculata two long, branched
*tentacles, armed with *colloblasts, emerge
from a deep canal in the *epidermis on the
*aboral side of the body and are used to
catch prey. Members of the class *Nuda lack
tentacles. Many ctenophores are spherical
or ovoid, and 1–5 cm in diameter, but some
are conical, cylindrical, or strap-like and
one species of the genus Cestum grows to
more than 1 m in length. Ctenophores have
no definitive fossil record, but their body
plan is similar to that of a *medusa and they
are believed to be descended from a medusoid cnidarian. There are two classes (Tentaculata with four orders, and Nuda with
one order) altogether comprising about
50 species, some of which occur in coastal
waters throughout the world, and others of
which are oceanic.

**Ctenostomata (subphylum *Ecto-
procta, class *Gymnolaemata)** An order
of bryozoans in which new individuals grow
by budding from a slender, creeping tube
(the stolon). The aperture of each *zooid is
closed by a fold of the body wall which bears
bristle-like *setae. The body wall is not calcified. The order first occurred in the *Ordovician, and is still extant.

Ctenostomatida *See* ODONTOSTOMA-
TIDA.

cubitus vein One of the main groups of veins in an insect wing. In the *Lepidoptera it tends to be reduced or lost in the more advanced species.

cuboidal cell One of the three types of epithelial (*see* EPITHELIUM) cells (*see also* COLUMNAR CELL; SQUAMOUS CELL) that is cubic in shape.

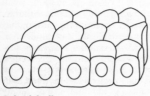

Cuboidal cell

cuboid bone The name sometimes given to each of the fourth and fifth distal *tarsal bones.

Cubomedusae (class *Scyphozoa, subclass *Scyphomedusae) An order of free *medusae, whose members have a cuboid or quadrangular shape with four *tentacles, or groups of tentacles, at the corners. They are inhabitants of warmer waters of the oceans, and also swarm in harbours and bays. The order includes *Chironex fleckeri* (sea wasp) whose stings are so virulent as to be very dangerous to humans (more so than those of the Portuguese man-of-war, *Physalia physalis*).

cuchia (*Amphipnous* (*Monopterus*) *cuchia*) *See* AMPHIPNOIDAE.

cuckoo bees *See* ANTHOPHORIDAE.

cuckoo-roller *See* LEPTOSOMATIDAE.

cuckoos *See* CUCULIDAE; CUCULIFORMES.

cuckoo-shrikes *See* CAMPEPHAGIDAE.

cuckoo-spit insect *See* CERCOPIDAE.

Cuculidae (cuckoos, anis, couas, coucals; class *Aves, order *Cuculiformes) A family of small to medium, long-tailed, generally short-legged birds most of which have a stout, slightly *decurved bill, and toes that are *zygodactylous or semi-zygodactylous. The 27 species of coucals (*Centropus*) of Africa, southern Asia, and Australasia have heavy, hooked bills. Most members of the family are arboreal, but some are terrestrial (e.g. *Geococcyx californianus*, greater roadrunner, and *Centropus*). They are mainly insectivorous and many species are brood-parasitic (*Clamator glandarius*, great spotted cuckoo, one of four species in its genus, all of which are migratory, parasitizes crows and starlings; each of the 13 *Cuculus* species parasitizes a variety of hosts). Most non-parasitic species build open nests in trees but *Centropus* builds an enclosed nest with a side entrance, close to the ground. They are mainly solitary, but *Crotophaga* species (anis) are gregarious and build communal nests. The eight *Coccyzus* species of America are secretive, dwelling in woodland and nesting in trees. There are 35 genera in the family, with 130 species, found world-wide. The call of *Cuculus canorus* (common cuckoo) gives rise to the family name.

Cuculiformes (cuckoos, turacos; class *Aves) An order that comprises two families of birds, *Cuculidae and *Musophagidae, both of which have loose-webbed plumage and *zygodactylous feet. Some authorities define them as separate orders.

cucullus In *Ricinulei, a hinged plate attached to the *carapace and covering the mouth-parts.

Cuculus (cuckoos) *See* CUCULIDAE.

Cuculus canorus (common cuckoo) *See* CUCULIDAE.

cucumberfish *See* CHLOROPHTHALMIDAE.

Culex *See* CULICIDAE.

Culicidae (mosquitoes, gnats; order *Diptera, suborder *Nematocera) Family of very slender flies, which generally have an elongate, piercing proboscis, and stiff *palps. The legs are long, the wings fringed with scales on the hind margins and along veins. The antennae of females are *pilose, those of males *plumose. Eggs are laid on or near the surface of water, singly or in compact masses as floating rafts. The transparent larva has a well developed, mobile head, with dense tufts of bristles which are used to filter from the water the planktonic organisms and detritus on which the larva feeds, and with an anal respiratory siphon which it uses to renew its oxygen supply at the water surface after a period of submersion. The *pupa is comma-shaped, very active, and able to swim beneath the water surface as a shadow passes across the surface. It breathes through a respiratory siphon held above the surface. Adult mosquitoes are well known as carriers of disease, especially those of the genera *Aëdes*, *Culex*, and *Anopheles*. Males generally feed upon plant juices. Females have stronger, piercing mouth-parts and feed upon the blood of vertebrates, acting as intermediary hosts for malaria, yellow fever, filariasis, dengue, and several other disease organisms. *Culex (pipiens) fatigans*, distributed throughout the tropics, is the carrier of *Wucheria bancrofti*, the nematode responsible for elephantiasis in humans. Adults of this species have a characteristic resting attitude, with the body almost parallel with the substrate. *Aëdes aegypti* is the yellowfever mosquito of the tropics and subtropics, a species that rarely breeds far from human habitations, laying its eggs in small water-containing receptacles such as tree holes, old tins, and bottles, or in broken coconut shells. Adult *Anopheles* (the malarial mosquito) can be distinguished by their resting attitude: the body makes an acute angle with the substrate as the body is held in line with the proboscis. *A. maculipennis* is the most important malarial vector in Europe. There are 3000 described species of Culicidae.

( ) SEE WEB LINKS
• Diseases transmitted by Culicidae.

culmen A ridge along the upper *mandible of a bird, from the tip of the bill to the forehead.

culpeos *See* CANIDAE.

cultural Applied to the transmission of information from one generation of animals to another by non-genetic means (e.g. by *imprinting or *imitation). Behaviour may be modified culturally, so that while individuals usually behave in a traditional way, traditional behaviour in the same species may vary widely from one population to another.

cultural evolution Evolution that is based on the transmission of information other than genetically (i.e. culturally). In humans, such transmission embraces customs, beliefs, and the acquisition and communication of knowledge. Adaptation by cultural change can be far more rapid than by genetic alteration.

culture 1. A population of micro-organisms or of the dissociated cells of a tissue grown, for experiment, in a nutrient medium; they multiply by asexual division. **2.** The transfer of behavioural traits between individuals in a non-genetic manner (i.e. the traits are not inherited genetically although they may be passed from parent to offspring by verbal or visual *communication). Culture is most developed in *Primates, particularly humans, but may also occur in other organisms such as social insects.

Cuon alpinus (cuon, dhole) *See* CANIDAE.

cupula *See* NEUROMAST.

cupulate Cup-shaped.

curassows *See* CRACIDAE.

Curculionidae (weevils, snout beetles; subclass *Pterygota, order *Coleoptera) Family of robust beetles, 1–50 mm long, in which the *elytra are toughened, and often highly sculptured, with a pattern of coloured or metallic scales. Some are flightless, with fused elytra. The head is produced into a *rostrum bearing *mandibles at the tip; some are short and stout, others long and narrow, up to three times the length of the body. The antennae are elbowed and clubbed. Larvae are legless, grub-like, and are usually found inside a plant, or underground, at the roots. Many are crop pests (e.g. *Anthonomus grandis*, the cotton boll weevil, and *Sitona lineatus*, the bean weevil or pea weevil). There are 60 000 species, making it the largest beetle family.

curlews 1. *See* SCOLOPACIDAE. **2. (stone-curlew)** *See* BURHINIDAE.

currawongs *See* CRACTICIDAE.

current competition Competition in which species restrict one another to *niches that are smaller than those they would occupy were competitors absent. *See* COMPETITIVE RELEASE.

cursorial Adapted or specialized for running.

cuscus *See* PHALANGERIDAE.

cusk-eel *See* OPHIDIIDAE.

cusp Of a tooth, the biting point.

cuspidate Having a sharp tip or point.

cutaneous respiration Breathing through the skin; in some vertebrates the body surface has become highly vascularized for gaseous exchange. Such exchange is of particular importance in the class *Amphibia, where mucous glands in the skin maintain a moist respiratory surface; and in the soft-shelled turtles (family *Trionychidae).

cuticle (*adj.* **cuticular)** A layer covering, and secreted by, the *epidermis. In invertebrates, it is mainly protective against mechanical or (in endoparasites, *see* PARASITISM) chemical damage. *See* EXOSKELETON.

cuticular *See* CUTICLE.

cuticulin A *lipoprotein secreted by an arthropod during *ecdysis that protects the animal and forms the outermost layer of the new *exoskeleton.

cutlassfish *See* TRICHIURIDAE.

cut-throat eel *See* SYNAPHOBRANCHIDAE.

cuttlefish *See* SEPIOIDEA.

cutworms *See* NOCTUIDAE.

Cuvier, Georges Léopold Chrétien Frédéric Dagobert, Baron (1769–1832) A French naturalist who was one of the founders of the disciplines of comparative anatomy and palaeontology. His studies of marine fauna led to detailed work on the structure and classification of *Mollusca and the anatomy of fishes. Later he turned his attention to fossil vertebrates and then to living members of those groups, eventually describing the structure of all known animal groups. He believed each species and each organ was created for a particular purpose and extinctions were caused by catastrophes, the affected areas being recolonized by immigration.

((⊕)) SEE WEB LINKS
• Biography of Cuvier.

cyanobacteria A large and varied group of *bacteria which possess chlorophyll and carry out photosynthesis in the presence of light and air with concomitant production of oxygen. They do not have chloroplasts. Cyanobacteria were formerly regarded as algae (division Cyanophyta) and were called 'blue-green algae'. Fossil cyanobacteria have been found in rocks almost 3000 Ma old and they are common as *stromatolite colonies in rocks 2300 Ma old. They are believed to have been the first oxygen-producing organisms and to have been responsible for generating the oxygen in the atmosphere, thus profoundly influencing the subsequent course of evolution. The organisms may be single-celled or filamentous, and may or may not be colonial. Some are capable of a gliding motility when in contact with a solid surface. Many species can carry out the fixation of atmospheric nitrogen.

cybernetics The study of communications systems and of system control in animals and machines. In the life sciences, it also includes the study of feedback controls in *homoeostasis.

Cyclarhidae (pepper-shrikes) *See* VIRE-ONIDAE.

cyclic AMP *See* C-AMP.

cyclic-GMP *See* GUANOSINE PHOSPHATE.

Cycliophora (phylum *Bryozoa) A bryozoan discovered by P. Funch and R. M. Kristensen in 1995 in the N. Sea and Kattegat, and so far known by only one species, *Symbion pandora*, which lives in the bristles around the mouth of the Norway lobster (*Nephropidae). *S. pandora* is less than 1 mm long, *acoelomate, *bilaterally symmetrical, sac-like, and attaches by an adhesive disc. It has a ring of *cilia around the mouth and an *anus just outside this ring. Sexes are separate; males are much smaller than females. It can also reproduce asexually, by *budding. The adult, feeding stage alternates with a non-feeding, free-swimming, larval stage.

cycloid scale In bony fish, a type of scale that lacks tooth-like processes, the edge being smooth all round.

cyclomorphosis Seasonal change in body shape found in rotifers (phylum *Rotifera), and in cladoceran *Crustacea (phylum *Arthropoda). In cladocerans, e.g. *Daphnia* species, the changes in shape involve the head, which is rounded from midsummer to spring and then progressively becomes helmet-shaped from spring to summer, reverting to the rounded shape by midsummer. The process is poorly understood and may be the result of genetic factors interacting with external conditions, e.g. temperature or day length, or, as in rotifers, the result of internal factors alone.

Cyclophyllidea (Taenioidea; phylum *Platyhelminthes, class *Cestoda) An order of parasitic worms which possess a relatively complex, suckered *scolex. Genitals are often repeated. Two hosts are inhabited during development.

Cyclopteridae (lumpsucker, sea snail; subclass *Actinopterygii, order *Scorpaeniformes) A large family of marine, cold-water fish that have thickset, almost rotund, short bodies with scaleless skin that bears many rows of bony thorns. Both second *dorsal and *anal fins are short. The presence of a sucking disc formed by modifications of the ventral fins is characteristic of the family. *Cyclopterus lumpus* (Atlantic lumpsucker) is found on both sides of the N. Atlantic, the larger males sometimes exceeding 35 cm in length. Treated with salt and black pigment, the roe is at times sold as cheap caviar. Atlantic snailfish, e.g. *Liparis liparis*, also have a scaleless skin, but no bony thorns or denticles. The dorsal and anal fins are much longer than those of the lumpsucker, but snailfish are also bottom-dwelling. There are about 140 species, found in the colder waters of the N. Atlantic and the Antarctic.

Cyclorhagida (phylum *Aschelminthes, class *Kinorhyncha) An order of worms in which the head is covered on retraction by platelets on the second segment of the body.

Cyclorrapha (subclass *Pterygota, order *Diptera) Suborder of true flies which is normally divided into two sections, Aschiza and Schizophora, comprising insects whose antennae have less than six segments and have, or lack, a ptilinal suture (*see* PTILINUM). The antennae usually have three segments, with a terminal *arista. The wing venation includes a closed discal cell with the *cubitus closed or contracted. The larvae have a vestigial head, and pupate within a puparium. The suborder includes 57 families.

cyclosis A type of *cytoplasmic streaming in which the organelles circulate within the cell *cytoplasm (e.g. in the passage of *food vacuoles in *Ciliatea).

Cyclostomata 1. (subphylum *Ectopro-cta, class *Gymnolaemata) An order of bryozoans in which the colony is made up of thin-walled, finely perforate, calcareous tubes. The aperture of each individual is rounded, without any covering. Reproduction and larval development take place in a single, enlarged individual called an 'ovicell'. The order appeared in the *Ordovician and exists at the present day. **2.** The group that comprises the jawless fish, the most primitive of living vertebrates, characterized by the absence of jaws and paired fins, by a series of gill pouches rather than gill slits, by a cartilaginous skeleton without proper vertebrae, and by never having more than two *semicircular canals. The taxonomic status of the term 'Cyclostomata' is somewhat confused: at various times it has been ranked as an order, superorder, subclass, and class. The extant orders usually included are the lampreys (Petromyzoniformes) and the hagfish (Myxiniformes) but there is now considerable doubt whether they are closely related to each other; the hagfish may be a sister group to other vertebrates.

Cydippida *See* CYDIPPIDEA.

Cydippidea (Cydippida; phylum *Ctenophora, class *Tentaculata) An order of ctenophorids which possess a body that may be slightly compressed. There are two retractable *tentacles at the opposite end of the body to the mouth. Only one species adopts an endoparasitic (*see* PARASITISM) mode of life.

cydippid larva A free-swimming, larval stage of a ctenophorid (*Ctenophora) which resembles adults of the order *Cydippidea.

Cygnus (swans) *See* ANATIDAE.

Cygnus atratus (black swan) *See* ANATIDAE.

Cygnus olor (mute swan) *See* ANATIDAE.

Cylindrachetidae (order *Orthoptera, suborder *Caelifera) Family of medium-sized, wholly subterranean caeliferans which bear a superficial resemblance to *Gryllotalpidae (mole crickets). Wings are absent, and the general body and leg structure is highly modified for burrowing. Only seven species have been described. Five

occur in Australia, one in New Guinea, and one in Patagonia.

Cymbalophus *See* HYRACOTHERIUM.

Cynipidae (suborder *Apocrita, superfamily *Cynipoidea) Family of insects: it is divided into several subfamilies, but the great majority of its members belong to the subfamily *Cynipinae (gall wasps). Gall wasps cause galls on a variety of plants, particularly *Quercus* (oak). Each species forms a characteristic gall on its host plant, the larva or larvae developing inside. Galls may be *unilocular or *multilocular. The life cycle is often complex and *heterogynous. In the past, adults produced from the two different generations of the same species were described as different species owing to the apparent lack of similarity. Other members of the Cynipidae are dipteran parasites, hyperparasites of aphidiine *Braconidae, and *inquilines in the galls of other gall wasps. Adults are usually shiny, black to brown, with straight antennae, and a laterally compressed abdomen. They are 2–8 mm long.

Cynipinae (gall wasps; superfamily *Cynipoidea, family *Cynipidae) Subfamily of insects whose larvae induce gall formation on many different species of plant. Galls arising from cynipid attack have names such as oak apple, oak marble, spangle gall, artichoke gall, and Robin's pincushion. Plant galls and the way in which insects are able to produce species-specific and often highly complex structures are little understood.

Cynipoidea (order *Hymenoptera, suborder *Apocrita)** A superfamily of small or very small insects, in which the *gaster is laterally compressed and the wing venation distinctive. Some are yellow or reddish-brown, but most are dark in colour. The most important family is the *Cynipidae (gall wasps).

Cynocephalidae (flying lemurs, colugos; cohort *Unguiculata, order *Dermoptera) A family of nocturnal, herbivorous, arboreal animals whose limbs support a *patagium extending from the sides of the body to the paws and from the neck to the tip of the tail, by means of which they glide. The digits are flattened, and the soles of the paws form sucking discs for adhesion. The teeth are sharp, the lower *incisors *procumbent and each divided into many prongs, like a comb. The upper incisors are positioned at the sides of the jaw, leaving the front of the jaw toothless. The *canines resemble *premolars. There is one genus (*Cynocephalus*), and two species, distributed throughout southeast Asia.

Cynocephalus *See* CYNOCEPHALIDAE.

Cynodontidae (subclass *Actinopterygii, order *Cypriniformes) A small family of freshwater fish that have an elongate body, large, oblique mouth, and small *dorsal and *adipose fins. They are found in the rivers of S. America.

Cynoglossidae (tonguefish, tonguesole; subclass *Actinopterygii, order *Pleuronectiformes) A small family of marine, warm-water, left-eyed flatfish whose *dorsal and *anal fins are continuous with the pointed tail fin. Most species do not exceed 20 cm in length, but some are of commercial importance. There are about 85 species, found in tropical and subtropical waters.

Cynomys (prairie dog) *See* SCIURIDAE.

Cyprinidae (carp, minnow; superorder Ostariophysi, order *Cypriniformes) In terms of numbers of species, by far the largest family of fish. It comprises freshwater fish that typically have a fairly elongate body, a single *dorsal fin without true spines, a forked tail fin, ventral fins in the abdominal position, and no teeth on the jaws (although pharyngeal teeth may be present). Cyprinidae range in size from *Rasbora maculata* (dwarf rasbora), measuring 3 cm, to *Barbus tor* (mahseer), which reaches a length of 1.6 m. Many cyprinids are herbivorous, feeding on plants or algae. Many are also of considerable

commercial importance as food fish (e.g. carp) or hobby fish (e.g. goldfish). *Tinca tinca* (tench) and *Rutilus rutilus* (roach), of Europe, are popular with anglers. The culture of *Cyprinus carpio* (carp) in ponds was established in parts of Europe and Asia several centuries ago. Many species are very popular with aquarium hobbyists. There are about 1600 species, occurring worldwide, but absent from S. and Central America, Madagascar, Australia, New Guinea, and New Zealand.

Cypriniformes (subclass *Actinopterygii, superorder *Ostariophysi) An order of carp-like fish. Typical freshwater inhabitants, the fins lack proper spines, the single *dorsal fin is usually half-way down the back opposite the ventral fins in the abdominal position, and an *adipose fin is present in some families. The order includes some 26 families, e.g. the *Characidae (tetras), the *Cyprinidae (carps), and the *Catostomidae (suckers).

Cyprinodontidae (killifish, toothcarp; subclass *Actinopterygii, order *Atheriniformes) A large family of small fresh- to brackish-water fish, that have a single *dorsal fin opposite the *anal fin, the slender body ending in a truncate or rounded tail fin. There is no *lateral line. There are teeth in the jaws. Several species are known to enter brackish water or even sea water; these include *Fundulus heteroclitus* (killifish). There are about 300 species, found in most tropical freshwater systems of the world.

Cyprinus carpio (carp) *See* CYPRINIDAE.

cypris larva Non-feeding larval stage in barnacles (class *Cirripedia), which follows the six planktonic *nauplius*instars. The cypris larva is the settling stage, and attaches to the substrate by means of cement secreted by glands in the first antennae. The body is enveloped in a bivalved carapace and the pair of *compound eyes is sessile; there are six pairs of thoracic appendages.

cyrtoconic Applied to the shell of a cephalopod (*Cephalopoda) when it is a curved, tapering cone.

Cyrtocrinida (class *Crinoidea, subclass *Articulata) An order of crinoids in which the *calyx consists of large radial plates and a single dorsal plate which is probably a modified column. The column is short or absent. The arms are short and stout and may completely cover the *aboral surface of the animal when withdrawn. The order is restricted to hard substrata (e.g. hardgrounds and reefs), and is known from the Early *Jurassic to the present.

Cyrtophorina (subclass *Holotrichia, order *Gymnostomatida) A suborder of ciliate *Protozoa in which the *cytostome is located on the ventral side of the body. The organisms feed on diatoms, algae, and bacteria, which are ingested through the non-expansible cytostome.

cyst A closed sac that is separate from the surrounding tissue.

cysteine An aliphatic, polar, *alpha amino acid that contains a *sulphydryl group.

cysticercus The larval form of a tapeworm (*Taenia* spp.), consisting of a single *scolex enclosed in a bladder-like *cyst filled with fluid.

cystine An *amino-acid *dimer formed from two *cysteine residues linked by a *sulphydryl bond.

cytidine The *nucleoside formed when *cytosine is linked to *ribose sugar.

cytochrome One of a group of haemoproteins, which are classified into four groups designated a, b, c, and d. They function as electron carriers in a variety of *redox reactions in virtually all aerobic organisms.

cytochrome oxidase An *enzyme, containing copper and iron, that, because of its ability (unique among the *cytochromes) to reduce molecular oxygen to water, catalyses the terminal reaction in *oxidative phosphorylation.

cytogenetics The scientific discipline that combines *cytology with *genetics. This usually involves microscopic studies of *chromosomes.

cytokines Molecules of *proteins, *peptides, or *glycoproteins that are secreted by cells of the immune system and that carry signals between cells.

cytokinesis (cell division) During the division of a *cell into two daughter cells, the separation of the *cytoplasm as distinct from the division of the *nucleus.

cytology The study of the structure, function, and life history of the *cell.

cytomatrix (cytoplasmic matrix) A fluid component of the *cytoplasm in many eukaryotic cells (*see* EUKARYOTE), consisting of a solution of *amino acids, *glucose, carbon dioxide, and oxygen, that fills the interstices of the *cytoskeleton.

cyton (soma) The body of a *neuron.

cytopharynx In some protozoan cells, a short, passage-like region through which food passes after being ingested at the *cytostome.

cytoplasm The part of a cell that is enclosed by the *cell membrane, but excluding the *nucleus.

cytoplasmic inheritance Non-Mendelian (extra-chromosomal) inheritance via *genes contained in cytoplasmic *organelles (e.g. *mitochondria) and viruses.

cytoplasmic matrix *See* CYTOMATRIX.

cytoplasmic streaming The continuous, often rapid, movement of *cytoplasm within a cell. It is a process requiring expenditure of energy by the cell and is thought to involve *microfilament and *microtubular activity.

cytosine A *pyrimidine base that occurs in both DNA and *RNA.

cytosis The *evagination, *invagination, *budding, or fusion of a *cell membrane. *See* ENDOCYTOSIS; EXOCYTOSIS; PHAGOCYTOSIS.

cytoskeleton The network of filaments, made from *microtubules and larger than those of the *microtrabecular lattice, that fill many eukaryotic cells (*see* EUKARYOTE). The cytoskeleton helps maintain the shape of the cell and assists in the movement of *organelles during cell division.

cytosol (cell sap) The fluid that fills a cell and surrounds the *organelles. It contains soluble *enzymes, free *ribosomes, and other substances dissolved in water.

cytostome The fixed site in some protozoan cells (e.g. in *Ciliatea) at which food is ingested; the cell's 'mouth'.

dabbling ducks (*Anas*) See ANATIDAE.

dab lizards See AGAMIDAE.

***Dacelo novaeguineae* (kookaburra)** See ALCEDINIDAE.

Dactylochirotida (class *Holothuroidea*, subclass *Dendrochirotacea*) An order of holothuroids that have digitate or feebly branched *tentacles.

Dactylopius coccus See COCHINEAL.

dactylopodite The *distal segment of the walking limbs in *Decapoda, and of certain limbs in other arthropods.

Dactylopteridae (flying gurnard; subclass *Actinopterygii*, order *Dactylopteriformes*) A small family of marine, tropical to subtropical fish which have a somewhat bizarre profile. The very large *pectoral fins are wing-like, the fin tips reaching the tail fin when folded. Two free spines preceding the first *dorsal fin are well forward on the back. The *pelvic fins are used by the fish to 'walk' on the sea-bottom, and the pectoral 'wings' are used to glide over it. There are about four species, widely distributed in coastal areas. The fairly common *Dactylopterus volitans* grows to 35 cm.

Dactylopteriformes (class *Osteichthyes*, subclass *Actinopterygii*) An order of marine fish that includes only the family *Dactylopteridae.

***Dactylopterus volitans* (flying gurnard)** See DACTYLOPTERIDAE.

Dactyloscopidae (sand stargazer; subclass *Actinopterygii*, order *Perciformes*) A family of mainly marine, tropical fish that have a somewhat bulky head, a large and almost vertical mouth, eyes positioned on top of the head, and long *dorsal and *anal fins, the anal fin originating well forward under the spinous dorsal. The *lateral line is well developed. The fish tend to bury themselves in the sand with only part of the head free, awaiting the arrival of unsuspecting prey. There are about 20 Atlantic and Pacific species.

dactylozooid In some colonial coelenterates, a defensive or protective *polyp. The mouth, *tentacles, and *enteron are reduced or lost, so that the more extremely specialized dactylozooids have the form of a club bearing many *nematocysts.

daddy-long-legs See TIPULIDAE.

Dalradian The last, or youngest, stratigraphic unit of the *Precambrian of Scotland and Ireland.

***Dama* (fallow deer)** See CERVIDAE.

damselflies See AMPHIPTERYGIDAE; CALOPTERYGIDAE; CHLOROCYPHIDAE; COENAGRIONIDAE; DICTERIASTIDAE; EUPHAEIDAE; HEMIPHLEBIIDAE; LESTIDAE; MEGAPODAGRIONIDAE; ODONATA; PLATYCNEMIDIDAE; PLATYSTICTIDAE; POLYTHORIDAE; PROTONEURIDAE; PSEUDOLESTIDAE; PSEUDOSTIGMATIDAE; ZYGOPTERA.

Danainae (tiger butterflies, crow butterflies; order *Lepidoptera*, family *Nymphalidae*) Subfamily of medium to large butterflies, usually orange and marked with black and yellow. The larvae have at least two pairs of fleshy filaments dorsally. Both adults and larvae have bright warning colours, and are distasteful and usually poisonous. They are distributed widely, but are found mainly in tropical Africa and south-east Asia.

dance flies *See* EMPIDIDAE.

dance language (bee dance) The ritualized behaviour by which a returned worker honey-bee (*Apis* species) communicates the quality and source of food to other workers. There are two dances, both performed in the darkness of the nest or hive, on the vertical wax comb. The 'round' dance conveys the presence and quality of a food source close to the nest; no directional information is given. It comprises runs in a small circle, with regular changes of direction. The more frequent the change of direction, the greater the profitability of the food source (i.e. the greater its calorific value). Workers of foraging age pay particular attention to the pollen carried by the dancer and constantly *antennate her. Presumably they identify the source of the pollen by its characteristic scent and are then recruited to exploit the nearby source of food. The 'waggle' dance conveys a wider range of information, relating to the distance and direction of the food source, as well as to its quality. It refers to more distant sources of food than does the round dance, and often recruits workers to exploit flowers several kilometres from the nest. The waggle dance comprises runs in a flattened figure-of-eight pattern, with the waggle action performed in the straight run between the two rounds of the figure. The duration and vigour of the waggle are scaled to indicate distance, as is the loudness and duration of the buzzing associated with the waggle. Direction is indicated by the angle by which the straight run deviates from the vertical, the angle being equal to the angle between the direction of the food source and the sun as seen at that time from the nest entrance. Workers recruited by this dance will fly on a course that maintains this angle until they reach the food source, which they recognize by olfactory cues picked up earlier by antennation of the dancing worker in the hive. The dance language can be regarded as a ritualized enactment or 'charade' of the foraging flight made by the dancing worker. The language varies slightly between one geographic race of honey-bees and another,

and thus resembles bird-song and human languages in having local dialects.

Danian The oldest of the *Cenozoic stages, about 65.5–61.7 Ma ago, which in Denmark is characterized by chalky limestone rich in reef-dwelling organisms. It is a problematic stage, once referred to the *Cretaceous but now placed at the base of the *Palaeocene.

Daphaenosittidae (Neosittidae; sittellas; class *Aves, order *Passeriformes) A family of small, grey-brown birds that have black and white markings on their heads, white-tipped tails, pale rumps, and coloured wing patches. They have thin, slightly hooked bills, long wings, short tails, and short legs with long toes. They are found in forests where they climb up and down trees, feed on insects, and nest in trees. There are two genera, with three species, found in Australia and New Guinea.

Daphnia (water flea) *See* CLADOCERA; CYCLOMORPHOSIS.

Daption (cape petrel, cape pigeon) *See* PROCELLARIIDAE.

dark-green whip snake (*Coluber viridiflavus*) *See* COLUBRIDAE.

darkling beetle *See* TENEBRIONIDAE.

darters *See* ANHINGIDAE.

darwin A measure of evolutionary rate (introduced by J. B. S. Haldane in 1949), given in units of change per unit time.

Darwin, Charles Robert (1809–82) An English naturalist who is remembered mainly for his theory of evolution, which he based largely on observations made in 1832–6 during a voyage around the world on HMS *Beagle*, which was engaged on a mapping survey. In 1858, in collaboration with Alfred Russel Wallace (who had reached similar conclusions), he published through the Linnean Society a short paper, 'A Theory of Evolution by Natural Selection'; and in 1859 he published a longer account in his book, *On the Origin of Species by Means of*

Natural Selection. In this he presented powerful evidence suggesting that change (evolution) has occurred among species, and proposed natural selection as the mechanism by which it occurs. The theory may be summarized as follows: (*a*) The individuals of a species show variation. (*b*) On average, more offspring are produced than are needed to replace their parents. (*c*) Populations cannot expand indefinitely and, on average, population sizes remain stable. (*d*) Therefore there must be competition for survival. (*e*) Therefore the best-adapted variants (the fittest) survive. Since environmental conditions change over long periods of time, a process of natural selection occurs which favours the emergence of different variants and ultimately of new species (the 'origin of species'). This theory is known as Darwinism. The subsequent discovery of *chromosomes and *genes, and the development of the science of genetics, have led to a better understanding of the ways in which variation may be caused. Modified by this modern knowledge, Darwin's theory is called 'neo-Darwinism'.

Darwinian fitness *See* ADAPTIVE VALUE.

Darwinism The theory of *evolution by *natural selection, often used incorrectly as a synonym for the theory of evolution itself. The term 'neo-Darwinism' is often used to denote the 'new synthesis' (i.e. *synthetic theory).

Darwin's finches Fourteen species of Geospizinae that are endemic to the Galápagos Islands. There are only six species of all other passerine birds and one species of cuckoo on the islands. Thus it is inferred that an ancestor of the finches arrived on the islands before other birds and then underwent *adaptive radiation. Each species has evolved a distinctive beak type, and feeds on food not eaten by the other species.

( ) SEE WEB LINKS
- Description of Darwin's finches and their evolutionary significance.

Darwin's frog (*Rhinoderma darwinii*) *See* LEPTODACTYLIDAE.

Darwin's rhea (*Pterocnemia pennatus*) *See* RHEIDAE.

dassies *See* HYRACOIDEA.

Dasyatidae (stingrays; subclass *Elasmobranchii, order *Rajiformes) A family of rays of temperate to warm waters. Most species are marine but several are found in fresh water. Typically, they have one or two stout, serrated spines half-way down the back of the tail. Wounds inflicted by these spines can be very painful. *Dorsal fins are usually absent; the head profile is continuous with that of the *pectoral fins. Some species are found in trawl-nets at depths of more than 100 fathoms (183 m), others are estuarine carnivores (and unpopular with oyster farmers). One of the larger stingrays is *Dasyatis brevicaudata* (smooth stingray) of eastern Australia (3.5 m). The butterfly rays, which have a very short tail and a very wide body disc, are often included in this family. There are about 65 species, distributed world-wide.

Dasyatis brevicaudata (smooth stingray) *See* DASYATIDAE.

Dasymutilla occidentalis (cow killer, mule killer) *See* MUTILLIDAE.

Dasypeltis *See* EGG-EATING SNAKES.

Dasypodidae *See* DASYPODOIDEA.

Dasypodoidea (armadillos; suborder *Xenarthra, infra-order *Cingulata) A superfamily that comprises the single family Dasypodidae (armadillos), the oldest known of all S. American edentates and the only types known there from the *Palaeocene and early *Eocene. Armadillos are terrestrial, burrowing, omnivorous animals in which the head is flattened dorsoventrally, the tongue is elongated, and the ears are prominent. In different genera there are usually 14–50 *homodont teeth, but up to 90 have been recorded. There are seven cervical vertebrae, some fused together. There are five digits on the hind limbs and the fore limbs possess three to five claws used in digging. The tail is short, and usually encased in rings of bone. The upper and lower body surfaces are encased in transverse bands of bony *scutes beneath horny plates. There is flexible skin between the plates, permitting some species to roll into a ball. Hair is present on the lower surface and between plates. Armadillos are very successful animals, distributed widely through S. and Central America, and now moving into N. America. There are nine genera, with 21 species.

Dasyproctidae (agoutis, pacas; order *Rodentia, suborder *Hystricomorpha) A family of rodents whose claws (fore limb: four; hind limb: three) resemble hoofs. They are long-legged animals with a running gait. The tail is short or absent. Cheek teeth are *hypsodont. These rodents are distributed throughout the northern part of S. America and Central America. There are four genera, with 11 species.

Dasyuridae (antechinus, *dunnarts, *quolls, Tasmanian devils; order *Dasyuromorphia, formerly part of *Marsupialia) A family of mainly nocturnal, marsupial carnivores, formerly known as 'marsupial rats', 'marsupial mice', 'native cats', etc., that show convergence with placental carnivores, especially in their dentition. The *incisors are small and pointed, the *canines large, the cheek teeth modified for shearing (and resembling *carnassials). Dasyurids are distributed widely from New Guinea to Tasmania, although *Sarcophilus harrisi* (Tasmanian devil), a nocturnal predator and scavenger, is now confined to Tasmania (*see* DEVIL FACIAL TUMOUR DISEASE). There are about 20 genera, with at least 50 species; new species are still being discovered, and in 1975 a completely new genus, *Ningaui*, was discovered.

Dasyuroidea (infraclass *Metatheria, order *Marsupialia) A superfamily that comprises the families *Dasyuridae, *Thylacinidae, and *Myrmecobiidae, marsupials in which the *marsupium, if present, opens toward the rear. The second and third digits of the hind feet are *didactylous, the first is often vestigial or absent and if present is clawless and not opposable. There are three

or more pairs of *incisors of equal size in upper and lower jaws.

Dasyuromorphia In classifications that divide the *Marsupialia into a number of orders, one of those orders, containing only *Dasyuroidea.

Daubentoniidae (aye-ayes; suborder *Strepsirhini (or *Prosimii), infra-order *Lemuriformes (or *Chiromyiformes)) A family of nocturnal primates in which all the digits except the big toe have claws rather than nails, and in which the third digit of the hand is elongated. There is one pair of *incisors, which are enlarged, chisel-like, reminiscent of rodent incisors, and used for gnawing. The ears are large and movable. Aye-ayes are found only in Madagascar. There is one genus (*Daubentonia*), with two species.

daughter cells The two cells that result from the division of a single cell, usually by *mitosis. *See also* DAUGHTER NUCLEI.

daughter nuclei The two nuclei that result from the division of a single *nucleus, usually by *mitosis. *See also* DAUGHTER CELLS.

dealfish *See* TRACHIPTERIDAE.

deamination The removal of an amino group (NH$_2$) from an organic compound.

death The permanent cessation of living functions within an organ or organism.

death feigning *See* THANATOSIS.

death-watch beetle (*Xestobium rufovillosum*) *See* ANOBIIDAE.

Decapoda (crabs, crayfish, lobsters, shrimps; subphylum *Crustacea, class *Malacostraca) Order of crustaceans, most of which are marine, but some of which inhabit freshwater or terrestrial habitats, or are amphibious. The first three pairs of thoracic appendages are modified; and the remaining five pairs, from which the order derives its name (from the Greek *deka*, 'ten', and *pod-*, 'foot') are legs. The first pair of legs is usually large, heavy, and *chelate (the 'chelipeds'): these legs are used in prey capture, defence, and sexual signalling. The head and *thorax are fused dorsally, lateral expansions of the *carapace enclosing the gills in branchial chambers. Gills may be *dendritic, filamentous, or lamellar, and ventilating currents are passed over them by means of gill bailers. The blood contains *haemocyanin and flows through the lamellae of the gills. Land crabs have a reduced gill number, and so conserve water. Most decapods are predacious or scavenging (the prey including echinoderms, bivalves, polychaetes, and other crustaceans), but most freshwater and terrestrial decapods are herbivorous. There is considerable sexual *dimorphism and the complex courtship displays have visual, acoustic, and *pheromonal components. Fertilization is internal in the true crabs (infra-order Brachyura) but occurs at the moment of egg-laying in most others. Decapods are very diverse in form, habit, and modes of locomotion, and are very important both in the marine ecosystem and to humans, for food and by-products. The most successful members of the order are the true crabs, which have evolved a specialized, short body form, the abdomen fitting tightly under the *cephalothorax, which permits rapid locomotion (1.6 m/s in *Ocypode*, a genus of ghost crab). Decapoda is the largest crustacean order, comprising some 8500 species, or about one-third of all known crustacean species.

decarboxylase An *enzyme that facilitates the removal of a molecule of carbon dioxide from the carboxyl group of an organic compound.

decarboxylation A chemical reaction in which a carboxyl group (COOH) is separated from a compound as carbon dioxide (CO_2).

decay index In *cladistic analysis, the number of additional steps required to dissolve a given *clade.

decay theory A theory to explain why animals forget learned behaviour, which holds that memories fade (decay) unless they are continually refreshed. *Compare* INTERFERENCE THEORY.

deception The deliberate provision of false information in order to benefit the deceiver. For example, if a new male takes over a group of langurs (*Cercopithecidae), a female that is already pregnant may undergo a false oestrus (*see* OESTRUS CYCLE) and mate with the new male in order to persuade him to accept the infant as his own rather than killing it as the offspring of a rival. Many prey species seek to deceive predators, for example by *Batesian mimicry or a *broken wing display.

deciduous Applied to parts of an animal that are shed once (e.g. molar teeth in *Primates), seasonally (e.g. deer antlers), or readily (e.g. the scales of many fish).

decomposer A *heterotroph which feeds on *detritus, breaking it down into its constituent nutrients, some of which it utilizes and some of which it releases to be recycled in the *ecosystem.

decurved Curved or bent downwards, e.g. as in the bill of a bird.

deep-sea corals *See* AHERMATYPIC.

deep-sea devil *See* CERATIIDAE.

deep-sea smelts *See* BATHYLAGIDAE.

deep-sea trevalla (*Hyperoglyphe porosa*) *See* CENTROLOPHIDAE.

deer *See* CERVIDAE.

deerfly *See* TABANIDAE.

deer mouse (*Peromyscus leucopus*) *See* CRICETIDAE.

degenerate Applied to parts of the body, or stages in the life cycle of an organism, that have become greatly reduced in size, or have disappeared entirely, in the course of evolution.

degenerate code Sections (triplets) of the *genetic code that differ from one another but encode similar *amino acids.

degus *See* OCTODONTIDAE.

dehydrogenase An *enzyme which catalyses the removal of hydrogen from a *substrate.

deimatic behaviour Intimidating behaviour in animals which serves to warn off potential predators. It may be bluff (e.g. the inflation of the lungs in some toads, which increases the animal's size) or may precede an attack (e.g. when skunks rise on their fore legs prior to spraying an evil-smelling liquid).

Deinotheriidae (order *Proboscidea, suborder *Deinotherioidea) An extinct family of what are often called 'hoe-tuskers' because they had large, down-turned tusks on their lower jaws. These proboscideans lived from the *Miocene until the end of the *Pleistocene, when they disappeared in a wave of extinctions which affected the larger mammals of all continents.

Deinotherioidea (superorder *Paenungulata, order *Proboscidea) An extinct suborder (sometimes classed as a full order) of elephant-like animals which lived in the *Miocene and *Pliocene of Eurasia and survived, in tropical Africa, into the *Pleistocene. Early forms were smaller, but later forms larger, than modern elephants. The proboscis was long, upper tusks were absent, the lower tusks were large, curving downward and sometimes to the rear. The suborder is considered to be divergent from the evolutionary path leading to modern elephants.

deletion The loss of a chromosomal segment from a *chromosome set. The size of the deletion may vary from a single *nucleotide to sections containing several *genes. If the deletion is from the end of a chromosome, it is called 'terminal'; if it is from elsewhere, it is termed 'intercalary'.

Delichon (martins) *See* HIRUNDINIDAE.

Delphacidae (plant hoppers; order *Hemiptera, suborder *Homoptera) Family of homopterans in which the hind *tibia is triangular in section and bears a mobile spur at the apex. There are thousands of species, distributed throughout the world, some of which are serious pests of sugar-cane and rice.

Delphinidae (suborder *Odontoceti, superfamily *Delphinoidea) A family that comprises the dolphins, killer whales, and pilot whales; relatively small cetaceans (1.5–9.4 m), the killer whale (*Orcinus orca*) being the largest member of the family. The dorsal fin is prominent except in Lissodelphinae (the subfamily of southern-right dolphins), in which it is absent. The first two or four, or all, cervical vertebrae are fused. Delphinids are distributed throughout all oceans. *Tursiops truncatus* (the bottle-nosed dolphin) is distributed widely in the northern Atlantic, and in Britain is prone to stranding. *Globicephala melaena* (pilot whale) is also prone to stranding, sometimes in large schools. The family comprises about 14 genera, and about 32 species.

Delphinoidea (order *Cetacea, suborder *Odontoceti) A superfamily that comprises the families *Delphinidae, *Phocoenidae, *Stenidae, and *Monodontidae, a large group of small cetaceans with teeth in both jaws and usually a prominent dorsal fin. Most are active predators, but *Sousa* (the white or long-backed dolphin), of the coasts and rivers of the Old World tropics, is herbivorous.

Delphinus (dolphin) *See* DELPHINIDAE.

Deltatheridium A small, probably carnivorous mammal from the Late *Cretaceous that was once thought to be an insectivore, but the possession of certain dental characters indicates that it was a likely ancestor or sister group to the *Creodonta and *Carnivora.

delthyrium A subtriangular opening or slit-like notch beneath the apex of the pedicle valve of some *Brachiopoda, for the passage of the *pedicle.

deltidial plate A calcareous deposit at the side of the *delthyrium which serves to constrict or close the opening.

deltoid muscle The major shoulder muscle in *Tetrapoda; in humans this triangular (deltoid, i.e. delta-shaped) muscle gives the shoulder its rounded shape.

deltoid ridge *See* DELTOID TUBEROSITY.

deltoid tuberosity (deltoid ridge) A rough, triangular (i.e. delta-shaped) region, forming a ridge on the upper part of the *humerus to which the *deltoid muscle is attached.

deme A spatially discrete, interbreeding group of organisms with definable genetic or cytological characters (i.e. a subpopulation of a *species). There is very restricted genetic exchange, if any, with other demes, although demes are usually contiguous with one another, unlike *subspecies or *races, which are often isolated by some geographical or *habitat barrier. All possible male and female pairings within a deme have an equal chance of forming, for one breeding season at least. Populations which fulfil only one of the two key criteria (i.e. very occasional or no *cross-breeding and free pairing) are also referred to as demes by some authors.

demersal Applied to fish that live close to the sea floor, e.g. the cod (*Gadus morrhua*), hake (*Merluccius merluccius*), and saithe (*Pollachius virens*).

demibranch The V-shaped structure of gills that is common to *Bivalvia. It consists of two *lamellae containing gill filaments; inner and outer demibranchs are joined by the *ctenidial axis, giving the gill a typical W shape.

demography The statistical study of the size and structure (e.g. with regard to age or sex distribution) of populations and of changes within them. The word is derived from the Greek *demos*, 'people', and *graphe*, 'writing'.

Demospongea *See* DEMOSPONGIAE.

Demospongiae (Demospongea; phylum *Porifera) A class of sponges which occur in both marine and freshwater environments. They have a *leucon body structure. Both siliceous *spicules and organic fibres may form a skeleton, but primitive types may lack skeletal support. This is the largest group of sponges and it first appeared in the *Cambrian.

den Subapical portion of the springing organ (*furcula) of *Collembola. A pair of dens arises from the basal *manubrium and each den terminates in lamellate *mucrones.

denaturation Reversible or irreversible alterations in the biological activity of *proteins or *nucleic acids, brought about by changes in structure other than the breaking of the primary bonds between *amino acids or *nucleotides in the chain. This may be accomplished by changes of solvent, *pH, or temperature, or through the physical abuse of the molecules.

dendrite (dendron) A many-branched process, arising from the cell body of a *neuron, that receives impulses from other neurons.

dendritic Branching many times, like the branches of a tree.

dendritic cells In mammals, cells that form part of the immune system. They are present in small numbers in the linings of the nose, *lungs, *stomach, and *intestines, but mainly as a specialized type (Langerhans cells) in the skin. When activated they process *antigen material and migrate to the lymphatic system where they interact with *B cells and *T cells. Their name refers to the branched processes (dendrites) the cells grow during their development.

***Dendroaspis* (mambas)** *See* ELAPIDAE.

***Dendrobates* (arrow-poison frogs)** *See* RANIDAE.

Dendroceratida (class *Demospongiae, subclass *Ceractinomorpha) An order of sponges which lack *spicules. Some also lack skeletons composed of organic fibres. When fibres do occur they emanate from a basal sponge plate in a branching fashion. Generally they do not incorporate *detritus.

Dendrochirotacea (subphylum *Echinozoa, class *Holothuroidea) A subclass of sea cucumbers that are characterized by the ability to retract the anterior part of the body wall within the oral aperture by use of muscles connecting it to a ring of calcite plates surrounding the oesophagus. The fossil record dates back to the *Ordovician.

Dendrochirotida (class *Holothuroidea, subclass *Dendrochirotacea) An order of sea cucumbers in which the *tentacles are enlarged to form branching, tree-like structures, ciliated and adapted to collect food from the *plankton.

Dendrocolaptidae (woodcreepers; class *Aves, order *Passeriformes) A family of brown birds, often streaked, that have stiff, spiny tails. Their bills vary from short and upturned to stout and straight or very long and *decurved. They are found in forests, probing for insects on tree surfaces and nesting in tree cavities. The 14 species of *Xiphorhynchus* are typical. There are 13 genera, with 52 species found from southern Mexico to northern Argentina.

Dendrocopos See PICIDAE.

***Dendrocygna* (whistling ducks)** See ANATIDAE.

***Dendroica* (warblers)** See PARULIDAE.

***Dendroica kirtlandii* (Kirtland's warbler)** See PARULIDAE.

***Dendroica petrechia* (yellow warbler)** See PARULIDAE.

dendroid Tree-shaped (from the Greek *dendron*, 'tree').

dendroid colony 1. In corals, a colony formed by the irregular branching of *corallites. The individual corallites are somewhat separated from one another, but may be joined by connecting tubules. **2.** In *Graptolithina, a bushy colony for-med by irregular branching of the *stipes.

dendron See DENDRITE.

Denisovan A species of human (*Homo*) that lived approximately 41 000 years ago. It was first identified from a fragment of bone from the fifth (little) finger of a juvenile female, discovered in 2008 in the Denisova Cave, in the Altai Mountains, Siberia. A toe bone and tooth from other individuals were found in the cave later. *Mitochondrial-DNA from the finger bone was found to be distinct from that of Neanderthals (*H. neanderthalensis*) and modern humans (*H. sapiens*). Nuclear DNA revealed that the Denisovans shared an ancestor with Neanderthals, lived among and occasionally interbred with modern humans, and were once distributed from Siberia to south-east Asia.

density dependence The regulation of the size of a population by mechanisms that are themselves controlled by the size of that population (e.g. the availability of resources) and whose effectiveness increases as population size increases. See also S-SHAPED GROWTH CURVE.

density independence See J-SHAPED GROWTH CURVE.

dentary The anterior bone of the lower jaw which bears the teeth. In mammals it forms the whole of the lower jaw.

dentate Toothed or serrated.

Denticipitidae (denticle herring) See CLUPEIFORMES.

denticle In many fish, a scale composed of *dentine, with a *pulp cavity, which resembles a tooth. In *Elasmobranchii denticles cover the entire body.

denticle herring See CLUPEIFORMES.

denticulate Having very small teeth or serrations.

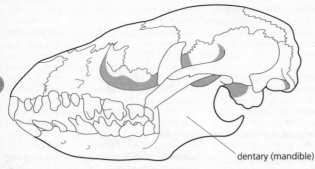

dentary (mandible)

Dentary

dentine A bone-like substance, lacking cell bodies and consisting mainly of calcium phosphate in a fibrous matrix.

deoxyribonucleic acid (DNA) A *nucleic acid, characterized by the presence of the sugar deoxyribose, the *pyrimidine bases cytosine and thymine, and the *purine bases adenine and guanine. It is the genetic material of organisms, its sequence of paired bases constituting the genetic code. *See also* WATSON–CRICK MODEL.

dependent rank In an animal society where rank is determined by alliances, the rank an individual possesses in a one-to-one encounter when some of its allies are present; its social rank then depends on the support of its allies. Dependent rank was first described in Japanese macaques (*Macaca fuscata*). *Compare* BASIC RANK.

depensation In some animal species (e.g. whales) an increase in *parasitism and *predation that occurs when the size of a population falls below a threshold. This further depletes the population, reducing the likelihood that it will recover.

deprivation studies A series of experiments conducted between 1957 and 1963 by Harry F. *Harlow who investigated the importance to infants of maternal bonding. Harlow separated macaque monkeys from their mothers 6–12 hours after birth and

raised them with two 'surrogate mothers', one made from wire mesh, the other covered in foam rubber and terry cloth; both were warmed by a light bulb. Only one 'mother' supplied milk (from a concealed feeding bottle). The infants spent more time with the cloth-covered 'mother' even when food was available only from the wire one. When a new object was introduced to the cage or an infant with its cloth 'mother' was taken outside, the monkey clung to the surrogate until it felt safe enough to explore. In later life the 'motherless' monkeys showed behavioural abnormalities and when they became mothers they were negligent or abusive.

Derichthyidae (long-neck eels; subclass *Actinopterygii, order *Anguilliformes) A small family of deep-water, tropical eels, 30 cm in length, that have a scaleless body and large eyes. There are two genera and three species, found in the Atlantic and Pacific.

derived (evol.) *See* APOMORPHHIC.

dermal Pertaining to the skin (dermis).

dermal denticles *See* PLACOID SCALE.

Dermaptera (earwigs; class *Insecta, subclass *Pterygota) Order of elongate, flattened, *exopterygote insects whose major diagnostic features include *cerci which are modified as a pair of forceps on the end of the mobile, telescopic abdomen; and fore wings which are reduced as short *tegmina, beneath which the large, membranaceous hind wings are intricately folded, although many species are wingless. Earwigs are interesting in that they show parental care of offspring. They are found in crevices, particularly among plant debris, are nocturnal, and feed on living and dead plant and animal material. Some species are pests, damaging flowers and fruit. About 1200 species have been described, most of them occurring in the tropics and warm-temperate regions.

Dermatemydidae (Tabasco turtles; order *Chelonia, suborder *Cryptodira) A monospecific family (*Dermatemys mawi*) of turtles in which the shell is smooth and fully ossified. Tabasco turtles are nocturnal, bottom-dwelling animals that seldom emerge from the water. Eggs are laid on the river bank. These turtles are herbivorous on water plants. The Tabasco turtle is found in coastal rivers of Mexico and Guatemala (Tabasco is a state of south-east Mexico).

Dermatemys mawi (Tabasco turtle) See DERMATEMYDIDAE.

dermatocranium See CRANIUM.

dermatome See SOMITE.

Dermestes lardarius (bacon beetle, larder beetle) See DERMESTIDAE.

Dermestidae (skin beetles, hide beetles, tallow beetles, museum beetles; class *Insecta, order *Coleoptera) Family of black or brown, small beetles, often with a pattern of fine hairs or scales, in which the *pronotum is strongly narrowed toward the head, and the antennae are clubbed. Dermestidae occur in flowers, feeding on pollen. The larvae are subcylindrical, and covered with long hair. *Anthrenus larvae* (woolly bears or carpet beetles) attack woollen carpets, furs, and dried museum specimens. *Dermestes lardarius* larvae (bacon beetles or larder beetles) attack furs, hides, and stored meat and cheese. There are 731 species, occurring world-wide.

dermis See INTEGUMENT.

Dermochelidae (leathery turtle, leatherback; order *Chelonia, suborder *Cryptodira) A monospecific family (*Dermochelys coriacea*) comprising the largest marine turtle (up to 600 kg and 2 m long). The shell is not fused to the ribs or vertebrae, being reduced to dermal ossicles in a thick, leathery skin. The *carapace is ridged, the flippers clawless. It occurs in all warm oceans.

Dermochelys coriacea See DERMO-CHELIDAE.

Dermoptera (colugo, flying lemur; infraclass *Eutheria, cohort *Unguiculata) An order which comprises the family *Cyno-cephalidae and a number of extinct forms. They are herbivores possessing a *patagium, and are the most highly developed of all mammalian gliding forms. There is controversy about their relationships—whether they are related to bats (perhaps especially to *Megachiroptera), or to *Primates, or are isolated among mammals. The fossil record for the group is poor, but what seem to have been early representatives are known from the late *Palaeocene and lower *Eocene of N. America.

Derwent whitebait (*Lovettia sealii*) See APLOCHITONIDAE.

desert locust (*Schistocerca gregaria*) See LOCUST.

desert mining bees See FIDELIIDAE.

desman See TALPIDAE.

desmodont Applied to a type of hinge condition found in certain *Bivalvia in which the teeth are very small or lacking, and ridges may have replaced them. The ligament may be supported by a *chondrophore.

Desmodontidae (vampire bats; order *Chiroptera, suborder *Microchiroptera) A family of bats (sometimes included in the *Phyllostomatidae) in which the *incisors are adapted for piercing and the

d

digestive system is modified for a diet consisting exclusively of blood. The nose has a fleshy pad, the ears are well separated, the tail is absent. The limbs are sufficiently strong to support the weight of the animal on the ground, enabling it to run and jump with agility. Vampire bats are distributed throughout tropical America and subtropical S. America. There are three genera (*Desmodus*, *Diaemus*, and *Diphylla*), with three species.

Desmodus *See* DESMODONTIDAE.

desmognathous In birds, applied to a structurally distinct form of the palate, in which the maxillopalatine bones are fused. It is found in *Anseriformes, *Pelecaniformes, *Psittaciformes, and some other groups.

desmosome An oval or circular junction between two cells where their respective *cell membranes run parallel to one another 30–50 nm apart. Immediately within each membrane there is a plaque of electron-dense material and behind this a less dense area of *microfilaments. The space between the membranes contains a *glycoprotein. The function of such a junction appears to be that of mechanical attachment.

Desmostylia An extinct order of subungulates known from *Miocene and *Pliocene rocks that resembled the hippopotamus in their general proportions and probably followed, like the hippopotamus, an amphibious way of life, though in shallow coastal waters. It seems that they shared a common descent with the *Sirenia and *Proboscidea, from early African subungulates.

Desmostylus hesperus A species of subungulate mammals from the mid- *Tertiary (upper *Miocene to lower *Pliocene) of western N. America and eastern Asia that lived in shallow coastal waters. It was an early sirenian-like animal, although the body was large (of hippopotamus size) with massive limbs. The skull was elongate, and there were 'tusks' on both upper and lower jaws.

detorsion The untwisting of the viscera during development in certain *Gastropoda. *Compare* TORSION.

detrital pathway (detritus food chain) A *food chain in which the living primary producers (green plants) are not consumed by grazing herbivores but eventually form *detritus on which *detritivores feed, with subsequent energy transfer to *carnivores (e.g. the pathway: leaf- litter→earthworm→ blackbird→sparrowhawk). Detritus from organisms at higher *trophic levels than green plants may also form the basis of a detrital pathway, but the key distinction between detrital and *grazing pathways lies in the fate of the primary producers.

detritivore (detritus feeder) *Heterotroph that feeds on dead material (*detritus). The detritus most typically is of plant origin, but may include the dead remains of small animals. Since this material may also be digested by decomposer organisms (fungi and bacteria) and forms the *habitat for other organisms (e.g. *nematode worms and small insects), these too will form part of the typical detritivore diet. Animals (e.g. hyena) that feed mainly on other dead animals, or that feed mainly on the products (e.g. exuviae, dung) of larger animals, are termed scavengers. *See also* FOOD CHAIN; *compare* CARNIVORE; HERBIVORE.

detritus Litter formed from fragments of dead material (e.g. leaf-litter, dung, moulted feathers, and corpses). In aquatic habitats, detritus provides habitats equivalent to those which occur in soil *humus.

detritus feeder *See* DETRITIVORE.

detritus food chain *See* DETRITAL PATHWAY.

Deuterostomia The name sometimes given to four animal phyla (*Chordata, *Hemichordata, *Echinodermata, and *Xenoturbellida) thought by many to share a common ancestry. The name indicates the mode of formation of the mouth opening in embryogeny: the first embryonic opening (the *blastopore) becomes the *anus, and the second opening becomes the mouth. *Compare* PROTOSTOMIA.

deuterostomy The condition in which the *blastopore forms the *anus of the adult animal. *Compare* PROTOSTOMY.

Devensian (Weichselian, Würm) The last glacial advance in N. Europe, approximately 70 000–10 000 years bp. It is approximately synchronous with the Wisconsinian glaciation in N. America.

devil facial tumour disease (DFTD) An aggressive form of cancer affecting the faces of Tasmanian devils (*Dasyuridae). The tumour begins close to the mouth, spreads across the muzzle, and eventually may cover the whole body. Growth of the tumour makes eating increasingly difficult and victims often die from starvation. The tumours are readily transmitted by living cancer cells; these are acquired when individuals bite each other's faces as they fight over food. The cancer is believed to have originated in Schwann cells (*see* MYELIN SHEATH). By 2010 approximately 80% of wild devils were infected and the population had decreased by 70% since 1996.

devilfish 1. *See* PLESIOPIDAE. 2. *See* OCTOPODA.

devil ray *See* MOBULIDAE.

devil's coach horse (*Staphylinus olens*) *See* STAPHYLINIDAE.

devil's darning needles *See* ODONATA.

Devonian The fourth of the six periods of the *Palaeozoic Era, about 416–359.2 Ma ago, and the first period of the Upper Palaeozoic Sub-Era. In Europe there are both marine and continental facies present, the latter being commonly known as the Old Red Sandstone. Although originally described from the type area in Devon, the marine Devonian is subdivided stratigraphically into stages established in the exceptionally fossiliferous deposits of the Ardennes in Belgium. These stages are the Gedinnian (416–411.2 Ma ago), Siegennian

(411.2–407 Ma ago), and Emsian (407–397.5 Ma ago) of the Early Devonian, the Eifelian (also called the Couvinian, 397.5–391.8 Ma ago) and *Givetian (391.8–385.3 Ma ago) of the Middle Devonian, and the *Frasnian (385.3–374.5 Ma ago) and Fammenian (370–359.2 Ma ago) of the Late Devonian. The subdivision of the marine deposits is based on lithologies and the presence of an abundant invertebrate fauna including goniatites (*Ammonoidea) and spiriferid brachiopods (Spiriferida). The continental Old Red Sandstone deposits contain a fauna of jawless fish and plants belonging to the primitive psilophyte group. As a result of the Caledonian orogeny (mountain-building episode) of late *Silurian times, much of the British Isles was covered with continental red-bed facies.

dewlap A fold of loose skin hanging below the throat.

dextrorse Growing or arising in a right-handed or clockwise spiral from the point of view of an observer.

dextrose *See* GLUCOSE.

D-fructose *See* FRUCTOSE.

DFTD *See* DEVIL FACIAL TUMOUR DISEASE.

dhole (*Cuon alpinus*) *See* CANIDAE.

diabetes (diabetes mellitus) A disease of mammals in which either the body fails to produce sufficient *insulin or cells respond inadequately to the insulin available. The effect is to raise blood-sugar levels, causing weight loss, increased appetite, increased thirst, and increased urination. Untreated, diabetes leads to complications arising from damage to blood vessels. Diabetes is most common in humans, other *Primates, and domesticated dogs and cats.

Diadematacea (class *Echinoidea, subclass *Euechinoidea) A superorder of echinoids that have perforate primary *tubercles on the rigid or flexible *test. The *perignathic girdle is complete, and the dentition *aulodont.

diadematoid In echinoids, applied to a compound ambulacral plate (*see* AMBULACRUM) that bears three pairs of pores. The middle, primary pore is the largest.

Diadematoida (subclass *Euechinoidea, superorder *Diadematacea) An order of regular echinoids in which the *ambulacra are composed of simple or compound *diadematoid plates. The gill slits conspicuously notch the *peristomal margin. Diadematoids first appeared in the Late *Triassic.

diadromous Applied to fish that regularly migrate between the sea and freshwater systems. *Compare* AMPHIDROMOUS. *See also* ANADROMOUS; CATADROMOUS; POTAMODROMOUS.

Diaemus *See* DESMODONTIDAE.

diakinesis *See* MEIOSIS.

dialect *Vocalizations among a *population of animals that differ from those of another population of the same species. There are many dialects in bird-song.

diallelic Applied to a *polyploid individual that has more than two sets of *chromosomes in which two different *alleles exist at a particular gene *locus.

dialysis The separation of dissolved crystalloids of low molecular weight from colloidal macromolecules of high molecular weight by means of a *semipermeable membrane that allows the passage of the former but not of the latter, separation being due to the difference in molecular weight of the substances.

diamond fish *See* MONODACTYLIDAE.

Dianthidium *See* MASON BEE.

diapause A temporary cessation that occurs in the growth and development of an insect or a mammal. Insects can enter the diapause state as eggs, *larvae, *pupae, or as an adult; mammals, only as *blastocysts. Diapause is frequently associated with seasonal environments, the insect entering it during the adverse period, and breaking from it when more favourable conditions return.

diapedesis The movement of blood cells through intact *capillary walls.

diaphragm In mammals, a transverse partition that separates the thoracic and abdominal cavities. When at rest the diaphragm is arched up into the thorax; its flattening increases the volume of the thorax, thus allowing the lungs to expand and so playing an important part in breathing. It is composed partly of muscle and partly of tendon, and is covered by a *serous membrane.

diaphysis The main part of a bone. In a long bone it is the shaft; in a vertebra the *centrum.

diapsid Applied to a type of skull that has two temporal openings behind the eye. This is typical of *Archosauria, *Lepidosauria, and *Rhynchocephalia. *See also* DIAPSIDA.

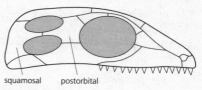

squamosal   postorbital

Diapsid

Diapsida The name formerly given to a subclass of reptiles whose skulls are of the *diapsid type. Today that classification has been abandoned as the more primitive and more advanced forms are not clearly related to one another and the diapsids are divided into two subclasses: the more primitive *Lepidosauria (the scaly reptiles), and the more advanced *Archosauria (ruling reptiles). The earliest known diapsids date from the end of the *Palaeozoic. They occur in southern Africa in Late *Permian and Early *Triassic rocks. Birds are closely related to the archosaurs, and although their skulls have only one temporal opening, this is apparently derived from the fusion of the two diapsid apertures.

Diarthrognathus broomi A species of mammal-like reptiles, recorded from the *Triassic, that is one of the closest to mammalian ancestry. It is assigned to the Ictidosauria and is characterized by a number of advanced cranial features, including the co-occurrence of the older reptilian (*quadrate-*articular) and newer mammalian (*squamosal-*dentary) jaw joints. The skull is only 4–5 cm long and its dentition is not as specialized as that of certain other mammal-like reptiles from the Late Triassic.

diarthrosis *See* SYNOVIAL JOINT.

diastema A naturally occurring (i.e. not resulting from the removal of a tooth) gap in the tooth row, most commonly between the *incisors and first *premolar.

diastole The period when the heart fills with blood, following *systole.

dibranchiate Applied to the condition (in *Mollusca) in which the ctenidia (*see* CTENIDIUM) used for respiration are paired.

Dicaeidae (flowerpeckers, pardalotes; class *Aves, order *Passeriformes) A family of small, often brightly coloured, short-tailed birds, whose bills vary from being thin and curved to stout, some being serrated. The birds are arboreal and often gregarious, feeding on insects, nectar, and fruit, and nesting in trees (flowerpeckers) or in holes in trees or banks (pardalotes). The 36 species of *Dicaeum* (flowerpeckers) build domed, pendant nests and many act as disseminators of seeds (e.g. *D. hirundinaceum* (mistletoe-bird) scatters mistletoe seeds). There are seven genera in the family, with 58 species, found from southern Asia to Australia.

Dicaeum (flowerpeckers) *See* DICAEIDAE.

Dicaeum hirundinaceum (mistletoe-bird) *See* DICAEIDAE.

dicentric Applied to a *chromosome or *chromatid that has two *centromeres.

Diceratiidae (subclass *Actinopterygii, order Lophiiformes) A small family of thickset, almost rotund, deep-sea fish that have a large mouth, short fins, small eyes, and a very long *dorsal-fin spine projecting over the mouth. There are about four species, occurring world-wide.

Dicerorhinus sumatrensis (Sumatran rhinoceros) *See* RHINOCEROTIDAE.

Diceros bicornis (black rhinoceros) *See* RHINOCEROTIDAE.

dice snake (*Natrix tessellata*) *See* COLUBRIDAE.

Dichobunidae A group, formerly classed as a family, of small, primitive, *Eocene and early *Oligocene artiodactyls. They show considerable variation, however, and more recently they have been regarded as a distinct artiodactyl group of subordinal status, the *Palaeodonta. They were neither true swine nor true ruminants.

dichthadiiform Of a hymenopteran insect, permanently wingless, with a greatly enlarged *gaster.

dicondylar With a double articulation.

Dicruridae (drongos; class *Aves, order *Passeriformes) A family of small to medium, glossy black birds that have stout, hooked bills and *rictal bristles, some being crested. Their tails are medium to long, often having distinctive curled or racquet-shaped feathers. They are aggressive and inhabit forests, flycatching for insects and building open or semi-pendant nests. There are two genera, with 20 species, 19 of them in the genus *Dicrurus*. (*D. hottentottus* (spangled drongo) has at least 31 subspecies and is one of the most variable birds.) The *monotypic genus *Chaetorhynchus papuensis* (Papuan mountain drongo) has 12, not 10, tail feathers, and is endemic to New Guinea.

Dicrurus (drongos) *See* DICRURIDAE.

Dicrurus hottentottus (spangled drongo) *See* DICRURIDAE.

Dicteriastidae (Heliocharitidae; damselflies; order *Odonata, suborder *Zygoptera) Family of damselflies in which there are many secondary *antenodals in the wing *venation, and the second and third branch of the radius arches towards the first soon after its origin. The family is neotropical and 12 extant species have been described.

Dictyoceratida (class *Demospongiae, subclass *Ceractinomorpha) An order of sponges which have no *spicules. The skeleton is composed entirely of organic fibres, which may enclose detritus. A tough dermis is present. The fibre structure is quite complex. The order includes the commercial sponges used by humans.

Dictyoptera (class *Insecta) Order of insects that includes the cockroaches and mantids. This classification is little used, the taxon having been split into the orders *Blattodea (cockroaches) and *Mantodea (mantids).

dictyosome A stack of flat, membranous *cisternae which, with the *vesicles, make up the *Golgi apparatus.

Dicyemida (Rhombozoa, lozenge animals) A phylum of protostomes (*Protostomia) that comprises three families, Conocyemidae, Dicyemidae, and Kantharellidae of small (0.7–7.0 mm when adult), worm-like parasites with elongated bodies that inhabit the kidneys of *Cephalopoda. Their life cycle is complicated.

Dicyemidae *See* DICYEMIDA.

didactylous In *Marsupialia, applied to the condition in which the second and third hind digits are not united within a common sheath.

Didelphidae *See* DIDELPHOIDEA.

Didelphimorphia *See* DIDELPHOIDEA.

Didelphoidea (infraclass *Metatheria, order *Marsupialia) A superfamily (or full order, Didelphimorphia) that comprises the opossums (family Didelphidae) and extinct relatives known since the Late *Cretaceous. Possibly they are the least modified of all therian mammals. They are arboreal (except *Chironectes*, an otter-like, semi-aquatic form), mainly nocturnal, insectivorous or omnivorous marsupials, in which the *marsupium may be well developed or (more usually) absent. There are more than three pairs of *incisors in both upper and lower jaws. The second and third digits of the hind feet are *didactylous. Didelphoids are distributed throughout tropical and subtropical America. There are twelve genera, and about 66 species.

Didolodontidae (superorder *Protoungulata, order *Condylarthra) An extinct family of ungulates which lived from the *Palaeocene to the *Miocene in S. America and which may have given rise to later ungulates characteristic of S. America.

diencephalon (thalamencephalon) The most anterior region of the *brain stem, part of the *prosencephalon. In *amniotes, the diencephalon forms a major relay between the sensory areas and higher brain centres via the *thalamus. Optic nerves develop from this region. The base of the diencephalon forms the *hypothalamus and a ventral outgrowth forms the *pituitary gland. The diencephalon is thought to have evolved from the median photoreceptor of *plesiomorphic vertebrates.

differentially permeable membrane A membrane that allows the passage of small molecules but not of large molecules.

differentiation The occurrence of changes in the structure and function of groups of cells due to increased specialization in an organism.

diffusion The movement of molecules from a region of higher to one of lower solute concentration as a result of their random movement. It is an important means of transport within cells.

Digenea (flukes; phylum *Platyhelminthes) A class, or in some classifications an order of the class *Trematoda, of parasitic worms all of which have a gut and an anterior (oral) sucker. Most have an intestine, a pharynx, and a ventral sucker. Mature individuals lack *cilia. In contrast to *Monogenea, Digenea have two to four hosts. The initial hosts are certain molluscs; the fluke then leaves and develops in a vertebrate host. All flukes are endoparasitic (*see* PARASITISM).

digger wasps *See* AMPULICIDAE; HUNTING WASP; SPHECIDAE; SPHECOIDEA.

digitalis An *alkaloid that is used for heart stimulation; it is derived from foxgloves (*Digitalis*). *See* CARDIAC GLYCOSIDE.

digitigrade Applied to a *gait in which only the digits make contact with the ground, the hind part of the foot remaining raised (e.g. in cats and dogs).

dihyodonty The condition in which an animal has two sets of teeth during its lifetime. Most *Mammalia are dihyodontic.

dimer A *protein that is made up of two *polypeptide chains or subunits paired together. If the subunits are identical in aminoacid sequence the protein is said to be homomeric; if they are different, it is heteromeric. Dimeric proteins may be detected by *electrophoresis. In monomeric *enzymes, the *isoenzyme pattern of the *heterozygote will represent a simple mixture of the two forms occurring by themselves in each of the corresponding *homozygotes. In dimeric enzymes, there are homomeric forms representing the two homozygotes, but the heterozygote occurs in heteromeric form: when stained after electrophoresis, this results in three bands (instead of two bands as with a monomer). An example of a dimeric enzyme is glucose–phosphate isomerase. Some enzymes are trimeric (comprising three components) or tetrameric (with four components).

Dimetrodon angelensis (subclass *Synapsida, order *Pelycosauria) A large, specialized, meat-eating reptile from the Early *Permian of N. America. *Dimetrodon* grew to over 3 m in length and is noted for the presence of a large sail on its back. The skull of this carnivore was high and narrow, with the teeth differentiated and well adapted to predatory habits.

dimidiate Divided into two.

dimorphism The presence of one or more morphological differences that divide a species into two groups. Many examples come from sexual differences of particular traits, such as body size (males are often larger than females), or plumage (male birds are usually more colourful than females). These result from sex-linkage of the genes coding for the particular trait. However, some dimorphisms, such as the colour-phases of some birds, may not be sex-linked.

dimyarian Applied to a condition in *Bivalvia in which two *adductor muscles are present, one anterior and one posterior.

Dinantian *See* CARBONIFEROUS.

dingo (*Canis dingo*) *See* CANIDAE.

Dinilysia A genus of snakes that is known from the Late *Cretaceous of Patagonia. Snakes are poorly known in strata of Cretaceous age; *Dinilysia* was a 2 m long booid (*Boidae).

Dinocerata (uintatheres; superorder *Paenungulata, order *Amblypoda) An extinct suborder of mainly N. American ungulates of the late *Palaeocene to *Eocene. *Uintatherium* was a grotesque animal resembling a rhinoceros in size and general appearance, ponderously built, with pairs of horn-like bony swellings on the upper side of the nasal, maxillary, and parietal bones of the skull.

dinoflagellate *See* DINOFLAGELLIDA.

Dinoflagellida (Dinophyceae, Pyrrophyta, dinoflagellates; subphylum *Sarcomastigophora, superclass *Mastigophora) An order of protozoons that are *heterotrophs but closely allied to brown algae and diatoms (they are sometimes classified as algae). Many have brown or yellow chromoplasts containing *xanthophyll and chlorophylls a and c; others are colourless. Typically, dinoflagellates have two flagella (*see* FLAGELLUM), one, propelling water to the rear and providing forward motion, attached just behind the centre of the body and directed posteriorly, the other, causing the body to rotate and move forwards, forming a transverse ring or spiral of several turns around the centre of the body. Some dinoflagellates are naked, others are covered with a membrane or plates of cellulose. Many species are capable of emitting light, and these are the main contributors to bioluminescence in the sea. Most are planktonic, some in freshwater but most in marine environments, and some live in *symbiosis with animals (e.g. the flatworm *Amphiscolops*, sea anemones, and corals) with which they exchange nutrients. Some are colonial. There are many species.

Dinogamasus *See* ACARINUM.

Dinomyidae (pacarana; order *Rodentia, suborder *Hystricomorpha) A monospecific family (*Dinomys branickii*) of ground-dwelling rodents marked by white stripes along the back and spots on the flanks. They are confined to the western part of tropical S. America.

Dinomys branickii (pacarana) *See* DINOMYIDAE.

Dinophyceae *See* DINOFLAGELLIDA.

Dinornithidae (moas; class *Aves, order Dinornithiformes) An extinct family of large, emu-like *ratites, the largest being 3 m tall. They had no wings, and were herbivorous and ground-nesting. They became extinct during the last few hundred years, some possibly surviving into the 19th century. There were at least six genera, with 19 species, confined to New Zealand.

Dinosauria Literally, 'terrible lizards' (from the Greek *deinos*, 'terrible', and *sauros*, 'lizard'), but in fact the dinosaurs were not

lizards. They were *diapsid reptiles whose closest living relatives are the crocodilians and birds. Dinosaurs first appeared in the *Triassic, became the dominant group of land animals by the Middle *Jurassic, and produced an astonishing array of different types and sizes before becoming extinct at the end of the *Cretaceous. The two groups of dinosaurs, *Saurischia and *Ornithischia, are not usually thought to be more closely related to each other than to other archosaurs (*Archosauria), so the concept of 'dinosaur' is a heterogeneous one.

Diodontidae (porcupine fish; subclass *Actinopterygii, order *Tetraodontiformes) A small family of marine, warm- to temperate-water fish in which the short, rotund body is covered by numerous hard spines. The teeth are fused in each jaw, resembling the beak of a parrot. Porcupine fish are well known for their peculiar habit of distending themselves with water until the body assumes a nearly spherical shape. The inflated, dried skin has been used to make lampshades. There are at least 15 species, inhabiting the coastal regions of the oceans.

dioecious Possessing male and female reproductive organs in separate, unisexual, individual animals. *Compare* MONOECIOUS.

dioestrus The period between two *oestrus cycles in a female mammal.

Diomedea exulans (wandering albatross) *See* DIOMEDEIDAE.

Diomedeidae (albatrosses; class *Aves, order *Procellariiformes) A family of large sea-birds that have long, narrow wings, a large head, and a long, strong bill with a hooked tip and tubular nostrils. They have short legs, webbed feet, and a reduced or absent *hallux. They are *pelagic, often gregarious, and have a powerful, gliding flight. They feed on fish, squid, and other marine animals, coming to land only to nest. *Diomedea exulans* (wandering albatross) has the largest wing-span of any bird, measuring 3.5 m. There are two genera, with 13 species, found in southern oceans and in the N. Pacific, occasionally wandering into other regions.

Diopsidae (stalk-eyed flies; order *Diptera, suborder *Cyclorrapha) Small family of flies whose eyes are borne on stalks. The head is usually sparsely haired, with *vibrissae absent. The anterior femora (*see* FEMUR) are stout, with ventral spines. The larvae are saprophagic or phytophagous. Adult males have lost *tergites seven and eight, and the seventh *sternite forms a complete ventral band. There are some 150 described species, found mainly in Africa and Asia, but also in New Guinea and N. America.

dipeptide A *protein comprising two *amino acid molecules linked by a single *peptide bond.

diphosphopyridine nucleotide (DPN) *See* NICOTINAMIDE ADENINE DINUCLEOTIDE.

diphycercal tail Possibly the original type of tail fin in fish, in which the body axis divides the fin into equal dorsal and ventral sections.

Diphylla *See* DESMODONTIDAE.

Diphyllobothridea (phylum *Platyhelminthes, class *Cestoda) An order, or in some classifications family, of parasitic worms which possess a simple *scolex. Most inhabit three hosts during their development.

diphyodont Applied to vertebrates in which a set of deciduous teeth (milk teeth) is shed and replaced by a second set of (permanent) teeth. *Compare* MONOPHYODONT, POLYPHYODONT.

dipleurula larva A type of echinoderm (*Echinodermata) larva that incorporates features common to all types of echinoderm larvae and is hypothetically the ancestral type. It has *bilateral symmetry, three coelomic sacs (*see* COELOM), longitudinal ciliated bands for locomotion, and *adoral ciliated bands for feeding. Some authorities claim similarities between this and the *tornaria larva indicate affinities between Echinodermata and *Chordata.

diploblastic Applied to animals whose body is derived from only two embryonic cell layers (*endoderm and *ectoderm) separated by a gelatinous *mesogloea.

diploid Applied to a *cell that has two *chromosome sets, or an individual with two chromosome sets in each cell (excluding the *sex chromosomes which may or may not be represented twice, according to the sex of the individual). A diploid state is written as 2n to distinguish it from the haploid state of n. Almost all animal cells (except *gametes) are diploid.

diploidy The diploid condition.

Diplomonadida (superclass *Mastigophora, class *Zoomastigophorea) An order of *Protozoa which are bilaterally symmetrical (*see* BILATERAL SYMMETRY) and have 2–8 *flagella. Most species are parasitic in vertebrates and some can cause disease.

Diplopoda (millipedes; phylum *Arthropoda, subphylum *Atelocerata) A class of arthropods in which the body is segmented, each segment of the trunk being formed from two fused somites. Each resulting diplosegment thus bears two pairs of legs. Millipedes do not move rapidly but are able to force their way through soil and similar loose material, and many are able to climb smooth surfaces. Many roll into a ball when threatened. Most millipedes are herbivorous, although some ingest soil from which they digest organic matter and a few species are carnivorous. More than 7500 species have been described. They occur in all continents but are most common in the tropics.

Diplostraca (clam shrimps, water fleas; subphylum *Crustacea, class *Branchiopoda) Order of branchiopods in which the body is laterally compressed and at least partially enclosed in a bivalved *carapace. The order comprises two suborders: *Conchostraca (clam shrimps) and *Cladocera (water fleas). There are 605 species.

diplotene *See* PROPHASE.

Diplozonina (starfish; subclass *Asteroidea, order *Paxillosida) A suborder of starfish in which the marginal frame consists of an upper and a lower pair of *ossicles. The *aboral skeleton consists of *paxillae set in a flexible membrane. The *tube feet are pointed, and without terminal discs. The suborder first appeared in the *Jurassic and persists to the present day.

Diplura (subphylum *Hexapoda, class *Insecta) A subclass of small (mostly 6 mm long or smaller), elongate, slender, whitish, blind, *entognathous insects. They have long, *moniliform antennae, legs that are five-segmented, and *abdomens that are ten-segmented and bear rudimentary limbs. There are two main families: the vegetarian Campodeidae, in which the abdomen bears long, moniliform *cerci; and the carnivorous Japygidae, in which the cerci are short and forcep-like. Post-embryonic development is *epimorphic. The Diplura form a cosmopolitan order of some 659 species, found world-wide, whose members are the most insect-like of the entognathous classes and live in damp soil, under stones or logs.

Dipneusti (Dipnoi; class *Osteichthyes) Often ranked as a subclass, the group that includes the extant lungfish and their fossil relatives. The lungfish are an order of bony fish (Osteichthyes) in the subclass *Choanichthyes (Sarcopterygii), or fleshy-finned fish. They first appear in early *Devonian rocks and were common in freshwater habitats in the late *Palaeozoic and *Triassic. Thereafter their fossil remains are very sparse and now there are just three surviving types, all in the tropics. *See also* CERATODIFORMES.

Dipnoi *See* DIPNEUSTI.

Dipodidae (jerboas; order *Rodentia, suborder *Myomorpha) A family of nocturnal or *crepuscular *saltatory rodents in which the hind limbs are much elongated, the tail is long with a tuft at the tip, and in most species the eyes and ears are large. Apart from the specialization of their limbs, dipodids are rather primitive. They are distributed throughout arid and desert regions of the Old World. There are 10 genera, with about 27 species.

dipole Applied to a molecule that has an uneven charge distribution, one pole having a net negative charge, the other a net positive charge.

dipole moment *See* POLAR MOLECULE.

dippers *See* CINCLIDAE.

diprotodont In *Marsupialia, applied to the condition in which the first *incisor teeth are enlarged.

Diprotodontia (subclass *Theria, infraclass *Metatheria) An order (or suborder) of marsupials, restricted to Australia and New Guinea, that are characterized by a pair of enlarged, *procumbent, lower *incisors. The order includes three superfamilies: *Macropodoidea; *Phalangeroidea; and *Vombatoidea.

Diprotodontidae (order *Marsupialia (or *Diprotodontia), superfamily *Vombatoidea) A family of extinct marsupials, most of which were of giant size. The largest, the rhinoceros-sized *Diprotodon*, survived until the late *Pleistocene.

Diptera (two-winged flies, true flies; class *Insecta, subclass *Pterygota) Order of insects in which the adults have a single pair of membranous wings, the hind wings having been modified into *halteres. The mouth-parts are generally adapted for sucking, and modified into a proboscis, often adapted for piercing. *Mandibles are absent in many families. *Metamorphosis is complete. Larvae lack true legs and are *eruciform, with up to 12 abdominal segments. Larval habits range from phytophagy to parasitism. The number of larval *instars ranges from three (*Cyclorrapha) to eight (*Brachycera). The order is one of the largest within the class, having more than 85 000 described species. Most authors recognize three suborders: *Nematocera (21 families), Brachycera (15 families), and Cyclorrapha (57 families).

direct flight muscle Muscle which attaches directly to the wing of an insect. In most insects flight is powered by indirect flight muscles, while trimming of the wing movement for steering and other flight adjustments is brought about by the direct flight muscles. Dragonflies are unusual in using the direct flight muscles to power flight.

directional evolution *See* ARISTO-GENESIS.

directional selection Selection that changes the frequency of an *allele in a particular, constant direction, either towards or away from fixation of that allele. It is often used in agriculture and horticulture to produce a shift in the population mean of a trait desired by humans. For example, the breeder might select for cows that yield more milk. *Compare* DISRUPTIVE SELECTION; STABILIZING SELECTION.

disaccharidase One of several *enzymes that break down *disaccharides into *monosaccharides.

disaccharide A *sugar composed by the joining of two *monosaccharides with the loss of one molecule of water. *Sucrose and *lactose are disaccharides.

disassortative mating Mating between individuals of unlike phenotype. *See also* ASSORTATIVE MATING.

Discinacea (order *Acrotretida, suborder *Acrotretidina) A superfamily of inarticulate brachiopods, in which growth of the *pedicle valve occurs all around the edge and the valve is usually conical to subconical. The phosphatic shell is homogeneous, finely punctate, and not in alternating organic and phosphate layers. All recent species are shallow-water forms, except one which is abyssal in habitat. The Discinacea first appeared in the *Ordovician.

Discoglossidae (class *Amphibia, order *Anura) A family of toads which have disc-like tongues. The lower jaws are toothless, there are eight *opisthocoelous vertebrae, the *sternum is three-pronged, and the ribs are retained at *metamorphosis. There are 15 species, in Europe and Asia. *Alytes obstetricans* (midwife toad) of western and central Europe lives in hilly terrain, hiding in crevices and under stones. It has a squat grey body with a whitish belly, is up to 5.7 cm long, has no vocal sacs, and the male winds a band of eggs round his back legs while mating and carries them until hatching occurs. *Bombina bombina* (fire-bellied toad) of central and eastern Europe is essentially aquatic, and grey with a red, speckled belly; the male has internal vocal sacs which inflate the throat, and *nuptial pads which develop on the lower arm and first two digits.

Discomedusae *See* SEMAEOSTOMATIDA.

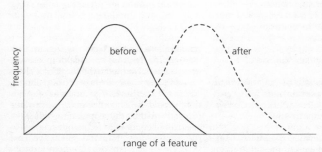

Directional selection

discrimination Differential responsiveness to stimuli when several stimuli are presented simultaneously.

disc-winged bats *See* THYROPTERIDAE.

disjunction The separation of *homologous chromosomes at the *anaphase stage of *mitosis and *meiosis, and movement towards the poles of the nuclear *spindle.

dispersal The tendency of an organism to move away, either from its birth site (natal dispersal) or breeding site (breeding dispersal): the opposite of *philopatry. Rates of regional dispersal depend on the interaction of several factors, notably the size and shape of the source area, the dispersal ability of the organisms, and the influence of such other environmental factors as winds or ocean currents. Dispersal may be passive (e.g. of ballooning spiderlings), active (e.g. of many mammals), passive but involving an active agent (e.g. seeds carried on the coats of mammals), or clonal; in practice these categories are difficult to define precisely. Mathematical modelling using these factors has practical applications in the design of nature reserves, and provides an insight into the present distribution of organisms.

dispersal barrier (ecological barrier) An area of unfavourable *habitat that separates two areas of favourable habitat, e.g. oceans in the case of terrestrial organisms, or a cereal monoculture in the case of woodland organisms.

dispersal biogeography The name now given to the traditional school of *biogeography in which organisms were regarded as having arisen in centres of origin and to have spread out by stages, utilizing *corridors, *sweepstakes routes, etc.

dispersal mechanism Characteristic adaptation for *dispersal which forms part of the reproductive strategy of many slow-moving or *sessile organisms.

dispersion In statistics, the internal pattern of a population, i.e. its distribution about the mean value. In spatial statistics,

the pattern relative to some specific location, or of individuals relative to one another, e.g. clumped or random.

((⊕)) SEE WEB LINKS
• Information about dispersion.

dispersion coefficient A measure of the spread of data about the mean value, or with reference to some other theoretically important threshold or spatial location, e.g. the standard deviation. *See also* DISPERSION.

Dispholidus typus (boomslang) *See* COLUBRIDAE.

displacement activity Behaviour of an animal that appears to be irrelevant to the situation in which it occurs and that may interrupt other activity. It may reduce *conflict; or it may arise because the animal is prevented from attaining a goal, and the consequent *frustration causes the attention to be switched to another stimulus to which it then responds.

display Stereotyped behaviour, involved in communication, that is largely acquired genetically. It may be associated with *courtship, in which physical characteristics (plumage, antlers, etc.) are exhibited by an animal as a means of attracting and securing the co-operation of a sexual partner, *deimatic, a threat (e.g. in a bird establishing a *territory), or cryptic (i.e. making the animal more difficult to see).

disruptive coloration In an animal, a colour pattern that is thought to disrupt the perceived contour of the body or parts of the body, thereby making the animal more difficult to see.

disruptive selection Selection that changes the frequency of *alleles in a divergent manner, leading to the fixation of alternative alleles in members of the population. The result after several generations of selection should be two divergent phenotypic extremes within the population; this process has been thought to provide a possible mechanism for sympatric speciation (*see* SYMPATRIC EVOLUTION). *Compare* DIRECTIONAL SELECTION; STABLIZING SELECTION.

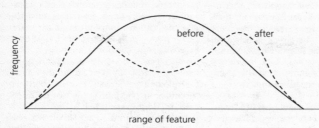

Disruptive selection

dissepiment 1. In corals, small, horizontal, domed plates which form cyst-like enclosures around the edge of a *corallite and which do not extend right across the corallite. **2.** In graptolites, a strand of chitinous material which connects adjacent branches in a *dendroid colony.

distal Applied to that part of an organ or structure which is furthest from its point of attachment to the body, or from the centre of the body. *Compare* PROXIMAL.

Distichodontidae (subclass *Actinopterygii, order *Cypriniformes) A family of African freshwater fish that have a deep body, a rather produced, 'pointy' snout, a soft-rayed *dorsal and an *adipose fin, and a forked tail fin. A few genera (e.g. *Citharinus*) are also placed in a separate family, the *Citharinidae.

distichous In two ranks.

disulphide bridge A covalent bond formed between two sulphur atoms. It is a particular feature of *peptides and *proteins, where it is formed between the *sulphydryl groups of two *cysteine residues, helping to stabilize the tertiary structure of these compounds.

Ditrysia (subclass *Pterygota, order *Lepidoptera) The suborder to which 95% of all moths and butterflies belong. The female bears an egg pore on abdominal segment 9/10, and a separate copulatory opening on segment 8 (in contrast with the *monotrysian condition). The venation of the fore wing differs from that of the hind wing.

diurnal 1. During daytime (as opposed to nocturnal), as applied to events that occur only during daylight hours, or to species that are active only in daylight **2.** At daily intervals, as applied to such daily rhythms as the normal pattern of waking and sleeping, or the characteristic rise and fall of temperature associated with the hours of light and darkness. *See also* CIRCADIAN RHYTHM.

divergence Genetic segregation and differentiation within a taxon to the extent that distinct derivative taxa result. The divergence may be at the species, genus, family, order, or higher level. Thus it is possible to refer for instance to the divergence of reptiles and mammals from a stem group, to the divergence of mammal orders, and to the divergence of a breeding population into two related species.

divergent evolution The situation in which descendants of an ancestral group of organisms split into two or more groups that become increasingly different as time passes. Genetic separation and differentiation occur to such an extent that distinct derivative taxa (*see* TAXON) may result. Divergence may be at the *species, genus, family, order, or higher level.

diversification Increase in the diversity of distinct types in one or more taxonomic categories, i.e. species, genus, etc.

*Phanerozoic, well-skeletonized marine invertebrates provide an illustration: their diversity at phylum level remains much the same throughout, whereas at family level there is a peak at the mid-*Palaeozoic, and a trough at the Permo-*Triassic boundary; after this there is a steady increase to a second, higher peak in the *Cenozoic.

diversity The species-richness of a *community or area. The concept provides a more useful measure of community characteristics when it is combined with an assessment of the relative abundance of species present.

diversity index A simultaneous mathematical expression of the number of species and number of individuals of each species that are present in a given *habitat. Various formulae are used, one of the most common being the *Shannon–Wiener index. A different and specialized case of a diversity index is the 'biotic index' used in water-pollution studies.

diversivore See OMNIVORE.

diverticulum A tube, closed at the end, that forms a side-branch or recess leading from a body cavity.

diving beetle See DYTISCIDAE.

diving petrels See PELECANOIDIDAE.

division furrow See CLEAVAGE FURROW.

divisive method A system of *hierarchical classification that proceeds by subdividing the whole into successively smaller and more homogeneous units.

dizygotic twins Fraternal twins produced by two separate ova that have been fertilized simultaneously by separate sperms. Such twins are genetically no more alike than *siblings. Compare MONOZYGOTIC TWINS.

DNA See DEOXYRIBONUCLEIC ACID.

DNA–DNA hybridization A technique that is used to compare DNA from two different species, to locate or identify *nucleotide sequences, and to establish the effective *in vitro transfer of nuclear material to a new host. A single strand of DNA from one source is bound to a special filter to which is added a single strand of radioactively labelled DNA from a different source. Complementary *base pairing between *homologous sections of the two DNAs results in double-stranded hybrid sections that remain bound to the filter, whereas single-strand sections are washed away. The amount of radioactivity remaining on the filter, compared with the amount washed away, gives a measure of the number of nucleotide sequences that the radioactive DNA and the original DNA share in common.

DNA fingerprinting A technique that is used to compare *DNA from different sources or to identify a *genome by comparison with a known standard. *Mini-satellite DNA is partially broken down by *restriction enzymes and the sections are separated by *electrophoresis. The fragments are then transferred to a nitrocellulose filter by the *Southern blotting technique and the filters washed with a *probe consisting of radioactive DNA. The probe becomes attached to a piece of mini-satellite DNA and shows up as a dark band on an *autoradiograph of the nitrocellulose filter. The resulting patterns are unique to each individual.

Dobson flies See CORYDALIDAE.

Docodonta (class *Mammalia, subclass *Prototheria) An extinct order of primitive mammals, known only from teeth found in Late *Jurassic rocks. Their brains may have been larger than those of reptiles; probably their young were fed with milk; possibly they were *homoiotherms; and perhaps they were insectivorous, arboreal, and nocturnal. They were about the size of, and must have resembled, shrews.

doctorfish (surgeonfish) See ACANTHURIDAE.

dodos See RAPHIDAE.

dog (Canis familiaris) See CANIDAE.

dog-bears *See* AMPHICYONIDAE.

dogfish (*Mustelus mustelus*) *See* CARCHARHINIDAE; SQUALIDAE.

dog louse (*Trichodectes canis*) *See* TRICHODECTIDAE.

Dolichoderinae (suborder *Apocrita, family *Formicidae) Subfamily of ants which have a single, flattened scale on the *petiole, often overhung by the first gastral segment. Venom is extruded through a ventral, transverse slit at the tip of the *gaster. There are about 500 species, with a worldwide distribution.

Dolichopodidae (long-legged flies; order *Diptera, suborder *Brachycera) Large family of flies which are generally small, bristly, and metallic green or blue-green. Their antennae have a dorsal or terminal *arista, and they have a short, fleshy proboscis. The *tarsus has two *pulvilli, and a straight or poorly lobiform *empodium. Males have secondary sexual features that may affect almost any part of the body. As adults, the flies are thought to be predacious on tiny, soft-bodied insects, which are chewed before being sucked dry. The flies are also found on flowers, feeding on nectar and other plant juices. Most larvae are carnivorous within their particular habitat, which may be aquatic, or terrestrial in soil or leaf-mould. There are more than 2000 described species.

Dolicochoerus loudonensis A middle *Oligocene peccary known from Europe. Living relatives include *Tayassu* from the New World.

doliform Barrel-shaped or jar-shaped.

doliolaria In *Holothuroidea, a larval form in which the ciliated band forms rings around the body. It develops from the *auricularia larva.

Dollo's law Evolutionary irreversibility: once regarded as inevitable, but now considered to apply mainly in special cases. The potential for further useful mutation may well be very limited in highly specialized organisms, since only those mutations that will allow the organism to continue in its narrow niche will normally be functionally possible. In such cases there is therefore a self-perpetuating, almost irreversible, evolutionary trend, so much so that it is regarded virtually as a law, called 'Dollo's law' after the palaeontologist Louis Dollo (1857–1931). The trend results from steady directional selective pressure, or orthoselection reinforced by specialization, or developmental *canalization.

(((∰))) SEE WEB LINKS
• Biography of Louis Dollo, of Dollo's law.

dolphin (*Delphinus*) *See* DELPHINIDAE.

dolphinfish *See* CORYPHAENIDAE.

domain The highest taxonomic category in a classification system based on comparisons of ribosomal RNA. There are three domains: *Archaea; *Eubacteria; and *Eukarya. Each domain comprises organisms that are not closely related genetically to members of either of the other domains. *See also* KINGDOM.

domestication The selective breeding of species by humans in order to accommodate human needs. Domestication also requires considerable modification of natural *ecosystems to ensure the survival of, and optimum production from, the domesticated species.

domestic horse (*Equus caballus*) *See* EQUIDAE.

dominance 1. The possession of high social status within an animal group that exhibits social organization; it is often achieved and sustained by *aggression toward inferior individuals. *See* PECKING ORDER. 2. *See* DOMINANT GENE.

dominant gene In *diploid organisms, a *gene that produces the same phenotypic character when its *alleles are present in a single dose (heterozygous) per nucleus as it does in double dose (homozygous). For example, if A is dominant over a, then AA (the homozygote) and Aa (the heterozygote) have the same *phenotype. A gene that is masked in the presence of its dominant allele in the heterozygote state is said to be recessive to that dominant. Such a dominant–recessive relationship is common between two alleles, the gene most frequently present at a given *locus being usually dominant to its alleles.

donor See GRAFT.

Doradidae (thorny catfish; subclass *Actinopterygii, order *Siluriformes) A large family of small catfish that have rows of thick, overlapping plates along each side of the body, each plate provided with a thorny process. The *dorsal and *pectoral fins each bear a strong spine. Doradidae usually have three pairs of *barbels around the mouth. Several species are known to aquarium hobbyists, including *Amblydoras hancocki* (grunting thorny catfish), which can make audible grunting noises. There are about 130 species, found in S. American rivers.

dorado See CORYPHAENIDAE.

dor beetle See GEOTRUPIDAE.

dories See PROTRUSIBLE MOUTH; ZEIDAE.

dormice 1. See GLIRIDAE. 2. (Chinese pygmy dormouse, *Typhlomys*; spiny dormouse, *Platacanthomys*) See PLATACANTHOMYIDAE.

***Dorosoma cepedianum* (gizzard shad)** See CLUPEIDAE.

dorsal Nearest to the back; in vertebrates to the spinal column.

dorsal aorta The main artery conveying blood to the posterior part of the body.

dorsal fin The unpaired fin located on the back (dorsal surface) of both bony fish and sharks. It may be single and soft-rayed, as in trout, or double with the anterior dorsal fin supported by fin spines, as in perch. In some species (e.g. eels) the dorsal fin is confluent with the tail fin.

dorsiventral With upper and lower sides differing in structure.

dorsum The inner margin of the shell or *carapace of an invertebrate animal; the *dorsal side of the body.

Dorylinae (army ant, driver ant; suborder *Apocrita, family *Formicidae) Subfamily of tropical ants which have one or two petiolar segments, and in which the eyes are reduced or absent. Colonies contain several million individuals, produced by the single *dichthadiiform queen. They are characterized by their *legionary foraging activity, and nomadic behaviour (there are frequent changes in nest site). The bivouac (nest) is formed from the bodies of living ants, protecting the queen and brood within. There are about 200 species.

dosage compensation A genetic process that compensates for *genes that exist in two doses in the homogametic sex and one in the heterogametic sex as a result of their location on the *X-chromosome. The process is not universal among species. For example, in *Drosophila*, the heterogametic male has one X-chromosome which doubles its effect, while in humans, in the homogametic female only one of the two X-chromosomes is functional.

dottyback See PSEUDOCHROMIDAE; PSEUDOGRAMMIDAE.

douc langur (*Pygathrix nemaeus*) See CERCOPITHECIDAE.

dove milk See PIGEON MILK.

doves 1. See COLUMBIDAE. 2. See HAWK-DOVE STRATEGIES.

Down's syndrome (formerly 'Mongolism') The condition whereby one of the small *autosomes (chromosome 21 in humans) is represented three times (trisomy) instead of twice. Individuals exhibiting the syndrome have characteristic *phenotypic signs. It is known only in humans, chimpanzees, and orang-utans. Presumably, trisomy of larger chromosomes is incompatible with survival.

DPN (diphosphopyridine nucleotide) *See* NICOTINAMIDE ADENINE DINUCLEOTIDE.

***Draco volans* (flying dragon, flying lizard)** *See* AGAMIDAE.

dragonet *See* CALLIONYMIDAE.

dragonfish 1. (*Bathydraco scotiae*) *See* BATHYDRACONIDAE; BATHYPELAGIC FISH. **2.** *See* IDIACANTHIDAE. **3.** *See* MELANOSTOMIATIDAE. **4.** *See* PEGASIDAE. **5.** *See* STOMIATIDAE.

dragonflies *See* AESHNIDAE; ANISOPTERA; ANISOZYGOPTERA; CORDULEGASTERIDAE; CORDULIIDAE; EPIOPHLEBIIDAE; GOMPHIDAE; LIBELLULIDAE; MEGANISOPTERA; ODONATA; PETALURIDAE; SYNTHEMIDAE.

Drepanididae (Hawaiian honeycreepers; class *Aves, order *Passeriformes) A diverse family of orange, yellow, green, brown, grey, or black birds, that have bills varying from long, thin, and *decurved to stout and hooked. They are arboreal, feeding on nectar, fruit, and seeds, and nesting in trees and other vegetation. The two extant species of *Loxioides* (three species are extinct) are often placed in the genus *Psittirostra*. There are 8–11 genera, with 16 species, confined to Hawaii.

drey The nest of a squirrel.

dried-fruit beetle *See* NITIDULIDAE.

drift *See* GENETIC DRIFT.

driftfish *See* NOMEIDAE.

drive An outdated term used formerly to describe a type of motivation in animals, the psychological 'force' that leads to physical action. The term was abandoned because 'force' in the sense of 'physical energy' plays no direct part in psychological processes, the attempt to attribute a drive to each

aspect of behaviour led to an uncontrollable proliferation of drives, and, ultimately, because the concept lacks explanatory power.

driver ant *See* DORYLINAE.

Dromadidae (crab plover; class *Aves, order *Charadriiformes) A *monotypic family (*Dromas ardeola*) which is a white, plover-like bird with a black back and black *remiges and *wing coverts. It has a medium-length, stout, black bill, and its legs are long with partially webbed toes and a *pectinate middle claw. It is found on coastal shores, feeding on crabs, *Crustacea, and molluscs, and nesting in a burrow, and occurs along Indian Ocean coasts from the Red Sea to Burma, and on Madagascar and other islands.

dromaeosaurid A member of the *Coelurosauria branch of the theropod dinosaurs which developed from the genus *Dromaeosaurus* (the 'emu reptile'), known from the early Late *Cretaceous of Canada. Dromaeosaurids had relatively large brain cases and may have been among the most intelligent reptiles ever to have lived.

Dromaiidae (emu; class *Aves, order *Casuariiformes) A monotypic family (*Dromaius novaehollandiae*) of large, *ratite birds that have grey-brown, hair-like plumage, a stout bill, vestigial wings, and no tail feathers. Emus have long legs with three toes and short claws. They inhabit dry, open country, feeding on vegetable matter and fruit, and nesting in a hollow in the ground. They are confined to Australia; a second species *D. diemenianus* (dwarf emu) is extinct and sometimes regarded as a subspecies. Emus are closely related to the *Casuariidae.

***Dromaius diemenianus* (dwarf emu)** *See* DROMAIIDAE.

***Dromaius novaehollandiae* (emu)** *See* DROMAIIDAE.

***Dromas ardeola* (crab plover)** *See* DROMADIDAE.

dromedary (*Camelus dromedarius*) *See* CAMELIDAE.

Dromiciops See DROMICIOPSIA.

Dromiciopsia (subclass *Theria, infra-class *Metatheria) The name sometimes given to an order of marsupials that comprises the single family Microbiotheriidae, with several fossil genera and one extant genus, *Dromiciops*, which occurs in S. America. Both its anatomy and blood proteins show this group to be more closely related to Australian marsupials than to other American marsupials.

drone The male of ants, bees, and wasps, whose only function is to mate with fertile females: the drone contributes nothing to the maintenance of the colony.

drone flies See SYRPHIDAE.

drongos See DICRURIDAE.

Drosophila See DROSOPHILIDAE.

Drosophilidae (fruit flies; order *Diptera, suborder *Cyclorrapha) Large family of flies which are of great economic importance as pests of fruit and other plant tissue. The adults are usually small, with light-red eyes. Eggs are laid on or near the fermenting fruit, and the larvae feed in a semiliquid soup of ripe fruit. Species of the genus *Drosophila*, of which more than 2000 are recognized, include some of the animals about whose genetics most is known. Hawaii is a site of active *Drosophila* speciation.

drum See SCIAENIDAE.

drying oil An oil that hardens to form a film on exposure to air. Drying oils are used in paints and varnishes.

Dryinidae (suborder *Apocrita, superfamily Bethyloidea) Family of small (1.5–10 mm long), black, parasitic wasps in which the fore *tarsi of the female are often *chelate and used to grasp the host adult or nymph. The host is stung and becomes insensible, whereupon the female wasp lays her egg or eggs. The larvae are mostly ectoparasitic on the abdomens of adult or nymphal plant hoppers, leaf hoppers, and

tree hoppers. The position of the parasite on the host varies; more than one may be present; and many species exhibit *polyembryony, with up to 60 individuals resulting from a single fertilized egg. The family is relatively uncommon but there are about 1400 species.

dry-matter production The expression of plant or animal productivity in terms of the dry weight of material produced per unit area during a specified time period. It is a more easily achieved, though technically less accurate, measure of organic production than are *calorific values (in which the inorganic (ash) component can be separated).

Dryopidae (class *Insecta, order *Coleoptera) Family of tiny, *pubescent, black to reddish beetles, less than 5 mm long. The antennae are very short and thick, with six or more segments forming a *pectinate club. They live in or near water, crawling on aquatic vegetation. Larvae often have tufts of external gills at the end of the abdomen, or a mobile *operculum, and are found under stones in streams or in permanently waterlogged fields. There are 250 species.

Dublin Bay prawn (*Nephrops norvegicus*) See NEPHROPIDAE.

duck-billed dinosaurs See HADROSAURIDAE.

duck-billed platypus (*Ornithorhynchus anatinus*) See ORNITHORHYNCHIDAE.

duckbill eel See NETTASTOMIDAE.

duck mole (*Ornithorhynchus anastinus*) See ORNITHORHYNCHIDAE

ducks See ANATIDAE.

duck virus hepatitis Disease, chiefly affecting ducklings, which involves haemorrhagic necrosis of the liver. Mortality rates are high. The disease is caused by an enterovirus.

ductless gland See ENDOCRINE GLAND.

ductus pneumaticus See PHYSOSTOMOUS.

Ducula (imperial pigeons) See COLUMBIDAE.

duetting A co-ordinated vocal *display performed by a monogamous (see MONOGAMY) pair of birds or primates in which the individuals vocalize alternately or simultaneously. Duetting occurs frequently after the pair have bonded and throughout the duration of the pair bond. The behaviour has been documented in more than 200 species of birds and most Old World primates.

duffers See NYMPHALIDAE.

Dufour's gland An abdominal gland found in the females of nearly all *Apocrita. It empties at the base of the *ovipositor or sting, and is thought in many groups to lubricate the valves of the ovipositor during egg-laying. In worker ants it secretes either alarm or trail-making *pheromones. It is massively developed in most solitary mining bees, and produces secretions to make a waterproof, fungus-resistant lining for brood cells. In at least two bee families (*Anthophoridae and *Megachilidae) the Dufour's gland secretion is also a dietary supplement for the developing larva, and is added to the pollen and nectar stored by the female before egg-laying.

dugong See DUGONGIDAE.

Dugong dugon (dugong) See DUGONGIDAE.

Dugongidae (dugong; superorder *Paenungulata, order *Sirenia) A monospecific family (*Dugong dugon*) of large (up to 5.8 m long), exclusively marine 'sea cows' which have notched tails and moderately cleft upper lips. The functional teeth are reduced to a pair of upper *incisors (tusks in the males) and there are four to six other, peg-like teeth, the lower jaw having a horny pad. The stomach is complex, the intestine very long. Dugongs are distributed throughout coastal waters from the Red Sea to Taiwan and northern Australia.

Dulidae (palmchat; class *Aves, order *Passeriformes) A *monotypic family (*Dulus dominicus*) which is an olive-brown bird with buff-white under-parts, streaked with brown, a stout bill, and stout feet. It inhabits open country with palms, feeding on berries and blossom. Its nest is a large communal structure with separate entrances and nesting compartments, built high in a palm. It is found in the West Indies, confined to Hispaniola and Gonâve Island, Haiti.

dulosis (*adj.* **dulotic**) The slave-making behaviour of certain parasitic ant species, in which the workers raid nests of other species and remove pupae. These are then reared as slaves in the nest of the dulotic ant species, and pass on foraged food to their captors.

Dulus dominicus (palmchat) See DULIDAE.

dun See EPHEMEROPTERA; SUB-IMAGO.

dung beetle See GEOTRUPIDAE.

dung roller See SCARABAEIDAE.

dunnart The preferred vernacular name for members of the dasyurid (*Dasyuridae) genus *Sminthopsis*, formerly known as 'marsupial mice'. These are voracious, mouse-sized carnivores found mainly in arid and grassland areas of Australia and southern New Guinea.

dunnock (*Prunella modularis*) See PRUNELLIDAE.

duodenum In vertebrates, the section of the *alimentary canal that lies immediately behind the *stomach, forming the start of the intestine.

duplication A chromosomal aberration in which more than one copy of a particular chromosomal segment is produced within a *chromosome set.

Duplicidentata See LAGOMORPHA.

duplicidentate In *Lagomorpha, applied to the condition in which the first *incisors (i.e. those at the centre of the front of each jaw) are large and have *persistent pulps, and the second are small and peg-like.

duplivincular Applied to a ligament formed from alternate bands of hard and soft tissue.

dura mater *See* MENINGES.

durophagic Adapted to the eating of hard materials, such as the diet of many benthic dwellers, comprising shelled invertebrates. Skates and rays, for instance, have a tough dentition and protrusible mouth capable of powerful suction which may be used to dislodge shellfish from rock faces.

Dusicyon (South American fox) *See* CANIDAE.

dwarf crocodiles *See* CROCODYLIDAE.

dwarf emu (*Dromaius diemenianus*) *See* DROMAIIDAE.

dwarf lemur (*Cheirogaleus*) *See* CHEIROGALEIDAE.

dwarf rasbora (*Rasbora maculata*) *See* CYPRINIDAE.

dyad In genetics, one of the products of the disjunction of the *tetrads at the first meiotic division, contained in the nuclei of secondary *oocytes and *spermatocytes.

dynein A *protein that binds at regular intervals to the *axonemes of the tubule doublets of *microtubules. Dynein forms projections that attach to one tubule of the doublet and it splits ATP (*adenosine triphosphate).

Dysdercus (cotton stainers) *See* PYRRHOCORIDAE.

dysgenic Genetically deleterious.

dysodont Applied to a type of hinge dentition, found in certain *Bivalvia, where teeth are simple, small, and situated very close to the dorsal margins of the valves.

Dysommidae (arrowtooth eels; subclass *Actinopterygii, order *Anguilliformes) A small family of eel-like fish which have a scaleless body, small *pectoral fins, and a row of strong teeth on the upper jaw. Possibly they live near the edge of the continental shelves.

Dytiscidae (true water beetles, diving beetles; class *Insecta, order *Coleoptera) Family of large, shiny beetles up to 38 mm long, in which the dorsal and ventral surfaces are equally convex. Air is trapped between the *elytra and *abdomen for breathing under water. The hind legs are flattened and paddle-like, and specialized for swimming. Segments of the male fore *tarsi are expanded to form suckers which hold the female during mating. The antennae are thread-like. Larvae are aquatic, with an elongate body, well-developed legs, and large, sickle-shaped *mandibles. Adults and larvae are predatory on small fish, snails, and tadpoles. There are 4000 species.

dzhalman (*Selevinia betpakdalensis*) *See* SELEVINIIDAE.

eagle owl (*Bubo*) *See* STRIGIDAE.

eagle ray *See* MYLIOBATIDAE.

eagles *See* ACCIPITRIDAE.

ear-stone (otolith) A generally oval-shaped, solid, calcareous structure that is enclosed in membranaceous sacs in the inner-ear region of the brain of vertebrates. Although there are usually three ear-stones on each side of the brain, the middle one, the sagitta, is often the largest. Because the ear-stone shows growth zones, or annual rings, it is used by fish biologists to determine the age of fish.

earth measurers *See* GEOMETRIDAE.

earwigs *See* DERMAPTERA; PYGIDICRANIDAE; LABIDURIDAE; FORFICULIDAE; APACHYIDAE; LABIIDAE; ARIXENIIDAE; HEMIMERIDAE.

eastern box turtle (*Terrapene carolina*) *See* EMYDIDAE.

Ecardines *See* INARTICULATA.

eccrine gland (sweat gland) In *Mammalia, one of the glands located in the skin that opens directly to the exterior (*compare* APOCRINE GLAND) and secretes a dilute solution, mainly of salty water, the evaporation of which cools the body surface. Its function is thermoregulation.

ecdysis The periodic shedding of the *exoskeleton by some invertebrates, or of the outer skin by some *Amphibia and *Reptilia.

ecdysone A *hormone, produced by the prothoracid glands, which brings about *moulting in an insect. The form of the

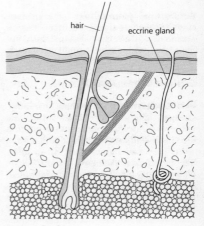

Eccrine gland

animal after each moult is determined by the *juvenile hormone. Ecdysone acts mainly on the epithelial cells.

Echeneidae (remora; subclass *Actinopterygii, order *Perciformes) A family of marine fish that have a flat, disc-like structure on top of the head. Actually a highly modified spinous *dorsal fin, this is used by the fish as a sucking disc; by means of it the remora attaches itself to the bodies of larger fish (e.g. sharks and marlins). In the past, fishermen reputedly used remoras with lines fastened to their tails in order to catch turtles and large fish. One of the larger species, *Remora remora*, may reach a length of 90 cm.

echidna (*Tachyglossus*) *See* TACHYGLOSSIDAE.

Echimyidae (spiny rats; order *Rodentia, suborder *Hystricomorpha) A family of ground-dwelling, burrowing rodents, some of which are arboreal and others semi-aquatic. The legs are long, the first digit of the fore limb is rudimentary, the hind limb is *pentadactyl. The hair tends to be spiny. Echimyidae are distributed throughout tropical S. America. There are 14 genera, with about 43 species.

Echinacea (subphylum *Echinozoa, class *Echinoidea) A superorder of regular echinoids that have rigid *tests, gill slits, complete *perignathic girdles, and keeled teeth. They first appeared in the Late *Triassic.

Echinodera *See* KINORHYNCHA.

Echinodermata A phylum of 'spiny-skinned', invertebrate animals which are entirely marine. The name is derived from the Greek *ekhinos*, 'hedgehog', and *derma*, 'skin'. They are characterized by an internal, mesodermal skeleton of porous calcite plates, a *pentameral symmetry (although a *bilateral symmetry is often superimposed upon this radial plan, especially in many modern echinoids), and the presence of a water-based vascular system, a complex internal apparatus of fluid-containing tubes and bladders which pass through pores in the skeleton and are seen from the outside as *tube feet. The larvae are bilaterally symmetrical and indicate affinities with the *Chordata (*see* DIPLEURULA LARVA). The phylum is varied, and includes *Ophiuroidea (brittle-stars), *Asteroidea (starfish), *Echinoidea (sea urchins), *Holothuroidea (sea cucumbers), and *Crinoidea (sea lilies); and the extinct Edrioasteroidea, Blastoidea, Carpoidea, and Cystoidea. *Tribrachidium* from the late *Precambrian of Australia is probably an echinoderm, close to the stock from which the other groups evolved in the *Cambrian and *Ordovician. Echinoderms first appeared in the Early Cambrian, but of the 20 classes known in the *Palaeozoic, only six survive into the *Mesozoic and on to the present day.

Echinoida (class *Echinoidea, superorder *Echinacea) An order of regular echinoids that have imperforate, non-crenulate *tubercles, solid spines, and shallow gill slits. They first appeared in the mid-*Cretaceous.

Echinoidea (sea urchins, sand-dollars, heart urchins; phylum *Echinodermata, subphylum *Echinozoa) A class of free-living echinoderms in which the body is enclosed in a globular, cushion-shaped, discoidal, or heart-shaped *test built of meridionally arranged columns of interlocking, calcareous plates, which bear movable appendages (spines which have small 'ball-and-socket' joints, *pedicellariae, and *spheridia). The test is composed of 20 vertical rows of plates arranged in five double rows of perforate (ambulacral) plates and five double rows of imperforate (interambulacral) plates. *Tube feet, connected to the internal water-based vascular system, emerge through the pores of the *ambulacra. The apical system on the upper surface consists of five *ocular plates and up to five *genital plates. In all regular echinoids the *anus is enclosed within the apical system, but in many irregular echinoids it is in the posterior interambulacrum. The mouth is always on the lower surface and may be central or anterior in position; in most species, the mouth is ringed with five teeth that form part of *Aristotle's lantern. The class first appeared in the *Ordovician. It underwent a great *adaptive radiation in early Upper *Palaeozoic times, when rigid tests were evolved, experienced a marked reduction in the *Permian and *Triassic, and thereafter resumed its diversification, which has continued until the present day. Fossils of the inner skeletons of these echinoderms are common in both *Mesozoic and *Cenozoic sediments. There are about 125 Palaeozoic, 3670 Mesozoic, 3250 Cenozoic, and more than 900 extant species.

Echinophthiriidae (order *Phthiraptera, suborder *Anoplura) Family of sucking lice which are parasitic on seals and otters. Species differ from other Anoplura in the lack of eyes or ocular points; the lack of abdominal *sclerites; the thick covering of *setae; and the long, tubular, spiracular chamber (*see* SPIRACLE). They are the only lice to parasitize marine mammals, and some species are known to obtain oxygen from sea water by cutaneous respiration. This is possible because the *exoskeleton is very thin, and such lice are thickly covered in protective, flattened scales. Some species of seal lice live in the nasal passages of the host, while others burrow a short distance into the skin. Many seals come to land only once a year, to breed; the life cycle of the lice is synchronized to this, so they breed at the same time as the seals. It is then that lice are able to move from seal to seal, and to colonize young seals. There are five genera, with 12 species.

(((⊕))) SEE WEB LINKS

• Information about the Echinophthiriidae.

echinopluteus larva In *Echinodermata, a planktonic and feeding larval form that possesses limb-like outgrowths bearing ciliated bands.

Echinosorex gymnurus (gymnure, Malayan moon rat) See ERINACEOIDEA.

Echinothurioida (subclass *Euechinoidea, superorder *Diadematacea) An order of regular echinoids that have flexible *tests and compound ambulacral plates bearing three pairs of pores. The *tubercles are non-crenulate. At the present day there are many abyssal forms. They first appeared in the Late *Jurassic.

Echinozoa (phylum *Echinodermata) A subphylum whose members have radial symmetry of a globoid body without arms or projecting rays. It contains seven classes in the *Palaeozoic, but only two of these (*Echinoidea and *Holothuroidea) survive to the present day.

echinulate Covered with small points or spines.

Echiura (spoon worms) A group of coelomate (see COELOM), marine, worm-like animals in which mature individuals lack *metameric segmentation. The body consists of a sausage-shaped, cylindrical trunk and an anterior, muscular *proboscis that cannot be retracted (unlike the *introvert

of *Sipunculida which Echiura otherwise resemble in size and shape). *Parapodia are absent in most forms. All have a pharynx. The nervous system is similar to that of *Annelida, but there are no special sense organs. Males and females occur. The 135–165 species are exclusively marine, and most are burrowers and *detritus feeders. Spoon worms were formerly classed as a family (Echiuridae) of the Annelida, but were then reclassified as a phylum (Echiura); probably the original classification was more reliable, and most authorities now regard them as an annelid family.

Echiuridae See ECHIURA.

echiurus larva In *Sipunculida, a *trochophore larva.

echo-location The detection of an object by means of reflected sound. It is used by bats, by some cetaceans, and by other animals, for purposes of orientation and to find prey.

eclosion The emergence of an insect larva from an egg, or of an adult insect from a pupa. The word is from the French *faire éclore*, meaning 'to hatch'.

eco- A prefix derived from the Greek *oikos*, 'house' or 'dwelling place'.

ecological barrier See DISPERSAL BARRIER.

ecological efficiency (trophic level assimilation efficiency) The ratio between the energy assimilated at one *trophic level and that assimilated at the immediately preceding level, usually expressed as a percentage; i.e. $(A_2/A_1) \times 100$, where A_1 is the lower trophic level and A_2 the higher trophic level.

ecological energetics The study of energy transformations within *ecosystems.

ecological factor See LIMITING FACTOR.

ecological genetics The study of genetics with particular reference to variation on a global and local geographic scale.

• Website of the Ecological Genetics Research Unit at the University of Helsinki.

ecological indicator Any organism or group of organisms indicative of a particular environment or set of environmental conditions. *See also* BIOTIC INDICES; INDICATOR SPECIES.

ecological isolation Separation of groups of organism as a result of changes in their *ecology or in the environment in which they live. This is one of the processes leading to *speciation, since there will be a restriction in the movement of *genes between groups thus separated, and changes in *gene frequencies may occur due to local selection or drift until eventually the groups may be so divergent that reproductive barriers exist.

ecological niche *See* NICHE.

ecological pyramid (Eltonian pyramid) Graphical representation of the trophic structure and function of an *ecosystem. The first *trophic level, of producer organisms (usually green plants), forms the base of the pyramid, with succeeding levels added above to the apex. There are three types of pyramids: of numbers, of biomass, and of energy. The alternative name for the concept is taken from the name of Dr Charles *Elton, FRS, the British ecologist who devised it.

ecological system *See* ECOSYSTEM.

ecology The scientific study of the interrelationships among organisms and between organisms, and all aspects, living and non-living, of their environment. Ernst Heinrich Haeckel is usually credited with having coined the word 'ecology' in 1869 (as the German word *Ökologie*, from the Greek *oikos*, 'house', and *logos*, 'discourse').

ecomorph A small difference in form or colour that distinguishes *populations of a *species that have become reproductively isolated fairly recently. Ecomorphs represent adaptations to very local environmental variations (e.g. among *Anolis* lizards

(*Iguanidae), all of the species *A. sagrei*, descended from populations that were introduced experimentally to 14 very small islands in the Bahamas in the 1970s, which differ in the length of their hind legs according to the structures on which they rest).

ecophysiology The scientific study of the physiological adaptation of *species to their environments.

ecosphere *See* BIOSPHERE.

ecosystem (ecological system) A term first used by A. G. Tansley (1871–1955) in 1935 to describe a natural unit that consists of living and non-living parts, interacting to form a stable system. Fundamental concepts include the flow of energy via food chains and food webs, and the cycling of nutrients biogeochemically. Ecosystem principles can be applied at all scales—thus principles that apply e.g. to an ephemeral pond apply equally to a lake, an ocean, or the whole planet. In Russian and central European literature '*biogeocoenosis' describes the same concept.

ecotone A narrow and fairly sharply defined transition zone between two or more different *communities. Such edge communities are typically species-rich. Ecotones arise naturally, e.g. at land–water interfaces; but elsewhere may often reflect human intervention, e.g. the agricultural clearance of formerly forested areas.

ecotope The habitat component of a *biogeocoenosis.

ecotype A locally adapted *population of a widespread species. Such populations show minor changes of morphology and/or physiology, which are related to *habitat and are genetically induced. Nevertheless they can still reproduce with other ecotypes of the same species.

ecto- A prefix meaning 'outside', from the Greek *ekto*, 'outside'.

ectocrine (environmental hormone, exocrine) A chemical substance, released by an organism into the environment during decomposition processes, which influences the activity of another organism.

ectoderm In a *triploblastic *embryo, the outer layer of cells which eventually give rise to the *epidermis, the most anterior and posterior portions of the digestive tract, and the majority of the nervous system.

Ectognatha In some classifications, a superorder of insects comprising the *ectognathous groups Pterygota and Thysanura.

ectognathous Of insect mouth-parts, well developed and projecting outwards from the head rather than being contained in an eversible pocket.

ectoparasite *See* PARASITISM.

Ectopistes migratorius **(passenger pigeon)** *See* COLUMBIDAE.

ectoplasm In plant cells and some *Protozoa, the outer, *gel-like layer of the cell *cytoplasm, which lies immediately beneath the *cell membrane and contains packed layers of *microtubules.

Ectoprocta *See* BRYOZOA.

ectotherm An animal that maintains its body temperature within fairly narrow limits by behavioural means (e.g. basking or seeking shade). Terrestrial reptiles are ectotherms. *Compare* POIKILOTHERM.

edaphic Of the soil, or influenced by the soil. Edaphic factors that influence soil organisms are derived from the development of soils and are both physical and biological (e.g. mineral and humus content, and *pH).

Edentata (infraclass *Eutheria, cohort *Unguiculata) An order that comprises two suborders, *Palaeanodonta (of ancestral forms) and *Xenarthra (the S. American ant-eaters, armadillos, and sloths). The teeth are reduced in number, have a simple peg-like form, and lack enamel; in the anteaters they are absent entirely, reflecting adaptation to a diet of invertebrates. The brain

case is low, the brain small, the olfactory region large. There are extra articulations between the lumbar vertebrae. The feet have well-developed claws, used for digging or hanging, and some edentates walk on the outsides of the feet. Body temperatures tend to fluctuate widely with ambient air temperature. Edentates are considered to have departed little from the general eutherian form, their superficial features being adaptations to particular ways of life.

edentulous **1.** Applied to a condition found in some *Bivalvia in which hinge teeth are absent. **2.** In mammals, toothless, either naturally or as a result of tooth loss.

Ediacaran The last (i.e. youngest) faunal stage of the *Proterozoic; it is marked by the presence, at localities all over the world, of the distinctive Ediacaran fauna, named after the Ediacara Hills, a site in South Australia. It lasted from 600 Ma ago to 542 Ma ago.

edible frog A first-generation hybrid between the marsh frog (*Rana ridibunda*) and pond frog (*R. lessoneae*). The parental chromosome sets are inherited as a whole, so e.g. the offspring of edible × marsh frogs are 50% marsh and 50% edible; there are no F_2 hybrids.

eelblennies *See* CONGROGADIDAE.

eel gobies *See* GOBIOIDIDAE.

eelpout (*Zoarces viviparus*) *See* ZOARCIDAE.

eels **1. (arrowtooth eel)** *See* DYSOMMIDAE. **2. (common eel)** *See* ANGUILLIDAE. **3. (conger eel)** *See* CONGRIDAE. **4. (cusk-eel)** *See* OPHIDIIDAE. **5. (cutthroat eel)** *See* SYNAPHOBRANCHIDAE. **6. (deep-sea eel)** *See* SERRIVOMERIDAE. **7. (duckbill eel)** *See* NETTASTOMIDAE. **8. (electric eel, *Electrophorus electricus*)** *See* ELECTRIC FISH; ELECTROPHORIDAE. **9. (false moray eel)** *See* XENOCONGRIDAE. **10. (knife eel)** *See* GYMNOTIDAE. **11. (long neck eel)** *See* DERICHTHYIDAE. **12. (long-tailed eel)** *See* MURAENIDAE. **13. (moray eel)** *See* MURAENIDAE.

14. (pikecongereel) See MURAENESOCIDAE.
15. (reef-eel) See MURAENIDAE. **16. (snake eel)** See OPHICHTHIDAE. **17. (snipe eel)** See NEMICHTHYIDAE. **18. (spiny eel)** See MASTACEMBELIDAE; NOTACANTHIDAE.
19. (swamp eel) See SYNBRANCHIDAE.
20. (witch eel) See NETTASTOMIDAE.
21. (worm eel) See MORINGUIDAE.

eelworms See NEMATODA.

effect hypothesis A model proposed in 1980 by the palaeontologist Elisabeth Vrba to account for *evolutionary trends. She proposed that a *species, occupying a restricted ecological *niche, would continually give rise to daughter species by *punctuated equilibrium. These new species would have a variety of character-istics; but because of the features of the particular ecological niche, only species that possessed a particular suite of char-acters would survive; the surviving species would speciate in their turn, with the same result, and at each level the lineage appears to be 'pushed' further and further in a given direction.

effective population size The aver-age number of individuals in a population that actually contribute *genes to succeed-ing generations. This number is generally rather lower than the observed, censused, population size, being reduced by the fol-lowing factors: (*a*) a higher proportion of one sex may mate; (*b*) some individuals will pass on more genes by having more offspring in a lifetime than others; (*c*) any severe past reduction in population size may result in the random loss of particular *genotypes.

effector Any structure (e.g. gland, muscle, electric organ, stinging cell, or pigment cell) that responds to a stimulus directly or in-directly. The nervous system controls most effectors.

efferent fibres In the *autonomic nerv-ous system, fibres that convey information from the *central nervous system to pe-ripheral organs, e.g. to control heart rate, contraction and dilatation of blood vessels,

and contraction and relaxation of smooth *muscle.

effused Loosely or irregularly spreading.

egestion The expulsion by an organism of waste products that have never been part of the cell constituents.

eggars See LASIOCAMPIDAE.

egg burster Spine or tooth on the surface of the *exoskeleton of the first larval *instar of an insect, which allows it to break out of the egg.

egg-eating snakes Two genera of snakes (*Colubridae), *Dasypeltis* from Africa and *Elachistodon* from India, which feed almost exclusively on eggs. Elongate projections from the neck ver-tebrae pierce the throat, and with throat-muscle action break the eggshell, which is regurgitated.

egg-raft See CULICIDAE.

egrets (*Egretta*) See ARDEIDAE.

***Egretta* (egrets, herons)** See ARDEIDAE.

eiders (*Somateria*) See ANATIDAE.

Eifelian See DEVONIAN.

eigenvalue (latent root λ) The compo-nents (latent roots) derived from the data which represent that variation in the origi-nal data accounted for by each new compo-nent or axis.

(((⊕))) SEE WEB LINKS
• Explanation of eigenvalue.

Eimeriina (subclass *Coccidia, order *Eucoccidia) A suborder of *Protozoa in which *syzygy does not occur. They are para-sitic, growing inside the cells of a wide range of vertebrate and invertebrate hosts, and are commonly parasitic in domestic animals. The suborder includes some pathogens (*see* COCCIDIOSIS). There are several families, and many species.

ejaculation The process of expulsion of *semen through the male sexual organ thereby releasing *spermatazoa from the *testis. In invertebrates, the seminal fluid is passed through the vas deferens to the outside, in vertebrates the vas deferens leads to the urethra which leads to the outside. *Compare* OVULATION.

Elachistodon See EGG-EATING SNAKE.

Elaenia (elaenias) See TYRANNIDAE.

Elagatis bipinnulatus (rainbow runner) See CARANGIDAE.

eland (*Tragelaphus* (formerly *Taurotragus*) *oryx*,) See BOVIDAE.

Elanus (kites) See ACCIPITRIDAE.

Elaphe longissima (Aesculapian snake) See COLUBRIDAE.

Elaphe quatuorlineata (four-lined snake) See COLUBRIDAE.

Elaphe situla (leopard snake) See COLUBRIDAE.

Elapidae (cobras, kraits, taipans, mambas, coral snakes; order *Squamata, suborder *Serpentes) A family of venomous snakes which have short, firm, grooved fangs at the front of the mouth. The venom is mainly neurotoxic (i.e. a nerve poison). Generally these are elongate, agile snakes, active at dusk or dawn. Their prey consists mainly of small vertebrates. *Bungarus* (kraits) of south-east Asia grow up to 2 m long; they are nocturnal and feed almost exclusively on other snakes. The genus *Dendroaspis* (mambas) of Africa includes the black mamba which grows up to 4.25 m long and has been recorded moving at 11 km/h across country, and the smaller (up to 2.7 m), arboreal green mamba. *Ophiophagus hannah* (king cobra) of southern Asia, the longest of all venomous snakes, sometimes exceeds 5 m; its neck ribs can be spread to form a small hood. There are about 200 species, found on every continent except Europe.

Elasipodida (class *Holothuroidea, subclass *Aspidochirotacea) An order of deep-water holothuroids which lack *respiratory trees, gas exchange occurring through the body surface.

Elasmobranchii (class *Chondrichthyes) A subclass of shark-like fish that have five to seven gill slits, fairly rigid fins, *placoid scales, a spiracular opening behind the jaws, numerous teeth, and *claspers on the ventral fins of the male. Modern elasmobranchs have a cartilaginous skeleton. Apart from modern sharks and rays, the subclass also includes many fossil species.

elastic fibre A fibre that gives *connective tissue elasticity and resilience. It consists of an extracellular *macromolecule with a core of *elastin surrounded by microfibrils (*see* FIBRIL) rich in *fibrillin.

elastin A *protein present in *connective tissue including skin that allows the tissue to resume its original shape after having stretched or contracted.

Elateridae (click-beetles, skip-jacks, snapping beetles; subclass *Pterygota, order *Coleoptera) Family of elongate, usually black or brown beetles, 3–54 mm long, with long, often comb-like antennae. If the beetle falls on its dorsal surface, the body jack-knifes with a pronounced click, throwing it into the air, and thus turning it over, so that it may land on its ventral surface. The larvae (wire-worms) are long, cylindrical, and tough-skinned; they can be serious crop pests, feeding on roots. Others are carnivorous or wood-feeders. *Pyrophorus* species (fireflies) occur in the neotropics: all stages are luminescent (*compare* LAMPYRIDAE). There are 7000 species, with very wide distribution.

electric catfish (*Malapterurus electricus*) See MALAPTERURIDAE.

electric eel (*Electrophorus electricus*) See ELECTROPHORIDAE; ELECTRIC FISH.

electric fish Fish that are capable of generating electric discharges. A number of species, belonging to different families, have this capability. Some (e.g. *Electrophorus electricus* (electric eel) and the electric rays, such as the Atlantic *Torpedo nobiliana*) can produce strong, stunning currents capable of immobilizing potential prey. The electric organs are derived from muscle tissue which is organized as a series of units, each one acting as a type of battery. The electric eel (which is related to the carp, not to the true eel) also uses the electrical pulses as a navigational aid, as does the African elephant fish (*Mormyridae).

electric ray *See* TORPEDINIDAE.

electrolocation The detection of an object by its distortion of a weak electrical field generated and sensed by the animal. *See* ELECTRIC FISH.

electromagnetic location The detection of an object by its distortion of the Earth's electromagnetic field. *See* AMPULLAE OF LORENZINI; ELECTROMAGNETIC SENSE.

electromagnetic sense Mechanism by which some fish (e.g. *Scyliorhinus*, dogfish) and many invertebrates are able to orientate themselves within an electric field that they themselves generate, or by disturbances that they detect in the Earth's electrical or magnetic fields. In some fish the electric pulses are altered by individuals to prevent interference with neighbours, and are used as a means of communication.

electron carrier A compound that functions as an acceptor and donor of electrons and/or protons in an electron-transport system.

electron-transport chain (respiratory chain) A system of *electron carriers, present in *mitochondria and the *cell membranes of *prokaryotes, which sequentially transport electrons and/or protons previously removed from metabolites in *glycolysis, the *citric-acid cycle, and other metabolic reactions.

electrophoresis The migration, under the influence of an electric field, of charged particles within a stationary liquid. The liquid may be a normal solution or held upon a porous medium (e.g. starch, acrylamide gel, or cellulose acetate). The rate at which migration occurs varies according to the charge on the particle and also its size and shape. The phenomenon is exploited in a variety of analytical and preparative techniques employed in studies of macromolecules.

Electrophoridae (electric eel; subclass *Actinopterygii, order *Cypriniformes) A monospecific family (*Electrophorus electricus*, electric eel) which has an elongate, eel-like body with a long *anal fin confluent with the tail fin. There are no scales, and the *dorsal fin is absent. The long tail region contains the specialized muscles that form the electric organ, which is capable of producing about 600 volts per discharge. It is found only in the Amazon and Orinoco rivers of S. America. *See also* ELECTRIC FISH.

***Electrophorus electricus* (electric eel)** *See* ELECTROPHORIDAE; ELECTRIC FISH.

electroreceptor A type of *receptor cell found in weakly electrical *teleost fish, which allows the detection of electrical discharge. There are two types of electroreceptor: *ampullary organs and *tuberous organs.

elementary particle *See* CRISTA.

***Eleops saurus* (ladyfish)** *See* ELOPIDAE.

Eleotridae (gudgeon, sleeper; subclass *Actinopterygii, order *Perciformes) A large family of small fish found in marine, brackish, and freshwater environments. Like the related gobies, they have a fairly round body, a blunt head with the eyes located almost above the jaws, two *dorsal fins, and a rounded tail fin. Unlike gobies, the *ventral fins of the gudgeons are close to one another but do not form a sucking disc. One of the largest species is *Gobiomorus dormitor* (bigmouth sleeper) which grows to 60 cm; most do not exceed 20 cm. There are about 150 species, distributed world-wide.

elephant *See* ELEPHANTIDAE; PROBOSCIDEA.

elephant beetle *See* SCARABAEIDAE.

elephant birds *See* AEPYORNITHIDAE.

elephant fish *See* CALLORHYNCHIDAE; MORMYRIDAE.

Elephantidae (order *Proboscidea, suborder Gomphotherioidea) A family than comprises the ancestral and modern elephants. They can be traced back to the *Miocene (*Stegolophodon*) and are distinct from the superficially similar mastodons (*Mammutidea). In living species the *incisors grow to tusks in the upper jaw only; other incisors and *canines are absent. Three milk *premolars and three *molars are present. Only one tooth in each half-jaw is used at a time; it is shed when worn and replaced by the tooth behind it, which moves forward. The skull is short, the nasal aperture high in the face, the proboscis long and muscular, with nostrils at its end. The limbs end in short hoofs; the gait is *digitigrade. The family is distributed throughout Africa south of the Sahara, and southern Asia. The family includes the extinct *Mammuthus* (mammoth) and three surviving species: *Elephas maximus* (Asian elephant), *Loxodonta africana* (African elephant), and *Loxodonta cyclotis* (African forest elephant).

elephant louse (*Haematomyzus elephantis*) *See* RHYNCOPHTHIRINA.

elephant shrew *See* MACROSCELIDEA.

Elephas maximus (Asian elephant) *See* ELEPHANTIDAE.

Eleutherozoa (phylum Echinodermata) Ranked previously as a subphylum, a term that is now used only informally, to denote free-living echinoderms. These include the subphyla *Asterozoa and *Echinozoa.

elk 1. N. American form of *Cervus elaphus* (red deer, wapiti). **2. (moose, *Alces alces*)** *See* CERVIDAE.

elm-bark beetle (*Scolytus scolytus*) *See* SCOLYTIDAE.

Elmidae (subclass *Pterygota, order *Coleoptera) Family of small beetles, 2–5 mm long, which have punctured *elytra and raised lines on the *thorax. The antennae are *filiform. The legs are long and slender, with an elongate terminal tarsal segment. They live in running water. Some breathe under water, using an air film trapped by hairs as a physical gill. Larvae are woodlouse-shaped, with terminal external gills, and are found clinging to stones or weeds in water. There are 300 species.

Elopidae (giant herring, ladyfish; superorder *Elopomorpha, order *Elopiformes) A family of marine to estuarine fish that have a very slender, silvery body, a single *dorsal fin, small *pectoral and *pelvic fins, and a large, forked tail fin. Small schools of *Elops saurus* (ladyfish), 90 cm, often enter estuaries from the western Atlantic. Different species may exist in the Indo-Pacific region.

Elopiformes (subclass *Actinopterygii, superorder *Elopomorpha) An order of marine bony fish, which have the *pelvic fins in the abdominal position, opposite the single *dorsal fin, wide gill openings, and a strongly forked tail fin. The order includes the families *Megalopidae (tarpon) and *Elopidae (ladyfish).

Elopomorpha (class *Osteichthyes, subclass *Actinopterygii) A superorder of bony fish that includes the orders *Elopiformes, *Anguilliformes, and *Notacanthiformes.

Elops saurus (ladyfish) *See* ELOPIDAE.

Elton, Charles Sutherland (1900–91) A British zoologist who studied animal communities and made major contributions to the development of ecological studies in general and of animal ecology in particular. One of his early books, *Animal Ecology* (first published in 1927), was very influential. He emphasized the importance of conserving species and habitats. He proposed the existence of *niches, occupied by species at particular points in a *food chain. From 1932 until 1967 he was director of the Bureau of

Animal Population at the University of Oxford. Elton was elected a Fellow of the Royal Society in 1953 and was awarded many scientific prizes.

Eltonian pyramid *See* ECOLOGICAL PYRAMID.

elver A juvenile eel (*Anguillidae).

elytron (*pl.* **elytra**) Hard, often coloured fore wing of a beetle or earwig, which gives the animal its characteristic appearance (e.g. the colour and spots of a ladybird) when it is not flying.

emarginate Having a notch or notches at the edge.

Emballonuridae (sac-winged bats; order *Chiroptera, suborder *Microchiroptera) A family of bats, in many species of which a glandular sac is present in the *propatagium. The third digit is elongated and folds twice (into an N shape). The ears are separate, with a small *tragus, the small tail projects from a loose sheath on the dorsal side of the *uropatagium, and the nose is simple and pointed in many species. Sac-winged bats often roost in the open or in well-lit cave entrances. They are distributed throughout the New and Old World tropics. There are about 13 genera, with about 40 species.

Embden–Meyerhof pathway *See* GLYCOLYSIS.

Emberiza striolata (house bunting) *See* EMBERIZIDAE.

Emberizidae (buntings, finches, longspurs, sparrows; class *Aves, order *Passeriformes) A family of small, mainly brown, streaked birds with medium-length feet, longish tails, and short conical bills. They inhabit open areas, grassland, and woodland, feeding mainly on seeds and building either an open or dome-shaped nest in trees, bushes, or on the ground. *Emberiza striolata* (house bunting) prefers to occupy buildings. The colourful plumage of the six species of *Passerina* makes them popular cage birds. There are about 70 gen-

era in the family, with about 280 species, found in N., Central, and S. America, Africa, Europe, and Asia.

Embiotocidae (surfperch; subclass *Actinopterygii, order *Perciformes) A family of coastal marine fish that have oval-shaped, compressed bodies, small mouths, and joined, spinous, soft-rayed *dorsal fins. Many species are known to be live-bearing. *Rhacochilus toxotes* (rubberlip sea-perch) reaches a length of about 40 cm: the other 22 species are smaller. All are found in the northern Atlantic.

Embrithopoda (cohort *Ferungulata, superorder *Paenungulata) An extinct order of subungulates of uncertain ancestry and with no known descendants, known by one form, *Arsinoitherium*, from the lower *Oligocene of Egypt. The animal was the size of a rhinoceros (about 3.4 m long); possibly it dwelt in marshes. It had short massive limbs, one pair of large horns (fused at their bases) growing from its nasal bone, and a second, much smaller pair of horns (behind the large pair) from the frontal bone.

embryo A young animal that is developing from a sexually fertilized or parthenogenetically activated ovum and that is contained within egg membranes or within the maternal body. The embryonic stage ends at the hatching or birth of the young animal.

Emmelichthyidae (bonnetfish, redbait; subclass *Actinopterygii, order *Perciformes) A family of marine fish that have a slender body, a large and oblique mouth, toothless jaws, a deeply notched *dorsal fin, and a forked tail fin. Several species are of some commercial importance. There are at least 17 species, distributed world-wide in tropical to temperate waters.

emperor *See* LETHRINIDAE.

emperor angelfish (*Pomecanthus imperator*) *See* CHAETODONTIDAE.

emperor butterflies *See* NYMPHALIDAE.

emperor moths *See* SATURNIIDAE.

Empididae (dance flies; order *Diptera, suborder *Brachycera) Large family of medium to small flies, which have body bristles and a horny proboscis used to pierce prey and suck it dry, the prey usually consisting of other Diptera. The males often exhibit secondary sexual characters on their legs, as adaptations for holding prey and the female during mating. Mating behaviour also involves 'dancing' up and down by the males, hence the common name. Males often catch items of prey and offer them to females as part of the courtship procedure. Larvae are *amphipneustic, with pseudopods to aid locomotion. The known larvae are carnivorous on other insects in soil and leaf-litter. There are more than 3000 described species, with a cosmopolitan distribution.

Empidonax (flycatchers) *See* TYRANNIDAE.

empodium Usually, the median bristle or pulvilliform structure that arises between the *unques on the feet of *Diptera; but the term is sometimes used to describe the terminal and sometimes claw-like structures on the collembolan *tibiotarsus.

Emsian *See* DEVONIAN.

emus *See* CASUARIIFORMES; DROMAIIDAE.

emu wrens (*Stipiturus*) *See* MALURIDAE.

Emydidae (freshwater turtles, terrapins; order *Chelonia, suborder *Cryptodira) A family of turtles that bask on land but enter slow-moving water to feed. Usually there is no reduction of the shell. The limbs are flattened with webbed, clawed toes. Emydidae are mainly carnivorous. *Emys orbicularis* (European pond terrapin or European pond tortoise) is native to central and southern Europe, south-western Asia, and north-western Africa; it was present in Britain during the last (Ipswichian) interglacial. *Terrapene carolina* (Carolina box turtle or eastern box turtle) of the eastern USA is omnivorous and lives on land; its *plastron is hinged allowing the shell to be closed. There are about 80 species, distributed widely in temperate zones, except southern Africa and Australia.

Emys orbicularis **(European pond terrapin, European pond tortoise)** *See* EMYDIDAE.

enamel Crystals of a calcium phosphate-carbonate salt, containing 2–4% of organic matter, which are formed from the *epithelium of the mouth and which provide a hard outer coating to *denticles and to the exposed part of teeth.

encephalitis Inflammation of the brain, which may be due, for example, to virus infection.

Encyrtidae (suborder *Apocrita, superfamily *Chalcidoidea) Family of minute (0.5–7.0 mm long), usually black or brown, parasitic wasps which have convex *mesopleura and *mesonota, and middle legs modified for jumping. The adults have a large apical spur on the front *tibia, and some are wingless. The larvae are mostly parasitic on scale insects and whiteflies, but some parasitize psyllids, nymphal ticks, potato moths, etc. One group forms galls. Many are *polyembryonic, producing from 10 to more than 1000 young from a single egg.

endangered *See* RARITY.

endemic *See* ENDEMISM.

endemism The situation in which a species or other taxonomic group is restricted to a particular geographic region, due to factors such as isolation or response to soil or climatic conditions. Such a taxon is said to be endemic to that region. The size of the region in this context will usually depend on the status of the taxon: thus a family will be endemic to a much larger area than a species, all other things being equal.

endo- A prefix meaning 'internal', derived from the Greek *endon*, 'within'.

endobiotic Growing within a living organism.

endobyssate Applied to the habit of specific bivalves (*Bivalvia) that live in sediment. In contrast to *epibyssate forms, the byssus (*see* BYSSATE) of these animals is used to anchor the animal within a burrow or boring.

endochondrial ossification *See* OS-SIFICATION.

endocone In some *Nautiloidea, apically pointed, conical layers of calcareous material in the *siphuncle, building forward from the rear and sometimes filling it.

endocrine disruption Interference with the *endocrine system that can result in changes in development and reproduction, and in the neurological and immune systems. A wide variety of chemical substances are known or believed to be endocrine disruptors in humans and non-humans, including pharmaceutical products, dioxin and dioxin-like compounds, polychlorinated biphenyls, plasticizers (e.g. bisphenol-A), and some insecticides.

endocrine gland (ductless gland) A *gland that secretes its *hormone product into the bloodstream or surrounding tissue directly, rather than through a duct. *Compare* EXOCRINE GLAND.

endocrine system In vertebrates, the system of ductless glands which secrete into the blood stream *hormones which act on a target elsewhere in the body. *See also* PITUITARY GLAND.

(SEE WEB LINKS)
• Description of the endocrine system.

endocuticle *See* EXOSKELETON.

endocytosis The mechanism by which a *cell engulfs large material by the invagination of the *cell membrane to form a *vesicle or *vacuole. *See also* PHAGOCYTOSIS; PINOCYTOSIS.

endoderm In an *embryo, the innermost layer of cells that gives rise to the lining of the *archenteron (digestive cavity) and its associated glands. *See also* GASTRULATION.

endogenous infection Infection with a micro-organism that is normally resident in the body so infected.

endolithic Growing between stones.

endolymph The fluid contained in the *membranous labyrinth of the inner ear of vertebrates.

endomitosis The doubling of the *chromosomes within a *nucleus that does not divide. The doubling may be repeated a number of times in a single nucleus. It occurs mainly in insects, where different adult tissues have specific *polyploid numbers; it also occurs in some vertebrate tissues.

endomixis In certain ciliated (*see* CILIUM) *Protozoa, the periodic division and reorganization of the cell nucleus. It was first observed in 1914 in *Paramecium aurelia*.

Endomychidae (subclass *Pterygota, order *Coleoptera) Family of rounded, convex, hairy beetles, 2–8 mm long, with an expanded *pronotum covering most of the head. Many are brilliantly coloured. The antennae are conspicuously clubbed. Normally the beetles are associated with fungi on forest trees. Larvae are oval, with sides concealed by expanded *tergites. Temperate species are found in decaying vegetation, dung, and stored products. There are 1100 species, mostly tropical.

endomysium A sheath of *connective tissue that encloses a fibre of *striated muscle.

endoneurial channel *See* ENDONEURIUM.

endoneurial sheath *See* ENDONEURIUM.

endoneurial tube *See* ENDONEURIUM.

endoneurium (endoneurial channel, endoneurial sheath, endoneurial tube) A sheath of *connective tissue that encloses the *myelin sheath of each of the nerve fibres of the *spinal cord.

endoparasite *See* PARASITISM.

endopeptidase *See* PROTEASE.

endophallus The inner wall of the *aedeagus of an insect, derived embryologically by the fusion of paired genital papillae to form a medial tube. The endophallus is flexible and eversible, is the true intromittent organ, and may be armed with backward-directed teeth or spines which gain purchase on the vaginal walls of the female.

endophragmal skeleton In *Decapoda, a complex internal structure composed of *arthrophragms, or from the fusion of *apodemes, that provides a framework for muscle attachment.

endoplasm In some *Protozoa, the inner layer of *cytoplasm within which are embedded the principal cell *organelles.

endoplasmic reticulum A complex network of cytoplasmic membranaceous sacs and tubules which appears to be continuous with both the nuclear and *cell membranes. It occurs in two forms: that bearing *ribosomes is termed 'rough ER'; that without, 'smooth ER'. Both are involved in the synthesis, transport, and storage of cell products.

endopod In *Crustacea, the inner *ramus of the *biramous limb.

endopodite *See* PROTOPODITE.

Endopterygota (class *Insecta, subclass *Pterygota) A division of insects in which the wings develop internally until the final moult, when they undergo a pupal stage and complete *metamorphosis.

endorphins *Peptides, peculiar to vertebrates, manufactured by the *pituitary gland, which have analgesic properties similar to opiates (e.g. morphine), because of an affinity for the same receptors in the *brain.

endoskeleton A skeleton that is contained within the body. In vertebrates, the endoskeleton comprises the *axial skeleton and the *appendicular skeleton. In *Echinodermata, the skeleton lies beneath the *epidermis and technically it is therefore an endoskeleton. *Compare* EXOSKELETON.

endosteum A thin layer of *connective tissue that lines the inner surface of the hollow centre of long bones.

endostracum A layer of generally *nacreous calcium carbonate which is present on the internal surface of the shells of certain *Bivalvia and other molluscs.

endostyle 1. A ciliated groove on the mid-ventral surface of the pharynx of *ammocoetes. Food particles are trapped by mucus produced by the pharynx, and then transported toward the *oesophagus. The endostyle is also able to take up iodine, a capacity it retains after it is transformed into the thyroid gland of the adult lamprey, when *metamorphosis is complete. 2. A *mucus-secreting groove in the *pharynx of *Urochordata and *Cephalochordata, and a comparable organ in *Hemichordata; it is *analogous to the thyroid gland in *Chordata.

endosymbiosis *Symbiosis in which one symbiont lives within the other. *Compare* EXOSYMBIOSIS.

endosymbiotic hypothesis The hypothesis that the plastid and *mitochondrial *organelles evolved from *prokaryotic endosymbionts within *eukaryotic cells early in eukaryotic evolution. At least two endosymbiotic events are postulated, one each for plastid and mitochondrial origins. It is contentious whether two independent events could account for the mitochondria that occur in plants and those that occur in animals, the present-day mitochondria of the two kingdoms being quite distinct from each other genetically.

endothelium The thin layer of tissue that lines the interior surfaces of blood and lymphatic vessels.

endotherm An animal that is able to maintain a body temperature that varies only within narrow limits by means of internal mechanisms (e.g. dilation or contraction of blood vessels, sweating, panting, and shivering). Birds and mammals are endotherms. *Compare* HOMOIOTHERM.

endozoochory Dispersal of spores or seeds by animals after passage through the gut.

energid *See* PROTOPLAST.

energy budget A comparison between the amount of energy that enters the body of an animal, or a particular *trophic level, and the amount of energy that leaves that animal or level.

energy of activation *See* ACTIVATION ENERGY.

engram A memory trace stored in the brain.

Engraulidae (anchovy; subclass *Actinopterygii, order *Clupeiformes) A family of small, slender, silvery, marine fish related to the herrings, which have a long snout projecting over the lower jaw, a large mouth, and a single, soft-rayed, *dorsal fin. They travel in huge schools near the shore. Many species, including *Engraulis encrasicolus* (European anchovy), are of considerable importance as food fish or as bait fish, the annual catch in Europe amounting to some 300 000 tonnes. There are about 110 species, distributed world-wide.

Engraulis encrasicolus (anchovy) *See* ENGRAULIDAE.

enigmatic taxon A genus or higher *taxon of restricted diversity, comprising only one or a few species with affinities that are poorly understood. Many species belonging to enigmatic taxa are known from very few specimens; some have not been seen since they were first described.

Enopla (phylum *Nemertini) A class of worms in which the *proboscis is divided into distinct regions, and may or may not bear *stylets. The mouth is either in front of or below the brain.

Enoplosidae (old wife; subclass *Actinopterygii, order *Perciformes) A monospecific family (*Enoplosus armatus*, old wife) of marine fish; an attractive, deep-bodied fish with a small mouth, long, spinous, soft-rayed *dorsal fins, and large *pelvic fins. It is found in coastal regions of eastern Indonesia and eastern Australia.

Enoplosus armatus (old wife) *See* ENOPLOSIDAE.

Ensifera (subclass *Pterygota, order *Orthoptera) Suborder of orthopterans in which the antennae have more than 30 segments, usually many more, and in some groups are several times the length of the body. When present, auditory organs are found on the fore tibia, and *stridulation is produced by friction between modified veins on the fore wings. When present, the modified *ovipositor is cylindrical or sword-shaped (ensiform— hence 'Ensifera'). The suborder includes the families *Cooloolidae, *Gryllacrididae, *Gryllidae, *Gryllotalpidae, *Myrmecophilidae, *Prophalanopsidae, *Rhaphidophoridae, *Schizodactylidae, *Stenopelmatidae, and *Tettigoniidae.

ensiform Sword-shaped.

ensign flies *See* SEPSIDAE.

ensign wasps *See* EVANIIDAE.

Entamoeba histolytica The *amoeba that causes amoebic dysentery in humans. *See* AMOEBOZOA.

entelechy An outmoded theory holding that evolution proceeds by the realization of that which was always potential. The word is from the Greek *entelekheia*, 'become perfect'. *See also* ARISTOGENESIS; ORTHOGENESIS.

entellus langur (*Semnopithecus* or *Presbytis entellus*) *See* CERCOPITHECIDAE.

Entelodontidae So-called giant pigs, which diverged from palaeodont stock in the *Eocene. They flourished in the *Oligocene but disappeared in the *Miocene, probably as a result of displacement by the true pigs and peccaries. Although they grew to the size of bison, to judge from their long limbs they were fast runners.

enteroceptor A sense organ that registers the internal state of an animal's body. In mammals, enteroceptors monitor body temperature. *Compare* EXTEROCEPTOR; PROPRIOCEPTOR.

enteron *See* COELENTERON.

Enteropneusta (acorn worms) A phylum of free-living, elongate, worm-like, burrowing marine animals, formerly included in the phylum *Hemichordata, in which the body is clearly divided into a proboscis (protosoma), collar (mesosoma), and trunk (metasoma). The *stomochord extends into the proboscis. The branchial apparatus is well developed, consisting of a long, double row of pores strengthened by a cuticular skeleton (thought to be *homologous with the pharyngeal gill clefts of *Chordata). The nervous system is *notoneural. There are many living genera. No fossil examples are known.

enterotoxin A type of *toxin, produced by certain bacteria, which affects the function of the intestinal mucosa, causing diarrhoea, gastroenteritis, etc.

entocoel In some corals, the space between two *mesenteries where the pleated muscle blocks face one another. *Compare* EXOCOEL.

Entodiniomorphida (class *Ciliatea, subclass *Spirotrichia) An order of ciliate *Protozoa which occur as commensals in the rumen of cattle, sheep, and other *ruminants, and in the gut of various other herbivores (e.g. horses). The organisms are anaerobic and can digest cellulose. There are two families, and many genera.

Entognatha In some classifications, a superorder of insects comprising the *entognathous groups *Collembola, *Protura, and *Diplura.

entognathous Eversible and contained within a small pocket. In members of the *Entognatha, the mouth-parts are enclosed by the head capsule.

entomochory Dispersal of spores or seeds by insects.

entomopathogenic Capable of causing disease in insects.

Entoprocta (goblet worms) A phylum of freshwater animals which entirely lack a mineralized skeleton. The *lophophore surrounds both the anus and the mouth. Many fossil forms are known, but the phylum is known only from the *Palaeogene. *Compare* ECTOPROCTA.

environment The external surroundings within which an organism lives.

environmental hormone *See* ECTOCRINE.

environmental resistance factors Those aspects of an *environment which constrain the growth of a *population and establish the maximum number of individuals that can be sustained. Such factors include the availability of essential resources (e.g. food and water), predation, disease, the accumulation of toxic metabolic wastes, and, in some species, behavioural changes due to stress caused by overcrowding.

environmental variance The portion of *phenotypic variance that is due to differences in the environments to which the individuals in a population have been exposed. The total amount of variance observed amongst individuals in a population will be made up of an environmental component, determined by environmental variation, and a genetic component, determined by the variation that is inherited.

enzootic Applied to a disease of animals that is restricted to a given geographical locality.

enzyme A molecule, wholly or largely *protein, produced by a living cell, that acts

e

as a biological catalyst. Enzymes are present in all living organisms, and through their high degree of specificity exert close control over cellular metabolism.

Eocene A *Palaeogene epoch, about 55.8–33.9 Ma ago, which began at the end of the *Palaeocene and ended at the beginning of the *Oligocene. It is noted for the expansion of mammalian stocks (horses, bats, and whales appeared during this epoch), and the local abundance of nummulites (marine protozoans of the *Foraminiferida). The name means 'dawn of the new' and is derived from the Greek *eos*, 'dawn', and *kainos*, 'new'.

Eodelphis A genus of early opossums, marsupials with long, prehensile tails, several species of which are known from the Milk River Formation (Late *Cretaceous) of Alberta. Their teeth were sharply cusped and somewhat primitive in form. Opossums are arboreal in habit and are thought to represent the ideal stem stock from which evolved the whole marsupial group.

Eognathosomata The name given to a group of irregular echinoids (*Echinoidea) comprising the order Pygasteroida, and the suborders Holectypina and Echinoneina of the order *Holectypoida.

Eohippus *See* HYRACOTHERIUM.

Eosimias A fossil primate, known mainly by jaws, teeth, and foot bones, from the middle *Eocene of southern China, described in 1994 by C. K. Beard, Tao Qi, M. R. Dawson, Banyue Wang, and Chuankuei Li. According to its describers, it has the features of a very primitive member of the *Simiiformes and may stand at the base of this group.

eosinophils In vertebrates, white blood cells, produced in the *bone marrow, that have coarse granules within their *cytoplasm. They migrate to tissues throughout the body and help combat invasion by foreign substances and certain infections.

Eozostrodon One of the earliest recorded mammals, known from the *Triassic of Europe and probably China. It was a small animal resembling a tree shrew.

Epallaginidae (Euphaeidae, damselflies; order *Odonata, suborder *Zygoptera) Family of damselflies, which are often metallic-coloured, or iridescent. The nymphs are thickset, with a short abdomen. They occur in the Old World tropics, and 47 extant species have been described.

ephemeral Short-lived, or of brief duration (e.g. the life of a mayfly, *Ephemeroptera). The word is derived from the Greek *ephemeros*, 'lasting for only a day'.

Ephemerellidae (mayflies; subclass *Pterygota, order *Ephemeroptera) Family of insects whose nymphs are strikingly and cryptically coloured, and favour rapid, cool, clear streams or small lakes. The family has an almost world-wide distribution.

Ephemeridae (mayflies; subclass *Pterygota, order *Ephemeroptera) Family of insects whose nymphs occur in ponds, lakes, and large rivers, and have a burrowing habit. The head has a frontal process and long, tusk-like *mandibles for loosening the silt, which is pushed sideways by the front legs and out of the burrow by the hind legs. The feathery gills are in continuous motion during digging, and the nymph feeds on detritus excavated from the burrow. The family contains some of the largest mayflies. Many have clear wings, although a few have spotted wings. Members of a particular species often emerge in large numbers.

Ephemeroptera (mayflies; subclass *Pterygota, infraclass *Palaeoptera) Order of insects in which the adults are short-lived, surviving from a few hours to a few days (*see* EPHEMERAL). The order is somewhat primitive, its members having two pairs of membranous wings with a simple network of veins, resembling the venation of dragonflies. The eyes are large; the antennae are short; the mouth-parts are unsclerotized and vestigial (the adults do not feed); the *thorax is strongly developed for flight; and the *abdomen possesses three long tail filaments. The adults are small to medium-sized, soft-bodied insects, with large, triangular fore wings, and small, rounded, hind wings. The wings are held rigidly upright above the body at rest. The nymphs are aquatic and herbivorous, although some have sharp, modified *mandibles and are partly predacious. The nymphal stage may last several years, but the majority are *univoltine. The *metamorphosis is simple and there is no pupal stage, the first winged stage being known as the sub-imago, or dun. The sub-imago resembles the adult, but is often a dull-brown colour. The sub-imaginal stage may last as little as five minutes before the skin is shed and the brightly coloured adult (spinner) emerges. The males fly in swarms with a dancing, vertical flight, and have several mating adaptations (e.g. reversible fore-leg tarsal joints) and modified eyes. Ephemeroptera are unique in that they undergo a moult after wings have become full-sized and functional. The 2100 species are an important part of the diet of freshwater fish.

Ephippidae (spadefish, batfish; subclass *Actinopterygii, order *Perciformes) A small family of marine fish in which the compressed, flattened body is usually nearly as deep as it is long. The small mouth, steep forehead, and the large second *dorsal and *anal fins are also typical. Fairly slow-moving fish, they are found near rocky outcrops or coral reefs. There are about 15 species, occurring world-wide in warm to temperate waters.

Ephthianuridae (Australian chats; class *Aves, order *Passeriformes) A family of small, brightly coloured, short-tailed birds that have fairly short, slightly *decurved bills. They inhabit marshy ground and dry plains, and are mainly terrestrial, running along the ground to catch insects. They build an open nest in bushes near the ground. There are two genera, with five species, found in Australia. This family is sometimes placed with the *Maluridae.

ephyrae The immature *medusae of *Scyphozoa which develop from *scyphistomae.

They are almost microscopically small and have incompletely developed adult structures. They feed mainly on protozoons.

epi- A prefix derived from the Greek preposition *epi*, 'upon', 'above', or 'in addition to'.

epibenthos The organisms living on the surface of the sea-bed or bed of a lake.

epibiotic Growing on the surface of a living organism.

epiblast The outer layer of a *blastula; following *gastrulation it gives rise to the *ectoderm.

epibyssate Applied to the habit of specific bivalves (*Bivalvia) that live in sediment. In contrast to *endobyssate forms, the byssus (*see* BYSSATE) of these animals is used to anchor the animal to rock or seaweed.

epicardial tube In *Urochordata, a perivisceral space formed by the outgrowth of the *pharynx.

epicuticle *See* EXOSKELETON.

epidemiology The scientific study of the causes, distribution patterns, and effects of diseases and other health-related events.

epidermis The outermost layer or layers of cells in an animal. It is one cell thick in many invertebrates, but many cells thick in vertebrates. In land-dwelling vertebrates its surface layer is formed from dead, hardened (keratinized) cells. *See also* INTEGUMENT.

epifauna Benthic organisms (*see* BENTHOS) that live on the surface of the sea-bed, either attached to objects on the bottom or free-moving. They are characteristic of the intertidal zone. *Compare* INFAUNA.

epigamic Character of an animal that serves to attract or stimulate members of the opposite sex during courtship. Examples are the distinctive coloration of some male birds and fish, the colourful crest of the male crested newt, and the song of birds. Such characters may also have other functions such as discouraging potential competitors for mates or food resources.

epigenesis The hypothesis that an organism develops by the new appearance of structures and functions. An alternative hypothesis (termed 'preformation') is that development of an organism occurs by the unfolding and growth of characters already present in the egg at the beginning of development.

epigenetics The study of the mechanisms by which *genes bring about their phenotypic effects.

epiglottis In mammals, a flap of *cartilage and mucous membrane on the ventral wall of the *larynx, at the base of the tongue. During swallowing the entrance to the *trachea is pushed against it and so closed.

epigyne In *Arthropoda, a sexual opening on the ventral surface of the *abdomen.

Epilachna varivestis (Mexican bean beetle) See COCCINELLIDAE.

Epimorpha See CHILOPODA.

epimorphic 1. In insects, applied to post-embryonic development in which little change occurs beyond the addition of antennal segments and the development of the *chaetotaxy towards the mature condition. **2.** See EPIMORPHOSIS.

epimorphosis (epimorphic growth) Condition in some *Chilopoda (centipedes) in which the young possess the full complement of adult body segments on hatching. *Compare* ANAMORPHOSIS.

epimysium *Connective tissue that encloses an entire *striated muscle.

epinephrine See ADRENALIN.

epineurium The outer layer of *connective tissue around a peripheral nerve (*see* PERIPHERAL NERVOUS SYSTEM).

epineustic See ZOONEUSTON.

epineuston The organisms living in the upper part of the surface film of water. *Compare* HYPONEUSTON; NEUSTON.

Epiophlebiidae (dragonflies; order *Odonata, suborder *Anisozygoptera) The only family in its suborder, comprising two known species, found in India and Japan.

epiphragm A temporary structure, made from *mucus, with which many terrestrial snails (*Gastropoda) seal the aperture of their shells when they are inactive (e.g. during the day or during dry weather). Some freshwater snails also produce an epiphragm if their pools dry out. It prevents loss of moisture and offers some protection against predation.

epiphysial eye *See* PINEAL EYE.

epiphysis 1. In the limbs and vertebrae of mammals, a cap of bone at the joints; during growth it is separated from the main bone by a cartilaginous plate. It then ossifies separately, the whole fusing when growth ceases. The extent of the fusion can sometimes be used to estimate the age of an animal. **2.** *See* PINEAL BODY.

epipodite In some *Crustacea, an accessory appendage on the basal segment of the limb.

epipodium *See* APPENDICULAR SKELETON.

Epipolasida (class *Demospongiae, subclass *Tetractinomorpha) An order of sponges in which the body has a radial structure. Organic fibres and *spicules form the skeleton. The *cortex is well developed. Most spicules have four axes. The group first appeared in the *Cambrian.

epipubic bone In male and female *Marsupialia and *Monotremata, a rod-like bone which projects forward from the pelvis. In females it helps support the *marsupium.

episodic evolution The overall picture of evolution that arises from the fact that the fossil record is characterized by extinction events and succeeding phases of rapid evolutionary innovation. The term has recently acquired other connotations, and tends to be linked with *punctuated equilibrium.

epistasis The situation in which an *allele of one gene (called the epistatic gene) prevents the expression of all allelic alternatives of another gene. For example, the recessive gene apterus (ap) in *Drosophila* produces wingless homozygotes: in such individuals any other recessive gene affecting wing morphology will have its expression masked. The apterus gene is then epistatic to genes like curly wing; conversely these are hypostatic to it.

epistome 1. In *Bryozoa and *Phoronida, a lobe or ridge which overhangs the mouth. 2. In some *Crustacea, a ventral plate which lies in front of the mouth.

epithelium A type of animal tissue which consists of a sheet or tube of tightly packed cells with the minimum of intercellular material. Epithelia cover the exposed surfaces of the body, often line tubes and cavities, and are frequently secretory in function.

epizoic Applied to non-parasitic animals that attach themselves to the outer surface of other animals.

epizootic Outbreak of disease (an epidemic) in a population of (non-human) animals.

equatorial division The division of each *chromosome during the *metaphase of *mitosis or *meiosis into two equal, longitudinal halves which are then incorporated into separate *daughter nuclei.

equatorial plate An arrangement of the *centromeres of *chromosomes in which they lie approximately in one plane, at the equator of the *spindle. This is seen during *metaphase of *mitosis and *meiosis. In mitosis the centromeres lie exactly on the equatorial plate; in the first stage of meiosis the centromeres of *homologous chromosomes lie on opposite sides of the equatorial plate.

Equidae (horses; order *Perissodactyla, suborder *Hippomorpha) A family that includes the modern horses, asses, and zebras (all of which are placed in a single genus *Equus*, divided into six or seven species, depending on the classification used) and many extinct forms, the earliest being known from the late *Palaeocene. Sufficient of these are believed to be ancestral to modern equids for the evolution of the family to have been traced in considerable detail. The domestic horse (*E. caballus*) is probably not descended from the only true living wild horse, Przewalski's horse (*E. ferus przewalskii*), but from a progenitor closely related to it. Modern equids are adapted for rapid movement: they have a short *humerus and *femur, and a long *radius and *tibia; the *ulna is reduced and fused to the radius, and the *fibula is also reduced. Only the third toe is developed. All the *incisors are present, *canines are present in males, and the cheek teeth are *hypsodont, with complex grinding surfaces. The brain is large and senses of sight, smell, touch, and hearing are highly developed. Equids are migratory grassland animals, graze mainly on grasses, and have complex social organization. Wild species are found in Africa and parts of central and western Asia. The horses are first represented in the fossil record by *Hyracotherium*, which diverged from a condylarth predecessor in the Palaeocene. Numerous evolutionary lines subsequently appeared from this fox-sized prototype. Most of the evolutionary advances occurred in the New World, although the Equidae were to survive the *Pleistocene only in the Old World. The *conquistadores* reintroduced horses to the Americas in the 16th century.

equids *See* EQUIDAE.

equilibrium species A species in which competitive ability, rather than dispersal ability or reproductive rate, is the chief survival strategy: competition is the typical response to stable environmental resources. In unstable or extreme environments (e.g. deserts) equilibrium species survive unfavourable periods by living on stored food resources and reducing life processes to a minimum.

Equus (horse) *See* EQUIDAE.

Equus caballus (domestic horse) *See* EQUIDAE.

Equus ferus przewalskii (Przewalski's horse) *See* EQUIDAE.

Eremophila alpestris (shore lark, horned lark) *See* ALAUDIDAE.

Erethizon dorsatum (North American porcupine) *See* ERETHIZONTIDAE.

Erethizontidae (New World porcupines; order *Rodentia, suborder *Hystricomorpha) A family of arboreal rodents in which some hairs are modified to form barbed spines. The body is thickset, the legs short, and long claws are borne on each of the four digits of all limbs. First digits are reduced. In the tree porcupines (*Coendou*) the tail is *prehensile. Porcupines are distributed widely throughout N. America and tropical S. America. There are four genera with 17 species; the North American porcupine is *Erethizon dorsatum*).

Eretmochelys imbricata (hawksbill turtle) *See* CHELONIIDAE.

ergatogyne A form intermediate between a worker and a queen which occurs in some ant species. The *gaster is larger than that of a worker, but the head and *thorax are worker-like. It can function as an additional reproductive, or replace the queen.

Erinaceidae *See* ERINACEOIDEA.

Erinaceoidea (hedgehogs, moon rats; order *Insectivora, suborder *Lipotyphla) A superfamily that comprises several extinct families (e.g. Adapiscoricidae and the Dimylidae) and the extant family Erinaceidae, with its two subfamilies, Erinaceinae (hedgehogs) and Echinosoricinae (moon rats). Erinaceoidea are insectivores, in the more advanced of which (hedgehogs) the hairs of the back are modified to form spines. The senses of sight and hearing are more acute than in many insectivores. The limbs have five digits (except in the genus

Atelerix which has four on the hind limbs). Hedgehogs have some immunity to toxins, including snake bite, and chew toxic substances to make a froth with which they anoint their spines. Nocturnal animals feeding on small invertebrates, they are distributed throughout the Old World. In the *Oligocene and *Miocene, hedgehogs and moon rats were equally common in Europe; today the moon rats (e.g. *Echinosorex gymnurus*, the gymnure or Malayan moon rat) are confined to south-east Asia. There are about 10 genera, with 15 species.

Eriocraniidae (subclass *Pterygota, order *Lepidoptera) Family of small, primitive moths with a broad distribution but few species. The venation of the fore wing and hind wing is similar. *Mandibles are reduced but distinct, and there is a short proboscis. Adults are day fliers. Larvae are leaf-miners and often legless. Pupae have large, curved mandibles, and pupation occurs in the soil in a strong cocoon.

Erithacus *See* TURDIDAE.

ermine 1. *See* ARCTIIDAE. 2. *See* MUSTELIDAE.

Erotylidae (pleasing fungus beetles; subclass *Pterygota, order *Coleoptera) Family of shiny, oval beetles, 2–23 mm long, and often brightly coloured. The head is *prognathous, with a three-segmented, flattened club. Larvae are pale and elongate, with toughened spiny plates. Legs are well developed, antennae three-segmented. The beetles are found in trees, where they feed on fungal material. There are 1500 species.

Errantia (scaleworms, sea mice; phylum *Annelida, class *Polychaeta) A subclass of polychaete worms which have a large number of body segments, those of the head and posterior differing from the rest. The mouth often has several paired jaws. Most are *vagile predators, others adopt a burrowing mode of life. They are first recorded from the *Cambrian. There is no acceptable ordinal classification for the Polychaeta; the class is divided into Errantia and *Sedentaria for convenience but genera belonging to each subclass may not be related.

eruciform Resembling a caterpillar; i.e. having an approximately cylindrical body with *prolegs in the hind region and true thoracic legs. The word comes from the Latin *eruca*, 'caterpillar'.

Erythrinidae (subclass *Actinopterygii, order *Cypriniformes) A family of freshwater fish that have a large gape, a single *dorsal fin opposite the *pelvic fins, and a rounded tail fin. There are about five species, occurring in S. America.

erythrism In mammals, the possession of red hair, caused most commonly by a lack of black pigment (eumelanin) which allows the red pigment (phaeomelanin) to dominate.

erythrocyte Red blood cell, a body consisting mainly of *haemoglobin that conveys almost all the oxygen carried in the blood. It is incapable of independent motion. In adult mammals (but not *embryos) the cell has no *nucleus; in other species there is a nucleus but it does not synthesize *RNA.

escape reaction A form of defensive behaviour that occurs when an animal detects a predator or the predator attempts to capture it. Many animals have specialized forms of escape (e.g. flying fish disappear from the view of the predator by leaping from the water).

Eschrichtiidae (order *Cetacea, suborder *Mysticeti) A monospecific family (*Eschrichtius gibbosus*, the grey whale) of baleen whales in which the baleen plates are short and narrow and the left and right rows are separated at the front of the mouth. The snout is blunt, typically there are two grooves on the throat, and the body is slender. Grey whales are mostly grey in colour but with white mottling; they grow to a length of 11–15 m. They are seasonal migrants within the northern Pacific and feed on *benthic *Amphipoda.

Eschrichtius (grey whale) *See* ESCHRICH-TIIDAE.

escutcheon A depressed area, variable in shape, which is found to the posterior of the beaks of certain *Bivalvia.

Esocidae (pike, muskellunge; superorder *Protacanthopterygii, order *Salmoniformes) A small family of freshwater fish that have a streamlined body, a produced, duck-like snout, and a large mouth containing many sharp teeth. The *anal and single *dorsal fins are located to the rear of the body, near the tail. Adult pike tend to be aggressive, solitary individuals. Restricted to the northern hemisphere, *Esox lucius* (European pike) ranges from Ireland, across Europe to eastern Siberia, and right across N. America. Other species, including *E. masquinongy* (muskellunge), are found only in N. America. Female European pike can grow to a length of 1.5 m. There are about five species.

Esox lucius (European pike) *See* ESOCIDAE.

Esox masquinongy (muskellunge) *See* ESOCIDAE.

ESS *See* EVOLUTIONARY STABLE STRATEGY.

estivation *See* AESTIVATION.

Estrildidae (waxbills, mannikins, munias, silverbills; class *Aves, order *Passeriformes) A varied family of small, brightly coloured finches that have rounded wings, medium to short tails, small to large, short, conical bills, and strong feet. They are gregarious, inhabit forests, grassland, and desert, feed on seeds and insects, and build an enclosed nest with a side entrance, usually in trees. *Nestlings have decorated palates and tongues. *Lonchura striata* (Bengalese finch), commonly kept in captivity, is bred in a variety of plumages. *Amandava amandava* (red munia) is the only estrildid with a dull non-breeding plumage. There are 27 genera, with 138 species, most of which are popular cage birds, found in Africa, southern Asia, Australasia, and Pacific Ocean islands.

estrogen *See* OESTROGEN.

estrus cycle *See* OESTRUS CYCLE.

estuarine crocodile (*Crocodylus po-rosus*) *See* CROCODYLIDAE.

Ethiopian faunal region Often called the Afrotropical faunal region, an area that corresponds with sub-Saharan Africa, although it is not completely separated from the neighbouring *faunal regions; generally it is taken to include the south-west corner of the Arabian peninsula. Varied game animals are found there; and there are a number of *endemic families, including various rodents and shrews, giraffe, hippopotamus, aardvark, tenrec, and lemurs.

ethocline A progressive change in the pattern of behaviour of a group of closely related organisms, the more complicated patterns often being associated with more recently evolved members of the group, which evolved from members exhibiting a more primitive pattern. For example, variations in *courtship behaviour among *species of *Hilaria* flies (*Empididae) suggest an evolutionary lineage: (*a*) a male searches for a female and courts her in isolation; (*b*) a male captures a fly, presents it to a female, and she consumes it before mating; (*c*) a male captures a fly, joins a swarm of males, presents the fly to a female, and she consumes it before mating; (*d*) as (*c*), but the male adds a few strands of silk to the fly before joining the swarm; (*e*) as (*d*), but the male covers the fly in silk; (*f*) as (*e*), but the male eats the edible portion of the fly, leaving the husk wrapped in silk; (*g*) males of non-carnivorous species use dried insect fragments or petals, around which they weave a silken balloon before joining the swarm; (*h*) the male makes only a balloon (i.e. containing nothing) before joining the swarm.

ethogram A detailed record of an animal's behaviour.

ethological isolation The failure of individuals of *species with recent common ancestry (i.e. related species or semispecies) to produce hybrid offspring because differences in their behaviour prevent successful mating taking place.

ethology (animal behaviour) The scientific study of the behaviour of animals in their normal *environment, including all the processes, both internal and external, by which they respond to changes in their environment.

***Etmopterus spinax* (Atlantic velvet belly)** *See* SQUALIDAE.

eu- From the Greek *eu*, 'well', 'good', or 'easily', a prefix with the same meaning, usually attached to words of Greek derivation, and used in *ecology to denote, in particular, enrichment or abundance, e.g. 'eutrophic', nutrient-rich; 'euphotic', light-rich.

Eubacteria In the widely used five-*kingdom system of *taxonomy, one of the subkingdoms in the kingdom *Bacteria, containing organisms that need to be distinguished from members of the other subkingdom, the *Archaea. The Eubacteria comprises all of the 'true' bacteria. In the three-*domain classification system, one of the three domains, containing the single kingdom Bacteria.

***Eubalaena* (right whales)** *See* BALAENIDAE.

Eucaryota A name formerly given to the *Eukarya.

euchromatin *See* HETEROCHROMATIN.

Eucoccidia (class *Telosporea, subclass *Coccidia) An order of *Protozoa which live parasitically inside the cells of an animal host. *Schizogony occurs, and *syzygy occurs in some species. There are three suborders, and the genera include some important disease-causing organisms.

Euechinoidea (subphylum *Echinozoa, class *Echinoidea) A subclass of sea urchins in which the normally rigid *test is composed of *ambulacra and interambulacra, made up of two columns of plates. *Spheridia are present. The subclass includes both regular and irregular types of echinoid. They first appeared in the Late *Triassic.

eugenics A postulated method of improving humanity by altering its genetic composition. It would involve encouraging the breeding of those presumed to have desirable genes (positive eugenics), and discouraging the breeding of those presumed to have undesirable genes (negative eugenics). It relies upon the notion of 'desirable' and 'undesirable' genotypes and assumes that selection for these can be particularly applied. Neither concept is generally accepted as valid, the former because it is often a value judgement that will vary between individuals, societies, and circumstances; the latter because genotypes are often present in the heterozygous state and so are not readily detected. Every human is heterozygous for several different deleterious recessive genes, so that if these were detectable, *no one* would be allowed to breed under such a programme.

Eugregarinida (class *Telosporea, subclass *Gregarinia) An order of *Protozoa which are common as parasites in *Annelida and *Arthropoda. *Schizogony does not occur; *syzygy does occur. There are two suborders, several families, and many species.

Eukarya The superkingdom or domain comprising all *eukaryotes. These form the kingdoms Plantae (plants), *Animalia (animals), Fungi, and *Protoctista.

(⊕) SEE WEB LINKS
• Description of the Eukarya.

eukaryote An organism comprising cells that have a distinct nucleus enveloped by a double membrane, and other features including double-membraned *mitochondria and 80S *ribosomes in the fluid of the *cytoplasm (i.e. all protoctists, fungi, plants, and animals). The first eukaryotes were almost certainly green algae, and what appear to be their microscopic remains appear in *Precambrian sediments dating from a little less than 1500 Ma ago.

eulamellibranchiate Applied to a specialized type of gill structure, common in *Bivalvia, in which the gill filaments are united by cross-members, forming sheets of tissue, and partitioning the gill space into vertical tubes through which water passes. Blood vessels are arranged vertically along the tissue junctions.

Eulemur (brown lemur) *See* LEMURIDAE.

eulittoral Applied to the *habitat formed on the lower shore of an aquatic *ecosystem, below the *littoral zone. The marine eulittoral zone is marked by the presence of barnacles (*Balanus* and *Chthamalus* species).

Eulophidae (suborder *Apocrita, superfamily *Chalcidoidea) Large and common family of parasitic insects, most of which are 1–3 mm long, and black or brightly metallic. They are unique among the Chalcidoidea in that some species, on approaching maturity, spin silken cocoons on or close to the host larvae. Some eulophids are parasites of leaf-miners, beetle larvae, *Lepidoptera, *Diptera, aphids, scale insects, and whiteflies. Others are *hyperparasites of *Braconidae.

Eumastacidae (order *Orthoptera, superfamily Acridoidea) Family of primitive, slender, wingless or *brachypterous, short-horned grasshoppers, which includes the monkey and morabine grasshoppers. Most of the 120 or so species are tropical.

eumelanin *See* MELANIN.

Eumenidae (mason wasps, potter wasps; suborder *Apocrita, superfamily Vespoidea) Family of solitary, predatory wasps, whose adults make flask-shaped nests of clay and small pebbles glued together with saliva. Most adults are 10–20 mm long, marked with black, yellow, or white, and have elongate, knife-like *mandibles. A large and widely distributed group, it is sometimes regarded as a subfamily of the *Vespidae. All species provision their nests with paralysed lepidopteran, symphytan, or coleopteran larvae.

Eumeninae *See* HUNTING WASPS.

Eunectes murinus (anaconda) *See* BOIDAE.

e

Eupantotheria (subclass *Theria, infraclass *Pantotheria) An order of extinct mammals, known from the *Jurassic of N. America and Europe and in its earliest, Middle Jurassic, form (*Amphitherium*) only from the lower jaw. The jaw identifies it as mammalian, and possibly ancestral to later pantotheres.

Euparkeria See THECODONTIA.

Eupetes macrocercus (rail-babbler) See TIMALIIDAE.

Euphaeidae See EPALLAGINIDAE.

Eupharynx pelecanoides (gulper) See EURYPHARYNGIDAE.

Euphausiacea (krill; subphylum *Crustacea, class *Malacostraca) Order of shrimp-like, marine crustaceans, dense swarms of which occur in ocean waters. They feed on diatoms and themselves comprise the main food of filter-feeding whales. Krill are up to 5 cm in length and are found in both surface and bottom waters. There are two families: Euphausiidae with 10 genera and 85 species, and Bentheuphausiidae with one species.

Euphonia (euphonias) See THRAUPIDAE.

euphotic zone The upper, illuminated zone of an aquatic *ecosystem.

Euplectes (bishops, whydahs) See PLOCEIDAE.

Eupleridae (order *Carnivora, suborder *Feliformia) A family of cat-like carnivores found only in Madagascar, the best-known member being *Cryptodonta ferox* (fossa). The euplerids are believed to have evolved from a common ancestor which entered Madagascar 24–18 Ma ago and they are the only carnivorous mammals native to Madagascar. There are two subfamilies. Euplerinae contains three species (including *C. ferox*) that were formerly placed in the *Viverridae. Galidiinae contains five species of mongooses formerly placed in the Herpestidae.

euploid Applied to a *cell that has any multiple number of complete *chromosome sets, or an individual composed of such sets. It is thus a *polyploid with a chromosome number that is an exact multiple of the basic number for the species from which it originated.

Eurglossinae See COLLETIDAE.

European angler (*Lophius piscatorius***)** See LOPHIIDAE.

European army worm See SCIARIDAE.

European bats See VESPERTILIONIDAE.

European chameleon (*Chamaeleo chamaeleon***)** See CHAMAELEONTIDAE.

European earwig (*Forficula auricularia***)** See FORFICULIDAE.

European freshwater pearl mussel See GLOCHIDIUM.

European frog (*Rana temporaria***)** See RANIDAE.

European glass lizard (Ophisaurus apodus) See ANGUIDAE.

European pike (*Esox lucius***)** See ESOCIDAE.

European pond terrapin (*Emys orbicularis***)** See EMYDIDAE.

European pond tortoise (*Emys orbicularis***)** See EMYDIDAE.

European toad (*Bufo bufo***)** See BUFONIDAE.

European tree frog (*Hyla arborea***)** See HYLIDAE.

European turbot (*Scophthalmus maximus***)** See BOTHIDAE.

European whip snake (*Coluber viridiflavus***)** See COLUBRIDAE.

eury- From the Greek *eurus*, 'wide', a prefix with the same meaning, usually attached to words of Greek derivation, and used in *ecology to describe species with a wide range of tolerance for a given environmental factor. *Compare* STENO-.

euryapsid Applied to a skull that has a single temporal opening. It is found only in members of the *Euryapsida.

Euryapsida (Synaptosauria; class *Reptilia) A subclass of reptiles that have a single, upper temporal opening behind the eye. The subclass includes the *Plesiosauroidea, *Nothosauria, and *Placodontidae. The *Ichthyosauria had a similarly placed opening but there were key differences in the arrangement of the bones forming its margins. For this and other reasons the ichthyosaurs are placed in the subclass *Ichthyopterygia. Euryapsids were typically aquatic especially from the mid-*Triassic onwards. They first appeared in the *Permian and became extinct at the end of the *Mesozoic.

Euryarchaeota *See* ARCHAEA.

euryecious Having a wide range of habitats.

Euryhalina (brittle-stars; subclass *Ophiuroidea, order *Phrynophiurida) A suborder of brittle-stars in which the small disc lacks plates. Both the arms and the disc are completely covered with a thick skin. Vertical enrolling of the branched arms may result in a basket-like form. The suborder is known since the *Devonian.

euryhaline Able to tolerate a wide range of degrees of salinity. *Compare* STENOHALINE.

Eurylaimidae (broadbills; class *Aves, order *Passeriformes) A family of small to medium-sized, brightly coloured birds, most of which have a whitish dorsal patch. *Eurylaimus steerei* (wattled broadbill) has a large blue *wattle round its eye. Their bills are large, broad, flattened, hooked, and often covered by a short crest. They have large heads, rounded wings, and short to long tails. They inhabit open wooded country, and feed on insects, fruit, seeds, and small vertebrates. Their nests, which are suspended over water, are pear-shaped with a side entrance. There are eight genera, with 14 species, found in Africa, southern Asia, Indonesia, and the Philippines.

Eurylaimus steerei (wattled broadbill) *See* EURYLAIMIDAE.

Eurymylidae (cohort *Glires, order *Lagomorpha) An extinct family of rabbit-like mammals, known from the upper *Palaeocene of east Asia.

euryoxic Able to tolerate a wide range of concentrations of oxygen.

euryphagic Using a wide range of types of food.

Eurypharyngidae (gulper, pelican eel, umbrella mouth gulper; superorder *Elopomorpha, order *Anguilliformes) A monotypic family of deep-sea fish that have an enormous mouth capable of accommodating very large prey; the floor of the mouth is a distendible pouch, resembling that of a pelican and hence its common name. The body itself tapers down toward a very thin tail region. As gulpers live at depths of more than 2000 m, the large mouth presumably is an adaptation to scarcity of food in that region, although its small teeth and analyses of its stomach contents suggest it feeds mainly on small crustaceans. The relatively common *Eupharynx pelecanoides* (also known as *Eupharynx richardi*, *Gastrostomus pacificus*, *G. bairdii*, *Leptocephalus pseudolatissimus*, and *Macropharynx longicaudatus*,) is not a large fish, rarely exceeding 60 cm in length.

Eupharynx pelecanoides *See* EURY-PHARYNGIDAE.

Eupharynx richardi *See* EURYPHARYN-GIDAE.

eurypterid *See* MEROSTOMATA; PTERYGO-TUS DEEPKILLENSIS.

Eurypyga helias (sunbittern) *See* EU-RYPYGIDAE.

Eurypygidae (sunbittern; class *Aves, order *Gruiformes) A monotypic family (*Eurypyga helias*) which is a grey, brown, and white, barred and mottled bird with a large head, a long slender neck, long bill, legs, and toes, and broad wings that show a central orange area when spread. It is found by forest streams and ponds, feeds on insects and crustaceans, and nests in a tree or bush. It is found in Central and S. America.

eurythermal Able to tolerate a wide temperature range.

eurytopic Able to tolerate a wide range of a number of factors.

Euscorpius flavicaudis See SCORPIONES.

Euscorpius germanus See SCORPIONES.

eusociality A social structure in which only one female produces offspring, all other individuals attending to her needs, caring for the young, and performing other tasks necessary to the provisioning, defence, and welfare of the community. Eusociality occurs in many members of the *Hymenoptera and is known in one mammal species, the naked mole rat (**Heterocephalus glaber*).

eustachian canal In tetrapods, a tube that connects the *pharynx with the *middle ear. It permits the equalization of air pressure to either side of the *tympanic membrane and so prevents distortion of the membrane.

Eusthenopteron **(class *Osteichthyes, subclass *Crossopterygii)** A genus of lobe-finned fish, known from the *Devonian of Europe and N. America, that exhibited a number of advanced characters (e.g. lungs and strong muscular fins), suggesting that it was closely linked with the evolution of the *Amphibia. It was related to the lungfish and coelacanths.

Eutamias **(chipmunk)** See SCIURIDAE.

Eutheria (class *Mammalia, subclass *Theria) The infraclass that includes all of the placental mammals and which probably arose during the *Cretaceous. The *blastocyst develops an outer layer of cells which surrounds a small inner group from which the *embryo develops. This is retained in the uterus, nourished by means of an *allantoic placenta, and born in an advanced stage of development. In adults there is no *marsupium and the pelvic bones form an *os innominatum. The *tympanic bone may be ring-like or may form a *bulla, but the *alisphenoid bone never forms part of the bulla. There is no in-turned angle of the jaw.

eutrophication The nutrient enrichment (usually by nitrates and phosphates) of an aquatic ecosystem, such that the productivity of the system ceases to be limited by the availability of nutrients. An increase in photosynthetic activity is often followed by a depletion of dissolved oxygen as plants die and are decomposed by *aerobic organisms. Deoxygenation has an adverse effect on the aquatic animal life.

evagination 1. Turning inside out. **2.** The release of the contents of membranaceous *vesicles to the exterior.

Evaniidae (ensign wasps; suborder *Symphyta, superfamily Evanioidea) Family of black symphytans, 3–15 mm long, which are somewhat spider-like in appearance. The small *abdomen is carried on a long petiole, the *thorax is very stout, and the *coxae are grooved to receive the long *trochanters. Adults are found in association with *Blattodea (cockroaches) whose *oothecae they parasitize. There are about 20 genera and more than 400 species.

event recorder An instrument that is used to record animal behaviour. The operator has a keyboard, similar to that of a small electronic organ, which permits a very large number of signals to be produced by the pressing of keys or combinations of keys. The signals are transmitted to a recording tape from which they can be read by a suitably programmed computer.

Evermanellidae (sabretooth; superorder *Scopelomorpha, order *Myctophiformes)

A small family of marine fish that have a relatively large head with well-developed teeth in the upper and lower jaws, eyes located above the jaws, and a single *dorsal and an *adipose fin. They are small (about 16 cm), rather fragile fish. There are about six species.

eversible Capable of being everted (turned inside out).

Evo *See* MIMETIDAE.

evo-devo (evolutionary development biology) The study of the way differences in the way genes are activated during the development of the *embryo lead to the formation of a range of structures from similar genes. This explains why arms differ from legs, arms and legs from wings and flippers, and one species from another. The study reveals the relatedness between species and the stages leading to the emergence of novel features.

(((●))) SEE WEB LINKS
• 'What Is Evo Devo?'

evolute Unfolded or turned back. *Compare* INVOLUTE.

evolution Change, with continuity in successive generations of organisms (i.e. 'descent with modification' as *Darwin called it). The phenomenon is amply demonstrated by the fossil record, for the changes over geological time are sufficient to recognize distinct eras, for the most part with very different plants and animals.

evolutionary allometry *See* ALLOMETRY.

evolutionary determinism Change in gene frequencies by directed or deterministic processes, in contrast with change due to random or stochastic processes. The relative importance of the two kinds of change in evolutionary development is uncertain.

evolutionary lineage Line of descent of a taxon from its ancestral taxon. A lineage ultimately extends back through the various taxonomic levels, from the species to the genus, from the genus to the family, from the family to the order, etc.

evolutionary rate Amount of evolutionary change that occurs in a given unit of time. This is often difficult to determine, for several reasons: e.g. should the unit of time be geological or biological (the number of generations)? How should morphological change in unrelated groups be compared? In practice it is necessary to adopt a pragmatic approach, such as the number of new genera per million years. *See also* DARWIN.

evolutionary species *See* CHRONOSPECIES.

evolutionary species concept A suggestion made by G. G. *Simpson for adapting the *biological species concept for a palaeontological context: 'a species is a lineage with its own evolutionary role and tendencies.'

evolutionary stable strategy (ESS) In the application of *game theory to evolutionary studies, moves in the game that cannot be beaten; i.e. traits or combinations of traits cannot be replaced by any invading mutant. Evolutionary stable strategy theory has proved very useful in the analysis of certain types of animal behaviour.

(((●))) SEE WEB LINKS
• Information about John Maynard Smith, proposer of the evolutionary stable strategy.

evolutionary trend Steady change in a given adaptive direction, either in an evolutionary lineage or in a particular attribute, e.g. dentition. Such trends are often apparent in unrelated taxa. Formerly they were attributed to orthogenesis; now orthoselection or the contending theory of species selection (*effect hypothesis) are invoked.

exafference The stimulation of an animal as a result of factors external to the animal itself. *Compare* REAFFERENCE.

exaptation A characteristic that opens up a previously unavailable *niche to its possessor. The characteristic may have originated as an *adaptation to some other niche (e.g. it is proposed that feathers were an adaptation to thermoregulation, but opened up the possibility of flight to their possessors), or as a neutral mutation.

exarate In some insects, applied to a *pupa in which the appendages are free and able to move.

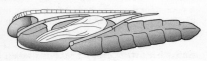

Exarate

excess baggage hypothesis The proposition that organs for which an organism no longer has any use (e.g. functional eyes in some cave-dwelling fish and excess brain tissue in some vertebrates) will be selected against because the energy cost of maintaining them reduces *fitness (i.e. they are 'excess baggage'). The hypothesis also applies to genetically engineered organisms. The cost they incur by carrying inserted genes may reduce their fitness if they are released into the environment, so reducing the risk that release will prove environmentally harmful.

exchange pairing The type of pairing of *homologous chromosomes that allows genetic *crossing-over to take place.

exclusion principle See COMPETITIVE-EXCLUSION PRINCIPLE.

exconjugant One of the two cells that results from *conjugation.

exoccipitals Bones situated to either side of the *foramen magnum of the skull, in mammals forming the *condyles.

exocoel In some corals, the space between pairs of *mesenteries where pleated muscle blocks face away from one another. *Compare* ENTOCOEL.

Exocoetidae (flying fish, halfbeak; subclass *Actinopterygii, order *Atheriniformes) A large family of marine and freshwater fish, all of which have streamlined bodies and spineless fins. The single *dorsal and *anal fins are set far back on the body, close to the forked tail fin. Typically, all flying fish have greatly expanded *pectoral fins, the tips of which may extend to the tail base. One of the more widely distributed species is the cosmopolitan *Exocoetus volitans* (flying fish), reaching a length of 25 cm. The halfbeaks have no expanded pectoral fins, but can be recognized by the elongated lower jaw, which in several species is at least three times the length of the upper jaw. The lower jaw is thought to scoop up minute animals as the halfbeak skims along the surface. One of the largest species, *Hemirhamphus far*, may reach 60 cm. There are about 108 species, distributed world-wide.

Exocoetus volitans (flying fish) See EXOCOETIDAE.

exocrine See ECTOCRINE.

exocrine gland A *gland that discharges its product through a duct. *Compare* ENDOCRINE GLAND.

exocuticle See EXOSKELETON.

exocytosis The process by which a membrane-bound *vacuole fuses with the *cell membrane and thus discharges its contents outside the cell. As well as a mechanism for the removal of wastes, it is commonly employed for the secretion of cell products.

exoenzyme An *enzyme that is discharged outside the cell.

exon Part of the *DNA of a *gene that encodes the information for the *amino-acid sequence in a polypeptide. Many *eukaryote genes contain a series of exons alternating with *introns; following *transcription, the introns are excised and the exon sequences are joined together to form m-RNA, which is used in protein synthesis.

exopeptidase *See* PROTEASE.

exopod In *Crustacea, the outer *ramus of the *biramous limb.

exopodite *See* PROTOPODITE.

exopterygote Applied to an insect in which the wings develop externally and gradually. The insect undergoes no pupal stage or rapid *metamorphosis, and its young are called 'nymphs'.

exoskeleton (cuticle) The horny skeleton that encloses the body of all *Arthropoda. It is secreted by an underlying cellular layer (the hypodermis). The structure is complex. Immediately above the epidermis is the *Schmidt layer; above that an endocuticle of *chitin; and above that an exocuticle of tanned chitin which is frequently dark in colour. There is a thin epicuticle (the outermost layer) covered with wax or grease. Internal to the epicuticle is the procuticle, comprising an outer exocuticle and an inner endocuticle. Both are composed of a complex glycoprotein formed from the binding of *chitin with protein; the molecular structure of the exocuticle is further stabilized by phenolic tanning and additional cross-linkages. The precise nature of these layers—whether they are tanned, lamellar, etc.—varies from species to species, and also from place to place in the exoskeleton of an individual insect, according to its species. In *Crustacea, the procuticle is also impregnated with mineral salts, calcium carbonate, and calcium phosphate. Cuticle is relatively impermeable to water and has a high strength-to-weight ratio, two characteristics important for the successful conquest of the land by the arthropods. Ingrowths of the cuticle, called 'apodemes', provide insertion sites for muscles. The division of the cuticle into separate plates facilitates movement: the plates are connected by the so-called arthrodial membrane, which comprises thin, flexible, untanned cuticle. Periodic moulting of the cuticle (*ecdysis) permits body growth.

exosymbiosis A form of *symbiosis in which one *symbiont lives on the

exterior of the body of the other. *Compare* ENDOSYMBIOSIS.

exotherm *See* POIKILOTHERM.

exotic species Introduced, non-native species.

exotoxin Toxin that is secreted by a living organism.

exozoochory Dispersal of spores or seeds by being carried on the surface of an animal.

experimental error Error that arises because of variation between experimental samples. It may be attributed to differences in materials and/or techniques, rather than to real differences (e.g. in growth or behaviour). Experimental error must be monitored in any statistical comparison of experimental data.

exploitation competition *Competition that occurs where two *species require the same resource and that resource is in short supply. Whichever of the two species is more efficient in accessing the resource is more likely to succeed. The alternative to exploitation competition is *interference competition.

exploratory behaviour A form of *appetitive behaviour that may be goal-orientated (e.g. the search for food or nesting material) or concerned with the examination of areas or articles with which an animal is unfamiliar, in which case the behaviour often exhibits signs of *conflict.

exploratory learning A form of *latent learning in which an animal learns by exploring new surroundings, without apparent reward or punishment.

explosive evolution Sudden diversification or adaptive radiation of a group. The term is an old one, rarely used nowadays. Phases of explosive evolution have occurred in all the higher taxonomic groups, i.e. in genera, families, orders, and classes. An example at the ordinal level is the diversification of the mammals at the dawn of the *Cenozoic Era.

exponential Applied to a rate of change (increase or decrease) that is calculated as a fixed percentage of the starting value, so that the amount of change in a particular period is calculated as the starting value plus the interest accrued in preceding periods, multiplied by the rate of change.

expressivity The degree to which a particular *genotype is expressed in the *phenotype.

extant Applied to a taxon some of whose members are living at the present time. *Compare* EXTINCT.

exteroceptor A sense organ that registers the state of the environment outside an animal's body. Exteroceptors effect the human senses of sight, hearing, taste, smell, and touch. *Compare* ENTEROCEPTOR; PROPRIOCEPTOR.

extinct Applied to a taxon no member of which is living at the present time. *Compare* EXTANT.

extinction 1. The elimination of a taxon. **2.** The loss of learned patterns of behaviour from the repertoire of an animal to which they have become irrelevant.

extracellular matrix A mesh of insoluble *proteins and *carbohydrates that is produced by cells and fills most of the space between cells, like mortar in a wall.

exumbrella The upper surface of the *umbrella of a jellyfish (*medusa).

exuvium In *Arthropoda, the cast *exoskeleton left behind after moulting.

eye An organ, sensitive to light, that is developed in many animals. In its simplest form, it consists of a light receptor (ocellus) capable of distinguishing light from shade. In *Arthropoda there are compound eyes, each consisting of many separate receptors (ommatidia) each of which contains light-sensitive cells. In *Cephalopoda and vertebrates each eye is a single organ, possessing an iris which controls the size of aperture (pupil) through which light passes to a lens, which in turn focuses the light on to a *retina.

eyebrow fish (*Psenes arafurensis*) *See* NOMEIDAE.

eyed lizard (*Lacerta lepida*) *See* LACERTIDAE.

eyespot A dark spot, resembling an eye, on an expendable part of an insect's body that protects the insect from predators either by deflecting attacks away from vital areas of the body, or by evincing a *startle response. Many butterflies have small eyespots along the edges of their wings, others have large eyespots on the hind wings that can be exposed suddenly by moving the forewings, and some have both.

F A notation, introduced by the geneticist Sewall Wright (1889–1988), for the inbreeding coefficient. *See also* COEFFICIENT OF INBREEDING.

F_1 The first filial generation of animals or plants, produced by crossing two parental lines (which are referred to as *P*).

F_2 The second filial generation of animals or plants, produced by selfing or intercrossing of F_1 individuals.

facet On the side of the *centrum of a dorsal vertebra, a cup-shaped depression to which the head of a rib is attached; if the 'cup' is incompletely hemispherical (i.e. 'half a cup') it is called a demi-facet or half-facet. A vertebra may have one or more facets or demi-facets.

facilitated diffusion A carrier-mediated membrane-transport mechanism, driven by a diffusion gradient, that is important in the transport of materials into and out of cells.

facilitation 1. The increase in the responsiveness of a nerve cell or *effector cell that is caused by the summation of impulses. **2.** The intensification of a behaviour that is caused by the presence of another animal of the same species.

facing-away display A part of the *courtship *display in some birds (e.g. black-headed gull, *Larus ridibundus*) in which the pair stand side by side facing the same direction with their necks extended, and their heads turned away from each other.

facultative Applied to organisms that are able to adopt an alternative mode of living. For example, a facultative anaerobe is an aerobic organism that can also grow under anaerobic conditions.

facultative mutualism *See* PROTOCO-OPERATION.

facultative parasite *See* PARASITISM.

FAD *See* FLAVIN ADENINE DINUCLEOTIDE.

faeces (feces) Waste material, consisting of the indigestible residues of food, bacteria, and dead cells from the lining of the *alimentary canal, that is expelled from the body through the *anus.

fairy-bluebirds *See* IRENIDAE.

fairyflies *See* MYMARIDAE.

fairy shrimps *See* ANOSTRACA.

fairy warblers *See* ACANTHIZIDAE.

fairy wrens (*Malurus*) *See* MALURIDAE.

falcate Curved, as a sickle.

***Falco* (falcons)** *See* FALCONIDAE.

Falconidae (falcons, caracaras; class *Aves, order *Falconiformes) A family of mainly grey or brown, long-winged birds that have long, usually barred tails. Falcons are fast fliers, catching animal prey with their feet; caracaras are long-legged, slow fliers that feed mainly on carrion. Falconids inhabit mainly forest and open country, nesting on the ground, on ledges, or in trees, but only caracaras build nests of their own. The largest genus is *Falco*, comprising typical falcons, many of them migratory, and there are five species of *Micrastur* (forest falcons) that inhabit dense forest in Central and S. America. There are 10 genera in the family, with about 62 species, found world-wide.

Falconiformes (eagles, falcons, hawks, osprey, secretary bird, vultures; class *Aves) An order of diurnal birds that have sharp claws and strong hooked bills with a basal *cere. They have powerful wings, medium to long tails, and are excellent fliers. They are mainly carnivorous, feeding on a wide variety of animal prey. There are five families: *Accipitridae, *Cathartidae, *Falconidae, *Pandionidae, and *Sagittariidae. They are found world-wide.

falcons *See* FALCONIDAE.

Fallopian tube In female mammals, the duct, with a funnel-shaped opening, that connects the uterus and the peritoneal cavity. The muscular contraction and ciliary action of the tube moves eggs from the ovary to the uterus, and sperms from the uterus to the upper part of the tube, where fertilization occurs.

fallow deer (*Dama*) *See* CERVIDAE.

false gharial (*Tomistoma schlegeli*) *See* CROCODYLIDAE; GAVIALIDAE.

false legs *See* PROLEGS.

false moray *See* XENOCONGRIDAE.

false roundhead *See* PSEUDOPLESIOPIDAE.

false sunbirds (*Neodrepanis*) *See* PHILEPITTIDAE.

false vampire *See* MEGADERMATIDAE.

Famennian *See* DEVONIAN.

fantails *See* MUSCICAPIDAE.

fanworms *See* SEDENTARIA.

farinose Floury or powdery in appearance or texture.

fascicle (fasciculus) A small bundle, e.g. of muscle fibres or nerve fibres.

fasciole In some *Echinodermata, a groove on the *test that bears no *tubercles or large spines; very small spines are present and are covered in *cilia. The cilia create water currents that move mud and other extraneous material from the surface of the test.

fat *See* LIPID.

father lasher (*Myoxocephalus scorpius*) *See* COTTIDAE.

fatigue *See* ACCOMMODATION.

fatty acid A long-chained, predominantly unbranched, carboxylic acid: it may be saturated or unsaturated. The side chains usually have an odd number (commonly 15 or 17) of carbon atoms to which hydrogen atoms may be attached; if all available hydrogen sites are occupied the fatty acid is said to be saturated, if one or more sites is vacant it is unsaturated, and if two or more sites are vacant it is polyunsaturated. Fatty acids have the general formula R—COOH, where R is hydrogen or a group of hydrogen and carbon atoms; saturated fatty acids have the general formula $C_nH_{2n+1}COOH$.

fauna (*adj.* faunal, faunistic) The animal life of a region or geological period.

faunal *See* FAUNA.

faunal province *See* FAUNAL REGION.

faunal region (faunal province, faunal realm, faunal zoogeographic kingdom) A biological division of the Earth's surface (i.e. a large geographical area) that contains a fauna more or less peculiar to it. The degree of distinctiveness varies with the region concerned and reflects partly climate and partly the existence of barriers to migration. The number of regions recognized varies from one authority to another, but a minimum of six are recognized: *Australian, *Ethiopian, *Nearctic, *Neotropical, *Oriental, and *Palaearctic. In defining a region, greatest emphasis is given to mammals.

faunal zone *See* ASSEMBLAGE ZONE.

faunal zoogeographic kingdom *See* FAUNAL REGION.

faunistic *See* FAUNA.

faunizone *See* ASSEMBLAGE ZONE.

fawns *See* NYMPHALIDAE.

feather A keratinous outgrowth of the skin of birds that is highly modified for the purposes of flight, insulation, and display. Feathers can be divided into distinct types: contour feathers, down feathers, intermediate feathers, filoplumes, powder down, and bristles. They are pigmented, iridescent colours being due to scattered light from specially structured feathers. Worn feathers are annually renewed by moulting.

Feather

featherback *See* NOTOPTERIDAE.

feather stars *See* COMATULIDA; CRINOIDEA.

feces *See* FAECES.

fecundity *See* FERTILITY.

feedback regulation A process by which the product of a metabolic pathway influences its own production by controlling the amount and/or activity of one or more *enzymes involved in the pathway. This influence is commonly inhibitory and is called negative feedback; this is a common form of homoeostatic control (*see* HOMOEOSTASIS). Positive feedback (e.g. in courtship rituals) leads to an increase in the amount and/or activity of enzymes.

feeding All behaviour that involves the obtaining, manipulation, and ingestion of food. *Compare* FORAGING.

Felidae (cats; suborder *Fissipedia, superfamily *Feloidea) A family that comprises the extant and extinct cats, the most specialized of all carnivorous mammals. The brain is large, with large olfactory centres and cerebral hemispheres which overlap the *cerebellum. The jaws are powerful and cannot be rotated (as for chewing). The *incisors are in a straight line across the jaws, the *canines are long, and the cheek teeth are reduced in number, the *carnassials being well developed for shearing. The skeleton is specialized for leaping. There are five digits in the fore limbs, four in the hind limbs, all clawed. The claws are retractile in most species, but not in *Acinonyx jubatus* (cheetah). The gait is *digitigrade. The cat family had diverged from the stem family of the *Carnivora, the *Miacidae, in *Eocene times and cats were more or less modern in appearance by the *Oligocene. Until the *Pliocene, 'sabretooth' and 'false sabretooth' forms are known, and probably it was from the latter that the present-day cat, with much reduced canines, was derived during the Pliocene. Cats are divided into several genera: *Acinonyx, Felis, Panthera* (or *Leo*), and *Neofelis* (*N. nebulosa* is the clouded leopard) are generally recognized, but some authorities separate *Leopardus* (ocelot and margay), *Herpailurus* (jaguarundi), *Prionailurus* (leopard-cat and fishing-cat), *Puma* (puma), and others as genera distinct from *Felis* in which they are otherwise included. Felids are terrestrial and/or arboreal, most feeding on higher vertebrates but some on invertebrates, fish, and fruit. They are distributed throughout the world except for some oceanic islands, Australasia, and Madagascar. There is controversy whether *Cryptoprocta* (fossa), of Madagascar, should also be placed in the Felidae; it is customarily assigned to the *Viverridae.

Feliformia In some classifications, a suborder of *Carnivora that comprises the *Feloidea and *Miacoidea.

***Felis** See* FELIDAE.

Feloidea (order *Carnivora, suborder *Fissipedia (or *Feliformia) A superfamily that comprises the families *Viverridae (civets), *Herpestidae (mongooses), *Hyaenidae (hyenas), and *Felidae (cats). These are the most specialized of all mammalian carnivores, the specialization being most marked in the Felidae. The feloids range from the civets, small animals occupying a position in tropical environments similar to that occupied by the weasels in higher latitudes, to the lions and tigers, and include the hyenas, in which the teeth are sufficiently massive to crush bones, permitting hyenas to subsist partly by scavenging.

femora *See* FEMUR.

femur (*pl.* **femora) 1.** In tetrapods, the upper bone of the hind limb. **2.** In *Insecta, the third segment in the leg, which articulates with the *trochanter proximally, and with the *tibia distally. The femur is the largest and most robust segment of the leg in most adult insects. Movement with respect to the trochanter is usually limited.

fenestrated Perforated with small openings or transparent areas.

fennec (*Fennecus* **(or** *Vulpes***)** *zerda***)** *See* CANIDAE.

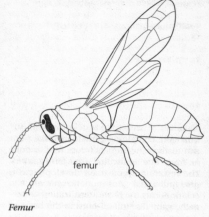
femur

Femur

Ferae (infraclass *Eutheria, cohort *Ferungulata) A superorder that comprises the extinct order *Creodonta and the extant order *Carnivora, thus including all the terrestrial mammalian carnivores, as well as the seals and sea lions (suborder *Pinnipedia).

feral From the Latin *ferus*, 'wild', an adjective applied to a wild or undomesticated organism. In particular, the term is applied to wild strains of an otherwise domesticated species or to an organism that has reverted to a wild condition following escape from captivity, etc. Some authors make these distinctions: wild species, subject to natural selection only; domestic species, subject to selection by humans; and feral species, formerly domestic species which are now, as escapees, subject once again to natural selection.

ferruginous Resembling rust, containing rust, or rust-coloured.

fertility (fecundity, fruitfulness) The reproductive capacity of an organism, i.e. the number of eggs that develop in a mated female over a specified period. It is usually calculated at the stage when this number is readily observable (i.e. in *oviparous animals when eggs are laid and in *viviparous animals when young are born), although strictly speaking it applies from the time that fertilization occurs. Sometimes the term 'fertility' is applied only to the production of fertilized eggs (ova), while 'fecundity' is used for the production of offspring, so excluding those *embryos which fail to develop.

fertilization The union of two *gametes to produce a *zygote, which occurs during sexual reproduction. Fertilization involves the fusion of two haploid nuclei (*see* HAPLOID NUMBER), containing genetic material from two distinct individuals (cross-fertilization) or from one individual (self-fertilization). The resulting zygote then develops into a new individual. Most aquatic animals, e.g. echinoderms, achieve fertilization externally, gametes uniting outside the body of the parents. Some other animals, particularly terrestrial species, have internal fertilization, with the union of gametes inside the female.

Ferungulata (subclass *Theria, infraclass *Eutheria) A cohort, proposed on palaeontological grounds but regarded by some authorities as artificial, that includes the *Carnivora (all of the modern mammalian carnivores), the primitive ungulates (including the elephants and sea cows), the *Perissodactyla (tapirs, rhinoceroses, horses, etc.), and the *Artiodactyla (pigs, camels, cattle, etc.), which are believed to have arisen from a common population in the *Palaeocene. At that time all eutherians were rather undifferentiated and it is supposed that the Ferungulata began to diverge rapidly following their break from the ancestral stock, so that today the cohort comprises 15 orders whose members are far more diversified than those of other cohorts.

fetus *See* FOETUS.

fever flies *See* BIBIONIDAE.

Feylina *See* FEYLINIIDAE.

Feyliniidae (order *Squamata, suborder *Sauria) A family of lizards which resemble limbless skinks but are not closely related to them. The head is flattened, with tiny eyes under transparent scales, and there is no ear opening. The body is cylindrical. The lizards feed mainly on termites. There are four species in a single genus, *Feylina*, occurring in Equatorial Africa and Madagascar.

fibril A small fibre or thread-like structure.

fibrillin A *glycoprotein secreted by *fibroblasts and released into the *extracellular matrix where it forms *microfibrils.

fibrin An insoluble *protein found in blood clots, which it helps to form. It is formed at the site of a wound from a soluble precursor, *fibrinogen, by the action of the enzyme *thrombin. *See also* BLOOD CLOTTING.

fibrinogen A soluble blood *protein that is acted upon by the enzyme *thrombin during *blood clotting to give the insoluble protein *fibrin. *See also* BLOOD PLASMA.

fibroblast A *connective-tissue cell that manufactures and secretes *collagen and that is extensively involved in the process of wound healing.

fibrocartilage A type of *cartilage in which fibres are embedded.

fibula In tetrapods, the posterior of the two bones of the lower part of the hind limb.

fibulare In tetrapods, the posterior (i.e. closest to and articulating with the *fibula) of the three tarsal (ankle) bones closest to the centre of the body (i.e. proximal).

Ficedula (flycatchers) *See* MUSCICAPIDAE.

Fick, Adolf Eugen (1829–1921) A German physiologist, born in Kassel, who qualified in medicine at the University of Marburg in 1851. In 1855 he proposed *Fick's laws, in 1870 he devised Fick's principle, which allows the measurement of cardiac output, and in 1887 he made the first successful contact lens. Fick died at Blankenberge, Belgium.

Fick's laws Two laws that describe the rate at which particles diffuse (*see* DIFFUSION) along a concentration gradient. The first law applies when the concentration of the diffusing particles remains constant over time, and the second law applies to situations where the concentration changes. Fick's first law states that $J = -D\nabla f$, where J is the rate at which particles of a specified type diffuse across a surface (the use of bold type indicating a vector quantity), D is the diffusion coefficient, which is a function of the type of diffusion medium and its temperature, ∇ is the concentration gradient, and f is the number of diffusing particles per unit volume. Fick's second law states that $\delta\varphi/\delta t = D (\delta^2\varphi/\delta x^2)$, where t is time, φ is the concentration, and x is the diffusion length.

((⊕)) SEE WEB LINKS
• Explanation of Fick's laws.

Fidelia *See* FIDELIIDAE.

Fideliidae (order *Hymenoptera, suborder *Apocrita) Small family of desert mining bees, the sister group of the *Megachilidae—members of the Fideliidae resemble those of the Megachilidae in having an abdominal pollen *scopa, but differ from them in having scopae also on the hind legs, and in possessing certain primitive characters: three submarginal cells in the fore wing, and pygidial (*see* PYGIDIUM) and basitibial plates in the female. The brood cells are unlined. The family has a disjunct distribution, with one genus, *Neofidelia*, found in the southern Atacama Desert, Chile; and two genera, *Fidelia* and *Parafidelia*, in southern Africa. One species of *Fidelia* has recently been found in N. Africa. There are fewer than 20 species in three genera.

field cricket *See* GRYLLIDAE.

field mouse (wood mouse; *Apodemus sylvaticus*) *See* MURIDAE.

fight or flight reaction (acute stress response, hyperarousal) A set of physiological responses to *stress, generated by the *autonomic nervous system, that prime an animal to flee from a perceived threat from another animal, or to confront the danger and if necessary fight the aggressor.

The responses include rapid breathing, increased heart rate, inhibition of stomach and intestinal movement, and sweating. The phenomenon was first described in 1915 by the American physiologist Walter Bradford Cannon (1871–1945). It is now recognized as the first stage of the *general-adaptation syndrome.

Figitidae (suborder *Apocrita, superfamily Cynipoidea) Family of wasps, generally small (3–6 mm long) and shiny black, whose larvae are pupal parasites of *Diptera, scale insects, and aphids. One subfamily parasitizes only Hemerobiidae (lacewings). The family is not common.

fig wasps *See* AGAONIDAE.

filament 1. A thin strand (e.g. of a feather or gill). **2.** One of the strands of *protein, variously grouped according to diameter (in the range 4–15 nm), found in many types of cell. Their functional significance is incompletely understood, but since they are largely composed of the contractile proteins actin and/or myosin it is presumed that the motility of the cell or its contents forms part of their role.

filefish 1. *See* BALISTIDAE. **2.** *See* MONACANTHIDAE.

file snakes *See* ACROCHORDIDAE.

filibranchiate Applied to a type of gill, common to *Bivalvia, that is composed of sheets of filaments forming a W shape.

filiform Thread-like; long and slender.

filoplume A fine, hairlike *feather with a thin *rachis and few *barbs present in circles around the contour feathers or down feathers.

filopodium A type of *pseudopodium that is extremely slender and tapers to a fine point.

filter route A term introduced by the American palaeontologist G. G. *Simpson in 1940 to describe a faunal migration route across which the spread of some animals is very likely but the spread of others is correspondingly improbable. The route thus filters out part of the fauna, but permits the rest to pass. Deserts and mountain ranges provide examples of filter routes.

fimbriate With the margin bearing a fringe, usually of hairs (fimbriae).

fin An appendage of fish and fish-like aquatic animals used for locomotion, steering, and balancing of the body. The skin fold forming the fin membrane is supported by cartilaginous, horny, or bony *fin rays, which can be soft and flexible (soft rays) or hard and inflexible (fin spines). The median unpaired fins of fish include the dorsal fin on the back, the caudal fin (tail fin), and the anal fin behind the vent (anus). Usually there are also two pairs of lateral fins: the pectoral fins and the pelvic (also called ventral or abdominal) fins. *See also* APPENDICULAR SKELETON.

finches 1. *See* EMBERIZIDAE; FRINGILLIDAE. **2.** (cardinals) *See* CARDINALIDAE. **3.** (mannikins, munias, waxbills) *See* ESTRILDIDAE. **4.** (plush-capped finch, *Catamblyrhynchus diadema*) *See* CATAMBLYRHYNCHIDAE.

finfoots *See* HELIORNITHIDAE.

finlet A small, isolated *fin formed either by a separate spine with its own membrane (e.g. in the stickleback, *Gasterosteus aculeatus*) or by a membrane without a spine (e.g. in tuna).

fin ray A rod-like support of the fin membrane, composed of *cartilage in lampreys, horny fibres in sharks, and *bone in bony fish.

fin whale (*Balaenoptera physalus*) *See* BALAENOPTERIDAE.

fire-bellied toad (*Bombina bombina*) *See* DISCOGLOSSIDAE.

firebrat (*Lepismodes inquilinus*) *See* LEPISMATOIDEA.

firebug (*Pyrrhocoris apterus*) *See* PYRRHOCORIDAE.

firefly 1. (*Pyrophorus*) *See* ELATERIDAE. **2.** *See* LAMPYRIDAE.

fire salamander (*Salamandra salamandra*) *See* SALAMANDRIDAE.

fish Broadly speaking, any *poikilothermic, legless, aquatic vertebrate that possesses a series of *gills on each side of the pharynx, a two-chambered heart, no internal nostrils, and at least a median *fin as well as a tail fin. If the *Agnatha (lampreys and hagfish) are excluded, the fish (*Pisces) would still include the *Chondrichthyes (sharks and rays), in which the skeleton is cartilaginous, as well as the *Osteichthyes (bony fish). In addition to the criteria mentioned, these two classes also possess well-developed *gill arches, a pair of *pectoral fins, and a pair of *pelvic fins. Some consider, however, that only the bony fish (Osteichthyes) should be classed as real fish.

fish eagles (*Haliaeetus*) *See* ACCIPITRIDAE.

fish-eating bat (*Noctilio*) *See* NOCTILIONIDAE.

fishflies *See* CORYDALIDAE.

fish lice Members of two orders of *Copepoda, the Caligoida and Lernaeopodida, which are highly modified ectoparasites of marine and freshwater fish and whales.

Fissipedia The name formerly given to the non-marine *Carnivora.

Fistulariidae (flutemouth, cornetfish; subclass *Actinopterygii, order *Syngnathiformes) A small family of tropical to subtropical, marine fish that have an elongate, scaleless body, a long, tubular snout terminating in a tiny mouth, and a small, forked tail fin with a central long ray trailing behind it. Some species may grow to a length of 2 m. There are about four species, distributed world-wide.

fitness **1.** In ecology, the extent to which an organism is well adapted to its environment. The fitness of an individual animal is a measure of its ability, relative to others, to leave viable offspring. **2. (Darwinian fitness)** *See* ADAPTIVE VALUE.

fixation **1.** The condition in which a gene occurs at 100% frequency in a population; i.e. there is no polymorphism, all members of a population being *homozygous for a particular *allele at a given *locus. **2.** The first step in making permanent preparations of tissues for microscopic study, by killing cells and preventing subsequent decay with as little distortion of structure as possible. Examples of fixatives are formaldehyde and osmium tetroxide, often used as mixtures.

fixed-action pattern Apparently stereotyped behaviour, exhibited or capable of being exhibited by all members of a species or higher taxonomic group, which may be used to achieve more than one objective, which may be innate or learned, and whose acquisition may be affected by environmental factors. Examples include the calls of certain birds: these are influenced by sounds heard by the birds early in their lives, but once acquired the songs are performed in a stereotyped way.

flabby whalefish *See* CETOMIMIDAE.

flabelliform Shaped like a fan.

flabellum A fan-shaped structure, e.g. on the tongue of some bees.

flacherie Disease of the silkworm (*Bombyx mori*), which is apparently caused by a virus.

flagella Plural of *flagellum.

flagellomere A segment of the *flagellum of an insect antenna. The number of flagellomeres varies greatly and is often of diagnostic importance when identifying species.

flagellum **1.** A thread-like *organelle which usually functions in locomotion.

Flagella are found on a range of *eukaryotic cells, including those of certain *Protozoa and *spermatozoa. Bacterial and eukaryotic flagella are very different both in structure and mode of operation and are not homologous. Bacterial flagella rotate; eukaryotic flagella undulate. The eukaryotic flagellum has the more complex structure, and there are two types: the whiplash type, which is smooth and whip-like, and the tinsel type, which bears numerous fine, hair-like projections along its length. **2.** In insects, the antennal segments distal to the *scape.

flagtail *See* KUHLIIDAE.

flame cell Excretory apparatus, similar in structure and function to a *solenocyte, that has a tuft of *cilia (rather than a single flagellum as in a solenocyte) at the blind end of the cell.

flamingo milk A secretion from the lining of the *crop in both male and female flamingos (*Phoenicopteridae) that the birds regurgitate to feed their young. Production of the milk is stimulated by *prolactin, and it consists of 8–9% protein, 15% fat, and 1% red blood cells. *Compare* PIGEON MILK.

flamingos *See* PHOENICOPTERIDAE.

Flandrian The present interglacial. Evidence suggests that the so-called postglacial period, the warm phase following the last (*Devensian) ice advance or cold phase, is more appropriately treated as another interglacial of the *Quaternary (or late *Cenozoic) Ice Age. In Europe, the warmest Flandrian stage occurred during Atlantic times, about 6000 BP (the Hypsithermal is the equivalent N. American climatic optimum). No consensus view exists as to when the ice advance or extreme cold conditions will prevail once again in high mid-latitudes, nor as to how quickly these conditions will arise. The Flandrian is sometimes referred to alternatively as the *Holocene interglacial.

flap-flight display In bird *courtship, a behaviour similar to *circle flight, but with more exaggerated wing beats, perhaps indicating an increase in the intensity of the ritual.

flathead *See* PLATYCEPHALIDAE.

flat-headed borer *See* BUPRESTIDAE.

Flatidae (order *Hemiptera, suborder *Homoptera) Family of bugs in which the *tegmina are broad and brightly coloured, and have a broad *costal cell with many transverse veins. Some species occur in two different colour forms, and when adults of these two forms cluster together on a twig they give the effect of a spike of coloured flowers with green bracts. There are about 1000 species occurring mainly in the tropics.

flatworms *See* PLATYHELMINTHES.

flavin adenine dinucleotide (FAD) A *coenzyme derivative of the vitamin *riboflavin, which participates in dehydrogenation reactions mediated by *flavoproteins. It is an important intermediate in *oxidative phosphorylation.

flavoprotein A *conjugated protein in which the *prosthetic group is a flavin nucleotide coenzyme such as flavin adenine dinucleotide (FAD) or riboflavin mononucleotide (FMN); some flavoproteins contain either haem or metal ions in addition. They are widely distributed in cells of all types and serve as electron transport agents.

flea beetle *See* CHRYSOMELIDAE.

fledge 1. To reach the stage in the growth of a young bird at which its flight muscles and feathers have developed sufficiently for it to be capable of flying. 2. To raise a young bird until it is fully grown. 3. To reach the stage in its development at which a young bird leaves the nest, whether or not it can fly. 4. To raise a young bird to a stage at which it can leave its parents and live independently.

flehmen Behaviour, probably with a sexual significance, exhibited by felids, goats, and some other mammals when examining a *scent mark. Having sniffed the mark intensively, the animal raises its head with its mouth partly open and upper lip slightly drawn back, stares fixedly to its front, and breathes slowly. Sniffing followed by flehmen may occur more than once at a single encounter with a scent mark.

flesh flies *See* SARCOPHAGIDAE.

flexuous Wavy; bending in a zigzag manner.

flies *See* DIPTERA.

floccose Having a loose, woolly or cottony surface.

flocking Among birds, the formation of a group with a social organization. In some species flocking is accompanied by communal nesting and roosting, in others it occurs only outside the breeding season. About half of all known species of birds exhibit flocking behaviour at some stage in their life cycles.

flounder (*Platichthys flesus*) *See* PLEURONECTIDAE.

flower flies *See* SYRPHIDAE.

flowerpeckers *See* DICAEIDAE.

fluid mosaic model A description of the structure of *cell membranes as consisting of a heterogeneous set of *globular protein molecules arranged with their polar groups (*see* POLAR MOLECULE) protruding into the aqueous phase and their nonpolar groups largely buried in the *hydrophobic interior of the membrane. The molecules are partially embedded in a matrix of *phospholipid. The model was first proposed in 1972 by S. J. Singer and Garth L. Nicolson ('The fluid mosaic model of the structure of cell membranes', *Science*, 175, 4023, 720–731).

flukes *See* DIGENEA; TREMATODA.

fluorescent-antibody technique A method for detecting the location of a specific *antigen in a cell by staining a section of

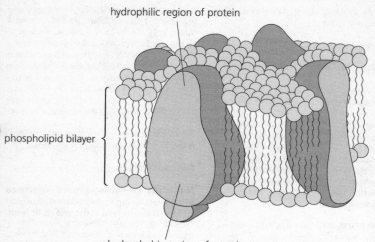

Fluid mosaic model

the tissue with *antibody (specific to that antigen) that is combined with a fluorochrome (a substance that fluoresces in ultraviolet light). The antigen is located wherever fluorescence is observed.

flutemouth 1. (trumpetfish) *See* AULO-STOMIDAE. 2. *See* FISTULARIIDAE.

flycatchers 1. *See* MUSCICAPIDAE. 2. (phoebes, tyrant flycatchers) *See* TYRANNIDAE.

flying dragon (*Draco volans*) *See* AGAMIDAE.

flying fish *See* EXOCOETIDAE.

flying fox *See* MEGACHIROPTERA.

flying frogs *See* RHACOPHORIDAE.

flying gurnard (*Dactylopterus volitans*) *See* DACTYLOPTERIDAE.

flying lemur (*Cynocephalus*) *See* CYNOCEPHALIDAE.

flying lizard (*Draco volans*) *See* AGAMIDAE.

flying squirrel (*Petaurista alborufa*) *See* SCIURIDAE.

foetus (fetus) A mammalian *embryo from the stage of its development where its main adult features can be recognized until its birth.

folic acid (pteroglutamic acid) Vitamin B$_9$ (*see* VITAMIN), a deficiency of which results, among other symptoms, in reduced growth and anaemia. *Coenzymes derived from folic acid participate in the production of the nitrogenous bases adenine and thymine, which are essential for DNA synthesis.

Folin–Ciocalteau reaction The basis of a colorimetric method for the quantitative determination of proteins. It is dependent upon the production of a blue colour as a result of the reaction of *tyrosine in the sample with a phosphomolybdotungstic-acid reagent.

folivorous Leaf-eating.

follicle A small sac, cavity, or gland. *See also* GRAAFIAN FOLLICLE.

follicle-stimulating hormone (FSH) A proteinaceous, gonadotropic hormone (*see* GONADOTROPIN) secreted by the anterior *pituitary (adenohypophysis) that stimulates spermatogenesis in the testis and the growth of follicles in the ovary.

fontanelle 1. In a vertebrate *foetus or infant, a soft region between the incompletely ossified bones of the *cranium. Fontanelles allow the cranium to be slightly flexible, easing the passage through the birth canal. **2.** In most species of termites (*Isoptera), a depression between the eyes that contains the *frontal gland.

food begging Juvenile behaviour that may endure into adulthood (e.g. forming part of some *courtship rituals), in which young that are dependent on their parents for food stimulate adults to feed them.

food chain The transfer of energy from the primary producers (green plants) through a series of organisms that eat and are eaten, assuming that each organism feeds on only one other type of organism (e.g. earthworm →blackbird→sparrowhawk). At each stage much energy is lost as heat, a fact that usually limits the number of *trophic levels in the chain to four or five. Two basic types of food chain are recognized: the *grazing and *detrital pathways. In practice chains interact to give a complex *food web.

food-chain efficiency The ratio between the energy value (the nutritional value, discounting indigestible parts such as hair or feathers) of prey consumed by a predator and the energy value of the food eaten by that prey. Maximum food-chain efficiency (gross ecological efficiency) occurs when the yield of prey to the predator is such that the surviving prey just consume all the available food: this implies that the food of the prey is being exploited to the best advantage by the predator.

food vacuole (gastriole) A small *vesicle that is formed inside a protozoan cell and within which food particles ingested by the cell are contained and subsequently digested.

food web A diagram that represents the feeding relationships of organisms within an *ecosystem. It consists of a series of interconnecting *food chains. Only some of the many possible relationships can be shown in such a diagram and it is usual to include only one or two carnivores at the highest levels.

football fish *See* HIMANTOLOPHIDAE.

footmen *See* ARCTIIDAE.

foraging All behaviour that is associated with the obtaining and consumption of food for which the animal must search or hunt. *Compare* FEEDING.

foram *See* FORAMINIFERIDA.

foramen An opening through which pass nerves, blood vessels, muscles, etc.

foramen magnum The opening at the posterior end of the skull through which the spinal cord passes.

foraminifer *See* FORAMINIFERIDA.

Foraminifera *See* FORAMINIFERIDA.

foraminiferan *See* FORAMINIFERIDA.

foraminiferid *See* FORAMINIFERIDA.

Foraminiferida (foram, foraminifer (*pl.* foraminifera), foraminiferan, foraminiferid; superclass *Sarcodina, class *Rhizopoda) An order (or in some classifications a subclass, Foraminifera) of amoeboid (*see* AMOEBA) *Protozoa in which the cell is protected by a *test, consisting of one to many chambers, whose composition is of great importance in classification. The three main types are: (*a*) most primitively, a test wall composed of *tectin (which also forms an underlying layer in the other two types); (*b*) a test formed from agglutinated sedimentary particles; and (*c*) a fully mineralized test composed of secreted calcareous or siliceous minerals of which *aragonite and *calcite are the most common. The arrangement of the chambers may be linear, spiral, cone-like, etc. Numerous fossil foraminifera are known, usually less than 1 mm across; though some, like the fusilinids (*Carboniferous to *Permian) and nummulitids (*Eocene to *Oligocene) were appreciably larger: some measured up to 100 mm in diameter. Most species live in marine environments. Agglutinated forms predominated in the *Cambrian and *Ordovician, presumably derived from a tectinous ancestor, while forms with fully mineralized tests appeared in the Ordovician and diversified greatly in the *Devonian. The Foraminiferida are important zonal fossils, and some planktonic varieties can be used for stratigraphic correlation on virtually a world-wide scale. Accumulations of their tests make up a substantial part of certain geological formations (e.g. chalk deposits of the *Cretaceous and *Globigerina ooze).

forceps Modified insect *cerci which are found in the *Dermaptera (earwigs) and the family Japygidae (order *Diplura). In the earwigs they are used to grasp prey, and in offence and defence. They may also assist in the folding of the membranaceous hind wings beneath the short *tegmina, after flight.

forcipate Forked, like pincers.

Forcipulatida (starfish; class *Stelleroidea, subclass *Asteroidea) An order of starfish characterized by the nature of the skeletal framework that surrounds the mouth, the *mouth angle plates originating as modified ambulacral ossicles. In addition, the *pedicellariae are *furcipulate.

forester moths See ZYGAENIDAE.

forest falcons See FALCONIDAE.

***Forficula auricularia* (European earwig)** See FORFICULIDAE.

Forficulidae (subclass *Pterygota, order *Dermaptera) Large, cosmopolitan family of typical earwigs which includes

Forficula auricularia (European earwig), an occasional pest of vegetable and fruit crops, and of ornamental plants.

Formicariidae (antbirds, antpittas, antshrikes; class *Aves, order *Passeriformes) A family of small to medium-sized, black, grey, or brown birds, sometimes streaked or barred, that have short, rounded wings, and short to long tails; their legs are short (arboreal) to long (terrestrial), and their bill is strong and slightly to very hooked. Antshrikes (18 species of *Thamnophilus*) superficially resemble shrikes, and have stout, hooked bills. They inhabit forest and scrub, where they are arboreal or terrestrial, building nests in tree holes, hanging from branches, or on the ground. They are insectivorous, often feeding on and associating with ants. There are 53 genera, with about 229 species, found in Central and S. America.

Formicidae (Formicoidea; ants; order *Hymenoptera, suborder *Apocrita) Family of insects in which the *petiole is composed of one or two narrow segments. All the known species are social. Males and females are usually *alate. After the nuptial flight, inseminated females shed their wings and found colonies by laying eggs and rearing larvae, often using the reserves in the now useless wing muscles. The female worker caste is always wingless, usually sterile, and often shows *polyethism. Some species are dulotic (see DULOSIS). Most species are scavengers of animal remains; others are seed-eaters, active predators, or fungus feeders. Many species obtain honeydew from *Homoptera. Nest sites range from cavities in wood or soil to carton or leaf constructions. Some species are symbiotic with plants. About 8000 species have been described, but the total number is probably around 13 000 species.

Formicinae (suborder *Apocrita, family *Formicidae) Subfamily of ants which have a single *petiole segment. The tip of the *gaster is armed with an acidopore through which venom containing formic acid is ejected. The subfamily includes the weaver ants. There are about 5000 species, with a

world-wide distribution. *See also* OECO-
PHYLLA LEAKEYI.

Formicoidea *See* FORMICIDAE.

forward display A behaviour by which
a male gull threatens a rival that persists
in an attempt to invade its nesting site.
Males arrive at the nesting ground before
females. Having taken up position, the
male repeatedly utters a loud call, known
as the long call. If another male then approaches, the bird raises its head, holding its bill in a position from where it can
strike downward. This is the upright display. If this fails and the rival persists, the
bird holds its body parallel to the ground
with its bill forward, pointing directly at its
rival. This is the forward display and the
bird will often follow it with a charge and
attempt to stab its rival and strike it with
a wing.

fossa 1. A depressed area, groove, pit, or
hollow, usually in a bone. **2.** *Cryptoprocta
ferox. See* EUPLERIDAE; VIVERRIDAE.

fossil The remains or traces of a once-living
organism. These include skeletons, tracks,
impressions, trails, borings, and casts. At
one time the term was used only of material dating from before the end of the most
recent glacial period, so fossils were more
than 10 000 years old; remains and traces
younger than 10 000 years were known as
'subfossils'. This restriction has now been
abandoned and all ancient remains are
called 'fossils' regardless of their age. Fossils are usually found in hard rock, but not
always: e.g. woolly mammoths living 20 000
years ago were recovered from the frozen
tundra of Siberia.

fossorial Burrowing.

founder effect (peripatric speciation)
The derivation of a new population (e.g. on
an oceanic island) from a single individual
or a limited number of immigrants. The
founder(s) represent a very small sample of
the *gene pool to which it or they formerly
belonged. *Natural selection operating on
this more restricted genetic variety yields

gene combinations different from those
found in the ancestral population.

founder lineage In *phylogenetics, an
ancestral lineage, often still extant, from
which other lineages have risen. The term
is usually applied to intraspecific studies of
*populations and used to describe opera-
tional taxonomic units that occur at internal
nodes of a *cladogram.

foureye butterfly fish (*Chaetodon
capistratus*) *See* CHAETODONTIDAE.

four-eyed fish *See* ANABLEPIDAE.

four-lined snake (*Elaphe quatuorlineata*) *See* COLUBRIDAE.

fovea (*pl.* **foveae**) **1.** A small depression
or pit. **2.** In the *retina of most diurnal vertebrates, a shallow depression that contains
*cones but no *rods, and has no blood ves-
sels or layer of nerve fibres between the cones
and the incoming light. Incoming light can
be focused on to the fovea to produce a sharp
image of the object being observed. *Compare*
MACULA.

foveate Marked with *foveae.

fowl cholera An acute, infectious, often
fatal disease of fowl caused by the bacterium
Pasteurella multocida.

fowl pest A disease of birds, which may
be either fowl plague or Newcastle disease.

fowl plague A disease of birds (including fowl, ducks, and turkeys) which is of
considerable commercial importance. Wild
birds, particularly waterfowl, are naturally
infected but do not normally show symptoms. In domestic fowl the disease affects
the respiratory tract, but the central nervous
system may also be involved. The disease is
caused by an avian influenza virus.

fox (*Alopex, Cerdocyon, Dusicyon,
Vulpes*) *See* CANIDAE.

francolins (*Francolinus*) *See* PHASIAN-
IDAE.

Francolinus (francolins) *See* PHASIAN-IDAE.

Fraser Darling effect Acceleration and synchronization of the breeding cycle that results from the visual and auditory stimulation experienced by individual members of large breeding colonies of birds. It tends to shorten the breeding season and to present predators with a sudden flush of eggs and young, thus increasing the likelihood that any given pair of adults will breed successfully. The effect is named after Sir Frank Fraser Darling (1903–79), who proposed its existence in 1938.

Frasnian A stage of the *Devonian Period, about 385.3–374.5 Ma ago, that is noted for the culmination of the first major radiation of the single-celled *Foraminiferida. The stage is the penultimate of seven. Three important groups of goniatites, the prolecanitids, goniatitids, and clymeniids, arose during this stage.

frass The fine powdery material to which phytophagous insects reduce plant parts.

Fratercula (puffins) *See* ALCIDAE.

fratricide The killing of siblings.

free-living 1. Living independently, i.e. not parasitic on, or symbiotic with, any other organism. *See* PARASITISM; SYMBIOSIS. 2. Not attached to a *substrate.

free-martin *See* INTERSEX.

free-tailed bat *See* MOLOSSIDAE.

freeze-etching *See* FREEZE-FRACTURING.

freeze-fracturing (freeze-etching) A technique for studying membrane *proteins that involves freezing the membrane and fracturing it longitudinally through the lamellae of the lipid bilayer of the membrane. When examined by electron microscopy, the protein molecules appear as protrusions on the undersurfaces of the lipid bilayer.

Fregata (frigatebirds) *See* FREGATIDAE.

Fregatidae (frigatebirds; class *Aves, order *Pelecaniformes) A family that comprises one genus, *Fregata*, of large, long-winged, slender-bodied, mainly black sea-birds that have long, deeply forked tails, long, slender, hooked bills, small feet with partial webbing, and strong claws, the middle one *pectinate. Breeding males have a large, inflatable, bright-red throat pouch. They are good fliers, feeding on fish and other marine animals taken from other birds or from the surface of the sea. Their plumage is not waterproof and they do not land on water; they alight rarely on land, on which they walk clumsily. They roost and build their nests in bushes. There are five species, found in the tropical regions of the Pacific, Indian, and Atlantic oceans.

frenulum Spine or hair found on the hind wing of an insect: it engages a hook to bring about the coupling of the fore and hind wings. *See also* RETINACULUM; WING COUPLING.

freshwater butterfly fish (*Pantodon buchholzi*) *See* PANTODONTIDAE.

freshwater fish Fish that live only in fresh water. Many families and orders of fish are considered to include only primary freshwater species, i.e. species that evolved without contact with a marine environment. Some families have a world-wide distribution, e.g. the *Cyprinidae are found in freshwater basins all over the world, except for S. America and the Australian region. Others, e.g. the *Cobitidae (loaches), are found only in the Eurasian region. A few families have a very restricted distribution, e.g. the *Amiidae (bowfin) are found only in the eastern half of the USA.

freshwater mullet (*Myxus* (*Trachlystoma*) *petardi*) *See* MUGILIDAE.

freshwater turtles *See* EMYDIDAE.

friarbird *See* MELIPHAGIDAE.

frigatebirds *See* FREGATIDAE.

fright substance *See* ALARM SUBSTANCE.

frilled shark (*Chlamydoselachus an-guineus*) *See* CHLAMYDOSELACHIDAE.

Fringilla (chaffinches, brambling) *See* FRINGILLIDAE.

Fringilla montifringilla (brambling) *See* FRINGILLIDAE.

Fringillidae (finches, brambling, bull-finches, canaries, chaffinches, crossbills, goldfinches, greenfinches, grosbeaks, hawfinches, seedeaters, siskins; class *Aves, order *Passeriformes) A family of small to medium-sized birds that have large conical bills. *Loxia* species (crossbills) have crossed *mandibles, an adaptation that assists them in opening pine cones, and they sometimes migrate in large numbers when cone crops fail. The nine species of *Coccothraustes* (hawfinches) have massive conical bills and well-developed jaw muscles that enable them to crack fruit stones. Fringillids have rounded to pointed wings with nine *primaries (the inner primaries of *Coccothraustes coccothraustes* are notched and curled), short to long tails, and medium-length legs. The three species of *Fringilla* (chaffinches and *F. montifringilla*, the highly migratory brambling) differ from other members of the family in having no crop. Fringillids are found in forests, scrub, and open country, where they feed on seeds, buds, and insects, and build open nests in trees, bushes, or on the ground. Bullfinches (six species of *Pyrrhula*) occur in Europe, Asia, and the Azores. The 32 species of *Serinus* (canaries and seedeaters) are noted for their song and some (e.g. *S. canaria*) are popular cage birds. The males of the 21 species of rosefinches (*Carpodacus*) are red and purple, the females brown or grey. There are 18 genera in the family, with 122 species, many migratory, found in Europe, Asia, Africa, and N., Central, and S. America. *Cardinalidae and part of *Emberizidae are often placed in this family.

Frisch, Karl von (1886–1982) An Austrian zoologist who was joint winner (with Konrad *Lorenz and Nikolaas *Tinbergen) of the 1973 Nobel Prize for Physiology or Medicine for their studies of animal behaviour.

Von Frisch specialized in insect behaviour and in the 1940s showed that bees can navigate by the sun and perform dances to communicate to other members of their hive the location of food sources. He also discovered that fish have an acute sense of hearing.

(((●))) **SEE WEB LINKS**
• Biography of Karl von Frisch.

frit flies *See* CHLOROPIDAE.

fritillaries *See* NYMPHALIDAE.

frogfish (anglerfish) *See* ANTENNARIIDAE.

frog-hopper *See* CERCOPIDAE.

frogmouths *See* PODARGIDAE.

frogs 1. *See* ANURA; ASCAPHIDAE; ATELO-PODIDAE; CENTROLENIDAE; LEIOPELMATI-DAE; LEPTODACTYLIDAE; MICROHYLIDAE; PHRYNOMERIDAE; PSEUDIDAE; RHACOPHO-RIDAE. **2. (horned frogs)** *See* PELOBATI-DAE. **3. (tree frogs)** *See* HYLIDAE. **4. ('true' frogs)** *See* RANIDAE.

frons In an invertebrate animal, the front part of the head capsule. *See* CLYPEUS.

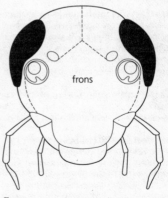

Frons

frontal bone One of the pair of bones that covers the anterior part of the brain; in mammals it includes air cavities (frontal sinuses) linked to the *nasal cavity.

frontal gland In most species of termites (*Isoptera), a sac-like structure between the eyes that in *soldiers secretes a toxic, repellant, or sticky substance into a reservoir. In *Reticulitermes* species the reservoir contains enough of this chemical to kill up to several hundred ant intruders.

frontal lobe In mammals, the part of the brain located at the front of each of the cerebral hemispheres (*see* CEREBRUM). It lies in front of the parietal lobe and in front and above the temporal lobe.

fructose (D-fructose, levulose, $C_6H_{12}O_6$) A sugar that occurs abundantly in nature as the free form; but also, with *glucose, in the form of the disaccharide sucrose.

frugivore An organism that feeds on fruit.

fruit bat *See* MEGACHIROPTERA.

fruit doves (*Ptilinopus*) *See* COLUMBIDAE.

fruit flies *See* DROSOPHILIDAE.

fruit moths *See* TORTRICIDAE.

frustration The motivational state which results when the consequences of behaviour are less satisfying than previous experience had led an animal to expect. Often it leads to more determined attempts to obtain satisfaction, but it may lead to *irrelevant behaviour.

FSH *See* FOLLICLE-STIMULATING HORMONE.

fugacious Short-lived; soon disappearing.

Fulgoridae (lantern flies; order *Hemiptera, suborder *Homoptera) Family of bugs, many of which are large and brightly coloured. In a number of genera the front of the head is produced into a hollow horn which was once believed erroneously to be luminous in some species, giving rise to the common name for the group. There are a few hundred species, most of which occur in the tropics.

Fulica (coots) *See* RALLIDAE.

fuliginous Sooty-coloured; dusky.

full-forward display An aggressive display with which some male birds (e.g. green heron, *Butorides virescens*) challenge another bird that seeks to invade its territory by landing on the tree where it intends to build its nest. The displaying bird holds its body horizontally, with its feathers fully erect, its eyes bulging, its mouth open, and its tail moving rapidly up and down. The bird then lunges at the intruder, waving its wings and uttering a harsh cry.

fulmars (*Fulmarus*) *See* PROCELLARIIDAE.

Fulmarus (fulmars) *See* PROCELLARIIDAE.

fulvettas *See* TIMALIIDAE.

fulvic acid A mixture of uncoloured organic acids that remains soluble in weak acid, alcohol, or water after its extraction from soil.

fulvous Reddish-brown or reddish-yellow in colour; tawny-coloured.

fumaric acid A dicarboxylic acid that is an important intermediate in the *citric-acid cycle.

fundamental niche The *niche a viable population of a species occupies in the absence of competition from other species. *Compare* REALIZED NICHE.

fundatrix (*pl.* fundatrices) The *parthenogenetic founder of a *population, for example of aphids (*see* APHIDIDAE).

Fundulus heteroclitus (killifish) *See* CYPRINODONTIDAE.

fungus ant *See* ATTINI.

fungus beetle *See* MYCETOPHAGIDAE.

fungus garden A 'garden' made by parasol ants (*Attini; *see also* MYRMICINAE) in which the ants cultivate fungi as a source of food.

fungus gnats *See* MYCETOPHILIDAE.

funiculus Any cordlike structure, e.g. the *umbilical cord or a *nerve cord.

funnel-web spiders *See* MYGALOMORPHAE.

furca *See* FURCULA.

furcipulate Pincer-like.

furcula **1. (furca)** The springing organ of *Collembola, which arises from the fourth or fifth abdominal segment. The furcula comprises a single basal portion, the *manubrium, which gives rise to two *dens, which terminate in lamellate mucrones (*see* MUCRO). It is bent forwards and beneath the body to engage on a ventral tube or *retinaculum, which acts as a catch. When the furcula is released, the energy stored propels the animal upwards and forwards, sometimes in a tumbling fashion. **2.** The wishbone of birds and some *Therapoda, supposedly formed by the fusion of *clavicles and *interclavicle.

furfuraceous Scurfy; covered with small, bran-like scales.

Furipteridae **(thumbless bats; order *Chiroptera, suborder *Microchiroptera)** A family of bats in which the first digit is much reduced. The nose is simple, with no leaf. The ears are simple, the *tragus small but distinct. The eyes are very small. The tail is within the *uropatagium but does not reach its edge. There are two genera, *Furipterus* and *Amorphochilus*, each with one species, found in the American tropics.

Furipterus *See* FURIPTERIDAE.

Furnariidae **(ovenbirds, spinetails; class *Aves, order *Passeriformes)** A very diverse family of small to medium-sized, brown and rufous birds, some streaked, that have a short to long, and straight to curved bill, wings that are short and rounded to long and pointed, a tail that is short and rounded to long and forked, and short to medium legs. Spinetails (about 30 species of *Synallaxis*) have medium to long, graduated tails with projecting shafts. Ovenbirds inhabit forests, open country, deserts, sea-shores, and cliffs (spinetails inhabit forests, bush, and scrub), and feed mainly on insects but also on some seeds and crustaceans. They vary from arboreal tree-climbers to terrestrial running species. Some build the domed mud nests which give the family its name, some nest in holes in trees or banks, others nest on the ground. There are some 34–53 genera, with about 218 species, found in Central and S. America.

furniture beetle *See* ANOBIIDAE.

furrowing *See* CLEAVAGE.

fuscous Sombre in colour; dark-coloured or black.

fusiform Spindle-shaped; elongated with tapering ends.

gadflies *See* TABANIDAE.

Gadidae (cod, haddock; superorder *Paracanthopterygii, order *Gadiformes) A family of marine fish found in cold to temperate waters. All members of the family have spineless fins. Some species have three *dorsal and two *anal fins. Typically, the *pelvic fins are located far forward, ahead of the *pectorals, and include more than five soft fin rays. Usually there is a single *barbel on the chin. Several species are of considerable commercial importance (e.g. the European *Gadus morrhua* (cod), which was already an important staple product to the Scandinavian Vikings). The omnivorous cod is a prolific species: large (10 kg) females may release some four million eggs. Other important species include *Melanogrammus aeglefinus* (haddock), *Molva molva* (ling), and *Lota lota* (burbot), a freshwater representative. There are about 55 species, occurring in the northern Atlantic and Pacific.

Gadiformes (subclass *Actinopterygii, superorder *Paracanthopterygii) An order of cod-like and predominantly marine fish that have elongate, tapering bodies and usually long *dorsal and *anal fins. All *fins are supported by soft rays only. There are two or three dorsal fins, and some species have two anal fins. The *pelvic fins, if present, are anterior to the *pectoral fins. The order comprises about 10 families, e.g. the *Gadidae (cod), *Merlucciidae (hake), and *Macruridae (whiptail).

Gadopsidae (river blackfish; subclass *Actinopterygii, order *Perciformes) A monospecific family (*Gadopsis marmoratus*, river blackfish), which is endemic to south-eastern Australia. A fairly slender fish, it has a long, low *dorsal fin and an *anal fin partly enveloped by thick skin. The *pectoral fins each consist of a single divided ray, and the tail fin is rounded. Found in fresh water only, it grows to 60 cm.

Gadopsis marmoratus (river blackfish) *See* GADOPSIDAE.

Gadus morrhua (cod) *See* DEMERSAL FISH; GADIDAE.

Gaia hypothesis The hypothesis, formulated by James E. Lovelock and Lynn Margulis, that the presence of living organisms on a planet leads to major modifications of the physical and chemical conditions pertaining on the planet, and that subsequent to the establishment of life the climate and major biogeochemical cycles are mediated by living organisms themselves.

gait Manner of walking.

galactose An aldohexose monosaccharide that is not normally found naturally in the free form, but usually as a unit in a larger molecule, e.g. a di- or polysaccharide.

galactosidase *See* BETA-GALACTOSIDASE.

galactoside *See* BETA-GALACTOSIDE.

Galago (bush-baby) *See* LORIDAE.

Galápagos cormorant (*Nannopterum harrisi*) *See* PHALACROCORACIDAE.

Galápagos giant tortoise (*Testudo* (or *Geochelone*) *elephantropus*) *See* TESTUDINIDAE.

Galápagos Islands A group of oceanic islands, about 970 km from the west coast of S. America, which *Darwin visited in 1835 and where he encountered a number of *endemic species that were to prove influential in the development of his ideas on evolution.

galaxias See GALAXIIDAE.

Galaxias maculatus See GALAXIIDAE.

Galaxiidae (galaxias, jollytail; subclass *Actinopterygii, order *Salmoniformes) A family of mainly freshwater fish that have a slender, almost tubular, scaleless body. The single *dorsal fin is placed well back, close to the rounded tail fin, but the ventrals are half-way down the belly. Most galaxids live and reproduce in fresh water, but some species migrate to estuaries to spawn. Of the 20 species known, 15 are found only in Tasmania. The family is distributed only in the cooler waters of the southern hemisphere, including Australia, New Zealand (where *G. maculatus* is the basis of the whitebait fishery), S. America, and the Falkland Islands.

Galbulidae (jacamars; class *Aves, order *Piciformes) A family of small to medium-sized birds that have metallic green and rufous or dull, brown-black plumage. They have long, thin, pointed bills, short, rounded wings, long tails, and four toes, two directed backwards, except for *Jacamaralcyon tridactyla* (three-toed jacamar). The four *Brachygalba* species are smaller and shorter-tailed than other jacamars. Jacamars inhabit forest edges, feed on insects, and nest in burrows dug in a bank. There are five genera with 15–17 species, found in Central and S. America.

Galeocerda cuvieri (tiger shark) See CARCHARHINIDAE.

Galerida (larks) See ALAUDIDAE.

gall bladder In many vertebrates, a small pouch in or near to the *liver in which *bile is stored. The presence of food (especially *lipids) in the intestine causes the gall bladder to contract, expelling its contents.

gall gnat See CECIDOMYIDAE.

Galliformes (curassows, hoatzin, megapodes, pheasants; class *Aves) An order of small to large, fowl-like birds, that have short, stout bills, short, rounded wings, and short legs. They are mainly terrestrial, and most are good runners. Important game-birds, many are domesticated and have been introduced world-wide. Many are endangered due to hunting and to habitat destruction. There are six families: *Megapodiidae, *Cracidae, *Tetraonidae, *Phasianidae, *Meleagrididae, and *Opisthocomidae. They are found world-wide.

Gallinago (snipe) See SCOLOPACIDAE.

Gallinula (gallinules, moorhens) See RALLIDAE.

gallinules See RALLIDAE.

gall midge See CECIDOMYIDAE.

Gallus (junglefowl) See PHASIANIDAE.

gall wasp See CYNIPINAE.

galvanotaxis A change in direction of locomotion in a *motile organism or cell that is made in response to an electrical stimulus.

game cropping The culling or husbanding of game animals for meat and other products, usually for human consumption. Game cropping is often advocated as an ecologically sound method of farming the African savannah and similar environments: it is argued that the wide variety of game animals, with their different food preferences, utilize the primary production more efficiently than cattle, and are generally better adapted to the environment. This gives added reason for conservation of the wild *ungulate population. Other social, economic, and political difficulties mean that game cropping is practised less widely than may seem desirable.

gamete A specialized haploid (see HAPLOID NUMBER) cell (sometimes called a sex cell) whose *nucleus and often *cytoplasm fuses with that of another gamete (from the opposite sex or mating type) in the process of fertilization, thus forming a *diploid *zygote. In some animals (e.g. mammals) the gametes are differentiated: the male is a motile sperm with reduced cytoplasm, and the female is immobile with a large amount of cytoplasm called the egg or ovum, which develops when stimulated. Usually there are many small male gametes, but only a few or one female gamete.

game theory The theory that relationships within a community (of organisms or of traits possessed by those organisms) can be regarded as a contest (i.e. a game) in which each participant seeks to secure some advantage. Numerical values can be attached to the gains and losses involved, allowing the contest to be simulated mathematically, usually by computer modelling. The application of game theory has produced many insights into ecological relationships and the significance of particular aspects of animal behaviour. *See also* EVOLUTIONARY STABLE STRATEGY.

gametic equilibrium (linkage equilibrium) The condition in which the frequency of the *gametes, formed by the association of *alleles at different *loci, is equal to the product of the frequencies of the alleles that constitute them. *Compare* LINKAGE DISEQUILIBRIUM.

gametocyst In some *Protozoa, a shell-like covering around *gametes.

gametocyte A cell that will undergo *meiosis to form *gametes. A cell giving rise to a male gamete is termed a spermatocyte; a cell giving rise to a female gamete is termed an oocyte.

gametogenesis The formation of *gametes from *gametocytes.

gamma globulin A specific fraction of *serum proteins that includes the immunoglobins.

Ganges dolphin (*Platanista*) *See* PLATANISTOIDEA.

ganglion (*pl.* ganglia) A mass of nervous tissue, usually surrounded by a layer of *connective tissue, that contains many nerve cells and *synapses. In vertebrates, most ganglia occur outside the central nervous system. In invertebrates, they occur at intervals along the nerve cord; the anterior pair of ganglia is analogous to the vertebrate brain.

gannets *See* SULIDAE.

ganoid In some fossil as well as extant bony fish, applied to a type of scale that has a rhomboid shape and consists of a superficial layer of enamel-like ganoine, a middle layer of *dentine, and a basal layer of vascular bony tissue.

gap analysis A technique, first performed in 1978 by the American ecologist Michael Scott, (1941–) for identifying *ecosystems in need of conservation. Ranges are mapped for a variety of rare or endangered *species. The maps are laid one above another and, when all of them are overlaid on a map showing the location of reserves and protected areas, gaps are revealed in which valuable ecosystems remain unprotected. Gap analysis identifies many more ecosystems deserving protection than the alternative *hot spot technique.

(⊕) SEE WEB LINKS
• Full explanation of gap analysis at the website of the GAP Analysis Program of the US Geological Survey.

gape In animals, the mouth opening between upper and lower *mandibles.

gap junction A communicating junction between cells, which provides a rapid passage for ions and small molecules such that they need not cross the bilipid layer of the opposed membranes. At the junction these membranes are only 2–3 nm apart, and within the intervening space there are arrays of closely packed cylinders, approximately 7 nm in diameter, each of which is pierced by a central channel. These junctions are a particular feature of cells that are in direct electrical coupling with one another (e.g. nerve and heart muscle cells, and perhaps embryonic cells).

gar *See* LEPISOSTEIDAE.

garden spider *See* ARANEIDAE.

garfish *See* BELONIDAE.

garlic toad (*Pelobates fuscus*) *See* PELO-
BATIDAE.

***Garrulax*(laughing thrushes)** *See* TIMAL-
IIDAE.

***Garrulus*(jays)** *See* CORVIDAE.

***Garrulus glandarius*(jay)** *See* CORVIDAE.

***Garrulus lidthi*(Lidth's jay)** *See* CORVIDAE.

Garstang, Walter (1868–1949) A comparative anatomist and embryologist who
was the first person to propose, in 1928, that
vertebrates arose from a neotenous ascidian
(*neoteny, *Ascidiacea) larva. Somewhat
diffident about his more radical evolutionary hypotheses, he propounded them in
anonymous comic verses that he would
leave scattered about the laboratory for his
students to find.

GAS *See* GENERAL-ADAPTATION SYNDROME.

gas bladder *See* SWIM-BLADDER.

gas gland A glandular structure found in
the wall of the *swim-bladder of bony fish.
Richly supplied with capillary blood vessels,
it is capable of secreting gas (mainly oxygen)
into the swim-bladder, thereby increasing
the internal pressure. This helps the fish in
downward migration, as it compensates for
the increase in external pressure.

gaster In *Hymenoptera, the *abdomen,
with the exception of the first abdominal
segment, which has become included in the
*thorax and is separated from the gaster by a
narrow neck (*petiole).

**Gasteropelecidae (hatchetfish; sub-
class *Actinopterygii, order *Cypriniformes)** A small family of freshwater fish
that have a compressed body, rather
straight dorsal profile, and a strongly curved

belly, terminating in a narrow tail (hence
'hatchetfish'). The large *pectoral fins are
operated by relatively huge shoulder muscles and can move so fast that the fish actually leaves the water and flies over the
surface, as a real flying fish. Hobbyists keeping *Gasteropelecus sternicla* (keeled belly),
5 cm, in an aquarium should keep a lid on the
tank! There are about nine species, found in
S. America.

***Gasteropelecus sternicla* (keeled
belly)** *See* GASTEROPELECIDAE.

**Gasterophilidae (warble fly, bot fly;
order *Diptera, suborder *Cyclorrapha)**
Anomalous group of true flies which are
sometimes classified under the broad family name *Oestridae.

**Gasterosteidae (stickleback; subclass
*Actinopterygii, order *Gasterostei-
formes)** A small family of marine, brackish, and freshwater fish that tend to have
torpedo-shaped bodies, a small mouth,
small fins, and a series of separate sharp
spines on the back preceding the *dorsal
fin. The body is scaleless, but may possess
rows of *scutes at the side. The well-known
Gasterosteus aculeatus (three-spined stickleback) grows to 8 cm and can live in fresh
as well as salt water. Gasterosteidae are
found in temperate waters of the northern
hemisphere.

**Gasterosteiformes (class *Osteich-
thyes, subclass *Actinopterygii)** An
order of fish that are found in both fresh
and salt water. The *dorsal fin is preceded
by several sharp spines, as are the *anal and
*pelvic fins. The pelvic fins are located just
behind the *pectorals. Gasterosteids have
rather elongate bodies with a narrow tail region. The order includes the families *Gasterosteidae and *Aulorhynchidae.

***Gasterosteus aculeatus* (three-spined
stickleback)** *See* GASTEROSTEIDAE.

**Gasteruptiidae (suborder *Apocrita,
superfamily Evanioidea)** Family of parasitic wasps which are very similar to the
*Ichneumonidae but have the head set on

a slender neck and the *abdomen set high above the hind *coxae. The adults are thin, black, and 13–20 mm long, with short antennae and *ovipositors often as long as the body. The larvae are parasitic on solitary bees and wasps nesting in wood, rotten logs, or in the ground. Adults are fairly common and usually found on flowers. There are 1500–2000 species.

gastric mucosa In vertebrates, the *mucous membrane (mucosa) that lines the *pylorus; it secretes *gastrin.

gastrin A *peptide *hormone, produced by the gastric mucosa, that stimulates secretion of gastric juices. Secretion of the hormone is initiated as a result of the taste, smell, and thought of food, as well as by mechanical stimulation following the entry of food into the stomach.

gastriole See FOOD VACUOLE.

gastro-enteritis Inflammation of the stomach and intestine, usually with vomiting and diarrhoea. There are various possible causes (e.g. food poisoning).

gastrolith A stone that is ingested into the stomach and acts as a masticatory agent.

Gastropoda (gastropods; phylum *Mollusca) A class of molluscs, of asymmetrical form, including snails and slugs, which have a true head, an unsegmented body, and a broad, flat foot. When present, the shell is in one piece and spirally coiled, at least in young stages. The *mantle cavity and visceral mass have undergone *torsion, at least in the developmental stage. There is usually a well-developed *radula. Gastropods inhabit a wide range of aquatic and terrestrial environments, and range from the *Cambrian to the present. Fossils of coiled gastropod shells are common in marine rocks, especially those of the *Cenozoic. All fossil gastropods and most modern ones have a coiled shell, which is all that remains for the identification of fossil forms. The identification of modern species is based largely on soft body parts.

gastropods See GASTROPODA.

Gastrostomus bairdii See EURYPHARYNGIDAE.

Gastrostomus pacificus See EURYPHARYNGIDAE.

Gastrotheca marsupiatum (marsupial frog) See HYLIDAE.

Gastrotricha (phylum Aschelminthes) A small phylum of microscopic, pseudocoelomate worms in which an excretory system may be present. Some body areas are ciliated and the *cuticle may have bristles. The worms occur in aquatic environments. Superficially they resemble *Rotifera. There are about 150 known species.

gastrovascular cavity A body cavity that has elaborated into a highly branched canal system with digestive and circulatory functions (e.g. the *coelenteron of *Cnidaria and *Turbellaria).

gastrozooid In polymorphic colonial coelenterates, a feeding *polyp.

gastrula The stage of embryonic development of an animal, following the *blastula stage, at which distinct *germ layers are present, the *archenteron opens to the exterior by the *blastopore, and *gastrulation movements occur.

gastrulation In the embryogenesis of an animal, the next stage after the *blastula, in which the three *germ layers differentiate to form the *triploblastic *embryo and the basic body plan of the vertebrate is laid down. During this stage there is a complicated movement of cells from various positions in the blastula to form the germ layers in an approximation of their final positions. This process usually occurs at the end of the *cleavage period.

(((🌐))) SEE WEB LINKS
• Description of gastrulation in a starfish.

Gault Clay A glutinous marine deposit found in south-eastern England and

France that contains abundant fossil *Bivalvia, *Gastropoda, *Ammonoidea, and vertebrates. It is Early *Cretaceous (*Albian) in age.

Gause, Georgyi Frantsevich (1910–86) A Russian biologist and ecologist, best known for his *competitive-exclusion principle, which he derived from his studies of competition among protists and described in a monograph, *The Struggle for Existence* (1934). Gause studied biology at Moscow University. In 1942 he and his wife, Maria Georgyevna Brazhnikova, isolated a strain of *Bacillus brevis* from which an antibiotic substance was produced and used later to treat infected wounds. For this Gause was awarded the Stalin Prize in 1946. From 1960 until his death he was director of the institute of antibiotics he and his wife had founded.

Gause principle *See* COMPETITIVE-EXCLUSION PRINCIPLE.

Gavia (divers) *See* GAVIIDAE.

Gavialidae (gharial, gavial; subclass *Archosauria, order *Crocodilia) A monospecific family (*Gavialis gangeticus*) comprising the Indian gharial (gavial) which is found in the Ganges, Indus, and Brahmaputra river systems. It is a fish-eater, with an elongate, narrow snout widening at the nostrils. Its teeth are all similar in shape and size. *Tomistoma schlegeli* (false gharial) is sometimes included in this family, but it is more probably a member of the *Crocodylidae which has developed a long, narrow snout.

Gaviidae (divers; class *Aves, order *Gaviiformes) A family of diving birds, brown and black above and white below, that have characteristic neck patterns. They are long-necked, with long, pointed bills, and short legs positioned at the rear of the body (which facilitates diving). They are very clumsy on land. They breed on inland lakes, wintering along sea coasts, feed mainly on fish and aquatic invertebrates, and nest on the ground. There is one genus, *Gavia*, and four species, which are migratory, and found in the northern *Holarctic.

Gaviiformes (divers; class *Aves) An order that comprises the single family *Gaviidae.

gazelle (*Gazella, Procapra*) *See* BOVIDAE.

GDP (guanosine diphosphate) *See* GUANOSINE PHOSPHATE.

Gecarcinidae (land crabs; class *Malacostraca, order *Decapoda) Family of omnivorous land crabs, which spend most of their lives on land, returning to the ocean to breed. They can survive in brackish or fresh water, and are very tolerant of high levels of nitrogenous wastes within their bodies. They occur in tropical America, W. Africa, and the Indo-Pacific region. There are three genera and four species.

gecko *See* GEKKONIDAE.

Gedinnian *See* DEVONIAN.

geese *See* ANATIDAE.

Gekko gecko (tokay gecko, great house gecko) *See* GEKKONIDAE.

Gekkonidae (geckos; order *Squamata, suborder *Sauria) A family of insectivorous, mainly nocturnal lizards in which the eye has a vertical pupil and spectacle (no eyelids). The skin is soft with sparse horny tubercles. The toes usually have transverse rows of hooked lamellae for adhesion. The tail is used as a fat store, and is capable of *autotomy. *Gekko gecko* (tokay gecko or great house gecko) of south-east Asia and Indonesia is one of the largest geckos, growing up to 35 cm long, and is strong and aggressive, feeding on small reptiles, birds, and mammals as well as insects; its common name refers to its call of 'tokay'. There are more than 400 species of geckos, distributed widely in warm latitudes. One species, the common house gecko (*Hemidactylus bibronii*), has been spread all around the tropical world by human agency, mainly as an insect-eater, but also simply for its attractiveness.

gel A rigid or jelly-like material, in which molecules form a loosely linked network, formed by the coagulation of a colloid.

geladas (*Theropithecus*) *See* CERCOPITHECIDAE.

gelechid moths *See* GELECHIIDAE.

Gelechiidae (gelechid moths, twirler moths; subclass *Pterygota, order *Lepidoptera*) Large, widespread family of small moths, whose larvae feed among leaves, which they spin together, or on seeds, or internally in stems. Occasionally they are leaf-miners. Pupation occurs in the larval shelter or on the ground. Many species are pests. There are more than 500 genera and more than 4500 species.

gel filtration A column-chromatographic technique that normally employs as a stationary phase polymeric carbohydrate-gel beads of controlled size and porosity. Mixture components are separated on the basis of their sizes and rates of diffusion into the beads. Smaller molecules tend to diffuse more rapidly into the beads, thereby leaving the mainstream of solvent and so becoming retarded with respect to larger molecules. This method can also be used to determine the molecular weight of an unknown substance.

Gelocidae *See* TRAGULOIDEA.

gemmation *See* BUDDING.

Gempylidae (snake mackerel, gemfish; subclass *Actinopterygii, order *Perciformes*) A small family of marine, coastal to oceanic fish that have a very elongate, sometimes ribbon-like body, a large mouth provided with strong teeth, and long, low first *dorsal and short second dorsal fins. Secondary small *finlets may occur anterior to the forked tail fin. Several species are of commercial importance. There are about 20 species, found in tropical to temperate waters, world-wide.

gena (*pl.* genae) **1.** The cheek. **2.** The side of an insect's head, below the eyes.

3. In birds, the feathered side of the lower *mandible.

gene The fundamental physical unit of heredity. It occupies a fixed chromosomal *locus, and when transcribed has a specific effect upon the *phenotype. It may mutate, and so yield various allelic forms. A gene comprises a segment of DNA (in some viruses it is *RNA) coding for one function or several related functions. The DNA is usually situated in thread-like *chromosomes, together with *protein, within the *nucleus; in bacteria and viruses, though, the chromosomes comprise simply a long thread of DNA. The part of a gene that functions as one unit is called a *cistron.

gene conversion A process whereby one member of a *gene family acts as a blueprint for the correction of the others. This can result in either the suppression of a new *mutation, or its lateral spread in the *genome.

gene duplication A process in *evolution where a *gene is copied twice; the two copies lie side by side along the same *chromosome.

gene family A group of similar or identical *genes, usually along the same *chromosome, that originate by *gene duplication of a single original gene. Some members of the family may work in concert, others may be silenced and become *pseudogenes. For example, each of the a- and b-*haemoglobin families of humans includes members that duplicate each other's functions, act at different stages of the life cycle, or may have lost their function altogether.

gene flow The movement of *genes within an interbreeding group that results from mating and gene exchange with immigrant individuals. Such an exchange of genes may occur in one direction or both.

gene frequency The number of *loci at which a particular *allele is found, divided by the total number of loci at which it could occur, for a given population, expressed as a proportion or percentage.

gene library *See* LIBRARY.

gene pool The total number of *genes, or the amount of genetic information, that is possessed by all the reproductive members of a population of sexually reproducing organisms.

general adaptation Adaptation that fits an organism for life in some broad environmental zone, as opposed to 'special adaptations' which are specializations for a particular way of life. Thus the wing of a bird is a general adaptation, while the particular kind of bill is a special adaptation. Major groups of organisms are differentiated very largely on the basis of general adaptations.

general-adaptation syndrome (GAS) A range of abnormal physiological systems reflecting pathological social *stress, and serving to regulate population growth, where no external resource limitation exists. Essentially it is a behavioural regulation of population growth. Mechanisms include the supression of the oestrus cycle, inadequate lactation, enlarged *adrenal glands, increased aggression, etc. It has been demonstrated experimentally with overcrowded laboratory rats. GAS was first described in 1939 by the Austrian-born Canadian endocrinologist Hans Selye (1907–82).

(((●))) SEE WEB LINKS
• Description of effects of general-adaptation syndrome in humans.

generalization The evocation of a learned pattern of behaviour by stimuli other than those to which an animal was trained to respond.

generation time The time required for a cell to complete one full growth cycle. If every cell in the population is capable of forming two daughter cells, has the same average generation time, and is not lost through lysis, the doubling time of the cell number in a population will equal the generation time.

generic cycles, theory of A theory that envisages a life cycle for a species or genus analogous to that of an individual. The first

stage is characterized by vigorous spread; the second by maximum phyletic activity giving rise to new forms; the third marks a phase of decline in area due to competition within the offspring species; and the fourth and final stage sees the extinction of the species. Although in many ways a useful analogy, there is no evidence that species become senile and die out spontaneously.

gene sharing The acquisition by a *gene of a secondary function without the loss of its primary function.

gene substitution The replacement of one *gene by an *allele while all other relevant genes remain unchanged.

genet A genetic individual; the product of a *zygote.

gene therapy The use of *genes to prevent or treat disease. At present experimental, in the future the technique may involve replacing a harmful mutated gene with a healthy copy, inactivating a mutated gene that is functioning incorrectly, or introducing a new gene into the body.

genetic Pertaining to the origin or common ancestor or ancestral type and, by extension, pertaining to *genes or heredity. The term is widely used outside the life sciences. *See also* GENETICS.

genetic code The set of correspondences between base (*nucleotide pair) triplets in DNA and *amino acids in *protein. These base triplets carry the genetic information for protein synthesis. For example, the triplet CAA (cytosine, adenine, adenine) codes for valine. The code is universal, but degenerate, in that certain *amino acids are coded for by more than one *codon, and the codon bias (the preferred codon for a particular amino acid from the several possible choices) differs somewhat in different organisms and is different in nuclear, *mitochondrial, and chlorophyll DNA.

genetic distance A measure of genetic similarity and evolutionary relationship (e.g. between two races of the same species), based on the frequency of a number of *genes, or on the number of base-pair differences in a DNA sequence, which can be easily detected and scored. A number of different indices of genetic distance are in use.

genetic drift The random fluctuations of *gene frequencies in a population such that the *genes amongst offspring are not a perfectly representative sampling of the parental genes. Although drift occurs in all populations its effects are most marked in very small, isolated populations, in which it gives rise to the random fixation of alternative *alleles so that the variation originally present within single (ancestral) populations comes to appear as variation between reproductively isolated populations.

genetic engineering (genetic manipulation) The manipulation of *DNA using *restriction enzymes which can split the DNA molecule and then rejoin it to form a hybrid molecule, i.e. a new combination of non-homologous DNA (so-called recombinant DNA). The technique allows the by-passing of all the biological restraints to genetic exchange and mixing, and may even permit the combination of genes from widely differing species. Genetic engineering developed in the early 1970s, and is now one of the most fertile areas of genetics.

genetic equilibrium An equilibrium in which the frequencies of two alleles at a given locus are maintained at the same values generation after generation. A tendency for the *population to equilibrate its genetic composition and resist sudden change is called genetic homeostasis.

genetic hitchhiking A process whereby a *gene with neutral value achieves a high value, or even *fixation, within a *population because it is closely linked to a gene that is being selected for.

genetic homeostasis See GENETIC EQUILIBRIUM.

genetic load The average number of *lethal mutations per individual in a population. Three main kinds: (a) input load, in

which inferior *alleles are introduced into the *gene pool of a population either by *mutation or immigration; (b) balanced load, which is created by *selection favouring allelic or genetic combinations that, by segregation and *recombination, form inferior *genotypes every generation; and (c) substitutional load, which is generated by selection favouring the replacement of an existing gene by a new allele. Originally called the 'cost of *natural selection' by the geneticist J. B. S. Haldane, substitutional load is the genetic load associated with transient *polymorphism. The term 'genetic load' was originally coined by H. J. Muller (1890–1967) in 1950 to convey the burden that deleterious mutations provide, but it is probably better recognized as a measure of the amount of natural selection associated with a certain amount of genetic variability, which provides the raw material for continued adaptation and evolution.

genetic manipulation See GENETIC ENGINEERING.

genetic map The linear arrangement of mutable sites on a *chromosome, deduced from genetic-recombination experiments. The percentage of recombinants is used as a quantitative index of the distance between two gene pairs, and this distance, together with those between other gene pairs, provides a map of their arrangement on the chromosome. One 'genetic map unit' is defined as the distance between gene pairs for which one product of *meiosis out of one hundred is recombinant (i.e. it equals a recombination frequency of 1%).

genetic polymorphism The occurrence in a *population of two or more *genotypes in frequencies that cannot be accounted for by recurrent *mutation. Such occurrences are generally long-term. Genetic polymorphisms may be balanced (such that *allele frequencies are in equilibrium with one another at a given *locus), or transient (such that a mutation is spreading through the population in a constant direction). In the former case, the different alleles may be maintained by different environmental conditions (in space or time), one being favoured under one set of circumstances and another under another set; or a heterozygous genotype (*see* HETEROZYGOSITY) may be in some way superior to the genotypes that are *homozygous at that locus (called a 'heterozygous advantage'). It has been pointed out, however, that in many populations polymorphism is so high that to account for it all by *natural selection would entail an impossible *genetic load; thus, a good deal of polymorphism would be due to chance increase or decrease of neutral alleles.

genetics The scientific study of *genes and heredity.

genetic system The organization of genetic material in a given species, and its method of transmission from the parental generation to its filial generations.

genetic variance The portion of *phenotypic variance that results from the varying *genotypes of the individuals in a population. Together with the environmental variance, it adds up to the total phenotypic variance observed amongst individuals in a population.

geniculate 1. Resembling a knee or capable of bending as a knee bends. The word is derived from the Latin *geniculare*, 'to bend' (like a knee). **2.** *See* FLAGELLUM (2).

genital capsule Those organs, and all their associated *sclerites, that are used for the transfer of sperm in a male insect. The genital capsule is made up of a number of abdominal segments.

genital lock The temporary inability of copulating partners to separate once intromission is achieved (e.g. in dogs the end of the penis is enlarged and a vaginal sphincter muscle closes around this enlargement).

genital plate In *Echinoidea, one of the five large plates, aligned with the *interambulacral areas, that surround the *periproct in regular echinoids and the *aboral centre in irregular echinoids. Each genital plate

bears a *gonopore, and one is porous and serves as the *madreporite. *See also* OCULAR PLATE.

genital ridge *See* MESODERM.

genome The total genetic information an individual inherits from its parents. In *eukaryotes, this will include chromosomal DNA and non-chromosomal DNA (e.g. DNA found in *mitochondria and chloroplasts).

genomic compartmentalization The presence within a cell of independent replication *genomes. In an animal cell the genomes include those of the cell *nucleus and the mitochondria (*see* MITOCHONDRION).

genotype 1. *noun*. The genetic constitution of an organism, as opposed to its physical appearance (phenotype). Usually this refers to the specific allelic composition of a particular *gene or set of genes in each cell of an organism, but it may also refer to the entire *genome. **2.** *verb*. To determine a genotype by means of an assay.

genotypic adaptation *See* ADAPTATION.

gentle A common name for the larva (maggot) of a housefly (*Musca domestica*). *See* MUSCIDAE.

gentle lemur (*Hapalemur*) *See* LEMURIDAE.

Geochelone elephantropus (Galápagos giant tortoise) *See* TESTUDINIDAE.

Geochelone gigantea (Seychelles giant tortoise) *See* TESTUDINIDAE.

Geococcyx californianus (greater roadrunner) *See* CUCULIDAE.

geometers *See* GEOMETRIDAE.

Geometridae (geometers, loopers, earth measurers, inch worms, carpet moths, waves, pugs; subclass *Pterygota, order *Lepidoptera) Huge family of small to medium-sized, often cryptically coloured moths, with a slender body and relatively broad wings. In the females of some species the wings are reduced. All or some abdominal 'legs' are absent from the larvae, except for the last pair. The larvae frequently resemble twigs and progress in a looping fashion. There are about 12 000 species. The family has a world-wide distribution.

Geomyidae (pocket gophers; order *Rodentia, suborder *Sciuromorpha) A family of colonial, burrowing rodents in which the body is thickset. There are five clawed digits on each limb, the claws of the fore limbs being especially strong. The tail is short, with sparse hair. There are many species, confined to N. and Central America.

Geophilomorpha See CHILOPODA.

geotaxis A change in the direction of locomotion in a motile organism or cell, made in response to the stimulus of gravitational attraction. The movement may be downward (positive geotaxis) or upward (negative geotaxis).

***Geothlypis* (yellowthroats, warblers)** See PARULIDAE.

Geotrupidae (dung beetles, dor beetles; subclass *Pterygota, order *Coleoptera) Family of robust, shiny, black or brown beetles, 7–27 mm long. The head and *thorax of the male often have projections, those of females much less commonly so. The antennae are clubbed. The broad, toothed *tibiae are used for digging. Eggs are laid in tunnels beneath patches of dung, and dung is carried down to feed larvae; in some species the sexes co-operate. Larvae are white and fleshy, with little reduction of legs. Both adults and larvae *stridulate. Some species may feed on underground fungi, others on plants. There are 300 species.

***Geotrygon* (quail doves)** See COLUMBIDAE.

gerbil See CRICETIDAE.

germ cell See GERMINAL SELECTION.

germinal disc See BLASTODISC.

germinal selection 1. The selection (by people) of germ cells (sex cells from which *gametes are derived) that possess qualities recognized as superior for use in the production of a subsequent generation. This method is used in the breeding of domesticated animals; and it is proposed by those supporting *eugenics that the method should also be applied to humans. In the latter case, germ cells would be stored, frozen, in germ banks, and would be available for couples who wished to use them rather than their own germ cells to generate a family. **2.** Selection during *gametogenesis against induced *mutations that retard the spread of mutant cells.

germ layers In an *embryo at the *gastrula stage, the layers of cells that will develop to form the organs of the body. See GASTRULATION.

germ line Cells from which *gametes are derived, and which therefore bridge the gaps between generations (unlike *somatic cells) in the body of an organism.

germ plasm The hereditary material transmitted to offspring through the germ cells, and giving rise in each individual to the *somatic cells.

Gerreidae (mojarra, silver biddy; subclass *Actinopterygii, order *Perciformes) A family of small marine fish that have an oblong to deep, silvery body, a small, *protractile mouth with tiny teeth, a long *dorsal fin, a forked tail fin, and long, pointed *pectorals. They inhabit coastal tropical to subtropical waters. There are about 40 species, found world-wide.

Gerridae (pond skaters; order *Hemiptera, suborder *Heteroptera) Family of bugs which live on the surface of lakes, ponds, slow rivers, and even the open sea. The legs and body are water-repellent, and so unwettable, and the insects are supported by surface tension. Their food consists mainly of drowning insects. There are about 400 species.

gestalt The perception of a pattern or structure as a whole, not as a sum of its constituent parts. The German *Gestalt* means 'form' or 'shape'. *See also* JIZZ.

gestation period Length of time from conception to birth in a *viviparous animal; it is usually a relatively fixed period for a particular species.

gharials 1. *See* GAVIALIDAE. **2. (*Tomistoma schlegeli*, false gharial)** *See* CROCODYLIDAE; GAVIALIDAE.

ghost-faced bats *See* MORMOOPIDAE.

ghostfish *See* CHIMAERIDAE.

ghost flathead *See* HOPLICHTHYIDAE.

ghost larva (*Chaoborus crystallinus*) *See* CHAOBORIDAE.

ghost moths *See* HEPIALIDAE.

ghost pipefish *See* SOLENOSTOMIDAE.

giant barracuda (*Sphyraena barracuda*) *See* SPHYRAENIDAE.

giant girdled lizard (*Cordylus giganteus*) *See* CORDYLIDAE.

giant gourami *See* OSPHRONEMIDAE.

giant herring *See* ELOPIDAE.

giant manta (*Manta birostris*) *See* MOBULIDAE.

giant maori wrasse (*Cheilinus undulatus*) *See* LABRIDAE.

giant panda (*Ailuropoda melanoleuca*) *See* AILUROPODA.

giant salamanders *See* CRYPTOBRANCHIDAE.

gibbon (*Hylobates*) *See* HYLOBATIDAE.

Gigantactinidae (subclass *Actinopterygii, order *Lophiiformes) A small family of deep-sea anglerfish that have unusually shaped, elongate bodies and large tail fins. The very long *illicium of the female is a conspicuous feature. There are about five species, living at depths exceeding 1000 m.

Giganthuridae (subclass *Actinopterygii, order *Salmoniformes) A small family of deep-sea fish, all of which have a large mouth with depressible teeth, tubular, telescopic eyes, a scaleless body, high *pectoral fins, and a long lower lobe to the forked tail fin. The small giganthurids (15 cm) occur at depths exceeding 2000 m. There are five species, found world-wide.

gila monster (*Heloderma suspectum*) *See* HELODERMATIDAE.

gill The respiratory organ of an aquatic animal, often complex in form, consisting of an outgrowth from the body surface or an internal layer of modified gut, past which water flows. It provides a large surface area, well supplied with blood vessels, across whose walls oxygen and carbon dioxide diffuse. It is the chief organ of respiration in fish, in which the gills usually are supported by four or more pairs of bony or cartilaginous gill arches. In sharks each gill is more or less surrounded by a separate gill chamber, but in bony fish the gills share a common gill (branchial) chamber, one on each side of the head. Each gill arch bears rows of plate-like filaments, in turn carrying secondary lamellae, which are capable of absorbing oxygen from the water flowing over the surface.

gill arch (branchial arch) *See* GILL.

gill book Gills composed of a series of plates, resembling the pages of a book.

gill cover (operculum) A hard but flexible cover which, in bony fish, forms the outer wall of the gill chamber. It protects the *gills and also plays a major role in the pumping mechanism that regulates the continuous flow of water over them.

gill-footed shrimps *See* BRANCHIOPODA.

gill pouch Part of the respiratory system of lampreys and hagfish, which have a row of gill pouches on each side of the head.

A gill pouch consists of a muscular wall with many respiratory lamellae lodged on its inner side. In lampreys the gill pouches each have a separate opening to the exterior.

gill raker In most bony fish, one of a set of fairly stiff, tooth-like processes, located on the inner side of the gill arch, which strain the water flowing past the *gills. In some fish (e.g. mullet and herring) the gill rakers are long and closely set, thereby acting as a sieve capable of retaining food particles.

ginglymoidy An articulation, such as a hinge joint, which allows movement in one plane. In *Cryptodira, the cervical vertebrae are articulated such that vertical rotation is permitted between the sixth and seventh, and seventh and eighth vertebrae, so that an S shape is formed which allows the retraction of the head in one plane.

Ginglymodi An older name for the American gars (*Lepisosteus*). See LEPISOSTEIDAE.

giraffe (*Giraffa camelopardalis*) See GIRAFFIDAE.

Giraffidae (giraffe, okapi; infra-order *Pecora, superfamily *Cervoidea) An artiodactyl family, now represented by only the giraffe and okapi of Africa, which diverged from the *Cervidae during the *Miocene. Relatively short-necked and short-legged forms like the okapi are known as fossils from the *Pliocene of Eurasia, and in the *Pleistocene of Africa and southern Eurasia there were very large, heavily built giraffids with branched horns, the best-known example being *Sivatherium*. The okapi is almost indistinguishable from the Pliocene genus *Palaeotragus*. Both sexes bear short horns covered by skin, but extinct forms grew larger horns (e.g. *Samotherium* and *Sivatherium*). Giraffidae are long-legged, and in the giraffe (*Giraffa camelopardalis*) the weight of the body is balanced on the fore limbs, the number of ribs is reduced, and the hind part of the spinal column functions as part of the hind limbs, which provide propulsion. The gait is 'rocking', both limbs of one side of the body moving together. There are mechanisms that regulate blood flow when the head is raised. There

are two genera and species (*Giraffa* and *Okapia johnstoni*), found in Africa south of the Sahara.

girdle 1. In *Polyplacophora (chitons), an extension of the mantle beyond the *shell plates; it is in contact with the *substrate below and often bears *spicules on its upper surface. **2.** In vertebrates, a group of bones attaching limbs to the trunk: the pectoral girdle attaches the fore limbs, the pelvic girdle the hind limbs.

girdled lizard giant (*Cordylus giganteus*) See CORDYLIDAE.

girdle-tailed lizards See CORDYLIDAE.

Givetian An age in the Middle *Devonian Epoch (about 391.8–385.3 Ma ago), preceded by the Eifelian and followed by the *Frasnian, that is zoned on goniatites (*Ammonoidea) and Spiriferida.

gizzard In many animals, a part of the *alimentary canal where food is broken into small particles before the main digestive processes commence. The walls of the gizzard are very muscular and may be equipped with a horny lining (e.g. in earthworms), sharp projections (e.g. in cockroaches), or may contain stones or grit swallowed by the animal (e.g. in birds).

gizzard shad (*Dorosoma cepedianum*) See CLUPEIDAE.

glabella The part of the head that is posterior to or above the eyes, and that is more or less smooth and hairless (i.e. in humans the bony swelling above the root of the nose).

glabrous Smooth, lacking hairs.

gland A cell or group of cells that is specialized for the secretion of a particular substance.

Glareolidae (pratincoles, coursers; class *Aves, order *Charadriiformes) A family of mainly brown and grey birds that have white under-parts and distinctive facial and wing markings. Pratincoles have four toes, short bills and legs, long, pointed wings, and forked tails. Coursers have three toes, longer bills, short, broad wings and tails, and long legs. Most have their middle toe *pectinate. They inhabit bare, stony areas, coursers being terrestrial and pratincoles more aerial, feed on insects, and nest on the ground. There are five genera, with 16 species, found in Europe, Africa, Asia, and Australia.

glassfish *See* SALANGIDAE.

glass snake (*Ophisaurus apodus***)** *See* ANGUIDAE.

Glaucidium **(pygmy owls, owlets)** *See* STRIGIDAE.

Glaucosomidae (pearl perch; subclass *Actinopterygii*, order *Perciformes*) A small family of marine coastal fish that have a robust, deep body, large mouth, and large eyes. There are about five species, found in the western Pacific region.

glaucous Sea-green or bluish-green in colour, or having a waxy, bluish-green bloom (e.g. the colour of the carpet formed on the surface of the open sea by swarms of sea 'lizards', *Glaucus eucharis*, members of the *Nudibranchia). Originally the word meant bright, sparkling, greyish, like the eyes of Pallas Athene.

glenoid Pertaining to a socket. The word is derived from the Greek *glene*, 'socket'.

glenoid cavity In tetrapods, the socket in the *pectoral girdle into which the head of the *humerus fits.

glenoid fossa 1. The smooth depression on the ventral side of the skull into which the *condyle of the jaw bone fits. **2.** Similar structures in the *scapula.

glial cell (neuroglia) A type of cell that occurs in vast numbers in the central nervous system, surrounding and separating *neurons and their processes. Apparently, the role of glial cells is to act as mediators of normal neuronal metabolism and/or as insulators between neurons to prevent unwanted

interactions. They may also be involved in the regeneration of damaged neurons.

Glires (subclass *Theria*, infraclass *Eutheria*) A cohort that comprises mammals possessing *incisors that are 'rootless', i.e. they continue to grow throughout life and are worn away by gnawing. The cohort includes the orders *Rodentia and *Lagomorpha, presumed to have diverged from the insectivorous eutherian stock during the *Cretaceous and subsequently differentiated into rodent and lagomorph forms; but similarities between the two orders may be superficial, implying no close relationship, so their inclusion in a single cohort is somewhat arbitrary.

Gliridae (dormice; order *Rodentia*, suborder *Myomorpha*) A family of small, mainly arboreal, nocturnal rodents in which the snout is often elongated, the eyes and ears are well developed, the tail is bushy, and the first digit of the fore limb is rudimentary. Dormice have lower population densities and lower breeding potential than most other rodents, and most *hibernate in winter. There are seven genera, with many species, distributed throughout the *Palaearctic region and Africa.

Globicephala melaena **(pilot whale)** *See* DELPHINIDAE.

Globigerina **(class *Rhizopoda*, order *Foraminiferida*)** A protozoan genus whose members are commonly *pelagic, unlike most foraminiferans which are *benthic. Globigerina species have more delicate shells than benthic foraminiferans and the shells often have spines.

Globigerina ooze Deep-sea ooze in which at least 30% of the sediment consists of planktonic *Foraminiferida, including chiefly *Globigerina. It is the most widespread deposit to form from the settling out of material from overlying waters, covering almost 50% of the deep-sea floor. Globigerina ooze covers most of the floor of the western Indian Ocean, the mid-Atlantic Ocean, and the Equatorial and S. Pacific.

globose Spherical.

globular protein A *protein in which at least one polypeptide chain is folded in a three-dimensional configuration. The stability of the structure is maintained by a number of intrachain bonds. Globular proteins have a variety of functions (e.g. acting as *enzymes, facilitating transport, and providing storage).

globulin One of a group of *globular, simple proteins, which are insoluble or only sparingly soluble in water, but soluble in dilute salt solutions. They occur in many animal tissues (e.g. *blood plasma) and have a variety of functions (e.g. as components of *antibodies).

glochidium The microscopic larva of a freshwater mussel (families Unionida (river mussels) and Margaritiferidae (European freshwater pearl mussels), order *Unionoida). It possesses sharp hooks by which it attaches itself to a fish; it then lives as a parasite (*see* PARASITISM) until it grows into the form of a juvenile mussel and detaches itself. The parasitic stage in its life cycle allows the fish host to distribute the mussel more widely than it could travel by itself.

Gloger's rule Individuals of many species of insects, birds, and mammals are darkly pigmented in humid climates and lightly coloured in dry ones. This may well be a camouflage adaptation—moist habitats are usually well vegetated and tend to lack pale colours. There are many exceptions to this so-called rule. The rule was proposed in 1833 by the German zoologist Constantin Wilhelm Lambert Gloger (1803–63). *See also* ALLEN'S RULE; BERGMANN'S RULE.

glomerulus A structure composed of a leaky arterial capillary bundle encapsulated in a Bowman's capsule at the end of a *nephron.

glossa The tongue. In insects, the median lobes (fused in some species) of the *labium.

Glossina *See* GLOSSINIDAE.

Glossinidae (tsetse fly; order *Diptera, suborder *Cyclorrapha) Family of flies which are well known as important carriers of disease organisms, e.g. the flagellate protozoon *Trypanosoma*, which causes sleeping sickness. The adults are moderately sized, with *aristae adorned with feathered hairs. The mouth-parts are produced as a needle-like proboscis which is protected by a sheath when not in use to pierce host skin. The *prosternum is membranous in the adult. Female tsetse flies feed on a wide range of animal blood, and produce mature larvae which develop within a 'uterus'. The deposited, full-grown larva buries itself in the ground and pupates. Different species of *Glossina* may carry specific forms of *Trypanosoma* which vary in their virulence and periodicity. Tsetse flies occur in Africa. Only 22 species have been described so far.

glossy starling (*Aplonis*) *See* STURNIDAE.

glottis In mammals, the vocal cords and the spaces between them.

glow-worm *See* LAMPYRIDAE.

glucagon A polypeptide *hormone secreted by the alpha-cells of the islets of Langerhans in the pancreas. By stimulating glycogenolysis (the breakdown of *glycogen to *glucose in the *liver), it increases the level of glucose in the blood. Thus its function is antagonistic to *insulin.

glucan (glucosan) A polysaccharide of glycosidically linked *glucose units. The types of linkage in a glucan chain may be mixed or may all be the same. Examples include cellulose, callose, and starch.

glucocorticoid hormones A group of *hormones that trigger the release of *glucose and guide processes linked to converting *sugar, *lipids, and stored *protein to usable energy, especially when energy is needed to deal with physical or emotional stress. *Cortisol is a glucocorticoid hormone.

gluconeogenesis The synthesis of *glucose from non-carbohydrate precursors.

Any compound that can be converted into one of the intermediates of *glycolysis is potentially glycogenic. These include *amino acids, *fatty acids, *glycerol, and intermediates of the *citric-acid cycle.

glucosan *See* GLUCAN.

glucose (dextrose) $C_6H_{12}O_6$. An aldohexose monosaccharide that is a major intermediate compound in cellular metabolism. *See also* GLYCOGEN.

glucoside A *glycoside formed from *glucose.

glutamic acid A dicarboxylic, polar *amino acid.

glutamine An *amino acid, the Y-amide of *glutamic acid.

glutathione The tripeptide of *glutamic acid, cysteine, and *glycine, which is universally found in animal tissue. It functions as a *coenzyme for some *enzymes, e.g. for *glyoxalase. Also it appears to act as an antioxidant, protecting the *sulphydryl groups of certain enzymes and proteins.

glutton (wolverine; *Gulo gulo*) *See* MUSTELIDAE.

glyc- Prefix, commonly with an added 'o' (i.e. glyco-), meaning 'pertaining to sugar', derived from the Greek *glukos*, 'sweet'.

glyceraldehyde An aldotriose sugar, a phosphorylated derivative of which is an important intermediate in the glycolytic pathway.

glyceride An ester formed from *glycerol and between one and three *fatty-acid molecules, respectively designated mono-, di-, or triglyceride. Glycerides serve variously as sources of energy, and triglycerides (lipids) also serve as thermal and mechanical insulators.

glycerol A three-carbon, linear, trihydroxy alcohol. Its fatty esters are a very important constituent of many *lipids, and some of its phosphorylated derivatives are intermediates in *glycolysis.

glycine (aminoacetic acid, aminoethanoic acid) The simplest *alpha amino acid, and the only one not to exhibit optical activity.

glycogen (animal starch) A highly branched homopolysaccharide composed of D-glucose (*see* GLUCOSE) units. It is the principal storage carbohydrate in animals.

glycolipid A *lipid attached to a *carbohydrate that serves to provide energy and as a marker for cell recognition. Glycolipids are found on the outer surface of all *eukaryotic cells.

glycolysis (Embden–Meyerhof pathway) The stepwise, anaerobic degradation of *glucose to produce as end-products either *lactic acid (in cells of animals) or ethanol and carbon dioxide (in those of fungi and plants). In animals, one mole of glucose yields two moles of lactic acid and the reaction sequence has a net yield of two moles of *ATP. However, in most cells under aerobic conditions, the pathway serves primarily to provide pyruvate, which is oxidized via the *citric-acid cycle, and intermediate compounds for biosynthetic processes.

glycoprotein A *conjugated protein that consists of a carbohydrate covalently linked to a protein (e.g. mucins and mucoids).

glycoside The product obtained when a sugar reacts with an alcohol or phenol.

glyoxalase An *enzyme that, in *liver and muscle, catalyses the intramolecular oxidation–reduction of methylglyoxal to *lactic acid.

glyoxysome A *microbody that occurs only in those cells which contain the *enzymes of the glyoxylate cycle, whereby lipids are converted into sugars. *See also* PEROXISOME.

Glyptodontoidea (order *Edentata, suborder *Xenarthra) An extinct superfamily whose earliest representatives are known from the late *Eocene and which flourished until the end of the *Pleistocene in N. and S. America. Glyptodons were armadillo-like animals, all larger than modern armadillos and some later forms growing to a length of 2.7 m. The armoured plates tended to be fused into a solid carapace, forming a mosaic of small plates, the skull was covered by a bony casque, the tail was armour-plated, some forms bearing tail spikes, and most of the vertebrae were fused to form three sections, cervical, dorsal, and a final section comprising the latter dorsal, lumbar, and sacral vertebrae.

GMP (guanosine monophosphate) *See* GUANOSINE PHOSPHATE.

gnatcatchers *See* POLIOPTILIDAE.

gnateaters *See* CONOPOPHAGIDAE.

Gnathifera A phylum or superphylum formed by combining the *Rotifera, *Acanthocephala, *Gastrotricha, and *Gnathostomulida. *Phylogenetic evidence indicates that the Gnathifera is monophyletic (*see* MONOPHYLY).

gnathobase In some *Arthropoda, an expanded process on the base segment of a limb, used for the manipulation of food.

Gnathobdellida (phylum *Annelida, class *Hirudinea) An order of aquatic or terrestrial, carnivorous, parasitic, or free-living worms in which a *pharynx is present. All members of the order have three pairs of jaws. There are two families.

gnathosoma *See* CAPITULUM.

Gnathostomata 1. A superclass of jawed vertebrates that includes the fishes, *Amphibia, *Reptilia, *Aves, and *Mammalia. Recently it has been proposed that Gnathostomata should comprise the *Elasmobranchii (sharks, rays, etc.) and the *Teleostomi (i.e. *Osteichthyes and *Sarcopterygii). In this classification *Pisces becomes redundant. 2. (phylum *Echinodermata, class *Echinoidea) A superorder of echinoids in which the *periproct is outside the *apical system, there are no compound ambulacral plates, an *Aristotle's lantern and girdle are present, and the teeth are keeled. It includes the orders *Holectypoida and *Clypeasteroida.

Gnathostomulida A *phylum of tiny, *acoelomate animals, discovered by Peter Ax in 1956; according to Ax it is the sister group of *Platyhelminthes, like which it has no *anus. Gnathostomulids are usually less than 1 mm long, elongate (often thread-like) and cylindrical, and externally ciliated (*see* CILIUM); they are *hermaphrodite, and there is a pair of lateral, toothed jaws. The 80 or so species live in fine sand in marine environments.

gnats *See* CHAOBORIDAE; CULICIDAE.

gnatwrens *See* POLIOPTILIDAE.

gnotobiotic Applied to a culture in which the exact composition of organisms is unknown, usually during the formation of experimental laboratory *ecosystems (*microcosms) from *axenic cultures.

goal The state that is presumed to exist within the brain of an animal and which corresponds to a state of affairs the animal seeks to achieve. It is inferred from observation of the stimuli that terminate a behaviour.

goat (*Capra*) *See* BOVIDAE.

goat antelopes *See* CAPRINAE.

goatfish *See* MULLIDAE.

goat moth (*Cossus cossus*) *See* COSSIDAE.

go-away birds *See* MUSOPHAGIDAE.

Gobiesocidae (clingfish; subclass *Actinopterygii, order *Gobiesociformes) A large family of small, marine, shallow-water fish, that have a powerful sucking disc formed by the *pelvic fins which are placed ahead of the *pectorals, a depressed head with a wide mouth, and a rather flat, scaleless body. Many species are found close inshore among kelp beds, but some occur in fresh water. There are about 100 species, found world-wide.

Gobiesociformes (Xenopterygii; class *Osteichthyes, subclass *Actinopterygii) An order of small, bottom-dwelling fish that have a depressed head and scaleless body. The order includes *Gobiesocidae (clingfish), *Callionymidae (dragonets), and Draconettidae. According to some authors the Callionymidae and Draconettidae belong to the order *Perciformes.

Gobiidae (goby; subclass *Actinopterygii, order *Perciformes) A very large family of coastal-water fish. Most species have the *pelvic fins united to form a sucking disc which is used to cling to hard objects on the sea floor. Unlike the clingfish (*Gobiesocidae) the gobies have two *dorsal fins, the first one being spiny. Gobies form an important part of the fish population of rocky shores, sea-grass beds, and coral reefs. Some species live in close association with other animals, e.g. sea urchins and shrimps. Most gobies are very small; *Gobius vittatus* (striped goby) of the Mediterranean does not exceed 4 cm. One of the largest gobies is *G. giuris* of the Indo-Pacific, which reaches a length of 50 cm. With about 800 species, Gobiidae is the largest family of marine fish, found in tropical to temperate waters world-wide.

Gobioididae (eel gobies; subclass *Actinopterygii, order *Perciformes) A small family of marine or freshwater fish, that have an eel-like body, very small eyes, very long *dorsal and *anal fins, and a rounded tail fin. The *pelvic fins usually form an adhesive disc under the head. There are about 18 species, found in the tropical regions of the Indo-Pacific, America, and W. Africa.

Gobiomorus dormitor (bigmouth sleeper) *See* ELEOTRIDAE.

Gobius vittatus (striped goby) *See* GOBIIDAE.

goblet worms *See* ENTOPROCTA.

goby *See* GOBIIDAE.

godwits (_Limosa_) *See* SCOLOPACIDAE.

golden bat (_Myzopoda_) *See* MYZOPODIDAE.

golden eagle (_Aquila chrysaetos_) *See* ACCIPITRIDAE.

golden hamster (_Mesocricetus auratus_) *See* CRICETIDAE.

golden mole *See* CHRYSOCHLORIDAE.

golden oriole (_Oriolus oriolus_) *See* ORIOLIDAE.

goldfinch *See* FRINGILLIDAE.

goldfish (_Carassius auratus_) *See* CYPRINIDAE.

Golgi, Camillo (1843–1926) An Italian physician who developed a method for staining nerve cells with silver nitrate and who first demonstrated the existence of the branching processes named after him (*Golgi body). In 1906 he was awarded (jointly with S. Ramón y Cajal) the Nobel Prize for Physiology or Medicine for his research into the structure of the nervous system.

( ) SEE WEB LINKS
• Biography of Camillo Golgi.

Golgi body A system of flattened, smooth-surfaced, membranaceous *cisternae, arranged in parallel 20–30 nm apart and surrounded by numerous *vesicles. A feature of almost all *eukaryotic cells, this structure is involved in the packaging of many products of cell metabolism. It is named after Camillo Golgi.

Goliath beetle (_Goliathus giganteus_) *See* SCARABAEIDAE.

Goliathus giganteus (Goliath beetle) *See* SCARABAEIDAE.

Gomphidae (dragonflies; order *Odonata, suborder *Anisoptera) Cosmopolitan family of swift-flying, slender dragonflies, which usually alight near running water. The nymphs are bottom-dwellers in streams, often burrowing in mud. Some possess respiratory tubes. More than 870 extant species have been described.

Gomphotheriidae (order *Proboscidea, suborder Gomphotherioidea) An extinct family of long-jawed mastodons, characterized by the development of multiple accessory tooth cusps and believed to be ancestral to the later mastodons, though not to modern elephants. The gomphotheres were one of three distinct proboscidean lines established by the *Miocene. They in turn diverged into a variety of descendent lines, several of which had highly specialized lower jaws. The family persisted into the *Pleistocene in both Old and New Worlds. Long snouts appeared in the earliest forms and in the Miocene *Gomphotherium* the lower jaw and *premaxilla were very long, both bearing tusks which may have been used for digging, and probably permitting only a short trunk. In later forms the face became shorter and the trunk presumably became longer. There were many genera and species. The evolutionary importance of the gomphotheres lies in the probable parallel between their development and that of the line which led to the true elephants.

Gomphotherium *See* Gomphotheriidae.

gonad A sex gland in which spermatozoa or ova are formed. *See* OVARY; OVUM; SPERMATOZOON; TESTIS.

gonadotropin A generic name for a group of *hormones secreted by the *pituitary, the *placenta, and the endometrium that stimulate the gonads. They regulate reproductive activity.

gonangulum In insects, a *sclerite formed from the ninth abdominal segment and forming part of the *ovipositor.

Gondwana A former supercontinent of the southern hemisphere from which S. America, Africa, India, Australia, and Ant-

arctica are derived. Their earlier connection explains why related groups of plants and animals are found in more than one of the now widely separated southern land masses; examples include the *Dipnoi (lungfish) common to S. America, Africa, and Australia, and the monkey-puzzle tree (*Araucaria*) common to S. America and Australia.

(((⊕))) SEE WEB LINKS
• Description of Gondwana.

goniatites *See* AMMONOIDEA.

gonochorist Applied to an organism, or an organism, that has either ovaries or testes. *Compare* HERMAPHRODITE.

gonocoel A body cavity that contains the *gonads.

gonopodium An *intromittent (copulatory) organ, formed by a modified *anal fin, that is present in adult male fish of the families *Poeciliidae (live-bearers) and *Goodeidae (goodeid topminnow).

gonopore An external opening of a reproductive system.

Gonorynchidae (sandfish; superorder *Ostariophysi, order *Gonorynchiformes) A family of marine fish that have an almost pencil-shaped, round body, a single, backward-placed *dorsal fin opposite the *ventral fins, and the *anal fin even closer to the tail fin. The small mouth is somewhat below the pointed snout. There are one or two species, found in the Indo-Pacific region.

Gonorynchiformes (subclass *Actinopterygii, superorder *Ostariophysi) An order of marine fish that have a small mouth, toothless jaws, and the single *dorsal fin and the *ventral fins located half-way down the body, or further backward. There are four families.

Gonostomatidae (bristlemouth; subclass *Actinopterygii, order *Salmoniformes) A family of small (about 7 cm long) deep-sea fish that have an elongate body, a large mouth, and a single *dorsal and a long *anal fin. The *pectoral and *pelvic fins are small. A series of *photophores is often present along the ventral side of the fish. Many species seem to be present in large numbers in the deep sea. There are about 60 species.

gonotheca In some colonial coelenterates, a cylindrical capsule containing a reproductive *polyp.

gonozooid In some colonial coelenterates, a reproductive *polyp.

Goodeidae (goodeid topminnow; subclass *Actinopterygii, order *Atheriniformes) A family of freshwater fish, including deep-bodied as well as elongate species, that have a single *dorsal fin opposite the *anal fin, and a fairly pointed snout. They are *live-bearing, the modified anal fin of the males acting as a primitive *gonopodium. There are about 24 species, found in western Mexico.

goodeid topminnow See GOODEIDAE.

goose See ANATIDAE.

goose barnacles See LEPADIDAE.

goosefish (*Lophius americanus*) See LOPHIIDAE.

gopher tortoise (*Gopherus polyphemus*) See TESTUDINIDAE.

***Gopherus polyphemus* (gopher tortoise)** See TESTUDINIDAE.

Gorgonacea (sea fans, horny corals; class *Anthozoa, subclass *Octocorallia) An order of colonial octocorals that are anchored by a holdfast and have a tree-like shape supported by a branching, central skeleton of horn-like, organic material (gorgonin) or calcareous material, or both. They first appeared in the *Cretaceous. The order contains two suborders, 18 families, and about 120 genera.

gorgonin See GORGONACEA.

gorilla (*Gorilla gorilla*) See HOMINIDAE.

goshawks See ACCIPITRIDAE.

Gould, John (1804–81) A British naturalist who described and depicted (although most of the paintings were executed by his wife) much of the Australian fauna, but whose main claim to fame is that he sorted out and analysed *Darwin's collection of finches brought from the Galápagos (*see* DARWIN'S FINCHES), opening Darwin's eyes to their geographic and ecological diversity. Before Gould's work, Darwin had not realized the true nature of this group of birds, which later became a keystone of his theory of evolution by natural selection. In 1838 Gould and his wife travelled to Tasmania and from there to some of the islands in the Bass Strait, then South Australia and New South Wales. They returned to England in 1840. Mrs Gould died in 1841, but he completed *Mammals of Australia*, which was published in 1863.

gourami See BELONTIIDAE.

gourami, giant See OSPHRONEMIDAE.

Graafian follicle A fluid-filled, spherical *vesicle in the ovary of a mammal, containing an *oocyte attached to its wall, and responsible for *oestrogen production in the ovary. A cavity first appears amongst follicle cells surrounding the oocyte, around the end of the period of cytoplasmic growth of the oocyte. In mammals with an oestrus cycle, it enlarges during the early part of the cycle, so that follicle cells separate into two layers, one around the oocyte and the other an external layer, until the follicle bursts on to the surface of the ovary, discharging the oocyte (ovulation). Most follicles, however, will not ovulate but will simply degenerate; the rest form the *corpus luteum in most mammals. Follicle growth is under the influence of the *pituitary gland.

Gracillariidae (subclass *Pterygota, order *Lepidoptera) Widespread family of very small moths, with narrow, strongly fringed, colourful wings. At rest the anterior part of the body is generally raised. Larvae are leaf-miners for all or part of their lives.

Pupation occurs within the mine or in a chamber formed when the larva rolls the edge of the leaf. There are about 151 genera and more than 2500 species.

grade 1. A group of things all of which have the same value. **2.** A distinctive functional or structural improvement in the organization of an organism. Thus fish, amphibians, reptiles, and mammals represent successive vertebrate grades. Grades may occur within a single lineage; or the same grade may be achieved independently in different ones (e.g. warm-bloodedness evolved independently in birds and mammals).

gradualism See PHYLETIC GRADUALISM.

graft 1. *noun.* A small piece of tissue that is implanted into an intact organism. **2.** *verb.* To transfer a part of an organism from its normal position to another position on the same organism (autograft), or to a different organism of the same species (homograft), or an organism of a different species (heterograft). The source of the part that is grafted is called the donor (in animals) and the organism to which it is united is called the host or recipient. A graft hybrid is an organism made up of two genetically distinct tissues due to fusion of host and donor after grafting.

Grallinidae (magpie larks, mudnest builders; class *Aves, order *Passeriformes) A family of black and white or grey birds whose features are varied; the mudnest builders are often separated into a separate family, Corcoracidae. They inhabit forest, woodland, and scrub, and build mud nests stuck to a horizontal branch. There are three genera, with four species, found in Australia and New Guinea.

Grammicolepididae (subclass *Actinopterygii, order *Zeiformes) A family of marine, deep-sea fish somewhat similar to dories, but which have a small, nearly vertical mouth, and long scales arranged in almost vertical rows across the body. They are rather rare fish; one species, *Vesposus egregius*, is based on a specimen brought up by a lava flow near Hawaii. They are widely distributed.

Grammidae (basslets; subclass *Actinopterygii, order *Perciformes) A small family of tropical, marine fish that have two *dorsal fins, thoracic *pelvic fins, and an interrupted or absent *lateral line. There are about nine species, occurring in the Indo-Pacific region and the western Atlantic.

Grammistidae (soapfish, six-lined perch; subclass *Actinopterygii, order *Perciformes) A small family of marine, deep-bodied fish in which all the fins have a rounded profile. Typically, members of the family are able when disturbed to release a thick mucus, which turns into a soapy froth. There are about 17 species, occurring in the Indo-Pacific and Atlantic, some among coral reefs.

Gram reaction A reaction obtained when *bacteria are subjected, in the laboratory, to a staining procedure called the Gram (or Gram's) stain (after the Danish scientist Christian Gram (1853–1938) who first devised the technique in 1884). The bacteria are killed and stained, for example with crystal violet; the stained cells are then treated with an organic solvent such as acetone or ethanol. Bacteria fall into two categories: those that readily decolorize under these conditions, and those that retain the stain. Cells that lose colour are said to be Gram-negative; those that retain colour are Gram-positive. The difference in reaction reflects a fundamental difference in the structure of the cell wall in the two types of bacteria.

granule See SECRETORY VESICLE.

granulocyte (polymorph) See LEUCOCYTE.

granulose Consisting of or covered with small granules or grains.

granulosis Disease of insects caused by certain DNA-containing viruses. The name of the disease derives from the granular inclusions seen in the cells of an infected insect. It affects mainly the larval stages of *Lepidoptera.

graptolites *See* GRAPTOLITHINA.

Graptolithina (graptolites; phylum *Hemichordata) A class of extinct, colonial, marine organisms which secreted a chitinous exoskeleton with characteristic growth bands and lines. The complete skeleton (colony) is referred to as a rhabdosome. They first appeared in the Middle *Cambrian and are last known from the lower *Carboniferous. The class comprises two principal orders: true (graptoloid) graptolites (Graptoloidea), which are important zonal fossils in *Ordovician and *Silurian rocks; and the dendroid graptolites (Dendroidea).

Graptoloidea *See* GRAPTOLITHINA.

graptoloid graptolites *See* GRAPTOLITHINA.

grass-cutters *See* THRYONOMYIDAE.

grass flies *See* CHLOROPIDAE.

grasshoppers *See* ACRIDIDAE; CAELIFERA; ENSIFERA; EUMASTACIDAE; GRYLLACRIDIDAE; LOCUST; ORTHOPTERA; PNEUMORIDAE; PROSCOPIIDAE; PYRGOMORPHIDAE; TETTIGONIIDAE; TETRIGIDAE.

grasshopper warblers (*Locustella*) *See* SYLVIIDAE.

grass snake (*Natrix natrix*) *See* COLUBRIDAE.

graviportal Applied to limbs that are short and massive.

grayling butterflies *See* SATYRINAE.

grazing food chain *See* GRAZING PATHWAY.

grazing pathway (grazing food chain) *Food chain in which the primary producers (green plants) are eaten by grazing *herbivores, with subsequent energy transfer to

various levels of *carnivore (e.g. plant (blackberry, *Rubus*)→herbivore (bank vole, *Clethrionomys glareolus*)→carnivore (tawny owl, *Strix aluco*). *Compare* DETRITAL PATHWAY.

greater glider (*Petauroides volans*) *See* PSEUDOCHEIRIDAE.

greater roadrunner (*Geococcyx californianus*) *See* CUCULIDAE.

great house gecko (*Gekko gecko*) *See* GEKKONIDAE.

great spotted cuckoo (*Clamator glandarius*) *See* CUCULIDAE.

great tegu (*Tupinambis teguixin*) *See* TEIIDAE.

great warty newt (*Triturus cristatus*) *See* SALAMANDRIDAE.

grebes *See* PODICIPEDIDAE.

Greek moray (*Muraena helena*) *See* MURAENIDAE.

Greek tortoise (*Testudo hermanni*) *See* TESTUDINIDAE.

greenbottles *See* CALLIPHORIDAE.

greenbuls *See* PYCNONOTIDAE.

greeneye *See* CHLOROPHTHALMIDAE.

greenfinch *See* FRINGILLIDAE.

greenfly *See* APHIDIDAE.

greenhead *See* TABANIDAE.

green iguana (*Iguana iguana*) *See* IGUANIDAE.

green lacewings *See* CHRYSOPIDAE.

Greenland halibut (*Reinhardtius hippoglossoides*) *See* PLEURONECTIDAE.

Greenland shark (*Somniosus microcephalus*) *See* SQUALIDAE.

greenling *See* HEXAGRAMMIDAE.

green lizard (*Lacerta viridis***)** *See* LACERTIDAE.

green pigeons (*Treron***)** *See* COLUMBIDAE.

green toad (*Bufo viridis***)** *See* BUFONIDAE.

green tree frog (*Hyla arborea***)** *See* HYLIDAE.

green turtle (*Chelonia mydas***)** *See* CHELONIIDAE.

gregaria *See* LOCUST.

Gregarinia (subphylum *Sporozoa, class *Telosporea) A subclass of *Protozoa which live parasitically in the digestive tracts of invertebrates, apparently causing little harm to their hosts. The anterior part sometimes possesses hooks, suckers, simple filaments, or a knob for anchoring the parasite to the cells of the host. The life cycle can be completed in a single host species.

gregariousness The tendency of animals to form groups which possess a social organization (e.g. schools of fish, herds of mammals, flocks of birds). *Compare* AGGREGATION.

grenadier *See* MACROURIDAE.

gressorial Applied to legs that are modified for running.

greybirds *See* CAMPEPHAGIDAE.

greylag goose (*Anser anser***)** *See* ANATIDAE.

grey matter Tissue of the central nervous system which contains a high density of nerve cell bodies. Grey matter occurs in the cortex of the *cerebrum and that of the *cerebellum, and forms an H in a cross-section of the spinal cord. *Compare* WHITE MATTER.

grey mullet *See* MUGILIDAE.

grey nurse shark (*Odontaspis taurus***)** *See* ODONTASPIDIDAE.

grey partridge (*Perdix perdix***)** *See* PHASIANIDAE.

grey ramus communicans *See* RAMUS COMMUNICANS.

grey whale (*Eschrichtius***)** *See* ESCHRICHTIIDAE.

grid analysis of pattern The detection of pattern using a contiguous grid of quadrats, rather than random sampling with a given quadrat size. By blocking adjacent quadrats in pairs, fours, eights, etc., the data may be analysed using increasingly larger quadrat sizes. This is important since the detection of pattern relates to the quadrat size used, with the most marked demonstration of contagion occurring when the quadrat size is approximately equal to the clump area (i.e. the size of the clumps). When the presence of contagion is not immediately obvious, detection of the scale of non-randomness (i.e. the quadrat size at which it is evident) provides very useful information as it may suggest the likely cause of the clumping.

grinner *See* SYNODONTIDAE.

Gromphadorhina *See* BLABERIDAE.

grooming Behaviour that is concerned with the care of the body surface, performed by an animal upon itself (autogrooming), or by one animal upon another of the same species (allogrooming). In some animals (e.g. rodents) the spreading of saliva on the fur during grooming may serve a role in thermoregulation. Autogrooming may serve to change the state of arousal of an animal; allogrooming strengthens social bonds and helps to provide all members of a group with a scent characteristic of the group, by means of which group members may identify one another.

grosbeaks *See* CARDINALIDAE; FRINGILLIDAE.

gross ecological efficiency *See* FOOD-CHAIN EFFICIENCY; LINDEMAN'S EFFICIENCY.

ground beetle *See* CARABIDAE.

ground cricket *See* GRYLLIDAE.

ground doves (*Columbina*) *See* COLUMBIDAE.

groundhog Another name for the woodchuck (*see* SCIURIDAE). In the USA, 'Groundhog Day' (2 February) is popularly supposed to be the day the groundhog emerges from his hole. The custom is celebrated in Punxsutawney, Pennsylvania. If 'Punxsutawney Phil' sees his shadow he considers this an omen of more bad weather and returns to his hole for a further six weeks; if he sees no shadow (the day is cloudy) he assumes spring is arriving and remains above ground.

groundhopper *See* TETRIGIDAE.

groundroller *See* BRACHYPTERACIIDAE.

ground sloth *See* MEGALONYCHOIDEA.

ground substance The amorphous, transparent, extracellular organic material, found in *connective tissue, which acts as a matrix within which a variety of fibres and cells are embedded. Most of it consists of acidic polysaccharides bound to proteins that endow it with substantial mechanical properties. These same properties also allow it to act as a barrier to bacterial infections.

grouper *See* SERRANIDAE.

group selection *Natural selection that works to the advantage of a population (the group) rather than individuals. There is considerable doubt about whether this actually occurs.

group translocation A mechanism widely utilized for the transportation of sugars across bacterial membranes and perhaps those of some higher cells. A donor compound is used to activate sugar molecules through the provision of a high-energy phosphate group. The activated sugar can thus more readily traverse the membrane. A number of membrane proteins have been implicated in the transfer process.

grouse *See* TETRAONIDAE.

grouse locust *See* TETRIGIDAE.

growth An increase in size of a cell, organ, or organism. This may occur by cell enlargement or by cell division.

growth form 1. The morphology of an animal, especially as it reflects its physiological adaptation to the environment. **2.** Shape of population growth, as expressed by a growth curve (e.g. *J-shaped, *S-shaped).

growth hormone (somatotropin) A proteinaceous *hormone secreted by the anterior lobe of the *pituitary (the adenohypophysis), which promotes growth of the body and has a stimulatory effect on many aspects of metabolism.

grubfish *See* MUGILOIDIDAE.

Gruidae (cranes; class *Aves, order *Gruiformes) A family of large, long-legged, long-necked, white, grey, or brown birds in which the head has areas of bare red skin or plumes, the bill is straight and medium to long, secondaries are elongated for display, the tail is short, and the wings are broad for soaring and gliding, the birds flying with the neck extended. Cranes inhabit plains and marshes, feed on animals and vegetable matter, and nest on the ground. There are four genera, with 14–15 species, many of which are endangered. They are found in N. America, Europe, Asia, and Australasia. The northern species are migratory. *Grus grus* (common crane), one of 10 *Grus* species, is found across Europe to eastern Asia, wintering in Africa and southern Asia.

Gruiformes (cranes, rails, bustards; class *Aves) A diverse order of small to large aquatic and terrestrial birds that comprises 12 families: *Aramidae, *Cariamidae, *Eurypygidae, *Gruidae, *Heliornithidae, *Mesitornithidae, *Otididae, *Pedionomidae, *Psophiidae, *Rallidae, *Rhynochetidae, and *Turnicidae. They are found world-wide.

grunt 1. *See* POMADASYIDAE. **2.** *See* THERAPONIDAE.

grunting thorny catfish (*Amblydoras hancocki*) See DORADIDAE.

Grus See GRUIDAE.

***Grus grus* (common crane)** See GRUIDAE.

Gryllacrididae (order *Orthoptera, suborder *Ensifera) Family of insects, most of which are fully winged, although in some species wings are reduced or absent. Antennae are much longer than the body. There are about 600 species listed, including leaf-rolling grasshoppers and a number of desert species.

Gryllidae (bush cricket, field cricket, ground cricket, house cricket, short-winged cricket, sword-tailed cricket, tree cricket; order *Orthoptera, suborder *Ensifera) Large family of true crickets, most of which are winged, with the fore wings box-like and bent down at the sides of the body. The *tarsi have no more than three segments, and the *ovipositor is cylindrical. Many species show well-developed auditory communication. Some are of economic importance as pasture pests. There are about 2000 species.

Grylloblattodea (rock crawlers; class *Insecta, subclass *Pterygota) Small order of *exopterygote insects, which are secondarily wingless. They are slender, about 2–3 cm long; and have long, *filiform antennae and *cerci, and a sword-shaped *ovipositor. They live in rock crevices, ice caves, and in rotting logs, and are seldom seen. Their optimal temperature is just above 0°C. They feed on moss, and are also predators and scavengers of other insects that have been immobilized by the cold. Rock crawlers occur in cold, wet, usually high-latitude areas of Japan, Siberia, and N. America. Only six species are known.

Gryllotalpidae (mole crickets; order *Orthoptera, suborder *Ensifera) Fairly cosmopolitan family of relatively large, burrowing mole crickets, which are greatly modified for digging. They have large, *fossorial fore legs, a strongly convex *prothorax, reduced fore wings, a cylindrical *abdomen, and no *ovipositor. Some of the 50 or so species are wingless, and all *stridulate.

GTP (guanosine triphosphate) See GUANOSINE PHOSPHATE.

guan- A prefix derived from the Quechua *huanu*, 'dung' (e.g. *guanine can be extracted from *guano).

guanaco (*Lama guanicoe*) See CAMELIDAE.

guanido group The basic-group configuration $C=(NH_2)_2$ found in arginine.

guanine A *purine base that occurs in both DNA and RNA.

guano The accumulated droppings of birds, bats, or seals, found at sites where large colonies of these animals occur. Guano is rich in plant nutrients (bird guano is richer than bat or seal guano).

guanosine The *nucleoside formed when *guanine is linked to *ribose sugar.

guanosine phosphate The *nucleotide of the *purine base *guanine. Guanosine phosphates are designated guanosine mono-, di-, and triphosphates (GMP, GDP, and GTP respectively). Cyclic-GMP is believed to be antagonistic in cells to *c-AMP, while GTP is involved as a high-energy compound in *peptide bond synthesis.

guans See CRACIDAE.

gudgeon See ELEOTRIDAE.

guenon (*Cercopithecus*) See CERCOPITHECIDAE.

guild A group of species that have similar ecological roles, because they require the same resources and obtain them by similar means.

guineafowl See NUMIDIDAE.

guinea-pig (*Cavia*) See CAVIIDAE.

guitarfish *See* RHINOBATIDAE.

gular plate In more primitive bony fish (*Osteichthyes), a bony plate that extends forward from the *gill covers over part of the throat and lower jaw.

gular pouch In some members of the *Pelecaniformes, but most conspicuously in *Pelecanidae (pelicans), the large, distensible pouch below the lower *mandible of the bill that is used in catching fish. As it flies low over the water, the bird opens its bill and dips the lower mandible into the water. The mandible distends as it fills with water and fish. As the bird raises its head and closes its bill, the pouch contracts, forcing out water but retaining the fish.

gulf wobbegong (*Orectolobus ornatus***)** *See* ORECTOLOBIDAE.

gullet *See* OESOPHAGUS.

gulls *See* CHARADRIIFORMES; LARIDAE.

Gulo gulo (carcajou, glutton, skunk bear, wolverine) *See* MUSTELIDAE.

gulper 1. (*Eupharynx pelecanoides***)** *See* EURYPHARYNGIDAE. **2. (whiptail)** *See* SACCOPHARYNGIDAE.

gundi *See* CTENODACTYLIDAE.

gunnel *See* PHOLIDIDAE.

guppy *See* POECILIIDAE.

gurnard *See* TRIGLIDAE.

gustation Tasting.

Guttera (guineafowl) *See* NUMIDIDAE.

Gymnarchidae (subclass *Actinopterygii, order *Mormyriformes) A *monotypic family that comprises the freshwater fish *Gymnarchus niloticus*. It has an eel-like body, small eyes, a very long *dorsal fin, and *pectoral fins; the *pelvic and *anal fins are absent. This fish grows to a length of at least 1.2 m, and can produce low-voltage electric currents using modified muscles in the tail region. It is found in tropical Africa.

Gymnoblastina (Anthomedusae, Athecata, Tubularina; class *Hydrozoa, order *Hydroida) A suborder of *Cnidaria, in which the *polyps and reproductive *zooids are not protected. *Medusae have gonads in the ectoderm of the *manubrium.

Gymnogyps californianus (California condor) *See* CATHARTIDAE.

Gymnolaemata (phylum *Bryozoa, subphylum *Ectoprocta) A class of bryozoans in which the circular *lophophore lacks a lip-like appendage covering the mouth. The body wall is non-muscular. The class extends from the *Ordovician to the present.

Gymnophiona *See* APODA.

Gymnorhina (Australian magpie) *See* CRACTICIDAE.

Gymnosomata (naked pteropods; phylum *Mollusca, class *Gastropoda) An order of *opisthobranch gastropods which typically have no shell or *mantle cavity. They have a planktonic mode of life, aided by small parapodial fins. They are carnivorous.

Gymnostomatida (class *Ciliatea, subclass *Holotrichia) An order of ciliate *Protozoa in which the body *cilia are uniformly arranged. The oral region lacks cilia, and the *cytostome opens directly to the exterior. There are two suborders. They are found in marine and freshwater aquatic environments.

Gymnotidae (knife eel; subclass *Actinopterygii, order *Cypriniformes) A small family of freshwater fish that have an eel-like body with no fins, except for a very long *anal fin and small *pectorals, a small head, small eyes, and powerful jaws. They can produce weak electric currents. There are three species, occurring in Central and S. America.

gymnure (Malayan moon rat; ***Echinosorex gymnurus*)** *See* ERINACEOIDEA.

gynandromorph An organism in which part of the body is male and the other part female.

gyne A reproductive female ant, bee, or wasp (*Apocrita) that will become a queen following mating.

***Gypohierax angolensis* (palm-nut vulture)** *See* ACCIPITRIDAE.

***Gyps* (vultures)** *See* ACCIPITRIDAE.

gypsy moth (*Lymantria dispar*) *See* LYMANTRIIDAE.

Gyrinidae (whirligig beetles; subclass *Pterygota, order *Coleoptera) Family of fast-moving, gregarious beetles which swim in small circles on the surface of ponds. They have dark, shiny *elytra. The antennae are short and stout. The eyes are divided for vision above and below the water surface. The middle and hind legs are paddle-like swimming structures, but the fore legs are modified for grasping prey. They feed on insects found on the water surface. Larvae are predatory and bottom-dwelling, with 10 pairs of feathery, external gills. There are 700 species.

Gyrinocheilidae (sucker loach; subclass *Actinopterygii, order *Cypriniformes) A small family of tropical freshwater fish that have a slender body, large eyes, and a single *dorsal fin placed slightly ahead of the *pelvic fins. The mouth is placed almost in line with the flat belly. The thickly built 'lips' enable the fish to attach itself firmly to rocks, but normally they are used to scrape algae from the surface of the rock. There are about three species, occurring in south-east Asia.

gyroconic Applied to the shell of a cephalopod (*Cephalopoda) when it coils relatively loosely.

Gyrocotylidea (phylum *Platyhelminthes, class *Monogenea) An order of parasitic worms all of which lack a gut. They are endoparasitic, living in certain fish.

Gyropidae (order *Phthiraptera, suborder *Amblycera) Family of chewing lice, which occur in S. America, where they are parasitic on *caviomorph rodents. Unlike most other mammal-infesting lice, which hold the host's hair between the *tarsal claw and the apex of the *tibia, Gyropidae clasp the hair between the *tarsus and the side of the *tibia or between the side of the tibia and the side of the *femur. The surfaces of the leg contacting the hair are ridged to provide greater purchase, a feature that distinguishes Gyropidae from other lice. There are eight genera, with 61 species.

gyrose Sinuous, curved; marked with curved lines or grooves.

habitat The living place of an organism or *community, characterized by its physical or biotic properties.

habitat fragmentation The dividing of an area of *habitat into a number of smaller areas separated by stretches supporting a radically different community of organisms. This occurs naturally, when geologic processes alter the environment, and when human activity changes the use of the land. Fragmentation tends to increase the amount of habitat edge while decreasing the amount of interior habitat, and to convert a contiguous area into a number of 'islands'. Animals that are unable to cross the intervening areas between these 'islands' risk isolation.

habitat selection The choice by an organism of a particular *habitat in preference to others (e.g. mayfly nymphs inhabit the underside of stones in fast-flowing streams and burrow into sediment in still water).

habituation A decrease in behavioural responsiveness that occurs when a stimulus is repeated frequently with neither reward nor punishment. The process involves learning to ignore insignificant stimuli and should not be confused with *accommodation.

haddock (*Melanogrammus aeglefinus*) *See* GADIDAE.

Hadean The interval of geologic time that lasted from the formation of the Earth 4567.17 Ma ago until the start of the *Archaean eon 3800 Ma ago. Divisions of geologic time are required to lie between firmly dated events. This is not possible with the Hadean and consequently its name is used only informally.

Hadromerida (class *Demospongiae, subclass *Tetractinomorpha) An order of sponges in which the body has a radial structure. *Spicules and organic fibres form the skeleton. The *cortex is simple. *Oviparous forms reproduce sexually. Some live in holes they have bored. They first appeared in the *Cambrian.

Hadrosauridae (duck-billed dinosaurs; subclass *Archosauria, order *Ornithischia) A family of dinosaurs of possibly amphibious habit which averaged about 9 m in body length. Several mummies of hadrosaur dinosaurs have been found, from which it was possible to establish that they were unarmoured and that the digits of the fore limbs were connected by a web of skin. Hadrosaurs flourished in the Late *Cretaceous.

Haeckel, Ernst Heinrich (1834–1919) A German zoologist who became Professor of Comparative Anatomy and Director of the Zoological Institute at Jena in 1862. Convinced of the validity of the theory of organic *evolution, he did much to promote it in Germany, and *Darwin attributed the success of his theory in that country to Haeckel's enthusiasm. Haeckel constructed genealogical trees for many living organisms and divided animals into Protozoa (single-celled) and Metazoa (multi-celled) groups. He was an accomplished field naturalist and in 1866 he coined the word '*ecology'.

haem (heme) A metal-*chelate complex in which iron is bonded to the four nitrogens of a *porphyrin ring structure. The word (and prefix) is derived from the Greek *haima*, 'blood'.

Haematomyzidae *See* RHYNCOPHTHIRINA.

Haematomyzus elephantis (elephant louse) *See* RHYNCOPHTHIRINA.

Haematomyzus hopkinsi (warthog louse) *See* RHYNCOPHTHIRINA.

Haematopinidae (order ***Phthiraptera, suborder *Anoplura**) Family of sucking lice which are parasitic on *Artiodactyla and *Perissodactyla. Members of this family lack eyes but possess very prominent *ocular points. There is only one genus, *Haematopinus*, with 22 species.

Haematopodidae (oystercatchers; class ***Aves, order *Charadriiformes**) A family of large, black, or black and white, waders, which have long, stout, orangered bills, and thick, orange legs with three webbed toes. They run and swim well, are very noisy, and inhabit coastal shores, where they feed on molluscs, crustaceans, worms, and insects. They nest on the ground. There is one genus, *Haematopus*, containing about eight species, found in Europe, Asia, N. and S. America, Australia, and New Zealand.

haemochorial In a eutherian *placenta, applied to the condition in which the relationship between the *embryo and its mother is very intimate, the embryonic *villi being bathed in the maternal bloodstream.

haemocoel A body cavity, containing blood, which has the same embryonic origins as the blood-vascular system (i.e. it is not a *coelom). Functionally, it replaces the coelom as the main body cavity and *hydrostatic skeleton but it does not contain *gametes.

haemocyanin (hemocyanin) A blue, copper-containing, respiratory pigment that is found free in the *blood plasma of molluscs and crustaceans. It is a non-haem compound that is a polymer of smaller monomeric units.

haemoglobin (hemoglobin) A *conjugated protein that binds molecular oxygen in a loose, easily reversible manner and is used as a respiratory pigment in many groups of animals, particularly the vertebrates.

Each haemoglobin molecule is a tetramer (*see* DIMER) of four *peptide chains, each bound to a haem group. In lower animals the haemoglobin occurs freely in the blood plasma, but in the vertebrates it is confined on the surface of specialized blood cells, the *erythrocytes.

haemolysis (hemolysis) The rupture of red blood cells and the dissolution of their contents.

haemopoesis (hemopoesis) The formation and development of red blood cells (*erythrocytes). In mammals and birds these cells arise from *reticular fibres (cells) in the bone marrow.

Haemosporina (subclass ***Coccidia, order *Eucoccidia**) A suborder of *Protozoa in which *gametocytes develop independently, i.e. *syzygy does not occur. More than one host species is required for the completion of the life cycle. The sexual stages occur within an insect host, while asexual reproduction occurs in a vertebrate host. There are three families, and several genera, including the malarial parasite *Plasmodium*.

hagfish *See* MYXINIDAE.

Haglidae *See* PROPHALANOPSIDAE.

hair-streaks *See* LYCAENIDAE.

hairtail *See* TRICHIURIDAE.

hairworms *See* NEMATOMORPHA.

hairy-nosed wombat (*Lasiorhinus latifrons*) *See* VOMBATIDAE.

hake (*Merluccius merluccius*) *See* DEMERSAL FISH; MERLUCCIIDAE.

Halcyon (kingfishers) *See* ALCEDINIDAE.

Halcyon chloris (white-collared kingfisher) *See* ALCEDINIDAE.

haldane A unit of evolutionary morphological change, equal to one standard deviation per generation, and named for John Burdon Sanderson *Haldane. *See also* DARWIN.

Haldane, John Burdon Sanderson (1892–1964) A Scottish physiologist and geneticist, who graduated from Oxford University in classics and philosophy, but who taught physiology and became interested in genetics. He investigated respiration and the effect of carbon dioxide in the blood, using himself as an experimental subject. Later he calculated the rate of *enzyme reactions, proving these obey the laws of thermodynamics. Finally, he turned to the mathematics of *natural selection. Haldane was born in Oxford, the son of John Scott Haldane, an eminent physiologist. In 1933 he was appointed professor of genetics at University College, London, where he later became professor of biometry. A committed atheist and Communist, he left the Communist Party because of the influence wielded by T. D. Lysenko. He migrated to India in 1957 in protest at the British and French invasion of Suez and died at Bhubaneshwar.

halfbeak *See* EXOCOETIDAE.

half-sib mating Mating between individuals that have one parent in common, i.e. between half-brother and half-sister.

halftooths *See* HEMIODONTIDAE.

***Haliaeetus* (fish eagles)** *See* ACCIPITRIDAE.

Halichondrida (class *Demospongiae, subclass *Ceractinomorpha) An order of sponges that have many skeletal *spicules, present in an unarranged, irregular fashion. A thin dermis occurs, and this is often strengthened by spicules. Organic fibres are also present.

Halictidae (sweat bees; order *Hymenoptera, suborder *Apocrita) Large, cosmopolitan family of mining bees, of great interest to behaviourists because the family encompasses all the grades of sociality. Like those of the *Colletidae, the brood cells are lined with macrocyclic lactones secreted by the *Dufour's gland. Small species of *Lasioglossum* are attracted to human perspiration and hence the name 'sweat bees'.

Haliplidae (subclass *Pterygota, order *Coleoptera) Family of small, yellow, convex water beetles, 2–5 mm long. The hind *coxae are expanded into large plates, the legs fringed with swimming hairs. Usually the beetles are found crawling on water plants. The aquatic larvae are noted for the long, fleshy processes of their body segments. Adults and larvae feed mainly on algae. There are 200 species.

hallux 1. The 'great toe'. On the *pentadactyl hind limb, the digit on the tibial side, which is often shorter than the other digits. 2. The first digit of the foot of a bird, forming the hind toe in many species. In some (e.g. *Coliidae) it is reversible, in others (e.g. *Apodidae) it points forward, but in some species it is reduced or absent.

***Halocynthia pyriformes* (sea peach)** *See* ASCIDIACEA.

Halosauridae (subclass *Actinopterygii, order *Notacanthiformes) A small family of very elongate deep-sea fish that have short *dorsal, *pectoral, and *pelvic fins, but a long *anal fin, a subterminal mouth, and relatively large scales. There are about 14 species, found (often near the edges of continental shelves) world-wide.

haltere In *Diptera, the modified hind wing, shaped like a drum stick and used as a balancing organ, enabling the fly to sense direction and movement in flight.

hamadryad (*Ophiophagus hannah***)** *See* VIPERIDAE.

hamadryas (baboon, *Papio***)** *See* CERCOPITHECIDAE.

Hamilton, William Donald (1936–2000) The evolutionary biologist who proposed the theory of kin selection to explain *altruism and who suggested the parasite theory to explain the evolution of sexual reproduction from asexual reproduction.

This theory holds that by inheriting *genes from two parents rather than one, sexually reproducing species present genomic (*see* GENOME) differences in each generation to which parasites (*see* PARASITISM) have to adapt; thus sexual reproduction gives organisms an evolutionary advantage over parasites. Hamilton was elected a fellow of the Royal Society in 1980 and from 1984 he was a Royal Society Research Professor in the zoology department of the University of Oxford.

hammerhead *See* SCOPIDAE.

hammerhead shark *See* SPHYRNIDAE.

hammerkop *See* SCOPIDAE.

hamster *See* CRICETIDAE.

hamula In some *Apterygota, a ventral abdominal appendage supporting the springing organ. *See also* COLLEMBOLA.

hamulus (*pl.* hamuli) In *Hymenoptera, a row of hooks along the costal margin of the hind wing which attach to a fold in the fore wing.

handicap principle An idea advanced by Amotz Zahavi (1928–) in 1975, which explains the existence of extravagant male traits (e.g. the tail of a peacock) by proposing that *sexual selection favours them because they indicate male prowess: the male demonstrates to females his ability to thrive despite the handicap (in this example of a heavy, cumbersome tail). Similarly, elaborate bird-song may demonstrate the success of the male in finding food quickly, allowing more time for song, and in avoiding predators, and elaborate plumage may demonstrate an ability to control skin and feather parasites. *Compare* RUNAWAY HYPOTHESIS.

Hapalemur (gentle lemur) *See* LEMURIDAE.

Haploactinidae *See* APLOACTINIDAE.

Haplochitonidae *See* APLOCHITONIDAE.

haplodiphoidy Genetic system in which males develop from unfertilized eggs and so are haploid, whereas females develop from fertilized eggs and are diploid. It is found in honey-bees, ants, wasps, spiders, and rotifers.

haploid number The number of *chromosomes in a *gamete, designated n. It is in contrast with the number of chromosomes in somatic cells, which is usually some multiple of this number, commonly $2n$ (diploid), but sometimes $3n$ (triploid), $4n$ (tetraploid), or many-n (polyploid). A haploid cell thus has only one chromosome set, and a haploid organism contains only haploid cells.

Haplomi Name for an order of bony fish, no longer recognized, that included the *Esocidae, *Umbridae, and *Galaxiidae.

Haplorrhini (cohort *Unguiculata, order *Primates) A suborder that comprises the anthropoids (suborder *Simiiformes, with the superfamilies *Ceboidea, *Cercopithecoidea, and *Hominoidea) and the tarsiers (infra-order Tarsiiformes), in all of which the upper lip is whole and the *placenta *haemochorial. On these and other grounds (including the proportion of proteins that are shared among primate groups) the tarsiers show closer affinities to the anthropoids than they do to the lemurs. The Haplorrhini are contrasted with the remaining primate groups, which are placed in the suborder *Strepsirrhini.

Haplosclerida (class *Demospongiae, subclass *Ceractinomorpha) An order of sponges that includes freshwater forms. If the skeleton is present it is either of siliceous *spicules, or organic fibres, or both. A dermis is not common, and where it does occur it is simple. The order first appeared in the *Cambrian.

Haplosporea (phylum *Protozoa, subphylum *Sporozoa) A class of protozoa that form spores with a resistant outer covering. There is one order. They are found as parasites in various invertebrates.

haplotype A set of closely linked genes that tend to be inherited together. The term is often applied to mitochondrial DNA (*see* MITOCHONDRION) in studies of animals, where the mitochondrial *genome is considered a haplotype due to the apparent lack of *recombination.

hapten A small molecule that by itself is non-allergenic, but that when combined with a larger carrier molecule forms a complex that reacts selectively with an *antibody.

Harderian gland A gland on the inner side of the *orbit in amphibians, reptiles, birds, and most mammals, that secretes a substance which lubricates the *nictitating membrane. The gland is absent or degenerate in *Primates. It was first described in 1693 by the Swiss anatomist Johann Jacob Harder (1656–1711).

Hardy–Weinberg law In an infinitely large, interbreeding population in which mating is random and in which there is no selection, migration, or mutation, *gene and *genotype frequencies will remain constant from generation to generation. In practice these conditions are rarely strictly present, but unless any departure is a marked one, there is no statistically significant movement away from equilibrium. Consider a single pair of *alleles, A and a, present in a diploid population with frequencies of p and q respectively. Three genotypes are possible, AA, Aa, and aa, and these will be present with frequencies of p^2, $2pq$, and q^2 respectively. The law was established in 1908 by G. H. Hardy and W. Weinberg.

(⊕) SEE WEB LINKS
• Explanation of the Hardy–Weinberg law.

hare-lipped bat (*Noctilio*) *See* NOCTILIONIDAE.

harem A group of female mammals with which a single male has exclusive mating rights.

harem group In some *Primates, a social group comprising one adult male together with several adult females and their offspring.

hares 1. *See* LEPORIDAE. **2.** (Cape jumping hare, *Pedetes*) *See* PEDETIDAE. **3.** (mouse hare, *Ochotona*) *See* OCHOTONIDAE. **4.** (Patagonian hare, *Dolichotis*) *See* CAVIIDAE.

harlequin chromosome *See* SISTER CHROMATID EXCHANGE.

Harlow, Harry Frederick (1905–81) An American psychologist who conducted *deprivation studies that demonstrated the importance of maternal contact during early development in *Primates. He was born Harry F. Israel in Fairfield, Iowa in 1905, and was educated at Reed College, Portland, Oregon, and Stanford University, graduating in psychology in 1927, and receiving his PhD in 1930; it was then that he changed his name to Harlow. In 1930 he was appointed an assistant professor in psychology at the University of Wisconsin, where he established a primate laboratory and it was there that he did his most important work. Harlow received many honours. After his retirement he lived in Tucson, Arizona, where he died in 1981.

(⊕) SEE WEB LINKS
• Biographical memoir of Harry F. Harlow.

Harpactes (trogons) *See* TROGONIDAE.

Harpadontidae (Bombay duck; subclass *Actinopterygii, order *Myctophiformes) A family of marine and brackish-water fish that have a torpedoshaped body, a compressed head with small eyes and a large mouth, a single *dorsal and an *adipose fin, and a trilobed tail fin. There are three species, of commercial value, in the Indo-Pacific region.

harriers *See* ACCIPITRIDAE.

harvester ants (*Messor*) *See* MYRMICINAE.

harvester termites (Hodotermitinae) *See* HODOTERMITIDAE.

harvestmen *See* ARACHNIDA.

hatchetfish See GASTEROPELECIDAE; STERNOPTYCHIDAE.

Haversian canal See OSSIFICATION.

Hawaiian goose (Branta sandvicensis) See ANATIDAE.

Hawaiian honeycreepers See DREPANIDIDAE.

hawfinches See FRINGILLIDAE.

hawk–dove strategies In *game theory, a contest in which only two *strategies are available, those of 'hawks' and those of 'doves' (the names do not refer to birds called hawks or doves). Hawks invariably fight aggressively; doves avoid fights and run away if they can. The purpose of the game is to calculate all the possible *payoffs to each strategy and thus determine which will prevail.

hawk eagles (Spizaetus) See ACCIPITRIDAE.

hawkfish See CIRRHITIDAE.

hawk moths See SPHINGIDAE.

hawks See ACCIPITRIDAE; FALCONIFORMES; HAWK–DOVE STRATEGIES.

hawksbill turtle (Eretmochelys imbricata) See CHELONIIDAE.

head louse (Pediculus capitis) See PEDICULIDAE.

head stander See ANOSTOMIDAE; CHILODONTIDAE.

head-tossing display 1. A *courtship display in some birds (e.g. gulls) in which a male and female walk side by side, repeatedly jerking their heads up and down, while uttering soft calls. 2. A courtship display in some birds (e.g. kittiwakes, Rissa tridactyla) in which the female moves her head up and down repeatedly; this causes the male to regurgitate food.

heart A muscular pump that causes blood to circulate. In invertebrates such as *Arthropoda, *Annelida, *Mollusca, and *Echinodermata, the heart is multichambered and occurs dorsally to the gut. In vertebrates, hearts are composed of cardiac muscle which is smooth in origin although striated in appearance, surrounded by the pericardium. There are three types of vertebrate heart structure: in fish there are two chambers, an atrium into which blood enters and a ventricle which expels blood into the circulatory system; in amphibians and most reptiles there are two atria and a single ventricle; in higher reptiles, such as *Crocodilia and above, there are two atria and two ventricles. Whilst there is a relatively low blood pressure in fish and a single circulatory system, all tetrapods have a double circulatory system. See also CARDIAC CYCLE.

heart urchins See ECHINOIDEA; SPATANGOIDA.

heath butterflies See SATYRINAE.

heavy-labelled water technique The use of water containing tritium (a radioactive isotope of hydrogen), such that the tritium replaces a stable (i.e. nonradioactive) atom in a compound, making the compound radioactive (e.g. tritium is readily incorporated into *DNA during DNA replication). The path taken by such a labelled compound in the biological system can be followed by the radiation it emits (e.g. on an *autoradiograph).

hectocotylized arm In male *Cephalopoda, a specialized *tentacle used as an *intromittent organ for the transfer of sperm. In some species of *Octopoda, during copulation the tentacle is inserted into the *mantle cavity and may break off; the detached hectocotylus was originally thought to be a parasitic worm, and was described as Hectocotylus (by *Cuvier).

Hectocotylus See HECTOCOTYLIZED ARM.

hedgehog See ERINACEOIDEA.

hedge rat See OCTODONTIDAE.

Helaletidae (order *Perissodactyla, suborder *Ceratomorpha) An *Eocene and *Oligocene family of animals that were related to the ancestral tapirs but also showed distinct affinities with the early rhinoceroses.

heliciform Coiled, like a snail shell.

helicotrema See SCALA.

Heliocharitidae See DICTERIASTIDAE.

Heliornithidae (sungrebes, finfoots; class *Aves, order *Gruiformes) A family of small to large, olive-brown or brown birds, which have a long, thin head, and broadly lobed toes. They inhabit the edges of rivers and lakes, feed on insects, molluscs, and small aquatic invertebrates, and nest low in bushes. They swim, dive, and run well, but rarely fly. There are three monotypic genera, found in Central and S. America, Africa, and southern Asia.

Heliozoa (superclass *Sarcodina, class *Actinopodea) A subclass of *Protozoa which have a spherical cell body from which radiate many slender *pseudopodia. In some species there is a porous *test, the pseudopodia extending through the pores. There are three orders, found in freshwater lakes, ponds, rivers, etc. Members of the subclass, but most commonly *Actinophrys sol*, are sometimes called 'sun animalcules'.

hellgramites See CORYDALIDAE.

helmeted guineafowl (*Numida meleagris*) See NUMIDIDAE.

helmetshrikes See LANIIDAE.

helminth A parasitic worm that lives inside its host; the study of helminths is called helminthology. The term is not taxonomic. There are three groups of helminths: cestodes (*Cestoda); nematodes (*Nematoda); and trematodes (*Trematoda).

Heloderma suspectum (gila monster) See HELODERMATIDAE.

Helodermatidae (beaded lizards, venomous lizards; order *Squamata, suborder *Sauria) A family that comprises the only poisonous lizards. The venom glands are at the rear of the lower jaw (unlike those of snakes). The body is cylindrical, with powerful limbs, and a thick, rounded tail. The upper surface is covered with large, bony scales. *Heloderma suspectum* (gila monster) is a sluggish, nocturnal lizard, mainly pink and brown with black shading, up to 60 cm long, that burrows in sand and preys principally on small rodents; its bite is seldom fatal to humans. There are two species in the family, occurring from Nevada to Mexico.

Helostoma temminckii (kissing gourami) See HELOSTOMATIDAE.

Helostomatidae (kissing gourami; subclass *Actinopterygii, order *Perciformes) A monospecific family (*Helostoma temminckii*) comprising a fish that is very popular with aquarium hobbyists because individuals often contact one another with their thick lips. They are relatively short but deep-bodied fish, with long *dorsal and *anal fins, and may reach a length of at least 25 cm. They occur in south-east Asia.

helotism A relationship between two organisms (e.g. some groups of ants, see DULOSIS) in which one 'enslaves' the other for its own benefit.

hemal arch See VERTEBRA.

hemal spine See VERTEBRA.

heme See HAEM.

Hemichordata (Stomochordata) A phylum first encountered in the Middle *Cambrian Burgess Shale, British Columbia, Canada. The hemichordates are related to the chordates (*Chordata), for although they lack a *notochord, the gill slits are very similar to those of primitive vertebrates and the dorsal collar nerve cord of hemichordates, which is sometimes hollow, is similar to the dorsal, hollow nerve cord of chordates. Hemichordates are exclusively marine. The body and *coelom are divided into three sections, the middle section bearing gill slits. Development is by a *tornaria larva, indicating affinities with the *Echinodermata; the link between echinoderms and chordates is supported by biochemical evidence, all three groups containing *creatine phosphate in the muscles whereas in all other invertebrates the phosphate is a compound of *arginine. The two classes, *Enteropneusta and *Pterobranchia, may not be descended from a common stock and nowadays are often ranked as separate phyla.

Hemicidaroida (class *Echinoidea, superorder *Echinacea) An order of regular echinoids, similar to the *Cidaroida but differing from them in having gill slits around the *peristome. They have perforate and crenulate primary *tubercles. They first appeared in the Late *Triassic.

Hemidactylus bibronii (common house gekko) See GEKKONIDAE.

hemielytron (_pl._ hemielytra) In *Heteroptera, a type of fore wing which is of a leathery texture, except at the tip, which is membranous.

Hemimeridae (earwigs; class *Insecta, order *Dermaptera) Family of earwigs which are well adapted as ectoparasites. They are flattened and streamlined, with specially modified tarsal segments with which they grip the fur of their hosts. They are wingless and blind, with rod-like *forceps. There are about 10 species, all of which live in the fur of a particular species of _Cricetomys_ (giant rat), in southern Africa.

hemimetabolous Applied to those insects in which the adult form is attained by a series of moults, and *metamorphosis is incomplete. All the immature stages are called nymphs.

Hemiodontidae (halftooths; subclass *Actinopterygii, order *Cypriniformes) A family of freshwater fish that have a slender body, rather flat belly, low-slung mouth, and single *dorsal and *adipose fins; the tail fin is strongly forked. Superficially these fish resemble some characid species, but they are distinguished by the absence of teeth in the lower jaw. They occur in tropical S. America. There are five genera with 29 species.

hemipenis In male *Squamata, one half of the paired, erectile, copulatory organ, protruded through the *cloaca. Only one hemipenis is inserted into the body of the female, which one being a matter of chance, and sperm is carried from the testes by ducts that open into the cloaca, whence it flows along a groove on the surface of the hemipenis.

Hemiphlebiidae (damselfies; order *Odonata, suborder *Zygoptera) Monospecific family whose only representative species is a tiny, metallic-coloured damselfly that occurs in Australia.

Hemiprocnidae (tree swifts; class *Aves, order *Apodiformes) A family of brown or grey swifts, many of which have a blue or green gloss to some feathers. The head is often crested, with white stripes above and below the eye. They have a non-reversible *hallux, and are less aerial than other swifts, perching in trees. They inhabit forest edges and clearings, feed on insects, and build their nests on the sides of branches. There is one genus with four species, found in India, south-east Asia, Indonesia, and New Guinea.

Hemiptera (true bugs; class *Insecta, subclass *Pterygota) Order of insects in whose life cycle there is no true pupal stage, and which never have an eleventh abdominal segment or *cerci. The mouth-parts are modified into a *rostrum or, in non-feeding stages of some species, are completely lacking. There are typically two pairs of wings, of which the anterior pair are usually wholly or partly toughened to protect the membranous posterior pair over which they are folded at rest. There are two suborders, *Homoptera and *Heteroptera.

hemocyanin See HAEMOCYANIN.

hemoglobin See HAEMOGLOBIN.

hemopoesis *See* HAEMOPOESIS.

Henlé, Friedrich Gustav Jakob (1809–55) A German anatomist, pathologist, and physician whose discoveries included the *loop of Henlé in the *kidney. He was one of the first scientists to propose a germ theory of disease. Henlé was born in 1809, in Fürth, Bavaria, and studied medicine at Heidelberg and then Bonn, where he graduated in 1832. He worked for six years as *prosector in Berlin, taught at Zürich in 1840, moved in 1844 to Heidelberg, and in 1852 to Göttingen. He died at Göttingen in 1885.

Hennig, Willi (1913–76) A German zoologist who originated *phylogenetic systematics. He was awarded his Ph.D. by the University of Leipzig in 1947 and conducted research on *Drosophila* larvae. His book, *Grundzüge einer Theorie der phylogenetischen Systematik* (1950), made no immediate impact, but when an English translation appeared in 1966, as *Phylogenetic Systematics* (with a 2nd edition in 1979), it suddenly achieved authoritative status (comparable, perhaps, to The Origin of Species!). Hennig argued that, as *taxonomy aims to depict relationships and the only objective meaning of 'related' means sharing a common ancestor, taxonomy must be based on *phylogeny. This struck an immediate chord. Hennig coined such terms as *apomorphic, *plesiomorphic, and *sister groups, and offered a redefinition of *monophyly, which he insisted must be paramount in taxonomy.

heparin A complex polysaccharide containing sulphur that is found in many tissues (particularly the lung), where it exhibits anticoagulant properties by preventing the conversion of *prothrombin to *thrombin.

Hepialidae (swift moths, ghost moths; subclass *Pterygota, order *Lepidoptera) Family of medium to very large moths, in which the wings are generally rounded apically; *venation is similar in fore wing and hind wing; and the wings are coupled in flight by a *jugum. Females scatter their eggs over vegetation; the larvae feed on roots or in wood. Hepialidae have a global distribution.

Hepsetidae (pike characin; subclass *Actinopterygii, order *Cypriniformes) A monospecific family (*Hepsetus odoe*) of freshwater fish that has an elongate, pike-like body, a large mouth, a forked tail fin, and the *dorsal and *anal fins placed well back. It is voracious and may grow to 1.8 kg. It occurs in tropical Africa.

Hepsetus odoe (pike characin) *See* HEPSETIDAE.

heptamerous Having seven parts.

Heptapsogasteridae (order *Phthiraptera, suborder *Ischnocera) Family of chewing lice, which occur in S. America where they parasitize birds of the order *Tinamiformes. Heptapsogasteridae differ from other Ischnocera in the fusion of the first three abdominal segments, the first pair of abdominal *spiracles appearing on the first visible segment. There are 15 genera, with about 130 species. This is a very high number of species for a host group of fewer than 50 species.

herbivore A *heterotroph that obtains energy by feeding on primary producers, most usually green plants. *Compare* CARNIVORE; DETRITIVORE. *See also* FOOD CHAIN.

Hercules beetle *See* SCARABAEIDAE.

herding The formation by mammals of groups of individuals that have a social organization. Usually the term is restricted to the behaviour of large herbivores.

hereditary disease A disease that is caused by a mutant gene and that may be transmitted from one generation to the next.

heritability The measure of the degree to which a *phenotype is genetically influenced and can be modified by *selection. It is represented by the symbol h^2: this equals V_a/V_p where V_a is the variance due to *genes with additive effects (known as the additive genetic variance) and V_p is the phenotypic variance. The variance V may also be written as s^2. Parent–offspring correlations, as used in studies of the heritability of human IQ, are estimates of familiarity and not of heritability: they cannot account for environmental correlations between relatives. This definition of heritability is a narrow one: heritability in the broad sense (represented by H^2) is the fraction of total phenotypic variance that remains after exclusion of the variance due to environmental effects.

Hermann's tortoise (*Testudo hermanni***)** See TESTUDINIDAE.

hermaphrodite An individual that possesses both male and female sex organs; i.e. it is *bisexual. *Compare* GONOCHORIST. *See also* SEQUENTIAL HERMAPHRODITE.

hermaphroditic fish Fish that, when mature, possess both male (testes) and female (ovary) sex glands at the same time. In such species cross-fertilization can occur during spawning. In other species (e.g. *Sparidae, sea bream), the testes develop first, these males turning into females as the fish grow larger and older. Most *Serranidae (sea perches) are females first or have both sets of glands equally developed.

hermatypic Applied to corals that contain *zooxanthellae and are reef-forming. Modern scleractinian hermatypic corals are characterized by the presence of vast numbers of symbiotic zooxanthellae in their endodermal tissue. They live in waters of normal marine salinity, at depths of up to 90 m, in temperatures above 18°C, and grow vigorously in strong sunlight. *Compare* AHERMATYPIC.

hermit crabs See PAGURIDAE.

hermits (*Phaethornis***)** See TROCHILIDAE.

herons 1. See ARDEIDAE. **2. (boat-billed heron)** See COCHLEARIIDAE.

Herpestidae (mongooses, meerkats (suricates)); order *Carnivora, suborder Feliformia) A family of small carnivores, closely related to the *Viverridae in which they were formerly placed, averaging about 230–750 mm in length and weighing from less than 1 kg to 6 kg. They have small heads with a long, flattened skull, pointed muzzles, and short, rounded ears. *Carnassials are well developed. Their claws are not retractile. Many possess *anal glands that secrete a foul-smelling substance. Males possess a *baculum. Most mongooses are predators of small mammals, birds, and reptiles, and also eat birds' eggs, insects, land crabs, and carrion. Mongooses are renowned for their speed and agility, which allows them to kill venomous snakes, including cobras and adders, despite having no immunity to haemotoxic snake venom, although they are immune to neurotoxins. Some species are solitary, but others travel in groups, called mongaggles, and share food. Meerkats, also known as suricates (*Suricata suricatta*), of southern Africa live in groups comprising two, three, or more families, each comprising an adult male, adult female, and up to five offspring. There are about 20 genera and 34 species, found mainly in Africa, but also in Asia and southern Europe. Mongooses are often kept as pets to control vermin, but they can be destructive, killing a wide variety of ground-dwelling animals.

herring See CLUPEIDAE.

herring smelts (argentines) See ARGENTINIDAE.

Hesperiidae (skippers; subclass *Pterygota, order *Lepidoptera) Large family of butterflies in which the wings are short compared with the large head and body. In contrast with other butterflies, the antennae are widely separated basally. The antennae are often clubbed, and extend into a short point terminally. The larvae are generally grass-feeders. The family has a wide distribution, but is absent from New Zealand.

Hesperornis See AVES.

Heteralocha acutirostris **(huia)** See CALLAEIDAE.

hetero- A prefix meaning 'different from', derived from the Greek *heteros*, 'other'. *Compare* HOMO-.

heteroalleles *See* HOMOALLELIC.

Heterobranchia (phylum *Mollusca, class *Gastropoda) The *clade that includes many of the land snails and slugs formerly included in the *Pulmonata, as well as the marine and freshwater gastropods.

***Heterocephalus glaber* (naked mole rat)** A *hypogaeous rodent of the family Bathyergidae and the only mammal known to exhibit *eusociality.

heterocercal tail In fish, a tail in which the tip of the vertebral column turns upward, extending into the dorsal lobe of the tail fin; the dorsal lobe is often larger than the ventral lobe. The heterocercal tail is present in many fossil fish, in the sharks (*Chondrichthyes), and in the more primitive bony fish, e.g. the families *Acipenseridae and *Polyodontidae.

heterochromatin *Chromosome material that accepts stains in the *interphase nucleus (unlike euchromatin). Such regions, particularly those containing the centromeric and nucleolus organizers, may adhere to form a chromocentre. Some chromosomes are composed primarily of heterochromatin: these are termed heterochromosomes. In many species, the *Y-chromosome is a heterochromosome.

heterochromosome *See* HETEROCHROMATIN.

heterochrony The dissociation, during development, of factors of shape, size, and maturity, so that organisms mature in these respects at earlier or later growth stages. This leads to *paedomorphosis or *recapitulation (*peramorphosis). *Compare* HETEROTOPY.

heterocoelus vertebra The condition in which the articulate surface of the *vertebra centrum is saddle-shaped, as is the case in birds. *Compare* AMPHICOELUS VERTEBRA.

heterodimer A *protein comprising paired polypeptides that differ in their *amino-acid sequences.

heterodont Possessing teeth that are differentiated into several forms, e.g. incisors, canines, premolars, molars.

Heterodonta (phylum *Mollusca, class *Bivalvia) A subclass of bivalves which have *heterodont dentition consisting of a hinge plate with distinct *cardinal teeth below the *umbo, and lateral teeth posterior to the cardinals. Some forms may be *desmodont. The ligament is *opisthodetic. Shell shape varies according to the mode of life of the organism, most having a crossed-lamellar, aragonitic shell structure. Heterodonta are mainly *infaunal siphon-feeders, the *siphons being well developed. They have *eulamellibranchiate gills and an *isomyarian musculature, and the *pallial line is entire or with a *sinus. They first appeared in the Middle *Ordovician.

Heterodontidae (Port Jackson shark, bullhead shark; subclass *Elasmobranchii, order *Heterodontiformes) A small family of marine sharks that have two *dorsal fins each preceded by a stout spine, and eyes protected by a dorsal ridge marking the otherwise blunt head profile. These sharks are bottom-dwelling and fairly slow-moving animals, feeding on molluscs, crabs, etc. There are about six species, occurring in the Indo-Pacific region.

Heterodontiformes (class *Chondrichthyes, subclass *Elasmobranchii) An order of marine sharks that includes only the family *Heterodontidae.

Heterodontosauridae (order *Ornithischia, suborder *Ornithopoda) A family of dinosaurs, known only from the Late *Triassic of S. Africa, whose name means 'different-toothed lizards'; they had tusk-like canines. Microscopic analysis of tooth wear in *Pegomastax africanus*, a heterodontosaur 60 cm tall, suggested that these animals used their fangs to nip and spar with rivals, rather than for eating meat.

heteroecious (heteroxenous) Applied to a parasitic organism in which different stages of the life cycle occur in different species of host organism.

heterogametic sex The sex that has *sex chromosomes that differ in morphology (e.g. XY), and that therefore produces two different kinds of *gametes with respect to the sex chromosomes. The term may also refer to the possession of an unpaired sex chromosome (a single X-chromosome). In mammals and most other animals, the male is the heterogametic sex, usually with equal numbers of different gametes X and Y, although some men, for example, have XXY or even XXXY. In birds, reptiles, some fish and amphibians, and Lepidoptera, the female is the heterogametic sex.

heterogamy 1. The alternation of reproduction by *parthenogenesis and bisexual reproduction. It is found in some aphids. **2.** See ANISOGAMY.

heterograft See GRAFT.

heterogynous Applied to species (e.g. many members of the *Cynipidae) whose life cycle involves the production of alternating sexual and *agamic generations. Among insect plant parasites, the alternating generations may occur on different parts of the host plant, and sometimes on different host plants.

heterokont Having *flagella of different lengths.

heterologous Applied to an *antigen and *antibody that do not correspond, so that one is said to be heterologous with respect to the other. The term may also be used to refer to a *graft originating from a donor of a species different from that of the host.

heteromeric See DIMER.

heteromerous 1. Composed of units (e.g. cells) of different types. **2.** Having unequal numbers of parts. **3.** Layered.

Heteromi A group of deep-sea fish that are now included in the order *Notacanthiformes.

heteromorphic Literally, differing in form, and applied to: (*a*) phases/stages of organisms in which there is alternation of generations; and (*b*) a *chromosome pair that have some homology but that differ in size or shape, e.g. the X and Y sex chromosomes.

Heteromyidae (kangaroo rats , pocket mice, spiny mice; order *Rodentia, suborder *Sciuromorpha) A family of nocturnal, burrowing rodents, most of which are adapted for jumping, having elongated hind feet and usually a long, haired tail. The limbs are *pentadactyl and the fore limbs bear powerful claws. Heteromyidae are distributed throughout the New World from British Columbia to the northern part of S. America. There are many species.

Heteronemertini (phylum *Nemertini, class *Anopla) An order of elongate worms which possess double-layered body muscles. A ciliated *epidermis and a highly developed dermis are present.

heterophagous Applied to an organism that feeds on a wide variety of items (i.e. one that can parasitize many different hosts).

Heteropneustidae (airsac catfish; subclass *Actinopterygii, order *Siluriformes) A family of freshwater fish that have four pairs of *barbels around the mouth, a single *dorsal fin, a very long *anal fin, a rounded tail fin, and (especially in the males) poisonous spines (which are dreaded by local fishermen) in the dorsal and *pectoral fins. The long air sac extending from the gill chamber is also typical, and gives rise to the alternative name, 'Saccobranchidae'. There is one genus (*Heteropneustes*) with four species.

Heteroptera (plant bugs, water bugs; subclass *Pterygota, order *Hemiptera) Suborder of bugs in which the head capsule is continuous ventrally behind the mouth-parts. Stink glands are almost always present.

heterosis (hybrid vigour) The increased vigour of growth, survival, and fertility of *hybrids, as compared with the two *homozygotes. It usually results from crosses between two genetically different, highly inbred lines. It is always associated with increased *heterozygosity.

Heterosomata An older name for the order *Pleuronectiformes (flatfish).

heterosomes *Chromosomes of dissimilar appearance that determine sex in some diploid organisms.

heterospecific Pertaining to individuals of different species. *Compare* CONSPECIFIC.

Heterostraci Alternative name for Pteraspida, an order of fossil jawless vertebrates. *See* PTERASPIDOMORPHA.

heterotopy An evolutionary change in the site at which a particular development occurs. The term was coined by E. H. *Haeckel in 1866 to complement *heterochrony.

Heterotrichida (class *Ciliatea, subclass *Spirotrichia) An order of ciliate *Protozoa in which the body *cilia are either uniformly arranged or absent. The cell body is often large (e.g. in *Spirostomum* it is up to 3 mm long). There are eight families, and many genera, found in freshwater and marine habitats, and as commensals (*see* COMMENSALISM) in the intestines of animals.

heterotroph An organism that is unable to manufacture its own food from simple chemical compounds and therefore consumes other organisms, living or dead, as its main or sole source of carbon. *Compare* AUTOTROPH.

heteroxenous *See* HETEROECIOUS.

heterozygosity The presence of different *alleles at a particular gene *locus. Heterozygosity provides a measure of the genetic variation, either in a population

(the frequency of individuals heterozygous at a particular locus), or in an individual (the proportion of gene loci that are heterozygous).

heterozygote A *diploid or *polyploid individual that has *alleles at at least one *locus. Its *phenotype is often identical to that of an individual that has one of these alleles in the *homozygous state. For example, the phenotype of A_1A_2 may be identical to that of A_1A_1, but different from that of A_2A_2. If so, that allele (A_1) is said to be dominant over the other (A_2), which is said to be recessive. When there exist only two alleles of a gene and when one of these is dominant over the other, one may use a lower-case letter for the recessive allele (e.g. one may write *Aa* rather than A_1A_2). However, alleles are not always dominant one over another; and often there are more than two alleles at a given locus, especially in polyploids. Because it has two or more different alleles at a given locus, a heterozygote does not breed true.

heterozygous advantage *See* GENETIC POLYMORPHISM.

hexacanth embryo *See* ONCHOSPHERE.

Hexacorallia *See* ZOANTHARIA.

Hexactinellida (Hyalospongea; phylum Porifera) A class of sponges whose skeleton is composed of mainly six-rayed, siliceous *spicules. Organic fibres do not occur. The body has a *leucon structure. They are marine, deep-water sponges and first appeared in the Early *Cambrian.

Hexagrammidae (greenling; subclass *Actinopterygii, order *Scorpaeniformes) A small family of marine fish that have a fairly robust body, a long and low *dorsal fin divided by a deep notch, a truncate tail fin, and rounded *pectorals. There are about 21 species, found in coastal regions of the northern Pacific.

Hexanchidae (cow shark; subclass *Elasmobranchii, order *Hexanchiformes) A small family of marine sharks, typically with a single, spineless *dorsal fin, an *anal fin, and six or seven gill slits which are not continuous across the throat. There are about five species, distributed world-wide.

Hexanchiformes (class *Chondrichthyes, subclass *Elasmobranchii) An order of marine sharks that have six or seven gill slits, one *dorsal fin, and an *anal fin. The order includes the families *Chlamydoselachidae and *Hexanchidae.

hexaploidy *See* POLYPLOIDY.

Hexapoda A subphylum that includes the classes *Insecta, *Collembola, *Diplura, and *Protura.

hexaster Applied to one of the small flesh *spicules of a sponge if it takes the shape of six rays emanating from a central point.

Hexasterophora (phylum *Porifera, class *Hexactinellida) A subclass of sponges in which the small, flesh *spicules occur as *hexasters.

hexose A monosaccharide sugar that contains six carbon atoms (e.g. glucose, fructose, and galactose).

hexose-monophosphate shunt (pentose-phosphate shunt) A metabolic pathway, alternative to that of *glycolysis, of carbohydrate interconversion: hexose-6-phosphate is converted into pentose-phosphate and carbon dioxide. The principal functions of the pathway are the production of deoxyribose and *ribose sugars for *nucleic-acid synthesis; the generation of reducing power in the form of NADPH for *fatty-acid and/or *steroid synthesis; and the interconversion of carbohydrates. In animals, the pathway occurs mainly in tissues that synthesize *steroids and fatty acids (e.g. *liver, *mammary glands, and *adrenal gland).

hibernal In the winter. The term is used with reference to the six-part division of the year used by ecologists, especially in relation to terrestrial and freshwater communities. *Compare* AESTIVAL; AUTUMNAL; PREVERNAL; SEROTINAL; VERNAL.

hibernation A strategy for surviving winter cold that is characteristic of some mammals. Metabolic rate is reduced to a minimum and the animal enters a deep sleep, surviving on food reserves stored in the body during the favourable summer period. *Compare* AESTIVATION. *See also* TORPOR.

hidden-necked turtles *See* CRYPTODIRA; PELOMEDUSIDAE.

hide beetle *See* DERMESTIDAE.

hierarchical and non-hierarchical classification methods The grouping of individuals by a series of subdivisions or agglomerations to form a characteristic 'family tree' (dendrogram) of groups. Alternatively, classification may be non-hierarchical, i.e. proceeding not by an organized series of progressive joinings or subdivisions, but instead achieving groupings by a series of simultaneous trial-and-error clusterings (successive approximation) until an optimum and stable pattern is found. A possible scheme, for example, is to select at random a number of starter individuals, equal to the number of groups required, and to which other individuals are added or from which they are removed according to their characteristics and the classificatory philosophy used (that of seeking to minimize internal group heterogeneity, or of seeking to maximize differences between groups). The chief advantage of hierarchical over non-hierarchical methods is the clarity with which the routes to the final groupings may be followed, facilitating explanation of those groups. However, since most hierarchical classifications either dichotomize or pair at each division or join, natural clustering patterns may be distorted or poorly represented. Hierarchical classifications are also more likely to be unduly affected by irrelevant background information.

hierarchy 1. A form of organization in which certain elements of a system regulate the activity of other elements. **2.** A form of social organization in which individuals, or groups of individuals, possess different degrees of status, affecting feeding, mating behaviour, etc.

higher categories In taxonomy, categories higher than *species, which are defined arbitrarily according to observed similarities among species, and which provide a useful hierarchical framework by which organisms may be described succinctly.

hillstream loach See HOMALOPTERIDAE.

Himantolophidae (football fish; superorder *Paracanthopterygii, order *Lophiiformes) A small family of deep-sea fish that have an almost rotund ('football') body, short fins, the *dorsal and *anal fins close to the tail, minute eyes, a large mouth, and a well-developed *illicium above the mouth. There are four species, distributed world-wide.

hinny The offspring of a female donkey and male horse (*Equidae), classed as an F_1 *hybrid. *Compare* MULE.

Hiodon tergisus (mooneye) See HIODONTIDAE.

Hiodontidae (mooneye; superorder *Osteoglossomorpha, order *Osteoglossiformes) A family comprising two species of freshwater, herring-like fish that have a single *dorsal fin positioned behind the *ventral fins. The eyes are large and the jaws bear small teeth ('toothed herring'). The larger of the two species is *Hiodon tergisus*(mooneye), 38 cm. Both species are found only in the eastern half of Canada and the USA.

Hippoboscidae (order *Diptera, suborder *Cyclorrapha) Family of aberrant flies which lack wings and are usually flattened dorsoventrally. The head is sunk into a groove in the *thorax, and the mouthparts are directed forwards, as a proboscis adapted for piercing. The antennae are inserted into a depression, and the eyes are round or oval. The adults are 'toughskinned', an adaptation to life as ectoparasites of large mammals or birds. Larvae are produced by the females in a mature state, and they pupate almost immediately. Some, e.g. *Lepoptena* species, are winged as a means of dispersal, but cast their wings once

a suitable host is found. More than 200 species are known so far.

hippocampus In vertebrates, a structure in the brain that is involved in the consolidation of new memories, generating emotional responses to events, spatial orientation, and navigation.

***Hippocampus hippocampus* (seahorse)** See SYNGNATHIDAE.

***Hippolais* (warblers)** See SYLVIIDAE.

Hippomorpha (superorder *Mesaxonia, order *Perissodactyla) A suborder of horse-like animals that comprises the extinct families *Palaeotheriidae and *Brontotheriidae and the extant family *Equidae.

Hippopotamidae (hippopotami; order *Artiodactyla, suborder *Suiformes) A family of large, amphibious artiodactyls, perhaps descended from the *Anthracotheriidae or from primitive *Tayassuidae, which appear in the fossil record for the first time in sediments of upper *Pliocene age. Now confined to Africa, during the *Pleistocene they were widespread in the warmer regions of the Old World. In modern forms the dentition is complete except for the outer *incisors. The eyes are small and set high on the skull, and the nostrils can be closed. Each limb possesses four digits, all of which are functional. There are two monotypic genera, *Hippopotamus amphibius* (common hippopotamus) and *Choeropsis liberiensis* (pygmy hippopotamus).

hippopotamus See HIPPOPOTAMIDAE.

***Hippopotamus amphibius* (common hippopotamus)** See HIPPOPOTAMIDAE.

Hipposideridae (Old World leaf-nosed bats; order *Chiroptera, suborder *Microchiroptera) A family of insectivorous bats in which the nose has a prominent leaf, the nostrils opening at the centre of a disc. The ears are pointed, and in some species joined; they are independently mobile, with an *antitragus but no *tragus. The tail extends to the edge of the *uropatagium. There are many species, distributed throughout the Old World tropics.

Hirudinea (leeches; phylum *Annelida) A class of worms in which segmentation is less obvious than in most annelids. Suckers are generally present at the anterior and posterior ends but *chaetae are normally absent. The nervous system is well developed. All are hermaphrodites. They occur in marine, freshwater, and terrestrial habitats, and some forms are bloodsuckers.

Hirundapus See APODIDAE.

hirundines See HIRUNDINIDAE.

Hirundinidae (swallows, martins, hirundines; class *Aves, order *Passeriformes) A family of small, brown, dark-blue, or dark-green birds, many with white under-parts, which have short bills with a broad gape, long and pointed wings, and a medium to long tail which is often strongly forked. The legs are short, and feathered in some species. Hirundines inhabit open areas and are aerial insectivores. They nest in cavities in trees and rocks, in burrows, or (e.g. *Hirundo rustica*, the swallow or barn swallow) in mud nests on ledges and buildings. *Delichon urbica* (house martin), one of the three species in its genus, is particularly associated with human habitations. Sand martins (four species of *Riparia*) nest in tunnels in banks. There are 17–19 genera in the family, with 77–81 species, found worldwide, many of them migratory.

Hirundo rustica (swallow, barn swallow) See HIRUNDINIDAE.

hispid Having short stiff hairs or bristles.

hist- A prefix, commonly with an added 'o' (i.e. histo-) meaning 'web-like', derived from the Greek *histos*, 'web'.

histamine A compound formed by the decarboxylation of *histidine and released principally by *mast cells during an allergic reaction, in mediating inflammation. It causes the dilation of blood vessels, the in-creased permeability of *capillaries, and the contraction of smooth muscle.

Histeridae (subclass *Pterygota, order *Coleoptera) Family of hard, shiny beetles, 1–6 mm long, with a convex ventral surface. They are generally black or metallic in colour. The last two abdominal segments are exposed. The antennae are clubbed and elbowed. The *tibiae are expanded and toothed. When the beetles are disturbed, the legs and antennae are withdrawn into grooves on the under-side. Adults and larvae live in dung and carrion, where they feed on insect larvae. Some species live in birds' nests, while others are associated with the nests of ants and termites. There are 2500 species.

histidine A basic, polar *amino acid that contains an imidazole group ($C_3H_4N_2$); it is the precursor of *histamine.

histiocyte A nonmotile *macrophage, derived from *bone marrow, that is found in *connective tissue in many organs.

Histiopteridae See PENTACEROTIDAE.

histochemistry The study of the chemistry of tissues and cells, using combined techniques from histology and biochemistry.

histocompatibility antigen See MAJOR HISTOCOMPATIBILITY COMPLEX.

histocompatibility genes See MAJOR HISTOCOMPATIBILITY COMPLEX.

histogenesis The formation of new tissues.

histology The microscopic study of the anatomy of cells and tissues.

histone One of a group of basic, *globular, simple *proteins that have a high content of the *amino acids *arginine and *lysine. Histones form part of the chromosomal material of *eukaryotic cells and appear to play an important, but as yet incompletely understood, role in gene regulation.

hive Artificial nest provided by beekeepers for honey-bees (*Apis mellifera* in the West and in Africa; *A. cerana* in parts of Asia), and for some stingless bees (*Melipona* species) in S. America.

HLA (human leucocyte antigen complex) *See* MAJOR HISTOCOMPATIBILITY COMPLEX.

hoarding Storage by an animal of some commodity, usually food, in a central cache or in specific locations throughout a *home range.

hoatzin *See* OPISTHOCOMIDAE.

Hodotermitidae (subclass *Pterygota, order *Isoptera) Family of lower termites. They feed on damp wood, except for the subfamily Hodotermitinae (harvester termites), which feed on vegetation. The Hodotermitidae forage above ground, sometimes in daylight; unlike other termites, they have well-developed eyes. There are 30 species, found mainly in the Old World tropics.

Hodotermitinae *See* HODOTERMITIDAE.

Holarctica A formerly unified, circumpolar, biogeographic region, embracing what are now N. America, Europe, and Asia (i.e. Laurasia). The legacy of this region is attested to by the great floral and faunal similarities between the three northern continents.

Holasteroida (subphylum *Echinozoa, class *Echinoidea) An order of irregular echinoids in which the *apical system is elongated, but not disjunct. In the *plastron, the labral plate is followed by a single, large plate, and a varying number of plates in a single column. The Holasteroida first appeared in the Early *Cretaceous.

Holaxonia (subclass *Octocorallia, order *Gorgonacea) A suborder of octocorals that have a distinct central axis of horny material either alone or more or less heavily permeated with a calcareous substance. The suborder contains 11 families, with about 100 genera.

Holectypoida (subphylum *Echinozoa, class *Echinoidea) An order of irregular echinoids in which the *ambulacra are narrower than the *interambulacra, there are lateral flanges on the teeth, and the *periproct is usually on the lower (sometimes the posterior) surface. They first appeared in the Early *Jurassic, and diversified in the *Cretaceous, but only two genera (*Echinoneus* and *Micropetalon*) have survived.

holistic (holological) Relating to the whole. In *ecology, the term is applied to studies which aim to understand *ecosystems as a whole (i.e. as entire systems), rather than examining their component parts. *Compare* MEROLOGICAL APPROACH.

Holling's disc equation A method for calculating the functional response of predators to increased prey density. The equation is based on laboratory experimental data simulating predation. A human predator gathers discs (prey) from boards with different 'prey' densities. In theory, the efficiency with which the predator consumes the prey should decline as the prey density increases, due to extra time spent handling the prey. Thus the relationship between prey density and numbers consumed by predators is not a straight line but a curve. This relationship was first summarized mathematically by C. S. Holling (1930–) in 1959 as $y = T_s ax/(1 - abx)$ where y is the number of discs removed, x is the disc density, T_s is the total experimental time, a is a constant describing the probability of finding a disc at a given density, and b is the time taken to pick up a disc. *Compare* DIVERSITY INDEX.

holo- A prefix meaning 'whole', from the Greek *holos*, 'whole' or 'entire'.

Holocene (Recent, Post-Glacial) The epoch that covers the last 11 000 years.

holocentric (polycentromic) Applied to *chromosomes with diffuse *centromeres such that the properties of the centromere are distributed over the entire chromosome.

Holocentridae (squirrelfish; subclass *Actinopterygii, order *Beryciformes) A large family of tropical- to temperate-water, marine fish, all species of which have large eyes, long *dorsal fins which are deeply notched, spiny edges on the *gill covers, and strongly developed *anal-fin spines. The tail fin is deeply forked. They are mostly nocturnal fish, and the body has a reddish colour. One of the more common western Atlantic species is *Holocentrus ascensionis* (squirrelfish), which may grow to about 35 cm. There are about 70 species, distributed world-wide.

Holocentrus ascensionis (squirrelfish) *See* HOLOCENTRIDAE.

Holocephali (class *Chondrichthyes) A group, often ranked as a subclass, of peculiar, shark-like fish, including both living and fossil species. They are characterized by a cartilaginous body, large eyes, large, wing-like *pectoral fins, a whip-like tail, a scale-less skin, and a *gill cover over the common gill chamber. *See also* CHIMAERIFORMES.

holocrine gland A *gland in which the entire secreting cell is destroyed on secretion. *Compare* APOCRINE GLAND.

holoenzyme An entire conjugated *enzyme, comprising a *protein component (apoenzyme) and its *prosthetic group.

hologamete In some *Protozoa a gamete formed by the fusion of two complete individuals, and therefore larger than either parent. *Compare* MEROGAMETE.

holological *See* HOLISTIC.

holometabolous Applied to a type of development in which distinct larval and adult forms occur.

holophyletic *See* HOLOPHYLY.

holophyly (*adj.* holophyletic) The condition of a group of taxa which not only are descended from a single ancestral species, but represent all the descendants of that ancestor. Holophyly represents a special case of *monophyly; some authors use the terms interchangeably.

holoplankton Zooplankton organisms that are planktonic throughout their life cycles. *Compare* MEROPLANKTON.

holoptic In a *compound eye, the condition in which the upper facets of the eye are larger than the lower ones.

Holostei (class *Osteichthyes) A group of marine and freshwater bony fish including many fossil species. Recently ranked as an infraclass, the Holostei includes the families Semionotidae, *Lepisosteidae, and *Amiidae.

holothuroid (sea cucumber) *See* HOLOTHUROIDEA.

Holothuroidea (sea cucumbers; phylum *Echinodermata, subphylum *Echinozoa) A class of worm-like echinoderms which may be free-living or attached. The mouth, surrounded by variably branched *tentacles, is at one end of the elongate body, the *anus at the other. The calcitic skeleton is not rigid, being reduced to small sclerites (or *spicules) of very variable shape: hooks, anchors, rings, and plates. Fossil spicules of sea cucumbers are first encountered in rocks of *Ordovician age.

Holotrichia (subphylum *Ciliophora, class *Ciliatea) A subclass of ciliate *Protozoa, which generally have a simple, uniform distribution of *cilia on their cells. The subclass includes a wide range of forms. There are seven orders, with numerous genera.

holotype *See* TYPE SPECIMEN.

holozoic Applied to the method of feeding in which nutrients are obtained by consuming other organisms (e.g. in most animals).

homalodotheres *See* NOTOUNGULATA.

Homalopteridae (hillstream loach; subclass *Actinopterygii, order *Cypriniformes) A family of freshwater fish that have a slender body, subterminal mouth, three or more pairs of *barbels around the mouth, a single *dorsal fin, and large, rounded *pectoral and *pelvic fins. They are bottom-dwelling fish inhabiting fast-flowing mountain streams. There are about 87 species, found in tropical Asia.

Homalorhagida (phylum *Aschelminthes, class *Kinorhyncha) An order of worms in which the head is covered on retraction by a series of body plates on the third segment.

homeobox (homoeobox) In an *embryo, a family of genes that cause the embryo to subdivide as it grows, to form groups of cells which will develop into particular organs and tissues.

homeobox gene (homoeobox gene) A gene that contains a conserved 180-base pair sequence called a homeobox. *See also* HOMEOTIC GENE.

homeomorph An organism which, as a result of *convergent evolution, comes to resemble another to which it may not be closely related.

homeostasis *See* HOMOEOSTASIS.

homeotherm *See* HOMOIOTHERM.

homeotic gene (homoeotic gene, Hox gene) A *homeobox gene in which mutations can transform part or whole of a body segment into the corresponding part of another segment. Homeotic genes are involved in the development of the basic body plan of *metamerically segmented animals.

home range The area within which an animal normally lives. The boundaries of the range may be marked (e.g. by *scent marking), and may or may not be defended, depending on species. *See* TERRITORY.

homing The return by an animal to a particular site that is used for breeding or sleeping. The term may apply to the return of an animal to its nest after foraging, or to seasonal migrations between breeding and feeding grounds.

Hominidae (suborder *Simiiformes, superfamily *Hominoidea) The family that, in old classifications, included humans and immediately ancestral forms now extinct; its members were distinguished from the apes (*Pongidae) by the possession of a much larger brain, in which the frontal and occipital lobes are especially well developed, allowing more complex behaviour including communication by speech; by a fully erect posture facilitated by the positioning of the *foramen magnum beneath the skull so that the head is held upright; by a bipedal gait; and by the slow rate of postnatal growth and development, which favours complex social organization and the emergence of distinct cultures. The family included the genera *Paranthropus*, *Australopithecus* (but *see* AUSTRALOPITHECINES), and *Homo*; formerly *Ramapithecus was also included in this family, but this is now rejected. There was said to be one surviving species, *Homo sapiens* (humans). There is now increasing evidence that gorillas and chimpanzees are more closely related to humans than to orang-utans, implying that all, or at least the African, apes should be included with humans in the Hominidae.

Hominoidea (order *Primates, suborder *Simiiformes) A superfamily that comprises the *Hylobatidae (gibbons), *Pongidae (great apes), and *Hominidae (humans). The latter two families are believed to be descended from a common stock of 'great apes' which diverged to form distinct Asian and African lines, the African line dividing again 4–6 million years ago into the African apes and the hominids. An increasing number of authorities hold that the hominids and African apes should be grouped together and the orang-utan separately; or else, that all great apes should be included in the Hominidae. The hominoids lack tails and cheek pouches; have *catarrhine nostrils and opposable thumbs (reduced in some species); and differ from the *Cercopithecidae in having less specialized dentition, a vermiform *appendix in the gut, and larger heads, longer limbs, and wider chests, which some authorities believe they inherited from ancestral *brachiating forms. Today only the Hylobatidae are specialized brachiators.

homo- A prefix from the Greek *homos*, 'same'. *Compare* HETERO-.

Homo *See* HOMINIDAE.

Homo africanus *See* AUSTRALOP-
ITHECINES.

homoallelic Applied to allelic mutants
of a *gene that have different mutations at
the same site; as opposed to heteroallelic
mutants, which have mutations at different sites with the one gene. Recombination
between heteroalleles can yield a functional
*cistron; recombination between homoal-
leles cannot.

homocercal tail In fish, a tail in which
the last vertebra adjoins modified bony elements (the *urostyle and *hypurals) which
support the *fin rays in such a manner that
the tail fin outwardly appears symmetrical. The homocercal tail fin is typical of the
higher bony fish.

homodimer A *protein that is made up of
two identical polypeptides paired together;
as opposed to a heterodimer, in which the
polypeptides are not identical.

homodont Possessing teeth all of which
are of the same form.

homoeobox *See* HOMEOBOX.

homoeobox gene *See* HOMEOBOX GENE.

homoeostasis (homeostasis) The tendency of a biological system to resist change
and to maintain itself in a state of stable
equilibrium.

homoeotic gene *See* HOMEOTIC GENE.

homogametic sex The sex that has *sex
chromosomes that are similar in morphology (e.g. XX); and hence the sex that produces only one kind of *gamete with respect
to the sex chromosomes. *Compare* HETERO-
GAMETIC SEX.

homograft *See* GRAFT.

homoiomerous Uniform in structure;
composed of units (e.g. cells) of all the same
type.

homoiotherm (homeotherm) An organism whose body temperature varies only
within narrow limits. It may be regulated
by internal mechanisms (i.e. in an *endotherm) or by behavioural means (i.e. in an
*ectotherm), or by some combination of
both (e.g. in humans, who light fires and
wear thick clothes to keep warm in cold
weather and wear light clothing to keep
cool in warm weather, but who are also
endothermic).

homologous **1.** Applied to an organ of
one animal that is thought to have the same
evolutionary origin as an organ of another
animal, although the functions of the two
organs may differ widely. Homology is generally deduced from similarity of structure
and/or position of the organ relative to other
organs, seen particularly during embryonic
development (e.g. the ear ossicles of a mammal, which are homologous with certain
bones involved in the articulation of the
jaw in fish). **2.** Applied to *chromosomes
that contain identical linear sequences of
*genes, and which pair during *meiosis.
Each homologue is therefore a duplicate of
one of the chromosomes contributed by one
of the parents; and each pair of homologous
chromosomes is normally identical in shape
and size. *Compare* HETEROLOGOUS.

homology The fundamental similarity of
a particular structure in different organisms,
which is assumed to be due to descent from
a common ancestor. Two organs sharing
a similar position, similar histological appearance, and a similar embryonic development are said to be homologous organs
irrespective of their superficial appearance
and function in the adult.

homomeric *See* DIMER.

homonym In nomenclature, one of two or
more identical names for different taxa (*see*
TAXON).

homoplasy In the course of *evolution,
the appearance of similar structures in different lineages (i.e. not by inheritance from
a common ancestor). The term includes
*convergence, *parallelism, and *reversal.

Homoptera (subclass *Pterygota, order *Hemiptera) Suborder of bugs, in which the head capsule is not continuous ventrally behind the mouth-parts. All homopterans are plant feeders. In some the fore wings are *tegmina.

Homosclerophorida (class *Demospongiae, subclass *Tetractinomorpha) An order of sponges that have a differentiated *leucon structure. Some lack skeletons and possess a thin dermis. A *cortex is present in more advanced forms. The larvae are *amphiblastulae.

homozygosity The presence of identical *alleles at one or more *loci in *homologous chromosomal segments.

homozygote An individual that has the same *genes at one or more loci. The *phenotype of the genes for a particular *locus will always be expressed since the two genes at the *homologous loci are identical. An individual will breed true at those homologous loci that are homozygous.

honest signal A communication from an animal that conveys truthful information.

honey Nectar collected by bees from flowers, and partly digested so that the complex sugars are broken down to simpler ones. It is concentrated by the evaporation of water, and also contains traces of gums, pollen, minerals, and enzymes. It is one of the principal foods of growing bee larvae, the other being protein-rich pollen. Honey is also eaten by adult bees, and is an energy-rich food which fuels their foraging trips.

honey badger (ratel; *Mellivora capensis*) See MUSTELIDAE.

honey-bee *Eusocial member of the family *Apidae, whose colonies are perennial. The term is usually restricted to the four species of *Apis* (*A. cerana*, *A. dorsata*, *A. indica*, and *A. mellifera*), but sometimes used to include the stingless bees, *Melipona* and *Trigona*,which also store large amounts of honey.

honey buzzards (*Pernis*) See ACCIPITRIDAE.

honeycomb Double-sided sheets of hexagonal wax cells built by *honey-bees (*Apis* species), in which pollen and honey are stored and the brood is reared.

honeydew Plant sap that has passed through the bodies of aphids (*Aphididae).

honeyeaters See MELIPHAGIDAE.

honeyguides See INDICATORIDAE; PICIFORMES.

Hooker, Sir Joseph Dalton (1817–1911) A British botanist, who in 1865 succeeded his father, Sir William Jackson Hooker, as director of the Royal Botanic Gardens, at Kew, London. He was a friend and champion of Charles *Darwin, who relied greatly on Hooker's botanical knowledge in his books on evolution.

hoolock (*Bunopithecus*) See HYLOBATIDAE.

hoopoe (*Upopa epops*) See UPUPIDAE; CORACIIFORMES.

hopeful monster An individual which carries a *macromutation that is of no benefit to that individual (i.e. the individual is a monster) but that may prove beneficial to one of its descendants if it undergoes further, but relatively minor, evolutionary change (i.e. the monster is hopeful). For example, the loss of the tail in descendants of *Archaeopteryx* was followed by the development of the tail feathers which stabilize the flight of modern birds (assuming there was a direct evolutionary line from *Archaeopteryx* to the birds). The hypothesis is advanced in order to explain the evolution of structures that appear to confer no benefit, or even disadvantages, until they reach their completed form. Many of the structures involved are explicable without invoking 'hopeful monsters' (e.g. partially evolved feathers may be useless for flight but provide thermal insulation; and an eye that works inefficiently may nevertheless be better than no eye at all). Macromutations that involve a major reorganization of the genes may effectively sterilize the animal and in any case an individual whose appearance is markedly different from that of other members of its species may have difficulty finding a mate. The hypothesis is now applied mainly to macromutations that affect regulatory genes (the genes that activate or deactivate genes which cause the synthesis of proteins); these may have major effects but involve no major, and probably sterilizing, genetic alteration.

Hoplichthyidae (ghost flathead; subclass *Actinopterygii, order *Scorpaeniformes) A small family of marine fish that resemble the true flatheads (*Platycephalidae) in that the head and anterior part of the body are very depressed and wide, the second *dorsal and *anal fins are very long, and the lower three of the *pectoral fin rays are free. There is a single genus (*Hoplichthys*) with 11 species, found in the Indo-Pacific region.

Hoplitis *See* MASON BEE.

Hoplocarida *See* STOMATOPODA.

Hoplonemertini (phylum *Nemertini, class *Anopla) An order of worms which have a relatively complex *proboscis. All have a straight intestine.

Hoplopleuridae (order *Phthiraptera, suborder *Anoplura) Family of sucking lice which are parasitic on *Rodentia, *Insectivora, and *Lagomorpha. They are very slender and hold on to single hairs of their host, so their detection and removal is difficult. There are five genera, with 132 species.

hormone A regulatory substance, active at low concentrations, that is produced in specialized cells but that exerts its effect either on distant cells or on all cells in the organism to which it is conveyed via tissue fluids.

horn A hardened outgrowth from the head of some members of the *Artiodactyla, most made from bone and used mainly for sparring in mating rituals. Rhinoceroses

(*Perissodactyla) have horns made from compressed hair.

hornbills *See* BUCEROTIDAE; CORACIIFORMES.

horned coot (*Fulica cornuta*) *See* RALLIDAE.

horned lark (*Eremophila alpestris*) *See* ALAUDIDAE.

horned lizards (*Phyrnosoma*) *See* IGUANIDAE.

horned viper (*Vipera ammodytes*) *See* VIPERIDAE.

hornet (*Vespa crabro*) *See* VESPIDAE.

horntails *See* SIRICIDAE.

horny corals *See* GORGONACEA.

horotely A normal or average rate of evolution per million years, of genera within a given taxonomic group. Thus slowly or rapidly evolving lines may be horotelic during certain episodes in their history. *Compare* BRADYTELY; TACHYTELY.

horse (*Equus*) *See* EQUIDAE.

horse-flies *See* TABANIDAE.

horseshoe bat *See* RHINOLOPHIDAE.

horseshoe crab *See* MEROSTOMATA.

horse-stingers *See* ODONATA.

horus swift (*Apus horus*) *See* APODIDAE.

host *See* GRAFT; PARASITISM.

hot spot An area which contains a large number of rare or endangered *species and for that reason is designated for protection. Identifying hot spots is the traditional technique by which sites are selected for protection, but it tends to concentrate that protection on a small number of areas, leaving others, and the many species in them, unprotected. Many conservation biologists prefer to identify conservation areas by *gap analysis.

house bunting (*Emberiza striolata*) *See* EMBERIZIDAE.

house cricket *See* GRYLLIDAE.

houseflies *See* MUSCIDAE.

house gekko (*Hemidactylus bibronii*) *See* GEKKONIDAE.

house martin (*Delichon urbica*) *See* HIRUNDINIDAE.

house mouse (*Mus musculus*) *See* MURIDAE.

house sparrow (*Passer domesticus*) *See* PLOCEIDAE.

house spider (*Tegenaria domestica*) *See* AGELENIDAE.

hoverflies *See* SYRPHIDAE.

howler monkey (*Alouatta*) *See* CEBIDAE.

Hox gene *See* HOMEOTIC GENE.

huia (*Heteralocha acutirostris*) *See* CALLAEIDAE.

human being *See* HOMINIDAE.

human body louse (*Pediculus humanus*) *See* PEDICULIDAE.

human head louse (*Pediculus capitis*) *See* PEDICULIDAE.

human leucocyte antigen complex (HLA) *See* MAJOR HISTOCOMPATIBILITY COMPLEX.

humerus In tetrapods, the upper bone of the fore limb.

humidity The amount of water vapour present in air, measured as the mass present in a given volume of air (absolute humidity), the ratio of the mass of water vapour present in a given mass of air to a unit mass of that air including the water vapour (specific humidity), the ratio of the mass of water vapour present in a given volume of air to a unit mass of dry air (mixing ratio or mass mixing ratio), or the amount present as a percentage of the amount needed to saturate the air at the prevailing temperature (relative humidity).

humming-birds *See* APODIFORMES; TROCHILIDAE.

humoral response An *antibody-mediated immunoresponse. Antibodies are secreted by *B cells that have been stimulated by helper *T cells. Antibodies bind to foreign *antigens which they recognize, to form an antibody–antigen complex. The resulting complex may then be consumed by a *macrophage. *Compare* CELL-MEDIATED RESPONSE.

hump-backed dolphin (*Sousa*) *See* STENIDAE.

humpback whale (*Megaptera novaengliae*) *See* BALAENOPTERIDAE.

hump-winged cricket *See* PROPHALANOPSIDAE.

humus The semi-decomposed organic matter in the soil; it provides nutrients for plant growth and increases the water-absorbing capacity of the soil.

hunger The desire for food that, in human experience, is generally an unpleasant, even painful sensation which may become so intense as to make the search for food dominate thought and action. Similar sensations in other animals are inferred from observation of their behaviour.

hunting dog (*Lycaon pictus*) *See* CANIDAE.

hunting spiders *See* LYCOSIDAE.

hunting wasp Any wasp species in which the females seek spider or insect prey that they sting and paralyse, and then transport to the nest as food for their larvae. The term is applied to the following families (division *Aculeata): Pompilidae (spider-hunting wasps), Vespidae subfamily Eumeninae (caterpillar- and beetle-hunting wasps), and Ampulicidae and Sphecidae (digger wasps). In primitive pompilids and ampulicids, prey is caught before a nest is prepared, and the nest is often no more than a shallow scrape in the soil, or a pre-existing cavity. In more advanced forms, a nest is excavated or built before the prey is caught.

hutia *See* CAPROMYIDAE.

Hyaena *See* HYAENIDAE.

Hyaenidae (hyenas; suborder *Fissipedia, superfamily *Feloidea) A family of carnivores whose ancestors evolved rapidly from the *Viverridae (civets) in late *Miocene times. They are the youngest feloids and displaced the last hyaenodont (*Hyaenodontidae) from the scavenging niche, and although today they are restricted to the Old World, one form entered N. America in the early *Pleistocene. Hyenas possess heavy, blunt teeth and strong jaws, allowing them to crush bones and so to live partly by scavenging. The aardwolf (*Proteles cristatus*) resembles a small hyena superficially, and is also descended from the viverrid group, but it feeds mainly on insects and has weaker jaws and fewer teeth. Hyaenids have hind limbs that are shorter than the fore limbs, non-retractile claws, a *digitigrade gait, and long skulls. They are distributed widely in Africa and southern Asia as far east as India. Apart from *Proteles* there are two or three genera: the spotted hyena (*Crocuta*) and the striped and brown hyenas (*Hyaena*); many authors now place the brown hyena in its own genus (*Parahyaena*).

Hyaenodontidae (superorder *Ferae, order *Creodonta) An extinct family of carnivorous mammals that appeared early in the *Eocene and flourished throughout most of the *Tertiary. They were slimly built, with elongated jaws; probably they possessed claws, and many had well-developed *carnassials, differing from those of modern carnivores in being formed from the first or second *molars of upper and lower jaws (rather than from the upper *premolar and lower molar). Some forms (e.g. *Machaeroides* and *Apataelurus*) had jaws similar to those of the later sabretoothed cats. Hyaenodonts ranged in size from animals the size of a weasel to some large enough to have preyed on large herbivores.

hyaloplasm The ground substance of cell *cytoplasm, within which the various subcellular *organelles and membranaceous components are embedded.

Hyalospongea *See* HEXACTINELLIDA.

hyaluronic acid A highly viscous acid mucopolysaccharide, composed of d-glucuronic acid and N-acetyl- d-glucosamine. It is widely distributed throughout the *connective tissues of animals. It is thought to have various functions, including maintaining the level of hydration in tissues, acting as a molecular sieve in restricting the movement of large molecules in tissues, and preventing the spread of pathogens and toxic substances.

hybrid 1. An individual animal that results from a cross between parents of differing *genotypes. Strictly, most individuals in an outbreeding population are hybrids, but the term is more usually reserved for cases in which the parents are individuals whose *genomes are sufficiently distinct for them to be recognized as different *species or *subspecies. A good example is the mule, produced by crossbreeding an ass and a horse (each of which can breed true as a species). Hybrids may be fertile or sterile depending on qualitative and/or quantitative differences in the genomes of the two parents. Hybrids like the mule, whose parents are of different species, are frequently sterile. **2.** By analogy, any *heterozygote. Each heterozygote represents dissimilar *alleles at a given *locus, and this difference results from a cross between parental *gametes possessing differing alleles at that locus. **3. (graft hybrid)** *See* GRAFT.

hybrid dysgenesis A complex of genetic abnormalities, which occurs in certain hybrids. The abnormalities may include sterility, enhanced rates of *gene*mutations, and chromosomal rearrangements. Hybrid dysgenesis occurs in the hybrid offspring of certain strains of *Drosophila*, in which it is thought to be due to mutations induced by *transposon-like elements.

hybrid swarm A continuous series of *hybrids that are morphologically distinct from one another, which results from the hybridization of two *species followed by the crossing and backcrossing of subsequent generations. The hybrids are very variable owing to segregation of *alleles at each *locus.

hybrid vigour *See* HETEROSIS.

hybrid zone A geographical zone in which the hybrids of two geographical races may be found.

Hydra *See* HYDROIDA.

Hydraenidae (subclass *Pterygota, order *Coleoptera) Family of tiny beetles, about 1.5 mm long, found in stagnant water. The ventral surface is flat, the upper surface very convex. The *elytra are generally black to metallic. The *maxillary palps are long, the antennae short, with five-segmented clubs. The *ocelli are well developed, possibly indicating an affinity with *Staphylinidae. Some larvae are aquatic, others live only partly in water. Adults and larvae are phytophagous, often feeding on dead material. There are 300 species.

Hydrobatidae (Oceanitidae; storm petrels; class *Aves, order *Procellariiformes) A family of small, brown-black or grey sea-birds which have white rumps or under-parts. The bill and legs are usually black. Storm petrels have long wings, and medium to short, square or forked tails. They are *pelagic and gregarious, feed on surface plankton, and breed in burrows and crevices on islands. The two species of *Oceanites* inhabit southern oceans, breeding on sub-Antarctic islands, but also occur north of the Equator, following ships, and *O. oceanicus* (Wilson's petrel) occurs in the N. Atlantic. The 11 species of *Oceanodroma* occur mainly in northern oceans and rarely follow ships. There are 2–8 genera in the family, with 20–22 species, found in all oceans.

hydrocarbon A naturally occurring compound that contains carbon and hydrogen.

Hydrochidae (subclass *Pterygota, order *Coleoptera) Family of small, black or bronze beetles about 3 mm long, which are generally found on water plants. The *maxillary palps are long. The antennae have three-segmented clubs. Formerly the family was included in the *Hydraenidae, but there are essential structural differences. Little is known of the larvae. There are 69 species.

Hydrochoeridae (capybara; order *Rodentia, suborder *Hystricomorpha) A monospecific family (*Hydrochoerus hydrochaeris*) that comprises the largest of all rodents, which grows to a weight of 50 kg or more. Capybaras are semi-aquatic, with digits partially webbed, and eyes and ears small and positioned high on the skull. The body is sparsely covered with hair, and the tail is vestigial. Capybaras are distributed throughout tropical S. America east of the Andes.

Hydrochoerus hydrochaeris (capybara) *See* HYDROCHOERIDAE.

Hydrocorallina (hydrocorals) A collective name for the cnidarian (*Cnidaria) orders *Milleporina and *Stylasterina.

hydrocorals *See* HYDROCORALLINA.

hydrofuge hair Water-repellent, cuticular hair on the body of an insect, which serves a variety of functions. Mosquito larvae use hydrofuge hairs to seal their respiratory siphons when they dive; some *Coleoptera use them to create a *plastron.

hydrogenase An *enzyme that catalyses reactions involving the reduction of a *substrate by molecular hydrogen.

hydrogen bond The force of attraction (the hydrogen force) that exists between *polar molecules containing hydrogen atoms, or between one part of a molecular chain and another part that contains bonded hydrogen. It occurs because the single electron in the hydrogen atom is held only weakly, so hydrogen readily forms ionic bonds, which allow a hydrogen atom to link other atoms. Some of the properties of water are due to the hydrogen bond that links water molecules. *DNA molecules are linked by hydrogen bonds, and hydrogen bonding is also important in linking other organic molecules.

hydroid See HYDROIDA; HYDROZOA.

Hydroida (Leptolinida, hydroids ; phylum *Cnidaria, class *Hydrozoa) An order of cnidarians that have a well-developed *polypoid generation. They are fixed, solitary or colonial, and many have a chitinous exoskeleton, although some (e.g. *Hydra*) are naked. There are free *medusae or aborted structures on the *polyps. The medusae have *ocelli or ectodermal *statocysts, or both.

hydrolase An *enzyme that catalyses reactions involving the *hydrolysis of a *substrate.

hydrolysis A reaction between a substance and water in which the substance is split into two or more products. At the points of cleavage the products react with the hydrogen or hydroxyl ions derived from water.

Hydrophiidae (sea snakes; order *Squamata, suborder *Serpentes) A family of marine descendants of the *Elapidae, comprising venomous, front-fanged snakes that have valvular nostrils and salt-secreting glands in the head, long bodies (generally with small ventral scales), and laterally compressed tails. Despite their salt-secreting glands, sea snakes drink only freshwater, which they obtain from rain. Most are *ovoviviparous. They prey mainly on small fish. There are about 50 species, occurring in the Indian and Pacific Oceans.

hydrophilic Applied to a molecule or surface that can become wetted or solvated by water. This ability is characteristic of polar compounds (see POLAR MOLECULE).

Hydrophilidae (water beetles; subclass *Pterygota, order *Coleoptera) Family of beetles, 1–40 mm long, most of which have a Y-shaped impressed line on the *vertex. The *maxillary palps are usually longer than the antennae, which, in aquatic forms, have a respiratory function. The aquatic species are dark, convex, and shiny. Females carry eggs beneath the abdomen in a silk cocoon. Adults are usually phytophagous; larvae are carnivorous. Land-living species are associated with damp places and dung, and are often paler than aquatics, with more heavily sculptured *elytra. There are 2000 species, and their distribution is more abundant in the tropics.

hydrophobic Applied to a molecule or surface that can resist wetting or solvation by water. This ability is characteristic of non-polar compounds (see NON-POLAR MOLECULE).

***Hydropotes* (Chinese water deer)** A small cervid (see CERVIDAE) from northern China and Korea that is characterized by its possession of long, tusk-like upper *canine teeth and its lack of antlers.

hydrorhiza The stem of a hydroid (*Hydroida) colony by which it is attached to the substratum; the name refers to its resemblance to a root (from the Greek *rhiza*, root).

hydrospire In Blastoidea (a class of *pelmatozoans that became extinct in the Early *Permian but are known as fossils), a system of pores and folds to either side of each *ambulacrum that may have served to circulate water into and from the *coelom, thus allowing gas exchange.

hydrostatic organ See SWIM-BLADDER.

hydrostatic skeleton In soft-bodied invertebrates, the *coelomic fluid, held under pressure; this maintains the shape of the animal and allows surrounding muscles to contract against it to provide locomotion.

hydrotaxis The locomotion of an organism in response to the stimulus of water.

hydroxyapatite A hydrated calcium phosphate mineral, which also contains fluoride, chloride, and carbonate calcium salts. It is often formed as a consequence of *biomineralization, producing hard structures, such as *bone.

Hydrozoa (hydroids; phylum *Cnidaria) A class of multicellular, mainly marine animals in which the cells are derived from two layers, epidermis and gastrodermis (endodermis), separated by a gelatinous *mesogloea. These enclose a continuous *gastrovascular cavity (coelenteron), which communicates directly with the exterior by a single aperture (mouth) and is lined by a gastrodermis. The gastrodermis lacks *nematocysts. Eggs and sperm are shed outside the animal and not into the gastrovascular cavity.

Hyemoschus (African chevrotains) *See* TRAGULIDAE.

hyena *See* HYAENIDAE.

hygienic strain Certain (but not all) strains of *honey-bees (*Apis* species) in which workers remove the rotting remains of larvae that have died, thus preventing the spread of infection and mite infestation. The workers are 15–20 days old (i.e. middle-aged and typically younger than foragers) and are able to detect a dead larva, uncap its cell, and remove the larva and throw it out of the hive.

(()) SEE WEB LINKS

• Information about hygienic strain and test for hygienic behaviour.

Hygrobiidae (screech beetles; subclass *Pterygota, order *Coleoptera) Family of highly convex water beetles, found in muddy ponds. They possess several primitive features—including, in the adult, swimming by means of alternate leg movements—and very prominent eyes. Adults produce high-pitched squeaks by rubbing the tip of the abdomen against the underside of the *elytra. Larvae are club-shaped, with a large head and thorax and a three-pronged tail, and feed on insect larvae. There are four species, with notably discontinuous distribution, occurring in Australia, Asia, and Europe.

hygrometer An instrument that is used to measure atmospheric *humidity.

Hyla arborea (European tree frog, green tree frog) *See* HYLIDAE.

Hylaeinae *See* COLLETIDAE.

Hylidae (tree frogs; class *Amphibia, order *Anura) A family of frogs whose feet have suction pads. The digits are elongated with an extra, cartilaginous element. Brood-care behaviour is often well developed. Females of *Gastrotheca marsupiatum* (marsupial frog), small (3 cm) S. American frogs, carry fertilized eggs in a dorsal pouch for more than 100 days before releasing the tadpoles into water. *Hyla arborea* (European tree frog or green tree frog), the only tree frog to occur in Europe, with a range extending into N. Africa and temperate Asia, can change colour from light green to brown; the male has a balloon-like vocal sac but does not develop *nuptial pads. There are 600 species in the family, widespread in tropical and temperate zones. They are mainly arboreal, but a few are secondarily ground-living, or aquatic.

Hylobates (gibbon) *See* HYLOBATIDAE.

Hylobatidae (lesser apes; suborder *Haplorrhini, superfamily *Hominoidea) A family of apes that includes gibbons and their relatives. They are small in size and specialized for *brachiation with long legs, extremely long arms, and a hook-like hand. The *canines are long and sharp in both sexes. Generally they are placed in one genus (*Hylobates*) with four subgenera: *Hylobates* (true gibbons), *Bunopithecus* (hoolock), *Symphalangus* (siamang), and *Nomascus* (concolors). Molecular evidence suggests, however, that these are nearly as distinct from each other as e.g. humans from chimpanzees. They are distributed throughout south-east Asia, from Assam to Java. There are about nine species.

Hylonomus lyelli (order *Cotylosauria, suborder *Captorhinomorpha) What is probably the oldest-known reptile, first found in fossilized tree stumps in the Coal Measures (lower Pennsylvanian, *see* CARBONIFEROUS) of Nova Scotia. The skull roof of this species was fully ossified and there were no openings behind the eye sockets. *Hylonomus* measured some 25 cm and possessed a long tail.

Hymenoptera (ants, bees, sawflies, wasps; class *Insecta, subclass *Pterygota) Very large, complex, and diverse order, whose members show a high degree of adaptive radiation. There are two suborders, the *Symphyta (sawflies), which do not have a constricted waist, and the *Apocrita (ants, bees, and wasps), of which there are about 105 000 species world-wide, in which the waist is constricted. Hymenoptera have two pairs of wings, the fore wings a little longer than the hind wings, and both pairs with relatively few veins. In flight, the wings can be linked together by means of *hamuli. Adults can be found in a large variety of habitats. They have mouth-parts that are either adapted for chewing or with the *maxillae and *labium modified to form a proboscis for sucking nectar, etc. The antennae are usually fairly long, with 10 or more segments. The *ovipositor may be modified for sawing, piercing, or stinging (for defence or the paralysis of prey and hosts). Members of the order have complete *metamorphosis, the pupae being *adecticous, and usually *exarate. Pupation normally takes place in a *cocoon, in a special cell, or in the body of the host (in the case of parasitic species). The larvae of many species are plant feeders; many others are parasitic, living in or upon the bodies of other insects; still others live in special nests, constructed by the adults, and feed on materials with which the nest is provisioned. Some Hymenoptera are social insects whose life histories and biology are well known. Many Hymenoptera are extremely useful in the biological control of pest species. Some are important pollina-

tors. Others are pests of cultivated plants, in some cases causing serious defoliation. The taxonomic characters used in the classification of the order include wing *venation, the legs, the antennae, the *pronotum, the thoracic structures, and the ovipositor.

Hymenostomatida (subphylum *Ciliophora, class *Ciliatea) An order of ciliate *Protozoa in which the mouth region is subterminal and located on the ventral surface of the cell. The arrangement of *cilia in the mouth region is complex. Beating of these cilia causes water currents to direct food toward the mouth. There are three suborders. Most species are free-living in freshwater habitats, a few are parasitic.

hynobiid Applied to the most primitive of living salamanders, the *Hynobiidae, and to amphibians that resemble them.

Hynobiidae (hynobiids, Asian land salamanders; class *Amphibia, order *Urodela) A family of amphibians in which *metamorphosis is complete, adults lacking larval teeth or gills. They grow to about 8–25 cm long. Some species are found in mountain streams at altitudes up to 400 m. The lungs are sometimes lost, presumably to reduce buoyancy in fast-flowing water. There are five genera, with about 30 species, all of which are Asiatic.

hyoid A bone or bones developed from the second *visceral arch: in vertebrates, it or they support the floor of the mouth. The tongue is supported by a hyoid.

Hyopsodontidae (superorder *Protoungulata, order *Condylarthra) An extinct family of *Palaeocene and *Eocene mammals, mainly N. American but with at least one S. American genus, whose members possessed *bunodont, quadritubercular cheek teeth, relating them to the ungulates, but had clawed digits. *Hyopsodus* was about 30 cm long, and possibly semi-arboreal. *Mioclaenus*, an earlier genus, had sharp-cusped, *creodont-like teeth.

hyostyly A type of jaw suspension in which the upper jaw is attached to the *cranium anteriorly only by means of ligaments and posteriorly by the hyomandibular. Hyostyly, found e.g. in sharks and rays, allows multiple jaw positioning. *Compare* AMPHISTYLY.

hyperarousal *See* FIGHT OR FLIGHT REACTION.

Hypermastigida (superclass *Mastigophora, class *Zoomastigophorea) An order of *Protozoa which has many *flagella. They inhabit the alimentary tracts of cockroaches, woodroaches, and termites, with which they appear to have a mutualistic relationship, the host depending on its flagellate for the digestion of wood.

hypermorphosis A type of *heterochrony in which growth is prolonged, so that the adult morphology of the descendant is produced by a prolongation of the growth trajectory of its ancestor. The organism reaches its adult size and form well before the attainment of sexual maturity, and continues to develop into a 'super-adult'.

Hyperoartii An older name for the order *Petromyzoniformes (lampreys).

Hyperoglyphe porosa **(deep-sea trevalla)** *See* CENTROLOPHIDAE.

Hyperoodon **(bottle-nosed whale)** *See* ZIPHIIDAE.

hyperosmotic In *osmosis, the side of the *semi-permeable membrane toward which water flows. The body fluids of the coelacanth (*see* COELACANTHIFORMES) and *Chondrichthyes are hyperosmotic, so water flows from the sea into the body. *Compare* HYPOSMOTIC.

Hyperotreti An older name for the order *Myxiniformes (hagfish).

hyperparasite A parasitic insect that lives in or on a host that is itself parasitic on another species of insect. If the parasite kills the host as a consequence of its own development it is more commonly called a 'parasitoid'.

hypersensitivity A condition in which an individual with a pre-sensitized immune system experiences an excessive response to what the immune system interprets as a foreign substance. The over-reaction produces undesirable symptoms that range from mild discomfort to death.

hypertonic Applied to a cell in which the *osmotic pressure is higher than that in the surrounding medium. *Compare* HYPOTONIC; ISOTONIC.

Hypertragulidae (order *Artiodactyla, superfamily *Traguloidea) A family of primitive ruminants that are often included in the *Pecora in the broadest sense. A few representatives are known from the late *Eocene and *Oligocene of Eurasia, but the family as a whole flourished in the Oligocene and early *Miocene of N. America.

hypobiosis Dormancy.

hypocercal tail In fish, a tail in which the lower lobe is more pronounced or larger than the upper lobe.

Hypocolidae (hypocolius) *See* BOMBYCILLIDAE.

Hypocolius *See* BOMBYCILLIDAE.

hypocone In the upper *molar tooth of mammals, the inside posterior cusp, which corresponds to the outer posterior cusp (hypoconid) of the lower molar.

Hypoderma **(warble flies)** *See* OESTRIDAE.

hypodermis *See* INTEGUMENT.

hypogaeous Living below ground.

hyponeustic *See* ZOONEUSTON.

hyponeuston The organisms living in the lower part of the surface film of water. *Compare* EPINEUSTON; NEUSTON.

hyponome In *Cephalopoda, the tube or funnel through which water is expelled from the *mantle cavity. The discharge of water provides the animal with jet propulsion.

hypopharynx 1. In insect mouthparts, a globular structure at the base of the *labium that aids swallowing. **2. (laryngeal pharynx)** in vertebrates, the lowest part of the *pharynx, connecting the throat to the *oesophagus.

hypophloeodal Growing or living beneath (or within) tree bark.

hypophysial sac (Rathke's pouch) A median pit in the roof of the mouth of vertebrates. In the *Agnatha (lampreys and hagfish) it is associated with the nasal pit to form the nasohypophysial sac. In all animals of the superclass *Gnathostomata it forms much of the *pituitary body.

hypophysis In vertebrate *embryos, a projection that develops from the *stomodaeum; it meets the *infundibulum, the two fuse, the hypophysis loses its connection with the stomodaeum, and the combined infundibulum and hypophysis form the *pituitary body.

hypopleural bristles In some *Diptera, two rows of bristles, one on each side of the *thorax, from below and in front of the *halteres to above the base of the hind legs.

hypoptile *See* AFTERSHAFT.

Hyposittidae (coral-billed nuthatch; class *Aves, order *Passeriformes) A monospecific family (*Hypositta corallirostris*), comprising a small, blue-grey bird that has a short, stout, hook-tipped, coral-red bill and a black face. It feeds on insects by climbing up trees. Sometimes it is included in the *Vangidae or the *Sittidae. It occurs only in Madagascar.

hyposmotic In *osmosis, the side of the *semi-permeable membrane away from which water flows. The body fluids of marine *teleosts and *lampreys are hyposmotic, so water flows from the body to the sea. *Compare* HYPEROSMOTIC. *See also* CHLORIDE CELLS.

hypostasis *See* EPISTASIS.

hypostome In *Hydrozoa, a conical mound at the oral end of the body: it contains the mouth.

hypothalamus In vertebrates, the basal part of the *diencephalon region of the *brain, located beneath the *thalamus. The hypothalamus controls the *autonomic nervous system and thus regulates many processes, including the regulation of body temperature, sleeping and waking, and feeding and digestion. Through its secretion of hormone-like substances it also influences the *pituitary.

hypotheres *See* NOTOUNGULATA.

hypothesis An idea or concept that can be tested by experimentation. In inductive or inferential statistics, the hypothesis is usually stated as the converse of the expected results, i.e. as a null hypothesis (H_0). This helps workers to avoid reaching a wrong conclusion, since the original hypothesis H_1 will be accepted only if the experimental data depart significantly from the values predicted by the null hypothesis. Working in this negative way carries the risk of rejecting a valid research hypothesis even though it is true (a problem with small data samples); but this is generally considered preferable to the acceptance of a false hypothesis, which would tend to be favoured by working in the positive way.

hypothesis-generating method A data-structuring technique, such as a *classification and ordination method which, by grouping and ranking data, suggests possible relationships with other factors (i.e. generates an *hypothesis). Appropriate data may then be collected to test the hypothesis statistically.

hypothesis testing The evaluation of an *hypothesis using an appropriate statistical method and significance test. *See also* STATISTICAL METHOD; NUMERICAL METHOD.

hypotonic Applied to the condition in a cell where the *osmotic pressure is lower than that in the surrounding medium. *Compare* HYPERTONIC; ISOTONIC.

Hypotremata 1. An alternative name for the superorder *Batoidimorpha (rays). **2.** An older name for the order *Rajiformes (rays). *Compare* PLEUROTREMATA.

hypotremate Applied to gill openings on the ventral surface beneath the pectoral fins (e.g. in skates and rays). *Compare* PLEUROTREMATE.

Hypotrichida (class *Ciliatea, subclass *Spirotrichia) An order of ciliate *Protozoa which are mostly ovoid and flattened, with few or no *cilia on the dorsal surface. *Cirri occur in groups on the ventral surface. There are three families, and many genera. They are free-living in fresh or brackish water, and feed mainly on bacteria.

Hypsilophodontidae (order *Ornithischia, suborder *Ornithopoda) A family of bipedal dinosaurs that ranged from the Late *Triassic to the Late *Cretaceous. *Hypsilophodon* itself, from the Wealden of the Early Cretaceous, is the most primitive of the ornithopod dinosaurs known to date, despite its relatively late appearance in the *Mesozoic fossil record.

Hypsipetes **(bulbuls)** *See* PYCNONOTIDAE.

Hypsithermal *See* FLANDRIAN.

hypsodont Applied to teeth in which the crowns are high and the roots are short.

hypurals Enlarged ventral ribs which, in the tail region of teleost fish (*Teleostei), support the *caudal fin.

Hyracodontidae (order *Perissodactyla, superfamily *Rhinocerotoidea) One of three families of rhinoceroses, known as the 'running rhinoceroses'. They were small, slender browsers, which first appeared in the *Eocene and became extinct in the lower *Miocene. They inhabited the northern continents, and were possibly displaced by ancestors of the horse. They had teeth similar to those of rhinoceroses, but differed from them in having long legs and three toes on each foot (similar to those of contemporary horses).

Hyracoidea (hyraxes , dassies , conies; cohort *Ferungulata, superorder *Paenungulata (or *Mesaxonia)) An order of primitive ungulates, containing the single family Procaviidae, which are small, gregarious, herbivorous, terrestrial or arboreal mammals, superficially resembling rabbits and occupying similar niches. The hyrax has short ears and no tail. The fore limb has four digits and the hind limb three. The second digit of the hind limb has a long claw, used possibly for grooming or for climbing, the remaining digits have hoof-like nails. The gait is *plantigrade. The single pair of upper *incisors grow continually and are used for defence; the lower incisors are comb-like and are used in grooming. The intestine has chambers containing *symbionts which digest cellulose. A variety of ancestors, some the size of small horses but others similar to modern forms, are encountered first in the lower *Oligocene of N. Africa: they lived throughout the *Tertiary, so modern hyraxes may be very similar to early small ungulates. There are about five species, in three genera: *Procavia* and *Heterohyrax* (rock hyraxes), and *Dendrohyrax* (tree hyrax). These are distributed throughout Africa, the Mediterranean region, and Arabia, some living in deserts, where they survive with very little water.

Hyracotherium **(suborder *Hippomorpha, family *Equidae)** Known formerly as *Eohippus* (the 'dawn horse'), the earliest known perissodactyl, an animal that was only 27 cm high, the size of a fox terrier. It was short-faced with low-crowned cheek teeth, and had four toes on the fore feet and three on the hind. Abundant in the early *Eocene of N. America and Europe, it was also discovered in *Palaeocene deposits in Mongolia. It was a browser dwelling in forest glades, and the likely ancestor for all the horses. Because of its small size, when the fossils were first found they were mistakenly associated with the African hyraxes, hence the name. In America the fossils were identified correctly and the name *Eohippus* was adopted, but according to the rules of taxonomic nomenclature, which require that the first name to be accepted must stand, *Hyracotherium* takes precedence. The modern tendency is to break up *Hyracotherium* into two or more genera, of which *Protorohippus* might be the ancestor of Equidae, while *Cymbalophus* would be close to the ancestry of the entire *Perissodactyla.

hyrax *See* HYRACOIDEA.

hysginous Red in colour.

Hystricidae (Old World porcupines; order *Rodentia, suborder *Hystricomorpha) A family of mainly vegetarian, burrowing rodents in which some body hairs are modified to form spines. The body is thickset, the legs are short with five digits, the first digit of the fore limbs being reduced. The tail is long in some species and bears modified spines. Old World porcupines are slow-moving, relying for defence on their colouring and on the noise made by the movement of their spines. The hystricoid arrangement of the jaw muscles probably evolved independently in Old and New World porcupines (*Erethizontidae), so these are not necessarily related closely. There are four genera, with about 16 species, distributed through Africa, Italy, and southern Asia.

Hystricomorpha (cohort *Glires, order *Rodentia) A suborder comprising many families of rodents, in which the medial *masseter muscles spread forward through the *infra-orbital foramen, while the lateral masseter is attached to the *zygomatic arch, as in primitive rodents. The suborder includes the porcupines of the Old and New Worlds, guinea-pigs, capybara, agoutis, vizcacha, mole rat, etc.

ianthinous Blue to purple in colour.

Ibaliidae (suborder *Apocrita, super-family *Cynipoidea) Family comprising a small and rare group of insects, parasitic on members of the *Siricidae, and containing some of the largest Cynipoidea (up to 15 mm long) and those with the most complex wing *venation. Adults are black and yellowish-brown, with an elongate and laterally compressed abdomen, the *gaster being almost sessile. The fore wing has brownish spots medially and apically. The abdomen is laterally flattened like a knife edge, and the *ovipositor is curled up beneath in a sheath. The sensitive antennae of the ibaliids seek out the tunnels made by the woodwasp, and eggs are laid in the body of the host. There are three genera with 20 species.

Iberomesornis The best known of several Early *Cretaceous birds, from Las Hoyas, central Spain. It is the first bird to show evidence of a perching foot.

Ibis **(wood storks)** *See* CICONIIDAE.

ibises *See* CICONIIFORMES; THRESKIORNITHIDAE.

icefish *See* CHANNICHTHYIDAE; SALANGIDAE.

Icelidae (sculpin; subclass *Actinopterygii, order *Scorpaeniformes) A small family of marine fish that have a broad head, narrow tail, large eyes placed on top of the head, and wide mouth in line with the flat belly. Closely related to the true sculpins (*Cottidae) and often placed with them, the icelid sculpins are sluggish, bottom-dwelling fish preferring a cold-water habitat. *Icelinus filamentosus* (threadfin sculpin), 27 cm, is one of the larger species of the N. Pacific. There are about ten species, found in the N. Atlantic and Pacific.

Icelinus filamentosus **(threadfin sculpin)** *See* ICELIDAE.

Ichneumonidae (ichneumon wasps; suborder *Apocrita, superfamily Ichneumonoidea) One of the largest insect families, comprising a very diverse and highly successful group, each subfamily or tribe of which is parasitic on a particular group of insects. Many species are important as biological control agents, parasitizing larvae and pupae of *endopterygote insects, and others parasitize spiders and the egg sacs of spiders and pseudoscorpions. Ichneumonids are medium to large (3–40 mm long) wasps, yellowish to black, or brightly patterned with black and brown or black and yellow. Most species have a long *ovipositor but do not sting, although they may make 'stinging' motions if handled. There are more than 60 000 species.

ichneumon wasps *See* ICHNEUMONIDAE.

ichnofossil (trace fossil) A structure formed in a sediment by the action of a living organism (e.g. a tube, burrow, footprint, or groove made by crawling across a surface) and preserved when the sediment becomes a sedimentary rock. Traces are most commonly found at the interfaces between different rock types (e.g. between sandstone and shale) and are classified in various ways, including their forms and the places of their occurrence.

ichnology (palichnology) A subdiscipline of *palaeontology or, more specifically, of *palaeoecology, which is concerned with the study of *ichnofossils and their value in the analysis of sedimentary sequences and the *morphology and behaviour of ancient organisms.

Ichthyoboridae (subclass *Actinopterygii, order *Cypriniformes) A family of freshwater fish that have a rather elongate body, a single *dorsal and an *adipose fin, a forked tail fin, large eyes, and a somewhat produced snout. There are many species, found in Africa.

ichthyology The study of fish. Originally this involved research on anatomy, classification, and the general biology of fish, but more recently research topics have included fish culture, fish diseases, conservation, and commercial fisheries.

Ichthyophiidae (ichthyophids; class *Amphibia, order *Apoda) A family of amphibians in which the adults are terrestrial and have functional eyes and a small tail. Eggs are laid on land. The gilled, aquatic, larval stage is of variable duration before *metamorphosis. Adults burrow but feed on the surface at night, eating earthworms, arthropods, and small vertebrates. There are 43 species, occurring in the New World tropics, southern Asia, and Australia.

Ichthyopterygia (class *Reptilia) A subclass that includes the one order *Ichthyosauria. The ichthyopterygids had a synapsid type of skull, but because of key differences in the arrangement of the bones forming the margins of the single temporal opening, and for other reasons, they are usually placed in a different subclass from the *Synapsida.

Ichthyosauria (ichthyosaurs; class *Reptilia, subclass *Ichthyopterygia) An order of so-called 'fish lizards', whose first members date from the *Triassic and were primitive in type. *Mixosaurus atavaus*, possibly the earliest of the mixosaurs, the first of the ichthyosaurs, is recorded from the Triassic of Germany. Unlike *Ichthyosaurus*, the mixosaur tail lacked the strong development of the fish-type of caudal fin. The typical, shark-like form, *Ichthyosaurus*,

appeared in the *Jurassic, when the group in general was especially common. Ichthyosaurs disappeared before the end of the *Cretaceous.

ichthyosaurs *See* Ichthyosauria.

Ichthyostega A genus of amphibians that, together with *Ichthyostegopsis* and *Acanthostega*, first appeared in the Late *Devonian. *Ichthyostega* shows refinements of the skull, and the development of strong limbs. It retained a long, rather fish-like tail, and links with the *Crossopterygii are confirmed by the presence of teeth with a labyrinthine infolding of the enamel, as well as a notochordal tunnel (*see* NOTOCHORD) underlying the brain case and a *lateral-line system. *Ichthyostega* grew to just under 1 m in length. An unexpected finding about *Ichthyostega* and its relatives, from recently discovered, more complete specimens, is that they had more than the five digits per limb that characterizes modern tetrapods: *Ichthyostega* had seven on the hind feet (the forefeet are still unknown), and *Acanthostega* had eight on both fore and hind feet.

Ichthyostegopsis *See* Ichthyostega.

Icosteidae (ragfish; subclass *Actinopterygii, order *Perciformes) A monospecific family (*Icosteus aenigmaticus*) of large (more than 2 m) marine fish that have a blunt snout, small eyes, and a distinct hump in front of the *dorsal fin. The dorsal and *anal fins are long, the *pectoral fins are absent in adults. There is a considerable difference between the juvenile and adult stages, juveniles possessing a very deep body. Icosteidae occur in the N. Pacific.

Icosteus aenigmaticus (ragfish) *See* Icosteidae.

Icteridae (New World blackbirds, orioles, cowbirds; class *Aves, order *Passeriformes) A diverse family of black, or black with brown, or orange, or red, or yellow birds, many of which have white wing markings. The bill is short and heavy to long and slender, the wings usually long and pointed, but some short and rounded, and the tail variable. Icteridae are arboreal or terrestrial, gregarious or solitary, and feed on fruit, insects, and small vertebrates. The cowbirds (five species of *Molothrus*) are *brood parasites, except for *M. badius* (bay-winged cowbird), which is parasitized by *M. rufoaxillaris* (screaming cowbird). Icterid nests vary from open and cup-shaped to domed, and are built on the ground, or suspended in trees, or in tree cavities. The 24 species of *Icterus* (orioles and troupials) build nests that are globular with side openings of open cups suspended between branches. Some icterids are colonial breeders, and many are migratory. There are 22 genera, with about 93 species, found in N., Central, and S. America, and the W. Indies.

Icterus (orioles, troupial) *See* ICTERIDAE.

ICZN *See* INTERNATIONAL CODE OF ZOOLOGICAL NOMENCLATURE (International Commission on Zoological Nomenclature).

-idae A standardized suffix used to indicate a family of animals in the recognized codes of classification. For example, 'Felidae' is the cat family (including lions, tigers, lynxes, pumas, etc.), and 'Canidae' the dog family (including foxes, jackals, and wolves).

ideal free distribution The distribution of *foraging organisms that results if all the consumers are ideal in their assessment of the profitability of patches and free to move from one patch to another.

Idiacanthidae (black dragonfish; subclass *Actinopterygii, order *Salmoniformes) A small family of deep-sea fish that have an elongate, eel-like body. The *dorsal and *anal fins are long, the *pectorals are absent in adults. The long chin *barbel of *Idiacanthus fasciola* (dragonfish) is very typical. There are about three species, living world-wide at depths of 100–200 m.

Idiacanthus fasciola (dragonfish) *See* IDIACANTHIDAE.

idiothetic Applied to information concerning its orientation in an environment that is obtained by an animal by reference to a previous orientation of its body, and without external spatial clues. *Compare* ALLOTHETIC.

-iformes A suffix often used, but not recommended by the *ICZN, when a name refers to an order of animals (e.g. 'Strigiformes', the owls).

Iguana iguana (common iguana, green iguana) *See* IGUANIDAE.

Iguania A group of lizards that includes the *Iguanidae and *Chamaeleontidae. *Compare* SCLEROGLOSSA.

Iguanidae (iguanas, basilisks, anoles; order *Squamata, suborder *Sauria) A family of lizards, with terrestrial, arboreal, burrowing, and semi-marine forms, which are the New World counterparts of the Old World *Agamidae, but distinguishable by their *pleurodont teeth. Typically, iguanids are long-limbed and agile. Dorsal crests, throat appendages, and 'helmets' are common. *Iguana iguana* (common iguana or green iguana) of S. American tropical forests is large (up to 1.8 m long), principally herbivorous, and has a short, fleshy tongue, serrated teeth, a throat dewlap, and a spiny crest along the back; it dives into water if startled. *Phyrnosoma* species (horned lizards) are flat and squat with short tails, their bodies covered with spiny tubercles, and they can burrow backwards into the sand; if disturbed they may squirt blood from the corners of their eyes. *Amblyrhynchus cristatus* (marine iguana) of the Galápagos Islands is the only extant marine lizard; it feeds on seaweeds and basks on rocks or shelters in fissures when not swimming. The 165 *Anolis* species (anoles) are small, slender, with long, whip-like tails, and have adhesive pads on their clawed toes; in some these pads have enlargements allowing the lizard to 'parachute' (i.e. to slow their rate of descent when falling, which allows them to jump to the ground from considerable heights). Anoles are adept at colour change. The arboreal *Basiliscus* (basilisks) of Mexico and Ecuador are excellent swimmers but also fast, bipedal runners on land, and over the surface of water! Adults develop a casque on the head, which is larger in males. They feed on fruit and small animals. There are more than 700 species in the family, found mainly in the New World, but also in Madagascar and some Pacific islands.

Iguanodontidae (order *Ornithischia, suborder *Ornithopoda) A family of *Jurassic–*Cretaceous, bipedal dinosaurs whose best-known representative is *Iguanodon*, remains of which have been found in Europe, Asia, and Africa.

ileum In *Mammalia, *Amphibia, *Reptilia, birds (*Aves), and some fish, the final part of the small intestine concerned with the absorption of digestive products not absorbed earlier. In insects, a narrow tube from the *pylorus to the *rectum, which together comprise the hind gut.

ilium In tetrapods, the dorsal section of the *pelvis, which articulates with one or more sacral vertebrae.

illicium In anglerfish, the 'fishing rod' situated above the snout. It consists of the first *fin ray of the *dorsal fin and bears a flap of skin, forming a lure, at the end.

imago The fully developed adult among pterygote insects.

imbricate With parts (e.g. plates) overlapping one another like tiles on a roof.

imbricate plates *See* IMBRICATE.

imino acid An acid derived from an imine in which the nitrogen of the imino group (==NH) and the carboxyl group (—COOH) are attached to the same carbon atom. Proline and hydroxyproline are imino acids, normally classified with *amino acids.

imitation The acquisition of patterns of behaviour by repeating exactly behaviour observed in others, not necessarily of the same species.

immigration In genetics, the movement or flow of *genes into a population, caused by immigrating individuals which interbreed with the residents. This is the usual source of new variation in a population, although the fundamental sources of all variation are gene mutation and recombination. *See also* MIGRATION.

immune deficiency A condition in which an individual's immune system is weakened or not functioning.

immunity The resistance of an organism to a pathogenic micro-organism or its products. Immunity may be active (*see* ACTIVE IMMUNITY) or passive (*see* PASSIVE IMMUNITY).

immunization The process by which an individual's immune system is prepared to act against an *immunogen. The process may be active or passive. Active immunization involves either the introduction of a foreign substance into the body, most commonly by vaccination, or natural exposure to an infective agent. Both stimulate the immune system to respond. Passive immunization involves the introduction of prepared elements of the immune system, e.g. *antibodies, that are able to combat an infection immediately. Active immunization confers *immunity for a prolonged period, in some cases for life; passive immunization lasts only until the prepared elements break down.

immunodiffusion A commonly used test for the existence of an *antibody– *antigen reaction. It involves the diffusion of known antibodies and suspected antigens through agar, and the identification of a zone of precipitation.

immuno-electrophoresis A technique for the differentiation of *proteins in solution, based on both their *electrophoretic and immunological properties. Initially the proteins are separated by gel electrophoresis: they are then reacted with specific *antibodies by double diffusion through the gel. The pattern of precipitating arcs thus formed can be used to identify the proteins.

SEE WEB LINKS
• Description of the immuno-electrophoresis technique.

immunogen A substance, often an *antigen, which stimulates an immunological response.

immunoglobin See ANTIBODY.

immunoglobulin An *antibody secreted by *B cells.

immunology The study of the immune system.

impala (*Aepyceros melampus*) See BOVIDAE.

imperforate nostrils See PERFORATE NOSTRILS.

imperial pigeons (*Ducula*) See COLUMBIDAE.

impermeable junction See TIGHT JUNCTION.

imprinting A general descriptive term applied to a form of *learning that only occurs early in a young animal's life. In it, the young animal learns to direct some of its social responses to a particular object, usually a parent. The phenomenon was first described in detail by Konrad *Lorenz in 1937 in respect of *precocial birds, which learn to follow a moving object and in which it is most strongly developed. Young mammals are often imprinted to the scent or vocalizations of the adult and some birds are imprinted to the vocalizations of their parents (e.g. wood ducks (*Aix sponsa*) nest in restricted spaces in tree hollows; prior to hatching the mother makes a characteristic call; later, when the call is repeated, the chicks respond by leaving the nest and when they are all assembled the mother moves away with the chicks following). In this case imprinting occurs only during a brief critical period, possibly of only a few hours, very early in the life cycle. Such imprinting is irreversible and influences subsequent behaviour patterns, most importantly sexual behaviour.

-ina A customary suffix, not specifically cited in the code of nomenclature, that indicates an animal subtribe.

-inae A standardized suffix used to indicate a subfamily of animals in the recognized code of classification.

Inarticulata (Ecardines; phylum *Brachiopoda) A class of brachiopods whose *chitinophosphatic or calcareous valves are held together only by muscles and the body wall, and not hinged. The valves may be *punctate or impunctate. There is no internal skeletal support for the *lophophore. The *pedicle is developed from the ventral *mantle. The alimentary canal has a functional anus. Inarticulata first appeared in the Early *Cambrian.

inborn error A genetically determined biochemical disorder that results in an *enzyme defect: this produces a metabolic block and has pathological consequences. Two examples in humans are phenylketonuria, in which tissue degeneration results from the accumulation of a toxic intermediate in the pathway of *tyrosine metabolism, and Wilson's disease, in which death results from poisoning by copper because of a blockage in the pathway of copper detoxification. In both cases a single *nucleotide substitution has produced a recessive *allele that interferes with a metabolic pathway and causes death of the *homozygotes.

inbreeding The interbreeding of closely related individuals (e.g. of *siblings, or cousins, or a parent and its offspring). The converse is *outbreeding. The inbreeding coefficient is the probability of *homozygosity due to a *zygote obtaining copies of the same *gene from parents of common ancestry. Inbreeding increases homozygosity: unfavourable recessive genes are therefore more likely to be expressed in the offspring. In general, the genetic variability of an inbred population declines; this is usually disadvantageous for its long-term survival and success. Typically, vigour is reduced in inbred stock. Many organisms, including humans, have physical or social mechanisms that tend to discourage inbreeding and promote outbreeding.

inbreeding depression The decreased vigour in terms of growth, survival, or fecundity that follows one or more generations of *inbreeding. This is the opposite outcome to that obtained by the crossing of two separate inbred lines. *Compare* HETEROSIS.

inch worms *See* GEOMETRIDAE.

incisiform Chisel-shaped, resembling an *incisor.

incisor In *Mammalia, one of the chisel-shaped teeth at the front of the mouth. In primitive forms there are five on each side of each jaw.

inclusive fitness The *adaptive value (fitness) of an individual, taking account not only of that individual's own success, but also of the success of all its kin (i.e. those bearing some portion of the same *genotype). J. B. S. *Haldane is reputed to have said that he would lay down his life for more than two brothers or sisters, eight cousins, 32 second cousins, etc., these numbers corresponding to the proportions of his own *genes shared by these relatives. *See also* KIN SELECTION.

incompatibility The relationship between patterns of behaviour that cannot be performed simultaneously because they are contradictory, because they require the simultaneous operation of reflexes that are mutually inhibitory, or because they are based on contradictory motivation.

incomplete dominance *See* PARTIAL DOMINANCE.

Incurvariidae (subclass *Pterygota, order *Lepidoptera) Family of small, primitive, drab or metallic moths. The antennae are very long in some species. The *ovipositor is extensible and used to lay eggs in leaves or stems. Larvae are leaf-miners or stem-borers, or live in galls or among detritus. Frequently they construct cases from leaf-litter or from excisions from leaves. They have a wide distribution.

incus In *Mammalia, the central auditory ossicle of the *middle ear, derived from the *quadrate bone of ancestral vertebrates (from the Latin *incus*, 'anvil').

independent assortment (random assortment) The random distribution in the *gametes of separate *genes. If an individual has one pair of *alleles A and a, and another pair B and b (this *genotype being represented as *AaBb*), then it should produce equal numbers of four types of gametes: *AB*, *Ab*, *aB*, and *ab*. The assortment of alleles of one gene occurs independently of that of the alleles of the other gene. This is found experimentally with many pairs of genes, and is asserted in *Mendel's second law (the law of independent assortment); in fact, though, it applies only to distantly linked or unlinked genes.

index fossil (zonal fossil, zone fossil) A *fossil whose presence is characteristic of a particular unit of rock (the *zone) in which it occurs and after which that zone is named.

Indian rhinoceros (*Rhinoceros unicornis*) *See* RHINOCEROTIDAE.

Indian variable lizard (*Calotes versicolor*) *See* AGAMIDAE.

Indicator (honeyguides) *See* INDICATORIDAE.

Indicatoridae (honeyguides; class *Aves, order *Piciformes) A family of grey, olive, brown, and white birds which have short, stout bills, and short legs with *zygodactylous toes. Their skin is thick and so protects them against insect stings. They are arboreal, and found in forest and woodland. They feed on beeswax and insects, and are nest parasites. Their common name refers to their habit of guiding animals to bee nests, where they feed on the remains left by the animals that have plundered them. The largest genus is *Indicator*, with 10 species. There are four genera, with 14 species, found in Africa and Asia.

indicator species A species that is of narrow ecological amplitude with respect to one or more environmental factors and that is, when present, therefore indicative of a particular environmental condition or set of conditions. For example, fish species and many aquatic invertebrates vary in the amount of dissolved oxygen they require and the species present in a body of water provide an indication of the extent to which the water is contaminated with organic material. If species are long-lived their performance represents an integration of the influence of the factor with time and may give a better assessment of its importance than can a more precise physical measurement taken on a particular day. *See also* INDUSTRIAL MELANISM.

indigenous *See* NATIVE.

indirect flight muscle The thoracic musculature that brings about the flight movements of the wings in most insects. It is not attached directly to the wings. The *thorax is a box to which the wings are joined. Longitudinal muscles compress the thoracic box and cause the wings to move down. Dorsoventral muscles pull the *tergites and *sternites of the thorax together, causing the wings to move upwards.

individual distance The distance from an individual at which the presence of another individual of the same species incites aggression or avoidance.

Indostomidae (subclass *Actinopterygii, order Indostomiformes) A monospecific family (*Indostomus paradoxus*) of very small (3 cm) freshwater fish that have a slender body, large eyes, and pointed snout. This fish, found in Burma, has been placed in an order (Indostomiformes) of its own.

Indostomiformes *See* INDOSTOMIDAE.

indri *See* INDRIDAE.

Indridae (formerly Indriidae; indri, sifaka; order *Primates, suborder *Strepsirrhini) A family of primitive primates that includes the largest of the living lemurs (up to 1 m long). Its members are distinguished from other lemurs by the absence of the lower *canine and by having one upper and one lower *premolar tooth. The tail is short in some species, and the ears may be concealed beneath long hair on the sides of the head. They are fully arboreal, although they visit the ground, and are exclusively vegetarian. There are six species (one discovered in 1988) in three genera: *Indri* (indri), *Propithecus* (sifaka), and *Avahi* (woolly indri). They are all confined to Madagascar.

induced-fit theory A variation of the *lock-and-key theory of enzymatic function. It is proposed that the *substrate causes a conformational change in the *enzyme such that the *active site achieves the exact configuration required for a reaction to occur. The overall effect would be a tighter binding for the substrate and enzyme.

industrial melanism The development of melanic forms of organisms in response to soot and sulphur dioxide pollution. Industrial melanism is especially associated with various moth species and was first noted early in the 19th century. In the 1950s, studies by H. B. D. Kettlewell (1907–99) on the moth *Biston betularia* showed that melanism is a simple inherited trait favoured by natural selection in soot-polluted environments, where dark colouring gives better camouflage and so better protection from predators. In time the proportion of melanic forms increases in such polluted environments. The reverse process—selection in favour of pale forms—occurs in pollution-free environments where the paler, speckled form is better concealed on the tree lichens on which it settles. With appropriate calibration, the extent of melanism in a population may thus be used as an indicator of industrial pollution. *See also* INDICATOR SPECIES.

infanticide The destruction of eggs or killing of pre-reproductive young by mature individuals of the same species. This occurs among many vertebrate and invertebrate species. Males that expel a rival to become dominant, with exclusive access to females, may kill any of their predecessor's young that are still dependent on their mothers;

females may kill some of their own young if there are insufficient resources to raise them.

infauna Benthic organisms (*see* BENTHOS) that dig into the sea-bed or construct tubes or burrows. They are most common in the subtidal and deeper zones. *Compare* EPIFAUNA.

inflammasome An *oligomer consisting of *protein units that activates *inflammation.

inflammation Part of the innate immune response (*see* INNATE IMMUNITY) to an infection or injury that marks the commencement of the healing process, involving irritation followed by swelling, and sometimes by suppuration. *Arterioles dilate, increasing blood flow, which causes reddening of the affected area and *capillaries allow fluid to pass into interstitial spaces.

information index *See* SHANNON–WIENER INDEX OF DIVERSITY.

information statistic A measure of disorder within a group; it is zero when all individuals within the group are identical. In information analysis (an agglomerative, *hierarchical classificatory technique, devised by W. L. T. Williams and others in 1966) a hierarchy is constructed by repeatedly joining together those individuals or groups that exhibit the smallest increase in heterogeneity (disorder), and therefore the smallest change in information.

informosome *See* MATERNAL MESSAGE.

infrabasals In some *Crinoidea, a circlet of plates at the base of the *calyx, adjacent to the top of the stem.

infra-orbital foramen A canal in the *maxilla, below and anterior to the *orbit, through which nerve fibres and blood vessels pass to the face.

infundibuliform Shaped like a funnel.

infundibulum 1. In *Ctenophora, a part of the *gastrovascular cavity from which the radial canals originate. **2.** In vertebrate *embryos, an outpushing from the floor of the brain which, together with the *hypophysis, develops into the *pituitary body. **3.** In *Equidae, a *lingual concavity, often called the 'cup', in the lower *incisors.

infuscate Brown in colour.

infusoria An old term for the teeming microscopic organisms, particularly *Protozoa, found in hay infusions (i.e. hay which has been left to soak in water).

inguinal Pertaining to the groin.

inhibition The complete abolition of, or the decrease in the extent or rate of, an action or process.

-ini In classification by the nomenclature code, the recommended, but not mandatory, suffix used to indicate a tribe of animals.

Iniomi An older name for a group of fish now placed in the order *Myctophiformes.

initiator *Transfer ribonucleic acid (t-RNA), that in *eukaryotes carries methionine and in *prokaryotes N-formylmethionine, which binds to the small unit of a *ribosome bearing *messenger ribonucleic acid (m-RNA), thus forming an initiation complex. This, in the presence of three protein initiation factors and GTP which is hydrolysed, enables the large ribosomal subunit to associate with the complex, and peptide-chain synthesis to proceed.

innate Applied to behaviour that is not learned (i.e. behaviour that is acquired genetically). In the 1950s and 1960s there was much debate concerning whether particular behaviour is innate or learned; today most ethologists hold that most behaviour has both innate and learned components and studies concentrate on the relative contribution made by each.

innate immunity *Immunity that is non-specific and represents the body's first line of defence against infection. It is conferred by barriers including the skin (*integument), chemicals in sweat, tears, and *saliva, and the microbial population of the skin, nasal passages, and intestinal tract. An invader that penetrates these barriers will trigger *inflammation, which is the second level of the innate response.

innate releasing mechanism (IRM) Special units within the nervous systems of young animals that detect neural messages generated by *sign stimuli and, upon receipt of an appropriate signal, activate motor cells, triggering a *fixed-action pattern of behaviour.

inositol A carbocyclic or sugar alcohol, widely distributed in both plants and animals. It is a constituent of *membranes, muscles, and nervous tissue and is necessary for growth. Inositol is often classed with the B vitamins as it has been reported to be essential in the diets of some organisms.

Inovicellata (order *Cheilostomata, suborder *Anasca) A division of bryozoans in which the slender *stolons have a creeping habit and give rise to separate, upstanding individuals which possess *opercula. The division occurs from the *Eocene to the present.

inquiline An animal which lives inside the body or nest of another animal without harming its host.

inquilism An intimate association between two animals in which one partner lives within the host, obtaining shelter and, perhaps, a share of the host's food.

Insecta (insects; phylum *Arthropoda, subphylum *Hexapoda) Class of arthropods that have three pairs of legs and, usually, two pairs of wings borne on the *thorax. Typically, there is a single pair of antennae and one pair of *compound eyes. Gas exchange takes place through a system of *tracheae and the gonoducts open at the posterior end of the body. The oldest fossil

insects occur in *Devonian rocks, and the first winged representatives are known from *Carboniferous rocks. Dragonflies and beetles were established before the end of the *Palaeozoic; social varieties such as ants and wasps are present in *Cretaceous sediments. The evolution of the flowering plants had a marked influence on insect development, so that many new forms appeared in the Cretaceous and *Tertiary Periods. More than 750 000 extant species of insects have been described. This is larger than the number of species belonging to all other animal classes combined.

Insectivora (infraclass *Eutheria, cohort *Unguiculata) An order, once recognized, that included the ancestors of all eutherian mammals, nowadays grouped in four orders: *Proteutheria, comprising many extinct forms; *Scandentia, the tree shrews; *Macroscelidea, the elephant shrews; and *Lipotyphla, the living hedgehogs, shrews, and moles. The order Insectivora is therefore disbanded. Insectivores are usually small, nocturnal animals showing many primitive features. The dentition is usually full; the teeth resemble those of the earliest mammals. The bony palate is incomplete. In many groups the tympanic cavity is open with no ossified *bulla, the *tympanic bone forming a partial ring. Limbs are *pentadactyl, the gait *plantigrade. The brain has large olfactory bulbs but the cerebral hemispheres are small and little convoluted. The stomach is simple. The snout is very sensitive, in some species being drawn into a small trunk. *Vibrissae are present on the snout and elsewhere on the body. The sense of hearing is acute, but the eyes are small and eyesight is poor. Some insectivores retain the *cloaca. The reproductive system has diverged from the early eutherian form. Fossil forms are known from the *Cretaceous and *Palaeocene.

insects *See* INSECTA.

insight learning The ability to respond correctly to a situation that is experienced for the first time and that is different from any experience encountered previously. This ability appears to involve some kind of mental reasoning. *See also* LATENT LEARN-ING; TRIAL-AND-ERROR LEARNING.

instar An insect larva that is between one moult (ecdysis) of its *exoskeleton and another or between the final ecdysis and its emergence in the adult form. Instars are numbered and there are usually several during larval development.

instinct A genetically acquired force that impels animals to behave in certain fixed ways in response to particular stimuli. The term is little used by modern ethologists because it is open to many of the same objections as the term '*drive', because it makes no allowance for environmental influences upon patterns of behaviour, and because behaviour that formerly was considered to be 'instinctive' is known now to result from several different categories of motivation.

instrumental conditioning *See* OPER-ANT CONDITIONING.

insulin A *protein *hormone that is secreted by the beta-cells of the *islets of Langerhans in the *pancreas. It stimulates the utilization of *glucose by the cells and thereby lowers the blood-sugar level. A deficiency of this hormone results in the condition known as *diabetes mellitus*.

integral protein A *protein molecule, or assembly of protein molecules, that is permanently attached to a biological *membrane.

integrin A receptor on a cell surface to which an adhesive *protein or *carbohydrate molecule binds.

integument The organ which covers the body including skin, glands, hair, scales, and feathers. The integument is the largest single organ of the vertebrate body, often making up 15–20% of the body weight. The skin is composed of three types of tissue: the epidermis (derived from embryonic *ectoderm); dermis (derived from *mesodermal tissue and *neural crest cells); and hypodermis (derived solely from mesoderm tissue). The dermis gives rise to most of the *axial skeleton.

intelligent design The assertion that certain features of the universe and of living organisms are best explained by supposing them to result from an intelligent cause rather than an undirected process such as *natural selection. Its proponents maintain that intelligent design is a scientific hypothesis, but that claim is rejected by the scientific community. The US National Academy of Sciences has stated that intelligent design and other claims of supernatural intervention in the origin of life are not science because they cannot be tested by experiment, do not generate any predictions, and propose no new hypotheses of their own. The American Association for the Advancement of Science (AAAS, publisher of the journal *Science*) has described intelligent design as pseudoscience. In fact, intelligent design is a modern version of the teleological argument for the existence of God, but with all references to its religious origin removed. *See also* CREATION 'SCIENCE'.

intensity scale A sequence of thresholds such that progressively stronger stimuli are needed before the appropriate patterns of behaviour result. These stimuli may be increasingly intensive repetitions of a single stimulus (e.g. a spider that is only slightly hungry does not respond to prey; when rather more hungry it turns toward the prey; when still more hungry it initiates an attack).

intention movement The first stage in a recognized sequence of behaviour; by itself it indicates to an observer the intention of an animal to perform the full sequence. For example, the swimming of a stickleback low over a small area of a sandy river bed during the breeding season may be an intention movement that heralds the digging of a pit and the building of a nest.

inter- Between or among.

interambulacral *See* INTERAMBU-LACRUM.

interambulacrum (*adj.* **interambulacral**) In *Echinodermata, the parts of the body that lie between the rows of *tube feet.

intercalary deletion *See* DELETION.

interclavicle In the *pectoral girdle of tetrapods, but absent from mammals except for the *Monotremata, a median bone on the ventral side between the *clavicles.

interference competition The *competition that occurs when two organisms demand the same resource and that resource is in short supply, and one of the organisms denies its competitor access to the resource. In essence, space is substituted for the resource as the prime object of competition and dominance of space provides an alternative to efficiency in resource exploitation. Territorial animals exhibit this type of competition. *Compare* EXPLOITATION COMPETITION.

interference theory A theory to explain why animals forget learned behaviour, which holds that new memories displace old memories. *Compare* DECAY THEORY.

interferon A group of *proteins that are produced in very small quantities, by *leucocytes and related cells, in response to viral infection. As a group they appear to induce the synthesis of a protein that interferes with a viral-specific function. This often confers a temporary resistance to subsequent infection by other viruses.

intergenic suppression *See* SUPPRESSOR MUTATION.

intermediate filament A filament or fibre, approximately 10 nm in diameter, that contributes to the cytoskeletal structure of *eukaryotic cells. There are several types, each composed of a different non-motile structural protein such as keratin, vimentin, or desmin.

interna A system of feeding 'roots' produced by certain parasitic barnacles (e.g. *Sacculina carcini*) that penetrates the entire body of the host.

International Code of Zoological Nomenclature (International Commission on Zoological Nomenclature, ICZN)
1. The regulations governing the scientific naming of animals. **2.** The international authority that draws up those regulations and that supervises their application.

((●)) SEE WEB LINKS
• The International Code of Zoological Nomenclature.

interneuron A *neuron that extracts information from a sensory neuron, or integrates information from different neural systems, or passes instructions to *motor neurons. Brain tissue consists entirely of interneurons. First order interneurons are attached to sensory neurons, second order interneurons are attached to first order interneurons, third order interneurons are attached to second order interneurons, and so through higher orders in a chain linking sensory and motor neurons.

internodal species concept The notion that a species exists between two branching points in a fossil lineage; it is based on a fiction, maintained as a convenience by Willi *Hennig, that a species ceases to exist as soon as it branches into two daughter species.

interparietal In the posterior part of the roof of the brain case of some mammals, a small bone lying between the *parietals.

interphase In the *cell cycle, the phase of the *nucleus between succeeding *mitoses, during which DNA is replicated and energy stored in preparation for nuclear division.

interradius In starfish (*Asteroidea) the space between two adjacent arms (radii).

intersegmental membrane Arthrodial membrane. *See* EXOSKELETON.

intersex A class of individuals that belong to a species in which two sexes occur but which possess sexual characteristics that are intermediate between the two sexes. The condition may result from the failure of the sex-determining mechanism of the genes or through hormonal or other influences during development (e.g. the condition in cattle called free-martin, in which the female member of a pair of oppositely sexed twins has been influenced, probably hormonally, by her twin brother, through the fusing of their placental circulations).

intersexual selection The version of *sexual selection in which members of one sex (usually females) select mates of the opposite sex. Compare INTRASEXUAL SELECTION.

interspecific Applied to phenomena occurring between different species. Compare INTRASPECIFIC.

interspecific competition See COMPETITION.

interstitial fauna Animals that inhabit the spaces between individual sand grains. The term is often used synonymously with *meiofauna, *mesofauna, and *microfauna.

interstitial fluid *Blood plasma minus the large proteins. Interstitial fluid is the basis of *lymph.

intervallum The space between the inner and outer walls of *Archaeocyatha.

intestine The part of the *alimentary canal, posterior to the *stomach, in which the digestion of food is completed, water is absorbed, and *faeces are produced.

intra- On the inside, or within.

intragenic suppression See SUPPRESSOR MUTATION.

intrasexual selection The version of *sexual selection in which members of the same sex (usually males) compete for access to mates of the opposite sex. Compare INTERSEXUAL SELECTION.

intraspecific Applied to phenomena occurring within a single species. Compare INTERSPECIFIC.

intraspecific competition See COMPETITION.

intra-uterine position effect (IUP effect) In mammals that give birth to litters of several young, the hormonal influence on foetuses of the sex of their neighbours. For example, a female *foetus developing between two males is exposed to *testosterone and acquires masculinized physical and behavioural traits, whereas a female developing between two females is born with more feminine traits.

intrinsic rate of natural increase See BIOTIC POTENTIAL.

introgression The incorporation of the genes of one species into the *gene pool of another. If the ranges of two species overlap and fertile hybrids are produced, the hybrids tend to backcross with the more abundant species. This results in a population in which most individuals resemble the more abundant parents but also possess some of the characters of the other parent species.

intromit To insert.

intromittent organ An organ that can be inserted, such as the copulatory organ in many species. See AEDEAGUS.

intron (silent DNA) In *eukaryotes and some *prokaryotes, part of the DNA of *genes that is not expressed in the polypeptide chains or in m-RNA. Compare EXON.

introvert In *Sipunculida, a slender, *proboscis-like structure that is regularly protruded from and withdrawn into the main part of the body.

invagination An infolding of one part of a structure within another part, e.g. of the *blastula during *gastrulation or the formation of a *typhlosole.

inversion A change in the arrangement of genetic material, involving the excision of a chromosomal segment that is then turned through 180° and reinserted at the same position in the *chromosome. The result is a reversal in the order of *genes in that segment of chromosome.

invertase *See* SUCRASE.

invertebrate drift The passive *dispersal of the *larvae of invertebrates living in rivers, which are carried by the flow of water from the sites where they hatched to sites where they can develop further.

in vitro Literally 'in glass', but applied more generally to studies on living material that are performed outside the living organism from which the material is derived. Examples include the use of a perfused organ, tissue cultures, cell homogenates, and subcellular fractions.

in vivo Applied to studies on whole, living organisms, on intact organ systems therein, or on populations of micro-organisms.

involute Having edges that roll under or inwards. Applied, for example, to coiled cephalopods (*Cephalopoda) in which the final *whorl envelops earlier ones.

ionic bond The bond formed when an electron is transferred from one atom to another. The atom that loses that electron becomes a positively charged ion and the atom that gains the electron becomes a negatively charged ion. A strong electrostatic force then bonds the two ions together. The bonding in sodium chloride crystal (NaCl) is ionic, the crystal lattice containing Na$^+$ ions and Cl$^-$ ions.

ioras *See* IRENIDAE.

Ipnopidae (subclass *Actinopterygii, order *Myctophiformes) A small family of deep-sea fish that have a very slender body, flat head with protruding lower jaw, and peculiar yellow eyespots directed dorsally. There are about five species, distributed world-wide, some at depths of 1000–2000 m.

Irenidae (leafbirds, ioras, fairy-bluebirds; class *Aves, order *Passeriformes) A family of bright green, green and black, white, or blue and black birds, which have fairly long, slightly curved bills, short to medium, rounded wings, and short to long tails. They are arboreal, and found in forests. They feed on fruit, berries, and insects, and

nest in trees. The eight species of *Chloropsis*, which are bright green, are popular cage birds. There are three genera in the family, with 14 species, found in southern Asia, Indonesia, and the Philippines.

Irian division The lowland and hill component of the New Guinea biota, including the rain forests of the Cape York Peninsula, Australia, nearly all of whose animals are identical at the species level with those of the lower-altitude forests of New Guinea. *See also* TUMBUNAN DIVISION.

iridescence A physical phenomenon in which fine colours are produced on a surface by the interference of light that is reflected from both the front and back of a thin film.

iris In the vertebrate eye, between the *cornea and the lens, a structure formed by the *choroids with a central aperture, the pupil, the size of which adjusts according to the light intensity. The iris is coloured.

Irish elk (*Megaceros giganteus*) *See* ALLOMETRY; CERVIDAE.

IRM *See* INNATE RELEASING MECHANISM.

Irregulares *See* ANTHOCYATHEA.

irrelevant behaviour Activity that interrupts a pattern of behaviour temporarily and that does not contribute to the function of that pattern.

irritability (sensitivity) The capacity of a cell, tissue, or organism to respond to a stimulus, usually in such a way as to enhance its survival.

irruption A sudden change or oscillation in the population density of an organism.

isabelline Greyish in colour, drab.

ischial callosity An area of hard or thickened skin on the buttocks.

ischiopodite (ischium) In the segmented walking limbs of *Crustacea, the third segment from the base.

ischium 1. In tetrapods, the part of the *pelvis that projects backward on the ventral side. In *Primates, it bears the weight of the sitting animal. **2.** *See* ISCHIOPODITE.

Ischnocera (subclass *Pterygota, order *Phthiraptera) Suborder of chewing lice characterized by the *pulvinus and the vertical articulation of the *mandibles. The suborder contains two families that are parasitic on placental mammals, and three that are parasitic on birds. Their food includes feathers, skin, sebaceous exudates, and blood; and some species are known to eat other lice, although these are probably not an important part of their diet. There are 120 genera, with about 1800 species.

Ischyromyidae (order *Rodentia, suborder *Sciuromorpha) An extinct family comprising the most ancient ancestors of rodents, which lived from the late *Palaeocene to the middle *Eocene, probably as scampering, squirrel-like animals. The skull and teeth retain many primitive features. There are more teeth than in later rodents, but the front upper *premolar is reduced and lost in some species. The jaw lacks modifications found in later rodents.

Isectolophidae (order *Perissodactyla, suborder *Ceratomorpha) One of several *Eocene families of ceratomorphs, a group represented today by the tapirs and rhinoceroses.

island biogeography Basically, the relationship between area and species number, as an equilibrium between immigration and extinction, on islands; a study that is possible because islands are numerous and their biotas are often small enough to be quantified. The subject also has great relevance to the continents, where plant and animal communities are effectively reduced to islands in a sea of cultivation or urbanization.

island biotas The plants and animals of oceanic islands; because of their isolation, such biotas normally include numerous *endemic taxa. Island biotas are generally

fragile, unbalanced in that they lack plants and animals with poor trans-oceanic dispersal capacities, and vulnerable to disturbance by human activities and introduced species.

island hopping The colonization of an island or islands by plants and animals that move from an adjacent island or islands. Birds are particularly likely to island-hop. Over geological time, islands drift away from their areas of origin. The descendant biotas maintain themselves in the ancestral environment by island hopping on to successively younger islands as these emerge.

Island mammal region A concept, proposed by the biogeographer Charles H. Smith (1950–), that combines the *Australian, *Madagascan, and W. Indian faunal subregions. The development of the characteristic mammalian faunas of these insular faunal regions is due at least as much to their isolation by the sea as to ecological and geographic factors.

islets of Langerhans In the *pancreas of jawed vertebrates, tissue that secretes the *hormones *insulin and *glucagon.

isocercal tail In fish, a tail in which there is an apparent symmetry between the dorsal and ventral regions, the *fin rays being supported by equal-sized arches of the penultimate vertebrae as well as by the reduced, plate-like, ultimate vertebra. It is found e.g. in cod.

Isocrinida (sea lilies; class *Crinoidea, subclass *Articulata) An order of sea lilies which have a long stem, characteristically five-lobed in cross-section, which bears whorls of *cirri at irregular intervals along its length. The small *calyx is dicyclic, with concealed *infrabasals. The arms are generally branched several times. The order is known from the *Triassic to the present.

isoenzyme (isozyme) A species of *enzyme that exists in two or more structural forms which are easily identified by *electrophoretic methods.

isogamy The fusion of *gametes that are morphologically alike. This is an uncommon condition, found in some *Protozoa.

isogeneic (syngeneic) Applied to a *graft involving a donor and a host that are genetically identical.

isokont Having *flagella of equal length.

isolecithal Applied to an egg in which the yolk is evenly distributed in the cytoplasm, e.g. an *oligolecithal egg. *Compare* TELOLECITHAL.

isomer Either or any of two or more compounds that have the same molecular composition but different molecular structure. Isomers differ from each other in their physical and chemical properties.

isomerase An *enzyme that catalyses a reaction involving the conversion of a compound into an *isomer of that compound.

isomyarian Applied to the condition in certain *Bivalvia in which the two *adductor muscles are of approximately the same size.

Isonidae (subclass *Actinopterygii, order *Atheriniformes) A small family of marine fish, closely related to the silversides (*Atherinidae), that are often found close to the surf zone. *Iso hawaiiensis*, a small (10 cm) fish with large eyes, a small mouth, and small first *dorsal and *pelvic fins, is typical of the family. There are about six species, in the Indo-Pacific region.

Isopoda (pill bugs, slaters, woodlice; class *Malacostraca, superorder *Pericarida) Large, diverse order of crustaceans, 5–15 mm long, forming a specialized subdivision of their superorder. Isopods occupy many habitats, from the deep sea to true terrestrial niches. Some are even parasitic. They are almost always dorsoventrally flattened and adapted for crawling, and lack a *carapace. The *pereon is made of up to seven somites, each bearing one pair of *uniramous walking legs. The leathery, flexible *exoskeleton may have brown or grey markings, and some species possess *chromatophores which to some degree enable the animal to match its colour to the substratum. The head bears unstalked, *compound eyes, and uniramous first antennae with sensory *setae. Most isopods are scavengers and omnivores, with compact mouth-parts and a well-developed cardiac (or triturating) stomach. The *pleopods, unlike those of most other Crustacea, have become developed for gaseous exchange. Burrowing is common in marine forms and some species can damage docks and wooden piling (e.g. *Sphaeroma* and the gribble, *Limnoria*). Characters such as internal fertilization, a marsupium (or brood pouch), and walking legs are preadaptations for the successful life-style of the terrestrial forms. There are 4000 known species, most of them marine.

isoprene (2-methyl butadiene) A five-carbon compound that forms the structural basis of many biologically important compounds, e.g. the *terpenes.

isoprenoid A compound that comprises two or more *isoprenes or their derivatives.

Isoptera (termites, white ants; class *Insecta, subclass *Pterygota) Order of *hemimetabolous insects, related to cockroaches, in which all species are social and *polymorphic, with primary and secondary reproductives, soldiers, and workers. Nests (termitaria) may be cavities in the ground or wood, earth mounds, or carton constructions. Termites forage for plant material, using odour trails, and rely on gut-dwelling, symbiotic micro-organisms (*Protozoa in lower termites, bacteria in Termitidae) for the digestion of cellulose. Anal *trophallaxis distributes these symbionts, and *pheromones involved in caste determination, among the colony members. There are about 2200 living species, most being restricted to the tropics.

Isospondyli An older name for a group of herring-like (soft-rayed) fish which comprise the four superorders: *Elopomorpha; *Osteoglossomorpha; *Clupeomorpha; and *Ostariophysi.

isotonic Applied to a cell in which the *osmotic pressure is equal to that in the surrounding medium. *Compare* HYPER-TONIC; HYPOTONIC.

isozyme *See* ISOENZYME.

Istiophoridae (marlin, sailfish, spear-fish; subclass *Actinopterygii, order *Perciformes) A small family of marine fish that have strongly streamlined bodies, a long bill (upper jaw), and a high *dorsal fin, which is even more accentuated in the sail-fish. When swimming, the long-based dorsal fin is folded down into a special groove at the back. The deeply forked tail fin has very long dorsal and ventral lobes. *Makaira nigricans* (blue marlin) can grow to about 5 m; *Istio-phorus platypterus* (sailfish) grows to about 2.5 m. There are about 10 species, distributed world-wide in tropical to subtropical seas.

Istiophorus platypterus **(sailfish)** *See* ISTIOPHORIDAE.

Isurus oxyrinchus **(mako shark)** *See* LAMNIDAE.

Italian wall lizard *(Podarcis sicula) See* LACERTIDAE.

iterative evolution The repeated evo-lution of similar or parallel structures in the development of the same main line. There are many examples of iterative evolution in the fossil record, spanning a wide range of groups. This evolutionary conservatism probably is due to the overriding morpho-genetic control exerted by certain *regula-tory genes.

iteroparity The condition of an organism that has more than one reproductive cycle during its lifetime. *Compare* SEMELPARITY.

IUP effect *See* INTRA-UTERINE POSITION EFFECT.

Jacamaralcyon tridactyla (three-toed jacamar) *See* GALBULIDAE.

jacamars *See* GALBULIDAE; PICIFORMES.

jacanas *See* JACANIDAE.

Jacanidae (jacanas; class *Aves, order *Charadriiformes) A family of rufous, greenish-brown, and black birds which have a medium-length bill, and broad wings. Their legs are long, with bare *tibiae and extremely long toes and claws. They inhabit the marshy shores of lakes and streams, walk on surface vegetation, feed on insects, aquatic animals, and vegetation, and nest on floating plants. There are six genera, with eight species, found in Central and S. America, Africa, southern Asia, and Australasia.

Jaccard's index (Jaccard's coefficient) In *biogeography, an index of faunal resemblance between two regions. It is calculated as C/N_1+N_2-C, where C is the number of taxa shared between a pair of regions and N_1 and N_2 are the number of species in each of the two regions.

jack *See* CARANGIDAE.

jackal (*Canis*) *See* CANIDAE.

Jacobson's organ *See* VOMERONASAL ORGAN.

jaguar (*Panthera onca*) *See* FELIDAE.

Japanese giant salamander (*Andrias japonicus*) *See* CRYPTOBRANCHIDAE.

Japygidae *See* DIPLURA.

Javan rhinoceros (*Rhinoceros sondaicus*) *See* RHINOCEROTIDAE.

javelinfish *See* POMADASYIDAE.

jawfish *See* OPISTHOGNATHIDAE.

jawless fish *See* AGNATHA.

jays (*Garrulus*) *See* CORVIDAE.

jejunum In most vertebrates, the middle section of the *intestine, between the *duodenum and the *ileum.

jellyfish *See* SCYPHOZOA.

Jenynsiidae (subclass *Actinopterygii, order *Atheriniformes) A small family of freshwater fish, of interest mainly because both males and females have the sex organs directed either to the left or to the right. A 'left' male can mate only with a 'right' female, and vice versa. There are about three species, none exceeding 12 cm in length, found in southern S. America.

jerboa *See* DIPODIDAE.

jerboa-rat (*Notomys*) *See* MURIDAE.

Jerusalem crickets *See* STENOPELMATIDAE.

jewel beetle *See* BUPRESTIDAE.

jewel wasps *See* PTEROMALIDAE.

jezebels *See* PIERIDAE.

jigger flea *See* SIPHONAPTERA.

jizz The term birdwatchers use to describe the *gestalt experience of instantly recognizing a bird.

John Dory (*Zeus faber*) See ZEIDAE.

joint capsule See ARTICULAR CAPSULE.

jollytail See GALAXIIDAE.

J-shaped growth curve A curve on a graph that records the situation in which, in a new environment, the population density of an organism increases rapidly in an exponential (logarithmic) form, but then stops abruptly as environmental resistance (e.g. seasonality) or some other factor (e.g. the end of the breeding phase) suddenly becomes effective. It may be summarized mathematically as: $dN/dT = r$ (with a definite limit on N) where N is the number of individuals in the population, T is time, and r is a constant representing the *biotic potential of the organism concerned. Population numbers typically show great fluctuation, giving the characteristic 'boom and bust' cycles of some insects, or as seen in algal blooms. This type of population growth is termed 'density-independent' as the regulation of growth rate is not tied to the population density until the final crash. *Compare* S-SHAPED GROWTH CURVE.

jugal bone (malar bone, zygomatic bone) The bone on the side of the face, beneath the eye.

jugular From the Latin *jugulum*, 'neck' or 'throat', pertaining to the neck or throat. In fish, applied to the position of *pelvic fins when these are located anterior to the *pectoral fins.

jugular vein In vertebrates, one of the blood vessels that carry blood from the head to the anterior *vena cava and thence to the heart.

jugum In some *Lepidoptera, an overlapping lobe of the fore wing that couples the fore and hind wing in flight.

jumping mouse See ZAPODIDAE.

jumping spiders See SALTICIDAE.

junglefowl (*Gallus*) See PHASIANIDAE.

Jurassic One of the three *Mesozoic periods, about 199.6–145.5 Ma ago, that followed the *Triassic and preceded the *Cretaceous. The Jurassic Period is subdivided into 11 stages, with clays, calcareous sandstones, and limestones being the most common rock types. *Brachiopoda, *Bivalvia, and *Ammonoidea were abundant fossils, along with many other invertebrate stocks. Reptiles flourished on land and in the sea, but mammals were relatively insignificant and are presumed to have been predominantly nocturnal. The first birds, including **Archaeopteryx*, appeared in the Late Jurassic.

((())) SEE WEB LINKS
• Detailed information about life in the Jurassic.

juvenile hormone (neotenin) A *hormone, released by the *corpora allata, which regulates the form of larvae after each moult. Juvenile hormone brings about the retention of juvenile characteristics in developing insects, but has other functions in adults (e.g. it may be required for the proper development of reproductive organs and for the deposition of yolk in eggs).

***Jynx* (wrynecks)** See PICIDAE.

***Jynx torquilla* (common wryneck)** See PICIDAE.

kagu (*Rhynochetos jubatus*) *See* RHYNOCHETIDAE.

Kainozoic *See* CENOZOIC.

Kalotermitidae (subclass *Pterygota, order *Isoptera) Family of lower termites which form small colonies and feed in dry, dead wood. *Pheromones secreted by the king and queen prevent *pseudergates from metamorphosing. When the queens die this inhibition is released, and the pseudergates are transformed into reproductives. There are about 300 species, with a world-wide distribution but found mainly in the tropics.

kangaroo *See* MACROPODIDAE.

kangaroo rat *See* HETEROMYIDAE.

Kantharellidae *See* DICYEMIDA.

karyogamy The fusion of two cell nuclei.

karyokinesis During *cytokinesis, the division of the cell *nucleus.

karyotype The entire chromosomal complement of an individual or cell, which may be observed during mitotic *metaphase.

katydid *See* TETTIGONIIDAE.

K cells *See* NATURAL KILLER CELLS.

keel A longitudinal ridge on a solid structure, e.g. a bone or shell. In birds, a keel (carina) along the *sternum is the site for the attachment of flight muscles.

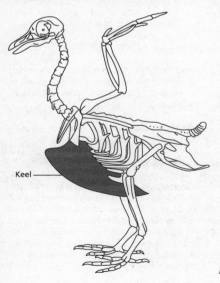

Keel

keeled belly (*Gasteropelecus sterni-cla*) *See* GASTEROPELECIDAE.

kelpfish *See* CHIRONEMIDAE.

kelp flies *See* COELOPIDAE.

kentron A hollow, dart-like structure produced by certain parasitic barnacles (e.g. *Sacculina carcini*) that pierces the *cuticle of the host.

keratin A group of fibrous scleroproteins that usually contain large amounts of sulphur as *cystine and which form the structural bases of hair, wool, nails, and other epidermal structures in animals.

ketone An organic compound that contains a ketone group, $\rangle$ C==O.

ketose A monosaccharide or its derivative that contains a *ketone group.

keystone species A *species that has a disproportionately strong influence within a particular *ecosystem, such that its removal results in severe destabilization of the ecosystem and can lead to further species losses.

kidney An organ that may have arisen in freshwater animals for the purpose of voiding excess water and that is concerned in modern forms with excretion and/or the retention of water.

killer cells *See* NATURAL KILLER CELLS.

killer whale (*Orcinus orca*) *See* DELPHINIDAE.

killifish (*Fundulus heteroclitus*) *See* CYPRINODONTIDAE.

Kimberella quadrata A fossil of *Precambrian age, known from Australia and the coast of the White Sea, Russia, that is believed to have been a bilaterally symmetrical *metazoan, possibly resembling a mollusc. It confirms that *triploblastic metazoans began to diversify in the Precambrian.

Kimmeridgian The penultimate stage of the *Jurassic, about 155.7–150.8 Ma ago.

Kimura, Motoo (1924–1994) A Japanese population geneticist who made important contributions to evolutionary theory by his mathematical modelling of evolutionary change and his proposal of the *neutrality theory of evolution.

kinaesthetic Applied to sensory receptors or organs that detect movement or changes in their own position.

kinaesthetic orientation The behaviour of an animal that moves through familiar terrain in the absence of sensory information (e.g. in total darkness) by the repetition of actions remembered from past experience of the terrain.

kinase An *enzyme that catalyses reactions involving the transfer of phosphates from a nucleoside triphosphate, particularly *ATP, to another *substrate.

kinesis The phenomenon in which a *motile organism or cell changes its rate of locomotion (or frequency of turning) in response to the intensity of a particular stimulus. The direction of locomotion remains random and is unrelated to the direction of the stimulus. *Compare* TAXIS.

kinetochore A dense, plaque-like area of the *centromere region of a *chromatid to which the *microtubules of the *spindle attach during cell division.

kinetonucleus Term formerly used to describe the *parabasal body.

kinetoplast *See* KINETOPLASTIDA.

Kinetoplastida (superclass *Mastigophora, class *Zoomastigophorea) An order of *Protozoa characterized by the possession of a kinetoplast (a region rich in *DNA) within the *mitochondrion of the cell. They may be free-living or parasitic. Some can cause important diseases of humans and domestic animals.

kinetosome *See* BASAL BODY.

kingbirds (*Tyrannus*) *See* TYRANNIDAE.

king cobra (*Ophiophagus hannah*) *See* ELAPIDAE.

king crabs *See* MEROSTOMATA.

king crickets *See* STENOPELMATIDAE.

kingdom In *taxonomy, one of the major groups into which organisms are placed. In the widely-used five-kingdom system of classification the kingdoms are: *Bacteria; *Protoctista; *Animalia; Fungi; and Plantae (plants). The kingdoms are grouped into two superkingdoms: Prokarya, containing the kingdom Bacteria; and *Eukarya, containing the remaining four kingdoms. In the three-domain classification system, kingdoms are ranked below the *domains.

kingfishers *See* ALCEDINIDAE; CORACIIFORMES.

kinglets (*Regulus*) *See* REGULIDAE.

king-of-the-salmon *See* TRACHIPTERIDAE.

Kinorhyncha (Echinodera; phylum *Aschelminthes) A phylum of very small, tube-shaped worms in which the *cuticle is divided into distinct segments (zonites) but there is no *metameric segmentation; there is a total lack of *cilia (unlike *Rotifera and *Gastrotricha). Most inhabit marine environments, and all adopt a *benthic mode of life. There are about 100 known species.

Kinosternidae (mud and musk turtles; order *Chelonia, suborder *Cryptodira) A family of freshwater turtles in which the smooth or keeled *carapace may be separate from the *plastron, which may be large and hinged to enclose the animal, or reduced. Some species produce evil-smelling secretions from cloacal glands. *Sternotherus odoratus* (stinkpot or common musk turtle) derives its common name from its cloacal secretions; it grows to 14 cm, lives on the bottom of ponds and sluggish streams, and is rarely seen on land. There are 20 species in the family, occurring in the New World.

kin selection A form of *natural selection in which the *altruism of an individual benefits its close relatives, and thereby helps to ensure the survival of at least some of its own genes. *See also* INCLUSIVE FITNESS.

Kirtland's warbler (*Dendroica kirtlandii*) *See* PARULIDAE.

kissing bug Name given to several American genera of *Reduviidae that feed on the blood of mammals and birds. Some species live in and around human dwellings, where they can transmit the blood parasite *Trypanosoma cruzi*, responsible for the debilitating *Chagas's disease.

kissing gourami (*Helostoma temminckii*) *See* HELOSTOMATIDAE.

kites *See* ACCIPITRIDAE.

Kitti's bat (*Craseonycteris thonglongyai*) *See* CRASEONYCTERIDAE.

kiwi (*Apteryx*) *See* APTERYGIDAE.

kleptoparasitism *Parasitism that is based on the theft of food from other organisms. Skuas (*Stercorariidae) practise this form of parasitism upon gulls and terns. Ants often attempt this parasitism upon carnivorous plants, robbing them of their insect prey; one estimate suggests that carnivorous plants may lose as much as 50 per cent of their prey to kleptoparasites.

klino-kinesis A change of direction of movement of an animal in response to a stimulus such that the rate at which the direction changes is proportional to the strength of the stimulus.

klino-taxis The movement of an animal in response to a stimulus; the animal compares the intensity of the stimulus to either side of its body and moves either towards or away from the stimulus, typically along a sinuous path with constant turning of the head from side to side.

klipfish (*Clinus superciliosus*) *See* CLINIDAE.

Kneriidae (subclass *Actinopterygii, order *Gonorynchiformes) A small family of freshwater fish that have a single *dorsal fin opposite the ventral fins, a slender body, and a low-slung mouth with a *protractile upper jaw. They have an *accessory respiratory organ and can climb over land, e.g. to negotiate waterfalls. There are about 12 species, found in tropical Africa.

knife eel *See* GYMNOTIDAE.

knifefish *See* NOTOPTERIDAE; RHAMPHICHTHYIDAE.

knight fish *See* MONOCENTRIDAE.

knuckle walking A style of locomotion, practised by gorillas and chimpanzees, in which an animal walks on all fours with the fingers of its forelimbs partially flexed, so its knuckles make contact with the ground. The backs of the intermediate *phalanges of the hand bear the weight and the metacarpophalangeal joints are hyperextended. Orang-utans (*Pongo pygmaeus*) practise a similar type of locomotion called fist walking, with the weight borne on the backs of the *proximal phalanges.

koala (*Phascolarctus cinereus*) *See* PHASCOLARCTIDAE.

Kogia *See* PHYSETERIDAE.

Kollikodon *See* MONOTREMATA.

Komodo dragon (*Varanus komodoensis*) *See* VARANIDAE.

kookaburra (*Dacelo novaeguineae*) *See* ALCEDINIDAE.

Kornberg enzyme The *enzyme DNA-polymerase, isolated from *Escherichia coli* in 1958 by A. Kornberg (1918–2007) and his colleagues, which functions in repair synthesis of damaged DNA.

Kraemeriidae (sand gobies; subclass *Actinopterygii, order *Perciformes) A small family of marine (rarely freshwater) fish that have a usually scaleless body, small eyes, and, typically, a strongly projecting lower jaw. Both *dorsal and *anal fins are long. Sand gobies are very small, burrowing fish, inhabiting sandy shallow waters. There are about 10 species, found in the Indo-Pacific region.

kraits (*Bungarus*) *See* ELAPIDAE.

Krebs's cycle *See* CITRIC-ACID CYCLE.

krill *See* EUPHAUSIACEA.

kronism The killing and eating of offspring.

krotovina (crotovina) An animal burrow that has been filled with organic or mineral material from another soil horizon.

K-selection Selection for maximizing competitive ability, the strategy of *equilibrium species. Most typically it is a response to stable environmental resources. This implies selection for low birth rates, high survival rates among offspring, and prolonged development. *K* represents the carrying capacity of the environment for species populations showing an *S-shaped population-growth curve. *See also* BET-HEDGING. *Compare* R-SELECTION.

Kuhliidae (flagtail; subclass *Actinopterygii, order *Perciformes) A small family of often silvery fish that have large eyes and a distinct notch between the two *dorsal fins. The forked tail fin may show oblique black bands ('flagtail'). There are about 12 species, found in the Indo-Pacific region, some in fresh water.

Kupffer cells *Macrophages located in the *liver that release a variety of compounds when activated by direct or indirect exposure to a toxic agent. They are also involved in normal physiology and the maintenance of homoeostasis. The cells were first observed in 1876 by Karl Wilhelm von *Kupffer.

Kupffer, Karl Wilhelm von (1829–1902) A German anatomist who discovered the *Kupffer cells. He was born in Courland, now in Latvia, and qualified in medicine in 1854 at the University of Tartu, where he worked from 1858 to 1865 as a *prosector. He was professor of anatomy at the University of Kiel, 1866–75, and at the University of Königsberg, 1875–80, and professor of *histology and curator of the anatomical museum at Ludwig-Maximilians-University of Munich from 1880 to 1901, when he retired. He died in Munich.

Kurtidae (nursery fish; subclass *Actinopterygii, order *Perciformes) A small family of mainly brackish-water fish that have a compressed, deep body, tapering sharply towards the tail, an oblique mouth, and a single *dorsal fin. The *anal fin is very long but low. Due to the steep head profile the male is able to carry the eggs delivered by the female. *Kurtus gulliveri* (nursery fish), 58 cm, will enter both freshwater and brackish-water regions of rivers in tropical Australia. There are about two species, found in the Indo-Pacific region.

Kyphosidae (sea chub; subclass *Actinopterygii, order *Perciformes) A family of marine fish that generally have oval-shaped bodies, long and low second *dorsal and *anal fins, relatively short heads with a small mouth, and forked to notched tail fins. Many species are predominantly plant-feeding. *Kyphosus incisor* (yellow chub) of the western Atlantic is one of the larger species (90 cm). There are at least 31 species, found world-wide in warm to temperate waters.

***Kyphosus incisor* (yellow chub)** *See* KYPHOSIDAE.

labella *See* LABELLUM.

labellum (*pl.* **labella**) In *Diptera, one of a pair of fleshy lobes into which the proboscis is expanded distally. It is often used for imbibing surface liquids.

labial mask *See* MASK.

labial palp 1. In some *Mollusca, one of a pair of flap-like folds at the end of each *tentacle by which food is transported to the mouth. **2.** One of the pair of jointed, sensory structures carried on the *labium of the mouth of an insect. They articulate with the part of the labium known as the prementum.

Labiduridae (earwigs; subclass *Pterygota, order *Dermaptera) Cosmopolitan family of rather primitive earwigs, some species of which are winged. Typical labidurids are the robust, red-brown, wingless species that are found in the rain forest of Australia and south-east Asia.

Labiidae (earwigs; subclass *Pterygota, order *Dermaptera) Cosmopolitan family of earwigs, all species of which are winged. Labiids are generally small (less than 1.5 cm long) and unremarkable in appearance.

labium Of an insect, the lower 'lip' of the mouth. *See also* LABIAL PALP; MASK.

Labracoglossidae (subclass *Actinopterygii, order Perciformes) A family of marine fish that have a single *dorsal fin (the soft-rayed part of the dorsal and *anal fins being covered in scales), and a *protractile mouth. *Labracoglossa argentiventris* is a common species in Japanese waters and grows to about 25 cm. There are about five species, found in the western Pacific.

labral plate In *Echinozoa, one of the plates of the *labrum.

Labridae (wrasse; subclass *Actinopterygii, order *Perciformes) A very large family of marine fish which are very diversified in shape and size. Usually a wrasse has a rather robust body, long and low *dorsal and *anal fins, and a tail fin that is rounded or notched, but never forked. The thick, *protrusible lips and well-developed canine-like teeth give the fish a distinctive profile. Most species are brightly coloured, but within a single species the colour pattern may differ according to age, sex, and season. In a number of species females turn into males as they grow older. Wrasses range in size from the 10 cm *Labroides dimidiatus* (cleaner wrasse) to the huge *Cheilinus undulatus* (giant maori wrasse) which grows to 2.3 m. There are at least 400 species, found world-wide in tropical to temperate waters.

labriform swimming A type of swimming practised by fish, in which movement is effected by a rowing motion involving only short, often pectoral, fins. *Compare* ANGUILLIFORM SWIMMING; BALLISTIFORM SWIMMING; CARANGIFORM SWIMMING; OSTRACIIFORM SWIMMING; RAJIFORM SWIMMING.

Labroides dimidiatus (cleaner wrasse) *See* CLEANER FISH; LABRIDAE.

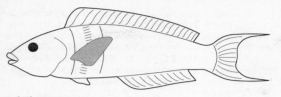

Labriform swimming

labrum 1. In *Echinozoa, a lip-like projection of the *peristome. **2.** A broad piece of *cuticle which forms the upper 'lip' in an insect. The labrum articulates with the *clypeus. It is anterior to the *mandibles, and can be moved away from them by a pair of muscles, or closed down on to them by another pair of muscles with a rocker-like action.

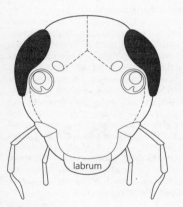

Labrum

labyrinth In the inner ear of *Gnathostomata, a structure consisting of a bony labyrinth that encloses a delicate membranous labyrinth. The labyrinth comprises three *semicircular canals.

labyrinth fish Fish that have *accessory respiratory organs in the *gill chambers, enabling them to utilize atmospheric oxygen when necessary.

Labyrinthodontia (class *Amphibia) A group (or subclass) comprising amphibians that were primitive in character and from a few centimetres to several metres long. The bones forming the palate often carried the folded teeth that give the subclass its name. They lived during the *Palaeozoic and *Triassic. Labyrinthodonts show affinities with *Crossopterygii and with reptiles, and are connected to other amphibians via such fossil groups as the Loxomammatids, Nectrideans, and *Temnospondyla. The modern tendency is to regard the labyrinthodonts as an informal assemblage of primitive amphibians, rather than as a formal taxonomic group.

Lacciferidae (order *Hemiptera, suborder *Homoptera) Family of bugs, related to the *Coccidae, in which the females are legless and have greatly reduced antennae.

lacebug *See* TINGIDAE.

Lacerta *See* LACERTIDAE.

Lacerta agilis (sand lizard) *See* LACERTIDAE.

Lacerta lepida (eyed lizard, ocellated lizard) *See* LACERTIDAE.

Lacerta viridis (green lizard) *See* LACERTIDAE.

Lacerta vivipara (common lizard, viviparous lizard) *See* LACERTIDAE.

Lacertidae ('true' lizards; order *Squamata, suborder *Sauria) A family of small, agile lizards, all of which have well-developed legs and long tails capable of *autotomy. The head plates are fused to the skull bones. The 32 species in the genus *Lacerta* are typical. They have tiny dorsal scales, increasing in size on the flanks, broad, smooth, ventral plates, a well-defined scaly collar, and in males of most species the *hemipenis has small, *sulcal lips. *L. vivipara* (common lizard or viviparous lizard) is *ovoviviparous; usually brown or grey-brown with darker and lighter spots and often a dark vertebral stripe, it grows to 18 cm. *L. agilis* (sand lizard), with large, brown, dorsal blotches and, in the male, green flanks, grows to 25 cm. In *L. viridis* (green lizard) the male is green, stippled with black, the female green or brown and sometimes striated; it is a large, stocky lizard that grows to 40 cm. The largest European lizard is *L. lepida* (eyed lizard or ocellated lizard) which grows up to 70 cm; the body is thick and has 13–24 blue spots on each flank. *Podarcis* species (wall lizards) of Europe, N. Africa, and the Near East, are similar to lacertids, their common name referring to their habit of basking on walls. *P. muralis* (wall lizard) occurs naturally in Jersey and colonies have been introduced in mainland Britain; *P. sicula* (Italian wall lizard or ruin lizard) is common in Italy. There are more than 200 species in the family, occurring in Europe, Asia, and Africa, some within the Arctic Circle.

Lacertilia *See* SAURIA.

lacewings 1. *See* NEUROPTERA. **2. (green lacewings)** *See* CHRYSOPIDAE.

lachrymal A small bone near the centre of the face, situated just within or just outside the *orbit, and often perforated to accommodate the lachrymal (tear) duct.

lacinia mobilis In *Peracarida, a mobile, tooth-like process on the *mandible.

laciniate Deeply cut, into irregular, narrow segments or lobes.

Lactariidae (false trevally; subclass *Actinopterygii, order *Perciformes) A small family of marine fish that have an oblong, compressed body and an oblique, large mouth with a prominent lower jaw. The *pectorals are long and pointed, the tail fin forked. *Lactarius lactarius* (24 cm), found in the Indo-Pacific region, is of some commercial importance in India and Malaysia. Two species are known.

lactase The *enzyme that metabolizes *lactose. It is present in the stomach of neonatal mammals, but usually ceases to be secreted after weaning, hence milk is indigestible to most mammalian species after this time. In humans, it is a *polymorphism: lactase is secreted throughout life in most people of north European and north Indian origin, but its frequency declines towards the tropics and most tropical, East Asian, Australian Aboriginal, Pacific, and Native American peoples cannot digest milk.

lactate fermentation The method by which muscles deprived of oxygen continue to produce ATP (adenosine triphosphate), but with the loss of 34 ATP molecules for every molecule of *glucose processed, and with the release of lactate (*lactic acid), which is toxic and must be excreted. The reduced nicotinamide adenine dinucleotide (NAD) produced by *glycolysis gives up its hydrogen to reduce pyruvate (*pyruvic acid), also produced by glycolysis, to yield lactate. The reaction, catalysed by lactate dehydrogenase, is: $C_6H_{12}O_6$ (glucose)$\rightarrow$ $2C_3H_6O_3$ (lactate). The net loss of ATP causes muscles to tire and the accumulation of lactate causes the pain sometimes associated with extreme exercise.

lactic acid A three-carbon hydroxy-acid, formed as the major metabolic product of certain bacteria, and also from *pyruvic acid in animal cells when *glycolysis occurs under anaerobic conditions.

lactogenic hormone *See* PROLACTIN.

lactose (milksugar) A disaccharide sugar comprising *glucose and *galactose; it is the principal sugar in milk.

lacunose Having a surface pitted with cavities or indentations.

LAD *See* LAST-APPEARANCE DATUM.

lady-beetle *See* COCCINELLIDAE.

ladybird *See* COCCINELLIDAE.

ladybug *See* COCCINELLIDAE.

ladyfish (*Elops saurus***)** *See* ELOPIDAE.

Laemobothriidae (order *Phthiraptera, suborder *Amblycera) Family of chewing lice which are parasitic on birds of the orders *Anseriformes and *Falconiformes, and on the hoatzin (*see* OPISTHOCOMIDAE). The family includes the largest of all lice, some being as long as 11 mm. There is only one genus in the family, *Laemobothrion*, the three subgenera of which each parasitize one of the three host groups. There are 14 species.

Lagomorpha (Duplicidentata; infraclass *Eutheria, cohort *Glires) An order that comprises the families *Eurymylidae (extinct forms), *Ochotonidae (pikas), and *Leporidae (rabbits, cottontails, and hares). The lagomorphs are believed to have diverged from a primitive eutherian stock at the same time as, or soon after, the rodents, so that similarities between lagomorphs and rodents may be very superficial. There is a single record from the late *Palaeocene of Mongolia, but the order is not encountered again until the late *Eocene, after which it spread widely in N. America and Eurasia. Ancestors of modern lagomorphs, very similar to modern forms, lived during the *Oligocene. The *incisors grow continually, but to either side of those of the upper jaw there is a small, peg-like incisor that gave the order its old name of Duplicidentata. There is a wide *diastema between the incisors and the cheek teeth. The *premolars and *molars have sharp edges and are used for cutting rather than grinding, the upper teeth biting outside the lower teeth. The nostrils can be closed and are surrounded by naked skin which may be covered by surrounding fur. There are four or five digits on the hind limbs and five on the fore limbs. The tail is reduced and may not be visible externally. Like some rodents, lagomorphs pass food twice through the alimentary canal (*caecotrophy).

Lagopus **(grouse)** *See* TETRAONIDAE.

Lalage sueurii **(white-winged triller)** *See* CAMPEPHAGIDAE.

Lama **(guanaco, llama)** *See* CAMELIDAE.

Lamarck, Jean Baptiste Pierre Antoine de Monet, Chevalier de (1744–1829) A French naturalist who, in 1809, proposed the first formal theory of evolution. He advanced the theory that evolutionary change may occur by the inheritance of characteristics acquired during the lifetime of the individual. For example, fossil evidence suggests that the ancestors of the giraffe had short necks: Lamarck proposed that competition for food encouraged them to stretch upward in order to browse among higher vegetation, causing their necks to lengthen, and that this lengthening was passed on to their descendants. Over millions of years the minute increases from each generation to the next culminated in the long-necked form. It is interesting to note that the theory of the inheritance of acquired characteristics did not hold a central position in Lamarck's own writings. His cardinal point was that evolution is a directional, creative process in which life climbs a ladder from simple to complex organisms. He believed the inheritance of acquired characteristics provided a mechanism for this evolution. Lamarck explained that this progress of life up the ladder of complexity is complicated by organisms being diverted by the requirements of local environments; thus cacti have reduced leaves (and giraffes have long necks). *Compare* DARWIN, CHARLES ROBERT.

Lamarckism The theory of evolution propounded by *Lamarck, based on the inheritance of *acquired characteristics.

lamella 1. A thin, plate-like structure (e.g. in the *gills of *Bivalvia and fish). **2.** A thin layer that forms part of the matrix of *bone.

lamellate Composed of, or resembling, lamellae (*see* LAMELLA). In birds, applied to a bill that has special plates adapted for sieving out food items. It is found in *Anatidae and *Phoenicopteridae.

Lamellibranchia *See* BIVALVIA.

lamina A flat, sheet-like structure.

laminate Comprising layers of material (laminae).

Lamnidae (mako, basking shark; subclass *Elasmobranchii, order *Lamniformes) A small family of marine sharks that have stout, streamlined bodies. The first *dorsal fin is placed midway on the back, but the second dorsal and *anal fins are small and near the tail fin. Distinct ridges (keels) are present on each side of the *caudal peduncle. Among the species considered dangerous are *Carcharodon carcharias* (white shark), 6.4 m, and *Isurus oxyrinchus* (mako shark), 3.6 m. The very large *Cetorhinus maximus* (basking shark), 10 m, is a harmless animal. Some workers place certain species in different families. There are about 14 species, distributed world-wide.

Lamniformes (class *Chondrichthyes, subclass *Elasmobranchii) An order of sharks (not recognized by all authorities) that have two *dorsal fins without spines and, typically, an *anal fin. There are five pairs of gill slits and a *spiracle is present in most species. The order includes a variety of sharks, e.g. *Orectolobidae (carpet sharks), *Lamnidae (mako sharks), and *Scyliorhinidae (catsharks). There are about 200 species, world-wide in distribution.

lampbrush chromosome One of the large *chromosomes found in the eggs (primary oocytes) of amphibians, with paired loops which extend from most of their *chromomeres giving a furry, brush-like appearance under the microscope. They are particularly obvious at the diplotene stage of *meiosis.

lamper eel *See* AMPHIUMIDAE.

lamprey *See* AMMOCOETE; PETROMYZONIDAE.

Lamprididae (opah, moonfish; subclass *Actinopterygii, order *Lampridiformes) A monospecific family (*Lampris guttatus*) comprising a species of large (1.8 m) marine fish that have a greatly compressed, and oval or almost round body, a small mouth, long first *dorsal and *pelvic fins, and a narrow tail ending in a forked tail fin. Despite the attention it draws because of its size and shape, little is known about its biology. It is distributed world-wide.

Lampridiformes (class *Osteichthyes, subclass *Actinopterygii) An order of marine fish of widely differing shape and form, ranging from the deep-bodied *Lamprididae (opah) to the eel-like *Regalecidae (oarfish). They have a unique type of protrusible jaw, fins without true spines, and a rather small mouth. Some families (e.g. Mirapinnidae) have also been placed in another order.

***Lampris guttatus* (opah, moonfish)** *See* LAMPRIDIDAE.

lamp shell *See* BRACHIOPODA.

Lampyridae (glow-worms, fireflies, lightning beetles; subclass *Pterygota, order *Coleoptera) Family of beetles in which the male *elytra are soft and *pubescent, and the eyes large. Females resemble larvae; the body is elongate, often without wings or elytra, and the eyes small. The antennae are *filiform. Larvae are tawny, the body segments are expanded, the head is small and concealed, the *mandibles are large. They feed on slugs and snails, which they inject with digestive juices. All stages are luminous, but this is most pronounced in adults; light is produced from the end of the abdomen to signal a mate (*compare* ELATERIDAE). There are 1700 species.

lancelet (amphioxus, *Branchiostoma lanceolatum*) *See* BRANCHIOSTOMIDAE.

land bridge A connection between two land masses, especially continents, that allows the migration of plants and animals from one land mass to the other. Before the widespread acceptance of continental drift, the existence of former land bridges was often invoked to explain faunal and floral similarities between continents that are now widely separated.

land crabs *See* GECARCINIDAE.

land-locked fish Marine fish which no longer migrate back to the sea, but have become established permanently in freshwater lakes and drainage basins. There are, for example, land-locked populations of *Salmo salar* (salmon) in parts of N. America and Sweden.

Langerhans, Paul (1847–88) A German physician and anatomist, Langerhans was born in Berlin in 1847, the son of a distinguished physician. He studied medicine at the universities of Jena and Berlin, qualifying in 1869. He developed new cell-staining techniques that allowed him to identify nerve endings (now called Langerhans cells) in the *Malpighian layer of the skin, and he also studied the *pancreas, mainly of rabbits, and was the first to describe structures (*islets of Langerhans) that, in 1893, G. E. Laguesse found to be endocrine cells that secrete *insulin. Langerhans served as a physician in the army during the Franco-Prussian War and after the war, in 1871, became a *prosector and later professor in pathological anatomy at the University of Freiburg im Breisgau. He contracted tuberculosis in 1874, possibly in the dissecting room, and had to retire. He then moved to Funchal, Madeira, where he took up the study of marine invertebrates as well as practising as a physician. In 1887 he suffered progressive renal failure and died at Funchal in 1888.

Langhian *See* BURDIGALIAN.

langur (*Presbytis, Semnopithecus, Pygathrix, Rhinopithecus*) *See* CERCOPITHECIDAE.

Laniidae (shrikes; class *Aves, order *Passeriformes) A family of mainly black, grey, or brown and white birds which have a strong, hooked bill, well-developed *rictal bristles, strong legs, and a long, narrow tail. They inhabit open areas and woodland edges, and feed on insects, reptiles, mammals, and birds, many species impaling their prey on thorns. Bushshrikes (six *Tchagra* species) of Africa often hunt on the ground by running. *Lanius* species (of which there are 65) feed from exposed perches, sometimes hovering to drop on their prey. Shrikes nest in bushes and trees. There are 12–13 genera, with 78–82 species, many migratory, found in Europe, Asia, Africa, and N. America. The helmetshrikes are often placed in a separate family, Prionopidae.

Lanius (shrikes) *See* LANIIDAE.

lantern-eye fish *See* ANOMALOPIDAE.

lanternfish *See* MYCTOPHIDAE.

lantern fly *See* FULGORIDAE.

Lanthanotidae (earless monitor; order *Squamata, suborder *Sauria) A monospecific family (*Lanthanotus borneensis*), from Borneo, comprising an elongate lizard with short limbs and a long tail with no *autotomy. It grows up to 43 cm long. The lower eyelid is transparent. The lizard feeds on fish and invertebrates. It burrows and swims. It may be a survivor of the lizard group ancestral to snakes.

Lanthanotus borneensis (earless monitor) *See* LANTHANOTIDAE.

lanugo The soft, downy hair that covers the body of some juvenile mammals (*Mammalia) and is replaced by hair of the adult type as the animal develops. The retention of lanugo in humans is regarded as evidence of *neoteny.

Lapparentophis defrennei One of the earliest snakes, which first appeared during the *Cretaceous Period (*Albian).

lappet moths *See* LASIOCAMPIDAE.

lapwings *See* CHARADRIIDAE.

larder beetle (*Dermestes lardarius***)** *See* DERMESTIDAE.

large-eye bream *See* PENTAPODIDAE.

largemouth bass (*Micropterus salmoides***)** *See* CENTRARCHIDAE.

large-nose fish *See* MEGALOMYCTERIDAE.

Laridae (gulls, terns; class *Aves, order *Charadriiformes) A family of small to fairly large birds, most of which are grey and white, although some have black on the head and wings. The bill is slender to heavy, the wings long and pointed, the tail square or forked, and the feet webbed. The 32 *Sterna* species have short tails, some being deeply forked with long tail streamers, and some have short crests. Terns are found on coasts, rivers, lakes, and marshes, where they feed on fish, crustaceans, molluscs, insects, and carrion. They nest on the ground, on cliffs, and in trees. The 38–45 *Larus* species are gregarious and nest colonially; they have yellow, red, or black bills and many have black or dark brown hoods (immature forms are darker). There are 11–17 genera in the family, with about 90 species, most migratory, and found world-wide.

larks 1. *See* ALAUDIDAE. **2. (magpie larks)** *See* GRALLINIDAE.

Larus **(gulls)** *See* LARIDAE.

larva The stage in the life cycle of an animal, during which it is motile and capable of feeding itself, that occurs after hatching from the egg, and prior to the reorganizations involved in becoming adult. The appearance of the larval form differs markedly from that of an adult of the same species. Larvae are not usually able to reproduce (but *see* NEOTENY; PAEDOMORPHOSIS). The term is applied loosely to fish, amphibians, and all *exopterygote and *endopterygote insects at this stage of growth and feeding, although the term 'nymph' is frequently applied to exopterygotes and to other invertebrates. *See also* PUPA.

larviparous Applied to animals (e.g. flesh flies, *Sarcophaga carnaria*, *see* SARCOPHAGIDAE) that reproduce by depositing larvae, rather than eggs.

laryngeal pharynx *See* HYPOPHARYNX.

larynx In tetrapods, a dilated region of the upper *trachea at the junction with the *pharynx, often possessing a folded membrane (vocal cords) which can be made to vibrate and so produce sound, the pitch of the sound being controlled by stretching or relaxing the cords.

Lasiocampidae (lappet moths, eggars; subclass *Pterygota, order *Lepidoptera) Family of medium-sized to large moths in which the body is stout and hairy, the proboscis absent, and the antennae *bipectinate. Generally females are larger and slower-flying than males. Larvae are densely hairy, with downward-directed tufts and a fringe. The *pupa is in a whitish, parchmentlike *cocoon. Distribution is world-wide, but the moths are absent from New Zealand.

Lasioglossum *See* HALICTIDAE.

Lasiorhinus krefftii **(***L. barnardi***, Queensland hairy-nosed wombat)** *See* VOMBATIDAE.

Lasiorhinus latifrons **(hairy-nosed wombat)** *See* VOMBATIDAE.

last-appearance datum (LAD) The last recorded occurrence of a key *taxon in biological history.

latent learning The formation of associations that appear to bring the animal neither reward nor punishment (although novel experiences may themselves constitute a reward). There is a delay between exposure to the learning situation and the performance of a behaviour pattern that demonstrates the effect of learning. For example, rats will explore a maze without apparent reward or punishment and will remember the information they obtain; later, when rewards are presented, those rats will perform better at finding their way through the maze than rats with no exploratory experience.

latent root *See* EIGENVALUE.

lateral fold lizards *See* ANGUIDAE.

lateral inhibition Where three *neurons are side by side, the inhibition of the middle neuron when both of the outer two are active.

lateralis system *See* ACOUSTICO-LATERALIS SYSTEM.

lateral line A system of receptors, often embedded in special grooves in the skin of an animal, that is capable of detecting vibrations (and therefore movements) in the water surrounding the animal. In most fish, the lateral line runs along the sides of the body, but usually it forms a number of branches on the head. It is found in all lower aquatic vertebrates and some *Amphibia.

lateral plate mesoderm *See* MESODERM.

laterotergite In an insect, one of a number of lateral flanges, each being a pair of *sclerites which flank each of the main abdominal *tergites. They are usually found on all the pregenital tergites.

Latimeria chalumnae (coelacanth) *See* COELACANTHIFORMES; CROSSOPTERYGII; LATIMERIIDAE.

Latimeriidae (coelacanth; subclass *Crossopterygii, order *Coelacanthiformes) A monospecific family (*Latimeria chalumnae*, coelacanth) which is a large, bulky fish characterized by peculiar stalk- or lobe-like supports for the second *dorsal, *pectoral, and *pelvic fins.

Latin American mammal region A concept that is considered by the biogeographer Charles H. Smith (1950–) to be more appropriate than the *Neotropical faunal region in respect of mammals. It extends northwards from Tierra del Fuego to just across the southern borders of the United States.

Latridae (trumpeter; subclass *Actinopterygii, order *Perciformes) A small family of marine fish that have a somewhat elongate body shape, a small mouth with sharp teeth, the *dorsal fin with a deep notch between the spinous and soft-rayed parts, and many rays (usually more than 30) in the second dorsal and *anal fins. The *pectoral and *pelvic fins are small, and the tail fin is forked. There are about 10 species, found in southern Australia, New Zealand, and Chile.

Latrodectus mactans (black widow spider) *See* THERIDIIDAE.

laughing frog (*Rana ridibunda*) *See* RANIDAE.

laughing thrushes (*Garrulax*) *See* TIMALIIDAE.

Laurasia The continent, formed when *Pangaea divided into two parts, that comprised what subsequently divided further to become N. America and Eurasia. *See also* HOLARCTICA.

Laurer's canal (Laurer–Stieda canal) In *Platyhelminthes, a short tube connecting the union of the *oviduct and yolk duct with a pore on the dorsal surface.

leaf beetle *See* CHRYSOMELIDAE.

leafbirds *See* IRENIDAE.

leafbug *See* MIRIDAE.

leaf-cutter ant *See* ATTINI.

leaf-cutter bee Any species of the cosmopolitan family *Megachilidae, the female of which cuts leaf pieces for use as material for lining or closing the nest. Sometimes the leaf fragments are augmented with mud or resin.

leaf-fish (*Monocirrhus polyacanthus*) *See* NANDIDAE.

leaf hopper *See* CICADELLIDAE.

leaf insects See PHASMATODEA; PHYLLI-IDAE.

leaf-mining flies See AGROMYZIDAE.

leaf-nosed bat 1. (Old World) See HIPP-OSIDERIDAE. **2. (New World)** See PHYLL-OSTOMATIDAE.

leaf-rollers See TORTRICIDAE.

leaf-rolling grasshoppers See GRYLL-ACRIDIDAE.

leaf warblers (*Phylloscopus*) See SYLV-IIDAE.

leafy sea dragon (*Phycodurus eques*) See SYNGNATHIDAE.

learning The acquisition of information or patterns of behaviour other than by genetic inheritance, or the modification of genetically acquired information or behaviour as a result of experience. *See also* CONDITIONING; CULTURAL; EXPLORATORY LEARNING; LATENT LEARNING; LEARNING, SPATIAL; OPERANT CONDITIONING; TRIAL-AND-ERROR LEARNING.

learning, spatial The acquisition and retention by an animal of information concerning its surroundings, including the location of potential food sources and other features. Spatial learning is particularly developed in birds that cache food stores and in many ants, bees, and wasps (*Apocrita).

leatherback See DERMOCHELIDAE.

leatherjackets 1. See MONACANTHIDAE. **2.** See TIPULIDAE.

leather-winged beetle See CANTHARIDAE.

leathery turtle See DERMOCHELIDAE.

Lebiasinidae (subclass *Actinopterygii, order *Cypriniformes) A family of freshwater fish that have a slender body, small mouth, large eyes, and the single *dorsal fin positioned half-way along the back. There are about 34 species, found in S. America.

lecithin One of a group of *phospholipid compounds (phosphatidyl cholines) found in higher plants and animals.

lecithotropic A mode of embryonic development in which the yolk of an egg provides all the nourishment. *Compare* MATROTROPHIC.

lectin A generic name for *proteins extracted from some molluscs, fish, and plants (especially legumes) that exhibit *antibody activity by effecting the agglutination of red blood cells. Some are *mitogenic, while others are capable of causing the preferential agglutination of cancer cells.

lectotype In taxonomy, one of a collection of *syntypes which, subsequent to the publication of the original description, is chosen and designated through published papers to serve as the *type specimen.

leeches See HIRUDINEA.

left-eye flounder See BOTHIDAE.

legionary In ants, applied to the type of foraging in which groups of workers seek food together.

legless lizards See AMPHISBAENIDAE; ANGUIDAE; ANNIELLIDAE; CORDYLIDAE; FEYLINIIDAE; PYGOPODIDAE; SCINCIDAE; TEIIDAE.

Leiognathidae (pony-fish, slipmouth; subclass *Actinopterygii, order *Perciformes) A small family of marine fish that have an oblong and strongly compressed body, large eyes, a small, highly *protrusible mouth, and long *dorsal and *anal fins covering at least half of the dorsal and ventral profiles. When handled, the skin may produce a soapy mucus. Pony-fish are fairly small, *Leiognathus equula* (common pony-fish) of the Indo-Australian region growing to a length of 25 cm. There are about 18 species, frequenting coastal waters of the Indo-Pacific region.

Leiognathus equula (common pony-fish) *See* LEIOGNATHIDAE.

Leiopelmatidae (class *Amphibia, order *Anura) Family of primitive frogs which have nine biconcave (*amphicoelous) vertebrae and free ribs. Their eggs, laid on damp ground, develop into froglets before hatching. There are three species, found in the uplands of New Zealand.

lek Territory that is held and defended against rivals by males of certain *species during the breeding season. The male displays within its lek in order to attract females into the lek for mating. Females move among the leks, mating with males to whose displays they respond. Consequently, for a local population of a species, leks are usually grouped together within a breeding area and dominant males tend to occupy the more central leks where their displays can be seen by the largest number of females.

lemming (*Lemmus*) *See* CRICETIDAE.

Lemur (ring-tailed lemurs) *See* LEMURIDAE.

Lemuridae (lemurs; suborder *Strepsirhini (or *Prosimii), infra-order *Lemuriformes) A family of primates which in some ways resemble the ancestors of modern monkeys, apes, and humans. The brain has small cerebral hemispheres and large olfactory regions, compared with other primates. The snout is long, the upper lip is cleft, and the *rhinarium is moist. The eyes are directed more to the side than in other primates and there is little binocular vision. In most species the external ears are large. The upper *incisors are small and the lower incisors and *canines are directed forward. Lemurs have marked breeding seasons. All modern lemurs are arboreal, herbivorous, and social. Lemurs were distributed widely throughout the warmer regions during the *Eocene. Today they survive only in Madagascar, possibly because the isolation of Madagascar during the *Tertiary left it with no native carnivores. There are four genera: *Lemur* (ring-tailed lemurs); *Varecia* (ruffed lemur); *Hapalemur* (gentle lemur); and the recently recognized *Eulemur* (brown lemur), with about 10 species. Formerly *Lepilemur* (sportive lemur) was included in this family but nowadays it is placed in the *Megaladapidae.

Lemuriformes (order *Primates, suborder *Strepsirhini (or *Prosimii) The infra-ordinal name sometimes given to the Malagasy representatives of the Strepsirhini, comprising the most primitive of living primates and their immediate ancestors, grouped into the families *Adapidae (extinct forms), *Cheirogaleidae (mouse lemurs), *Lemuridae (lemurs), *Indriidae (indris), and *Daubentoniidae (aye-ayes). An alternative recent classification separates the Daubentoniidae as *Chiromyiformes and the Adapidae as Adapiformes, but includes the *Loridae in the Lemuriformes. The primitive, insectivore-like features that they retain include a long face, eyes that are directed to the sides of the head, and a brain smaller than that of other primates. Lemur-like animals are known to have lived in both Old and New Worlds from the lower *Eocene. Some authorities now think that the 'lemuriform' characters are those of primitive Strepsirrhini, and that despite the fact that they all live on Madagascar the various families may not be closely related to each other.

Lemuroidea (cohort *Unguiculata, order *Primates) In some classifications, a superfamily comprising the lemurs and lorises, placed together in a separate suborder (*Strepsirrhini) of the primates. In other classifications lemurs and lorises are linked with the tarsiers into the suborder *Prosimii. The lemuroids appeared in the *Eocene and closely resembled their living descendants.

lemurs The *Strepsirrhini in general, or just those that live today in Madagascar. Early lemurs (family *Adapidae) lived in Europe and N. America during the *Eocene. *See also* LEMURIDAE.

lentic Applied to *habitats formed in very slow-moving or standing water (e.g. in ponds and lakes).

lenticular Shaped like a biconvex lens.

Leonardian The second of the four stages of the *Permian in N. America, 280–270.6 Ma ago. In some areas it is zoned by the use of fusulinid *Foraminiferida. It is the N. American equivalent of the Rotliegende. Red-bed localities of Wolfcampian–Leonardian age in Texas and Colorado have yielded many vertebrate fossils.

leopard frog (*Rana pipiens*) See RANI-DAE.

leopard snake (*Elaphe situla*) See COLU-BRIDAE.

Lepadidae (goose barnacles order *Thoracica, suborder Lepadomorpha) One of seven families of so-called goose barnacles, in which the *capitulum has five plates, though some of these may be missing or reduced. There are no articulated caudal appendages, and all known species are hermaphroditic. *Anelasma* has no capitular plates and is entirely parasitic; it sends root-like growths into the tissues of its dogfish hosts.

Lepidoblepharon See CITHARIDAE.

lepidoid Scaly.

Lepidoptera (moths and butterflies; class *Insecta, subclass *Pterygota) Major order of insects, characterized by wings with overlapping scales. The wingspan ranges from less than 5 mm to about 250 mm. Adults generally have a sucking proboscis; rarely they have chewing mouthparts (*Micropterigidae). The larvae have chewing mouth-parts and are nearly always plant feeders. The *pupae generally lack functional *mandibles. Lepidoptera are *endopterygotes, with an estimated 165 000 species.

Lepidosauria (scaly lizards; class *Reptilia) A subclass of reptiles that have a *diapsid skull. The subclass includes the *Rhynchocephalia (tuatara) and *Squamata (lizards and snakes).

Lepidosirenidae (South American lungfish; subclass *Dipneusti, order *Lepidosireniformes) A monospecific family (*Lepidosiren paradoxa*) of freshwater fish that have a long, cylindrical body with small scales, small eyes, confluent *dorsal and *anal fins, and reduced, almost filamentous, *pectoral and *pelvic fins. It can absorb atmospheric oxygen by utilizing a pair of lungs located beneath the gut. These lungs seem to be even more important for respiration than the gills. During the dry season the fish makes a burrow and becomes dormant. It may reach a length of 1.2 m. It is found in fresh water in S. America.

Lepidosireniformes (class *Osteichthyes, subclass *Dipneusti) An order of lungfish with a cylindrical body, paired lungs, and nearly filamentous *pectoral and *pelvic fins. The larvae have tadpole-like external gills. The order includes the African (*Protopteridae) and American (*Lepidosirenidae) lungfish, which live only in freshwater basins.

Lepidosiren paradoxa (South American lungfish) See AESTIVATION; LEPIDO-SIRENIDAE.

Lepidosteidae See LEPISOSTEIDAE.

Lepidosteus spatula (alligator gar) See LEPISOSTEIDAE.

Lepilemur (sportive lemur) See MEGA-LADAPIDAE.

Lepisma saccharina (silverfish) See LEPISMATIDAE.

Lepismatidae (suborder Zygentoma, superfamily *Lepismatoidea) A family of insects which includes some common and cosmopolitan household pests (e.g. *Lepisma saccharina*, silverfish), which feed on food scraps, carbohydrate and dextrin compounds, glues, and sizes. There are about 190 species.

Lepismatoidea (order *Thysanura, suborder Zygentoma) Superfamily whose members resemble the *Pterygota more closely than those of the other thysanuran superfamily, the *Machiloidea. Pterygota and Lepismatoidea have two mandibular articulations, a *gonangulum, and other morphological similarities. *Lepismodes inquilinus* (firebrat) occurs as a pest in human habitations. The superfamily occurs throughout the world.

Lepismodes inquilinus (firebrat) *See* LEPISMATOIDEA.

Lepisosteidae (Lepidosteidae; gars infraclass *Holostei, order *Semionotiformes) A small family of freshwater fish that have a slender, cigar-shaped body and long jaws bearing many teeth. The *anal and single *dorsal fins are located far back on the body, close to the rounded tail fin. Living in shallow, weedy waters, gars have a vascularized *swim-bladder that enables them to take in air when necessary. Found only in N. America, they range from Quebec Province to Costa Rica and, in the case of *Lepisosteus* (*Lepidosteus*) *spatula* (alligator gar), reach a length of 3 m. There are about seven species.

Lepisosteus spatula (alligator gar) *See* LEPISOSTEIDAE.

Leporidae (cohort *Glires, order *Lagomorpha) The family that includes the rabbits, cottontails, and hares. These are lagomorphs in which the tail is reduced, the hind legs are modified for jumping, and the ears are usually long. Rabbits are adapted for burrowing, and their young are born in burrows, naked and blind. Hares are born above ground, their eyes open, and fully furred. Cottontails do not burrow, but may use burrows dug by other animals. There are eight genera. They are distributed widely throughout the *Holarctic region, where they are highly successful (there are more than 30 species), but are less common in Africa (about eight species) and S. America (two species).

Lepospondyli (class *Amphibia) A group of small, *Palaeozoic amphibians in which the vertebrae have spool-shaped centra (*see* CENTRUM) below the *neural arches. The centra were each perforated lengthwise, providing a channel for the *notochord. The lepospondylous, or 'husk', vertebra also

occurs in living amphibians, but its evolutionary derivation is unclear.

leprose Consisting of or bearing powdery or scurfy granules.

Leptanillinae (suborder *Apocrita, family *Formicidae) Little-known subfamily of tiny, tropical ants, which have reduced eyes, and slender bodies adapted for foraging underground. They are believed to be *legionary since the queen is *dichthadiiform, as are queens in the *Dorylinae. There are about 30 species.

Leptobramidae (beach salmon; subclass *Actinopterygii, order *Perciformes) A monospecific family (*Leptobrama muelleri*) of marine and brackish-water fish that have a deep and compressed body 30 cm long, fairly small eyes, a short *dorsal fin placed above the *anal fin, and a large mouth. It is found only in the coastal waters of New Guinea and tropical Australia.

leptocephalus larva A type of larva, found only in certain species of marine bony fish, whose appearance is different from that of the adult. In the case of the eel (*Anguilla*), for example, the larva has a small head with tiny eyes, and the body is shaped like a willow leaf and is virtually transparent; the dorsal and ventral portions of the body divided by the *notochord and spinal cord are of approximately the same size. Only after *metamorphosis does the fish begin to look like an adult eel.

Leptocephalus pseudolatissimus *See* EURYPHARYNGIDAE.

Leptodactylidae (class *Amphibia, order *Anura) A varied family of frogs, including arboreal and burrowing forms, that are adapted to arid conditions. Metamorphosis is rapid and they have a well-developed capacity to retain water. The vertebrae are *procoelous. *Rhinoderma darwinii* (Darwin's frog or the mouth-breeding frog) of Chile and Argentina is small (up to 2.5 cm long), has a soft outgrowth from the tip of its snout, and its eggs develop into froglets inside the vocal pouch of the male. The family includes *Sminthillus limbatus,* the world's smallest frog (body length 1 cm), but many species are large and voracious. There are about 650 species occurring in Australia, tropical America, and southern Africa.

leptodermous Thin-skinned or thin-walled.

leptokurtic Applied to a distribution that is more peaked than a Gaussian distribution (i.e. a few points occur far from the origin, but most are very close to it).

Leptolinida See HYDROIDA.

Leptomedusae See CALYPTOBLASTINA.

Leptopoecile (titwarblers) See REGULIDAE.

Leptoptilos (storks) See CICONIIDAE.

Leptoscopidae (sandfish; subclass *Actinopterygii, order *Perciformes) A small family of marine fish that have very long *dorsal and *anal fins which are without spines. The small ventral fins are situated in front of the much larger *pectoral fins. Sandfish have a large mouth, with the lower jaw longer than the upper jaw. Found in very shallow water, *Crapatalus arenarius* (9 cm) is able to bury itself quickly in the sand. There are about three species, found in the coastal waters of Australia.

Leptosomatidae (coural, cuckoo-roller; class *Aves, order *Coraciiformes) A monospecific family (*Leptosomus discolor*), which is a large, mainly grey bird. The male has metallic green and purple wings, the female is barred and spotted brown and cream. *L. discolor* has a stout bill, a reversible fourth toe, and a forehead crest which curves forward, partly covering the bill. It inhabits forest and scrub, feeds on insects and lizards, and nests in a tree hole. It is confined to Madagascar and the Comoro Islands.

leptotene See MEIOSIS.

Leptotyphlopidae (slender blind snakes, thread snakes order *Squamata,

suborder *Serpentes) A family of snakes which are similar in appearance and habits to the *Typhlopidae but which have teeth only on the lower jaw and spurs behind the anus (the remnants of hind limbs). There are 40 species, occurring in Africa and tropical America.

lesser rorqual (minke whale; *Balaenoptera acutorostrata*) See BALAENOPTERIDAE.

Lestidae (damselflies; order *Odonata, suborder *Zygoptera) Ubiquitous family of damselflies, comprising insects which hold their wings partly open when they are at rest. The larvae have elongate labia (*see* LABIUM), and are typically found in vegetation in still water. More than 180 extant species have been described, with a cosmopolitan distribution.

Lestoideidae (Pseudolestidae, damselflies; order *Odonata, suborder *Zygoptera) Family of unusual damselflies, which shows many features common to other families, but whose members can be recognized by the reduced first and second anal wing vein, and from the third intercullary, and fourth branch of the radius, arising midway between the arculus and subnodus in the wing. The larvae are short, with saccoid or lamellar *caudal gills. There are 14 described extant species, found in the eastern Palaearctic, south-east Asia, Australasia, and Central and S. America.

lethal mutation A *gene *mutation whose expression results in the premature death of the organism carrying it. Dominant lethals kill *heterozygotes, recessive lethals kill *homozygotes only. See also VISIBLE.

Lethrinidae (emperor; subclass *Actinopterygii, order *Perciformes) A small family of marine, tropical fish that resemble the *Pomadasyidae, being rather deep-bodied fish, often brightly coloured, with a continuous *dorsal fin, a pronounced or pointed snout, scaleless cheeks, and strong 'canine' teeth in the jaws. *Lethrinus chrysostomus* (sweetlips emperor), 90 cm, among the largest species, is found in the

Indo-Australian region. There are about 21 species, found in the Indo-Pacific region and in W. Africa.

Lethrinus chrysostomus (sweetlips emperor) *See* LETHRINIDAE.

Leucettida (class *Calcarea, subclass *Calcinea) An order of sponges whose body possesses a *sycon or *leucon structure. The dermis is present and in some forms a *cortex may occur as well.

leucine An aliphatic, *non-polar, neutral *amino acid. Unlike most amino acids it is sparingly soluble in water.

leucocyte (leukocyte) A white blood cell, present in invertebrate and vertebrate animals. There are three types: lymphocytes, amoeboid, non-*phagocytic cells that produce or convey *antibodies; monocytes, phagocytic cells that ingest invading organisms, sometimes entering tissue to do so; and polymorphs (polymorphonuclear leucocytes, or granulocytes), also phagocytic cells, with granular *cytoplasm and lobed nuclei.

leucon Applied to the body structure of a sponge if it is complex and consists of many chambers.

Leucophaea maderae (Madeira cockroach) *See* BLABERIDAE.

Leucosolenida (class *Calcarea, subclass *Calcaronea) An order of sponges whose body possesses an *ascon structure.

Leucospidae (suborder *Apocrita, superfamily *Chalcidoidea) Family of wasps which are similar to the *Chalcidae but larger (8–15 mm long) and which have a somewhat hump-backed appearance. The family is sometimes included in the Chalcidae, but its members hold the wings folded longitudinally at rest through the retention of wing coupling. Adults are brightly coloured black and yellow, and the females have a curved *ovipositor which lies *anteriad along the dorsal surface of the abdomen. Leucospids are usually found on

flowers and are parasitic on various species of wasp and bee.

leukocyte *See* LEUCOCYTE.

levulose *See* FRUCTOSE.

LH *See* LUTEINIZING HORMONE.

Lias A name commonly given to the lowest stage of the *Jurassic, 199.6–175.6 Ma ago, with extensive outcrops of blue-grey shales and muddy limestones in Britain and France in which many fossils have been found.

Libellulidae (dragonflies; order *Odonata, suborder *Anisoptera) Ubiquitous family of dragonflies, of most recent and tropical origin, which contains about one-quarter of all the known species of living Odonata. These dragonflies are usually colourful, and are seen frequenting bodies of still water. The short, stout nymphs are found swimming in still water, where they hunt in the bottom mud. There are about 3750 species, with a cosmopolitan distribution.

library (gene library) A random collection of cloned (*see* CLONE) *DNA fragments in a number of *vectors that ideally includes all the genetic information of that species.

Libytheidae (snout butterflies; subclass *Pterygota, order *Lepidoptera) Small family of butterflies in which the palps project from the front of the head like a 'snout'. They have small, angular fore wings. The fore legs of males are reduced, but in females they are fully developed. They are closely related to the *Lycaenidae, and sometimes treated as a subfamily of the *Riodinidae. They are distributed world-wide.

lice *See* PHTHIRAPTERA.

Lidth's jay (*Garrulus lidthi*) *See* CORVIDAE.

life cycle The series of developmental changes undergone by the individuals comprising a population, including fertilization, reproduction, and the death of those individuals and their replacement by a new generation. The life 'cycle' in fact is linear with respect to individuals, but cyclical with respect to populations. In many animals there is a succession of individuals in the entire cycle with sexual or asexual production linking them. In vertebrates, the life cycle is confined to the period from fusing of the *gametes to the death of the resulting individuals.

life form The structure, form, habits, and life history of an organism.

life table A table that displays the mortality data of a population according to age groups.

ligament A band of tissue that holds together adjacent bones in a vertebrate and *valves in an invertebrate.

ligand An atom, ion, or molecule that acts as the electron-donor partner in one or more co-ordination bonds. A heterocyclic ring is formed if the ligand is an organic compound, and the product is termed a chelate.

ligase An enzyme that catalyses a reaction that joins two *substrates, using energy derived from the simultaneous *hydrolysis of a *nucleotide triphosphate.

lightning beetle *See* LAMPYRIDAE.

Ligiidae (rock slaters, sea slaters; class *Malacostraca, order *Isopoda) Family of isopods whose members live in middle- and upper-shore regions, hiding during the day in cracks and crevices (e.g. in harbour walls), and beneath stones and seaweed. They are particularly active at night, can run fast when disturbed, and are widely distributed and abundant.

limb borer *See* BOSTRICHIDAE.

limiting factor (ecological factor) Any environmental condition or set of conditions that approaches most nearly the *limits of tolerance (maximum or minimum) for a given organism. It was defined originally as the essential material that is available in an amount most closely approaching the critical minimum needed, but the term is now used more generally.

limits of tolerance The upper and lower limits to the range of particular environmental factors (e.g. light, temperature, availability of water) within which an organism can survive. Organisms with a wide range of tolerance are usually distributed widely, while those with a narrow range have a more restricted distribution. *See also* SHELFORD'S LAW OF TOLERANCE.

limnetic zone The area in more extensive and deeper freshwater *ecosystems that lies above the depth at which light penetration is reduced to a level where the amount of oxygen produced by photosynthesis balances the amount consumed by respiration, and beyond the *littoral zone. This zone is inhabited mainly by *plankton and *nekton with occasional *neuston species. The limnetic and littoral zones together comprise the *euphotic (well-illuminated) zone. In very small and shallow lakes or ponds the limnetic zone may be absent.

Limnognathia maerski *See* MICROGNATHOZOA.

limnology The study of freshwater ecosystems, especially lakes.

Limosa **(godwits)** *See* SCOLOPACIDAE.

limpet (*Patella vulgata*) *See* ARCHAEOGASTROPODA.

limpkin (*Aramus guarauna*) *See* ARAMIDAE.

Lincoln index A simple estimate of animal population density, based on data obtained by *mark–recapture techniques.

(((⊕))) SEE WEB LINKS
• Explanation of the Lincoln index technique.

Lindeman's efficiency (gross ecological efficiency) The ratio of energy assimilated at one *trophic level to that assimilated at the preceding trophic level; the ratio of energy intake at successive trophic levels. It is one of the earliest and most widely applied measures of *ecological efficiency.

lineage See EVOLUTIONARY LINEAGE.

line transect A tape or string laid along the ground in a straight line between two poles as a guide to a sampling method used to measure the distribution of organisms. Sampling is rigorously confined to organisms that are actually touching the line. *Compare* BELT TRANSECT.

ling (*Molva molva*) See GADIDAE.

lingua In insects, a fleshy, horny, or membranaceous, tongue-like structure forming the anterior part of the *labium.

lingual Of the tongue.

Lingulacea (class *Inarticulata, order *Lingulida) A superfamily of brachiopods that have shells made from calcium phosphate. The *pedicle valve, bearing the pedicle groove, is usually larger than the *brachial *valve. Lingulacea appeared in the Early *Cambrian.

Lingulida (phylum *Brachiopoda, class *Inarticulata) An order of brachiopods that have valves of calcium phosphate with some layers of organic material. The valves may be finely punctate or impunctate. The *pedicle emerges between the valves posteriorly. Lingulida are usually marine, but some are tolerant of reduced salinity. They first appeared in the Early *Cambrian. There are two superfamilies, comprising eight families, and about 50 genera.

linkage The association of *genes that results from their being on the same *chromosome. Linkage is detected by the greater association in inheritance of two or more non-allelic genes than would be expected from *independent assortment. The nearer such genes are to each other on a chromosome, the more closely linked they are, and the less often they are likely to be separated in future generations by *crossing-over. All

the genes in one chromosome form one linkage group.

linkage disequilibrium The non-random association of *alleles at different *gene loci (*see* LOCUS) in a population (e.g. when two loci occur close together on the same *chromosome and selection operates to keep the allele combinations together).

linkage equilibrium See GAMETIC EQUILIBRIUM.

linkage group See LINKAGE.

linkage map An abstract map of chromosomal *loci, based on experimentally determined recombinant frequencies, which shows the relative positions of the known genes on the *chromosomes of a particular species. The more frequently two given characters recombine, the further apart are the genes that determine them.

Linnaeus, Carolus (1707–78) The Swedish botanist and physician who became lecturer in botany at the University of Uppsala in 1730 and who, in 1735, published his *Systema naturae.* The 10th edition of this, published in 1758, is taken as the starting-point of the method of *binomial classification. As he wrote in Latin, the latinized form of his name has often been mistaken as his real name. In 1761 he was raised to the nobility (the honour was backdated to 1757), from which time he was known as Carl von Linné.

(((🌐))) SEE WEB LINKS
• Biography of Carolus Linnaeus.

Linognathidae (order *Phthiraptera, suborder *Anoplura) Family of sucking lice which are parasitic on members of the mammalian orders *Artiodactyla, *Perissodactyla, *Carnivora, and *Hyracoidea. Like *Haematopinidae, these lice lack eyes, but they also lack the prominent *ocular points characteristic of the other family. The family comprises 69 species in three genera, of which one, *Prolinognathus*, is sometimes placed in its own family. This genus contains all the species of *Anoplura* parasitic on hyraxes, other lice found on hyraxes belonging to the *Ischnocera (family *Trichodectidae).

linoleic acid A polyunsaturated *fatty acid containing two double bonds and comprising 18 carbon atoms. It is classed as an essential fatty acid since it is required by mammals but cannot be synthesized by them. It is a precursor of *prostaglandins.

linolenic acid $C_{17}H_{29}COOH$, an essential *fatty acid that forms a series with its three *isomers. It is found in linseed oil and all *drying oils, and can be synthesized from *linoleic acid.

Linophrynidae (monster anglerfish; superorder *Paracanthopterygii, order *Lophiiformes) A small family of deep-sea anglerfish that have a bizarre appearance. They have a short and almost round body, a very large mouth with needle-like teeth, and a strongly developed *illicium. The fins are small, with the exception of the larger, rounded tail fin. Females may possess a long tassel-like or branching *barbel under the chin. The mature male is tiny and remains attached to the body of the female until spawning. There are about 20 species, distributed world-wide.

Linyphiidae (money spiders; order *Araneae, suborder Araneomorphae) Family of small, uniformly black or brown spiders which construct sheet webs among vegetation. The webs do not have a retreat, as do those constructed by the *Agelenidae; the adult spider hangs on the underside of the web and runs over it to catch prey. The *chelicerae have many teeth, and the legs bear many strong *setae. The members of this large family are among the commonest and most abundant species in the northern hemisphere (40% of the British spider fauna belonging to the family). Many species travel long distances by 'ballooning' on long threads of silk.

lionfish (*Pterois volitans***)** See SCORPAENIDAE.

lipase An *enzyme that catalyzes the formation or *hydrolysis of *lipids.

Liphistiidae (order *Araneae, suborder Mesothelae) Family of spiders which have the normal *carapace covering the *prosoma, but in which the abdomen is distinctive in being clearly segmented. The spiders are 1–3.5 cm long, and have two pairs of *book lungs. They sit in tubes up to 60 cm deep, dug out from the substrate, and use their *chelicerae to hold shut a trapdoor. A few radial threads spread out from the tube to trip prey items and alert the concealed spider. The family occurs from south-east Asia to southern Japan.

lipid A member of a heterogeneous group of small organic molecules that are sparingly soluble in water, but soluble in organic solvents. Included in this classification are fats, oils, waxes, *terpenes, and *steroids. The functions of lipids are equally diverse and include roles as energy-storage compounds, as *hormones, as *vitamins, and as structural components of cells, particularly membranes.

lipofuscin (age pigment) The dark-pigmented material that accumulates in the *lysosomes of older animal cells. It may be partly responsible for the ageing of such cells.

lipoic acid (6–8-dithio-n-octanoic acid) A compound that functions as a *coenzyme in the oxidative decarboxylation of pyruvic acid to alpha-ketoglutaric acid and of the latter to succinic acid. It is sometimes classed as a *vitamin as it is an essential requirement for some micro-organisms.

lipoprotein A water-soluble, *conjugated protein in which the *prosthetic group is a *lipid. Lipoproteins transport lipids in the blood and lymph from the small intestine to the *liver and from the liver to fat deposits.

Lipotes vexillifer **(baiji)** See PLATANISTOIDEA.

Lipotyphla (cohort *Unguiculata, order *Insectivora) A suborder (or, more correctly, full order) that includes most extant insectivores, grouped into two superfamilies, *Erinaceoidea, the hedgehogs and moon rat, and *Soricoidea, the shrews and moles. Most are small, nocturnal animals showing many features, especially of the skull and teeth, similar to those of early fossil eutherians, for which reason the insectivores are held to be evolutionarily primitive.

Lissamphibia (class *Amphibia) A subclass that comprises all the extant amphibians which are thought to be descended from the *Lepospondyli. *See* URODELA; ANURA; APODA.

listroceline grasshoppers *See* TETTIGONIIDAE.

Lithisida (class *Demospongiae, subclass *Tetractinomorpha) An order of sponges which possess bulky, intertwined *spicules of silica. The order is rather artificial, containing members from several orders with this type of skeleton. They are found mainly in deep water below the *photic zone. They are first recorded from the *Cambrian.

litho- A prefix meaning 'pertaining to rock or stone', from the Greek *lithos*, 'stone'.

Lithobiomorpha *See* CHILOPODA.

Litopterna (cohort *Ferungulata, superorder *Protoungulata) An extinct order of S. American ungulates that lived from the *Eocene to the *Pleistocene and are considered to be descendants of the *Condylarthra. They were hoofed, and the central metapodials became elongated and the lateral ones reduced to leave three toes and, in *Thoatherium*, one. The *incisors were reduced in some forms, but in most dentition was complete. One line developed remarkable 'pseudo-horses' (e.g. *Diadiaphorus* and *Thoatherium*) which strongly resembled their perissodactyl contemporaries. The other major lineage culminated in *Macrauchenia*, which strongly resembled the N. American camels, apart from the possible presence of a short, tapir-like trunk. From the presence of a nasal opening on the top of the skull, however, other authorities have inferred that *Macrauchenia* followed a life in water, rather than one that involved a trunk.

litter layer The layer of organic material that lies on the surface of the soil.

littoral Applied to the shallow-water regions of aquatic *ecosystems in which rooted plants occur and light penetrates to the bed. In the marine littoral zone periodic exposure and submersion by tides is usual; in Britain it is defined as the upper limit of periwinkles (*Littorina*). *Compare* EULITTORAL.

littoral fish 1. (freshwater) Those fish that are found along the shores of a lake from the edge of the water down to the limits of rooted vegetation. **2. (marine)** Those fish that are found in the intertidal zone of the sea-shore, most of the fish moving in and out with the tide.

live-bearing fish Fish in which the fertilized eggs are retained initially within the body of the female, but are subsequently released as newly hatched larvae (e.g. some *Scorpaenidae) or born as free-swimming *postlarvae in which the *yolk has been fully utilized and the *yolk sac has disappeared (e.g. some sharks, and *Poeciliidae).

liver In vertebrates, a large gland, richly supplied with blood, that arises from the intestine. It is concerned with the detoxification of blood; with the storage of sugars, vitamins, and other food substances; with aiding digestion; with the production of *proteins and *antibodies; and with the removal of wastes.

living fossil A member of a living species of animals that are almost identical to species known from the fossil record (not the recent fossil record); i.e. that have changed very little over a long period. For example, *Latimeria chalumnae* (coelacanth) is close to Upper *Palaeozoic genera; *Xiphosura* species (horseshoe crab) are close to middle Palaeozoic genera; and *Tapirus pinchaque* (Andean tapir) is close to a *Miocene species. Note, however, that none of these species is identical to any fossil taxon; evolution has been slow, but not absent.

lizard *See* SAURIA.

lizardfish *See* SYNODONTIDAE.

llama (*Lama glama*) *See* CAMELIDAE.

loach goby *See* RHYACICHTHYIDAE.

Lobata (phylum *Ctenophora, class *Tentaculata) An order of ctenophorids which have laterally compressed bodies. The two main *tentacles are generally absent in mature forms. Small lateral tentacles are present.

Lobatocerebromorpha A *phylum of three species, discovered in 1980 and proposed as a phylum in 1991 by G. Haszprunar, R. M. Rieger, and P. Schuchert, who suggested it may be related to the *Annelida and *Sipunculida (peanut worms); the Lobatocerebromorpha is now classed as an unranked *clade and a sister *taxon to Annelida, *Mollusca, *Nemertea, and Sipuncula. Lobatocerebromorphs have a ciliated body (*see* CILIUM) with a *cuticle, *ventral nerve cords, segmentally arranged protonephridia (*see* NEPHRIDIUM), and a posterior *anus. The *parenchyma has very little intercellular matrix, but there is no true *coelom. The organisms are 1 mm long, worm-like, and *hermaphrodite, with no vascular system, but with an unexpectedly large brain.

Lobopodia A suggested phylum to contain the *Onychophora (walking worms), as well as the subphyla *Tardigrada and *Pentastomida, removing them from the *Arthropoda, which is now considered to be monophyletic (*see* MONOPHYLY). Fossil lobopodians occur only in the early *Cambrian.

(SEE WEB LINKS)
• Description of Lobopodian fossils

lobopodium A type of *pseudopodium which is broad with a blunt, rounded tip.

Lobotidae (tripletail; subclass *Actinopterygii, order *Perciformes) A small family of marine fish that have a large and heavy body. Characteristically, the rounded *dorsal, *anal, and tail fins are of similar size and give the fish the appearance of having a trilobed tail. The best-known species is *Lobotes surinamensis* (1 m), a sluggish fish found in coastal waters of tropical seas. Probably it is a cosmopolitan species. There are about four species, with some southeast Asian species found in brackish-water or freshwater habitats.

lobsters 1. *See* CRUSTACEA; DECAPODA; NEPHROPIDAE. **2.** (spiny lobsters) *See* PALINURIDAE.

loci *See* LOCUS.

lock-and-key theory A theory to explain the mechanism of enzymatic reactions, in which it is proposed that the *enzyme and *substrate(s) bind temporarily to form an enzyme–substrate complex. The binding site on the enzyme is known as the 'active site' and is structurally complementary to the substrate(s). Thus the enzyme and substrate(s) are said to fit together as do a lock and a key.

locus (*pl.* loci) The specific place on a *chromosome where a *gene is located. In *diploids, loci pair during *meiosis and unless there have been translocations, inversions, etc., the *homologous chromosomes contain identical sets of loci in the same linear order. At each locus is one gene; if that gene can take several forms (*alleles), only one of these will be present at a given locus.

locust (order *Orthoptera, family *Acrididae) Name given to several species of acridids which show density-related changes in their morphology and behaviour. At low population densities the insects develop as solitary, cryptically coloured grasshoppers (phase *solitaria*). At higher densities, such as may result from an abundance of food after rain, the insects develop into gregarious, brightly coloured individuals, which swarm and migrate, often causing great destruction to vegetation (phase *gregaria*). Major species include *Locusta migratoria* (migratory locust), *Schistocerca gregaria* (desert locust), and *Nomadacris septemfasciata* (red locust).

Locusta migratoria (migratory locust) *See* LOCUST.

Locustella (grasshopper warblers) *See* SYLVIIDAE.

loggerhead turtle (*Caretta caretta*) *See* CHELONIIDAE.

logistic equation (logistic model) A mathematical description of growth rates for a simple population in a confined space with limited resources. The equation summarizes the interaction of *biotic potential with environmental resources, as seen in populations showing the *S-shaped growth curve, as: $dN/dt=rN(K-N)/K$ where N is the number of individuals in the population, t is time, r is the biotic potential of the organism concerned, and K is the saturation value or *carrying capacity for that organism in that environment. The resulting growth rate or logistic curve is a parabola, while the graph for organism numbers over time is sigmoidal. *Compare* J-SHAPED GROWTH CURVE.

⊕ SEE WEB LINKS
• Detailed explanation of the logistic equation.

logistic model *See* LOGISTIC EQUATION.

Lonchura (mannikins, munias, silverbills) *See* ESTRILDIDAE.

Lonchura striata (Bengalese finch) *See* ESTRILDIDAE.

long call *See* FORWARD DISPLAY.

longevity The persistence of an individual for longer than most members of its *species, or of a genus or species over a prolonged period of geological time.

longhorn beetle *See* CERAMBYCIDAE.

long-horned grasshopper *See* TETTIGONIIDAE.

longicorn beetle *See* CERAMBYCIDAE.

Longisquama insignis One of the first archosaurs (*Archosauria) that is known to have been able to glide or parachute. Described from Soviet Kyrgyzstan in the 1970s, *Longisquama* is noted for the development of elongate paired scales along its back, the anterior (*nuchal) ones resembling feathers. It is recorded from sediments of *Triassic age.

long-legged bat (*Natalus*) *See* NATALIDAE.

long-legged flies *See* DOLICHOPODIDAE.

longneck eels *See* DERICHTHYIDAE.

longnose chimaera *See* RHINOCHIMAERIDAE.

long-nosed batfish (*Ogcocephalus vespertilio*) *See* OGCOCEPHALIDAE.

long-nosed viper (*Viper ammodytes*) *See* VIPERIDAE.

longspurs *See* EMBERIZIDAE.

long-tailed groundroller (*Uratelornis chimaera*) *See* BRACHYPTERACIIDAE.

long-tailed reef-eel (*Thyrsoidea macrura*) *See* MURAENIDAE.

long-tailed tits (*Aegithalos*) *See* AEGITHALIDAE.

long tom *See* BELONIDAE.

long-whiskered catfish *See* PIMELODIDAE.

Longworth trap A metal (usually aluminium) trap that is used to collect small mammals without injuring them. It consists of a nest-box, in which appropriate nesting material and food are placed, and a tunnel leading into it. As an animal enters it trips a door which falls shut so it cannot escape and must remain in the nest-box.

loopers *See* GEOMETRIDAE.

loop of Henlé Part of a *nephron that filters solutes. It comprises a descending arm connected to the *medulla of the *kidney and a thin and thick ascending arm connected to the *cortex. Water is filtered out in the descending arm and enters the interstitial fluid in the medulla. Salts are filtered out in the thin ascending arm, which is permeable to solute, especially sodium and chloride ions, but impermeable to water. The solute ions are transported along the thick ascending arm. This produces a region of high salt (urine) concentration in the medulla, and a concentration gradient. The kidney's collecting duct passes through this region. Water flows out of the collecting duct, so water is reabsorbed and urine is concentrated. The loop was discovered by Friedrich *Henlé. See COUNTER-CURRENT EXCHANGE.

loosejaw See MALACOSTEIDAE.

loph On molar teeth, a ridge connecting cusps.

Lophiidae (goosefish, angler; superorder *Paracanthopterygii, order *Lophiiformes) A small family of marine fish that have a huge, wide, flattened head, an enormous mouth whose lower jaw is longer than the upper, and eyes placed almost on top of the head. The fins are short and truncate. The goosefish are bottom-dwelling, and lure prospective prey by means of the movable first dorsal *fin ray, which is provided with an *illicium. Found at depths of 100–1000 m, *Lophius piscatorius* (European angler) can reach a length of 1.7 m. Encountered regularly by commercial fishermen, the annual catch is of the order of 30 000 tonnes. *Lophius americanus* (goosefish) of the western Atlantic is slightly smaller. There are at least 12 species, with world-wide distribution but absent from the eastern Pacific.

Lophiiformes (subclass *Actinopterygii, superorder *Paracanthopterygii) An order of marine, bottom-dwelling fish that have a relatively large head, wide mouth, short, tapering body and tail, and short, rounded fins. The *illicium is typical. The order includes 14 families.

Lophiodontidae (order *Perissodactyla, suborder *Ceratomorpha) A family of ceratomorphs that are known only from *Eocene sediments and were closely related to the tapirs.

Lophius americanus (goosefish) *See* LOPHIIDAE.

Lophius piscatorius (European angler) *See* LOPHIIDAE.

lophodont Applied to cheek teeth (*molars), found in some *Mammalia, in which the cusps are fused to form transverse ridges (lophs) that aid the mastication of plant material. *See also* BILOPHODONTY.

Lophophorata Coelomates that have two longitudinal coelomic compartments (*see* COELOM) separated by a *septum. There are three living lophophorate phyla, *Phoronida, *Brachiopoda, and *Bryozoa, and a number of extinct phyla.

lophophorate Applied to animals, such as *Brachiopoda, that possess *lophophores, used to collect food.

lophophore In invertebrates (e.g. *Brachiopoda and *Bryozoa), the circular or horseshoe-shaped feeding organ composed of ciliated *tentacles surrounding the mouth. Each tentacle is hollow and contains an outgrowth of the *coelom which keeps it rigid. The *cilia drive a current of water through the lophophore, and *plankton is collected in the process.

Lophotidae (crestfish; subclass *Actinopterygii, order *Lampridiformes) A small family of open-ocean fish that have an elongate, greatly compressed body, with the very long *dorsal fin commencing above the upper jaw and extending all the way to the tail. The *pelvic fins are minute. They have an ink sac containing a dark pigment which can be released through the *cloaca when required. *Lophotus capellei* (crestfish) of the eastern Pacific may reach a length of 80 cm. There are probably two species, distributed world-wide.

Lophotus capelei (crestfish) *See* LOPHOTIDAE.

Lord Derby's zonure *(Cordylus gigan-teus) See* CORDYLIDAE.

lordosis 1. In some mammals (*Mammalia), including cats (*Felidae) and mice (*Muridae), a sexual response in which the female bends her back downwards, which raises her pelvis to allow intercourse. **2.** In humans, curvature of the spine.

Lorenz, Konrad (1903–89) An Austrian zoologist who was joint winner (with Karl *von Frisch and Nikolaas *Tinbergen) of the 1973 Nobel Prize for Physiology or Medicine for their studies of animal behaviour. Lorenz worked for some years at a centre for behavioural physiology built for him by the Max Planck Institute at Seewiesen, Germany, and it was there he conducted the studies of *imprinting, especially with geese, for which he became widely known.

(((()) SEE WEB LINKS)
• Biography of Konrad Lorenz.

lorica A loose-fitting, shell-like covering which encloses the cells of some *Protozoa.

Loricariidae (armoured catfish; subclass *Actinopterygii, order *Siluriformes) A very large family of herbivorous, freshwater catfish that have the head and part of the body enclosed in bony plates; the posterior part of the body is very slender. The ventrally located mouth is modified into a sucking apparatus which is used by the fish to fasten on to rocks in fast-flowing mountain streams. There are about 410 species, occurring in Central and S. America.

Loricata *See* POLYPLACOPHORA.

Loricifera A *phylum, first described in 1983 by R. M. Kristensen (1948–), of animals only 0.25 mm long that live in marine gravel. The body bears a *cuticle of four plates (one *dorsal, one *ventral, two lateral) and has *recurved spines on the anterior end. The *anus is terminal. Sexes are separate; larvae resemble small adults. The phylum appears to be related to the *Priapulida (priapus worms) and *Kinorhyncha.

Loridae (formerly Lorisidae; lorises; order *Primates, suborder *Strepsirrhini) A family that includes the lorises, pottos, and galagos (bush-babies). They are related to Malagasy lemurs (*Lemuriformes), compared with which they have on average shorter muzzles, higher brain cases, and eyes that are directed more forward. They are sometimes included in a separate infraorder, *Loriformes (formerly Lorisiformes). In the African potto (*Perodictus potto*) and two species of angwantibo (*Arctocebus*) the index fingers are reduced in size, so providing a wider grip between the thumb and other digits. The lorises of Asia (eight species of slender lorises (*Loris*) and seven extant species of slow lorises (*Nycticebus*)) are slow-moving, arboreal, and nocturnal; they feed largely on invertebrates that are unpalatable to other predators, detecting them by smell. The bush-babies (three *Otolemur* spp., two *Euoticus* spp., and 15 *Galago* spp.) are quick-moving and feed on insects (some of which they catch in flight), small mammals, and some fruit and other plant material. The Loridae are distributed throughout the Old World tropics, apart from Madagascar, as far east as Indonesia. Formerly about 11 species were recognized, usually classed in six genera, but a number of new species of bush-babies have recently been described, and in 1966 J. H. Schwartz described a new genus and species from Cameroon, *Pseudopotto martini*, related to the potto and angwantibo.

lories *(Lorius) See* PSITTACIDAE.

Loriformes (formerly Lorisiformes; order *Primates, suborder *Strepsirrhini) The name sometimes given to a group comprising the single family Lorisidae, to contrast them with the Malagasy lemurs (*Lemuriformes).

loris *See* LORIDAE.

Lorisidae *See* LORIDAE.

Lorisiformes *See* LORIFORMES.

Lorius (lories) *See* PSITTACIDAE.

Lota lota (burbot) *See* GADIDAE.

Iotic Applied to habitats formed in running water (i.e. rivers and streams).

Lotka–Volterra equations Mathematical models of competition between resource-limited species living in the same space with the same environmental requirements. They have been modified subsequently to simulate simple predator–prey interactions. The competition model predicts that coexistence of such species populations is impossible; one is always eliminated, according to the *competitive-exclusion principle. The predation model predicts cyclic fluctuations of predator and prey populations. Reduction of predator numbers allows prey to recuperate, which in turn stimulates the population growth of the predator. Increasing predator numbers depress the prey population, leading eventually to a reduction in the predator population.

(⊕) SEE WEB LINKS
• Explanation of the Lotka–Volterra equations.

louse *See* PHTHIRAPTERA.

louvar (*Luvarus imperialis***)** *See* LUVARIDAE.

Lovettia sealii **(Derwent whitebait)** *See* APLOCHITONIDAE.

Loxia **(crossbills)** *See* FRINGILLIDAE.

Loxioides **(Hawaiian honeycreepers)** *See* DREPANIDIDAE.

Loxodonta africana **(African elephant)** *See* ELEPHANTIDAE.

Loxodonta cyclotis **(African forest elephant)** *See* ELEPHANTIDAE.

lubber grasshopper *See* ACRIDIDAE.

Lucanidae (stag beetles; subclass *Pterygota*, order *Coleoptera*) Family of large, shiny, brown or black beetles, 10–66 mm long, in which the *mandibles of males are often developed into 'antlers', which are probably used in fighting between males.

The antennae are elbowed, with clubs of flattened, expanded segments. There is a wide size range within species, females generally being smaller. The larvae are curved, and fleshy. Their antennae are well developed, their legs slender. They live in rotting wood, taking 3–6 years to develop. They *stridulate by rubbing the last two pairs of legs together. Adults feed on sap and liquids. There are 750 species.

Lucernarida *See* STAUROMEDUSIDA.

luciferase In the *bioluminescence reactions of animals such as the firefly, the enzyme that catalyses the oxidation of the substrate luciferin, with the consequent release of visible light.

luciferin *See* LUCIFERASE.

Luciocephalidae (pikehead; subclass *Actinopterygii*, order *Perciformes*) A monospecific family (*Luciocephalus pulcher*) of predatory, surface, freshwater fish, about 20 cm long, that have an elongate, nearly cylindrical body, a very pronounced or even pointed snout, and large eyes. The *dorsal and *anal fins are placed far back, with the anal fin showing a deep notch. The fish has an *accessory respiratory organ. It is found in Malaysia.

Luciocephalus pulcher **(pikehead)** *See* LUCIOCEPHALIDAE.

lugworms *See* SEDENTARIA.

lumbar vertebra *See* VERTEBRA.

lumen A cavity inside a cell or bodily structure.

lumpsucker *See* CYCLOPTERIDAE.

lunate Half-moon shaped.

lung 1. In terrestrial *Mollusca, a highly vascular part of the *mantle involved in respiration. **2.** In air-breathing vertebrates, the respiratory organ. It is probably derived from an *accessory respiratory organ of aquatic vertebrates inhabiting oxygen-depleted fresh water: this developed into the *swim-bladder of many fish. The lung is present embryonically as a diverticulum of the gut. It contains many *alveoli across whose walls oxygen and carbon dioxide diffuse.

lungbook See BOOK LUNG.

lungfish See AESTIVATION; LEPIDOSIRENI-DAE; LEPIDOSIRENIFORMES.

lungless salamanders See PLETHO-DONTIDAE.

lunule 1. In many sand-dollars (*Echinoidea), a perforation in the *test. **2.** In *Bivalvia, a depressed area along the hinge line, anterior to the *umbo.

Luscinia See TURDIDAE.

Luscinia megarhynchos (nightingale) See TURDIDAE.

luteinizing hormone (LH) A *gonadotropic protein hormone, secreted by the adenohypophysis (anterior *pituitary), that stimulates in the male the production of *testosterone and in the female the final ripening and rupture of the ovarian follicles. *See also* CORPUS LUTEUM.

Lutjanidae (snapper; subclass *Actinopterygii, order *Perciformes) A very large family of mainly marine fish that have rather deep bodies, a continuous, slightly notched *dorsal fin, and a slightly forked tail fin. The jaws bear large 'canine' teeth. Snappers are important food fish and popular with fishermen, but may sometimes carry *ciguatera poison. *Lutjanus argentimaculatus* (mangrove jack), 91 cm, sometimes enters estuaries along the coast of northern Australia. About 230 species are distributed throughout the tropical seas.

Lutjanus argentimaculatus (mangrove jack) See LUTJANIDAE.

Luvaridae (louvar; subclass *Actinopterygii, order *Perciformes) A monospecific family (*Luvarus imperialis*), living in the open seas, of large fish (up to 2 m) that

have a robust, deep body, blunt head, small mouth, long *dorsal and *anal fins, and a narrow tail with a lunate tail fin. The *anus is located between the minute *pelvic fins, a rather unusual position in fish. The louvar has world-wide distribution in tropical and subtropical seas.

Luvarus imperialis (louvar) See LUVARI-DAE.

lyase An *enzyme that catalyses non-hydrolytic reactions in which groups are either removed or added to a *substrate, thereby creating or eliminating a double bond, especially between carbon atoms ($C{=}C$) or between carbon and oxygen ($C{=}O$).

Lycaenidae (blues, hair-streaks, coppers; subclass *Pterygota, order *Lepidoptera) A very large family of small to medium-sized butterflies. Coloration is often brilliant metallic, and sexual *dimorphism is widespread. 'Tails' are often present on the hind wings. The fore legs of males are slightly reduced. The larvae are frequently associated with ants, and some species prey on aphids and scale insects. Distribution is widespread but mainly tropical and subtropical.

Lycaon pictus (Cape hunting dog) See CANIDAE.

Lycidae (net-winged beetles; subclass *Pterygota, order *Coleoptera) Family of brightly-coloured, distasteful beetles, 5–20 mm long, which are often mimicked by harmless insects. The *elytra have pronounced longitudinal ridges. The *pronotum is expanded and ridged, covering most of the head. The antennae are long and thick, often serrate or pectinate, and placed close together on the head. The eyes are protuberant. Some are sexually *dimorphic, with larviform females. Larvae are usually found in dead trees. Their antennae are much reduced. Both adults and larvae are predatory. There are 3000 species, mostly tropical.

Lycosidae (hunting spiders, wolf spiders; order *Araneae, suborder Araneomorphae) Family of medium to large spiders that run over the ground hunting for their prey. Some genera are burrowing (up to 1 m deep) and others make funnel-shaped webs. Females carry their egg cocoons, which can be quite large, attached to their circularly arranged *spinnerets.

Lyctidae (powder-post beetles; subclass *Pterygota, order *Coleoptera) Family of small, elongate, parallel-sided beetles, distinguished from other woodborers by having a smaller pronotum, which leaves the head visible. The antennae have two-jointed clubs. Larvae are white, fleshy, and C-shaped, with the head largely covered by the *prothorax. They attack the heartwood of deciduous trees, but never conifers, reducing wood to a powdery dust. There are 70 species.

Lydekker's line A line that defines the easternmost extension of oriental animals into the zone of mixing between the *Oriental and *Australian *faunal regions. The corresponding western limit of the zone is known as *Wallace's line, and marks the maximum extent of marsupials (*Marsupialia) in that direction. It is named after the British naturalist Richard Lydekker (1849–1915).

Lygaeidae (order *Hemiptera, suborder *Heteroptera) Family of bugs, many of which feed on ripe seeds, whereas others (e.g. the N. American chinch bug) destroy the shoots of grasses. Some are predators. Most species live on the surface of the ground and many are flightless. There are about 2000 species, distributed world-wide.

Lymantria dispar (gypsy moth) *See* LYMANTRIIDAE.

Lymantriidae (tussock moths; subclass *Pterygota, order *Lepidoptera) Family of generally medium-sized moths. The proboscis is usually absent, and the antennae are *bipectinate in males and usually so in females. Sometimes females are wingless. The larvae are hairy, often brightly coloured, and hairs are frequently woven into the coccoon. *Lymantria dispar* (gypsy moth) was supposed to be extinct in Britain

from about 1850, but several individuals have been identified since the late 1960s. There are about 2000 species. The family has a wide distribution.

lymph A colourless fluid, similar to *blood plasma, that consists mainly of salts and *proteins in water and suspended fats (whose presence and amount varies according to food intake). It drains from spaces between cells into a network of vessels (the lymphatic system) that convey it to the bloodstream which it enters close to the heart. Lymph nodes, occurring at intervals in the lymphatic system (in humans especially in the neck, armpits, and groin), filter out bacteria and particles of foreign matter and produce lymphocytes (*see* LEUCOCYTE).

lymphadenitis Inflammation of a lymph node due to e.g. bacterial infection.

lymphatic system *See* LYMPH.

lymph node *See* LYMPH.

lymphocyte *See* LEUCOCYTE.

lyrebirds *See* MENURIDAE.

Lyrurus (black grouse) *See* TETRAONIDAE.

Lysenko, Trofim Denisovich (1898–1976) A fanatic anti-geneticist, who attained extraordinary power in the USSR under Stalin, and retained it under Khrushchev. He sought successfully the suppression of research in genetics and was responsible for the imprisonment and execution of a number of noted Soviet geneticists. The failure of his own attempts, based on a kind of *Lamarckism, to increase grain production led to his own downfall and contributed to that of Khrushchev.

lysine An aliphatic, basic, polar *amino acid that is generally abundant in animal proteins, but is of limited occurrence in those of plants.

lysosome A membrane-bound *vesicle, commonly 0.1–0.5μm in diameter, in a cell that contains numerous acid hydrolases capable of digesting a wide variety of extra- and intracellular materials. The membranes of lysosomes apparently originate from the *Golgi body and the hydrolases are synthesized in the rough *endoplasmic reticulum. The functions of lysosomes are: (*a*) the digestion of material taken in by *endomitosis, usually by fusing with the material and discharging their contents into a *vacuole; (*b*) the destruction of redundant *organelles (*see* AUTOPHAGY); (*c*) the storage of indigestible residues (*see* RESIDUAL BODY); (*d*) the destruction of cellular materials during *autolysis; and (*e*) the release of *enzymes outside the cell (e.g. during the resculpturing of bone).

lysozyme A bacteriolytic *enzyme that is found in a number of biological fluids, including tears, saliva, and egg white. It facilitates the *hydrolysis of structural polysaccharides in the cell walls of bacteria.

Lytta vesicatoria (Spanish fly) *See* MELOIDAE.

Macaca (macaques) *See* CERCOPITHECIDAE.

macaques (*Macaca*) *See* CERCOPITHECIDAE.

macaw *See* PSITTACIDAE.

Machilidae (order *Archaeognatha, superfamily *Machiloidea)** One of the two present-day archaeognathan families, whose member species are active jumpers, usually found under bark and stones, or among rocks along the sea-shore, principally in the northern hemisphere. There are about 250 species.

Machiloidea (subclass *Apterygota, order *Archaeognatha)** Superfamily whose members are divided into two groups, one in the northern hemisphere and one in the southern.

mackerel *See* SCOMBRIDAE.

Macquarie perch *See* PERCICHTHYIDAE.

macro- A prefix meaning 'large' or 'long', from the Greek *makros*, 'large' or 'long'.

macroconsumer *See* CONSUMER.

Macrodasyida (phylum *Aschelminthes, class *Gastrotricha)** An order of microscopic, hermaphroditic worms with adhesive tubes on the body found in marine and brackish water.

macroevolution Evolution above the *species level (i.e. the development of new species, genera, families, orders, etc.). There is no agreement as to whether macroevolution results from the accumulation of small changes due to *microevolution, or whether macroevolution is uncoupled from microevolution.

macrofauna The larger animals; the term is sometimes used to include larger insects and earthworms in this category, but otherwise these form part of the *mesofauna. *Compare* MEIOFAUNA; MESOFAUNA; MICROFAUNA.

macrolecithal Applied to an egg that has a large yolk, typical of *Myxinidae, *Elasmobranchii, *Teleostei, *Chelonia, *Sauria, *Serpentes, *Crocodilia, *Aves, and *Monotremata. *Compare* MESOLECITHAL; OLIGOLECITHAL. *See also* ISOLECITHAL; TELOLECITHAL.

macromere *See* BLASTOMERE.

macromolecule A molecule that has a high molecular weight, often a polymer.

macromutation A *mutation that has very large phenotypic (*see* PHENOTYPE) effects (e.g. a mutation affecting early *ontogeny). Macromutations have been proposed as the leading mechanism of evolution, as in the '*hopeful monster' hypothesis.

macrophage A large, phagocytic (*see* PHAGOCYTE; PHAGOCYTOSIS), white blood cell, which occurs in large numbers at sites of infection where it is instrumental in removing foreign cells and cell debris. Macrophages are derived from monocytes (*see* LEUCOCYTE) and adopt different shapes according to their location.

macrophagy Feeding on items that are large compared with the size of the organism consuming them. *Compare* MICROPHAGY

Macropharyngea *See* NUDA.

Macropharynx longicaudatus *See* EURYPHARYNGIDAE.

Macropis *See* MELITTIDAE.

Macropodidae (kangaroos, wallabies; order *Marsupialia (or *Diprotodontia), superfamily *Macropodoidea) A family of herbivores in which the lower *incisors are directed forward and can be moved against one another like shears, the *molars are modified for intensive grinding, and the stomach has a sacculated, nonglandular, rumen-like chamber containing *symbionts. The *ilia and thigh muscles are modified for bipedal motion and the fourth metatarsal is elongated. Kangaroos and wallabies are mainly terrestrial, but in the semi-arboreal tree kangaroos the fore limbs are almost as long as the hind limbs. There are about 60 species, all native to Australia and New Guinea.

Macropodoidea (infraclass *Metatheria, order *Marsupialia (or *Diprotodontia) A superfamily of hopping marsupials that includes the families *Macropodidae and *Potoroidae.

Macrorhamphosidae (snipefish, bellowsfish; subclass *Actinopterygii, order *Syngnathiformes) A small family of marine tropical and subtropical fish that have a compressed and deep body. The long, stout second spine of the *dorsal fin and the long, tubular snout are characteristic. The fish have large eyes and short fins, and the body is partially enclosed in armour (consisting of bony plates). *Macrorhamphosus elevatus* (common bellowsfish), 15 cm, of S. Australia often moves in large schools at depths of 25–160 m. There are about 11 species, distributed world-wide.

***Macrorhamphosus elevatus* (common bellowsfish)** *See* MACRORHAMPHOSIDAE.

Macroscelidea (Macroscelididae; elephant shrews; cohort *Unguiculata, order *Insectivora) A suborder of omnivorous mammals (now generally classed as a separate order) in which the hind limbs are elongated, enabling the animals to jump well, and the snout is elongated. The *molars are large and squared, like those of ungulates. Macroscelids are known from the early *Oligocene and are believed to be an evolutionary offshoot from a primitive insectivore stock, confined to Africa and with no close living relatives. There are many species, found in N. Africa and south of the Sahara.

Macroscelididae *See* MACROSCELIDEA.

Macrotermitinae *See* TERMITIDAE.

Macrouridae (Macruridae; grenadier, whiptail; subclass *Actinopterygii, order *Gadiformes) A large family of marine mid-water fish that have very long *dorsal and *anal fins tapering down to the pointed tip of the tail, the tail fin being absent. The blunt head has large eyes and often shows a chin *barbel. The small *pelvic fins are usually located under or in front of the *pectorals; all fins are spineless. Most grenadiers live near the sea-bottom at depths of 200–1200 m. There are about 250 species, distributed world-wide.

***Macrozoarces americanus* (ocean pout)** *See* ZOARCIDAE.

Macruridae *See* MACROURIDAE.

macula In the *retina of many vertebrates that lack a *fovea, an area of relatively acute sensitivity to light stimulation and so of relatively acute vision. In the human retina the macula lutea is a similarly sensitive area, containing a yellow pigment, that surrounds the fovea.

Madagascan faunal subregion The strongly *endemic and insular fauna of Madagascar, that includes five orders found nowhere else. For example, four of the five families of lemurs survive only in Madagascar; and this is the only part of the African continent from which the more advanced apes and monkeys are entirely absent.

Madeira cockroach (*Leucophaea maderae*) *See* BLABERIDAE.

Madreporaria (Scleractinia; stony corals; class *Anthozoa, subclass *Zoantharia) An order of colonial (or, more rarely, solitary) corals, which always possess an external ectodermal, calcareous skeleton consisting essentially of radial *septa. Septa develop following the pattern of *mesenteries in cycles of 6, 12, 24, 48, etc. The order first appeared in the Middle *Triassic.

madreporite In *Echinodermata, a sieve-like (it may be perforated by up to 250 pores), button-shaped process on the *aboral surface of the body; through its opening the water-vascular system is connected to the water outside.

magpie goose (*Anseranas semipalmatis***)** See ANATIDAE.

magpie larks See GRALLINIDAE.

magpies 1. See CORVIDAE. **2. (Australian magpie, Gymnorhina tibicen)** See CRACTICIDAE.

Mahlanobis' D^2 A measure of generalized distance between samples based on the means, variances, and covariances of various properties of replicate samples in multivariate analysis. The larger D^2, the greater is the difference between the samples (or the properties measured).

mahseer (*Barbus tor***)** See CYPRINIDAE.

maintenance evolution See STABILIZING SELECTION.

Majidae (spider crabs; class *Malacostraca, order *Decapoda) Family comprising slow moving, marine crabs, whose carapace is longer than it is wide. The *chelipeds are usually not much longer than the other *pereopods, and the body and appendages are equipped with hooked *setae for the attachment of camouflage derived from pieces of seaweed, sponges, hydrozoans, bryozoans, and ascidian colonies. Some species lie among the spines of sea urchins and the tentacles of anthozoans, from which they derive a degree of protection. There are about 200 species.

major gene A *gene that has pronounced phenotypic effects, in contrast to a modifier

gene, which modifies the phenotypic expression of another gene.

major histocompatibility complex (MHC) A large multigene family that codes for histocompatibility *antigens. Present on the surface of nucleated cells, histocompatibility antigens bind to and present foreign antigens to *T cells which in turn may destroy infected cells or invading bodies, and use the histocompatibility antigens to recognize cells as self or foreign. The MHC tends to be highly *polymorphic even within a single population, allowing a high level of discrimination between foreign and self, consequently organs transplanted from one member of a population to another are likely to be rejected by the new host unless the two members are very closely related. In different species different symbols are assigned to the MHC, e.g. in humans HLA is used, standing for human leucocyte antigen complex.

***Makaira nigricans* (blue marlin)** See ISTIOPHORIDAE.

mako shark (*Isurus oxyrinchus***)** See LAMNIDAE.

Malacanthidae (tilefish; subclass *Actinopterygii, order *Perciformes) A family of marine fish, known formerly as Branchiostegidae, found mainly in tropical and temperate waters. The rather elongate body possesses long *dorsal and *anal fins and a truncate tail fin, the *pelvic fins being located under the *pectorals. Although many species are less than 40 cm long, *Caulolatilus princeps* (ocean whitefish) of the eastern Pacific reaches a length of more than one metre. There are about 24 species.

***Malacochersus tornieri* (pancake tortoise)** See TESTUDINIDAE.

Malacopterygii A name now little used, describing the group (superorder) comprising the 'lower' bony fish (*Osteichthyes), characterized by the possession of only soft-rayed fins (no spines), often single *dorsal and *adipose fins, both dorsal and *pelvic fins located roughly midway along the back and belly profile, and the *pectorals set low on the body. The 'higher' bony fish are usually grouped as the *Acanthopterygii.

Malacostega (order *Cheilostomata, suborder *Anasca) A division of bryozoans in which the individuals of the colony possess uncalcified (membranous) body walls anterior to the aperture. The division occurs from the *Cretaceous to the present.

Malacosteidae (loosejaw; subclass *Actinopterygii, order *Salmoniformes) A small family of marine deep-water fish that have a very large, low-slung mouth, eyes located far forward on the blunt head, and the *dorsal and *anal fins placed close to the small tail fin. There are about 10 species, distributed world-wide.

Malacostraca (phylum *Arthropoda, subphylum *Crustacea) Class which contains nearly 75% of known crustaceans, including the larger forms, e.g. shrimps, crabs, and lobsters. The typical malacostracan trunk comprises 14 segments and a *telson. The thorax is composed of eight segments, and the remaining six form the abdomen. The first antennae are often *biramous, and all the body segments bear appendages. The cutting edge of the *mandibles comprises a grinding molar process and a cutting incisor; the mandible usually bears a *palp. The carapace is lost in some orders and in most malacostracans one or more of the anterior thoracic limbs is upturned and modified into a *maxilliped. Primitively, the biramous thoracic limbs are all similar, with the endopodite better developed than the exopodite, and are used for crawling or grasping. The anterior abdominal limbs are *pleopods, and there are usually five pairs. They may be modified for crawling, for creating feeding and respiratory currents, or in the females for egg-carrying. In the males, the first two pairs are usually adapted for use as *intromittent organs. The terminal tail fin, used in swimming, comprises the dorsal *telson and a pair of ventral *uropods. In the female, the genital openings (gonopores) are on the sixth thoracic segment, while those of the male are on the eighth. The *nauplius stage

is passed within the egg. There are 16 orders and more than 25 000 species.

Malapteruridae (electric catfish; subclass *Actinopterygii, order *Siluriformes) A monospecific family (*Malapterurus electricus*) of fish found in freshwater basins of tropical Africa: there may be a second species in the Congo basin. The massive, stubby body shows an *adipose but no *dorsal fin. Both adipose and *anal fins are located close to the rounded tail fin. The eyes and spineless *pectoral and *pelvic fins are small. A large electric catfish can deliver shocks strong enough to stun or kill prey fish. Its electric organs are located under the skin along the whole length of the body.

Malapterurus electricus (electric catfish) *See* MALAPTERURIDAE.

malar bone *See* JUGAL BONE.

malarial mosquito (*Anopheles***)** *See* CULICIDAE.

Malayan moon rat (gymnure; *Echinosorex gymnurus*) *See* ERINACEOIDEA.

malic acid A dicarboxylic acid formed during the *citric-acid cycle by the reversible hydration of *fumaric acid.

mallard (*Anas platyrhynchos***)** *See* ANATIDAE.

malleolus 1. A small prominence on the *distal end of the *tibia and of the *fibula; in humans they form the 'knobs' of the ankle. **2.** *See* RACQUET ORGAN.

malleus In *Mammalia, the outer ossicle of the *middle ear, which is derived from the *articular bone of ancestral vertebrates.

Mallophaga (chewing lice, biting lice; subclass *Pterygota, order *Phthiraptera) Group of lice, comprising the suborders *Ischnocera, *Amblycera, and *Rhyncophthirina, all of which possess functional *mandibles. The common name 'CHEWING LICE' applied to this group is preferred to 'biting lice' because members of the fourth suborder of lice, the *Anoplura, also bite. The use of the Mallophaga as a taxonomic group is to be discouraged because the *Rhyncophthirina are more closely related to the *Anoplura than they are to the *Ischnocera and the *Amblycera, and the Ischnocera are more closely related to the Rhyncophthirina and Anoplura than they are to the Amblycera.

Malpighi, Marcello (1628–94) An Italian physiologist who is credited with having been the first person to study the structure of plants and animals by means of a microscope. He was the first person to observe directly the blood coursing through vessels on the surface of the lung, and he described the structure of secreting glands (see MALPIGHIAN TUBULES). He also attempted to study the fine structure of the brain.

Malpighian layer In the skin of mammals, the lowest layer of the *epidermis and the site of active cell division, some of the cells migrating outwards, becoming keratinized (see KERATIN) and joining the outermost layer of the epidermis. The Malpighian layer also contains most of the skin's *melanin. It was discovered by Marcello *Malpighi.

Malpighian tubules The excretory organs of insects. They are long and narrow, made up of transporting epithelial cells, which join the gut at the junction of the midand hind gut. When stimulated hormonally they produce copious amounts of isotonic urine. Their numbers vary greatly from species to species.

maltose A disaccharide composed of two alpha-glucose units linked by an μ-1,4 glycosidic bond.

Maluridae (Australian wren; class *Aves, order *Passeriformes) A family of small, long-tailed birds; *Malurus* (fairy wrens) have bright blue males, other genera are mainly brown. *Stipiturus* (emu wrens) have only six tail feathers. They are found in grass and scrub, feed on insects, and build a domed nest in grass. There are five or six genera, with 23–30 species, found in Australia and New Guinea. See also EPHTHIANURIDAE.

Malurus **(fairy wrens)** See MALURIDAE.

mambas (*Dendroaspis*) See ELAPIDAE.

Mammalia (phylum *Chordata, superclass *Gnathostomata) A class of *homoiothermic animals in which the head is supported by a flexible neck, typically with seven vertebrae, articulating through two *occipital condyles, the side wall of the skull is formed by the *alisphenoid bone, the lower jaw is formed from the *dentary bone and articulates with the *squamosal, the *quadrate and *articular bones form auditory ossicles, and the *angular bone forms the *tympanic bone. Typically teeth are present and are *thecodont, *heterodont, and *diphyodont; the mouth cavity is separated by a hard palate from the nasal cavity. *Epiphyses are present on many of the bones. The thorax and abdomen are separated by a *diaphragm, the heart is fourchambered, the right *aortic arch is absent, except in *Monotremata the egg is small and develops in the uterus, and the young are fed milk secreted by mammae (which give the class its name). The skin has at least a few hairs. Many mammalian features were present in therapsids (mammal-like reptiles) during the *Triassic. Mammals are believed to have first appeared toward the end of the Triassic and to have diversified rapidly from the end of the *Mesozoic, 100 Ma later, following the mass extinction which marks the *Mesozoic–Tertiary boundary 65 Ma ago.

mammal regions A biogeographic scheme that has been proposed by Charles H. Smith on the basis of statistical analyses. Four regions are recognized: *Holarctica (the only region consistent with traditional faunal regions); *Afro-Tethyan; *Island; and *Latin American.

mammary gland In female mammals, a gland on the ventral surface which produces milk. It is probably derived from modified sweat glands. It gives the *Mammalia its name.

mammoth See *MAMMUTHUS*; ELEPHANTIDAE.

Mammuthus (mammoth) A genus of *Pleistocene elephants that were adapted to steppe and tundra habitats. The tusks were elongated and strongly curved, and the skull was shorter and higher than that of other elephants. The woolly mammoth was adapted to Arctic environments. The largest mammoth, indeed the largest probiscidean of all time, was *Mammuthus armeniacus* of Eurasia, which stood about 4.5 m at the shoulder.

Mammutidae (Mastodontidae; order *Proboscidea, suborder *Mammutoidea) An extinct, monospecific family (*Mammut* or *Mastodon*) of mastodons, comprising elephant-like animals that diverged from the evolutionary line leading to the modern elephants. The genus was long-lived, extending from the lower *Miocene to the *Holocene, and it survived in Africa and the *Holarctic region at least until the end of the *Pleistocene. Mastodons were shorter and heavier in build than elephants. *Mammut* (or *Mastodon*) species had short, high skulls, longer jaws than elephants, and usually tusks in both upper and lower jaws, the upper tusks often being large and curving outward and upward. There were never more than two teeth in use at a time, never more than a vestige of the lower *incisors, and the *molars were low-crowned, simple, and lacked cement. It is believed that the evolution of the mastodons paralleled that of the *Gomphotheriidae.

Mammutoidea (superorder *Paenungulata, order *Proboscidea) A suborder that includes the families *Mammutidae (Mastodontidae) and *Stegodontidae, the mastodons.

manakins *See* PIPRIDAE.

manatee (*Trichechus*) *See* TRICHECHIDAE.

mandible 1. In vertebrates, the lower jaw. 2. In birds, specifically the lower jaw and bill but the term is also used to denote the two parts of the bill of a bird, as upper and lower mandibles. 3. In *Arthropoda, one of the pair of mouth-parts most commonly used for seizing and cutting food.

Mandibulata (phylum *Arthropoda) Arthropods that possess antennae. These were formerly classed as a subphylum, named for the fact that the first appendages behind the mouth are *mandibles, and included the insects, crustaceans, millipedes, and centipedes. The grouping is now held to be artificial.

mandrill (*Mandrillus*) *See* CERCOPITHECIDAE.

Mandrillus (mandrill) *See* CERCOPITHECIDAE.

maned wolf (*Chrysocyon brachyurus*) *See* CANIDAE.

mangos *See* TROCHILIDAE.

mangrove jack (*Lutjanus argentimaculatus*) *See* LUTJANIDAE.

Manidae (pangolins, scaly ant-eaters; cohort *Unguiculata, order *Pholidota) A family of animals which are similar superficially to the New World ant-eaters, to which they may be related distantly. Pangolins (or scaly ant-eaters) are nocturnal, insectivorous, terrestrial, or arboreal mammals up to 1.5 m long, lacking teeth, with elongated snouts, and tongues that are long, thin, and sticky. There are long claws on all five digits of each limb. The tail is long and *prehensile in arboreal species. The dorsal surface of the body is covered with overlapping epidermal scales, and a manid will roll itself into a ball when threatened. The eyes and ears are small, the stomach simple. The brain is very small, the hemispheres folded. There is one genus, *Manis*, and several species, distributed throughout the Old World tropics except for Madagascar and Australia.

mannan A polysaccharide, composed mostly of glycosidically linked *mannose units, that act as storage polymers in some species and perform a structural role in others. A great range of linkage has been reported and some mannans also contain other sugars.

mannikins *See* ESTRILDIDAE.

mannose An aldohexose monosaccharide, which occurs in a wide variety of organisms including vegetable gums, fungi, bacterial capsular layers, and some immunoglobulins in animals.

man-of-war fish (*Nomeus gronovius*) *See* NOMEIDAE.

manometer An instrument that is used to measure differences in pressure, usually by comparing the heights of two liquid columns. The simplest version comprises a U-shaped tube containing a liquid. One end of the tube is open and the other is connected to the container holding the pressure that is to be measured. The pressure is registered as the difference in the level of liquid on the two sides of the tube. *See also* RESPIROMETER.

manta *See* MOBULIDAE.

Manta birostris (giant manta) *See* MOBULIDAE.

Mantidae (praying mantises; subclass *Pterygota, order *Mantodea) Family of medium to large, cryptically coloured, predatory insects (e.g. *Mantis religiosa*, the praying mantis), which is by far the larger of the two families of praying mantises. The 1800 or so described species are found in all the warmer parts of the world.

mantid flies *See* MANTISPIDAE; NEUROPTERA.

mantids *See* MANTIDAE; MANTODEA.

Mantispidae (mantid flies; subclass *Pterygota, order *Neuroptera) Family of insects whose common name refers to their mantid-like appearance. The *prothorax is elongate, with raptorial fore legs arising from its anterior end. The predacious adults have large, transparent wings, up to 35 mm in span. The larvae are parasites on the egg sacs of wolf spiders, and one subfamily is parasitic in the nests of vespoid wasps. Mantispids are widely distributed, but are commoner in southern regions.

Mantis religiosa (praying mantis) *See* MANTIDAE.

mantis shrimps *See* STOMATOPODA.

mantle 1. (pallium) In *Mollusca and some *Brachiopoda (in which it is known as the mantle lobe), a fold of skin on the dorsal surface that encloses a space (the mantle cavity) containing the gills. The mantle is responsible for the secretion of the shell. **2.** In *Cirripedia, the name often given to the *carapace.

mantle cavity *See* MANTLE.

mantle lobe *See* BRACHIOPODA; MANTLE.

Mantodea (mantids, praying mantises; class *Insecta, subclass *Pterygota) Order of medium to large, terrestrial insects which are predacious, having large, raptorial fore legs used for seizing their insect prey. The *prothorax is usually narrow and elongate, and the head is typically triangular and freely movable on a slender neck. The mantids are *exopterygote but the wing buds do not reverse their orientation in later *instars. The fore wings are narrow, and may be hardened as *tegmina. The females of many species have reduced wings. Eggs are laid in an *ootheca. About 1800 species of mantids have been described, and are found in the warmer regions of the world.

manubrium 1. In coelenterates, a projection around the mouth. **2.** In *Collembola, the single, basal portion of the *furcula, which bears a pair of structures called dens.

manus The hand (the Latin for 'hand'). The term includes *homologous structures in other members of the *Tetrapoda.

many-plumed moths *See* ALUCITIDAE.

map butterflies *See* NYMPHALIDAE.

marbled murrelet (*Brachyramphus marmoratus*) *See* ALCIDAE.

marbled newt (*Triturus marmoratus*) *See* SALAMANDRIDAE.

marbled white butterflies *See* SATYRINAE.

March flies *See* BIBIONIDAE.

Margaritiferidae *See* GLOCHIDIUM.

marginal value theorem A mathematical rule, proposed by E. L. Charnov (1947–) in 1976, according to which the optimum time a *foraging animal remains in a patch is defined in terms of the rate at which the forager is extracting energy at the time it leaves (the marginal value of the patch). The optimum foraging strategy is to abandon each patch when the rate of energy extraction from it falls to a certain level, this level being the same for all patches. The theorem predicts that foragers will remain a shorter time in patches with little food than in patches with more food, patches will be abandoned more quickly when they are close together than when they are scattered, and patches will be abandoned more quickly in an area of abundance than in a poorer area.

(((⊕))) SEE WEB LINKS
• Detailed explanation of the marginal value theorem.

marine iguana (*Amblyrhynchus cristatus***)** *See* IGUANIDAE.

marine toad (*Bufo marinus***)** *See* BUFONIDAE.

marker A piece of genetic material that bears or produces a distinctive feature. It is usually a mutant *allele and can be dominant or recessive.

marker-assisted selection In *genetic engineering, the introduction of a *marker that confers antibiotic resistance into bacterial cells along with other *DNA. Offspring containing both antibiotic resistance and the new DNA will survive exposure to the antibiotic, while other offspring will be killed.

mark–recapture technique A technique for estimating the population density of more elusive or mobile animals. A sample of the population is captured, marked, and released. Assuming that these marked individuals become randomly distributed through the wild population, and that subsequent trapping is random, any new sample should contain a representative proportion of marked to unmarked individuals. From this, the size of the population may be estimated, most simply by multiplying the number in the first sample by the number in the second sample, and dividing the product by the number of marked individuals in the recaptured sample. This calculation is appropriate only when the population is fairly static or where changes (due to migration, natality, or mortality) are known. It may also be distorted if marked individuals become more vulnerable to predators and if particular individuals become 'trap-addicted' or 'trap-shy'. When a population is fluctuating rapidly, as is common in insect populations, more sophisticated indices, which allow for the probabilities of change, are preferable.

marlin *See* ISTIOPHORIDAE.

marmoset *See* CALLITRICHIDAE.

marmot **(woodchuck,** **groundhog;** *Marmota marmota*) *See* GROUNDHOG; SCIURIDAE.

***Marmota marmota* (marmot, woodchuck, groundhog)** *See* SCIURIDAE.

marsh A more or less permanently wet area of mineral soil (as opposed to a peaty area), typically found around the edges of a lake or on an undrained river flood-plain. Colloquially, 'marsh' is often used interchangeably with 'swamp' and 'bog'.

marsh frog (*Rana ridibunda***)** *See* RANIDAE.

marsh wrens (*Cistothorus***)** *See* TROGLODYTIDAE.

Marsipobranchii Alternative name for *Cyclostomata (Agnatha).

marsupial ant-eater *See* MYRMECOBIIDAE.

marsupial carnivores *See* DASYURIDAE.

marsupial frog (*Gastrotheca marsupiatum*) See HYLIDAE.

Marsupialia (subclass *Theria, infraclass *Metatheria) An order that comprises some 250 species of living marsupials and many extinct forms. In the 1960s it was divided into three suborders (Polyprotodonta, which includes the opossum-like insectivorous, carnivorous, and omnivorous forms; Diprotodontia, containing the phalangers, kangaroos, and other forms evolved from an opossum-like stock, but differing structurally from the polyprotodonts; and *Caenolestoidea (classed by others as a superfamily), containing a small group of 'opossum rats'), but nowadays it is usual to divide the marsupials into several orders: *Dasyuromorphia; Didelphimorphia (*see* DIDELPHOIDEA); *Dromiciopsia; *Notoryctemorphia (*see* SYNDACTYLIFORMES); *Paucituberculata; and *Peramelemorpha; as well as the extinct *Sparassodontia. These are often allocated to two cohorts: *Ameridelphia and *Australidelphia. In this scheme, the name Marsupialia would cease to be used formally. Marsupials are characterized principally by their method of reproduction. The egg is yolky and has a thin shell protecting it from maternal antigens. Placental development is usually very limited and except in the Peramelemorpha the *allantois serves no nutritional function, but uterine milk may be taken up by the *yolk sac. Within 10–12 days of the breaking of the shell, the *embryo (whose fore limbs and associated neural development, mouth, and olfactory system have developed precociously) is born. It crawls into the pouch (*marsupium) and attaches itself to a teat, its lips growing around the teat, which injects milk without choking the embryo. In the later stages of its development an offspring may receive high-fat, low-protein milk from one teat while a newer embryo receives high-protein, low-fat milk from another. Marsupials also differ from placentals in their dentition, in the possession of an inflected angular process to the jaw, and in the presence of two marsupial bones which articulate with the pubes. Marsupials and placental mammals apparently diverged from a common ancestor in the *Cretaceous. The first marsupials were similar in general form to the opossums of America. In Australia the marsupials radiated to produce a wide array of adaptive types, while in S. America they filled the insectivorous and carnivorous niches for much of the *Cenozoic, while placentals occupied the herbivorous niches.

marsupial mice *See* DUNNART.

marsupial mole *See* NOTORYCTIDAE.

marsupial pouch *See* MARSUPIUM.

marsupial rats *See* DASYURIDAE.

marsupium A pouch (from the Greek *marsupion*, 'pouch') on the surface of the body of an adult animal in which its young are held securely. Marsupia occur in some invertebrates (e.g. some *Isopoda), some *Anura, but most notably in the *Marsupialia, which derive their name from the possession of marsupia, although not all species have them. In those female marsupials that do, and also in echidnas (*see* TACHYGLOSSIDAE), the marsupium is a fold of skin, supported by *epipubic bones and containing the *mammary glands, into which the eggs of echidnas and young of marsupials are placed.

marten (*Martes*) *See* MUSTELIDAE.

Martes martes (pine marten) *See* MUSTELIDAE.

martin bug (*Oeciacus hirundinis*) *See* CIMICIDAE.

martins *See* HIRUNDINIDAE.

Mary Lyon effect The inactivation of one of the *X-chromosomes in female mammals; its discovery was predicted by the geneticist Mary Lyon (1925–), who observed that otherwise females would have considerably more genetic material than males, since in mammals the Y-chromosome is tiny whereas the X-chromosome is very large. In marsupials it is the paternally derived X that is inactivated, but in placentals paternal or maternal are inactivated at random in different cell lines.

Masaridae (suborder *Apocrita, super-family Vespoidea) Family of medium-sized (9–24 mm long), stout-bodied wasps, which are often strongly marked with orange. They are sometimes regarded as a subfamily of the *Vespidae, but should be excluded from that family on the basis of the relatively weak wing coupling, the possession of two submarginal wing cells (vespids have three), and other structural features. The antennae are clubbed, and the *mandibles do not cross over. Single nests or groups of nests are provisioned with pollen and nectar, and are constructed of mud or sand attached to rocks or twigs.

mask (labial mask, labium) Organ formed from the *labium, and found only in the nymphs of the insect order *Odonata: it is so named because it covers the mouth-parts, and sometimes the face. It is prehensile, and nymphs use it to capture prey by extending it at high speed and impaling the prey on the terminal spikes, or by trapping prey in a basket formed by the mask and its associated bristles.

mason bee Any member of the genera *Anthocopa*, *Chalicodoma*, *Dianthidium*, *Heriades*, *Hoplitis*, or *Osmia*, of the cosmopolitan family *Megachilidae. The female collects soft, malleable building materials which are shaped at the nest site into durable structures. The nests may be inside existing cavities in timber, under stones, or may be built on exposed surfaces, e.g. rocks, or leaves and branches of woody plants. According to the bee species, the collected materials may be mud, pebbles, quartz chips, resin, or a mastic of chewed leaves or petals, or a combination of two or three of these substances.

mason wasps *See* EUMENIDAE.

masseter muscle One of the muscles that raise the lower jaw.

massive hydrocorals *See* MILLEPORINA.

mass mixing ratio *See* HUMIDITY.

mass provisioning The provision of all the food required for larval development by

wasps and bees. The food, either insect or spider prey (for wasps) or honey and pollen (for bees) is stored in a specially prepared space (cell) within the nest. An egg is laid on or in the food, and the cell is sealed. Mass provisioning is very largely found in solitary (non-social) species, but is also practised by the highly social stingless bees.

Mastacembelidae (spiny eel; subclass *Actinopterygii, order *Perciformes) A family of freshwater fish that have very elongate, eel-like bodies. The *dorsal fin consists of a series of short spines followed by a long second dorsal fin matched by the *anal fin. Typically, the pointed snout terminates in long, fleshy, tubular nostrils, probably used by these nocturnal fish to sniff out food in darkness. There are about 50 species, occurring in tropical Africa and Asia.

mast cell A *connective tissue cell which, on becoming injured at a site of tissue damage, releases certain chemicals. Some of these cause a restriction of *lymph vessels, thus allowing localized swelling to occur. Another, called *histamine, increases capillary permeability, allowing *leucocytes to leave the circulatory system more readily and to congregate at the site of injury.

Mastigophora (phylum *Protozoa, subphylum *Sarcomastigophora) A superclass of protozoa, which employ one or more *flagella for locomotion. There are two classes, and 19 orders. The superclass includes a wide variety of organisms, many of which are classified alternatively with the algae.

mastodon (*Mammut*, *Mastodon*) *See* MAMMUTIDAE.

Mastodontidae *See* MAMMUTIDAE.

mastoid process In animals, a bony outgrowth behind the ear: it contains air spaces which communicate with the *middle ear.

Mastotermitidae (subclass *Pterygota, order *Isoptera) Family of 'lower' termites containing only one surviving species, *Mastotermes darwiniensis*, which inhabits northern Australia. Colonies may contain one million inhabitants, living in many underground nests connected by covered passages built along the surface of the ground. Their very catholic diet makes them extremely destructive.

matamata (*Chelus fimbriatus*) *See* CHELIDAE.

maternal message *Messenger-RNA that is synthesized during *oogenesis and deposited in the egg *cytoplasm, with *ribosomes, as an inactive complex often termed an 'informosome'. It is activated after fertilization and helps to control the earliest stages of development.

mating The union of two individuals of opposite sexual type to accomplish sexual reproduction.

mating system In *artificial selection, a procedure that is used to control the genetic constitution of offspring (usually the degree of *homozygosity due to inbreeding). Mating may be assortative (i.e. between similar *phenotypes), resulting in increased homozygosity; or disassortative (i.e. between unlike phenotypes), tending to maintain or increase *heterozygosity.

mating type The equivalent in lower organisms (particularly micro-organisms) of sexes in higher organisms. Micro-organisms may be subdivided into mating types on the basis of their physiology and mating behaviour. Different mating types are usually identical in physical form although individuals of one mating type possess on their surfaces proteins that will bind to complementary proteins or polysaccharides found only on the coats of individuals of the opposite mating type. In this way, only individuals of different mating types will undergo *conjugation.

matrilineage In a social group of females, those females which are related. Matrilineages are a common feature of *Primates.

matrix The gel-like interior of a *mitochondrion.

matrotrophic Applied to a mode of *embryo development in which the source of nourishment is not limited to the yolk, but is supplemented by nourishment derived from the mother. *See also* PLACENTOTROPHIC. *Compare* LECITHOTROPHIC.

Matuyama The penultimate *chron; during most of it the prevailing magnetization was reversed (i.e. the magnetic north and south poles were reversed from their present orientation). It lasted, with a few subchrons of normal polarity, from 2.4 to 0.71 Ma ago.

Mauthnerian system A neurolocomotory system, well developed in *teleosts, that gives rise to a startle response. Two giant nerves located to either side of the midline in the medulla oblongata which synapse with the acoustic nerve have giant axons which cross over and extend the length of the body, synapsing with motor neurons. Stimulation of the acoustic nerve from one side results in a rapid, forceful contraction of the body trunk on the other side, initiating rapid locomotion forward and away from the direction of stimulus.

maxilla 1. In vertebrates, the posterior bone of the upper jaw, which bears the teeth other than the *incisors. The term is sometimes applied to the whole of the upper jaw. **2.** In some *Arthropoda, one of a pair of mouth-parts used in eating. The maxillae are paired, limb-like structures located immediately behind the *mandibles, which articulate with the head capsule. *See also* MAXILLARY PALP.

maxillary gland In *Crustacea, one of a pair of antennal glands opening into the bases of the second antennae and/or the second pair of *maxillae, and used for the excretion of nitrogenous waste.

maxillary palp Jointed, sensory structure found on the *maxillae of an insect.

maxilliped In some *Arthropoda, a thoracic limb that is modified to filter food, to assist in respiration by filtering water before it enters the gills, or (in centipedes) to inject prey with poison.

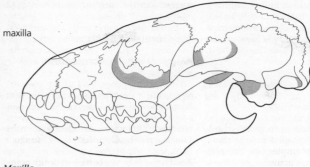

Maxilla

maxillopalatine Part of the upper jaw of some avian orders, which distinguishes them structurally from others. It consists of a plate formed by the fusion of the two maxillopalatine bones.

maximum sustained yield (MSY) *See* OPTIMUM YIELD.

maximum velocity of enzyme The maximum rate at which a given quantity of an *enzyme can catalyse a reaction under defined conditions. It is achieved only when the enzyme is saturated with *substrates and cosubstrates.

maybug (*Melolontha melolontha*) *See* SCARABAEIDAE.

mayflies *See* EPHEMERELLIDAE; EPHEMERIDAE; EPHEMEROPTERA.

meadow frog (*Rana pipiens*) *See* RANIDAE.

mealworm (*Tenebrio molitor*) *See* TENEBRIONIDAE.

mealybug *See* COCCIDAE.

mealy hairs Hairs which collectively have the consistency of meal and give the appearance of a mealy covering.

mean square (variance) The square of the mean variation of a set of observations around the sample mean.

mechanical senses Senses that detect mechanical stimuli (e.g. touch, hearing (detecting air vibration), and balance).

mechanoreceptor A *receptor cell which responds to mechanical deformation, such as those in the inner ear that sense touch and detect movement. *See also* CHEMORECEPTOR; RADIORECEPTOR.

Meckel's cartilage In mammalian *embryos, a cartilaginous rod derived from the first *visceral arch (mandibular arch). It runs along the *dorsal surface of the *mandible (otherwise a membrane bone) and parts of it ossify as the *malleus and *incus. Its particular interest is its duplication of the processes by which the mandibular and ear bones evolved during the transition from reptiles to mammals.

Mecoptera (scorpionflies; class *Insecta, subclass *Pterygota) An order of insects with slender bodies, often brightly coloured, mouth-parts extended to form a beak, and in many species the tip of the abdomen curled upward so it resembles that of a scorpion (although scorpionflies do not sting). Most species have two pairs of membranous wings, held horizontally over the back when the insect is at rest. Larvae dwell in burrows, feed at the soil surface, and pupate in their burrows, undergoing complete *metamorphosis. The *pupa is *exarate and moves to the surface prior to the emergence of the adult. Scorpionflies are mainly scavengers, as larvae and adults. Fossil mecopterans are known from the *Permian, making them the oldest *endopterygotes, and some extant species appear to have changed little since the Permian, making them *living fossils. There are about 300 species, widely distributed, in three families: Panorpidae; Bittacidae; and Boreidae (snowflies or snowfleas).

medaka (*Oryzias latipes*) See ORYZIATIDAE.

median eye See PINEAL EYE.

median fin In fish, one of the unpaired (i.e. *dorsal, *anal, and tail) fins.

median vein In insects, a vein that begins along the centre of the wing, then branches into anterior and posterior sections, each of which branches further.

Mediterranean chameleon (*Chamaeleo chamaeleon*) See CHAMAELEONTIDAE.

Mediterranean faunal subregion See AFRO-TETHYAN MAMMAL REGION.

medulla 1. The central part of an organ or bone (i.e. the marrow). **2.** Abbreviation for medulla oblongata. See MYELENCEPHALON.

medulla oblongata See MYELENCEPHALON.

medullary sheath See MYELIN SHEATH.

medusa A 'jellyfish'; in *Cnidaria, the free-swimming body type, resembling an umbrella or bell, that floats convex side uppermost. The mouth is located at the centre of the underside of the bell and *tentacles hang from its edge.

meerkat (*Suricata suricatta*) See HERPESTIDAE.

Megabraula See BRAULIDAE.

Megaceros giganteus (Irish elk) See ALLOMETRY; CERVIDAE.

Megachile rotundata (alfalfa leaf-cutter bee) See MEGACHILIDAE.

Megachilidae (mason bees, leaf-cutter bees, carder bees; order *Hymenoptera, suborder *Apocrita) Large, cosmopolitan family of solitary bees, in which the pollen *scopa consists entirely of tracts of stiff hairs on the abdominal sterna (*see* STERNUM). The *labrum is longer than it is broad, and *basitibial and pygidial (*see* PYGIDIUM) plates are absent. The family includes the mason bees and leaf-cutter bees, and also such genera as *Anthidium* and *Callanthidium*, the females of which tease out (card) plant hairs to use as nest-lining material—hence the common name 'carder bees'. Nests are usually made in pre-existing cavities, e.g. in beetle borings in dead wood or pithy stems. Some species specialize in using old snail shells; others build exposed nests on rocks, woody branches, etc. One species, *Megachile rotundata* (alfalfa leaf-cutter bee), is cultured commercially in the western USA as a pollinator of *Medicago sativa* (alfalfa or lucerne).

Megachiroptera (flying foxes, fruit bats; cohort *Unguiculata, order *Chiroptera) A suborder of large bats, with a wing-span of up to 1.5 m, that roost in colonies usually in trees (except in the only cave-dwelling genus, *Rousettus*), often feed in large groups, emit loud cries, and orientate by sight and smell, not by echo-location (except in one genus, *Rousettus*). Most feed on fruit. The snout is long, the teeth are flattened for grinding, the tail lies below the *uropatagium and is not part of it, and the second digit of each fore limb has three joints and a claw. Megachiropterans retain certain primitive features in the skull and are less specialized than microchiropterans. The suborder comprises only one family, Pteropodidae, containing about

45 genera, with more than 170 species, distributed throughout the Old World tropics and subtropics.

Megadermatidae (false vampires; order *Chiroptera, suborder *Microchiroptera) A family of bats in which the ears are large, joined extensively, and have a *tragus that is very large, the eyes are relatively large, and the nose has a well-developed leaf, often concealing the nostrils. The *uropatagium is not wrapped around the body when the animal sleeps. There is no tail. There are four genera, with five species, distributed widely throughout the Old World tropics.

megafauna 1. Animals that are large enough to be seen with the naked eye. **2.** Very large mammals (*Mammalia) and birds (*Aves), i.e. those exceeding 50 kg in body weight. Much controversy concerns the extinction of many members of the megafauna in the early part of the *Holocene, especially in Europe and N. America. Climatic change is a possible factor, but most of the *species involved had already survived previous interglacials. The distinctive aspect of the present interglacial is the rise to prominence of humans, and many megafaunal species became extinct at a time when hunting pressures were gaining strength.

megakaryocyte A cell in the *bone marrow that produces blood *platelets.

Megaladapidae (order *Primates, suborder *Strepsirrhini) A family of Malagasy lemurs, most of which were giants. They survived until a few centuries ago and were eliminated by human activity. One extant genus of small lemurs, *Lepilemur* (sportive lemurs), with seven species, used to be included in the *Lemuridae but is now recognized as a surviving member of the Megaladapidae.

***Megaloglossus* (echidna, spiny anteater)** *See* TACHYGLOSSIDAE.

Megalomycteridae (large-nose fish; subclass *Actinopterygii, order *Lampridiformes) A small family of deep-sea fish that have an elongate body, the spineless *dorsal and *anal fins located near the tail fin, and an unusually large olfactory organ ('large nose'). There are five species, with world-wide distribution.

Megalonychoidea (ground sloths; suborder *Xenarthra, infra-order *Pilosa) An extinct superfamily of ground-dwelling edentates which are known first from the *Oligocene but which thrived during the *Pleistocene in N., Central, and S. America, and which may be broadly ancestral to modern tree sloths. Early forms attained a length of little more than a metre; but later much larger forms appeared, including *Megatherium*, which was more than 6 m long, and more massive than an elephant. Ground sloths had claws on all five digits on each limb, the claws in some species being so large as to require the animal to walk with its feet turned on their sides, as do modern New World ant-eaters. The teeth were reduced and simple and some may have possessed horny plates used in cropping vegetation. Later ground sloths were contemporaries of early humans: a skeleton of *Nothrotherium* (an animal about the size of a tapir) has been found that shows signs of having been killed by humans. In the W. Indies ground sloths evolved into dwarf forms, some no larger than a cat.

Megalopidae (tarpon; subclass *Actinopterygii, order *Elopiformes) A small family of marine fish that have a streamlined body and silvery scales. The mouth is in a slightly oblique position with the lower jaw protruding. All the fins are soft-rayed. The *dorsal fin is located midway along the back, and the last ray is extended into a long, filamentous tip. The forked tail fin is indicative of a strong swimmer. There are two species, found in tropical and subtropical oceans.

megalops *See* POSTLARVA (2).

Megaloptera (alderflies; class *Insecta, subclass *Pterygota) Order of mandibulate endopterygotes, which have two dissimilar pairs of wings and whose larvae are aquatic. The larvae have well-developed mandibulate mouth-parts, functional legs, and lateral, abdominal gills. The *pupae are *adecticous and *exarate, pupation taking place in a silken *cocoon spun from the larval anus. The predacious larvae are an important part of the diet of freshwater fish, and are often used for bait. Many adults are large, *Corydalus* (American Dobson fly) having a wing-span of 16 cm. This small order comprises some 300 described species, most of which occur throughout the temperate regions, although a few occur in the tropics.

meganephrostomal In the *nephridia of *Oligochaeta, applied when the nephridial funnel is well developed and when its upper lip is shaped like a horseshoe. *Compare* MESONEPHROSTOMAL; MICRONEPHROSTOMAL.

Meganeura See MEGANISOPTERA.

Meganisoptera (Paralogidae, dragonflies; order Odonata) Extinct suborder of large, *Palaeozoic dragonflies that lacked the well-known venational characteristics of present-day Odonata. The suborder includes some of the largest insects known. *Meganeura gracilipes* had a wing-span of 70 cm and *M. monyi* had a wing-span of 60–70 cm.

Megapodagrionidae (damselflies; order *Odonata, suborder *Zygoptera) Family of damselflies which breed in bogholes and streams. The family occurs in the tropics. More than 180 extant species have been described.

megapodes See MEGAPODIIDAE.

Megapodiidae (megapodes, scrubfowl; class *Aves, order *Galliformes) A family of mainly brown or black, turkey-like birds, many of which have bare heads, although one species is crested. They have small to large bills, large, rounded wings, and very large, strong feet. They inhabit forests and scrub, and feed on insects, worms, seeds, and fruit. Their eggs are incubated in mounds of earth and vegetation, warmed by the sun and by fermentation. There are seven genera, with 12 species (including three species of *Megapodius*, scrubfowl),

found in the Philippines, Indonesia, Australasia, and Polynesia.

Megapodius **(scrubfowl)** See MEGAPODIIDAE.

Megaptera novaeangliae **(humpback whale)** See BALAENOPTERIDAE.

megasclere In *Porifera (sponges) a large, supporting *spicule. *Compare* MICROSCLERE.

Meinertellidae (rock bristletails; subclass *Apterygota, order *Archaeognatha) One of the two present-day archaeognathan families, whose members are distinguished from the *Machilidae by the possession of very small abdominal *sternites which protrude slightly between the coxal plates. The legs and *scape, and the *pedicel of the antennae, are not covered with scales. Principally a southern-hemisphere taxon, members of the family are found on coastal cliffs, in forests, and in rain forests. There are three monospecific genera.

meiofauna That part of the *microfauna which inhabits algae, rock fissures, and the superficial layers of the muddy sea-bottom. The term is often used synonymously with *interstitial fauna. *Compare* MACROFAUNA; MESOFAUNA; MICROFAUNA.

meiosis (reduction division) A form of nuclear division whereby: (*a*) each *gamete receives only one member of a *chromosome pair (this forms one of the bases of *Mendel's first law of genetic segregation); and (*b*) genetic material can be exchanged between *homologous chromosomes. Two successive divisions of the *nucleus occur (known as division I and division II), with corresponding cell divisions, following a single chromosomal duplication. Thus a single *diploid cell gives rise to four *haploid cells. This produces gametes (in animals) or sexual spores (in plants and some *Protozoa) that have one half of the genetic material or chromosome number of the original cell. This halving of the chromosome number ($2n$ to n) compensates for its doubling when the gametes $(n + n)$ unite to form a *zygote $(2n)$ during sexual reproduction. The process occurs during gamete formation in animals or during spore formation in plants and Protozoa. The first stage of the first division of meiosis is often called prophase I and for convenience it has been divided into the leptotene, zygotene, pachytene, diplotene, and diakinesis stages; these are not distinct and grade into each other. Chromosomes first appear in the first stage (leptotene) of meiosis, as single threads. The two homologous members of each chromosome pair associate side by side with corresponding *loci adhering together: this is called pairing. It occurs during the zygotene stage, each resulting pair being called a bivalent. Thus the *apparent* number of chromosome threads is half what it was before, being the number of bivalents rather than the number of

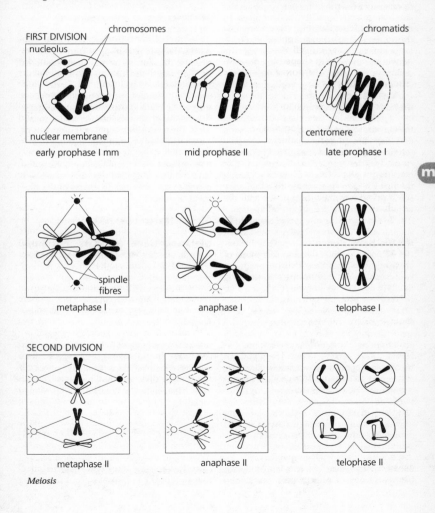

Meiosis

single chromosomes. During the pachytene stage, each bivalent separates into two sister *chromatids (except at the region of the *centromere), with some localized breakage and *crossing-over of genetic material of both maternal and paternal origin. There are now *n* groups of four chromatids lying parallel to each other and forming a *tetrad. During the diplotene stage, one pair of sister chromatids in each of the tetrads begins to separate from the other pair except at the sites where exchanges have taken place. In these regions the overlapping chromatids form a cross-shaped structure called a *chiasma and these chiasmata slip towards the ends of the chromatids so that their position no longer coincides with that of the original cross-overs. This process continues until, during diakinesis, all the chiasmata reach the ends of the tetrads and the homologues can separate during *anaphase. At diakinesis the chromosomes coil tightly, so shortening and thickening to form a group of compact tetrads which are well spaced out in the nucleus, and the *nucleolus disappears. This ends the prophase I stage of meiosis. During the first division (metaphase to telophase), the *nuclear envelope disappears, with the tetrads arranged at the equator of the *spindle. The chromatids of a tetrad separate in such a way that maternal chromosomal material is kept distinct from paternal material except at regions distal to the points of crossing-over. This first division produces two secondary gametocytes containing dyads (a dyad is half a tetrad) each of which becomes surrounded by a nuclear envelope. After a short interphase, the second division (prophase II) begins, during which the sister chromatids of a single chromosome are separated. The nuclear membrane disappears once more and the dyads arrange themselves upon the metaphase plate, the chromatids of each dyad being equivalent to one another (except for those regions distal to points of crossing-over). The centromere divides and so allows each chromosome to pass to a separate cell and the process is complete. Four cells result from the two divisions of meiosis. *Compare* MITOSIS.

Melamphaeidae (Melampheidae; bighead, bigscale; subclass *Actinop-

terygii, order *Beryciformes) A small family of deep-sea fish that have a relatively large head supported by thin bones, a large mouth, and small eyes. The body, not deeper than the head, is covered by thin and large scales. The *pectorals are as long as the single *dorsal fin and much larger than the *pelvic fins. Melamphaeidae are commonly found at depths of around 1000 m. There are about 20 species, distributed world-wide.

Melanerpes (woodpeckers) *See* PICIDAE.

melanin One of a group of dark pigments found in the skin and formed in melanoblast cells through the oxidation of *tyrosine, *phenylalanine, and other aromatic compounds. There are two chemically differing forms: the black-brown eumelanin and the red-yellow phaeomelanin. In many mammals, these alternate during the growth of a hair, resulting in a banded ('agouti') hair.

melanism In an animal population, the occurrence of individuals that are black owing to the presence in excess of the pigment *melanin.

Melanitta (scoters) *See* ANATIDAE.

Melanocetidae (subclass *Actinopterygii, order *Lophiiformes) A small family of deep-sea anglerfish that have a very large head, an *illicium, and a short body. The jaws of the huge mouth bear many sharp teeth. There are about nine species found at depths of 50–2000 m in all major oceans.

melanocytes Cells derived from *neural crest cells in the dermis and deeper *epidermis, which produce *melanin granules. Melanin is then injected by melanocytes into adjacent cells that do not produce melanin.

Melanogrammus aeglefinus (haddock) *See* GADIDAE.

Melanostigma pammelas (Pacific softpout) *See* ZOARCIDAE.

Melanostomiatidae (black dragon-fish; subclass *Actinopterygii, order *Salmoniformes) A fairly large family of deep-sea fish that have slender bodies, with a small head and small eyes, but a wide mouth. Both the *dorsal and *anal fins are located very close to the small tail fin. The fish are scaleless, but bear rows of small *photophores along the sides of the body. Many species possess a chin *barbel. There are at least 90 species, found in all oceans.

Melanotaeniidae (rainbowfish; subclass *Actinopterygii, order *Atheriniformes) A family of mainly freshwater fish that have a deep, compressed body (especially in the males), a pointed snout, and a small mouth. The base of the *anal fin is longer than that of the second *dorsal fin. There are about 19 species, found in tropical to subtropical waters around eastern Indonesia and Australia.

Meleagrididae (turkeys; class *Aves, order *Galliformes) A family of very large gamebirds which are dark, with a green and bronze sheen, and have a bare, blue or red head with *wattles. The wings are barred with white, the tail broad and rounded, the legs long and spurred. Turkeys are gregarious and terrestrial, and are found in wooded country. They feed on vegetable matter, and breed on the ground. *Meleagris gallopus* is the wild ancestor of the domestic turkey, with characteristic bare, red head wattles, and a 'gobbling' call. There are two genera, with two species, found in N. and Central America.

Meleagris (turkey) See MELEAGRIDIDAE.

Meles (badger) See MUSTELIDAE.

Meliphagidae (friarbird, honeyeaters; class *Aves, order *Passeriformes) A family of small to medium-sized, mainly green or grey-brown birds, some of which have patches of red, yellow, or white. Many have bare, *wattled skin on the head, the bill is slender and curved, and the tongue has a brush-like tip. (Friarbirds (17 species of *Philemon* found in Australia, New Guinea, and Indonesia) have on their heads areas of bare, black skin, with a black knob on the forehead.) The wings are long and pointed, the tail is medium to long, and the legs are strong. Honeyeaters inhabit mainly forests, feed on nectar, insects, and fruit, and build a domed, open, or cup-like pendant nest in trees. There are 37 genera, with about 170 species, found in S. Africa and Australasia.

Melipona (stingless bees) See APIDAE; HIVE; HONEY-BEE.

Melittidae (order *Hymenoptera, suborder *Apocrita) Family of short-tongued mining bees, which superficially resemble the Andrenidae, but differ from members of that family in that they lack subantennal plates and have the *scopa restricted to the posterior *tibia and *basitarsus. Females of *Macropis* collect plant oils as well as nectar.

melliferous Producing honey.

Mellisuga helenae (bee humming-bird) See TROCHILIDAE.

Mellivora capensis (honey badger, ratel) See MUSTELIDAE.

Meloidae (oil beetles, blister beetles; subclass *Pterygota, order *Coleoptera) Family of beetles, of which the ground-living species have a large, swollen *abdomen, reduced *elytra, and no wings. Others occur on flowers, are parallel-sided, have normal elytra, have warning coloration, and are fully winged. Meloidae contain cantharidin, a toxin producing skin blisters. Powdered elytra of *Lytta vesicatoria* (Spanish fly) is used in minute quantities as a supposed aphrodisiac. Larvae are minute and louse-like, with strong claws, and are insect parasites. Some attach to a bee host at a flower and are transported back to the nest, where they feed on host eggs and food stores. There are 2000 species.

Melolontha melolontha (maybug, cockchafer) See SCARABAEIDAE.

Melopsittacus undulatus (budgerigar) See PSITTACIDAE.

m

Membracidae (tree hoppers, thorn bugs; order *Hemiptera, suborder *Homoptera) Family of bugs in which the *pronotum is produced backwards into a spine; it is often also produced laterally, and may be elaborated into fantastic shapes. Some species live as small 'families', each comprising a mother and her offspring, and these are usually attended by ants. There are about 2500 species, occurring mainly in the warmer regions of the world.

membrane A sheet-like structure, 7–10 nm wide, that forms the boundary between a *cell and its environment and also between various compartments within the cell. It is composed mainly of lipids, proteins, and some carbohydrates, the structural arrangement of which is still subject to speculation. Membranes function as selective barriers and also as a structural base for enzymes, which may form an integral part of the membrane itself.

membranous labyrinth A structure in the inner ear of vertebrates that lies inside the *bony labyrinth and contains *endolymph.

memnonious Brownish black in colour.

memory The ability to store and recall the effects of experience. In non-human animals, memory may be studied by observing the persistence of learned responses.

Mendel, Gregor Johann (1822–84) An Austrian Augustinian monk who is credited with founding the science of *genetics, based on his experiments breeding peas in the monastery garden at Brünn (now Brno). These led him to formulate the laws of heredity (*see* MENDEL'S LAWS) which he presented to the Naturforschenden Verein (Natural History Society) in Brünn and, in 1866, published as an article ('Versuche über Pflanzenhybriden') in the Society's transactions. His work was not widely recognized until 1900, when similar results were obtained by other workers who, searching the literature, came across Mendel's original paper.

(⊕) SEE WEB LINKS
• Information about Gregor Mendel from the Masaryk University Mendel Museum.

Mendelian character A *character that follows the laws of inheritance formulated by Gregor *Mendel (*see* MENDEL'S LAWS).

Mendelian population An interbreeding group of organisms that share a common *gene pool.

Mendel's laws Two general laws of inheritance formulated by the Austrian monk Gregor Johann *Mendel. **1. (segregation)** The two members of a gene pair segregate from each other during *meiosis, each *gamete having an equal probability of obtaining either member of the gene pair. **2. (independent assortment)** Different segregating gene pairs behave independently. This second law is not universal as was originally thought, but applies only to unlinked or distantly linked pairs. At the time of Mendel, genes had not been identified as the units of inheritance: he considered factors of a pair of characters segregating and members of different pairs of factors assorting independently.

***Mene maculata* (moonfish)** *See* MENIDAE.

menhaden (*Brevoortia tyrannus*) *See* CLUPEIDAE.

Menidae (moonfish; subclass *Actinopterygii, order *Perciformes) A monospecific family (*Mene maculata*) of marine fish, about 20 cm long, that have an unusual body profile. The very deep body almost resembles a flat disc because of the strongly curved ventral contour. A very narrow tail ends in a forked tail fin. Both *dorsal and *anal fins are long and low; the first ray of the *pelvic fin is rather prolonged. Moonfish are found in the warm coastal waters of the Indo-Pacific region.

meninges Tissue that encloses the *central nervous system. In *Mammalia the meninges are the dura mater, arachnoid layer, and the pia mater.

meningitis An inflammation of the meninges (the membranes that cover the brain and spinal cord) which may be caused by any of a variety of bacteria or viruses.

Menippe mercenaria (stone crab) *See* XANTHIDAE.

Meniscotheriidae (superorder *Protoungulata, order *Condylarthra) An extinct family of archaic ungulates, which were distributed throughout N. America and Europe during the *Palaeocene and early *Eocene, being common locally. *Meniscotherium*, typical of the family, was an animal about the size of a rabbit. It had *molars of a *selenodont pattern, rather similar to those of later artiodactyls. Its limbs were short and stout and appear to have ended in claws rather than hoofs.

Menoponidae (order *Phthiraptera, suborder *Amblycera) Family of chewing lice which are parasitic on birds of many orders. One genus, *Piagetiella*, lives inside the *gular pouch of pelicans and cormorants, adult females passing through the nasal holes to lay eggs on the feathers. Another genus, *Actornithophilus*, contains species that live inside the quills of wing feathers, feeding on the quill material. Most species of Menoponidae, however, live among the feathers of the host. It is the largest family in the suborder, and includes about 650 species in 50 genera.

menstrual cycle The cycle of *ovulation that replaces the *oestrus cycle in most *Primates and follows an approximately monthly rhythm.

mentum 1. In insects, the *distal median plate of the *labium, bearing the *palps, *glossae, and *paraglossae or *ligula. **2.** In humans, the part of the chin that protrudes, especially in the *foetus. **3.** In some marine *Gastropoda, a projection of soft tissue below the mouth.

Menura (lyrebirds) *See* MENURIDAE.

Menuridae (lyrebirds; class *Aves, order *Passeriformes) A family of large, brown birds which have spectacular long tail feathers, most of them fine and filamentous, with two long central feathers, and two curved outer feathers in *Menura novaehollandia* (superb lyrebird). They inhabit forest and scrub, feed on small invertebrates, and build a large, domed nest on the ground. There are two species, confined to eastern Australia.

Mephitidae (skunks, stink badgers; superorder *Ferae, order *Carnivora) A family, formerly included in the *Mustelidae, of nocturnal mammals that inhabit fields and woods, but also occur close to human habitations, feeding on invertebrates which they dig from below ground, in urban areas augmenting this diet with garbage. They are renowned for their defensive use of a foul-smelling secretion they squirt with considerable force and accuracy, but only after giving ample warning of their intention. There are four genera and ten species of skunks, found throughout the United States, southern Canada, and Central and South America, and one genus and two species of stink badgers found in southern Asia and the Philippines.

mergansers *See* ANATIDAE.

Mergus (mergansers, sawbills) *See* ANATIDAE.

Merluccidae (hake; subclass *Actinopterygii, order *Gadiformes) A small family of marine fish that have a rather elongate body shape, a large head, and mouth with well-developed teeth. The first *dorsal fin is short and triangular, the second dorsal and *anal fins are long and may show a notch near the middle. Several species are important food fish. Often found at depths of more than 200 m, the large schools of hake may come closer inshore during summer. There are about 13 species, with world-wide distribution.

mermaid *See* SIRENIA.

merocenosis *See* MEROTOPE.

merocrine gland A gland whose secretions contain little or no solid material contributed by the disintegration of cells of the gland itself. *Compare* APOCRINE GLAND.

merogamete In *Protozoa, a *gamete formed by multiple division of the parent and so smaller than the parent. *Compare* HOLOGAMETE.

merological approach An approach to *ecosystem studies in which the component parts of an ecosystem are studied in detail in an attempt to compose a picture of the whole system (i.e. it assumes a system is the sum of its component parts). *Compare* HOLISTIC.

Meropidae (bee-eaters; class *Aves, order *Coraciiformes) A family of small to medium-sized, brightly coloured birds which have long, slender, *decurved bills, long, pointed wings, and *syndactylous feet. Many have elongated central tail feathers, others (e.g. *Merops* species) have square or forked tails; some have long throat feathers. They inhabit forest and open country, feed on insects, particularly bees and wasps, and nest colonially in burrows in a bank. There are three genera, with 24 species, many migratory, found in Europe, Africa, Asia, and Australasia. (*Merops apiaster* is the only species that breeds in Europe.)

meroplankton Temporary zooplankton (*see* PLANKTON), i.e. the larval stages of other organisms. *Compare* HOLOPLANKTON.

Merops (bee-eaters) *See* MEROPIDAE.

Merostomata (phylum *Arthropoda, subphylum *Crustacea) Class that includes the horseshoe crabs (king crabs) and extinct eurypterids. The eurypterids ('water scorpions'), which grew up to 3 m long, ranged from the *Ordovician to the end of the *Palaeozoic; the horseshoe crabs also appeared in the Lower Palaeozoic and have survived to the present.

merotope A microhabitat that forms part of a larger unit (e.g. a fruit or a pebble). Organisms colonizing the merotope form a merocenosis.

Merycoidodontidae A relatively small, extinct group of artiodactyls that appeared in the *Eocene, expanded in the *Oligocene, and dwindled to extinction in the *Pliocene.

Mesaxonia (infraclass *Eutheria, cohort *Ferungulata) A superorder that comprises the one order, *Perissodactyla, of those extant and extinct ungulates in which the third digit bears most of the weight of the animal, the other digits being reduced. The name is derived from the Greek *mesos*, 'middle', and *axon*, 'axis'. It has recently been proposed that the *Hyracoidea also belong to the Mesaxonia.

mesaxonic Applied to vertebrate limbs in which the weight of the animal passes through the third digit.

mesencephalon The mid-brain and second region of the *brain stem. The mesencephalon develops in conjunction with the eye, the roof (tectum) receiving input from the optic nerve.

mesenchyme cells In the developing vertebrate *embryo, cells that break away and become migratory, giving rise to various tissues such as cartilage, bone, smooth and voluntary muscle, blood cells and vessels, lymph vessels, and glands. Mesenchyme cells are derived from both the *mesoderm and *ectoderm germ layers of a *triploblastic embryo.

mesentery 1. In *Anthozoa, vertical partitions in the coelenteron (single body cavity). 2. In vertebrates, folds in the dorsal portion of the membrane lining the abdominal cavity (peritoneum) in which the abdominal organs are supported, so that these organs are slung from the wall rather than lying in the cavity proper. The folds supporting the *stomach, *liver, *duodenum, and *spleen are termed omenta (*sing.* omentum); those supporting the remainder of the *intestine are mesenteries.

mesepimeron *See* MESEPISTERNUM.

mesepisternum A pleural *sclerite on the *mesothorax of an insect, which is thought to be derived from the subcoxal elements of the ancestral legs, and is dissected by a suture from the other pleural sclerite, the mesepimeron.

mesites *See* MESITORNITHIDAE.

Mesitornithidae(mesites; class *Aves, order *Gruiformes) A family of medium-sized birds which are mainly brown above and plain, spotted, or barred below. They have long tails, short wings, and medium-length, straight or *decurved bills, and a well-developed *hallux. They are mainly terrestrial, rarely flying. They inhabit forests and scrub, feed on insects, fruit, and seeds, and nest in bushes and low trees. There are two genera, with three species, confined to Madagascar.

meso- From the Greek *mesos*, 'middle', a prefix with the same meaning (e.g. 'meso-trophic', neither nutrient-poor nor nutrient-rich; and '*mesofauna').

mesoblast A *mesoderm cell.

***Mesocricetus auratus* (golden hamster)** *See* CRICETIDAE.

mesoderm In a *triploblastic *embryo, the middle layer of cells, which form a hollow ring surrounding the *endoderm, giving rise to muscle, blood vessel, some nervous tissue, and the organs of the body within the mesoderm cavity. On development of the *notochord, the mesoderm develops bilaterally paired blocks on either side of the notochord, known as *somites, in which voluntary muscles, vertebrae, and deep portions of the skin develop. Somites are connected by undifferentiated mesoderm, known as lateral plate mesoderm. The ectodermal layer of the mesoderm is called the somatic mesoderm and the endodermal layer is called the splanchnic mesoderm layer. The mesoderm region between the somite and lateral plate mesoderm eventually separates and forms a nephric ridge which will develop into the kidney. A genital ridge later forms from the medial border of the kidneys which gives rise to the gonads. After differentiation of the mesoderm, *organogenesis occurs.

mesofauna Generally, animals of intermediate size, including small, invertebrate animals found in the soil, characteristically *Annelida, *Arthropoda, *Nematoda, and *Mollusca. These organisms are readily removed from a soil sample using a *Tullgren funnel or similar device. *Compare* MACROFAUNA; MEIOFAUNA; MICROFAUNA.

Mesogastropoda (class *Gastropoda, subclass *Prosobranchia) An order of prosobranch gastropods in which individuals possess one *pectinibranch gill and one *kidney. The shell is generally helical, the aperture being entire. The morphology of the *radula is variable, and the sexes are separate. This large order contains terrestrial, marine, and freshwater forms, and first appeared in the Early *Ordovician.

mesogloea In coelenterates and *Porifera, a gelatinous layer between the external and internal layers of the body wall. The mesogloea may range from a thin, non-cellular membrane to a thick, fibrous, jelly-like material, and may contain cells that have migrated from other areas.

mesolecithal Applied to an egg that has a yolk of intermediate size and strongly concentrated in one hemisphere, typical of *Petromyzonidae, *Acipenseridae, *Amiidae, *Lepidosirenidae, and *Amphibia. *Compare* MACROLECITHAL; OLIGOLECITHAL. *See also* ISOLECITHAL; TELOLECITHAL.

mesonephros In the development of an *amniote *embryo, the part of each *kidney that at first is concerned with excretion. In later development that role is assumed by the *metanephros, the mesonephros degenerating in females and in males becoming connected to the testes and associated with reproduction.

mesonephrostomal In the *nephridia of *Oligochaeta, applied when the upper lip of the nephridial funnel consists of only a few cells. *Compare* MEGANEPHROSTOMAL; MICRONEPHROSTOMAL.

mesonotum The dorsal, sclerotized *cuticle of the second thoracic segment of an insect. *See also* MESOTHORAX.

Mesonychidae (superorder *Protoungulata, order *Condylarthra) An extinct family of condylarths, formerly classified as *Creodonta, which are believed to show that modern ungulates and carnivores are descended from a common stock. They first appeared in the *Palaeocene, became more common in the *Eocene, and had disappeared by the middle of the *Oligocene. They had upper cheek teeth that were triangular, each with three blunt cusps, and lower *molars apparently adapted for shearing. The feet probably were hoofed. Some later forms were large. *Mesonyx* was about 1.5 m long, but judging by the comparative sizes of their skulls *Andrewsarchus*, which lived in Mongolia, may have attained a length of 4.5 m. The animals may have fed on tough vegetable material or carrion.

mesopleurum Lateral, sclerotized *cuticle of the second thoracic segment of an insect. *See also* MESOTHORAX.

mesopodium *See* APPENDICULAR SKELETON.

Mesosauria (class *Reptilia, subclass *Anapsida) An order of reptiles that comprises one family, the Mesosauridae. The mesosaurs are known only from late *Carboniferous or early *Permian rocks in S. America and S. Africa. They were adapted to life in fresh water, lightly built, and up to 1 m in length.

mesosoma 1. In invertebrates whose body is divided into sections (e.g. *Lophophorata and *Hemichordata), the 'collar' that lies between the *protosoma and the *metasoma. **2.** In *Hymenoptera, the combined *thorax and first abdominal segment. *See also* GASTER; METASOMA; SCORPIONES. **3.** *See* ENTEROPNEUSTA.

mesosternum Ventral, sclerotized *cuticle of the second thoracic segment of an insect. *See also* MESOTHORAX.

mesostracum A shell layer, composed of calcium carbonate, that occurs in some of the more advanced polyplacophorans. It is situated between the middle and outer layers of the shell.

mesothelium *See* SEROUS MEMBRANE.

mesothorax The second thoracic segment of the body of an insect, consisting of a cuticular box made up of a *mesonotum dorsally, *mesopleura laterally, and a *mesosternum ventrally. The mesothorax bears a pair of legs, and in addition may carry a pair of wings.

Mesoxaea *See* OXAEIDAE.

Mesozoa A phylum of small, marine, multicellular organisms which are simply ciliated and which adopt an endoparasitic mode of life within certain marine invertebrates.

Mesozoic The middle of three eras that constitute the *Phanerozoic period of time, about 251–65.5 Ma ago. The Mesozoic (literally 'middle life') was preceded by the *Palaeozoic Era and followed by the *Cenozoic Era. The Mesozoic comprises the *Triassic, *Jurassic, and *Cretaceous Periods.

messenger-RNA (m-RNA) A single-stranded *RNA molecule that is responsible for the transmission to the *ribosomes of the genetic information contained in the nuclear *DNA. It is synthesized during *transcription and its base sequencing exactly matches that of one of the strands of the double-stranded DNA molecule.

Messinian An important stage of the late *Miocene, about 7.2–5.3 Ma ago, marked by thick evaporite deposits in the Mediterranean, suggesting the Straits of Gibraltar were closed and the Mediterranean was reduced to a series of evaporite basins.

Messor (harvester ants) *See* MYRMICINAE.

metabolic pathway A sequential series of enzymatic reactions involving the synthesis, degradation, or transformation of a *metabolite. Such a pathway may be linear, branched, or cyclic, and directly or indirectly reversible.

metabolic rate The rate at which energy is used as an organism carries out its *metabolism. The metabolic rate is usually measured by the oxygen consumption of the organism, and calibrated using either the *basal metabolic rate or a *standard metabolic rate.

metabolism The total of all the chemical reactions that occur within a living organism (e.g. those involved in the digestion of food and the synthesis of compounds from *metabolites so obtained). The metabolism comprises anabolism, which includes synthetic reactions, and catabolism, which includes breakdown reactions.

metabolite Any compound that takes part in or is produced by a chemical reaction within a living organism.

metacarpal In the forelimb of a tetrapod, one of the rod-like bones that articulate proximally with the *carpus and distally with the phalanges. In humans, the metacarpals occupy the palm region of the hand.

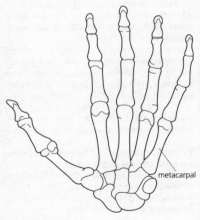

metacarpal

Metacarpal

metacarpus In the forelimb of a tetrapod, the *metacarpal bones.

metacentric Applied to a *chromosome that has its *centromere in the middle.

metachromism The widely accepted evolutionary theory which holds that the primitive colour of mammalian hair is *agouti. Subsequent evolution leads to saturation by one or other colour (i.e. eumelanic or phaeomelanic, *see* MELANIN and PHAEOMELANIN) owing to the elimination of the other colour, followed by bleaching, eventually to white. In the course of these changes the two primitive colours may be distributed about the body, producing distinctive patterns.

metacone In primitive, *triconodont, upper *molar teeth whose *cusps are in line, the posterior cusp. Where the cusps are not in line the metacone is the inside, posterior cusp.

metagenesis The occurrence, during the life cycle of an organism, of two types of individual, both *diploid, both of which reproduce, one sexually and the other asexually. The term is sometimes used loosely (but incorrectly) as a synonym for the *alternation of generations.

metallic bond The chemical bond that links the atoms in a solid metal. The atoms are ionized and electrons move fairly freely among them as an 'electron gas', the bond being between the electropositive atoms and the electrons. It is the free electrons that give metals their high electrical and thermal conductivities. *See also* COVALENT BOND; HYDROGEN BOND; IONIC BOND.

metamere *See* SOMITE.

Metameria Segmented organisms, including the annelids and arthropods, which radiated from a flatworm ancestor about 700 Ma ago.

metameric segmentation The repetition of organs and tissues at intervals along the body of an animal, thus dividing the body into a linear series of similar parts or segments (metameres). It is most strikingly seen in *Annelida. Essentially, metameric segmentation is an internal, *mesodermal phenomenon, the body musculature and *coelom being the primary segmental divisions; this internal segmentation imposes a corresponding segmentation on the nerves, blood vessels, and excretory organs. In some metameric animals the segmentation is visible externally but in others (e.g. *Chordata) external segmentation has been lost and internal segmentation is best seen in the *embryo. Metameric segmentation is thought to have arisen as an adaptation to more efficient locomotion.

metamorphosis Abrupt physical change. The word derives from the Greek *meta*, 'after', and *morphe*, 'form'. The term is used in a restricted sense for the transformation from the larval to the adult condition, as in the classes *Insecta, *Amphibia, etc.

metanauplius larva Later nauplius stages in *Crustacea. *See also* NAUPLIUS LARVA.

metanephridium *See* NEPHRIDIUM.

metanephros In the development of an *amniote *embryo, the part of each *kidney that develops later than and posterior to the *mesonephros and takes over functions associated with excretion to become the functional kidney in the later stages of embryonic development and throughout the life of the animal.

metanotum Dorsal, sclerotized *cuticle of the third thoracic segment of an insect. *See also* METATHORAX.

metaphase The stage of *mitosis or *meiosis at which the *chromosomes move about within the *spindle until they eventually arrange themselves in its equatorial region.

metapleuron The sclerotized *cuticle that makes up the sides of the third thoracic segment of an insect. *See* METATHORAX.

metapodium *See* APPENDICULAR SKELETON.

metapophysis In *Edentata, the additional, anterior, articulating facet on the posterior vertebrae.

metapopulation A group of *conspecific *populations that exist at the same time, but in different places. In other words, it is a population that is fragmented into numerous subpopulations.

metasepta In a *corallum, the plates that lie between the *prosepta.

metasoma 1. In invertebrates whose body is divided into sections (e.g. *Lophophorata and *Hemichordata), the 'trunk' that lies behind the *protosoma and *mesosoma. **2.** In *Hymenoptera, those segments that lie behind the *petiole (i.e. the abdomen from the second segment). *See also* GASTER; MESOSOMA; SCORPIONES.

metaspecies Ancestral species. They are not *monophyletic (in the strictest sense) so do not, and cannot, conform to the canon that taxa must always be monophyletic.

metastasis The act of dispersal by cancer cells.

metasternum (xiphoid process; xiphosternum) Ventral, sclerotized *cuticle of the third thoracic segment of an insect. *See also* METATHORAX.

metatarsal In the hind limb of a tetrapod, one of the rod-like bones that articulate proximally with the *tarsus and distally with the phalanges.

metatarsus In the hind limb of a tetrapod, the *metatarsal bones.

Metatheria (class *Mammalia, subclass *Theria) An infraclass that comprises the marsupials and their extinct relatives, sometimes included in the single order *Marsupialia and now more usually divided into several orders.

metathorax Third thoracic segment of the body of an insect, consisting of a box of sclerotized *cuticle (*see* METANOTUM; METAPLEURON; METASTERNUM). The metathorax bears a pair of legs, and in addition may carry a pair of wings.

Metazoa *See* ANIMALIA.

metazoan A multicellular animal. *See* ANIMALIA.

metencephalon The first region of the *rhombencephalon in the *brain. The metencephalon gives rise to a dorsal outgrowth, the *cerebellum, which constitutes a major brain region.

methionine A neutral, non-polar *amino acid that contains sulphur. It is an essential requirement in the diet of mammals.

methylation The introduction of a methyl group ($-CH_3$) into an organic compound.

Mexican bean beetle (*Epilachna varivestis***)** *See* COCCINELLIDAE.

Mexican burrowing toad (*Rhinophrynus dorsalis***)** *See* RHINOPHRYNIDAE.

MHC *See* MAJOR HISTOCOMPATIBILITY COMPLEX.

Miacidae (suborder *Fissipedia, superfamily *Miacoidea) An extinct mammalian family, known from the *Palaeocene and *Eocene, which is believed to be directly ancestral to the modern carnivores and which itself evolved directly from insectivore stock. Most Miacidae were small animals, typically much the size and shape of a modern weasel (with a long body and short legs), and were arboreal forest dwellers. They had larger brains than their creodont (*Creodonta) contemporaries and the *carnassials were well developed, formed by the elongated fourth *premolar and first *molar, the cusps forming ridges.

Miacoidea (order *Carnivora, suborder *Fissipedia) An extinct superfamily of early carnivores, comprising the single family *Miacidae and known from the *Palaeocene and *Eocene, during which it had a wide *Holarctic distribution.

Michaelis constant A kinetic constant (K_m) characteristic of each *enzyme, which is numerically equal to the concentration, in moles per litre, of the given *substrate that gives half the maximal velocity.

Micrastur (forest falcons) *See* FALCONIDAE.

micro- A prefix meaning 'small', from the Greek *mikros*, 'small'. Attached to SI units it means the unit$\times 10^{-6}$ (abbreviated as μ).

microbial genetics The study of the *genetics of micro-organisms. This is an important discipline of genetics, since micro-organisms may be bred in captivity and many generations can be obtained in short periods of time; their heredity can therefore be studied more readily than that of higher organisms.

microbiocoenosis The decomposer organisms that form part of the *biocoenosis in a *biogeocoenosis.

microbiology The study of micro-organisms and allied subjects.

Microbiotheriidae *See* DROMICIOPSIA.

microbivore An animal which feeds on micro-organisms.

microbody A term sometimes used to describe *peroxisomes and *glyoxysomes.

Microcebus (mouse lemur) *See* CHEIROGALEIDAE.

Microchiroptera (cohort *Unguiculata, order *Chiroptera) A suborder of bats that orientate themselves and locate prey principally by means of echo-location, aided by large, specialized ears and in many species by modifications of the nose which serve to beam the sound emissions. Typically, microchiropterans are insectivorous, but many have adapted to diets of fruit, fish, flesh, nectar, or blood, and although the suborder has a world-wide distribution many families within it have a very restricted range. In some families implantation is delayed until the spring following autumn copulation. Gestation is from 50 days to eight months, and the young of some species do not fly until they are 10 weeks old. Microchiropterans are known to have lived in the *Palaeocene and today they are among the most successful of all mammal groups. There are about 18 families, with some 700 species in all.

microclimate The *climate of a *microhabitat.

microconsumer See CONSUMER.

microcosm 1. A late 19th-century American term encompassing essentially the same ideas as the word '*ecosystem'. 2. (micro-ecosystem) A small-scale, simplified, experimental ecosystem, laboratory- or field-based, which may be: (a) derived directly from nature (e.g. when samples of pond water are maintained subsequently by the input of artificial light and gas-exchange); or (b) built up from *axenic cultures until the required conditions of organisms and environment are achieved.

Microcyprini An older name for a group of fish including the *Cyprinodontidae (toothcarps) and *Poeciliidae (live-bearers).

Microdesmidae (wormfish; subclass *Actinopterygii, order *Perciformes) A small family of mostly marine fish that have an elongate to eel-like body, small eyes, a pronounced lower jaw, and minute scales in the skin. The *dorsal fin extends along most of the body, and both dorsal and *anal fins are confluent with the tail fin. Wormfish are coastal fish living partially burrowed in the coarse sand of the sea-bottom. There are about 21 species, found in tropical waters world-wide.

micro-ecosystem See MICROCOSM.

microevolution Evolutionary change within *species, which results from the differential survival of the constituent individuals in response to *natural selection. The genetic variability on which selection operates arises from *mutation and sexual reshuffling of gene combinations in each generation.

microfauna The smallest animals (i.e. micro-organisms). Compare MACROFAUNA; MEIOFAUNA; MESOFAUNA.

microfibril One of many threadlike filaments formed in the *extracellular matrix from *fibrillin. Microfibrils form *elastic fibres.

microfilament A filament, 0.4–0.7 nm in diameter, containing the protein *actin. Such structures often occur in abundance immediately beneath the *cell membrane and play a role in cell motility and *cytokinesis (see MITOSIS).

microgamete The smaller of the two *gametes that fuse in the process of fertilization; it is usually the male gamete.

Micrognathozoa (phylum (or superphylum) *Gnathifera) A new class (or phylum) of microscopic invertebrates, first proposed in 2000, that is based on one species, Limnognathia maerski, found in a cold spring at Disko Island, West Greenland. L. maerski resembles *Aschelminthes. It is up to 0.1 mm long and has complex jaws, located in its *pharynx, which it can extend beyond its mouth while eating and when regurgitating indigestible items. The jaws are similar to those of some *Gnathostomulida and *Rotifera, with which the Micrognathozoa has been combined into the phylum Gnathifera. The body of L. maerski comprises a head, thorax, and abdomen, the dorsal and lateral *epidermis is covered by plates formed from the intercellular *matrix, but the epidermis is not syncytial (see SYNCYTIUM). The animal moves by means of two ventral rows of multiciliated cells, and it bears tactile bristles made from 1–3 cilia, and cilia around the head direct food particles towards the mouth. All the specimens so far discovered are female, suggesting that L. maerski reproduces by *parthenogenesis. It lays two types of egg: thin-walled eggs that hatch rapidly and thick-walled eggs that can withstand freezing, making it possible for them to overwinter.

microhabitat A precise location within a *habitat where an individual species is normally found (e.g. within a deciduous oak woodland habitat woodlice may be found in the microhabitat beneath the bark of rotting wood).

Microhylidae (narrow-mouthed frogs; class *Amphibia, order *Anura) A family of burrowing or nocturnal, arboreal frogs, most of which have an egg-shaped body, tapered head, and narrow mouth. The tadpoles may be free-swimming, or develop within the egg. *Breviceps adspersus* (rain frog or shorthead) of southern Africa lays its egg mass in a damp burrow and the froglets develop without an aquatic phase; large numbers of rain frogs appear after rain. There are many species, occurring in the Old and New World tropics.

micromere *See* BLASTOMERE.

micronephrostomal In the *nephridia of *Oligochaeta, applied when the upper lip of the nephridial funnel is absent. *Compare* MEGANEPHROSTOMAL; MESONEPHROSTOMAL.

micro-organism Literally, a microscopic organism; the term is usually taken to include only those organisms studied in microbiology (i.e. bacteria, fungi, microscopic algae, *Protozoa, and viruses), thus excluding other microscopic organisms such as eelworms and rotifers.

microphagy Feeding on items that are very much smaller than the organism consuming them. *Compare* MACROPHAGY.

Micropharyngea *See* TENTACULATA.

Micropsitta (pygmy parrots) *See* PSITTACIDAE.

Micropterigidae (class *Insecta, order *Lepidoptera) Family of very small, metallic-coloured moths believed to be the most primitive of Lepidoptera, and sometimes classified as a suborder (Zeugloptera). Adults are usually day-fliers, have functional *mandibles (rare in Lepidoptera), and feed on pollen. The venation is similar in fore wing and hind wing. Larvae may feed on detritus. The moths are widely distributed, but with few species. The family includes the oldest fossil lepidopteran, which was found in *Cretaceous amber.

Micropterus salmoides (largemouth bass) *See* CENTRARCHIDAE.

micropylar *See* MICROPYLE.

micropyle (*adj.* **micropylar)** A pore in the egg membrane of an insect *oocyte, which allows sperm to enter and fertilize the egg.

microsatellite DNA *See* SATELLITE DNA.

Microsauria A group of very small, *Carboniferous and *Permian amphibians that are thought to be ancestral to the modern *Apoda.

microsclere In *Porifera (sponges), a small, supporting *spicule. *Compare* MEGASCLERE.

microsome A small, membrane-bound *vesicle formed from the *endoplasmic reticulum in large numbers following the homogenization of cells. Microsomes may or may not bear *ribosomes, depending upon the type of endoplasmic reticulum from which they are derived.

Microsporidea (phylum *Protozoa, subphylum *Cnidospora) A class of protozoa which live parasitically inside the cells of a wide range of invertebrates and some lower vertebrates. There is one order, divided into two suborders. The class includes *Nosema*, species of which cause diseases of economic importance in silkworms and honey-bees.

microtine cycles The periodic, *density-dependent fluctuations in *population size, involving mass migrations, of certain animal *species, the best known of which is the Scandinavian lemming (*Myodes lemmus*) of the rodent subfamily Microtinae, after which the phenomenon is named. An abundance of nutritious food (phosphate being the limiting nutrient for lemmings) causes a rapid increase in numbers; their feeding reduces plant productivity until overcrowding triggers the migration. Cyclical *locust migrations can cause catastrophic damage to agricultural crops over large areas.

microtrabecular lattice The irregular lattice of slender, interconnected strands of fibre, anchored to the *cell membrane, that permeate the *cytoplasm in many eukaryotic cells (see EUKARYOTE).

microtubule A tubular structure, 15–25 nm in diameter and of indefinite length, that is composed of subunits of the protein tubulin. It occurs in large numbers in all *eukaryotic cells, either freely in the *cytoplasm or as a structural component of *organelles (e.g. *centrioles, *cilia, and *flagella). Microtubules appear to function in the motility of cells, the maintenance of cell shape, and the transport of materials within cells. In addition, they form part of the structure of the *mitotic spindle and have been implicated in sensory transduction in some receptor cells.

microtubule-organizing centre (MTOC) A diffuse zone of material around the *centriole from which *spindle*microtubules grow.

microvillus A small, finger-like projection. Microvilli are found in large numbers on the free surfaces of many cells. They serve to increase the surface-area-to-volume ratio of a cell and are a particular feature of those with an absorptive function, e.g. the epithelial cells of the intestine, and renal tubules.

microwhipscorpions See PALPIGRADI.

middle ear In tetrapods, except for *Urodela, *Apoda, and snakes (*Serpentes), the central chamber of the ear, derived from a *gill pouch, lying between the *tympanic membrane and the inner ear, linked to the *pharynx by the *eustachian canal, and containing the ear ossicles. In most mammals it is encased in the *bulla.

mid-water Applied to fish that swim clear of the sea-bottom.

midwife toad (*Alytes obstetricans*) See DISCOGLOSSIDAE.

migration The movement of animals from one area to another. Three cases may be distinguished: (*a*) emigration (outward only); (*b*) immigration (inward only); and (*c*) migration, which in this stricter sense implies periodic two-way movements to and from a given area, usually along well-defined routes. Such migratory movement is triggered by seasonal or other periodic factors (e.g. changing day length), and occurs in many animal groups.

migration route A link between two biogeographical regions that permits the interchange of plants and/or animals. Various types are recognized in the literature: e.g. G. G. *Simpson's '*corridors', '*filters', and '*sweepstakes routes' are widely referred to in connection with mammalian, and more recently reptilian, migrations.

migratory locust (*Locusta migratoria*) See LOCUST.

Milichiidae (order *Diptera, suborder *Cyclorrapha) Small family of flies, in which the adults are very small to small, and greyish-black or shining black. The larvae are basically scavengers in the nests of ants and birds, but a few species have larvae which feed on fungi. The family is distributed throughout the world, and 250 species have been described.

milkfish (*Chanos chanos*) See CHANIDAE.

milk sugar See LACTOSE.

millepores See MILLEPORINA.

Milleporina (massive hydrocorals, millepores; phylum *Cnidaria, class *Hydrozoa) An order of cnidarians whose members build massive, calcareous exoskeletons, which have pores through which the *polyps protrude. The *medusae are free but *degenerate and are formed in special cavities. Milleporina are known from the Late *Cretaceous to the present.

Millericrinida (sea lilies; class *Crinoidea, subclass *Articulata) An order of sea lilies which possess a stout *calyx consisting of basals and radials only (i.e. which is monocyclic). The long column is cylindrical in section and is distally attached to the substrate by a root-like structure or disc. The order is known from the Middle *Triassic to the present.

millipedes *See* DIPLOPODA.

milt A mass of spermatozoa and other secretions of the sperm ducts exuded by male fish during spawning, or during copulation in the case of the live-bearing species.

Mimetidae (order *Araneae, suborder Araneomorphae) Family of spiders which are predatory on other spiders, and have a characteristic row of large *setae on the first pair of legs. Species of the genus *Evo* hang under leaves and catch passing spiders by extension of their long legs. Other species invade spider webs, and some lure males of other species to their deaths by imitating the courtship display of the spider concerned.

mimicry *See* AUTOMIMICRY; BATESIAN MIMICRY; MÜLLERIAN MIMICRY.

Mimidae (mockingbirds, thrashers; class *Aves, order *Passeriformes) A family of blue-grey, grey, or brown birds, one species of which is entirely black (many of the nine *Mimus* species have a white *supercilium). Some have black-spotted under-parts. They have medium to long bills (slightly to very *decurved in thrashers, 10 species of *Toxostoma*), some have highly decurved, short, rounded wings, and they have long tails. They sing well, and many are mimics. They are arboreal and terrestrial, found in trees and dense scrub, feeding on insects, fruit, and seeds, and nesting in trees and bushes, or on the ground. There are 13 genera, with about 32 species, some migratory, found in N., Central, and S. America, and the W. Indies.

***Mimus* (mockingbirds)** *See* MIMIDAE.

mina (myna) *See* STURNIDAE.

mind, theory of The ability to attribute mental states to oneself and to other individuals and thereby to be able to predict the behaviour of others. There are standard tests to detect this ability in human infants. Research into the cognitive abilities of non-humans has centred mainly on monkeys and chimpanzees; it is not yet clear whether non-human primates possess a theory of mind.

mining bees 1. (desert mining bees) *See* FIDELIIDAE. **2. (solitary mining bees)** *See* ANDRENIDAE.

minisatellite DNA Short stretches of repetitive DNA that are distinctly different from the sequences classified as *satellite DNA.

minivets *See* CAMPEPHAGIDAE.

mink (*Mustela*) *See* MUSTELIDAE.

minke whale (lesser rorqual; *Balaenoptera acutorostrata*) *See* BALAENOPTERIDAE.

minnow *See* CYPRINIDAE.

Miocene The first of the two epochs of the *Neogene Period, about 23.03–5.332 Ma ago, extending from the end of the *Oligocene to the beginning of the *Pliocene. Many mammals with a more modern appearance evolved during this epoch, including deer, pigs, and several elephant stocks.

SEE WEB LINKS
• Description of the Miocene epoch of the Neogene period

***Mirafra* (bush larks)** *See* ALAUDIDAE.

mire Generally, a wetland area, especially a peaty area, and its associated *ecosystem.

Miridae (capsid bugs, leafbugs; order *Hemiptera, suborder *Heteroptera) Family of bugs which live on plants. Some are predators; others feed on plants. The family includes some destructive pests of both temperate and tropical crops. There are more than 5000 species, distributed world-wide.

Misgurnis fossilis (weatherfish) *See* COBITIDAE.

mis-sense mutant In genetics, a mutant in which a *codon has been altered by mutation so that it encodes a different *amino acid. The result is almost always the production of an inactive or possibly unstable *enzyme.

missing-plot technique Standard formula for the estimation of a missing datum observation (or, with suitable modification, the estimation of several missing values) in the analysis of the variance of data collected according to a recognized experimental design. The missing observation (x'_{ij}) may be estimated as:

$$x'_{ij} = (tT'_i + bB'_j - G')/(t-1)(b-1),$$

where T'_i, B'_j and G' are the treatment, block, and grand totals for the available observations, i is the ith treatment, j is the jth block, t is the number of treatments, and b is the number of blocks.

Mississippian *See* CARBONIFEROUS.

mistletoe-bird (*Dicaeum hirundinaceum*) *See* DICAEIDAE.

Misumena vatia *See* THOMISIDAE.

mites *See* ARACHNIDA.

mitochondrial-DNA (mt-DNA) Circular DNA that is found in mitochondria (*see* MITOCHONDRION). It is entirely independent of nuclear DNA and, with very few exceptions, is transmitted from females to their offspring with no contribution from the male parent. Mitochondrial-DNA codes for specific *RNA components of *ribosomes that are unique to those *organelles. It also codes for some of the respiratory enzymes found in mitochondria.

mitochondrion An oval, or occasionally round or thread-shaped, *organelle, whose length averages 2μm and that occurs in large numbers in the *cytoplasm of *eukaryotic cells. It is a double-membrane-bound structure in which the inner membrane is thrown into folds (the *cristae) that penetrate the inner matrix to varying depths. It is a semi-autonomous organelle containing its own DNA and 70S *ribosomes, and reproducing by *binary fission, partly under nuclear control. It is the major site of *ATP production and thus of oxygen consumption in cells, and houses the enzymes involved in the *citric-acid cycle and in *oxidative phosphorylation. Ribosomes have a *sedimentation factor; those with a factor of 70 are known as 70S ribosomes. Bacteria have 70S ribosomes, but the cytoplasmic ribosomes of eukaryotic cells are 80S. Endosymbiotic bacteria are believed to have evolved into mitochondria.

mitogenic Able to induce or to stimulate *mitosis.

mitosis The normal process of nuclear division by which two *daughter nuclei are produced, each identical to the parent *nucleus. Before mitosis begins each *chromosome replicates; these daughter chromosomes then separate during mitosis so that one duplicate goes into each daughter nucleus. The result is two daughter nuclei, each with an identical complement of chromosomes and hence of genes. Mitosis is generally divided into four phases: prophase, metaphase, anaphase, and telophase. During prophase, chromosomes become visible within the nucleus because they shorten, thicken, and coil up as a spiral. Each chromosome is longitudinally double except in the region of the *centromere, and each single strand of the chromosome is called a *chromatid. The *nucleolus and *nuclear envelope disappear. In metaphase the chromosomes move within the *spindle connecting the two *centrosomes (formed by division of the *centriole into two at the start of prophase, and the subsequent moving apart of these two daughter centrioles). The chromosomes finally arrange themselves along the equator of the spindle. During anaphase, the two chromatids making up each chromosome separate as the centromere becomes functionally double, and are thus converted to independent chromosomes moving to opposite poles. During telophase, the spindle disappears, a nuclear envelope

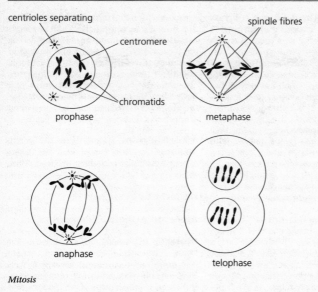

centrioles separating

centromere

chromatids

prophase

spindle fibres

metaphase

anaphase

telophase

Mitosis

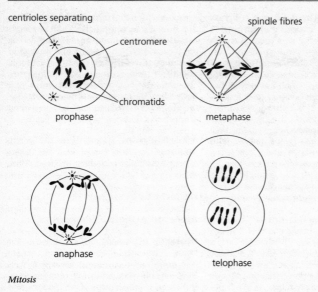

reappears around each of the two groups of daughter chromosomes, the chromosomes return to their extended state, and the nucleoli reappear. This is usually followed by cytoplasmic division (cytokinesis) into two parts by a cleavage furrow. The final result is two daughter cells with identical nuclear contents. *Compare* MEIOSIS.

mitotic spindle The spindle-shaped system of *microtubules that, during cell division, traverses the nuclear region of *eukaryotic cells. The *chromosomes become attached to it and it separates them into two sets, each of which can be enclosed in the envelope of a separate daughter nucleus.

mitral valve In the vertebrate heart, the valve that opens to allow blood to flow into the left *ventricle when the left *auricle contracts, and closes to prevent blood flowing in the opposite direction when the left ventricle contracts.

mixing ratio *See* HUMIDITY.

Mixosaurus atavus See ICHTHYOSAURIA.

mixotroph An organism in whose mode of nutrition both organic and inorganic compounds are used as sources of carbon and/or energy.

mnemon A *memory unit within the brain that is associated with a particular sensory system and with a particular set of behavioural responses to stimuli received by that system.

moas *See* DINORNITHIDAE.

mobbing The harassment of a predator by members of a prey species.

Mobilina (class *Ciliatea, order *Peritrichida) A suborder of ciliate *Protozoa which do not have stalks and are motile. The cell has a modified posterior region which forms a complex adhesive disc. Mobilina are found in association with aquatic animals.

Mobulidae (manta, devil ray; subclass *Elasmobranchii, order *Rajiformes) A small family of marine rays most of which are large, with very large *pectoral fins spread out like wings and pointed at the tips. A small *dorsal fin is located above the base of the whip-like tail, which has no tail fin. The two flap-like appendages extending forward on each side of the mouth, which assist in the uptake of food, are characteristic of the family. Perhaps the largest of the devil rays is *Manta birostris* (giant manta), with a 'wing-span' that may exceed 6.2 m. It is a non-aggressive, harmless species, feeding mainly on tiny planktonic animals. There are about 10 species, distributed worldwide in tropical and subtropical waters.

Mochokidae (upside-down catfish; subclass *Actinopterygii, order *Siluriformes) A large family of freshwater catfish that have, typically, three pairs of feathery *barbels around the mouth, an unusually large *adipose fin behind the single *dorsal fin, and a strongly forked tail fin. Several species, including *Synodon schall* (40 cm) and *S. nigriventris* (7 cm), often swim on their backs (upside-down), perhaps in search of insects at the water surface. There are about 150 species, found in Africa.

mockingbirds *See* MIMIDAE.

modern synthesis (neo-Darwinism) The fusion of Mendelian genetics and Darwin's *natural selection to produce a more comprehensive evolutionary theory than either can offer separately. *Darwin showed that evolution involves selection interacting with variation within populations, *Mendel that the bases of this variation are discrete units of heredity (*genes). A further synthesis (the *synthetic theory) has been achieved in recent years with the incorporation of knowledge of evolution at the molecular level.

modifier A *gene that modifies the phenotypic expression of another gene.

Moeritherioidea (superorder *Paenungulata, order *Proboscidea) An extinct suborder of subungulates, known from the upper *Eocene and lower *Oligocene in Africa, in which the beginning of tusk formation from the second *incisors can be recognized, and whose dentition was modified by the formation of a *diastema and the loss of the lateral lower incisor, the lower *canine, and the first *premolar. The animal was about the size of a modern tapir, with a long body, short legs, and a flexible snout. It may have inhabited marshes, and may have been amphibious.

Moeritherium lyonsi The first of the 'elephant' (proboscidean) line; an upper *Eocene mammal with short legs and a long trunk.

mojarra *See* GERREIDAE.

***Mola mola* (sunfish)** *See* MOLIDAE.

molar In mammals, one of the posterior teeth, commonly used for crushing, that are not preceded by milk teeth. Usually molars have several roots and their biting surface is formed from patterns of projections and ridges.

molariform row In mammals, a row of similarly formed *premolar and *molar cheek teeth all of which are used for crushing.

mole *See* TALPIDAE.

mole crickets *See* GRYLLOTALPIDAE.

molecular clock The idea that molecular evolution occurs at a constant rate, so that the degree of molecular difference between two *species can be used as a measure of the time elapsed since they diverged. Its accuracy depends on the validity of the *neutrality theory of evolution.

molecular drive The concept that changes within the *genome itself affect *evolution, quite irrespective of *natural selection. Apart from persistent directed *mutation, *gene conversion occurs so that whole *gene families are affected, and a mutation can creep through a population until a threshold is reached when large numbers of mutated individuals appear to arise all at once.

molecular evolution The substitution of one *amino acid for another in *protein synthesis as a result of *mutation of the genetic code. According to the *neutrality theory of evolution the variability at the molecular level which results from mutation is caused by random drift of the mutant genes rather than by selection.

molecular hybridization The artificial synthesis of double strands of polynucleotide molecules by binding together complementary strands of *RNA and/or *DNA. Such techniques permit the estimation of genetic similarity between DNA/DNA, DNA/RNA, and RNA/RNA polynucleotide chains. *See* COMPLEMENTARY BASE PAIRING.

mole rat *See* BATHYERGIDAE; SPALACIDAE.

mole salamanders *See* AMBYSTOMATIDAE.

Molidae (sunfish; subclass *Actinopterygii, order *Tetraodontiformes) A small family of fish of the open ocean that have a very peculiar body profile—they seem to be all head and no body. The head and body are very deep, the body short and ending in a tail fin that is no more than a narrow band without a *caudal peduncle. The *pectoral fins are small, the *pelvic fins are absent, but the *dorsal and *anal fins are long and provide the major locomotory thrust. The largest species, *Mola mola* (sunfish), may grow to 3 m, reaching a weight of some 1400 kg; the other species are smaller. Probably there are three species, occurring in tropical and subtropical seas.

Mollusca (molluscs) A phylum of coelomate (*see* COELOM) invertebrates comprising classes which are morphologically quite diverse, including the *Amphineura, *Bivalvia, *Cephalopoda, *Gastropoda, *Monoplacophora, and *Scaphopoda. They are fundamentally *bilaterally symmetrical, with *metameric segmentation almost or completely absent. Shell material is secreted by the *mantle covering the visceral hump. Calcium carbonate shells may be univalve, bivalve, or plated. Some groups have shells modified to serve as internal skeletons. Most have a well-differentiated head, with *radula and *salivary glands. A ventral muscular foot is very common. The *anus and *kidneys open into the *mantle cavity. The *alimentary canal usually has a buccal mass. A heart and an arterial and venous system are present, and there are paired *ctenidia for respiration. The nervous system is ganglionic (*see* GANGLION). Most Mollusca are aquatic. They first appeared in the Early *Cambrian. There are six extant and three extinct classes, with more than 80 000 species.

molluscs *See* MOLLUSCA.

Moloch horridus (thorny devil) *See* AGAMIDAE.

Molossidae (free-tailed bats; order *Chiroptera, suborder *Microchiroptera) A family of insectivorous bats in which the thick tail is surrounded by the *uropatagium which forms a loose sheath, and projects below the edge of the uropatagium when the animal is at rest. The nose is simple, with no leaf, and in most species overhangs the jaw. The ears are heavy-looking and deeply folded. Molossids often form very large colonies in caves and roofs. They are of world-wide tropical and subtemperate distribution. There are about 12 genera, with about 90 species.

Molothrus (cowbirds) *See* ICTERIDAE.

Molothrus badius (bay-winged cowbird) *See* ICTERIDAE.

Molothrus rufoaxillaris (screaming cowbird) *See* ICTERIDAE.

Molpadiida (class *Holothuroidea, subclass *Apodacea) An order of holothuroids which have *respiratory trees. A few *tube feet form a ring around the anus.

molybdeous Lead-coloured; drab grey.

Momotidae (motmots; class *Aves, order *Coraciiformes) A family of mainly bright green and blue birds which have black ear patches. They have strong, *decurved bills and *syndactylous feet, and in most the two, long, central tail feathers have bare shafts and racquet-shaped tips. Motmots are found in forests, feed on insects and fruit, and nest in holes in banks. There are six genera, with nine species, found in tropical Central and S. America.

Monacanthidae (filefish, leatherjacket; subclass *Actinopterygii, order *Tetraodontiformes) A family of deep-bodied marine to estuarine fish in which the first *dorsal fin can be locked in an upright position by the small second spine. Several species are valuable food fish. There are about 85 species, all of which recently have been accommodated in the family *Balistidae.

Monarcha (monarchs) *See* MUSCICAPIDAE.

monarchs (*Monarcha*) *See* MUSCICAPIDAE.

money spiders *See* LINYPHIIDAE.

mongaggle A group of mongooses that travel together and share food. *See also* HERPESTIDAE.

'Mongolism' *See* DOWN'S SYNDROME.

mongooses *See* HERPESTIDAE.

moniliform Resembling a string of beads.

monitors *See* VARANIDAE.

monkey *See* CATARRHINI; CEBOIDEA; CERCOPITHECIDAE; PLATYRRHINI.

monkey grasshopper *See* EUMASTACIDAE.

monkfish *See* SQUATINIDAE.

monoallelic In genetics, applied to a *polyploid in which all *alleles at a particular *locus are the same.

Monoblastozoa A *phylum proposed by R. Blackwelder (1909–2001) in 1963 to contain a single species, *Salinella salve*, which was found in water from a salt mine in Argentina and was described by J. Frenzel in 1892. The organism consists of a single layer of cells surrounding a gut with mouth and *anus; the mouth is surrounded by long *cilia and body cells are ciliated both internally and externally. Individuals conjugate (*conjugation) to produce cysts. *S. salve* has not been observed since the original find and may have been a misidentification.

Monocentridae (pinecone fish, knight fish; subclass *Beryciformes) A small family of marine fish that are short but deep-bodied and covered by heavy platelike scales. The eyes are large, but the fins are small, the first *dorsal fin consisting of at least four strong, membrane-free spines. The fish live in small schools, hiding in caves during the day. The better known species, *Cleidopus gloriamaris*, grows to 22 cm. There are two species, found in the Indo-Pacific region.

Monocirrhus polyacanthus (leaffish) *See* NANDIDAE.

monoclonal *See* MONOCLONAL ANTIBODY.

monoclonal antibody A particular *antibody produced by a *cell that is one of many identical cells each of which is a *clone of a single parent cell (i.e. each cell is monoclonal). The parent cell is formed by fusing a cell that produces the desired antibody (a lymphocyte) with a cell taken from a malignant lymphoid tumour in a mouse, yielding a hybrid that multiplies rapidly.

monocondylar With a single articulation.

monocyte *See* LEUCOCYTE.

monodactyl Applied to the condition (e.g. in *Thoatherium*, *see* LITOPTERNA) in which the lateral *metacarpals and *metatarsals are reduced, leaving only one functional digit.

Monodactylidae (diamond fish, moonfish; subclass *Actinopterygii, order *Perciformes) A small family of marine coastal fish that have a strongly compressed and deep body: the large *dorsal and *anal fins give the fish a diamond-shaped profile. Except for two dark bands across the head, these fish are a plain, silvery colour. About five species occur in the Indo-Pacific region.

Monodon (narwhal) *See* MONODONTI-DAE.

Monodontidae (suborder *Odontoceti, superfamily *Delphinoidea) A family comprising *Monodon monoceros* (narwhal) and *Delphinapterus leucas* (beluga, or white whale), whales in which the forehead is high and rounded, there is no dorsal fin, the flippers are rounded and short, and the tail fin is notched at the posterior margin. The beluga has 8–10 teeth in each jaw. The narwhal has two teeth. In the male, one (almost always the left) and rarely both of these may grow forward into a spiral tusk up to 2.5 m long. Narwhals are grey with black mottling, adult belugas white or near-white. Adults of both species grow to a length of about 6 m. Their diet consists of *demersal and *pelagic fish, and *Crustacea. They are distributed throughout Arctic waters and enter the larger rivers within the Arctic Circle.

monoecious Applied to an organism in which separate male and female organs occur on the same individual: e.g. to an *hermaphrodite animal. *Compare* DIOECIOUS.

monogamy Mating of a male with a female involving no extra individuals of either sex. Usually the bond operates through the breeding season and in some cases it may extend through the adult life of the two individuals.

Monogenea (phylum *Platyhelminthes) A class (or in some classifications an order of the class *Trematoda) of parasitic worms most of which have a gut with an intestine and *pharynx. An anterior sucker may be present, flanking the mouth. Mature individuals lack *cilia. Most are ectoparasitic; fish are the main hosts. Their name is derived from the fact that their life cycle

involves only one host (from the Greek *monos*, 'alone', and *genes*, 'of a particular kind'); they are distinguishable from *Digenea by their possession of a large, posterior, adhesive organ (an opisthaptor).

Monogononitida (phylum *Aschelminthes, class *Rotifera) An order of mainly freshwater rotifers most of which adopt a *vagile or *sessile mode of life. A degree of jointing is present in the cuticle. Sexes differ morphologically.

monogyne Applied to an ant, bee, or wasp (*Apocrita) nest that contains only one *queen. *Compare* POLYGYNE; *see also* GYNE.

monohybrid A cross between two individuals which are identically *heterozygous at one particular gene pair (e.g. $Aa \times Aa$).

monohybrid heterosis *See* HETEROSIS; OVERDOMINANCE.

monomer A molecule that may bind to others to form an *oligomer or *polymer.

Monomorium pharaonis (Pharaoh ant) *See* MYRMICINAE.

monomyarian Applied to the condition in certain *Bivalvia in which only the posterior adductor muscle is present.

monophagous Applied to an organism (e.g. many insect larvae) that feeds on only one type of plant or prey, or on species that are closely related.

monophyletic *See* MONOPHYLY.

monophyly The condition in which a group of taxa share a common ancestry, being ultimately derived from a single interbreeding (or *Mendelian) population, as opposed to a polyphyletic group, which is derived from many such populations. If the members of a given *taxon are descended from a common ancestor they are said to be 'monophyletic' (e.g. the families within a class would be monophyletic if they were all descended from the same family or lower taxonomic unit). Under the strictest definition they would all have to be descended from a single species. Monophyly is sometimes used interchangeably with *holophyly.

monophyodont Applied to animals in which the teeth are not replaced. *Compare* DIPHYODONT, POLYPHYODONT.

Monopisthocotylea (phylum *Platy-helminthes, class *Monogenea) An order of ectoparasitic worms which have a mouth, *pharynx, and intestine. The posterior end is ciliated.

Monoplacophora (phylum *Mollusca) A class of primitive, almost *bilaterally symmetrical, univalved molluscs, whose limpet-shaped shell of calcium carbonate consists of an external *periostracum, a prismatic layer, and an internal, *nacreous deposit. The circular foot and *radula are ventrally placed, and the internal organs appear to be *metamerically segmented, although some authorities question the evidence for metamerism, suggesting that monoplacophoran segmentation is not true metamerism but a secondary replication of structures such as has occurred in chitons and Nautilus. Typically, Monoplacophora have paired muscle scars. All are marine, *benthic filter-feeders. Fossil forms inhabited shallow water, but modern forms occur in deep water. They first appeared in the Early *Cambrian.

monopodial 1. A type of branching in which lateral branches arise from a definite main, central stem. **2.** With a single axis, an extension growth from the apex.

Monorhina An older name for the group of fish including the lampreys (*Petromyzonidae) and related jawless fossils which possess a single median nostril between the eyes.

monosaccharide The simplest *sugar, most monosaccharides having the formula $C_x(H_2O)_y$, where $x \geq 3$. *Glucose and *fructose are monosaccharides.

monosomic genome A *genome that is basically *diploid but that has only one copy of one particular *chromosome type so that its chromosome number is $2n - 1$. An XO human individual is monosomic for the *X-chromosome.

monothetic In numerical classification schemes, the use of a single criterion or attribute as the basis for each subdivision of a sample population, as for example in *association analysis. *Compare* POLYTHETIC.

Monotremata (class *Mammalia, sub-class *Prototheria) An order comprising the duck-billed platypus (*Ornithorhynchus anatinus*, *see* ORNITHORHYNCHIDAE) and the echidnas or spiny ant-eaters (*Tachyglossus* and *Zaglossus*, *see* TACHYGLOSSIDAE). There were some extinct forms, of which very few are known in detail, though of those some attained large sizes. The echidnas have no fossil record older than the *Pleistocene, but a fossil platypus, *Obdurodon*, is known from the *Miocene and in the early 1990s teeth of an undoubted platypus-like form, *Monotrematum sudamericanum*, were discovered in *Palaeocene deposits in Patagonia. Two *Cretaceous genera, *Steropodon* and *Kollikodon*, are also known. In view of their reptilian affinities they are thought to represent a separate and direct line of descent from the earliest *Mesozoic animals, possibly the *Docodonta, independent of the line leading to other mammals, but the dental features of *Steropodon* are considered by some to point to affinities with the *Eupantotheria. They retain many primitive features and are quite unlike marsupials or eutherians. The rectum and urinogenital system open to a common *cloaca (the name 'monotreme' is derived from the Greek *monos* meaning 'alone' and *trema* meaning 'hole', although the feature is shared by marsupials and some insectivores). The male is *heterogametic as in other mammals. The young are hatched from large, yolky eggs, incubated in a nest by the female platypus and in a pouch in the echidnas. The *embryos develop a *caruncle and egg-tooth. After hatching they are fed milk secreted by the female from specialized sweat glands which do not open through central nipples. The *diaphragm is fully developed, and the heart possesses a single left *aortic arch as in other mammals. The *larynx is developed, and monotremes make sounds. The *tarsus of the male has a grooved erectile spine to which (in the platypus) poison is fed from a gland in the thigh. The poison (said to be capable of killing a dog) may serve to immobilize the female during mating. Adults lack teeth but possess bills. The skull is specialized but retains primitive features. The jaw consists of a single bone. The cervical ribs are not fused, the shoulder girdle is reptilian in form, the pelvis is reminiscent of that of marsupials. The body is covered with hair, in the echidnas partly modified into spines on the back. The limbs are modified for digging (echidna) or swimming (platypus); it has recently been discovered that both platypus and echidna locate their prey by detecting weak electrical fields around the snout. Echidnas *hibernate, and in hot weather all monotremes shelter in burrows or caves. Monotremes are known only from Australia and New Guinea, where they may have survived because of their high degree of specialization and the isolation of that continent.

Monotrematum sudamericanum A *Palaeocene ornithorhynchid (*Ornithorhynchidae) from Patagonia, Argentina, known so far only by its *molar teeth, which closely resemble the distinctive molars of *Obdurodon and, to some extent, those of *Steropodon and the vestigial ones of *Ornithorhynchus. Its presence demonstrates that monotremes, like marsupials, were part of the *Gondwana fauna.

monotrysian The condition in which a single opening serves for mating and egg laying.

monotypic Applied to any *taxon that has only one immediately subordinate taxon. For example, a genus that contains only one species would be described as monotypic, as would a family containing only one genus. Compare POLYTYPIC.

monoxenous See AUTOECIOUS.

monozygotic twins Twins that are formed at some time during early development by the fission into two of the *embryo

derived from a single fertilized egg. Such twins are genetically identical and therefore also of the same sex. Compare DIZYGOTIC TWINS.

monster anglerfish See LINOPHRYNIDAE.

Monticola (rock thrushes) See TURDIDAE.

Montypythonoides riversleighensis The name given by Michael Archer to a species of fossil python he discovered and described from *Oligocene–Miocene deposits at Riversleigh, Queensland, Australia. This species is now known as Morelia riversleighensis.

mooneye (Hiodon tergisus) See HIODONTIDAE.

moonfish 1. (opah, Lampris guttatus) See LAMPRIDIDAE. **2.** (Mene maculata) See MENIDAE. **3.** See MONODACTYLIDAE.

moon moths See SATURNIIDAE.

moon rat See ERINACEOIDEA.

moor An acidic, usually upland area, commonly with peat development, dominated by low-growing heaths and heathers, with some areas dominated by grasses and sedges.

moorhens See RALLIDAE.

moose (elk; Alces alces) See CERVIDAE.

mor An acid *humus composed of several layers of organic matter in various stages of decomposition, in which animal and plant remains are visible, and there is little microbial activity except by fungi.

morabine grasshopper See EUMASTACIDAE.

moray eels See MURAENIDAE.

morbidity 1. The proportion of individuals in a population suffering from a particular disease. **2.** The state or condition of being diseased.

Morelia riversleighensis *See* MON-
TYPYTHONOIDES RIVERSLEIGHENSIS.

Morgan's canon The injunction, first
stated by Lloyd Morgan (1852–1936) in
1894, that animal behaviour should never
be attributed to a high mental function if
the behaviour can be explained in terms of
a simpler process.

Morganucodon A small, mouse-sized
mammal (sometimes considered identical
to *Eozostrodon*), from the Late *Triassic
to Early *Jurassic of Europe and Asia, that
is known from complete skeletons. It is the
earliest known mammal, transitional from
the mammal-like reptiles, with such features as a double jaw articulation (both the
primitive reptilian quadrate–articular joint
and the advanced mammalian squamosal–
dentary joint).

**Moridae (morid cod; subclass *Ac-
tinopterygii, order *Gadiformes)** A family of marine fish that have a fairly elongate
body, and sometimes a wide head and chin
*barbel. Although the first *dorsal fin is
short, the second dorsal and *anal fins are
long, but may be subdivided into sections.
The first few *pelvic fin rays are prolonged.
The Moridae are related to the Gadidae (true
cods): some species are found in deep water,
others may come closer inshore. There are
about 70 species, found world-wide in temperate waters.

morid cod *See* MORIDAE.

**Moringuidae (worm eel; subclass *Ac-
tinopterygii, order *Anguilliformes)**
A small family of eel-like marine fish that have
a very elongate, cylindrical body. The mouth
and eyes are rather small, and the *pectoral
fins minute. The low *dorsal and *anal fins
are near the tail and confluent with the tail
fin. There are about 10 species, found in
the Indo-Pacific region and the western
Atlantic.

**Mormoopidae (ghost-faced bats,
moustached bats, chin-leafed bats;
order *Chiroptera, suborder *Microchiroptera)** A family of insectivorous bats in
which there is no nose leaf, but the nostrils
form part of a fleshy pad extending on to
both lips and, in *Mormoops* species, continuing in foliar outgrowths on the chin. The
ears cover the sides of the head to below the
eyes, and the tail emerges dorsally from the
*uropatagium. There are two genera, *Mor-
moops* and *Pteronotus*, with about eight species, distributed through tropical Central
and S. America.

**Mormyridae (elephant fish; super-
order *Osteoglossomorpha, order
*Mormyriformes)** A fairly large family of
freshwater fish, characterized in most species by the presence of an elongate, trunk-
like snout. The eyes and mouth are small
and the tail fin is forked; the size of the *dorsal and *anal fins is variable. Several species
possess electric organs capable of emitting
series of low-voltage discharges, perhaps
as a means of finding the way in darkness,
since mormyrids are basically nocturnal.
There are about 100 species, found in tropical Africa.

**Mormyriformes (subclass *Actinop-
terygii, superorder *Osteoglossomorpha)** An order of primarily nocturnal
freshwater fish that have small eyes and
mouth and varying body profile. Several
genera (e.g. *Mormyrus* and *Gymnarchus*)
have electric organs capable of producing
weak electric discharges.

Morone saxitilis (striped bass) *See*
PERCICHTHYIDAE.

morpho butterflies *See* SATYRINAE.

morphology The form and structure of individual organisms. *Compare*
PHYSIOGNOMY.

morphometrics A technique of taxonomic analysis using measurements of the
form (*morphology) of organisms and typically involving multivariate statistics.

morphospecies A group of biological organisms that differs in some morphological
respect from all other groups.

morphotype In taxonomy, a specimen chosen to illustrate a morphological variation within a species population.

morula An *embryo at the stage where it consists of about 30 cells (*blastomeres) resembling a mulberry (Latin *morus*, hence the name).

morwong *See* CHEILODACTYLIDAE.

mosaic evolution Differential rates of development of various adaptive attributes that occur within the same *evolutionary lineage. For example, a particular *taxon might show greatly different rates of change with respect to the head, body, and limbs. This is a common phenomenon and makes the reconstruction of transitional fossil types very difficult.

mosaic heterochrony *Heterochrony in which a number of heterochronic processes occur simultaneously, so different parts of an organism develop at different rates. *Compare* ASTOGENETIC HETEROCHRONY; ONTOGENETIC HETEROCHRONY.

Mosasauridae (mosasaurs; order *Squamata, suborder *Sauria) A family of marine lizards, fossils of which have been found in Late *Cretaceous rocks throughout the world. They evolved into many types but all of them became extinct at the end of the Cretaceous. Mosasaurs were large, some reaching 9 m in length, with a long, slim body, short neck, and long head. The tail was used for swimming and the limbs for steering; the digits may have been webbed. The teeth were set in pits rather than being fused to the jaws. Most mosasaurs fed on fish, although some are believed to have fed on molluscs.

mosasaurs *See* MOSASAURIDAE.

Moschidae (musk-deer; infra-order *Pecora, superfamily *Cervoidea) A family of small ruminants, including *Moschus* (musk-deer) and its fossil relatives, that differ from *Cervidae, in which they were once included, in many anatomical features. Musk-deer are characterized by skeletal specializations for 'pogo-stick' jumping, and by possession in the male of deep *inguinal scent pouches and large, *recurved and slightly movable maxillary *canine teeth. The five or six living species have thick, quilly *pelage, large ears, and very long legs. They are found in the Himalayas, China, and Siberia, especially in mountainous country. Most species are threatened by hunting for the product of the males' scent glands, 'musk', that is used in perfumes.

***Moschus* (musk-deer)** *See* MOSCHIDAE; PALAEOMERYCIDAE.

mosquito *See* CULICIDAE.

moss-animals *See* BRYOZOA.

***Motacilla* (wagtails)** *See* MOTACILLIDAE.

***Motacilla flava* (yellow wagtail)** *See* MOTACILLIDAE.

Motacillidae (wagtails, pipits; class *Aves, order *Passeriformes) A family of medium-sized to small, black, grey, yellow, and brown birds, some of which are streaked. *Motacilla flava* (yellow wagtail), one of 10 species in its genus, exists in many subspecific forms, each having a different head colour. The bill is thin and slender, and the legs fairly long, with long hind claws. Motacillids have long bodies and long tails, usually edged with white or buff, which most species wag up and down. They inhabit grassland and open country, often near water. They feed mainly on insects, and nest on the ground. There are five genera, with 54 species, many migratory, and found world-wide.

moth fly *See* PSYCHODIDAE.

moths *See* LEPIDOPTERA.

motile Capable of independent locomotion.

motivation The cause for a spontaneous change in the behaviour of an animal that occurs independently of any outside stimulus, or of a change in the threshold of responsiveness of an animal to a stimulus, and that is not due to fatigue, *learning, or the maturation of the animal.

motivational conflict The simultaneous existence of two or more *motivations that lead to contradictory patterns of behaviour (e.g. to approach a human offering food in order to obtain the food, and to flee from the human).

motmots See CORACIIFORMES; MOMOTIDAE.

motor neuron A *neuron that is attached to a muscle. When a sufficient number of the motor neurons attached to a particular muscle become electrically active the muscle contracts.

moulting The periodic, often seasonal, shedding of hair or feathers by animals. In birds, it is the process by which feathers are periodically renewed, at least once a year, sometimes twice or (rarely) three times. Feather shedding is usually a gradual process and does not affect flight or other functions but *Anatidae lose their flight feathers simultaneously, becoming temporarily flightless. The usual sequence of moult begins with the loss of primaries, followed by secondaries, tail feathers, and then body feathers. See also ECDYSIS.

mountain beaver See APLODONTIDAE.

mouse 1. (Old World) See MURIDAE. **2. (New World)** See CRICETIDAE.

mousebird See COLIIDAE; COLIIFORMES.

mouse hare (*Ochotona*) See OCHOTONIDAE.

mouse lemur (*Microcebus*) See CHEIROGALEIDAE.

mouse-tailed bat (*Rhinopoma*) See RHINOPOMATIDAE.

moustached bats See MORMOOPIDAE.

mouth angle plates In *Asteroidea (starfish) plates that surround the mouth.

mouth-breeding fish See BUCCAL INCUBATION.

mouth-breeding frog (*Rhinoderma darwinii*) See LEPTODACTYLIDAE.

mouth brooding See BUCCAL INCUBATION.

m-RNA See MESSENGER-RNA.

MSY (maximum sustained yield) See OPTIMUM YIELD.

mt-DNA See MITOCHONDRIAL-DNA.

MTOC See MICROTUBULE-ORGANIZING CENTRE.

mucin A mucoprotein secreted by mucous cells.

mucoprotein A class of mucopolysaccharides in which large numbers of disaccharide units are bound to a protein chain.

mucosa See MUCOUS MEMBRANE.

mucous membrane (mucosa) In vertebrates, a moist membrane that consists of *epithelium (which is often ciliated) overlying *connective tissue; such membranes commonly secrete *mucus. The term is particularly applied to the membranes of the digestive and urinogenital systems.

mucro (*pl.* **mucrones**) A sharp point. Terminal portion of the springing organ (*furcula) of *Collembola, which is in contact with the substrate when the animal jumps.

mucronate Applied to an organ that ends in a sharp point.

mucrones See MUCRO.

mucus A secretion of mucous cells and glands located in epithelial structures. It is composed largely of a mixture of *mucin and water.

mud crabs See XANTHIDAE.

mud minnow *See* UMBRIDAE.

mudnest builders *See* GRALLINIDAE.

mudpuppy (*Necturus maculosus*) *See* PROTEIDAE.

mud turtles *See* KINOSTERNIDAE.

***Mugil cephalus* (common mullet)** *See* MUGILIDAE.

Mugilidae (grey mullet; subclass *Actinopterygii, order *Perciformes) A family of marine and brackish-water fish, which occasionally enter rivers. A mullet has a fairly thick, solid body with rather large scales, the mouth bearing rows of fine teeth. The *pelvic fins are located between the *pectoral and first *dorsal fins. The two widely separated dorsal fins are characteristic. Often congregating in large schools, most species feed on plankton, algae, or detritus. They are very tolerant of low salinity: some species, including *Myxus* (*Trachystoma*) petardi (freshwater mullet) of eastern Australia, are freshwater inhabitants. *Mugil cephalus* (striped or common mullet), 65 cm, is an example of the commercially important food fish in the family. There are about 70 species, distributed world-wide in tropical and temperate waters.

Mugiloididae (Parapercidae; sand perch, grubfish; subclass *Actinopterygii, order *Perciformes) A small family of marine fish that have an elongate body, small scales, a *protractile mouth, and a large head. The *dorsal and *anal fins are long; the *pelvic fins are located beneath the *pectoral fins. The dorsal position of the eyes suggests a bottom-dwelling way of life. One of the larger species is *Parapercis colias* (blue cod), 65 cm, a popular food fish in New Zealand. Generally they are inshore fish. There are about 25 species, confined to the southern hemisphere.

mule The offspring of a male donkey and female horse (*Equidae), classed as an F_1 *hybrid. *Compare* HINNY.

mule killer (*Dasymutilla occidentalis*) *See* MUTILLIDAE.

mull A well-aerated, fertile, surface soil that is rich in organic material mixed with mineral matter by the activity of animals (especially earthworms). Animal and vegetable material is not recognizable microscopically (*compare* MOR).

Müller, Fritz Johann Friedrich (1821–97) The German-born Brazilian zoologist who first recognized, in 1879, the type of mimicry named after him.

Müllerian mimicry The similarity in appearance of one species of animal to that of another, where both are distasteful to predators. Both gain from having the same warning coloration, since predators learn to avoid both species after tasting either one or the other. The phenomenon is named after Fritz MÜLLER who described it in relation to insects in S. America. *Compare* BATESIAN MIMICRY.

mullet *See* MUGILIDAE; MULLIDAE.

Mullidae (goatfish, red mullet; subclass *Actinopterygii, order *Perciformes) A family of marine fish that have an elongate body tapering towards the tail, but a flat belly, a steep forehead, and, characteristically, a pair of chin *barbels which are probably used as feelers. They are bottom-living fish, often with distinctive colour patterns. *Mullus surmuletus* (red mullet), 40 cm, of the eastern Atlantic has been a popular food fish at least since Roman times. There are about 55 species, mainly marine, distributed world-wide in warmer water.

***Mullus surmuletus* (red mullet)** *See* MULLIDAE.

multifactorial (polygenic) In genetics, a hypothesis to explain quantitative variation by assuming the interaction of a great number of genes (polygenes), each with a small additive effect on the character.

multilocular Containing more than one larva.

multiple allelism The existence of several known allelic forms of a *gene. *See* ALLELE.

multiseriate Arranged in many rows.

multistate character A *character that can occur in several *character states.

Multituberculata (class *Mammalia, subclass *Prototheria) An extinct order of rather rodent-like mammals, which first appeared in the Late *Jurassic in Europe, flourished during the *Cretaceous and *Palaeocene, but became extinct during the *Eocene. Probably they were the first herbivorous mammals, with skull and teeth analogous to those of rodents. Most were small, but some attained the size of modern woodchucks. The limbs sprawled more widely than those of most mammals. The olfactory bulbs were large, which suggests that the animals depended heavily on their sense of smell. The skull was massive, but unlike that of other mammalian groups, the multituberculates appear to have been a side branch from the main line of mammalian evolution, and they are believed not to be related closely to other groups.

munias *See* ESTRILDIDAE.

***Muntiacus* (muntjak)** *See* CERVIDAE.

muntjak (*Muntiacus*) *See* CERVIDAE.

***Muraena helena* (Greek moray)** *See* MURAENIDAE.

Muraenesocidae (pike conger; subclass *Actinopterygii, order *Anguilliformes) A small family of marine, eel-like fish that have a cylindrical body, scaleless skin, narrow head with large eyes, strong teeth, and the *dorsal fin starting above the well-developed *pectoral fins. Some species are known to enter brackish water. These rather aggressive fish are found throughout the world in tropical to subtropical seas, and some are quite large (e.g. *Muraenesox arabicus*, 2 m). There are about seven species.

Muraenidae (moray eel; subclass *Actinopterygii, order *Anguilliformes) A large family of marine, eel-like fish that have a slightly compressed, scaleless body, well-developed *dorsal and *anal fins, but no *pectoral or *pelvic fins. Many species exhibit colourful markings in the skin. Usually found in cracks and crevices between rocks near the sea-bottom, they can be very aggressive when handled. *Muraena helena* (Greek moray), 1.1 m, was a food fish well known to the Romans. Some species are very large, e.g. *Thyrsoidea macrura* (long-tailed reef-eel), 4.1 m, of northern Australia. There are about 100 species, distributed world-wide in tropical to temperate waters.

Muridae (order *Rodentia, suborder *Myomorpha) A family of Old World rats and mice that are perhaps the most successful of all mammalian families. They are small, terrestrial, arboreal, burrowing, or semi-aquatic animals. The tail is long and scaly, the limbs *pentadactyl and the first digit of the fore limb rudimentary. In some *saltatorial forms the hind limbs are modified. *Apodemus sylvaticus* is the European field mouse. *Mus musculus* is the house mouse, which originated in the drier areas of Eurasia but has spread throughout the world. The Muridae first appeared during the *Oligocene and radiated explosively during the *Pliocene. They reached Australasia (including New Guinea and the Solomon Islands) before the first humans, and radiated there into some 28 genera, including *Notomys* (jerboa-rat), which resembles the true jerboas (family *Dipodidae). There are at least 220 genera, with more than 1000 species.

muriform Resembling a brick wall; having both transverse and longitudinal cross-walls.

murine Of or like a mouse; mouse-coloured.

***Muscicapa* (flycatchers)** *See* MUSCICAPIDAE.

Muscicapidae (fantails, Old World flycatchers, monarchs, whistlers; class *Aves, order *Passeriformes) A diverse family of often highly coloured birds, some of which have crests or *wattles. The bill is narrow to broad, with *rictal bristles, the wings are short and rounded to long and pointed, the tail is usually short to medium, and the legs are short. Muscicapids are mainly arboreal and feed on insects caught in the air or on the ground. The nest is usually cup-shaped and made in a tree or bush, or in a hole in a tree or bank. Many are migratory. The flycatchers (27 species of *Ficedula* and 22 species of *Muscicapa*) and monarchs (30 species of *Monarcha*) are typical. Fantails (about 40 species of *Rhipidura*) inhabit dense forest and scrub in south-east Asia, Australasia, and some Pacific islands (but willie wagtail (*R. leucophrys*) is a common garden bird). Whistlers (28 species of *Pachycephala*, found in south-east Asia, Indonesia, New Guinea, Australia, and Pacific islands) have large heads and stout bills. There are about 60 genera in the family, with about 350 species, most of which are migratory, found in Europe, Africa, Asia, and Australasia. *Maluridae and *Epthianuridae are sometimes included in this family.

Muscidae (houseflies; order *Diptera, suborder *Cyclorrapha) Family of small to large flies, all of similar appearance, most of which can be distinguished from the similar *Tachinidae and *Calliphoridae families by their lack of *hypopleural bristles. *Musca domestica* is typical of the family. The female fly selects faeces or other organic refuse as a medium on which eggs are laid in masses of 100–150. The larvae feed on the substrate and pupate after eight days at 30°C. The speed of development and potential for increase make the housefly a formidable pest. The adult fly is a well-documented carrier of many human-disease organisms because of its preferred habitat and its association with human dwellings. The destruction of breeding sites and the fly-proofing of premises are the best means of control. Other species include several that are haematophagous (feeding on blood), e.g. of the genus *Stomoxys* (stable flies). More than 3800 species occur throughout the world.

muscle Tissue consisting of cells that form fibres, arranged as sheets or bundles, which are able to contract and thus produce tension. There are two types of muscle in vertebrates: smooth (involuntary) and striated (voluntary). Smooth muscle is derived from the splanchic *mesoderm, striated or skeletal muscles are derived from the myotomes of the mesoderm.

musculo-epithelial *See* MYOEPITHELIAL.

museum beetle *See* DERMESTIDAE.

Musiphagiformes *See* MUSOPHAGIDAE.

musk-deer (*Moschus*) *See* MOSCHIDAE; PALAEOMERYCIDAE.

muskellunge (*Esox masquinongy*) *See* ESOCIDAE.

musk gland A *gland, usually an *anal gland, present in many animals, that secretes musk, a substance with a penetrating odour. Different *species produce musks of varying composition. Originally, musk was obtained from the glands of musk-deer (*Moschidae), located between the stomach and genitals ('musk' is from Sanskrit *muská* meaning 'testicle').

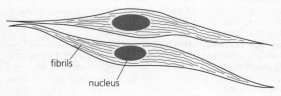

Muscle cell

musk-ox (*Ovibos moschatus*) *See* BOVI-DAE.

musk-rat (musquash; *Ondatra zibethicus*) *See* CRICETIDAE.

musk turtles *See* KINOSTERNIDAE.

***Mus musculus* (house mouse)** *See* MURIDAE.

Musophagidae (turacos, go-away birds, plantain-eaters; class *Aves, order *Cuculiformes) A family of medium to large birds, some of which are brightly coloured with green, purple, and red, others a duller grey or brown. They have crested heads, bare skin around the eyes, and short, stout bills. Their wings are short and rounded, and their tails long. The red pigments are copper-based and, together with the green pigments, are unique to turacos. They inhabit forests and bush, feed on fruit and berries, and nest in trees. The largest genus, *Turaco* (about 15 species), comprises bright green birds with black, blue, or green, glossed tails, bare areas of red or yellow skin around the eye, coloured crests, and crimson flight feathers. There are five genera in the family, with 18 species, often placed in their own order, Musiphagiformes. They are found in Africa.

musquash (musk-rat; *Ondatra zibethicus*) *See* CRICETIDAE.

mussels *See* PTERIOMORPHIA.

mussel shrimps *See* OSTRACODA.

***Mustela* (mink, stoat, weasel)** *See* MUSTELIDAE.

Mustelidae (weasels; suborder *Fissipedia (or *Caniformia), superfamily *Canoidea) A family of carnivores that are characteristic mainly of northern temperate latitudes, where they occupy niches similar to those occupied by the *Viverridae of lower latitudes. They are relatively primitive, with short, stocky legs. They have long, sharp *canine teeth and well-developed *carnassials but, in modern forms, never more than one

post-carnassial *molar in each jaw. Most are small, but a *Miocene American mustelid, *Megalictis*, was the size of a black bear. They had diverged from their miacid ancestors by the *Eocene–*Oligocene transition and later branched widely, one line giving rise to the seals (*Phocidae). The largest modern mustelid is *Gulo gulo* (wolverine, glutton, carcajou, or skunk bear) of Arctic regions, which grows to more than 1 m in length and a weight of more than 18 kg. Some (e.g. *Meles* and *Taxidea*, the Old and New World badgers) have adapted to an omnivorous diet, as has *Mellivora capensis* (honey badger or ratel) of Asia. The family also includes the martens (e.g. *Martes martes*, pine marten of Eurasia), and *Mustela* (weasel, stoat (or ermine), and mink). Otters (e.g. *Lutra*) have adapted to a semi-aquatic way of life and a diet consisting mainly of fish, with *Enhydra* becoming wholly marine. Mustelids are distributed throughout the world, except for Madagascar and Australasia. *Mephitis* (skunk), adapted to an omnivorous diet, was formerly classified in the Mustelidae, but on the basis of DNA comparisons this genus is now usually placed in its own family, *Mephitidae. Excluding *Mephitis*, there are 22 living mustelid genera, with 59 species.

***Mustelus mustelus* (dogfish, smooth hound)** *See* CARCHARHINIDAE.

musth In sexually mature male elephants, a physiological and behavioural condition associated with sexual activity. It lasts two to three months, during which *testosterone levels are high; the temporal glands become enlarged and temporin secretion becomes copious; urine is discharged continually; and behaviour becomes more aggressive. Musth occurs annually or biennially, but at any time of year rather than within a particular season. Its occurrence is not synchronized among groups of males.

mutagen An agent that increases the *mutation rate within an organism. Examples of mutagens are X-rays, gamma rays, neutrons, and certain chemicals such as carcinogens.

mutagenic Causing a *mutation.

mutant 1. A cell or organism th... gene *mutation. **2.** A *gene that ha... gone mutation.

mutation 1. A process by which a *gene or *chromosome set undergoes a structural change or a change in the amount of *DNA it contains. **2.** A gene or chromosome set that has undergone a structural change. The majority of mutations are changes within individual genes (e.g. the substitution of a different *nucleotide at some point in the *DNA so that the *amino acid sequence is altered), but some are gross structural changes of chromosomes (e.g. *inversion or *translocation) or changes in the number of chromosomes per nucleus (e.g. *polyploidy). Mutations are the raw material for evolution: they provide the source of all variation. For mutations to affect subsequent generations, though, they must occur in *gametes or in cells destined to be gametes, since only then will they be inherited. A mutation that occurs in a body cell is called a 'somatic' mutation: it is transmitted to all cells derived, by *mitosis, from that cell. Most mutations, however, are deleterious; evolution progresses through the few that are favourable. Some mutations are recurrent: they occur repeatedly within a population, or over long periods of time (as does haemophilia, for example).

mutation rate The number of *mutation events per *gene per unit of time (e.g. per cell generation). The term is also applied to the frequency with which mutation events occur in a given *species or to the frequency with which a specified mutation event or mutational class occurs in a given population. Normally mutations occur at a constant, very low, rate: this rate can be greatly increased by irradiation with X-rays, gamma rays, neutrons, etc., or by treatment with carcinogens.

mute swan (*Cygnus olor*) See ANATIDAE.

Mutica (class *Mammalia, infraclass *Eutheria) A cohort that comprises the single order *Cetacea, the whales. They are believed to have evolved from carnivorous or precarnivorous mammals which fed on

M... **ocrita, su...** wasps which are ...hed during the ing sphecoid and vesp... ...ural adtary bees. Adults are hairy and o... coloured, the males being winged an... females *apterous. The apterous females resemble ants but have a 'felt line' on the second segment of the *gaster, and can inflict painful stings (e.g. *Dasymutilla occidentalis* (cow killer or mule killer) which, despite its name, does not kill any large mammals).

mutual antagonism A relationship in which the effect of competition between two or more species (interspecific competition) exceeds that of competition within each species (intraspecific competition).

mutual interference Behavioural interactions among feeding organisms that reduce the time each individual spends obtaining food or the amount of food each individual consumes. It occurs most commonly where the amount of food is small or the number of animals feeding is high. When the searching efficiency of an individual consumer is calculated (*see* HOLLING'S DISC EQUATION) and plotted against consumer density on a graph, both logarithmically, the resulting negative slope is given a value, $-m$, and m is known as the coefficient of interference.

mutualism An interaction between members of two *species that benefits both. Strictly, the term may be confined to obligatory mutualism, in which neither species can survive under natural conditions without the other. Sometimes the term is used more generally to include *protocooperation.

mycetocyte In certain types of insect, a cell that contains symbiotic fungi or bacteria.

mycetome Special organ comprising a group of specialized gut or fat cells, containing symbiotic micro-organisms, e.g. as in

...ptera,
...nes appear
...intestinal crypts
...uc yeasts or bacteria,
Myc...believed to play a part in the
...olism of vitamins and other essential
substances.

Mycetophagidae (fungus beetles; subclass *Pterygota, order *Coleoptera) Family of tiny, oval, pubescent beetles, 1.5–3.0 mm long, with inconspicuous coloration. The head is triangular, the antennae clubbed. The eyes are many-faceted, giving a rough appearance. Larvae are tawny, subcylindrical, with well-developed legs. Adults and larvae are found in fungi and rotting material. There are 200 species, with world-wide distribution.

Mycetophilidae (fungus gnats; order *Diptera, suborder *Nematocera) Family of small flies which have long antennae, and *ocelli. The *tibiae have spurs, and the *coxae are elongate. The *thorax is more or less arched, often strikingly so. The common name refers to the usual feeding substrate of the larvae, which often live gregariously among decaying vegetable matter or in fungi. The larvae are usually pale in colour, and have a hard, sclerotized head, and 12 segments. Several species are economically important as pests of mushrooms, but others attack only wild fungi. Several species are luminous, the light being emitted from the distal part of the *Malpighian tubules. Although they are similar in many respects to the Mycetophilidae, the Sciaridae (Sciarinae) are now regarded as a separate family. There are more than 2000 described species, with a world-wide distribution.

Myctophidae (lanternfish; superorder Scopelomorpha, order *Myctophiformes) A large family of marine, mainly deep-water, fish, that have a slender and compressed body, a single *dorsal and *adipose fin, and a distinct *anal fin. The mouth and eyes are large. Generally small fish, they are probably the most abundant deep-sea fish, occurring in schools at depths exceeding 500 m during the day, but sometimes found near the surface at night. They have

rows of photophores along the lower part of the body. There are about 220 species, found world-wide.

Myctophiformes (subclass *Actinopterygii, superorder *Scopelomorpha) An order of marine fish that have a slender body, single *dorsal and often an *adipose fin, and forked tail fin. Some families (e.g. the *Synodontidae) are found in coastal waters. Others (e.g. the *Myctophidae and *Evermanellidae) are typical deep-sea fish.

myelencephalon (medulla oblongata) The most posterior region of the *brain stem, involved with the balance and hearing stimuli received from the ear and coordination of many involuntary functions.

myelin *See* MYELIN SHEATH.

myelin sheath (medullary sheath) A fatty sheath, composed of *lipids, *proteins, and *polysaccharides (myelin), that surrounds and insulates the *axons of *neurons and permits increased current flow of nerve impulses. The myelin is produced by Schwann cells. Constrictions which occur along the sheath (nodes of Ranvier) delineate adjacent Schwann cells.

Mygalomorphae (funnel-web spiders, tarantulas; class *Arachnida, order *Araneae) Suborder of spiders which construct funnel-shaped webs. Some species are poisonous to humans, and most species occur in Australia. Most large mygalomorphs are commonly called tarantulas. Many have hairs which produce an urticalike rash. Some species are venomous, but are less dangerous than is generally supposed, inflicting bites that in humans are no worse than wasp stings.

Myina (class *Bivalvia, order *Myoida) A suborder of bivalves which possess essentially the same characteristics as those of the order, except that forms with an internal *resilium have an external ligament developed on prominent *nymphs. The *pallial line has a *sinus. They first appeared in the *Permian.

Myliobatidae (eagle ray; subclass *Elasmobranchii, order *Rajiformes) A small family of marine rays that have large, wing-like, *pectoral fins with pointed tips. The elevated head, with its large eyes, stands clear of the disc formed by the pectorals. A small *dorsal fin, but no *caudal fin, is found on the whip-like tail; this is longer than the body and may bear a venomous spine just behind the dorsal fin. Although usually feeding near the bottom of the sea, eagle rays are active swimmers, sometimes making spectacular jumps out of the water. There are about 15 species, distributed world-wide.

Mymaridae (fairyflies; suborder *Apocrita, superfamily *Chalcidoidea) Family of *Hymenoptera, which are perhaps noteworthy for being among the smallest known insects, sometimes being less than 0.25 mm long. Adults are usually blackish or black and yellow, with long, thin legs and antennae, and very narrow, hair-fringed wings. The hind wing is very narrow or nearly linear, and both fore and hind wings have a characteristic, highly reduced venation. The *ovipositor is sometimes very long and brought forward in a loop beneath the *thorax. All members of the family are egg parasites of various insects. Their common name refers to their small size, delicate shape, and tiny, hair-fringed wings.

mynas *See* STURNIDAE.

Myocaster coypu (coypu) *See* CAPROMYIDAE.

myocoel A coelomic cavity (*see* COELOM) which forms in that part of the mesoderm (the germ layer of *triploblastic animal *embryos composed of cells that have moved from the surface to the interior) that will differentiate into *striated muscle.

myoepithelial (musculo-epithelial) In hydras, applied to the most conspicuous cells of the ectoderm. They are columnar in form, with contractile processes extending into the *mesogloea and broad outer bases which meet to form a continuous sheet. In some coelenterates (e.g. jellyfish) the epithelial part of the cells may be reduced.

myofibril A bundle of contractile filaments (myofilaments), 1–2μm in diameter, that are arranged in parallel groups in the *cytoplasm of *striated muscle cells.

myoglobin A haemoprotein, consisting of a single polypeptide chain surrounding a haem group, that is capable of reversibly binding oxygen. It has a higher affinity for oxygen than has *haemoglobin and releases it only when the oxygen supply becomes limiting. Myoglobin is found in the muscles of vertebrates and certain invertebrates, where it serves to store oxygen.

Myoida (phylum *Mollusca, class *Bivalvia) An order of bivalves which are not equilateral in shape and in which the valves may be equal or unequal. The shells are thin and never *nacreous. The dentition consists of a *cardinal tooth in each valve, and some have *edentulous valves. Musculature is *isomyarian or anisomyarian, and there is no *escutcheon or *lunule. *Siphons are well developed for a burrowing or boring way of life. They first appeared in the *Carboniferous.

myomere A block of striated muscle within a single *somite.

Myomorpha (cohort *Glires, order *Rodentia) A suborder that comprises nine families of rodents, in which the median portion of the *masseter muscle runs through the *infraorbital foramen and both superficial and median parts of the masseter are inserted on the face. If present, *premolars are small and are shed early in life. The families in the suborder probably are not closely related, the grouping being based on superficial characteristics. The suborder includes the rats, mice, voles, hamsters, dormice, etc.

myoseptum The partitioning *connective tissue that occurs between *myomeres.

myosin The predominant *protein of the *myofibrils of muscle cells. It has an unusual shape for a protein, having a globular head and a rod-like tail.

myotome *See* SOMITE.

Myoxocephalus scorpius (father lasher) *See* COTTIDAE.

Myriapoda (phylum *Arthropoda) A subphylum of nearly 13 000 species of arthropods that have a head and an elongated body comprising up to nearly 100 segments, each bearing a pair of *uniramous appendages ('myriapod' means 'many-legged'). There are four classes: *Chilopoda (centipedes), *Diplopoda (millipedes), *Pauropoda, and *Symphyla (the last two have no common name).

Myrmeciinae (bulldog ants; suborder *Apocrita, family *Formicidae) Subfamily of primitive ants, which have one or two petiolar nodes, and a powerful sting. The workers are large and predatory, and usually inhabit soil nests. There are 120 species, most of which occur only in Australia.

Myrmecobiidae (superfamily *Dasyuroidea, order *Dasyuromorphia) A monospecific family (*Myrmecobius fasciatus*), the numbat or banded ant-eater, which has up to 52 teeth, an elongated snout, no cheek pouch, and feeds on ants and termites. It is distributed in open forest and scrub habitats in western and southern Australia.

Myrmecobius fasciatus (banded ant-eater, numbat) *See* MYRMECOBIIDAE.

myrmecochory Dispersal of spores or seeds by ants.

Myrmecophagidae *See* MYRMECOPHA-GOIDEA.

Myrmecophagoidea (ant-eaters; suborder *Xenarthra, infra-order *Pilosa) A superfamily that comprises the one family Myrmecophagidae, in which the snout is elongated, the extent of the elongation depending on the overall body size of the animal, so that it is *Myrmecophaga* (giant ant-eater) that possesses the longest snout. There are no teeth. The tongue is long and sticky, the hard palate extended to the rear by the union of the *pterygoid bones. The limbs, of approximately equal length, bear four or five digits, with large claws used for digging and for defence. The tail is long. *Myrmecophaga* is terrestrial; *Cyclopes* and *Tamandua* are smaller, arboreal, and have *prehensile tails. There are three genera, with four species, distributed throughout tropical Central and S. America.

myrmecophile Applied to a species which relies on ants for food or protection in order to complete its life cycle. Food is obtained from the ants by direct parasitism, predation, stealing, or scavenging.

Myrmecophilidae (ant-loving crickets; order *Orthoptera, suborder *Ensifera) Family of small, ovoid, wingless crickets which live in ant nests. They feed on ant secretions and appear to be accepted by the ants as colony members. Representatives of the single myrmecophilid genus (*Myrmecophilus*) are found in Europe, Asia, America, and Australia.

Myrmeliontidae (subclass *Pterygota, order *Neuroptera) Family of insects in which the adults resemble damselflies, but have conspicuous, clubbed antennae as long as the combined head and *thorax, and are softer-bodied. Most species are large, with lightly patterned wings, and the abdomen is long and thin. The eggs are laid singly in dry soil or sand, and the larvae (ant lions) burrow into the substrate, forming a steep-sided, conical pit into which prey items fall. The larvae of some species have a very elongate *prothorax, but all have large, sickle-shaped jaws for capturing their prey.

Myrmicinae (parasol ants; suborder *Apocrita, family *Formicidae) Subfamily of ants in which the *petiole is composed of two node-like segments. The tip of the *gaster is equipped with a sting, supplied with venom from glands. *Messor* species (harvester ants) feed on grain and store seeds in chambers within the soil nest. In *Monomorium pharaonis* (Pharaoh ant), a *tramp species that is now a widespread pest, *budding has replaced reproduction by mating flights. The subfamily has a worldwide distribution and there are about 5000 species.

Mystacinidae (order *Chiroptera, suborder *Microchiroptera) A monospecific family (*Mystacina tuberculata*) of omnivorous bats that have thick fur, talons on all claws, and wings which tuck into pockets when the animals are at rest. Mystacinid bats are very active on the ground and may feed there on insects, fruit, pollen, flowers, and possibly on carrion. The ears are simple, long, and well developed. The nose is simple. The species is found only in New Zealand.

Mystacocarida (phylum *Arthropoda, subphylum *Crustacea) Class of crustaceans, first described in 1943, known from only three species in one genus, *Derocheilocharis*, and related to the copepods and barnacles. Mystacocarids are tiny filter feeders, 0.5 mm or less in length, and are adapted for life between intertidal sand grains. Although the elongate, cylindrical body is divided as that of the copepods, transverse constriction divides the head, and the thoracic segment bearing the *maxillipeds is only incompletely fused with the head. Four of the five thoracic segments bear *uniramous appendages, each reduced to a single, simple lamella. Both pairs of antennae are well developed, and the feeding appendages are *setose, and longer than those of the copepods. Only a *nauplius eye is present.

Mysticeti (baleen whales; cohort *Mutica, order *Cetacea) A suborder that comprises three families (*Eschrichtiidae, *Balaenidae, and *Balaenopteridae) of baleen whales. Teeth are absent in adults (vestigial teeth may be present in the foetus) and transverse sheets of *baleen extend like combs from the roof of the mouth into the mouth cavity: these are used to strain plankton from water. There is a pair of nasal openings.

Mytiloida (phylum *Mollusca, class *Bivalvia) An order of *epifaunal, *byssate bivalves which have an equivalve, but highly non-equilaterally shaped, shell with a prismatic-nacreous microstructure. They have a *dysodont dentition, the ligament is *parivincular and *opisthodetic, and the gills are *eulamellibranchiate and *filibranchiate. Mytiloida have an anisomyarian musculature. The *pallial line is complete, the *siphons poorly developed. They first appeared in the *Devonian.

Myxinidae (hagfish; class *Pteraspidomorphi, order *Myxiniformes) A small family of marine, jawless fish (or fish-like vertebrates) that have an elongate and slender body, a cartilaginous skeleton, but no paired fins or scales. The slit-like mouth is surrounded by *barbels and is supported by a powerful rasping tongue. There is considerable variation in the number of round gill openings. There are about 19 species, found in temperate to cold water.

Myxiniformes (superclass *Agnatha, class *Pteraspidomorphi) An order of jawless marine fish, or fish-like vertebrates, which comprises only the family *Myxinidae.

Myxosporidea (phylum *Protozoa, subphylum *Cnidospora) A class of protozoa, which parasitize invertebrates and lower vertebrates, particularly fish, often with fatal consequences for the host. The spores contain a coiled polar filament. When a spore is ingested by a new host the filament is extruded and attaches the spore to the gut wall of the host. There are three orders.

Myxozoa (phylum *Cnidaria) A taxonomically unranked group of highly modified cnidarians comprising more than 1300 species of aquatic parasites, many with a *life cycle involving two hosts, a fish, and an oligochaete (*Oligochaeta) or polychaete (*Polychaeta) worm or bryozoan (*Bryozoa). They may also be capable of infecting vertebrates. Myxozoans have a very reduced body, usually 0.01–0.02 mm long and consisting of several cells forming shell *valves and capsules bearing structures resembling *nematocysts with extrudible filaments.

(((•))) SEE WEB LINKS
• Description of Myxozoa

Myxus (*Trachystoma*) *petardi* (freshwater mullet) *See* MUGILIDAE.

Myzopodidae (golden bat, snake-footed bat; order *Chiroptera, suborder *Microchiroptera) A monospecific family (*Myzopoda aurita*) of bats in which the ears are large, simple, and separate, but have a near-circular process in the *antitragus. The nose is simple. There are adhesive suckers on all four limbs. This bat is found only in Madagascar.

nacreous Applied to the iridescent appearance of the pearly inner surface of some molluscan shells (mother-of-pearl) in which the mineral is laid down in thin, lustrous sheets, each deposited over a thin, organic matrix.

NAD *See* NICOTINAMIDE ADENINE DINUCLEOTIDE.

NADP *See* NICOTINAMIDE ADENINE DINUCLEOTIDE PHOSPHATE.

naked mole rat *See* HETEROCEPHALUS GLABER.

naked pteropods *See* GYMNOSOMATA.

Nandidae (leaf-fish; subclass **Actinopterygii, order *Perciformes) A small family of freshwater fish that have a strongly compressed, deep body, and a rather pointed snout, emphasizing the low-slung mouth. The *dorsal and *anal fins are long, the short tail ending in a rounded tail fin. Most species have irregular colour markings consisting of spots of different sizes and colours. *Monocirrhus polyacanthus* (leaf-fish), 8 cm, is a well-camouflaged, voracious predator. There are about 10 species, but the genera *Pristolepis Badis* have also been placed in different families (Pristolepidae and Badidae). They are found in W. Africa, south-east Asia, and S. America.

nannoplankton *See* PLANKTON.

Nannopterum harrisi (Galápagos cormorant) *See* PHALACROCORACIDAE.

nannygai (*Trachichthodes affinis*) *See* BERYCIDAE.

Narcomedusina (class *Hydrozoa, order *Trachylina) A suborder of cnidarians in which *tentacles arise some distance *aborally to the margin of the bell of the *medusa and in which gonads are situated on the *manubrium.

nares *See* NASAL CAVITY.

narial pads Naked, muscular pads surrounding the nostrils (e.g. in some *Microchiroptera).

narrow-mouthed frogs *See* MICROHYLIDAE.

narwhal (*Monoceros*) *See* MONODONTIDAE.

nasal bone One of a pair of bones that together form the roof of the *nasal cavity.

nasal cavity In tetrapods, a cavity in the head lined with mucous membrane and connected by passages (nares) to the mouth and to the exterior of the body. It contains the olfactory organ.

Nasalis larvatus (proboscis monkey) *See* CERCOPITHECIDAE.

Nasanov's gland *Pheromone-producing gland in the abdomen of a worker honeybee (*Apis* species). It opens at the base of the last *tergite via a partially *eversible membrane, and is functional only in bees of foraging age (10 days old). During emission, the bee adopts a characteristic stance, with abdomen raised, gland exposed, and the wings beating to waft away a plume of scent. The gland secretes a mixture of scents (geraniol, nerolic and geranic acids, and citral) which have several functions according to the context in which they are released. Thus in combination with 'queen substance' the Nasanov's gland scents recruit additional workers to a settled swarm; in the presence of disorientated bees the scent attracts them back to the hive. The scents are also emitted by foragers at a food source that does not have a characteristic scent of its own, and additional workers are recruited. Thus the Nasanov's gland scents may augment information conveyed by the round dance and waggle dance. The scents are also used to recruit workers to sources of water.

nasohypophysial sac In *Agnatha, a cavity above the mouth formed from the nasal pit and the *hypophysial sac, connected with the exterior by a single nostril, and performing an olfactory function.

nasopharynx The uppermost part of the *pharynx, extending from the base of the skull to the soft *palate.

nasus See NASUTE.

nasute A termite soldier that is equipped with a projection on the head (nasus). It squirts at enemies an adhesive or toxic fluid produced by the frontal gland.

natal dispersal See DISPERSAL.

Natalidae (long-legged bats; order *Chiroptera, suborder *Microchiroptera) A family of very agile, insectivorous bats in which the large *uropatagium is supported by very long legs, *calcars, and the tail which extends to its margin. The nose is simple, without a leaf, but has lateral tufts of hair. The ears are simple, separate, and very large. The bats are distributed throughout tropical N. and Central America and the W. Indies, and in tropical S. America east of the Andes. There is one genus, *Natalus*, with about eight species.

natant Floating entirely under water.

national park As defined internationally, an extensive area of land that has not been influenced significantly by human activities

and that is set aside in perpetuity to preserve its landscapes, species, or *ecosystems. British national parks are large areas, designated officially on the recommendation of Natural England, the Countryside Council for Wales, or Scottish Natural Heritage, that are preserved for the enjoyment of the public because of the beauty of their countryside and the opportunities they afford for outdoor recreation.

native (indigenous) Applied to a *species that occurs naturally in an area, and therefore one that has not been introduced by humans either accidentally or intentionally.

native cats See DASYURIDAE.

Natrix maura (viperine snake) See COLUBRIDAE.

Natrix natrix (grass snake, ringed snake) See COLUBRIDAE.

Natrix tessellata (dice snake, tessellated snake) See COLUBRIDAE.

natterjack toad (Bufo calamita) See BUFONIDAE.

natural immunity Immunity that has been inherited in contrast to one that has been acquired. If an individual acquires immunity by having the disease in question, as opposed to acquiring it as the result of vaccination, then that individual is said to have natural acquired immunity.

natural killer cells (K cells, killer cells, NK cells) A type of lymphocyte (white blood cell, see LEUCOCYTE) that is part of the *innate immune system and helps reject cancer cells and cells infected with *viruses. When close to a target, a natural killer cell releases compounds that open pores in the target cell, and *proteases that enter the target and induce *apoptosis.

SEE WEB LINKS
• Description of natural killer cell

natural selection ('survival of the fittest') A complex process in which the total *environment determines which members of a *species survive to reproduce and so pass on their *genes to the next generation. This need not necessarily involve a struggle between organisms.

nature and nurture Synonyms for heredity and environment as they affect a *character. Both may affect observed variation amongst individuals; only variations due to 'nature' are inherited, and it is these that form the subject of quantitative genetics.

nature reserve An area of land that is managed primarily to protect its flora, fauna, or physical features from harm that might result from a change in land use. Public access may be restricted or forbidden. Reserves may be owned privately or by a government agency. In the UK, reserves designated as national nature reserves are managed (but the land is not necessarily owned) by a government agency; those designated by local authorities, in consultation with the appropriate national agency, are known as local nature reserves. Voluntary conservation bodies often manage local nature reserves and have also established many reserves of their own.

nauplius eye A single median eye comprising three- or four-pigment cup *ocelli, sometimes with a lens, probably enabling the organism to determine the direction of a source of light for orientation. The median eye is typical of crustacean *nauplius larvae. It may degenerate as the larva develops, or it may persist into the adult form.

nauplius larva The first, free-swimming, planktonic larva of most marine and some freshwater crustaceans. It has no evident segmentation. There is a single, median, *nauplius eye at the front of the head. There are only three pairs of appendages, the first and second antennae, and the *mandibles. The second antennae and mandibles bear swimming *setae. Additional trunk segments and appendages appear with successive moults, the increments proceeding from anterior to posterior. The late nauplius stages are often called metanauplii. The term 'postlarva' is applied to all immature

crustaceans when the full complement of segments and appendages have developed. Some or all larval stages are absent in certain groups of *Crustacea, and in others they may be considerably modified. (Nauplius was the son of Poseidon, Greek god of the sea.)

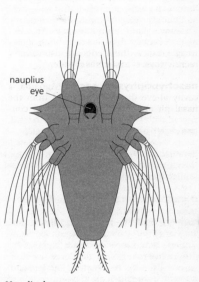

nauplius eye

Nauplius larva

nautiliconic Applied to the shell of a cephalopod (*Cephalopoda) when it is coiled and highly *involute.

Nautilida (class *Cephalopoda, subclass *Nautiloidea) The only present-day order of nautiloids, whose extant members are *involute nautilicones (*see* NAUTILICONIC). The majority of nautilids have a narrow *siphuncle in a subcentral position. Sutures vary from simple to very complex. *Nautilus* is the only extant genus. The order first appeared in the Middle *Devonian.

Nautiloidea (phylum *Mollusca, class *Cephalopoda) A subclass of cephalopods which possess a multi-chambered, external shell composed of calcium carbonate, which is siphunculate (*see* SIPHUNCLE) and may be coiled. The gill structure is tetrabranchiate. Simple sutures are produced by contact between the internal *septa and the shell wall. The subclass includes the oldest cephalopods, first recorded from Late *Cambrian rocks. They diversified and became common throughout the *Palaeozoic but were greatly reduced at the end of this era. Further diversification occurred in the *Mesozoic but the group dwindled again in the *Cenozoic. There are more than 300 extinct genera and one extant genus, *Nautilus*, which dates from the *Oligocene.

Nautilus *See* NAUTILIDA; NAUTILOIDEA.

navel *See* UMBILICUS.

navigation The *orientation of itself by an animal towards a destination, regardless of its direction, by means other than the recognition of landmarks. *Compare* COMPASS ORIENTATION; PILOTAGE.

Nearctic faunal region The fauna of N. America, south to Mexico. At the order and family level the fauna is essentially the same as that of the *Palaearctic region, but some genera and more especially species are distinctive to the Nearctic. Because of their strong affinities, the Nearctic and Palaearctic regions are often grouped into one unit, *Holarctica, reflecting their former connection via the Bering *land bridge.

necrology The scientific study of all the processes affecting dead animal and plant material, including decomposition and *fossilization.

necrophoresis The removal of dead individuals by carrying them away from places inhabited by living individuals.

necrosis The death of a circumscribed piece of tissue. Necrotic wounds in mammals are produced, for example, by the bite of *Loxosceles reclusa* (brown recluse spider).

necrotrophic Applied to a parasitic organism that obtains its nutrients from dead cells and tissues of its host organism.

Nectarinia (sunbirds) *See* NECTARINIIDAE.

Nectariniidae (sunbirds, spider-hunters; class *Aves, order *Passeriformes) A family of small birds, most of which have a long, narrow, pointed, *decurved bill, which is partly serrated. (*Anthreptes*, a genus of small sunbirds which are less specialized than most members of the family, have relatively shorter and straighter bills.) They have short legs, short, rounded wings, and a short to long, square, rounded, or graduated tail. Sunbirds have bright, metallic-coloured plumage, and are the Old World equivalent of humming-birds (e.g. the 75 species of *Nectarinia*, the males of which have bright, metallic blue, green, red, and purple plumage, although that of the females and some non-breeding males is dull). Spiderhunters are mainly brown. They inhabit forests, clearings, and bush, and feed on insects, spiders, nectar, and fruit. Their nests are built suspended from branches or leaves; some are sewn on to the undersides of leaves. There are five genera, with about 117 species, many of them kept as cage birds, found in Africa, Asia, the Philippines, Indonesia, and Australia.

nectocalyx In some *Siphonophora, a bladder-like swimming bell.

necton *See* NEKTON.

nectophore In some *Siphonophora (e.g. Stephalia and Nectalia), a modified, pulsating *medusa which functions as a swimming bell.

Necturus maculosus (mudpuppy, water dog) *See* PROTEIDAE.

needlefish *See* BELONIDAE.

nekton (necton) Aquatic organisms that swim actively (e.g. ciliated *plankton), rather than drifting passively.

Nematocera (subclass *Pterygota, order *Diptera) Suborder of flies in which the adults usually have long, thin antennae of not less than six segments, which are more or less uniform, and unfused. If the antennae are short, members of the suborder may be recognized by their pendulous, multi-segmented palpi. In many species the larvae have well-developed, biting *mandibles. The pupae are *obtect, and normally free. There are about 21 families with worldwide distribution.

nematocyst In *Cnidaria, part of a *cnidoblast; it is a pear-shaped sac with a lid (operculum), the outer end of the sac being extended to form a long, hollow thread which is stored coiled within the sac, immersed in a fluid. On stimulation of a sensory bristle (a cnidocil) the thread is expelled, as a weapon or to capture or hold prey, depending on the type of nematocyst. Some nematocysts are barbed and some inject venom.

Nematoda (eelworms, roundworms, threadworms) A phylum of worms which vary greatly in size up to about 5 cm, although most are microscopically small. The whole body is covered by a tough cuticle that has flanges, and may also have ridges or spines. Nematodes are spindle-shaped (the nearly perfect cylindrical shape is a diagnostic feature), have *bilateral symmetry, with strikingly radial or biradial arrangements of structures around the mouth (also a diagnostic feature), and are unsegmented. They lack true *cilia, have a *pseudocoelom, and the head is not distinct from the rest of the body. There are more than 15 000 known species, morphologically all similar, occurring as parasites in plants and animals, and as free-living forms. They are first known from rocks of *Carboniferous age.

(()) SEE WEB LINKS
• Introduction to the Nematoda

Nematognathi An older name for the order *Siluriformes (catfish).

Nematomorpha (hairworms) A small phylum of *pseudocoelomate, unsegmented worms, the vast majority of which are marine. Their narrow, hair-like bodies have a thick cuticle. All lack a true excretory system. The young are parasitic, and mature

individuals are free-living in fresh water. They are first recorded from the *Eocene.

Nemeobiidae *See* RIODINIDAE.

Nemertea (Rhyncocoela, Nemertinea, Nemertini, proboscis worms, ribbon worms) A phylum of unsegmented, non-parasitic worms which are bilaterally symmetrical and elongate. Externally they resemble *Platyhelminthes (flatworms) but are generally larger and more elongated. Unlike Platyhelminthes they have a mouth and *anus which allows ingestion and egestion to occur simultaneously. Adults are ciliated and possess a long, tubular *proboscis that can be thrust out from the body by a sudden contraction of the fluid-filled *rhynchocoel. The proboscis, which may bear piercing barbs, is used to capture prey and in defence. The mouth and brain are well developed, the rhynchocoel limited in extent. Most are marine, although freshwater and terrestrial forms occur. Most are not hermaphrodites. The first fossil forms occur in the Middle *Cambrian Burgess Shales of British Columbia, Canada.

Nemertinea *See* NEMERTEA.

Nemertini *See* NEMERTEA.

Nemichthyidae (snipe eel; subclass *Actinopterygii, order *Anguilliformes) A small family of deep-sea eels that have a very long, ribbon-like body and a thin, beak-like snout. One of the better-known species, *Nemichthys scolopaceus*, may reach a length of 1.2 m. There are about 10 species, distributed world-wide.

Nemipteridae (threadfin bream; subclass *Actinopterygii, order *Perciformes) A family of marine fish that have a compressed and slender body, somewhat produced snout, and a single, continuous *dorsal fin. The *pelvic fins are situated just behind the *pectorals. The forked tail fin may show a trailing filament off the upper lobe in some species. Nemipteridae are distributed throughout the Indo-Pacific region. There are many species.

ne-ne (*Branta sandvicensis*) *See* ANATIDAE.

Neobalaenidae (order *Cetacea, suborder *Mysticeti) A monotypic family of *baleen whales that comprises *Caperea marginata* (pygmy right whale). It grows to 6–5 m and inhabits the Southern Ocean.

Neoceratias spinifer See NEOCERATII-DAE.

Neoceratiidae (subclass *Actinopterygii, order *Lophiiformes) A monospecific family (*Neoceratias spinifer*) of small (6 cm long), deep-sea fish that have a large head, a large mouth, and small eyes. The jaws are equipped with many long and slender teeth. The single *dorsal, *caudal, and *anal fins have a rounded profile, while the *pectorals are small. No *illicium is present. Females are larger than males; males may form a parasitic relationship with their female sex partners. They are found in all oceans.

Neocyttus rhomboidalis (spiky dory) See OREOSOMATIDAE.

neo-Darwinism See DARWIN; DARWINISM; MODERN SYNTHESIS.

Neodrepanis (false sunbirds) See PHILEPITTIDAE.

Neofelis nebulosa (clouded leopard) See FELIDAE.

Neofidelia See FIDELIIDAE.

Neogastropoda (class *Gastropoda, subclass *Prosobranchia) An order of prosobranch gastropods which may have one *pectinibranch gill, or less commonly two. A proboscis develops and there is a large *osphradium. The nervous system is concentrated. Sexes are separate. The shell has an obvious siphonal notch or canal (i.e. in contrast to the entire aperture in *Mesogastropoda) and has an *operculum. Most neogastropods are carnivorous. They first appeared in the *Cretaceous.

Neogea A neotropical zoogeographical region that comprises Central and S. America. *Compare* ARCTOGEA; NOTOGEA. *See also* FAUNAL REGION.

Neogene The second period of the *Cenozoic Era, comprising the *Pliocene and *Pleistocene epochs, and lasting from 23.03 Ma ago to 1.81 Ma ago.

((●)) SEE WEB LINKS
• Description of the Neogene

Neognathostomata The name given to a group of irregular echinoids (*Echinoidea) comprising the orders Galeropygoida, *Cassiduloida, Oligopygoida, and *Clypeasteroida.

neo-Lamarckism Modern evolutionary theories that in some sense allow the possibility that *acquired characteristics may be inherited (as proposed by *Lamarck). For example, in 1980 E. J. Steele proposed that what in effect is Lamarckian evolution may occur by the insertion of new genetic material into the host *genome by a *retrovirus.

Neolampadoida (subphylum *Echinozoa, class *Echinoidea) A small and poorly known order of irregular echinoids, with non-*petaloid *ambulacra, which may be descended from the *Cassiduloida. They are known as fossils from the upper *Eocene, and are extant in tropical seas.

Neolinognathidae (order *Phthiraptera, suborder *Anoplura) Family of sucking lice which are parasitic on *Macroscelidea (elephant shrews). The species in this family can be distinguished from all other lice by their possession of a single pair of abdominal *spiracles, on segment 8 (most other lice have a pair of spiracles on each of segments 3 to 8, and those that have fewer than six pairs have lost those on segment 8). The family comprises a single genus, *Neolinognathus*, with two species.

Neoloricata (class *Amphineura, subclass *Polyplacophora) An order of polyplacophorans, some of which have the *articulamentum present as insertion plates. Advanced forms possess a *mesostracum layer to the valves. They first appeared in the *Carboniferous. The suborder Acanthochitonina includes the largest of all polyplacophorans, although some of its members are small.

Neomeniida (class *Amphineura, subclass *Aplacophora) An order of spiculose aplacophorans which possess a groove on the central surface containing a ciliated ridge. They have a simple *mantle cavity with internal folds for respiration. True gills do not occur. Individuals are bisexual. They are *Holocene in age.

neomycin An aminoglycoside antibiotic produced by *Streptomyces fradiae*, which functions by interfering with ribosomal activity and so causing errors in the reading of the m-RNA.

Neoophora (phylum *Platyhelminthes, class *Turbellaria) A subclass of worms, most of which occur in freshwater or marine habitats, but some of which are terrestrial. The ovary is divided.

neopallium In higher vertebrates, such as birds and mammals, that part of the *cerebrum occupied with senses other than the sense of smell. *See also* TELENCEPHALON.

Neopilina A genus of *Monoplacophora, 10 living specimens of which were dredged from a deep ocean trench off Costa Rica in 1952, since when specimens of seven different species have been collected from other oceans, at depths of 2000–7000 m.

(((●))) **SEE WEB LINKS**
• Description of *Neopilina* and its relatives

neoplastic cell A cell in the body of a multicelled *eukaryote that grows and multiplies outside the control of the organism as a whole and without reference to the organism's best interests.

Neoptera (class *Insecta, subclass *Pterygota) One of the two infraclasses (*compare* PALAEOPTERA) into which insects are placed according to the position in which they hold their wings when at rest. Members of the Neoptera fold their wings across their backs. This is the more advanced condition. The infraclass includes most modern insects. The distinction between the two infraclasses emerged in the *Carboniferous, very early in insect evolution, and is known from fossils of that period.

Neopterygii An older, but still current, name for a large group of bony fish (*Osteichthyes) including the infraclasses *Holostei (e.g. bow-fin, Amiidae) and *Teleostei (e.g. perch, Percidae).

Neorhabdocoela *See* RHABDOCOELA.

Neoscopelidae (subclass *Actinopterygii, order *Myctophiformes) A small family of deep-sea fish that resemble lanternfish (*Myctophidae) but have a somewhat pointed snout. A few photophores are present on the slender body. Generally small fish (20 cm), they have a single, short, *dorsal fin as well as an *adipose fin opposite the *anal fin; the *pelvic fins are just behind the dorsal fin. There are about five species, distributed world-wide.

Neosittidae *See* DAPHAENOSITTIDAE.

Neostethidae (subclass *Actinopterygii, order *Atheriniformes) A small family of brackish-water and freshwater fish that have a rather slender and almost transparent body. The *pelvic fins of the males are modified to form a complex copulatory apparatus (pelvic fins are absent in the females). The *anus is located far forward. Most species are very small (4 cm) and are found in coastal creeks in south-east Asia. There are about 16 species.

neotenin *See* JUVENILE HORMONE.

neoteny A form of *heterochrony that involves the slowing down in a descendant of part or all of its ancestor's rate of development, so that at least some aspects of the descendant resemble a (generally large-sized) juvenile stage of the ancestor. This may lead to *paedomorphosis. Since the juvenile stages of many organisms are less specialized than the corresponding adult stages, such shifts allow the organisms concerned to switch to new evolutionary pathways. The word comes from the Greek *neos* (meaning 'youthful'). Neoteny is common among *Urodela.

Neotropical faunal region The region which includes S. and Central America, including southern Mexico, the W. Indies, and the Galápagos Islands. Much of the area was isolated for the greater part of the *Tertiary Period, which explains the distinctiveness of the fauna and the survival of ancient forms of mammals (e.g. the *Marsupialia and *Edentata).

neotype In taxonomy, the specimen that is chosen to act as the 'type' material subsequent to a published original description: this occurs in cases where the original types have been lost, or where they have been suppressed by the *ICZN.

nephric ridge See MESODERM.

nephridiopore See NEPHRIDIUM.

nephridium In many invertebrates, an organ probably concerned with excretion and the regulation of the water content of the body. It consists of a simple or branched tubule, lined with *cilia. Excretory products diffuse into the tubule and are wafted by the cilia through a nephridiopore leading to the exterior of the body. There are several types of nephridia. A protonephridium is a tubule with a blind inner end bearing a *flagellum or tuft of cilia, found in all *acoelomates and *pseudocoelomates, and in some *coelomates. A metanephridium is an unbranched tubule open at the inner end through a funnel (the *nephrostome), found in some coelomates.

nephrocoel The coelomic cavity in a *nephrostome.

nephrocyte In most *Arthropoda, one of the large phagocytic cells that accumulate waste products.

nephromixium In many *Polychaeta, a composite structure formed from the attachment of a *nephridium to a coelomoduct (through which *gametes move to the exterior), so that excretory products and gametes reach the exterior through the same pore.

nephron One of the units in the vertebrate kidney that extracts metabolic wastes which are discharged as urine. The Bowman's capsule, at one end of the nephron, receives many blood constituents, and the loop of Henlé extracts water and some salts. See COUNTER-CURRENT EXCHANGE.

Nephropidae (lobsters; class *Malacostraca, order *Decapoda) Family of lobsters, which have large chelipeds and are similar in form to crayfish. Males are generally much larger than females. The family include *Nephrops norvegicus* (Dublin Bay prawn, Norway lobster, scampo) which lives off Atlantic and Mediterranean shores at depths of 40–80 m, usually on soft, sandy substrates.

***Nephrops norvegicus* (Dublin Bay prawn, Norway lobster, scampo)** See NEPHROPIDAE.

nephrostome 1. In *Polychaeta, the opening of the *nephridium into the *coelom. **2. (nephrocoelostome)** In the excretory system of vertebrate *embryos and in the vertebrate *kidney, the opening of an excretory tubule into the cavity of the Bowman's capsule.

nephrotome The part of the mesoderm (germ layer in triploblastic animals, composed of cells that have moved from the surface to the interior) that will differentiate to form *kidney and gonad tissues.

Nepidae (water scorpions; order *Hemiptera, suborder *Heteroptera) Family of aquatic bugs which live submerged, breathing through a bristle-like, tubular tail that reaches the surface. Being poor swimmers, they lie in wait for their prey, which they seize with the front pair of legs. They live in still or slow-moving waters. There are about 200 species, occurring on all continents.

Nepticulidae (subclass *Pterygota, order *Lepidoptera) Family of minute, primitive moths, with a wing-span sometimes less than 5 mm. Adults are metallic in colour or drab, and have much reduced wing venation, and a short proboscis. The antennal *scape is expanded into a prominent 'eye-cap'. The larvae are generally leaf-miners with legs and *prolegs hardly visible. Pupation occurs in a cocoon, usually in the soil. The family has a global distribution.

neritic Applied to that part of the ocean extending from the low-tide level to a depth of 200 m. Because light penetrates most of the shallow water, the neritic zone is the part of the ocean most densely populated by benthic organisms (*see* BENTHOS).

Nerophis ophidion (straight-nosed pipefish) *See* SYNGNATHIDAE.

nerve cord The bundle of nerve fibres that runs along the longitudinal axis of the body. In most invertebrates there are two, solid nerve cords; in *Chordata the nerve cord is hollow and in vertebrates it is the spinal cord.

nervous system In all multicellular animals except *Porifera, a network of nerve cells, linked together by thread-like processes, through which electrochemical impulses transmit information to and from sense organs, muscles, etc. This is the means by which the activities of the animal and its awareness of, and reaction to, events outside itself are co-ordinated.

nervure *See* VEIN.

nest-building frogs *See* RHACOPHORIDAE.

nestling 1. Young bird, before it leaves the nest. **2.** A mode of life adopted by certain *Bivalvia which live in crevices and depressions in hard substrata that have not been excavated by the bivalve itself.

nest parasite 1. A parasite (*see* PARASITISM) that lives in the nest of its host. **2.** *See* BROOD PARASITE.

net plankton *See* PLANKTON.

Nettastomidae (duckbill eel, witch eel; subclass *Actinopterygii, order *Anguilliformes) A small family of deepsea, eel-like fish that have a very slender, elongate body, long and thin jaws, and a large mouth with sharp teeth. *Pectoral fins are absent, but the *dorsal and *anal fins are long and low. Duckbill (or witch) eels are found only at great depth in the oceans. There are about 16 species.

net-winged beetle *See* LYCIDAE.

neural arch *See* VERTEBRA.

neuralation The formation of the nervous system in a developing *embryo.

neural crest cell A type of *mesenchyme cell, derived from the *ectoderm, that is unique to *Craniata. Derivatives of neural crest cells give rise to most of the *peripheral nervous system, some *endocrine glands, pigment cells, and the skeleton and *connective tissues of the head.

neural spine *See* VERTEBRA.

neural tube In a vertebrate *embryo, the *dorsal tubular structure that develops into the *spinal cord and *brain.

neurilemma (sheath of Schwann) A thin membrane that surrounds the *myelin sheath enclosing *neurons, especially those of the *peripheral nervous system.

neurobiology The scientific study of the biology of the nervous system, encompassing the structure and function of *neurons and the way they are organized into systems for transmitting and receiving signals. *Compare* NEUROSCIENCE.

neurocyte *See* NEURON.

neurofibril (neurofilament) One of the many fine filaments that runs in all directions through the *cytoplasm of *neurons, extending into the *axons and *dendrites, and forming part of the *cytoskeleton.

neurofilament *See* NEUROFIBRIL.

neuroglia *See* GLIAL CELL.

neurohypophysis In vertebrates, part of the *pituitary gland that is derived from the brain during the development of the *embryo. It is fused closely with the pars intermedia of the *adenohypophysis to form the posterior lobe of the pituitary.

neuromast In fish and some *Amphibia, the basic receptor units of the *acousticolateralis system. The receptor cells are surrounded by a clump of supporting cells, the units being covered by a gelatinous capsule (cupula). Physical changes in the environment are relayed to the brain by means of electrical discharges through special nerve fibres.

neuron (neurocyte, neurone) A nerve cell. It is a nucleated cell through which impulses are conducted, being transferred from one neuron to the next at a *synapse. Neurons are commonly long and threadlike in shape.

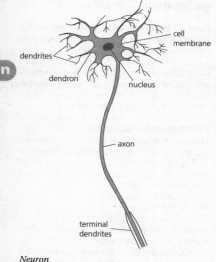

Neuron

neurone *See* NEURON.

neuropodium In *Polychaeta, the ventral division of the *parapodium.

Neuroptera (ant lions, lacewings, mantid flies, owlflies; class *Insecta, subclass

*Pterygota) Order of *endopterygote insects which have simple, biting mouthparts. The antennae are conspicuous and multi-segmented, and the two pairs of large, equal or subequal wings are lace-like, divided into many small cells by numerous cross-veins. The larvae are predacious or parasitic, with distinctive, sickle-shaped, sucking jaws. Pupation occurs in a silken cocoon. The silk is produced from modified excretory tubules opening into the hind gut of the larva and is extruded from the anus, rather than being produced by salivary glands. Many species are highly coloured and patterned, sometimes with dense hairs, and most can be recognized by the highly branched terminal portions of the main veins (end-twigging). With the exception of the *Hymenoptera, few other insect groups are more beneficial to humans, since adults and larvae are predacious on a vast range of sap-sucking insects (e.g. aphids and psyllids), and others on lepidopteran eggs and larvae, mites, and immature dipterans. The order is represented in all the major zoogeographical regions of the world.

neuroscience The scientific study of all aspects of the nervous system, comprising *neurobiology but also encompassing the chemistry and physics of the system, as well as psychology, and relevant areas of computer science and philosophy.

neurosecretion A secretion by nerve cells of chemical compounds, including *neurotransmitters and neurohormones.

neurotoxin A *toxin that affects the functioning of the *nervous system.

neurotransmitter A substance that functions in the transmission of nervous impulses. Although numerous substances have been implicated in neurotransmission, the two most widespread and best understood systems involve *acetylcholine and *noradrenalin, the so-called cholinergic and adrenergic systems respectively.

neuston Organisms resting or swimming on the surface of water. *Compare* EPINEUSTON; HYPONEUSTON.

neutralism The situation in which two *species populations coexist, with neither population being affected by its association with the other.

neutrality theory of evolution (neutral-mutation theory) A theory proposed by Motoo *Kimura that many genetic *mutations are adaptively equivalent (effectively neutral), and do not affect significantly the *adaptive value of the carrier. Thus they can become fixed in the *genome at a random rate. Changes in their frequencies are due more to chance than to *natural selection. The theory has been formally applied only to protein evolution, although there is no reason why it should not also work for aspects of gross morphology, and does not deny the role of natural selection in addition. *See also* MOLECULAR EVOLUTION.

neutral-mutation theory *See* NEUTRALITY THEORY OF EVOLUTION.

neutrophils The most abundant type of white blood cells, accounting for more than half of all white blood cells. They are the first cells of the immune system to arrive at a site of infection, and are *phagocytes. Neutrophils are short-lived and are destroyed instantly on ingesting a *pathogen, but they comprise most of the immune response.

Newcastle disease An acute, infectious disease which affects birds, including domestic fowl. Although not necessarily fatal, the disease is economically important since affected birds are less productive. The disease affects mainly the respiratory tract, but the nervous system may become involved. Mortality rates vary. It is caused by a paramyxovirus.

newt *See* SALAMANDRIDAE.

New World blackbirds *See* ICTERIDAE.

New World monkeys *See* CEBOIDEA.

New World vultures *See* CATHARTIDAE.

New World warblers *See* PARULIDAE.

New Zealand wrens *See* XENICIDAE.

niacin (nicotinic acid) Vitamin B_3. It is unique among the B *vitamins in that it is synthesized by animal tissues, from *tryptophan. A deficiency of this vitamin results in the disease pellagra.

nibbling In many bird *species, a *courtship behaviour, derived from *allogrooming, in which the pair nibble at each other's neck feathers or at each other's bills.

niche 1. (ecological niche) The functional position of an organism in its environment, comprising the *habitat in which the organism lives, the periods of time during which it occurs and is active there, and the resources it obtains there. **2. (evolutionary niche)** Way of life.

Nicolettidae *See* THYSANURA.

nicotinamide adenine dinucleotide (NAD, diphosphopyridine nucleotide, DPN) A derivative of the vitamin *niacin, which functions as a *coenzyme for various dehydrogenase enzymes involved in cellular respiration. It acts as an electron carrier in respiratory-chain phosphorylation.

nicotinamide adenine dinucleotide phosphate (NADP) A phosphorylated derivative of *NAD which functions as a *coenzyme in the reaction involving the reduction of ferrodoxin by NADP reductase.

nicotinic acid *See* NIACIN.

nictitating membrane In some vertebrates, a lid-like membrane that is attached to the anterior corner of the eye. When drawn across the eye it reduces illumination of the retina. It is common in *Reptilia and *Aves, and present in some *Mammalia and in many sharks (e.g. those of the family *Carcharhinidae).

nidicolous Applied to a young bird that stays in the nest until it is able to fly. Nidicolous young are usually naked and incapable of locomotion.

nidifugous Applied to a young bird that leaves the nest on hatching or soon after. Nidifugous young are down-covered and capable of locomotion.

night-hawks *See* CAPRIMULGIDAE; CAPRIMULGIFORMES.

night herons (*Nycticorax*) *See* ARDEIDAE.

nightingales (*Luscinia*) *See* TURDIDAE.

nightjars *See* CAPRIMULGIDAE; CAPRIMULGIFORMES.

night lizards *See* XANTUSIIDAE.

night monkey (*Aotus trivirgatus*) *See* CEBIDAE.

night shine *See* TAPETUM.

Nile bichir (*Polypterus bichir*) *See* POLYPTERIDAE.

Nile crocodile (*Crocodylus niloticus*) *See* CROCODYLIDAE.

Ningaui *See* DASYURIDAE.

ninhydrin (triketohydrindene hydrate) A compound which reacts with the free *alpha amino groups of amino acids, *peptides, and *proteins to yield coloured compounds, usually blue or purple. Ninhydrin is therefore used in the chromatographic detection and quantification of amino acids and peptides.

Nissl body (tigroid body) A large structure found in the *cytoplasm of *neurons and composed of granules of rough *endoplasmic reticulum that are the sites of *protein synthesis. The bodies were first described by the German neuropathologist Franz Nissl (1860–1919).

nit Name given to the eggs of lice (order *Phthiraptera), particularly to those of the human body and head lice, *Pediculus humanus* and *P. capitis*. Eggs are cemented to the hair or feathers of the host and are very difficult to dislodge. Before hatching they are brownish-yellow, but afterwards air enters and they become silvery-white; it is in the latter form that eggs of *P. capitis* are generally recognized. The hatched eggs of some bird lice are used by mites, which enter them to shed their skins.

Nitidulidae (pollen beetles, sap beetles, dried-fruit beetles; subclass *Pterygota, order *Coleoptera) Family of small, rounded, flattened beetles, 1–7 mm long, with *elytra that are *pubescent and short, exposing the terminal segments. The antennae have three-jointed clubs. The legs are often wide and flattened. Larvae are white and cylindrical, with well-developed legs. They feed on pollen and nectar, fermenting fruit, sap, and carcasses. Some are associated with fungi, others are predatory on insects. There are 2200 species.

NK cells *See* NATURAL KILLER CELLS.

Noctilionidae (bulldog bats, fish-eating bats, hare-lipped bats; order *Chiroptera, suborder *Microchiroptera) A family of large bats in which the legs are long, the feet very large and equipped with strong claws, the wings long and narrow. The tail passes through the *uropatagium. The ears are funnel-shaped, large, and separate, the nose simple, the lips full and divided by a fold of skin to give a 'hare lip' appearance. The bats are distributed throughout tropical America. There is one genus, *Noctilio*, and two species, one of which (*N. albiventris*) is insectivorous; the other (*N. leporinus*) feeds mainly on fish which it seizes from the water with its claws and transfers to its mouth in flight.

Noctuidae (noctuids, owlets, underwings, cutworms, army worms; subclass *Pterygota, order *Lepidoptera) Huge family of small to large moths, which are usually drab, sometimes with eyespots. The proboscis is usually well developed. Larvae are not hairy. Pupation occurs below ground. The family includes serious pests, e.g. cutworms, stem borers, and foliage strippers. There are about 20 000 species, with a world-wide distribution.

noctuids *See* NOCTUIDAE.

nocturnal During the night-time; applied to a type of *circadian rhythm in which the organism performs its main activities at night.

nocturnal ground beetle *See* TENEBRIONIDAE.

nodes of Ranvier *See* MYELIN SHEATH.

Noguerornis The earliest known bird after *Archaeopterix*, from the basal *Cretaceous of Montsec, northern Spain. It was the size of a finch and was the first bird to have a wing that was well developed, as indicated by the elongation of the distal portions of the fore limb and the rigid interlocking of the hand bones.

noisy scrub-bird (*Atrichornis clamosus*) *See* ATRICHORNITHIDAE.

Nomadacris septemfasciata **(red locust)** *See* LOCUST.

Nomascus **(concolors)** *See* HYLOBATIDAE.

Nomeidae (driftfish, man-of-war fish; subclass *Actinopterygii, order *Perciformes) A small family of marine fish (sometimes included in the *Stromateidae) that have oval bodies, large eyes, two separate *dorsal fins, and forked tail fins. Species such as *Nomeus gronovius* (man-of-war fish) and *Psenes arafurensis* (eyebrow fish) are sometimes found swimming among the tentacles of *Physalia physalis* (Portuguese man-of-war, *see* RHIZOPHYSALIINA). There are about 15 species, found in tropical and subtropical seas.

nomen abortivum In taxonomy, a name which contravened the *ICZN in operation at the time.

nomen ambiguum In taxonomy, a name which is ambiguous, because different authors apply it to different taxa.

nomen conservandum In taxonomy, a name, otherwise unacceptable under the

*ICZN rules of nomenclature, which is made available using specified procedures, with either the original or altered spelling.

nomen correctum In taxonomy, a name whose spelling is required or allowed to be intentionally altered under the *ICZN rules of nomenclature but which does not have to be transferred from one *taxon to another.

nomen dubium In taxonomy, a name which cannot be attached certainly to any particular taxon, and is therefore dubious.

nomen illegitimum In taxonomy, a name which must be rejected under the rules of the *ICZN and is therefore illegitimate.

nomen imperfectum In taxonomy, a name that, as originally published, meets all the mandatory requirements of the *ICZN, but that contains a defect needing correction (e.g. incorrect original spelling).

nomen invalidum In taxonomy, a name that has not been published properly, or is unavailable, and which is therefore invalid.

nomen inviolatum In taxonomy, a name that, as originally published, meets all the mandatory requirements of the *ICZN rules of nomenclature and is not subject to any sort of alteration.

nomen neglectum In taxonomy, a name that was published at some time in the past, but which has subsequently been overlooked.

nomen novem In taxonomy, a new name that is proposed as a replacement or substitute for an existing name.

nomen nudum In taxonomy, a name that, as originally published, fails to meet all of the mandatory requirements of the *ICZN rules of nomenclature and thus has no status in the nomenclature, even if corrected.

nomen oblitum In taxonomy, a forgotten name; i.e. the name of a senior *synonym that has not been used in the zoological literature for at least 50 years. Such names should not be used if they upset a better-known name, but should preferably be suppressed by the *ICZN.

nomen perfectum In taxonomy, a name that, as originally published, meets all of the requirements of the *ICZN, needing no correction of any kind, but which nevertheless is validly alterable by a change of ending.

nomen substitutum In taxonomy, a new, replacement name, published as a substitute for an invalid one (e.g. a junior *synonym).

nomen translatum In taxonomy, a name that is derived by the valid change of a previously published name as a result of a transfer from one taxonomic level to another within the group to which it belongs.

nomen triviale In taxonomy, a species (or trivial) name.

***Nomeus gronovius* (man-of-war fish)** *See* NOMEIDAE.

nomogenesis An evolutionary model holding that the direction of evolution operates to some degree by rules or laws, independently of *natural selection. For a long time it was regarded as an outmoded hypothesis, but recently it has been maintained that it corresponds rather well with observations of evolution in the fossil record, and that such mechanisms as *heterochrony and *molecular drive would produce nomogenetic effects. *See also* ARISTOGENESIS; ENTELECHY; ORTHOGENESIS.

non-biting midges *See* CHIRONOMIDAE.

non-competitive inhibition The irreversible inhibition of the activity of an *enzyme, brought about by the presence of an inhibitor that is generally structurally unrelated to the normal *substrate.

non-hierarchical classification method *See* HIERARCHICAL CLASSIFICATION METHOD.

non-parametric test *See* STATISTICAL METHODS.

non-polar molecule A molecule in which the electrons are shared equally between the nuclei. As a result the distribution of charge is even and the force of attraction between different molecules is small. Non-polar molecules show little reactivity.

nonsense codon A *codon which causes the termination of *translation. *Compare* SENSE CODON.

nonsense mutation A *mutation that alters a *gene so that a *nonsense codon is inserted. Such a codon is one for which no normal *t-RNA molecule exists; the codon therefore does not code for an *amino acid. Usually a nonsense codon causes the termination of translation (i.e. the end of the polypeptide chain). Three nonsense codons are recognized, and are called the amber, ochre, and opal codons.

noradrenalin (norepinephrine) A *catecholamine hormone, secreted by the adrenal medulla, which has effects similar to, but less pronounced than, those of *adrenalin. It elicits a wide variety of responses, including an increase in blood-sugar level due to its stimulation of the breakdown of *glycogen.

norepinephrine *See* NORADRENALIN.

normalizing selection *See* STABILIZING SELECTION.

northern viper (*Vipera berus*) *See* VIPERIDAE.

Norway lobster (*Nephrops norvegicus*) *See* NEPHROPIDAE.

no-see-um *See* CERATOPOGONIDAE.

nose-horned viper (*Vipera ammodytes*) *See* VIPERIDAE.

nose leaf In many *Microchiroptera, a fleshy process on the face, sometimes large and of complex shape, concerned with the emission of sounds used in *echo-location.

Notacanthidae (spiny eel; superorder *Elopomorpha*, order *Notacanthiformes*) A small family of deep-sea fish that have a blunt snout and ventrally located mouth. The posterior half of the body tapers down to a narrow tail ending in a tiny tail fin. Typically, there are at least six short and isolated *dorsal spines on the back. The long-based *anal fin also has short *fin rays. Spiny eels are bottom-living fish sometimes caught in trawl-nets. There are about nine species, distributed world-wide.

Notacanthiformes (subclass *Actinopterygii, superorder *Elopomorpha) An order of deep-sea fish that have elongate, tapering bodies terminating in a very small tail fin. The *pelvic fins are located on the abdomen slightly ahead of the origin of the *dorsal fin. All species have a long *anal fin and a ventrally positioned mouth. The order includes the spiny eels (*Notacanthidae) and other deep-sea fish.

Notaspidea (Pleurobranchomorpha; class *Gastropoda, subclass *Opisthobranchia) An order of opisthobranch gastropods most of which have flat bodies and most of which are similar to slugs. There is no *mantle cavity, and the shell is external or, if internal, reduced. One true gill is present. Most Notaspidea are grazing carnivores.

Noteridae (subclass *Pterygota, order *Coleoptera) Family of shiny, aquatic beetles with paddle-like swimming legs, found in slow-moving water. Formerly they were placed in *Dytiscidae, but they are distinguishable by their more flattened ventral surface and more convex dorsal surface. Each larva has a long body, walking legs, and large, curved jaws, and taps air spaces of aquatic plants for oxygen. Adults and larvae prey on soft-bodied larvae and snails.

Notharchus (puffbirds) *See* BUCCONIDAE.

Nothosauria (nothosaurs; subclass *Archosauria, order *Sauropterygia) A suborder of marine reptiles that flourished during the *Triassic. They possessed long necks and their limbs were adapted to swimming. *Nothosaurus* is one of several genera and one of the species referred to it was the last of the line. The nothosaurs were

replaced by the plesiosaurs (*Plesiosauroidea) in the Early *Jurassic.

notochord (chorda dorsalis) A somewhat flexible, rod-like structure, composed of disc-like, turgid cells, which extends virtually the entire length of the body of adult and/or larval members of the phylum *Chordata. Lying below the nerve cord, but dorsal to the intestine, the notochord provides a form of flexible support to the body. In vertebrates the notochord is replaced wholly or partly by the vertebral column, but it is retained throughout life in *Cephalochordata and *Agnatha.

Notogea The Australian *faunal region, which possesses a very distinctive fauna. It comprises Australia (including Tasmania), New Guinea, New Zealand, and the islands to the south and east of *Wallace's line. It can be subdivided into the Australian, Polynesian, and Hawaiian regions. *Compare* ARCTOGEA; NEOGEA.

Notograptidae (subclass *Actinopterygii, order *Perciformes) A small family of primarily marine fish that have an eel-like body, large eyes, a low-slung mouth, and rather reduced *pelvic fins. The very long *dorsal fin starts anterior to the *pectoral fins. Both dorsal and *anal fins are confluent with the tail fin. There are about three species, found in tropical Australia.

Notomyotina (starfish; subclass *Asteroidea, order *Paxillosida) A suborder of starfish which have long, flexible arms. The lower and upper rows of marginal plates alternate, and bear long, outwardly directed spines. The distal marginals *imbricate with each other. The suborder is known from the Late *Cretaceous to the present.

Notomys (jerboa-rat) *See* MURIDAE.

Notonectidae (backswimmers, water boatmen; order *Hemiptera, suborder *Heteroptera) Family of predatory, aquatic bugs, which swim in the water with their backs directed downwards, propelling themselves by means of the hair-fringed hind pair of legs; hence their common names. The front two pairs of legs are used to grasp prey. There are about 200 species, distributed world-wide.

notoneural nervous system In *Hemichordata, a nervous system that comprises a sheet of nerve cells and fibres lying in the *epidermis that, in the dorsal part of the collar, is rolled into a tube, which is open at both ends and hollow in some species.

notopodium In *Polychaeta, the dorsal division of the *parapodium.

Notopteridae (featherback, knife fish; superorder *Osteoglossomorpha, order *Osteoglossiformes) A small family of freshwater fish, that have a small, featherlike *dorsal fin half-way along the back, and a very long *anal fin starting near the *pectoral-fin base and merging with the small tail fin at the end of the tapering body. There are about six species found in south-east Asia and Africa.

Notoryctemorphia *See* SYNDACTYLIFORMES.

Notoryctidae (marsupial mole; order *Marsupialia (or *Syndactyliformes), superfamily Notoryctoidea) A monospecific family (*Notoryctes typhlops*) of small, burrowing marsupials in which the eyes are much reduced, there are no external ears, the cervical vertebrae are fused, and the strong fore limbs bear large, triangular claws on the third and fourth digits. The teeth are well separated, and often blunt. They are found in part of southern and north-western Australia.

Notoryctoidea *See* NOTORYCTIDAE.

Notostraca (tadpole shrimps; subphylum *Crustacea, class *Branchiopoda) Order of branchiopods in which the head and anterior trunk are enveloped by a large, shield-like *carapace. The first antennae are absent, and the second pair are vestigial. The sessile *compound eyes are close together beneath the *dorsum of the carapace. The highly flexible abdomen bears up to 70 paired appendages, and culminates in a pair of long processes (caudal furcae). Tadpole shrimps inhabit temporary pools in desert areas, and their resistant eggs may survive 20 years or more of being blown about in desert dusts. There are 15 species.

Notosudidae *See* SCOPELOSAURIDAE.

Nototheniidae (cod icefish; subclass *Actinopterygii, order *Perciformes) A small family of marine fish, most species of which have an elongate body, a large head, and a short first *dorsal fin; the second dorsal and *anal fins are long. Nototheniidae form the most numerous group of Antarctic fish, adapted to a life in extremely cold sea water. In some species the blood lacks red blood cells. *Notothenia coriiceps* (60 cm) feeds on a variety of bottom-dwelling invertebrates. *Pleurogramma antarcticus* (20 cm) is a *pelagic species, dominant among Antarctic mid-water fish. There are about 59 species, found in coastal Antarctica and S. America.

Notoungulata (cohort *Ferungulata, superorder *Protoungulata) An extinct order of ungulates that, apart from one lower *Eocene representative from N. America and a well-established representative from the late *Palaeocene of eastern Asia, are exclusively S. American. They flourished on that continent during the *Oligocene and survived to the *Pleistocene, evolving into many forms. Their dentition was complete and the tympanic *bulla was large, the structure of the ear being different from all other mammals. They were *mesaxonic, many of them possessing only three digits. Some had claws, but others appear to have possessed hoofs, though none achieved an *unguligrade gait. Some later forms became large. *Toxodon*, which may have been the most common large ungulate during the S. American Pleistocene, was about 2.75 m long, had a massive head, and hind legs that were much longer than the fore legs. In life it may have looked like a very large guinea-pig. In addition there were the homalodotheres, which resembled the *Ancylopoda (the so-called 'clawed horses'), and a range of large and small rodent-like animals referred to as hypotheres.

Notoxaea *See* OXAEIDAE.

notum In insects, the *dorsal part of a thoracic segment.

nuchal Of, or pertaining to, the nape of the neck.

Nucifraga (nutcrackers) *See* CORVIDAE.

nuclear envelope (nuclear membrane) The structure that separates the *nucleus of *eukaryotic cells from the *cytoplasm. It comprises two unit membranes each 10 nm thick separated by a perinuclear space of 10–40 nm. At intervals, the two membranes are fused around the edges of circular pores which allow for the selective passage of materials into and out of the nucleus.

nuclear membrane *See* NUCLEAR ENVELOPE.

nuclear-pore complex A complex formed from the nuclear pore of a *eukaryotic cell and the granular or fibrous annular material with which it is filled. The structure of the filling material, in particular, is still ill-defined, though it is thought to be a hollow cylinder. Its function is equally obscure, though probably it is concerned with regulating the transport of materials through the pores.

(((⊕))) SEE WEB LINKS
• 'The Nuclear Pore Complex'

nucleic acids *Nucleotide polymers, with high relative molecular mass, produced by living *cells and found both in the *nucleus and *cytoplasm of cells. They occur in two forms, designated DNA (*see* DEOXYRIBONUCLEIC ACID) and *RNA, and may be double- or single-stranded. DNA embodies the genetic code of a cell or *organelle, while various forms of RNA function in the transcriptional and translational aspects of protein synthesis.

nucleolus A clearly defined, often spherical area of the *eukaryotic *nucleus, composed of densely packed fibrils and granules. Its composition is similar to that of *chromatin, except that it is very rich in

RNA and protein. It is the site of origin of *ribosomes.

nucleoplasm *See* PROTOPLASM.

nucleoprotein A *conjugated protein that is composed of a *histone or *protamine bound to a *nucleic acid as the non-protein portion.

nucleoside A *glycoside, composed of *ribose or deoxyribose sugar bound to a *purine or *pyrimidine base.

nucleosome A particle, approximately 10 nm in diameter, found in large numbers in isolated *chromatin, where groups of *histone molecules combine with and strengthen DNA strands.

nucleotide A *nucleoside that is bound to a phosphate group through one of the hydroxyl groups of the sugar. It is the unit structure of *nucleic acids.

nucleus The double-membrane-bound *organelle containing the *chromosomes that is found in most non-dividing *eukaryotic cells; it is essential to their long-term survival. It is variously shaped, although normally spherical or ovoid. It disappears temporarily during cell division. It is absent from *viruses. The chromosomes, though probably intact, are not visible when the cell is in a resting stage (i.e. not dividing). The nucleus also contains nucleoli (*see* NUCLEOLUS).

Nuculoida (class *Bivalvia, subclass *Palaeotaxodonta) An order of bivalves which have a *taxodont hinge and equivalve, aragonitic shells. They are *isomyarian, the ligament generally *amphidetic, and the gills are *protobranchiate. The foot is grooved. Adults lack a *byssus. Nuculoids are small, *infaunal, *labialpalp feeders, which first appeared in the *Ordovician.

Nuda (Macropharyngea; phylum *Ctenophora) A class of ctenophorids which lack *tentacles. There is only one order, *Beroidea.

n

Nudibranchia (sea slugs; class *Gastropoda, subclass *Opisthobranchia) An order of molluscs in which the shell and mantle cavity are absent and the body is secondarily (*see* SECONDARY) *bilaterally symmetrical. Many nudibranchs have projections from the body surface (cerata). These occur in a variety of shapes and are often brightly and beautifully coloured. Lacking shells for defence, most nudibranchs swim well and many possess skin glands that secrete sulphuric acid. Some utilize *nematocysts from prey they have eaten; the ingested nematocysts are transported along ciliary tracts to *cnidosacs at the distal tips of the cerata, where they are used for defence, being replaced every few days.

null allele *See* SILENT ALLELE.

null hypothesis *See* HYPOTHESIS.

numbat (*Myrmecobius fasciatus*) *See* MYRMECOBIIDAE.

numbfish *See* TORPEDINIDAE.

Numenius (curlews, whimbrel) *See* SCOLOPACIDAE.

numerical method (numerical taxonomy) Any method for quantifying individuals or *communities, or for comparing such quantifications, that makes no assumptions about the data (e.g. about the sampling approach used, the distribution of the data, or the probability of particular patterns or frequencies). *Compare* STATISTICAL METHOD.

numerical taxonomy *See* NUMERICAL METHOD.

Numida meleagris (helmeted guineafowl) *See* NUMIDIDAE.

Numididae (guineafowl; class *Aves, order *Galliformes) A family of grey or black fowl, mostly spotted white, which have bare heads, some with *wattles or a crest. They have rounded wings, medium-length tails, and strong legs and feet, some with spurs. They inhabit forest and brush, feed on insects and seeds, and nest on the ground. The three species of *Guttera* are typical. There are five genera, with seven species, found in Africa. *Numida meleagris* (helmeted guineafowl) has been widely domesticated.

nun babblers *See* TIMALIIDAE.

nuptial pad In *Ranidae (frogs), one of the horny or thickened pads on each thumb of the male. They are especially prominent during the mating season, when they assist the male in grasping the female during the sexual embrace, in which the male extrudes sperm over the eggs as these are ejected by the female.

nursery fish *See* KURTIDAE.

nursery-web spiders *See* PISAURIDAE.

nurse shark *See* ORECTOLOBIDAE.

nutcrackers (*Nucifraga*) *See* CORVIDAE.

nuthatches 1. (*Sitta*) *See* SITTIDAE. **2. (coral-billed nuthatch, *Hypositta corallirostris*)** *See* HYPOSITTIDAE.

Nyctereutes procyonoides (raccoon dog) *See* CANIDAE.

Nycteribiidae (order *Diptera, suborder *Cyclorrapha) Small family of highly modified flies, which are ectoparasites of bats. Adults have the head folded back, to rest in a groove on the *thorax. The antennae are two-segmented, and have a terminal *arista. They have a *ctenidium on the front of the thorax. Wings are absent, and the legs are long and adapted for clinging. Larvae are produced in a mature state, and form puparia after falling from the host bat. There are more than 200 species, occurring most commonly in Asia, Europe, and Africa.

Nycteridae (slit-faced bats; order *Chiroptera, suborder *Microchiroptera) A family of bats in which the face is marked by a slit from the nose to between the eyes, formed from a nose leaf composed of fleshy pads. The ears are very large and separate and the *uropatagium is large, supported by long *calcars and by the tail which extends to its edge, where it ends in a T-shaped vertebra. The body is covered in long, loose fur. The bats are distributed through tropical and subtropical Africa, western Arabia, and part of the Mediterranean region, and in Malaysia and Indonesia. There is one genus, *Nycteris*, with about 10 species.

Nyctibiidae (pottoos; class *Aves, order *Caprimulgiformes) A family of grey, brown, and black, cryptically coloured birds which have long wings and tails, small bills, large gapes, and no *rictal bristles. They are nocturnal, inhabit open forest, feed on insects, and nest in tree crevices. There is one genus, *Nyctibius*, with five species, found in Central and S. America.

***Nyctibius* (pottoos)** *See* NYCTIBIIDAE.

***Nycticorax* (night herons)** *See* ARDEIDAE.

nymph 1. A crescent-shaped platform, present in certain bivalves, to which the ligament is attached. **2.** See LARVA.

Nymphalidae (admirals, brush-footed butterflies, duffers, emperors, fawns, fritillaries, map butterflies, rajah butter-flies, saturn butterflies, vanessas; subclass *Pterygota, order *Lepidoptera) Family of small to large, colourful butterflies. The fore legs are reduced, non-functional, and have long, hair-like scales. The egg is ribbed, the larvae are spined, and the pupae are not supported by a silken girdle. There are about 5000 species and the family has a worldwide distribution.

oak galls *See* CYNIPIDAE.

oarfish *See* REGALECIDAE.

***Obdurodon* (order *Monotremata, family *Ornithorhynchidae)** A genus of early *Miocene platypus, with two species, known from very well preserved material from such sites as Riversleigh, Queensland, Australia. They possessed well-developed, functional teeth, unlike the living platypus, and had skull bones that were less fused, from which it can be seen that monotremes, unlike other living mammals, retained *septomaxillae.

obligate Applied to an organism that can survive only if a particular environmental condition is satisfied. For example, an obligate aerobe can survive only in the presence of air, an obligate parasite only in association with its host.

obligate parasite *See* PARASITISM.

obtect Applied to the *pupae of those *endopterygote insects in which the appendages adhere to the body by means of a secretion produced at the last larval moult. Obtect pupae are found in some *Coleoptera, many *Chalcidoidea, and in all nematoceran *Diptera and higher *Lepidoptera.

Obtect

obturate foramen (obturator foramen) In the *pelvis of mammals, a large opening between the *pubis and *ischium, through which nerves and blood vessels pass.

Occam's razor (Ockham's razor) The axiom, proposed by William of Occam (William Ockham, *c.*1280–1349), that *pluralitas non est ponenda sine necessitate* ('multiplicity ought not to be posited without necessity'); i.e. when alternative hypotheses exist, the one requiring the fewest assumptions should be preferred.

occipital Pertaining to the posterior part of the *cranium.

occipital condyle At the back of the skull, a bony knob which articulates with the first vertebra. It is absent in fish, and double in amphibians and mammals.

occlusal Applied to the grinding or biting surfaces of a tooth.

oceanic Applied to the regions of the sea that lie beyond the continental shelf, with depths greater than 200 m.

***Oceanites* (storm petrels)** *See* HYDROBATIDAE.

Oceanitidae An alternative name for the *Hydrobatidae.

***Oceanodroma* (storm petrels)** *See* HYDROBATIDAE.

ocean pout (*Macrozoarces americanus*) *See* ZOARCIDAE.

ocean whitefish (*Caulolatilus princeps*) *See* MALANCANTHIDAE.

ocellar Applied to the region on the frons (*see* CLYPEUS) of the head of an insect that bears the ocelli (*see* OCELLUS). This includes an approximately triangular area (the ocellar triangle) and may bear ocellar bristles.

ocellated lizard (*Lacerta lepida*) *See* LACERTIDAE.

ocelli *See* OCELLUS.

ocellus (*pl.* ocelli) A simple *eye, which has a single, thickened, cuticular lens. The epithelial cells below the lens are transparent, and below them there are many light-sensitive nerve cells.

Ochotonidae (calling hares, conies, mouse hares, pikas; cohort *Glires, order *Lagomorpha) A family of lago-morphs in which the ears are short, the tail is not visible externally, and the hind limbs, bearing four digits, are only slightly longer than the fore limbs, which bear five digits. Ochotonidae live in mountainous regions throughout the northern hemisphere at altitudes higher than those available to most mammals. They shelter in burrows or crevices, some socially, and do not hibernate, feeding during winter on plant material that they have gathered and dried during the summer and autumn. There is one genus, *Ochotona*, with about 14 species.

Ockham's razor *See* OCCAM'S RAZOR.

octocoral *See* OCTOCORALLIA.

Octocorallia (Alcyonaria; phylum *Cnidaria, class *Anthozoa) A subclass of sedentary, colonial corals with *polyps that always have eight *tentacles which are almost invariably *pinnate. There are eight complete, unpaired *mesenteries and *septa, and a ventral groove on the *stomo-daeum Skeletal structures are usually made from calcareous *spicules.

Octodontidae (bush rats, hedge rats, degus; order *Rodentia, suborder *Hystricomorpha) A family of rat-like, ground-dwelling or burrowing rodents in which the tail is haired, the hairs growing longer closer to the tip and ending in a tuft. The limbs are short, the fore limb having five digits and the hind limb four. The hind digits have bristles which extend stiffly beyond the claws. Octodonts are confined to the west

coast of S. America. There are five genera, with about seven species.

octoploidy *See* POLYPLOIDY.

Octopoda (octopuses, devilfish; class *Cephalopoda, subclass *Coleoidea) An order of cephalopods which have eight long arms possessing suckers and a round, sack-like body. A few have a much reduced, internal shell. Probably they arose from a belemnite (*Belemnitida) forebear, some time in the early *Cretaceous. Since they lack hard parts they have a poor fossil record.

octopuses *See* OCTOPODA.

ocular plate In *Echinoidea, one of the small plates that alternate with the *genital plates and coincide with the *ambulacra.

ocular point A non-sensory conical projection from the side of the head at the position occupied by the eyes in other species. *See* HAEMATOPINIDAE.

Odacidae (rock whiting; subclass *Actinopterygii, order *Perciformes) A small family of marine fish that have an elongate body, long *dorsal fin, and straight-edged tail fin. The teeth are fused together, like those of the *Scaridae (parrotfish). Generally colourful fish, they tend to live in weedy patches in relatively shallow water. There are about eight species, found in the cooler waters of Australia and New Zealand.

Odobenidae (walruses; order *Carnivora, suborder *Pinnipedia (or *Caniformia)) A monospecific family (*Odobenus rosmarus*) of marine carnivores known since the *Miocene and related to the *Otariidae. There are no external ears or tail and the hind limbs are brought forward beneath the body when the animal is on land. The upper *canines form large tusks in both sexes but the remaining teeth are small and often lost, leaving only the heavy, peg-like *molars for crushing. Walruses feed exclusively on shellfish. They are found around the margins of the Arctic Ocean, in shallow waters and on ice floes.

Odonata (damselflies, devil's darning needles, dragonflies, horse-stingers; class *Insecta, subclass *Pterygota) Order of primitive insects, the earliest fossils of which are found in the upper *Carboniferous coal measures of Commentry (France). These slender-bodied insects have two pairs of large, glassy wings, and large, prominent eyes. They usually hunt on the wing, taking swarming flies and other flying insects. Odonates are unique in possessing two sets of genitalia (*see* ACCESSORY GENITHALIA). The nymphs are aquatic, and catch prey with a prehensile organ (called the mask) derived from the labia (*see* LABIUM). Despite their fearsome common names, these beautiful insects are not dangerous to large animals. About 5000 species are known, with a cosmopolitan but mainly tropical distribution.

Odontaspididae (Carchariidae; grey nurse shark, sand tiger; subclass *Elasmobranchii, order *Lamniformes) A small family of large sharks that have all the gill slits in front of the *pectoral fin, sharp, non-serrated teeth, and two *dorsal fins equal in size to the *anal fin. The fearsome *Odontaspis taurus* (grey nurse shark), growing to 5 m in length, is common along Australian and S. African beaches, but may be less dangerous than is often claimed. There are possibly seven species, distributed world-wide.

***Odontaspis taurus* (grey nurse shark)** *See* ODONTASPIDIDAE.

Odontoceti (toothed whales; cohort *Mutica, order *Cetacea) A suborder of marine mammals that comprises ancestral forms, known from the upper *Eocene, and the four superfamilies: *Squalodontoidea (extinct forms), *Platanistoidea (river dolphins), *Physeteroidea (large whales), and *Delphinoidea (smaller whales, dolphins, and porpoises). Teeth are always present, and may be numerous (e.g. squalodonts, which possessed up to 180 teeth, one porpoise which had 300, and the river porpoise *Platanista*, which has more than 50), or reduced to a single tooth (*Monoceros*). The teeth are simple and peg-like and are not differentiated into incisors, canines, etc.

There is a single nasal opening and the nasal bones do not form part of the roof of the nasal passage.

odontoid process A peg-like, bony process that projects forward from the second (axis) vertebra into the first (atlas) vertebra, but is formed from the *centrum of the atlas which has become detached and fused to the axis.

***Odontophorus* (wood quails)** *See* PHASIANIDAE.

Odontostomatida (class *Ciliatea, subclass *Spirotrichia) An order of ciliate *Protozoa whose cells are compressed and wedge-shaped. Body *cilia usually are confined to several rows at each end. There are two families, with several genera, found in freshwater habitats that contain organic material and that are deficient in oxygen.

***Oeciacus hirundinis* (martin bug)** *See* CIMICIDAE.

Oecophylla leakeyi Species of weaver ant (subfamily *Formicinae) whose early *Miocene remains, from Mfwangano Island (L. Victoria), illustrate the early development of ant society. Remains of 366 ants were discovered in one spot, with castes well differentiated and represented in the same proportions as in modern representatives of the genus. Larvae and pupae were discovered; and some workers were still attached to modern leaf fragments. Other *Oecophylla* species, preserved in amber, are known from the *Oligocene of northern Europe, but none in a colony deposit like the *O. leakeyi* discovery. At the present time species of *Oecophylla* live in Africa and tropical Asia.

Oedemeridae (subclass *Pterygota, order *Coleoptera) Family of soft-bodied, parallel-sided, finely pubescent beetles, 5–18 mm long, in which the *elytra may not fully cover the abdomen. Some species are toxic. Oedemeridae are found feeding at flowers, particularly on the coast. Larvae are found in old wood, preferring wood that is waterlogged. There are 1500 species.

Oegophiurida (class *Stelleroidea, subclass *Ophiuroidea) An order of primitive brittle-stars in which dorsal and ventral arm-plates are lacking, as are large plates of the disc (radial shields, etc.). The body cavity, containing gonads and digestive organs, extends into the base of the arms. The order is, though mainly *Palaeozoic, *Ordovician to *Holocene with one extant genus, *Ophiocanops*.

Oenanthe (wheatears) *See* TURDIDAE.

oesophagus (gullet) The part of the *alimentary canal that lies between the *pharynx and the *stomach.

Oestridae (warble flies, bot flies; order *Diptera, suborder *Cyclorrapha) Small family of stout, hairy flies, which are often bee-like. Their antennae are short, and partially sunken into facial grooves. The larvae are all internal parasites of large mammals, found beneath the skin, where they feed on exudations from the host tissues. The larvae of *Hypoderma* species (warble flies) form swellings ('warbles') on either side of the spine of their host after migrating through the subdermal tissues. The mature larvae bore through the hide of the host before pupating on the ground. The sheep-nostril fly, or bot fly, is *larviparous, the young larvae being deposited in the nostrils of sheep, from where they migrate to the sinuses. So far 60 species have been recognized. *See also* GASTEROPHILIDAE.

oestrogen (estrogen) A generic name for a group of C18 *steroids that act as female sex *hormones, having a wide range of physiological effects. The major oestrogens are oestrone and betaoestradiol.

oestrus cycle (estrus cycle) In female mammals (other than most *Primates, *compare* MENSTRUAL CYCLE) the hormonally controlled, regularly repeated stages by which the body is prepared for reproduction. In anoestrus the female reproductive apparatus is inactive; in pro-oestrus it becomes active; and in oestrus *ovulation usually occurs (in some species ovulation is triggered by copulation) and the female be-

comes receptive to males. Unless fertilization occurs, oestrus gives way to anoestrus as the cycle repeats.

Ogcocephalidae (batfish; subclass *Actinopterygii, order *Lophiiformes) A family of bizarre, marine, bottom-dwelling fish in which the body is usually much depressed and flattened sideways, with the eyes located on the upper surface. Awkward swimmers, the batfish use the widely separated *pectoral fins as legs to 'walk' over the bottom of the sea-bed. Often partly hidden in the sand, they use a long tentacle on the head to entice small fish to come close by, and then swallow them. The brownish *Ogcocephalus vespertilio* (long-nosed batfish) is one of the more common species of the Atlantic region. There are about 55 species, distributed world-wide.

Ogcocephalus vespertilio (long-nosed batfish) *See* OGCOCEPHALIDAE.

-oidea In taxonomy, a recommended, but not mandatory, suffix used to indicate a superfamily of animals (e.g. Chelonoidea, the sea turtles).

oil beetle *See* MELOIDAE.

oilbird (*Steatornis caripensis*) *See* STEATORNITHIDAE; CAPRIMULGIFORMES.

okapi (*Okapia*) *See* GIRAFFIDAE.

old wife (*Enoplosus armatus*) *See* ENOPLOSIDAE.

Old World flycatchers *See* MUSCICAPIDAE.

Old World monkeys *See* CERCOPITHECIDAE.

Old World orioles *See* ORIOLIDAE.

Old World vultures *See* ACCIPITRIDAE.

Old World warblers *See* SYLVIIDAE.

olecranon The bony prominence at the elbow, formed by the *apophysis of the *ulna. The word comes from the Greek *olene*, 'elbow', and *kranion*, 'head'.

olfaction Detecting by means of smell.

olfactometer A device used to determine the response of an animal to odours. The animal is placed inside the stem of a Y-shaped tube; equal air currents flow through each arm of the Y, one of them after crossing a source of the odour being tested, and a record is made of whether or not the animal moves through the arm carrying the odour.

olfactory centre One of the regions of the brain that are concerned with the sense of smell.

oligo- From the Greek *oligos*, 'little', a prefix meaning few or small; in *ecology it is often used to denote a lack, e.g. 'oligotrophic', nutrient-poor; 'oligomictic', subject to little mixing.

Oligocene The final epoch of the *Palaeogene Period, 33.9–23.03 Ma ago, following the *Eocene and preceding the *Miocene epochs.

(((⊕))) SEE WEB LINKS
• Description of the Oligocene

Oligochaeta (phylum Annelida) A class of worms that have very well-developed *metameric segmentation. The segments have only a few *chaetae (hence the name), but *parapodia are not present. They are all hermaphrodites; asexual reproduction is predominant in aquatic forms. Eyes and *tentacles are absent. A few marine forms occur but most are freshwater or terrestrial. There are 15 families, and the class is first recorded from the Late *Ordovician.

oligolecithal Applied to an egg that has a little yolk, typical of *Cephalochordata. *Compare* MACROLECITHAL; MESOLECITHAL. *See also* ISOLECITHAL; TELOLECITHAL.

oligolectic Applied to bee species that specialize in collecting pollen from one genus or species (or from only a few genera or species) of flowering plants.

oligomer A molecule that consists of a small number of *monomer units.

oligonucleotide A macromolecule that consists of a short chain of *nucleotides (e.g. a short length of DNA or *RNA).

oligopeptide A linear *peptide of 2–10 *amino acids. *Compare* POLYPEPTIDE.

oligophagous Applied to an organism that feeds on a few types of plants or prey. *Compare* MONOPHAGOUS; POLYPHAGOUS.

oligosaccharide A linear or branched carbohydrate of 2–10 monosaccharides.

Oligotrichida (class *Ciliatea, subclass *Spirotrichia) An order of ciliate *Protozoa in which body *cilia are sparse or absent. Oral cilia are conspicuous. They are found mostly in marine environments.

olm (*Proteus anguinus*) *See* PROTEIDAE.

olynthus A short-lived stage in the development of some calcareous *Porifera, formed when the colony becomes attached to a substrate. It consists of a hollow, vase-shaped body composed of an inner layer of *choanocytes and an outer dermal layer. Water enters through pores in the body and leaves through the 'mouth' (the osculum) of the 'vase'.

omasum (psalterium) In *Ruminantia, the third chamber of the stomach, in which water is absorbed.

omentum *See* MESENTERY.

ommatidium One of the individual, light-sensitive units that together form a *compound eye. Each ommatidium consists of a cuticular lens below which is the crystalline cone. Below this are six or seven elongated *retinula cells, which have close-packed *microvilli towards the centre forming a structure called the rhabdome. Pigment cells surround the retinula cells. *See also* EYE.

omnivore (diversivore) A heterotroph that feeds on both plants and animals, and thus operates at a range of *trophic levels.

onchosphere (hexacanth embryo) The first, *motile, six-hooked (hexacanth) larva of *Cyclophillidea.

oncogene A gene that has the ability to cause eukaryotic cells to grow in an un-regulated fashion, like that of a cancerous tumour. *Compare* APOPTOSIS.

oncogenic Capable of causing the formation of tumours.

***Ondatra zibethicus* (musk-rat, mus-quash)** *See* CRICETIDAE.

one-gene–one-polypeptide hypothesis The hypothesis that a large class of structural genes exists in which each gene encodes a single *polypeptide, which may function either independently or as a subunit of a more complex *protein. Originally it was thought that each gene encoded the whole of a single *enzyme, but it has since been found that some enzymes and other proteins derive from more than one polypeptide and hence from more than one gene.

Oneirodidae (subclass *Actinop-terygii, order *Lophiiformes) A family of marine fish that are related to the anglerfish. Typically, the head, mouth, and eyes are large, and the fins (apart from the tail fin) are much reduced in size. There are about 33 species, distributed world-wide.

ontogenetic allometry *See* ALLO-METRY.

ontogenetic heterochrony In colonial animals, *heterochrony that affects individuals, rather than the colony as a whole. *Compare* ASTOGENETIC HETEROCHRONY; MOSAIC HETEROCHRONY.

ontogeny The development of an individual from fertilization of the egg to adulthood.

Onychophora (velvet worms; *Peripatus) A phylum comprising animals that combine annelid and arthropod features and that may be an evolutionary link between the two. They first appeared in the *Cambrian (*see* AYSHEAIA PEDUNCULATA), since when they have changed little. The body is approximately cylindrical and slug-like, but covered with bands of tubercles covered by scales, and with 14 to 43 pairs of legs. The anterior bears antennae. The mouth is in a ventral position, flanked on either side by claw-like *mandibles and oral papillae. Reproduction is *oviparous, *ovoviviparous, or *viviparous. Many onychophorans are brightly coloured blue, green, or orange; others are black. There are about 70 species, found only in humid, tropical habitats.

oocyst In some parasitic *Protozoa, a *zygote formed by the fusion of *gametes, in which form the organism may be transmitted to a new host.

oocyte The cell that, following *meiosis, forms the *ovum. Most of the cytoplasmic growth involved in ovum formation is accomplished by the primary oocyte; this undergoes the first meiotic division, giving rise to two cells, the polar body, with very little *cytoplasm, and the secondary oocyte, which has a large amount of cytoplasm. This latter oocyte undergoes the second meiotic division to give rise to another polar body and an ovum. Fertilization often takes place without the formation of an ovum at one of the two oocyte stages.

oogamy Sexual reproduction in which the female *gamete (*ovum) is much larger than the male gamete (*spermatozoon).

oogenesis The formation of eggs or ova, including the meiotic division of the *oocytes and the formation of the *yolk and the egg membranes.

oogonium A cell that undergoes repeated *mitosis to give rise to *oocytes. It forms the cell of the animal *ovum.

ookinete The *motile *zygote of the malaria parasite (*Plasmodium*) that results from fertilization inside the body of a female *Anopheles* mosquito. It penetrates the mosquito's stomach lining and attaches itself to the stomach wall where it develops into an *oocyte that gives rise to *sporozoites.

ootheca Type of egg case produced by certain insects, best seen among the *Blattodea and *Mantodea. The ootheca encloses a cluster of eggs, and is secreted by accessory reproductive glands. On exposure to air during ovipositing, the frothy secretions tan and harden.

ootype In some *Platyhelminthes, the central structure of the female reproductive system, which receives eggs, sperm, and yolk cells from the ovovitelline duct. Shell forms around each egg in the ootype and the eggs leave through the uterus, to be expelled to the exterior.

oozoid In *Thaliacea and some *Cnidaria, an individual formed from an egg that buds asexually to produce sexually reproducing forms.

opah (*Lampris guttatus*) See LAMPRIDIDAE.

Opalinata (phylum *Protozoa, subphylum *Sarcomastigophora) A superclass of ciliated protozoa which have nuclei of one type only. Cysts are formed. Opalinata are found as parasites in the intestines of amphibians.

open population A population that is freely exposed to *gene flow. *Compare* CLOSED POPULATION.

open system A system in which energy and matter are exchanged between the system and its environment (e.g. a living organism, or an *ecosystem).

operant conditioning (instrumental conditioning) *Conditioning in which an animal forms an association between a particular behaviour and a result that *reinforces the behaviour, its behaviour being operant (or instrumental) in producing the result. For example, a bird that turns over dead leaves may find food beneath them, so it may come to associate turning over dead leaves with finding food. The process may be negative, as when an animal learns to associate a particular activity with an unpleasant result.

operator The region of DNA at one end of an *operon that acts as the binding site for a specific repressor protein, and so controls the functioning of adjacent *cistrons.

opercula See OPERCULUM.

opercular flap In *Holocephali, a skin flap attached to the *hyoid arch: it covers the gills.

operculum (*pl.* **opercula**) A little lid or cover. In *Prosobranchia, a rounded, horny or calcareous plate, carried on the foot, that closes the aperture when the animal withdraws into its shell. *See also* GILL COVER.

operon A set of adjacent *structural genes whose *messenger-RNA is synthesized in one piece, together with the adjacent regulatory genes that affect the *transcription of the structural genes. The operon is under the control of an *operator gene lying at one end of it. Operons are concerned with the control of gene transcription leading to the formation of particular *enzymes used in metabolic pathways. To date they have been found only in *prokaryotes (e.g. *Escherichia coli*), but a number of homologous systems exist in lower *eukaryotes such as fungi.

((⊕)) SEE WEB LINKS
• Detailed description of the *lac* operon.

Ophichthidae (snake eel; subclass *Actinopterygii, order *Anguilliformes) A large family of eel-like, marine fish, in most of which the *dorsal fin extends almost the full length of the body, originating behind the head and extending either to a spike-like tail or to a tiny tail fin. The tail is thought to be used for quick burrowing activity into the sea-bottom. Many species reach a length of 60 cm; others may grow to 1.7 m. There are about 270 species, distributed world-wide.

Ophiclinidae (snake blenny; subclass *Actinopterygii, order *Perciformes) A small family of marine coastal fish that have an elongate, almost snake-like body. Both *dorsal and *anal fins are long-based, but the dorsal consists of at least 40 spines with only one soft ray. They resemble blennies in that the *pelvic fins are reduced to a pair of free rays in a *jugular position. There are about five species, found in the temperate waters of Australia and New Zealand.

Ophidia *See* SERPENTES.

Ophidiidae (cusk-eel, brotula; subclass *Actinopterygii, order *Gadiformes) A fairly large family of marine fish that have long, compressed bodies with minute scales. The *dorsal, *caudal, and *anal fins are often united to form a continuous fin, although a few species possess a small, separate, tail fin. The *pelvic fins are reduced to slender, filamentous rays in the throat region, and probably are used as feelers. The Ophidiidae inhabit the seas at all depths, with those living at greater depths possessing rudimentary eyes. The family includes several species of commercial importance. There are about 190 species, distributed world-wide.

Ophiomyxina (brittle-stars; subclass *Ophiuroidea, order *Phrynophiurida) A suborder of modern brittle-stars which are completely enclosed in a thick, soft skin. They are known only from the *Holocene.

***Ophiophagus hannah* (king cobra)** *See* ELAPIDAE.

ophiopluteus In some *Ophiuroidea, applied to a larva that has four pairs of elongated arms supported by calcareous rods and bearing bands of *cilia.

***Ophisaurus apodus* (European glass lizard, glass snake)** *See* ANGUIDAE.

Ophiurida (brittle-stars; class *Stelleroidea, subclass *Ophiuroidea) An order of brittle-stars in which the *ambulacral groove is covered by the lateral arm-plates. The dorsal and ventral arm-plates are well developed. Large *radial shields, buccal shields, and genital plates are present in the disc. The order is known from the *Silurian to the present.

Ophiuroidea (brittle-stars; subphylum *Asterozoa, class *Stelleroidea) A subclass of brittle-stars in which a central, rounded disc is sharply demarcated from the long, slender arms. The body cavity of the arms is filled almost completely with axial skeleton. The subclass is known from the *Ordovician to the present.

Opiliones (harvestmen) *See* ARACHNIDA.

opioid A *neurotransmitter that blocks the sensation of pain. Morphine, heroin, and *endorphins are opioids.

opioid receptor A *receptor on one of the *neurons responsible for feelings of pain and pleasure that binds to *opioids.

opisthaptor *See* MONOGENEA.

Opisthobranchia (phylum *Mollusca, class *Gastropoda) A subclass of marine organisms, which are often 'naked', including the sea slugs and pteropods, ranging from the *Cretaceous to the present day. They have only one gill and are thought to represent an evolutionary *grade between the prosobranch gastropods (*Prosobranchia) and the pulmonate gastropods (*Pulmonata).

opisthocoelous Applied to vertebrae with a concavity on the posterior surface.

Opisthocomidae (hoatzin; class *Aves, order *Galliformes) A monospecific family (*Opisthocomus hoatzin*) which is a dark brown, rufous, and buff bird with a stout, short bill, a small head with long crest feathers, a long neck, and a long tail. It is a weak flier, normally clambering through trees. Its young possess claws on their wings and climb well. The hoatzin is arboreal, inhabiting forests, feeds on fruit and leaves, and nests in trees. It is confined to northern S. America.

***Opisthocomus hoatzin* (hoatzin)** *See* OPISTHOCOMIDAE.

opisthodetic Applied to the ligament of a bivalve when it is in a position posterior to the *shell beaks.

Opisthognathidae (jawfish; subclass *Actinopterygii, order *Perciformes) A small family of marine fish that have a long, low, *dorsal fin and a large mouth which enables them to swallow large prey. They tend to live in burrows in tropical to subtropical waters of the Atlantic and Indian Oceans. One of the larger species, *Opisthognathus aurifrons*, which grows to 12 cm, is found in the tropical western Atlantic region. There are about 30 species.

Opisthopora (phylum *Annelida, class *Oligochaeta) An order of mainly terrestrial worms all of which have paired testes. There are eight families, and many members are earthworms (e.g. the genera *Lumbricus*, *Glossoscolex*, and *Megascolex*).

Opisthoproctidae (barreleye, spookfish; subclass *Actinopterygii, order *Salmoniformes) A small family of somewhat bizarre fish: they are of variable form, but the eyes are usually tubular and directed upward. They tend to be deep-sea fish, rarely exceeding 15 cm in length, and have a cosmopolitan distribution. There are about 11 species.

opisthosoma *See* ARACHNIDA.

opistoglyphous Applied to snakes that have enlarged teeth towards the rear of the maxilla (upper jaw) and smaller teeth in front. The fangs may be solid or with a groove to allow saliva to enter wounded prey. *Compare* PROTEROGLYPHOUS; SOLENOGLYPHOUS.

Oplegnathidae (knifejaw; subclass *Actinopterygii, order *Perciformes) A small family of robust, deep-bodied, marine fish, typically with the teeth in both jaws fused to form a parrot-like beak with sharp cutting edges. The spines of the *dorsal, *anal, and *pelvic fins are very strong. There are about four species, found in the Indo-Pacific region.

Oporornis (warblers) *See* PARULIDAE.

opossum 1. *See* DIDELPHOIDEA. **2.** (possum) *See* PHALANGERIDAE.

opossum rat *See* CAENOLESTOIDEA.

opportunist pathogen A usually nonaggressive micro-organism, commonly a normal member of the population present in the body, which under certain conditions takes on the role of a *pathogen. Predisposing factors include surgery, antibiotic therapy, and suppression of the normal immune responses.

opposable Applied to a *pollex which can be turned so that its pad makes contact with the pad of each of the other digits on the same limb.

opsonification *See* OPSONIZATION.

opsonin An *antibody molecule that renders an invading cell susceptible to *phagocytosis.

opsonization (opsonification) The process by which an *opsonin molecule prepares a cell for *phagocytosis. The opsonin molecule binds to the surface of the foreign cell and attracts *phagocytes to the site. At the same time, the complex formed by the opsonin and the *receptor releases compounds on to the cell surface that facilitate the cell's destruction.

optimality theory The theory that behavioural strategies involve decisions which tend to maximize the efficiency of that behaviour.

(((●))) SEE WEB LINKS
• Deep Ethology Ecological Perspective: Optimality Theory

optimum foraging theory The theory that *foraging strategies may involve decisions which maximize the net rate of food intake, or of some other measure of foraging efficiency.

optimum yield (maximum sustained yield, MSY) The theoretical point at which the size of a population is such as to produce a maximum rate of increase, equal to half the *carrying capacity. The concept is of practical use in farming and has been applied widely to commercial fisheries. It forms the basis for models that predict the stocking density required to maintain optimum fish production, and the harvesting methods and food supply needed to maintain production at that level.

oral Pertaining to the mouth.

orange-tip butterflies *See* PIERIDAE.

orang-utan (*Pongo pygmaeus***)** *See* PONGIDAE.

orbicular Disc-shaped, circular, or globular.

orbit The bony socket of the eye.

orb-web spiders *See* ARANEIDAE.

***Orcinus orca* (killer whale)** *See* DELPHINIDAE.

ordination method Method for arranging individuals (or sometimes attributes) in order along one or more lines. The method is used, with many techniques, in the biological and earth sciences, and especially in biology. There is an extensive ecological literature that discusses ordination methods, their applicability to different situations, and the relative merits of ordination as opposed to *classification schemes.

Ordovician The second of six periods that constitute the *Palaeozoic Era, named after an ancient Celtic tribe, the Ordovices. It lasted from about 488.3 to 443.7 Ma ago. The Ordovician follows the *Cambrian and precedes the *Silurian. It is noted for the presence of various, rapidly evolving, graptolite genera (*Graptolithina) and of the earliest jawless fish.

(⊕) SEE WEB LINKS
• Detailed description of life in the Ordovician.

Orectolobidae (carpet shark, nurse shark; subclass *Elasmobranchii, order *Lamniformes) A small family of slow-moving, bottom-living sharks in which the two *dorsal fins are located far back on the body. Typically, the nostrils are connected with the mouth by a pair of oro-nostral grooves. The lips and sides of the head usually bear fringes or fleshy *barbels. An *anal fin, always smaller than the dorsal fins, is present. Perhaps the most colourful of the carpet sharks is *Orectolobus ornatus* (gulf wobbegong), of southern Australia, which grows to 3 m. Like its relatives, it may be dangerous only when provoked. There are about 25 species, distributed world-wide.

***Orectolobus ornatus* (gulf wobbegong)** *See* ORECTOLOBIDAE.

oreodont A member of an extinct N. American group of artiodactyls, which may have been related to the entelodonts, pigs, and peccaries, but which had *selenodont molars, suggesting that in fact they were ruminating swine. *See also* AGRIOCHOERIDAE; MERYCOIDODONTIDAE.

Oreosomatidae (subclass *Actinopterygii, order *Zeiformes) A small family of marine fish that have certain features in common with the Zeidae (John Dories), e.g. the large, distensible mouth, a deep, compressed body, large eyes, and a small tail fin. *Neocyttus rhomboidalis* (spiky dory), which is typical of the family, lives at depths of more than 600 m near the coasts of S. Africa and southern Australia. There are about seven species, distributed world-wide.

organ A group of *tissues that performs a specific function or group of functions. Animal organs include the *eye, *brain, *heart, *integument (skin), *kidneys, *lungs, *liver, *spleen, and *stomach.

organelle Within a cell, a persistent structure that has a specialized function.

organizational effects of hormones Permanent changes to the structure and function of the body triggered by sex *hormones, usually during critical developmental periods in foetal growth and puberty. These changes affect the development of the nervous system and primary and secondary sexual features, thereby leading to typically male and female *phenotypes. *Compare* ACTIVATIONAL EFFECTS OF HORMONES.

organogenesis A stage of embryogenesis in vertebrate *embryos that follows the differentiation of the *mesoderm and in which the development of organs occurs. Organs are derived from five basic types of tissues: epithelilial, connective, blood, muscular, and nervous. *See also* BLASTULA; GASTRULA; PHARANGULA.

organotroph An organism that obtains energy from the metabolism of organic compounds. The term is sometimes inaccurately used as a synonym of *heterotroph.

Oriental faunal region The area that encompasses India and Asia south of the Himalayan–Tibetan mountain barrier, and the Australasian archipelago, excluding New Guinea and Sulawesi. There are marked similarities with the *Ethiopian region (e.g. both have elephants and rhinoceroses) but there are *endemic groups (e.g. pandas and gibbons).

orientating response A spontaneous reaction of an animal to a stimulus in which the head and/or body are moved so that the source of the stimulus may be examined more thoroughly.

orientation A change of position by an animal in response to an external stimulus.

orioles *See* ICTERIDAE; ORIOLIDAE.

Oriolidae (Old World orioles; class *Aves, order *Passeriformes) A family of mainly yellow and black, red and black, or green birds which have stout, red, blue, or black bills. Many *Oriolus* species have yellow and black males and green females (the golden oriole is *O. oriolus*); orioles of the genus *Sphecotheres* have bare, red skin around the eye. Orioles are arboreal, feed on insects and fruit, and nest in trees. There are two genera, with 28 species (of which 24 belong to the genus *Oriolus*), many of them migratory, found in Europe, Africa, Asia, and Australasia.

Oriolus (Old World orioles) *See* ORIOLIDAE.

Oriolus oriolus (golden oriole) *See* ORIOLIDAE.

ornithine A non-protein *amino acid which, in the *liver of vertebrates, contributes to the conversion of ammonia and carbon dioxide into *urea.

ornithine cycle *See* UREA CYCLE.

Ornithischia (class *Reptilia, subclass *Archosauria) One of two orders of *Mesozoic dinosaurs, which is distinguished primarily on the basis of its bird-like pelvis. Ornithischian dinosaurs were exclusively vegetarian and produced both bipedal forms (the *Ornithopoda) and four-footed forms.

ornithomimid One of the *Coelurosauria, which were theropod dinosaurs from the Late *Cretaceous. They are referred to as the 'ostrich' dinosaurs because of their general build, and their name means 'bird imitators'.

ornithophily Pollination by birds.

Ornithopoda (subclass *Archosauria, order *Ornithischia) A suborder of dinosaurs which had an essentially bipedal gait. The suborder includes several families (e.g. *Iguanodontidae, *Pachycephalosauridae, and *Hadrosauridae). The ornithopods are regarded as the most primitive of the ornithischian suborders.

Ornithorhynchidae (platypus, duck mole; subclass *Prototheria, order *Monotremata) A monospecific family (*Ornithorhynchus anatinus*) of nocturnal monotremes that are highly adapted for aquatic life. They are now known to be related more closely to the marsupials than to the echidnas. The digits are webbed, the webs of the fore limbs being folded when the animal is not swimming; those of the hind limbs bear strong claws. The snout is covered in thick skin and is shaped like the bill of a duck. A spur on the inside of the hind leg carries venom. The animal feeds on aquatic invertebrates. Breeding takes place in spring, with one to three eggs being laid and incubated in a nest built in a burrow beside water. The animals are distributed throughout southern and eastern Australia and

Tasmania. Fossil genera are *Obdurodon*, from the early *Miocene of Queensland, and *Monotrematum*, from the *Palaeocene of Patagonia.

ornithosis A disease of birds caused by the bacterium *Chlamydia psittaci*. Humans can also be infected. Sometimes the term 'ornithosis' is used synonymously with 'psittacosis', and sometimes it is reserved for the infection in birds only, 'psittacosis' being reserved for the disease in humans.

Ortalis (chachalacas) *See* CRACIDAE.

ortho- A prefix meaning 'straight', 'rectangular', or 'correct', derived from the Greek *orthos*, 'straight'.

orthoconic Applied to the shell of a cephalopod (*Cephalopoda) when it is a straight, tapering cone.

orthogenesis Evolutionary trends that remain fairly constant over long periods of time and so appear to lead directly from ancestor organisms to their descendants. This was once explained as the result of some internal directing force or 'need' within the organisms themselves. Such metaphysical interpretations have been displaced by the concepts of *orthoselection and species selection. *See also* ARISTOGENESIS; ENTELECHY; NOMOGENESIS.

ortho-kinesis The movement of an animal in response to a stimulus, such that the speed of movement is proportional to the strength of the stimulus.

Orthonectida A phylum of small, elongate mesozoans, whose life cycle is simpler than that of the *Dicyemida. They are endoparasitic within some marine invertebrates.

Orthoptera (crickets, grasshoppers, locusts; class *Insecta, subclass *Pterygota) Order of medium to large, terrestrial insects, whose major diagnostic features include hind legs that are usually modified for jumping (saltatorial), and a *pronotum with large, external flanges. Orthopterans are *exopterygote, and the external wing buds reverse their orientation in later *instars. The fore wings are modified as *tegmina. Wings are reduced or absent in some members of all the larger families. *Stridulation is commonly exhibited, with the mechanisms involved and the nature of the sounds produced varying widely. Many species live in association with plants, while others are found on the ground among debris, or under rocks. A number of groups burrow in the ground, and some are cave-dwelling. One family even lives in ant nests. Orthopterans may be nocturnal or diurnal, cryptically or brightly coloured, solitary or gregarious. Most species are herbivorous, but some are omnivorous, and a few are predacious. There are more than 20 000 species, some of economic significance.

orthoselection A primary selective pressure of a directional kind, which results in a self-perpetuating evolutionary trend. Species selection, via the *effect hypothesis, has been advanced as an alternative explanation for such trends. *See also* DOLLO'S LAW; SPECIESselection.

Orussidae (parasitic woodwasps; suborder *Symphyta, superfamily Siricoidea) Small and relatively rare family of symphytans which differ from other groups in having a long, slender, curved *ovipositor which is spirally twisted within the body of the female. The adults resemble horntails but are much smaller and behave like chalcid wasps. The antennae arise below the eyes and just above the mouth, and the eyes are hairy (a feature unique among the Symphyta). Little is known about the biology of the adults, but they may feed on *frass and fungi in the tunnels of wood-boring beetles. The *apodous larvae are ectoparasitic on the larvae of cerambycid and buprestid wood-boring beetles. The family is regarded as representing an intermediate stage between the Symphyta and the *Apocrita.

Orycteropodidae (aardvark; superorder *Protoungulata, order *Tubulidentata) A family of animals, known from the *Miocene in Africa and the *Oligocene in Europe, that are isolated from all other mammal groups and that may be similar to *Condylarthra. An orycteropodid is about the size of a small pig. The teeth are peg-like, consisting only of *premolars and *molars in the form of columns of *dentine with no *enamel, and grow continuously. The snout is long, the mouth round, the tongue long and worm-like. The animal is able to bury its snout in earth and locate and swallow prey while continuing to breathe. The limbs are specialized for digging, and digging is used as a means of escape from danger as well as to obtain food. The olfactory region of the brain is highly developed. The ears are long and the sense of hearing acute. There is only one extant species, *Orycteropus afer* (aardvark), distributed throughout Africa south of the Sahara.

Orycteropus afer (aardvark) *See* ORYCTEROPODIDAE.

Oryzias latipes (medaka) *See* ORYZIATIDAE.

Oryziatidae (medaka; subclass *Actinopterygii, order *Atheriniformes) A small family of egg-laying toothcarps that are small, mainly surface-living fish inhabiting freshwater floodplains and coastal, brackish-water marshes. They are closely related to the true toothcarps (*Cyprinodontidae). The slim-bodied *Oryzias latipes* (medaka), which grows to 4 cm, has a flattened head with a single *dorsal fin placed near the tail. There are about seven species, found in the river systems of eastern Asia.

os coxa *See* OS INNOMINATUM.

osculum In sponges (*Porifera), an excretory structure consisting of a large opening through which the sponge expels water carrying metabolic wastes. *See* PARAGASTER.

os innominatum (os coxa) In adult *Reptilia, *Aves, and *Mammalia, in which the *pubis, *ilium, and *ischium bones of the *pelvis are fused into a single bone, one of the two lateral halves of the resulting bone.

Osmeridae (smelt; superorder *Protacanthopterygii, order *Salmoniformes) A small family of fish, with both freshwater and marine representatives, that, like the salmon, have a single, soft-rayed, *dorsal fin as well as an *adipose fin on the back, a forked tail, and the *pelvic fins located in a mid-ventral position. Most species are fairly small and inhabit the cooler waters of the northern hemisphere. Many marine species are *anadromous. There are at least 10 species.

Osmia *See* MASON BEE.

osmoregulation The process whereby an organism maintains control over its internal *osmotic pressure irrespective of variations in the environment.

osmosis The movement of water or of another solvent from a region of low solute concentration to one of higher concentration through a *semi-permeable membrane. *See also* HYPEROSMOTIC; HYPOSMOTIC.

osmotic potential (solute potential) The part of the *water potential of a tissue that results from the presence of solute particles. It is equivalent to *osmotic pressure in concept but opposite in sign.

osmotic pressure The pressure needed to prevent the passage of water or other pure solvent through a *semi-permeable membrane separating the solvent from the solution. Osmotic pressure rises with an increase in concentration of the solution. Where two solutions of different substances or concentrations are separated by a semi-permeable membrane, the solvent will move to equalize osmotic pressure within the system.

osmotrophic Applied to an organism that absorbs nutrients from solution, as opposed to ingesting particulate matter.

osphradium A water-sampling organ, common to all *Neogastropoda and also found in many other gastropods, that can detect very small amounts of chemicals in solution. Scenting for prey is its primary use. It consists of a patch of sensory epithelium on the posterior margin of afferent gill membranes which acts as a chemoreceptor and determines the amount of sediment in the water being inhaled.

Osphronemidae (giant gourami; subclass *Actinopterygii, order *Perciformes) A monospecific family (*Osphronemus goramy*); a fish that inhabits swamps and rivers, and is cultured in ponds as a food fish in southern Asia. Somewhat oval in outline, the giant gourami will grow into a rather heavy, robust fish reaching a length of about 65 cm. It is easily recognizable by the very long pelvic *fin rays, and a rather large *anal fin. It is closely related to other gouramis and labyrinth fish of the family *Belontiidae.

osprey *See* FALCONIFORMES; PANDIONIDAE.

ossein *Collagen that is the main organic constituent of *bone.

osseous labyrinth *See* BONY LABYRINTH.

ossicles 1. In invertebrates (e.g. *Echinodermata), irregularly fenestrated calcareous plates, rods, and crosses arranged to form a lattice and bound together by connective tissue; together they comprise the skeleton. **2.** Small bones.

ossification The formation of *bone tissue, which is ultimately derived from *neural crest cells. The process is unique to vertebrates and differs from *dentine formation in that it allows the continued existence of the mineralizing cells (osteoblasts) which reside in the lacunae of bone connected to each other by canalicula. Ossification is carried out by osteoblasts, which become arranged concentric to blood vessels (Haversian canals) to form cylindrical Haversian systems. The resulting bone, together with its series of Haversian systems, is encased in outer lamellae of bone plates and is then known as lamellar bone. Bone tissue may originate from the dermis of the *integument giving rise to dermal bone, or occur by endochondrial ossification, which is the invasion and replacement of cartilaginous tissue with osteoblasts and subsequently bone.

Ostariophysi A long-established name for a group of fish possessing *Weberian ossicles and including the carps and catfish. The name has been reintroduced recently for a superorder of bony fish including the orders *Cypriniformes, *Gonorynchiformes, and *Siluriformes.

Osteichthyes The class that comprises the bony fish; with more than 25 000 species it is the largest class of vertebrate animals. Typically, bony fish have a bony skeleton, gill chambers covered by a *gill cover, and usually a *swim-bladder; there are usually many flat, bony scales embedded in the skin, with epidermis over them. The class comprises the *Actinopterygii (all living jawed fish except the *Dipneusti and the coelacanth). *See also* TELEOSTOMI.

osteoblast *See* OSSIFICATION.

osteoclast A large, multinucleate cell that breaks down *bone and is responsible for bone resorption. Osteoporosis is due to excessive osteoclast activity.

osteocyte The most abundant type of cell in *bone. It is approximately star-shaped and long-lived, sometimes lasting the entire lifetime of the animal.

osteoderm A bony plate embedded in the skin; osteoderms occur in many reptiles.

Osteoglossidae (bony tongue, spotted barramundi; superorder *Osteoglossomorpha, order *Osteoglossiformes) A small family of freshwater fish that have an elongate, robust body. The single *dorsal fin is located close to the small, rounded tail fin and the *ventral fins are placed in a mid-ventral (abdominal) position. It is considered a fairly primitive family. Representatives are found in S. America, W. Africa, northern Australia, and south-east Asia. Some species practise *buccal incubation. The S. American *Arapaima gigas* grows to a length of 4.5 m and is one of the world's largest freshwater fish. There are about six species.

Osteoglossiformes (subclass *Actinopterygii, superorder *Osteoglossomorpha) An order of freshwater fish that includes several families with fossil relatives. All have a single, soft-rayed, *dorsal fin, low-placed *pectoral fins, small, mid-ventral, *pelvic fins, and a long-based *anal fin. The order includes the *Notopteridae (featherback), *Hiodontidae (mooneye), *Osteoglossidae (bony tongue), and *Pantodontidae (butterfly fish).

Osteoglossomorpha (class *Osteichthyes, subclass *Actinopterygii) A superorder of bony fish including the orders *Osteoglossiformes and *Mormyriformes. The name is derived from the Greek *osteon*, 'bone', *glossa*, 'tongue', and *morphe*, 'form'.

Osteostraci An order of fossil, fish-like, jawless vertebrates that had a flattened head covered by a bony shield with a series of small gill openings on the side. Usually small (30 cm), they ranged from the *Silurian to the *Devonian Periods. The internal structures of the head resemble those of extant *Petromyzonidae (lampreys).

ostium An opening, derived from the Latin word for a door or opening. In *Porifera, one of the openings through which water is drawn into the body. In *Arthropoda, one of the openings in the wall of the heart into the pericardial (blood-filled) *sinus.

ostraciiform swimming A type of swimming practised by inflexible fish in which undulation is restricted to the *caudal fin. *Compare* ANGUILLIFORM SWIMMING; BALLISTIFORM SWIMMING; CARANGIFORM SWIMMING; LABRIFORM SWIMMING; RAJIFORM SWIMMING.

Ostracion lentiginosum (boxfish) *See* OSTRACIONTIDAE.

Ostraciontidae (boxfish, cowfish; subclass *Actinopterygii, order *Tetraodontiformes) A family of marine fish in which the body is enclosed in a hard, bony shell or carapace. Special openings are provided to accommodate the various fins. Rather slow-moving fish, they are well represented in

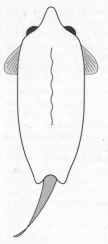

Ostraciiform swimming

tropical and subtropical waters, with many species entering estuaries. *Ostracion lentiginosum* (boxfish), which grows to 23 cm, and many other species are thought to discharge a toxic substance capable of killing other fish kept in the same aquarium. There are about 24 species, distributed world-wide.

Ostracoda (mussel shrimps, seed shrimps; phylum *Arthropoda, subphylum *Crustacea) Class of widely distributed marine and freshwater crustaceans in which the entire body is enclosed in a rounded or elliptical, bivalved *carapace, the outer layer of which is impregnated with calcium carbonate. A cluster of transverse *adductor muscle fibres closes the *valves, the muscles being inserted near the centre of each valve. The trunk is greatly reduced, so that the head accounts for much of the body. Trunk appendages are usually reduced to two pairs and there is no external segmentation of the trunk. By contrast, the head appendages, especially the antennules and antennae, are well developed and are the means of locomotion, even in the remarkable terrestrial species of *Mesocypris*, which plough through damp humus in the forests of southern Africa and New Zealand.

Ostracods show a wide range of feeding types, and the group includes filter feeders, carnivores, herbivores, and scavengers. The sexes are separate, and *parthenogenesis is common in freshwater forms. There are more than 2000 living species and the group is known from more than 10 000 fossil species, ranging from the early *Cambrian to the present.

Ostracodermi (ostracoderms) The name often used in older textbooks for the fossil, armoured, jawless, agnathan fish of the *Ordovician to early *Carboniferous Periods. Probably the name indicates only a grade of development; some ostracoderms are close to the line of ancestry of the living *Agnatha (lamprey and hagfish), others to the origin of the jawed vertebrates. The informal version of the name survives. *See also* CEPHALASPIDOMORPHI; PTERASPIDOMORPHI.

ostracoderms *See* OSTRACODERMI.

ostracum The calcified part of the shell of an invertebrate. In life it is covered by layers of protein forming a periostracum. This disappears after death.

Ostreina (oysters; class *Bivalvia, order *Pterioda) A suborder of *epifaunal, cemented or free-living bivalves in which the adults are non-*byssate, and the foot is absent. The shell is composed of foliated calcite with an aragonitic area below the adductor muscle. In most cases oysters live cemented to a hard substrate by the left valve, in a *pleurothetic attitude. They have a *monomyarian musculature; the ligament is *alivincular and divided into three parts, the central section being the resilium. The hinge margin is *edentulous. The gills are *eulamellibranchiate, and the *pallial line lacks a *sinus. Oysters first appeared in the Late *Triassic.

ostrich (*Struthio camelus*) *See* STRUTHIONIDAE.

Otariidae (sea lions; order *Carnivora, suborder *Pinnipedia (or *Caniformia) A family of marine mammals that have small external ears, a short tail, and hind limbs that can be turned forward to aid movement on land. They are less fully adapted to aquatic life and are considered to be evolutionarily more primitive than the *Phocidae, to which they may not be related closely, having, according to some authorities, arisen independently from canid stock. They may have first appeared on the shores of the Pacific during the lower *Miocene. Today they are distributed on Pacific and S. Atlantic coasts and on many islands in the southern hemisphere. There are six genera and about 14 species.

otic Pertaining to the ear.

otic capsule The part of the skull that encloses the inner ear.

Otididae (bustards; class *Aves, order *Gruiformes) A family of medium to large birds which have grey or brown upper-parts, buff or white under-parts, with many having black and white head and neck markings. They have broad wings, long necks and legs, and three-toed feet. They inhabit open plains, where they are mainly terrestrial, flying occasionally. They are omnivorous, and nest on the ground. There are 10 genera, with 25 species, many endangered, found in Europe, Africa, Asia, and Australia.

otolith *See* EAR-STONE.

otters *See* MUSTELIDAE.

otter shrew (*Potamogola*) *See* TENRECIDAE.

***Otus* (scops owls, screech owls)** *See* STRIGIDAE.

oubain A *cardiac glycoside that inhibits the transport of sodium and potassium ions across *cell membranes.

outbreeding The crossing of plants or animals that are not closely related genetically, in contrast to *inbreeding, in which the individuals are closely related. *See also* CROSS-BREEDING.

outgroup In *phylogenetics, a *species which is the least related to the species under analysis. The inclusion of a known outgroup allows the identification of *plesiomorphic and *apomorphic *character states, which might otherwise remain unclear, a situation that can give rise to topological errors (*see* TOPOLOGY). A tree that includes an outgroup is said to be rooted.

ova *See* OVUM.

ovary The female gonad that is responsible for the production of ova. The ovary is made from relatively undifferentiated tissue (stroma) is which *endocrine follicles are embedded. *See* also GRAAFIAN FOLLICLE; CORPUS LUTEUM; TESTIS.

ovenbirds *See* FURNARIIDAE.

overcrowding effect The regulation of population growth by behavioural abnormalities that are caused by malfunction of the *endocrine glands, which themselves occur in response to overcrowding.

overdominance (superdominance) In genetics, the phenomenon in which the character of the *heterozygotes is expressed more markedly in the *phenotype than in that of either *homozygote. Usually the heterozygote is fitter than the two homozygotes: this can give rise to monohybrid *heterosis, the hybrid vigour obtained by crossing parents differing in a single specified pair of allelic genes. *See also* BALANCED POLYMORPHISM.

overspecialization An old theory which held that straight-line evolution or orthogenetic trends (*see* ORTHOGENESIS) might proceed to the point at which the lineage was at an adaptive disadvantage. Overspecialization was therefore considered as one of the causes of extinction. There is no reason to believe, however, that *natural selection would permit evolution to proceed beyond maximum adaptation. More recently the term has been applied to highly specialized organisms which have proved incapable of responding to environmental change and so have become extinct.

Ovibos moschatus (musk-ox) *See* BOVIDAE.

ovicell *See* CYCLOSTOMATA.

oviduct A tube that carries eggs from the ovary to the *uterus; or from the *coelom, into which ova are shed, to the exterior.

oviger In *Pycnogonida, one of a pair of legs, located behind the *palps and in front of the walking legs, used for grooming and, in the male, for carrying eggs. In females of some species the ovigers are reduced or absent.

ovipary The method of reproduction in which eggs are laid and *embryos develop outside the mother's body, each egg eventually hatching into a young animal. Little or no development occurs within the mother's body. Most invertebrates and many vertebrates reproduce in this way. *Compare* OVOVIVIPARY; VIVIPARY.

oviposition The act of depositing eggs (i.e. ovipositing).

ovipositor Specialized egg-laying organ which in most insects is formed from outgrowths of the eighth and ninth abdominal segments. There are six outgrowths in all, known as valves, and these may be modified to form a variety of structures. In some insects the valves are short and in others they are drawn out to form a long, stylet-like tube. Extreme forms of ovipositor occur in the sawflies, ichneumons, and wood wasps, where it may be longer than the entire body, and equipped apically with saw teeth to aid penetration of hard plant tissues. Worker bees and sterile female wasps have the ovipositor modified to form a stinging organ and poison sac.

Ovis (sheep) *See* BOVIDAE.

ovovivipary The method of reproduction in which young develop from eggs retained within the mother's body but separated from it by the egg membranes. The eggs contain considerable *yolk which provides nourishment for the developing *embryo.

Many insect groups, fish, and reptiles reproduce in this way. *Compare* OVIPARY; VIVIPARY.

ovulation The release of a ripe egg on bursting from the ovarian follicle of a mammal, frequently stimulated by a pituitary hormone. The egg is discharged on to the surface of the ovary and then passes into the *oviduct. See also GRAAFIAN FOLLICLE. Compare EJACULATION.

ovum (*pl.* **ova**) An unfertilized egg cell. In animals it is often the product of the division of a secondary *oocyte. Fertilized by the sperm, an ovum develops as an *embryo.

owlet-nightjars *See* AEGOTHELIDAE; CAPRIMULGIFORMES.

owlets 1. *See* NOCTUIDAE. **2.** (*Glaucidium*) *See* STRIGIDAE.

owlflies *See* NEUROPTERA.

owls 1. *See* STRIGIDAE; STRIGIFORMES. **2.** (**barn owls, bay owls**) *See* TYTONIDAE.

ox *See* BOVIDAE.

Oxaea *See* OXAEIDAE.

Oxaeidae (**order *Hymenoptera, suborder *Apocrita**) Small, neotropical family of fast-flying, ground-nesting bees, a few of which are found also in the southern USA. Oxaeids were once considered a subfamily of the *Andrenidae, which they resemble in having subantennal plates and pollen *scopa that extend from the *coxa to the *basitarsus of the hind leg. They differ from andrenids in that the fore wing lacks a *stigma, and has an elongate marginal cell. The hind femora (*femur) of the females are enlarged and form a plate. The family includes the genera *Mesoxaea*, *Notoxaea*, *Oxaea*, and *Protoxaea*.

oxaloacetic acid A dicarboxylic acid; a key intermediate in the *citric-acid cycle, where it condenses with *acetyl coenzyme A to form citric acid and coenzyme A, thus initiating the cycle.

oxidase An *enzyme that catalyses reactions involving the oxidation of a *substrate using molecular oxygen as an electron acceptor.

oxidation A reaction in which atoms or molecules gain oxygen or lose hydrogen or electrons.

oxidative phosphorylation The action of a system, coupled to the electron-transport chain of *mitochondria, that is responsible for the production of *ATP from *ADP and inorganic phosphate.

oxidative potential The potential to cause a *redox reaction to occur through loss of electrons.

oxidoreductase An *enzyme that catalyses *oxidation–*reduction reactions. Included within this group are the *dehydrogenases and *oxidases.

Oxyaena (**order *Creodonta, family *Oxyaenidae**) A lower *Eocene, wolverine-like carnivore, which was one of the most powerful predators of its time and probably was the ancestor of *Patriofelis*, which was the size of a bear, and of *Sarkastodon*, which was much larger even than that. It is known from N. America and E. Africa.

Oxyaenidae (**superorder *Ferae, order *Creodonta**) An extinct family of carnivores that appeared during the late *Palaeocene and flourished during the *Eocene, but were extinct by its end. The large, deep jaw was mounted on a broad skull and in most forms *carnassials were developed from the first and second *molars. Most oxyaenids were rather like mustelids in shape, although *Paleonictis* resembled a cat.

oxygen debt The situation that arises in an aerobic animal when insufficient oxygen is available for necessary metabolic functions (e.g. during strenuous physical exertion). Energy is supplied to muscles by the anaerobic conversion of pyruvate (ionized *pyruvic acid) with the production of *lactic acid which is stored in the muscles (and whose accumulation leads to the sensation of fatigue). When the exertion ceases and an adequate amount of oxygen is once more available, the lactic acid is conveyed to the *liver and oxidized.

oxygen-dissociation curve A graphical representation of percentage of saturation of *haemoglobin with oxygen plotted against the oxygen tension. The resulting curve is S-shaped, indicating the efficiency with which haemoglobin takes up oxygen.

Oxyruncidae (sharpbill; class *Aves, order *Passeriformes) A monospecific family (*Oxyruncus cristatus*) which is a small, olive-green bird with black, spotted under-parts, orange-red crown stripe, and a short, sharply pointed bill. It inhabits forest in Central and S. America, and feeds on fruit.

***Oxyruncus cristatus* (sharpbill)** *See* OXYRUNCIDAE.

oxytocin A *peptide hormone, secreted by the mammalian *neurohypophysis, which causes the contraction of smooth muscle (particularly that of the uterus) and the secretion of milk.

***Oxyura* (stifftails)** *See* ANATIDAE.

oystercatchers *See* HAEMATOPODIDAE.

oysters *See* PTERIOMORPHIA.

P *See* PARENTAL GENERATION.

pacarana (*Dinomys*) *See* DINOMYIDAE.

pacas *See* DASYPROCTIDAE.

Pachycephala (whistlers) *See* MUSCICAPIDAE.

Pachycephalosauridae (order *Ornithischia, suborder *Ornithopoda) A family of 'bone-headed' dinosaurs of the Late *Cretaceous, which apparently may have lived in herds in upland regions.

pachydont Applied to a type of hinge dentition, found in certain cemented *Bivalvia, in which the teeth are heavy, very large, and generally few in number.

pachyostosis Condition (e.g. in *Sirenia) in which bones have a solid structure, with little or no marrow.

Pachyptila (prions) *See* PROCELLARIIDAE.

pachytene *See* MEIOSIS; PROPHASE.

Pacific sandfish (*Trichodon trichodon*) *See* TRICHODONTIDAE.

Pacific saury (*Cololabis saira*) *See* SCOMBERESOCIDAE.

Pacific softpout (*Melanostigma pammelas*) *See* ZOARCIDAE.

paddlefish *See* POLYODONTIDAE.

paedogenesis Reproduction by larval or other immature forms.

paedomorphosis Evolutionary change that results in the retention of juvenile characters into adult life. It may be the result of *neoteny, of *progenesis, or of *postdisplacement. It permits an 'escape' from specialization, and has been invoked to account for the origin of many taxa, from subspecies to phyla.

Paenungulata (infraclass *Eutheria, cohort *Ferungulata) A superorder that comprises herbivorous animals which diverged from the primitive *Palaeocene ungulate stock to become 'subungulates' or 'near ungulates'. The superorder includes the orders *Hyracoidea (hyraxes), *Proboscidea (elephants, their ancestors, and other elephant-like forms), *Pantodonta, *Dinocerata, *Pyrotheria, *Embrithopoda, and *Sirenia (sea cows or mermaids). It has been proposed recently, however, that the Hyracoidea belong to the *Mesaxonia; some consider the Paenungulata to be an entirely artificial group and combine only the Proboscidea and Sirenia into a superorder, Tethytheria. In all of them the *ulna and *fibula are complete, and the upper bones of the limbs are long. The digits bear nails but, in most cases, not hoofs. *Molars are specialized for grinding, and the *incisors and *canines are often reduced and modified to form tusks.

Paguridae (hermit crabs; class *Malacostraca, order *Decapoda) A family of crustaceans which use empty gastropod shells as cover for their soft abdomens. The abdomen spirals to the right, and the shell is held by tubercles on the *uropods, by abdominal contraction, and by the larger, left uropod which hooks round the *columella of the shell. As the crab grows, it selects new shells with great care, the colour contrast, weight, and fit being gauged before the crab takes up occupancy. In several species the shells become overgrown by sponges or bryozoan colonies. The *chelae are subequal, or sometimes with the right one larger and modified as an *operculum to close the shell. *Pleopods are reduced except on the left side, where they function for respiratory current production and for egg attachment. Hermit crabs are very active inhabitants of coastal and deep waters.

painted snipe *See* ROSTRATULIDAE.

pala The hair-fringed scoop found on the *tarsus of the fore legs of filter-feeding *Corixidae, used as a diagnostic characteristic.

Palaeanodonta (cohort *Unguiculata, order *Edentata) An extinct suborder that comprises three known genera of primitive edentates which lived in N. America from the late *Palaeocene to the *Oligocene, and which may resemble the animals from which later edentates evolved although they themselves occurred too late to be considered ancestral. The *incisors were absent, the cheek teeth were pegs with no *enamel, reduced in number and in some forms probably replaced functionally by pads of horny material, but the *canines were large. The vertebrae had not been modified in ways found in later edentates. Claws were well developed and there was a tendency for the wrist to be twisted. The most complete fossils are those of *Metacheiromys*, an animal about 45 cm long which probably fed on small invertebrates which it dug from the ground.

Palaearctic faunal region The region that includes Europe, Asia north of the Himalayan–Tibetan physical barrier, N. Africa, and much of Arabia. Despite its size and great range of habitats, the faunal variety in it is far less than that of the *Ethiopian and *Oriental faunal regions to the south of it (e.g. monkeys are nearly absent and reptiles few in number). The region is similar at the family level, and rather less so at the generic level, to the *Nearctic region: the Bering *land bridge connected the two for much of the *Cenozoic Era, and they are often combined into one region, *Holarctica.

Palaeocene (Paleocene) The lowest epoch of the *Palaeogene Period, about 65.5–55.8 Ma ago. The name is derived from the Greek *palaios* 'ancient', *eos* 'dawn', and *kainos* 'new', and means 'the old part of the *Eocene' (the subsequent epoch).

(⊕) SEE WEB LINKS
• Detailed description of life in the Palaeocene.

Palaeodonta (superorder *Paraxonia, order *Artiodactyla) An extinct suborder of small animals which lived during the *Eocene and early *Oligocene, mainly in N. America, and are classed as artiodactyls because of the modification of the *astragalus to typical artiodactyl form. Otherwise they had departed little from the original eutherian stock. Most were short-legged, had four digits on each limb, and were neither *ruminants nor entirely pig-like. The earlier forms may be the true ancestors of later artiodactyls, but the later forms left no descendants.

palaeoecology The application of ecological concepts to fossil and sedimentary evidence to study the interactions of Earth surface, atmosphere, and *biosphere in former times.

Palaeogene The first period of the *Cenozoic Era, 65.5–23.03 Ma ago, comprising the *Palaeocene, *Eocene, and *Oligocene epochs.

Palaeoheterodonta (phylum *Mollusca, class *Bivalvia) A subclass of equivalve bivalves which have prismatic-nacreous, aragonitic shells, and a small number of teeth that diverge from the *umbo. The external ligament is *opisthodetic or *amphidetic, and *parivincular. The hinge plate is present in some forms, and the musculature is variable. The subclass first appeared in the Middle *Cambrian, and contains one extinct and two extant orders.

Palaeomerycidae (infra-order *Pecora, superfamily *Cervoidea) A family of pecorans that appear to be ancestral to the deer and giraffes, and of which the musk-deer (*Moschus*) may be a surviving representative. The limbs and teeth resembled those of later groups. Most forms lacked horns, and some had enlarged upper *canines which formed tusks. Where horns existed they appear not to have been shed and probably they were covered with skin, as in modern giraffes or in the short pedicel at the base of deer antlers. The family had a wide *Holarctic distribution during the late *Oligocene and *Miocene, and some genera survived into the *Pliocene.

Palaeonemertini (phylum *Nemertea, class *Anopla) An order of elongate proboscis worms in which the muscles of the body are layered. The epidermis is ciliated, and the dermal layer is not well developed.

palaeontology The study of *fossil flora and fauna. Information thus gained may be used to establish ancient *environments and evolutionary lineages.

Palaeoptera (class *Insecta, subclass *Pterygota) One of the two infraclasses (*compare* NEOPTERA) into which insects are placed according to the position in which they hold their wings when at rest. Members of the Palaeoptera are unable to fold their wings: when the insect rests, its wings are held out as they are when it is flying; this is the more primitive condition. The infraclass includes the mayflies (*Ephemeroptera) and dragonflies (*Odonata). The distinction between the two infraclasses emerged in the *Carboniferous, very early in insect evolution, and is known from fossils of that period.

Palaeopterygii An older name for a supposed subclass including some of the more primitive fish. They were bony fish with many fossil relatives, including e.g. the families *Acipenseridae, *Polyodontidae, and *Polypteridae.

palaeospecies (chronospecies) A group of biological organisms, known only from *fossils, which differs in some respect from all other groups.

Palaeotaxodonta (phylum *Mollusca, class *Bivalvia) A subclass of bivalves which contains only one order, *Nuculoida.

Palaeotheriidae (order *Perissodactyla, suborder *Hippomorpha) An extinct family of horse-like animals which lived in Europe during the *Eocene and *Oligocene. They appear to have diverged from the main line of equid evolution and in some respects anticipated developments that were to occur later among the true horses. Several grew to large size, one (*Palaeotherium magnum*) to about the size of a rhinoceros but they had only three toes on each foot, and had molarized *premolars. One form had slim lateral metapodials and *cement on the crowns of its molars. The family had disappeared by the end of the Oligocene.

Palaeozoic (Paleozoic) The first of the three eras of the *Phanerozoic, about 542–251 Ma ago. The *Cambrian, *Ordovician, and *Silurian periods together form the Lower Palaeozoic Sub-Era; the *Devonian, *Carboniferous, and *Permian the Upper Palaeozoic Sub-Era. The faunas of the Palaeozoic are noted for the presence of many invertebrate organisms including *Trilobitomorpha, *Graptolithina, *Brachiopoda, *Cephalopoda, and corals. By the end of the era, amphibians and reptiles were major components of various communities and giant tree-ferns, horse-tails, and cycads gave rise to extensive forests.

palatal teeth Teeth that are borne on one or more of the bones of the *palate.

palate The roof of the mouth; it separates the nasal and mouth cavities. In *Mammalia and *Crocodilia bones project inward from the jaws to form a false palate below the original palate, so extending the nasal cavity and moving its opening to a position in the throat. In mammals the bony part of the (false) hard palate is extended at the rear of the mouth by membrane and *connective tissue which form the soft palate.

Paleocene *See* PALAEOCENE.

Paleozoic *See* PALAEOZOIC.

palichnology *See* ICHNOLOGY.

Palinuridae (spiny lobsters; class *Malacostraca, order *Decapoda) Family of large, edible decapods which have a more or less cylindrical *cephalothorax and a well-developed, dorsoventrally flattened abdomen. The head bears large, spiny antennae, and in most species the first pair of legs is not enlarged. The family occurs in tropical waters.

Pallas's sandgrouse (*Syrrhaptes paradoxus*) See PTEROCLIDAE.

pallette In some insects, a fleshy lobe in front of the mouth.

pallial line In most *Bivalvia, a linear feature which runs between the two adductor muscles on the inner surface of the shell, and close to the margin. It delimits the extent of the inner calcareous shell layer, and marks the line of marginal *mantle muscles.

pallium 1. The part of the *cerebrum that is concerned with olfaction. **2.** See MANTLE.

palmate newt (*Triturus helveticus*) See SALAMANDRIDAE.

palmchat See DULIDAE.

palm-nut vulture (*Gypohierax angolensis*) See ACCIPITRIDAE.

palp A sensory appendage situated near the mouth of many invertebrates. In *Arthropoda, it is a jointed, sensory structure that articulates with the *labium or *maxillae. *See also* LABIAL PALP; MAXILLARY PALP.

Palpigradi (microwhipscorpions; subphylum *Chelicerata, class *Arachnida) Order of minute colourless, thin-skinned arachnids that are probably closely related to the uropygids and amblypygids. The abdomen is joined by a stalk, and has a large *mesosoma and a short *metasoma, with a long, multisegmented flagellum. The *chelicerae are thin, three-segmented, and *chelate, with a movable lateral finger. The 50 or so species favour habitats of high humidity (under stones and soil), and although they

have a world-wide distribution most occur in southern Europe, southern USA, and S. America.

Paludicellea (class *Gymnolaemata, order *Ctenostomata) A suborder of bryozoans whose individuals possess elongated, thin, proximal ends. They are known only from the *Holocene.

panbiogeography The name given to a synthesis of the sciences of plant and animal distribution. The main features are that consistencies recur in distribution patterns and that analysis of these produces 'tracks' (joining areas of common floras and faunas) and 'nodes' (where different tracks meet). The name, and the associated method of analysis, are due to the Venezuelan botanist Leon Croizat (1894–1982). These ideas form the basis of the new school of *vicariance biogeography.

((⊕)) SEE WEB LINKS
• 'Never a Serious Scientist: The Life of Leon Croizat'

pancake tortoise (*Malacochersus tornieri*) See TESTUDINIDAE.

pancreas In vertebrates, a gland lying between the *duodenum and *spleen, that secretes *trypsin and other digestive *enzymes, and that contains the *islets of Langerhans.

panda 1. (giant panda, *Ailuropoda melanoleuca*) See AILUROPODA; URSIDAE. **2.** (red panda, *Ailurus fulgens*) See AILURIDAE; PROCYONIDAE.

Pandinus imperator See SCORPIONES.

Pandion haliaetus (osprey) See PANDIONIDAE.

Pandionidae (osprey; class *Aves, order *Falconiformes) A monospecific family (*Pandion haliaetus*) which is a large, brownish-black and white bird of prey. Its bill is short and strongly hooked, its wings are long and pointed, and its feet have long toes and claws, the outer toe being reversible, and the soles having sharp scales for gripping prey. It inhabits coastal areas and inland fresh water, and feeds on fish caught with its feet. It nests in a tree or on the ground. It is found world-wide.

panendemic distribution See COSMOPOLITAN DISTRIBUTION.

Pangaea A single supercontinent which came into being in late *Permian times and persisted for about 40 Ma before it began to break up at the end of the *Triassic Period. It was surrounded by the universal ocean of Panthalassa.

Pangasiidae (subclass *Actinopterygii, order *Siluriformes) A family of freshwater catfish that have a high *dorsal fin, a small *adipose fin, one pair of chin *barbels, a long *anal fin, and a forked tail fin. Some species, e.g. *Pangasianodon gigas*, can exceed a length of 2 m in their native south-east Asian waters. There are about 25 species. The members of this family are sometimes included in the family *Schilbeidae.

pangolin See MANIDAE.

panmictic unit A local population in which mating is completely random.

panmixia The process of *panmixis.

panmixis The random mating of individuals. *Compare* ASSORTATIVE MATING.

panniculus adiposus (subcutaneous fat) The layer of fat that lies immediately below the *dermis.

Pannonian The land-mammal stage of the late *Miocene in Europe and western Asia, beginning 12 Ma ago.

Panorpidae See MECOPTERA.

Pan paniscus See HOMINIDAE; PONGIDAE.

Panthalassa See PANGAEA.

Pantodon buchholzi **(freshwater butterfly fish)** See PANTODONTIDAE.

Pantodonta (superorder *Paenungulata, order *Amblypoda) An extinct suborder of ungulates which lived during the *Palaeocene and *Eocene, mainly in N. America, but survived into the *Oligocene in Asia. Early forms were small (e.g. *Pantolambda* was the size of a large dog), had short limbs, and large, broad feet. Later forms grew much larger (e.g. *Barylambda*, which was about 2.4 m long, and the hippopotamus-like *Corphodon*). Some developed claws rather than hoofs and in some the *canines became enlarged, probably enabling them to feed on roots. The name Pantodonta is sometimes used synonymously with Amblypoda (in which case it is regarded as an order).

Pantodontidae (freshwater butterfly fish; superorder *Osteoglossomorpha, order *Osteoglossiformes) A monospecific family (*Pantodon buchholzi*) of small, freshwater fish that have a straight dorsal outline, keel-shaped belly, large scales, and a large, obliquely orientated mouth. The species is of interest because of its large, wing-like, *pectoral fins, supported by special muscles capable of moving the fins up and down. Kept in aquaria, these fish have been known to jump out of the tank if no cover was provided. In their native W. African rivers they can jump and glide above the water, presumably in pursuit of fast-moving prey.

Pantopoda See PYCNOGONIDA.

pantothenic acid Vitamin B_5, a water-soluble *vitamin synthesized by green plants and micro-organisms, but not by animals, for which it is an essential dietary requirement. It forms part of the structure of the key metabolic compound *coenzyme A.

Pantotheria (class *Mammalia, subclass *Theria) An extinct infraclass of primitive Middle and Late *Jurassic mammals, known from N. America, Europe, and E. Africa. They were egg-laying insectivores of shrew-like appearance, and are believed, on the basis of the structure of the jaw and teeth, to include the ancestors of all later placental and marsupial mammalian groups. They were distributed widely by the later Jurassic, were numerous, and were varied in form.

Pan troglodytes (chimpanzee) *See* HOMINIDAE; PONGIDAE.

Panura biarmicus (bearded reedling) *See* PARADOXORNITHIDAE.

paper wasps (superfamily Vespoidea, family *Vespidae) Social wasps which build their nests from chewed wood pulp. Most belong to the subfamilies Vespinae, Polistinae, and Polybiinae, and the name is sometimes restricted to the Polistinae. The nest structure is variable, with or without outer coverings, but is usually spherical and contains a large number of cells. Nest construction is started by the queen in spring and the nest is subsequently enlarged by many generations of workers. Nests sometimes attain 35 cm or more in diameter. In *Polistes dominula* (European paper wasp), the brighter the yellow and black abdominal stripes, the larger the insect's poison gland, adding an additional advertisement to the *aposematic coloration as a warning to potential predators.

Papilionidae (Apollo butterflies, birdwing butterflies, swallowtails, swordtails; subclass *Pterygota, order *Lepidoptera) Family of spectacular butterflies, medium to large in size, and frequently brightly coloured. In most species the hind wings are extended into 'tails'. All the legs are functional. Adults and larvae are usually distasteful. Protrusible, odiferous, forked sacs are present on the thorax of the larva. There are about 530 species. The family is widely distributed but mostly tropical.

papilla A nipple or nipple-like structure.

Papio (baboons) *See* CERCOPITHECIDAE.

Papuan mountain drongo (*Chaetorhynchus papuensis*) *See* DICRURIDAE.

papulae *See* ASTEROIDEA.

parabasal body In certain flagellate *Protozoa, a small, granular mass located

close to the *blepharoplast. It was known formerly as the kinetonucleus, but apparently serves no function associated with movement (kinesis) or with the nucleus.

Paracanthopterygii (class *Osteichthyes, subclass *Actinopterygii) A recently recognized superorder of bony fish, that includes the orders *Batrachoidiformes, *Gadiformes, *Lophiiformes, *Percopsiformes, and *Polymixiiformes.

paraclade A group of *evolutionary lineages; paraclades may be *paraphyletic or *monophyletic, but not *polyphyletic.

paracone In primitive, *triconodont, upper, *molar teeth, either the anterior *cusp (where the three cusps form a straight line) or the outside anterior cusp (where they are not in line).

Paradisaeidae (birds of paradise; class *Aves, order *Passeriformes) A family of small to medium-sized birds which have brightly coloured, iridescent plumage and elaborate plumes on the head, back, wings, and tail; some also have *wattles. Females tend to be dull-coloured. The bill is short to long, straight and slender, and highly *decurved or hooked. The wings are short and rounded, the tail short and square to long and graduated. The birds are arboreal, feed on insects, frogs, lizards, fruit, and seeds, and nest in a cavity or tree fork. Their plumes are used in highly elaborate displays which also involve unusual vocal sounds. There are 20 genera, with 43 species, found mainly in New Guinea, but also in Australia and the Moluccas.

Paradoxornis (parrotbills) *See* PARADOXORNITHIDAE.

Paradoxornithidae (bearded tits, parrotbills; class *Aves, order *Passeriformes) A family of small to medium-sized, long-tailed birds which have short, deep, laterally compressed, parrot-like bills (yellow or orange in *Paradoxornis* species). Their plumage is soft and mainly reddish-brown. They inhabit forest, scrub, and reedbeds, feed on insects and seeds, and nest in trees, bushes, and reeds. The largest genus, *Paradoxornis*, with 18 species found only in Asia, is typical. *Panurus biarmicus* (bearded reedling) is the only parrotbill in Europe. There are three genera, with about 20 species, found in Europe and Asia.

Parafidelia *See* FIDELIIDAE.

paragaster The central cavity of a sponge (*Porifera), which opens to the surface through the *osculum.

paraglossa In insects, one of a pair of small appendages attached to either side of the *lingua or *labium.

Paralepididae (barracudina; subclass *Actinopterygii, order *Myctophiformes) A fairly large family of marine, *mid-water fish that have slender, streamlined bodies, pointed heads, and a single *dorsal fin located half-way along the back. They are fairly common deep-sea fish, with few species exceeding 32 cm in length. There are about 50 species, distributed world-wide.

Paralichthys californicus **(California halibut)** *See* BOTHIDAE.

parallel evolution Similar evolutionary development that occurs in lineages of common ancestry. Thus the descendants are as alike as were their ancestors. The nature of the ancestry imposes or directly influences the development of the parallelism.

parametric **test** *See* STATISTICAL METHOD.

paramylum In some *phytoflagellate *Protozoa, a granular carbohydrate food reserve.

paramyosin A form of *tropomyosin that is found in the *adductor and retractor muscles of *Bivalvia. It is a water-insoluble protein and forms filaments 100 nm thick.

Paranthropus A genus recognized by some authorities to describe the robust *australopithecines (often classified as *Australopithecus robustus* and *A. boisei*). They were early protohominids, the sister group of *Homo*. Members of the genus, often called 'nutcracker man', were characterized by enormous *molars and *premolars, and reduced front teeth. They were common at such sites as Olduvai and Koobi Fora, in the late *Pliocene and early *Pleistocene.

parapatric Applied to *species whose *habitats are separate but adjoining. *Compare* ALLOPATRIC; SYMPATRIC.

parapatric **speciation** Speciation that occurs regardless of minor *gene flow between *demes. In many species selective pressures are sufficiently strong on the whole to prevent homogenization of the immigrant genes by interbreeding.

Parapercidi Alternative name for the fish family *Mugiloididae.

Parapercis *colias* **(blue cod)** *See* MUGILOIDIDAE.

paraphyletic Of a *taxon, including some but not all descendants of the common ancestor (i.e. not *holophyletic).

Parapithecidae (order *Primates, suborder *Simiiformes) An extinct family of primates which lived during the *Eocene and *Oligocene in Egypt; Eocene fossils from Burma are sometimes included in the family in addition. They showed certain similarities in dentition to *Condylarthra, but had short faces and jaws shaped like those of tarsiers. They are usually placed in a superfamily, Parapithecoidea, equally related to *Ceboidea and *Cercopithecoidea plus *Hominoidea.

parapodia 1. In *Polychaeta, usually *biramous, muscular, lateral projections of the body, paired and extending from the body segments, more or less compressed laterally, and bearing *setae; in free-swimming forms they are used for locomotion, and in sessile forms to fan water. 2. In *opisthobranch *Mollusca, lateral projections from the side of the foot which form fins used in swimming.

paraproct In some insects, one of a pair of lobes or *sclerites near the *anus.

parapsid A variation of the euryapsid (*see* EURYAPSIDA) skull, which has a single opening in the upper temple region. The parapsid skull, found in *Ichthyosauria, has a single temporal opening bordered by the *parietal, postfrontal, and supratemporal bones.

parasite *See* PARASITISM.

Parasitica One of the two customary divisions of the suborder *Apocrita (order *Hymenoptera). The *Aculeata are the stinging forms (ants, bees, and wasps) and the Parasitica are those whose larvae parasitize other insects. There are four principal superfamilies: Ichneumonoidea, Chalcidoidea, Cynipoidea, and Proctotrupoidea.

parasitic barnacles *See* RHIZOCEPHALA.

parasitic catfish *See* TRICHOMYCTERIDAE.

parasitic flies *See* TACHINIDAE.

parasitic woodwasps *See* ORUSSIDAE.

parasitism An interaction of *species populations in which one (typically small) organism (the parasite) lives in or on another (the host), from which it obtains food (when the parasite may be called a biotroph), shelter, or other requirements. Whereas a predator kills its host (i.e. lives on the capital of its food resource) a parasite does not (i.e. lives on the income). Parasitism usually implies that some harm is done to the host, but this interpretation must be qualified. Effects on the host range from almost none to severe illness and eventual death, but even where such obvious immediate harm accrues to the individual host it does not follow that the relationship is harmful to the host species in the long term or in an evolutionary context (e.g. it might favour beneficial adaptation in the host species population). Obligate parasites can live only parasitically. Facultative parasites may live as parasites or as independent *saprotrophs. Partial parasites are facultative parasites that live more successfully as parasites

than they do independently. Ectoparasites live externally on the host. Endoparasites live inside the body of the host. *Compare* COMMENSALISM; MUTUALISM; NEUTRALISM. *See also* HYPERPARASITE; NECROTROPHIC.

parasitoid An organism that spends part of its life as a parasite and part as a predator (e.g. many wasps that are parasites during their larval stages and predators when mature).

parasitology The study of small organisms (parasites) living on or in other organisms (hosts), regardless of whether the effect on the hosts is beneficial, neutral, or harmful. The study uses the term 'parasite' in a wider sense than is usually associated with *parasitism.

parasol ants *See* ATTINI; MYRMICINAE.

parasympathetic nervous system That part of the *autonomic nervous system which generally causes bodily functions to become appropriate for calm conditions, for instance by slowing down the heart rate and promoting digestion. Nerves of the parasympathetic nervous system tend to form ganglia (*ganglion) at the effector organ such that preganglionic fibres from the spinal cord are long, post-ganglionic fibres are short. The nerve endings of the parasympathetic nerve system are *cholinergic.

parataxon In taxonomy, an artificial classification that is suggested for certain common organisms of doubtful affinities, or as yet unknown origins (e.g. fossil spores, dinosaur footprints).

parathyroid hormone *See* PARATHYROIDS.

parathyroids In some vertebrates, two pairs of *endocrine glands, embedded within the *thyroid, that secrete parathyroid hormone, which regulates the concentration of calcium in the blood.

paratype In taxonomy, a specimen, other than a *type specimen, that is used by an author at the time of the original description, and designated as such by the author.

Paraxonia (infraclass *Eutheria, cohort *Ferungulata) A superorder that comprises the extinct and extant members of the order *Artiodactyla.

paraxonic Applied to the condition in which the axis of the weight of an animal passes between the third and fourth digits.

Parazoa (kingdom *Animalia) A subkingdom of aquatic animals comprising the phyla *Placozoa and *Porifera. Parazoans have differentiated cells, but no tissues and they are asymmetrical. There are about 5000 species, of which about 150 inhabit fresh water.

pardalotes *See* DICAEIDAE.

Pareiasauridae (subclass *Anapsida, order *Cotylosauria) A family that comprises the largest (up to 3 m in length) of the stem reptiles (cotylosaurs), found in Middle and Late *Permian rocks in Europe and Africa. They were herbivorous, and perhaps lived semi-aquatically in swamps. They were distinctive among cotylosaurs in that the limbs were rotated in towards the body, thus supporting the body in a more upright fashion.

parenchyma In *Platyhelminthes, the tissue, composed of cells and intercellular spaces, that fills the interior of the body. In other animals, the essential cells of an organ, as opposed to blood cells, cells of connective tissue, etc.

parenchymula A sponge larva in which the area of non-flagellate cells is very small.

parental care All activities that are directed by an animal towards the protection and maintenance of its own offspring or those of a near relative.

parental generation (P) The generation comprising the immediate parents of the *F1* generation. The symbols P_2 and P_3 may be used to designate grandparental and great-grandparental generations respectively.

Paridae (tits, chickadees; class *Aves, order *Passeriformes) A family of small, active birds which are blue, olive, brown, or grey above, and white, yellow, or buff below. They have small, stout bills, and many have crests. Their wings are short and rounded, their tails short and square-ended. They are arboreal, inhabiting woodland and gardens, feeding on insects, nuts, and seeds, and nesting in holes in trees and banks, and in nest-boxes. The largest genus is *Parus*, with 47 species. There are three genera in the family, with 49 species, found in Europe, Africa, Asia, and N. and Central America.

parietal 1. One of a pair of bones that form much of the posterior part of the roof of the brain case. **2.** Applied to the lining on the inside of the body wall.

parietal pleura *See* PLEURA.

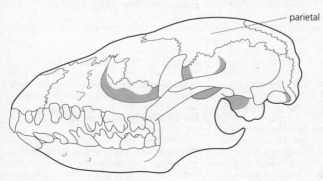

parietal

Parietal

parivincular In *Bivalvia, applied to a type of *ligament that is elongate and cylindrical in shape. It occurs in a position posterior to the *shell beaks.

paroccipital process At the rear of the base of the skull, a downward-projecting spur to which is attached the muscle used in opening the lower jaw.

Parodontidae (superorder *Ostariophysi, order *Cypriniformes) A small family of freshwater fish that have a single, soft-rayed, mid-dorsal fin, small *adipose fin, and a ventrally positioned mouth. They are found in S. American rivers.

Paromomyidae (order *Primates, suborder *Plesiadapiformes) An extinct family that comprises the earliest known primates, which may have originated during the late *Cretaceous (when they were contemporaries of dinosaurs). They flourished in the *Palaeocene, and died out in the middle *Eocene. Most are known only from their teeth, but increasing numbers of skulls and limb bones are now known. Probably they were the size of modern shrews, herbivorous, and lived on or close to the ground.

parotoid gland One of a pair of glands, resembling warts, that occur behind the eyes or on the neck or shoulders of some toads. In some species the parotoid glands secrete toxins.

parr The juvenile, freshwater stage of salmon. Parr bearing dark spots or bars tend to stay in the rivers for periods of at least two years before turning into *smolt and migrating to the sea.

parrotbills See PARADOXORNITHIDAE.

parrotfish See SCARIDAE.

parrots See PSITTACIDAE; PSITTACIFORMES.

parsimony In *cladistic analysis, the convention whereby the simplest explanation is preferred, because it requires the fewest conjectures, although the most parsimonious explanation is not always the correct one. See OCCAM'S RAZOR.

pars intercerebralis In insects, a cleft along the anterior midline of the brain, rich in neurosecretory cells.

parthenogenesis The development of an individual from an egg without that egg undergoing fertilization. It occurs in some groups of animals (e.g. flatworms, leeches, aphids, rotifers), in which males may be absent. The eggs (ova) that develop in this way are usually *diploid, so all offspring are genetically identical with the parent. Commonly, parthenogenesis with only females in the population alternates with ordinary sexual reproduction, which allows the recombination of genetic material and presents a need for males. This alternation is called heterogamy.

partial dominance (incomplete dominance, semidominance) In genetics, the production of an intermediate *phenotype in individuals that are heterozygous for the *gene concerned (i.e. the *heterozygote shares a phenotype that is quantitatively intermediate between those of the corresponding *homozygotes). The impact of a gene at the level of the phenotype depends on its dominance relations, but also on the conditions of the rest of the *genome and on the conditions of the environment. See also PENETRANCE.

partial parasite See PARASITISM.

partial random sample See STRATIFIED RANDOM SAMPLE.

partial refuge A feeding patch in which the density of prey is low and, therefore, prey individuals have a relatively low risk of being attacked. See also AGGREGATIVE RESPONSE.

partridges See PHASIANIDAE.

Parulidae (New World warblers; class *Aves, order *Passeriformes) A family of small birds that are mainly olive and grey, but which are boldly patterned with other colours (e.g. the 11 species of *Vermivora*, which are green or brown with areas of yellow, orange, red, black, and blue on their wings, head, or throat). They have slender, pointed bills, medium-sized, pointed wings with nine *primaries, and medium-length tails. They are mainly arboreal, some climbing trees; others are terrestrial. Many are migratory. They inhabit forests and brush, feed on insects, nectar, and fruit, and nest in bushes, trees, and on the ground. The 21 species of *Basileuterus* (wood-warblers) are insectivores that build domed nests with a side entrance on or near the ground. *Dendroica petrechia* (yellow warbler) exists in a number of distinct racial forms and throughout the Americas, while *D. kirtlandii* (Kirtland's warbler) has a breeding range confined to a small area of jack pine in Michigan. Some of the 14 species of *Geothlypis* (yellowthroats and warblers) are placed in a separate genus, *Oporornis*. There are about 25 genera in the family, with 126 species, found in N., Central, and S. America, and the W. Indies.

Parus (tits, chickadees) *See* PARIDAE.

pasang (*Capra aegagrus*) *See* BOVIDAE.

Passalidae (subclass *Pterygota, order *Coleoptera) Family of large shiny, flattened beetles, 20–60 mm long, which have an elongate, parallel-sided body, which is usually black or brown. The antennae are curved and comb-like. They are found under bark in tropical forests. They regurgitate chewed wood to feed larvae. It is claimed that adults of both sexes tend larvae to maturity. Larvae are curved and fleshy, with much reduced back legs. Adults and larvae *stridulate. There are 490 species, none occurring in Europe.

passenger pigeon (*Ectopistes migratorius*) *See* COLUMBIDAE.

Passer (sparrows) *See* PLOCEIDAE.

Passer domesticus (house sparrow) *See* PLOCEIDAE.

Passeriformes (perching birds, passerines; class *Aves) An order of small to medium-sized birds which have widely varied plumage and shape. The bill is modified for a range of feeding habits. The feet have three toes pointing forward and one backward which assists perching. Nests can be simple or elaborate. Young are usually naked at hatching. Passerines can be terrestrial, aerial, or arboreal. The order is divided into 82 families, and contains more than 5000 species, 60% of the class. They are found world-wide.

Passerina (buntings) *See* EMBERIZIDAE.

passerine Any member of the order *Passeriformes; all other avian orders can be called non-passerines.

passive dispersal *See* DISPERSAL.

passive immunity Immunity against a given disease that is acquired by the injection into the host of *serum containing *antibodies that have been formed by a donor organism itself possessing active immunity to the disease.

passive transport The movement of atoms or molecules by *diffusion, *facilitated diffusion, filtration, or across a *semipermeable *membrane by *osmosis.

Pataecidae (prowfish; subclass *Actinopterygii, order *Scorpaeniformes) A small family of marine fish that have a scaleless, smooth body, and a very long *dorsal fin extending from the head to the tip of the tail. The body is deep in the head region, but tapers down sharply toward the tail. Occurring in temperate Australian waters, several species are caught occasionally by anglers or in trawl-nets. There are at least four species.

patagium In *Mammalia, a fold of skin between the fore and hind limbs; it is used in gliding or flying.

Patagonian hare (*Dolichotis*) *See* CAVIIDAE.

patch dynamics The study, and mathematical modelling, of feeding areas in which patches are disturbed by competitive interactions and by such factors as burrowing, predation, severe weather, and extreme wave action.

(((●))) SEE WEB LINKS
• Example of the application of patch dynamics.

patch residence time The length of time an animal spends *foraging in a particular area (i.e. a patch) of its *habitat before diminishing returns compel it to move elsewhere.

patella The knee-cap. In most *Mammalia, and in some *Aves and *Reptilia, a bone located in front of the knee joint in the hind limb and set in a tendon of the muscles which straighten the limb.

Patella vulgata (limpet) See ARCHAEOGASTROPODA.

pathogen Any micro-organism that causes disease. Pathogens may be ecologically important in controlling the distribution of *species and interspecific and intraspecific *competition.

pathogenesis The process of disease development.

pathognomonic Applied to symptoms that are characteristic of a particular disease and that may therefore be useful in diagnosis.

Patterus Former name for *Eulemur* (brown lemur). See LEMURIDAE.

Paucituberculata The name given to the *Caenolestoidea in classifications that regard them as a separate order.

Pauropoda (phylum *Arthropoda, subphylum *Atelocerata) A class of small (0.5–2.0 mm long), soft-bodied animals resembling millipedes. There are usually 11 segments of which nine bear legs, the first and last segments and *telson being legless. Some of the dorsal plates are large and overlap adjacent segments. The head bears two-branched antennae and a disc-like sensory organ on either side. Pauropods occur widely in temperate and tropical regions, often in forest litter, and about 380 species are known.

Pavlov, Ivan Petrovich (1849–1936) A Russian physiologist who is renowned for his research into blood circulation, digestion (for which he was awarded the 1904 Nobel Prize), and conditioned behaviour in animals. From 1928 until his death he concentrated on the application of the principles of conditioning to psychiatric therapy. His approach to his behavioural work was based on the application of objective experiment to mental processes and the rejection of the mind–body dualism characteristic of earlier psychological studies.

(((●))) SEE WEB LINKS
• Biography of Ivan Pavlov.

paxillae In starfish, small, peg-like plates.

Paxillosida (starfish; class *Stelleroidea, subclass *Asteroidea) An order of starfish in which the *mouth angle plates are large and derived from fused ambulacral *ossicles. Marginal ossicles are not invariably developed, but commonly bear dorsoventral, intermarginal *fascioles. The marginals are separated from the mouth frame by small, oral, intermediate ossicles. The order is known from the *Ordovician to the present.

payoff In *game theory, the points won or lost by each contestant.

PCR See POLYMERASE CHAIN REACTION.

peafowl See PHASIANIDAE.

peanut worms See SIPUNCULIDA.

pearleye See SCOPELARCHIDAE.

pearlfish See CARAPIDAE.

pea weevil (Sitona lineatus) See CURCULIONIDAE; BRUCHIDAE.

pebrine Disease of the silkworm (*Bombyx mori*), caused by the protozoon *Nosema bombycis*. The disease is usually fatal.

peccary (*Tayassu*) *See* TAYASSUIDAE.

pecking order The name given to the hierarchical social organization found in some insect species (e.g. ants and bees) and in many vertebrate species. It is so called because the phenomenon was first described for chickens. Based on *dominance, it allows each member of the group to threaten (or actually peck) the individual immediately subordinate to it and so gain prior access to food or other resources.

Pecora (cattle; order *Artiodactyla, suborder *Ruminantia) An infra-order that comprises the three superfamilies *Traguloidea (chevrotains), *Cervoidea (deer and giraffes), and *Bovoidea (antelopes, cattle, goats, sheep, etc.). These form the most recent ungulate group, known since the *Miocene, and are the most advanced of ruminants. The stomach is four-chambered, the upper *incisors and in some forms the upper *canines are absent, and the *molars are *selenodont. Only two digits are functional on each limb. Most members possess horns: these may be present only in the male, and used principally for intraspecific fighting; they are developed in some forms for ritualized contests that seldom cause injury. Tragulids, musk-deer (*Moschidae), and some deer possess enlarged upper canines which are used in defence.

pectinate Resembling a comb in arrangement or shape.

pectine In *Scorpiones, a sensory appendage composed of three chitinous plates forming an axis attached by its centre to the second abdominal segment near the ventral mid-line, below which tooth-like processes are suspended, so that the whole structure resembles a comb.

pectinibranch Applied to the type of gill structure, common in the order *Mesogastropoda, in which only one (very efficient) gill is present.

pectinolytic Capable of digesting pectin, the material that acts as a kind of cement between the cells of higher plants.

pectocaulus In some colonial invertebrates (e.g. *Rhabdopleurida), applied to the condition in which each *zooid is attached to the *stolon by a contractile stalk.

pectoral fin In fish, one of the pair of fins that are situated one on each side of the fish just behind the gills ('breast fins'). Normally they are used for balancing and braking, but in some species (e.g. *Exocoetidae, flying fish) the extra-large fins are used for jumping and for gliding over the water surface.

pectoral girdle In vertebrates, the skeletal structure that provides support for the fore limbs or fins.

pectoralis major In vertebrates, the major chest muscle. In humans the two pectoralis major muscles cover most of the chest. In birds these muscles account for about 20% of the total body weight and move the wings downwards; the much smaller pectoralis minor muscles raise the wings upwards.

pectoralis minor *See* PECTORALIS MAJOR.

pectoral ridge A ridge on the upper *ventral part of the *humerus, below the greater *tubercle, to which the *pectoralis major muscle inserts.

***Pedetes capensis* (Cape jumping hare)** *See* PEDETIDAE.

Pedetidae (Cape jumping hare; order *Rodentia, suborder *Sciuromorpha) A monospecific family (*Pedetes capensis*) of rabbit-like, burrowing or terrestrial, nocturnal rodents which move by hopping or leaping. The ears are long and can be closed by rolling the *pinna during digging, the hind limbs are long and modified for jumping, the fore limbs are short and possess long claws. There are five digits on the fore limbs and four on the hind limbs. The tail is long, and covered with long hair. Pedetidae are distributed through central and southern Africa.

pedicel 1. In *Araneae, the short, narrow portion of the body that connects the *prosoma with the abdomen. **2.** In *Insecta, the second segment of the antenna. **3.** In *Apocrita, the *petiole.

pedicellariae In some invertebrates (e.g. some *Echinodermata and Asteriadina, asteroid starfish), minute, stalked organs, each consisting of a stem with a movable head built of three jaw-like valves and attached to small *tubercles on the *tests. They are used for grasping, defence, and scavenging. In many species they inject venom.

pedicle The muscular stalk by which *sessile, aquatic invertebrates are attached to their *substrate.

Pediculati An older name for the group of fish now placed in the order *Lophiiformes.

Pediculidae (order *Phthiraptera, suborder *Anoplura) Family of sucking lice, which are parasitic on some primates, including humans. The single genus, *Pediculus*, contains four species, two parasitic on humans, one on the chimpanzee, and one on spider monkeys (*Ateles* species). The two species parasitic on humans, *P. humanus* and *P. capitis*, have sometimes been considered two subspecies of a single species, but they differ in the relative lengths of parts of their legs, and in the environment they use. *P. humanus* lives and lays its eggs on human clothing, moving on to the body to feed; whilst *P. capitis* lives and lays its eggs on the hair of the scalp, and can cause mechanized dandruff. *P. humanus* is a vector of typhus, European relapsing fever, and trench fever, and because of this has been responsible for millions of human deaths. *P. capitis* is not known to transmit any diseases in natural conditions.

Pediculus (human louse) *See* PEDICULIDAE.

Pedinoida (class *Echinoidea, superorder *Diademataceae) An order of regular echinoids, of subconical to depressed hemispherical shape, which have simple or compounded, *diadematoid, ambulacral plates, and shallow slits notching the *peristomal margin for the passage of gills. They first appeared in the *Triassic.

Pedionomidae (plains-wanderer; class *Aves, order *Gruiformes) A monospecific family (*Pedionomus torquatus*) which is a small, quail-like bird, mainly brown, with loose plumage. It has a neck longer than those of quails, and tends to stand erect. Its feet have a hind toe. A terrestrial species, rarely flying, it is found in open grassland, feeds on insects and seeds, and nests on the ground. It is confined to Australia.

Pedionomus torquatus (plains-wanderer) *See* PEDIONOMIDAE.

Pedipalpi (subphylum *Chelicerata, class *Arachnida) Arachnid order in which the *Uropygi (including Schizomida) and *Amblypygi are often combined as suborders because of their similarities. All members are strikingly flattened (an adaptation for living in confined spaces) and have the first walking leg modified as a slender whip. The seventh body segment is constricted to form a *pedicel, and there are two pairs of *book lungs. There are nevertheless two distinct groups, the scorpion-like Uropygi which have a terminal flagellum, and the spider-like, tail-less whipscorpions (amblypygids).

pedipalps In *Arachnida, the second of the six pairs of appendages possessed by the *prosoma. They have become walking legs, or true palps, in those arachnids that have large *chelicerae (*Araneae, *Palpigradi, *Solifugae, and many *Opiliones and mites). In orders with small chelicerae (*Scorpiones, *Amblypygi, *Uropygi, and *Pseudoscorpiones) the pedipalp has become modified to form a large raptorial organ. The pedipalps are universal tools for the killing and manipulation of prey, as well as for defence and digging. In all arachnids they also serve as tactile and olfactory organs. In the spiders, the pedipalps guide food into the mouth, and the pedipalpal tarsus has become modified in males to form a complex copulatory structure (often used by taxonomists as a means of species identification).

peduncle 1. The stalk by which an invertebrate is attached to the substrate. **2.** In goose barnacles (*Lepadidae), the stalk by which the main part of the animal (the *capitulum) is attached to the substratum. **3.** *See* CAUDAL PEDUNCLE.

Pegasidae (dragonfish, sea moth; subclass *Actinopterygii, order *Pegasiformes) A small family of marine fish that have oddly shaped bodies encased by a bony armour, large, wing-like, *pectoral fins, and a produced snout. *Pegasus volitans*, growing to 15 cm, is a typical member of the family and is widely distributed in warmer waters of the Indo-Pacific region. There are about five species.

Pegasiformes (class *Osteichthyes, subclass *Actinopterygii) An order of bony fish including only the family *Pegasidae.

pelage The hair covering the body of a mammal.

pelagic 1. In marine ecology, applied to the organisms that inhabit open water, i.e. *plankton, *nekton, and *neuston (although neuston are fairly unimportant in such environments). **2.** In ornithology, applied to sea-birds that come to land only to breed, and that spend the major part of their lives far out to sea.

Pelecanidae (pelicans; class *Aves, order *Pelecaniformes) A family of large, white, brown, or grey birds which have long, broad wings, short tails, and a long, broad bill, the lower *mandible of which has a large, distensible pouch. The feet are *totipalmate. Pelicans inhabit estuaries and inland and coastal waters, feed on fish and crustaceans, and nest colonially in trees or on the ground. There is one genus, *Pelecanus*, with eight species, found in southern Europe, Africa, Asia, Australia, and N., Central, and S. America.

Pelecaniformes (tropicbirds, pelicans, gannets, cormorants, frigatebirds, anhingas; class *Aves) A diverse order of mainly marine birds, most of which have large bills and bare skin on the throat, which is distensible in some. They have short legs, the feet being *totipalmate (but only basally so in *Fregata* species). They are good fliers and swimmers, diving for food from the air or water surface, and feeding on fish and crustaceans. Most nest on cliffs or in trees. There are six families: *Anhingidae, *Fregatidae, *Pelecanidae, *Phaethontidae, *Phalacrocoracidae, and *Sulidae. They are found world-wide.

Pelecanoides **(diving petrels)** *See* PELECANOIDIDAE.

Pelecanoididae (diving petrels; class *Aves, order *Procellariiformes) A family of stocky, short-necked, short-winged petrels which have short bills and nostrils open at the top. They have a whirring flight, can dive into the sea, and can swim well under water to catch crustaceans. They breed in burrows on sub-Antarctic islands, and are less *pelagic than most petrels. There is one genus, *Pelecanoides*, with four species, found in the southern oceans.

Pelecanus **(pelicans)** *See* PELECANIDAE.

Pelecypoda *See* BIVALVIA.

pelican eel (*Eupharynx pelecanoides*) *See* EURYPHARYNGIDAE.

pelicans *See* PELECANIDAE; PELECANIFORMES.

pellicle The living, proteinaceous, layered structure which surrounds the cells in many types of *Protozoa. It is immediately below the *cell membrane and surrounds the *cytoplasm (it is not extracellular, like the cell wall in a plant). In *Ciliatea it may be very complex and contain pellicular, bottle-shaped *organelles (*trichocysts).

pelma *See* PELMATOZOAN.

pelmatozoan Applied to primitive echinoderms (*Crinozoa) that are attached to the substratum by a stalk (pelma). Formerly the pelmatozoans were ranked as a subphylum (Pelmatozoa) but the term is now used only informally.

Pelobates fuscus (spadefoot toad, garlic toad) *See* PELOBATIDAE.

Pelobatidae (spadefoot toads, horned frogs; class *Amphibia, order *Anura) A family of amphibians which have no ribs, minute teeth, and digging tubercles developed on their hind feet. *Pelobates fuscus* (spadefoot toad or garlic toad) of eastern Europe and western Asia has large, sharp, horny growths for digging; females are up to 8 cm long, males smaller. Pelobatids are adapted to dry habitats, appearing on the surface only after rain. Metamorphosis is rapid. There are 54 species, occurring in N. America, Eurasia, and N. Africa.

Pelomedusidae (hidden-necked turtles; order *Chelonia, suborder *Pleurodira) A family of freshwater turtles in which there is some vertical bending of the neck in withdrawal before the head is laid to the side. Pelomedusids are omnivores; they live in shallow water, but may travel over land. There are 14 species, occurring in S. America, Africa, and Madagascar.

Peloridiidae (subclass *Pterygota, order *Hemiptera) Family of bugs which have characteristics of both hemipteran suborders. They are said to feed on mosses. The fore wings have reticulate venation, and both they and the hind wings are almost always reduced to a condition where flight is impossible. They are restricted to temperate regions of the southern hemisphere.

pelta A shield or shield-like structure. The word is derived from *pelte*, a small, light shield used by Greek and Roman troops.

peltate Shield-shaped; having a central rather than a lateral stalk. *See* PELTA.

pelvic fin (ventral fin) One of the pair of fins positioned on the underside of the body of a fish. Depending on the species, the pelvic fins can be found in a mid-ventral (abdominal) position, underneath or just behind the *pectoral fins (thoracic position), or in front of the pectorals in the throat region (jugular position).

pelvis 1. In vertebrates, part of the *appendicular skeleton that is fused to the sacral *vertebrae and provides support for the hind limbs or *fins. **2.** The part of the abdomen that is surrounded by the bony pelvis.

Pelycosauria (class *Reptilia, subclass *Synapsida) An order of reptiles, dating from the upper *Carboniferous and Early *Permian, several of which sported large, sail-like, dorsal fins. There were both carnivorous and herbivorous types. They gave rise to and were replaced by the mammal-like reptiles (*Therapsida).

Pempheridae (sweeper, bullseye; subclass *Actinopterygii, order *Perciformes) A family of marine fish that have a fairly deep body, a short *dorsal fin, a long-based *anal fin, and large eyes. They are mostly schooling, shallow-water species found near caves, rock ledges, etc., and are well known to divers. Most species have an orange-red colour. There are about 20 species, distributed world-wide.

penduline tits *See* REMIZIDAE.

Penelope (guans) *See* CRACIDAE.

Penelope albipennis (white-winged guan) *See* CRACIDAE.

penetrance The proportion of individuals of a specified *genotype who manifest that genotype in the *phenotype under a defined set of environmental conditions. If all individuals carrying a *lethal mutation die prematurely, then the mutant gene is said to show complete penetrance. An organism may not express the phenotype normally associated with its genotype because of the presence of modifiers, epistatic genes (*see* EPISTASIS), or suppressors in the rest of the *genome; or because of a modifying effect of the environment. *Expressivity describes the degree or extent to which a given genotype is expressed phenotypically in an individual.

penguins *See* SPHENISCIDAE.

Peniculina (class *Ciliatea, order *Hymenostomatida) A suborder of ciliate *Protozoa in which characteristic bands of *cilia (comprising the peniculus) occur in the cytopharyngeal region (*see* CYTOPHARYNX). Body cilia are arranged uniformly. There are several families, including the well-known genus *Paramecium*, found in aquatic environments.

peniculus *See* PENICULINA.

Pennatulacea (sea pens; class *Anthozoa, subclass *Octocorallia) An order of colonial octocorals that have an elongate primary *polyp embedded in mud on the sea-bottom. The distal end of the primary polyp bears secondary polyps, usually on lateral branches. The polyps are supported by a calcareous or horny skeleton. They are first known as fossils from *Cretaceous and *Cenozoic strata, and with uncertainty from the *Silurian. The order contains two suborders, 14 families, and about 40 genera.

pennibrachium A modification of the forelimb into a wing-like structure bearing very long feathers found in the dinosaurs (*Dinosauria) Oviraptosauria (oviraptors) and Ornithomimosauria (ornithomimids). Pennibrachia may have been used in *display and *courtship.

Pennsylvanian *See* CARBONIFEROUS.

Pentacerotidae (Histiopteridae; boar fish; subclass *Actinopterygii, order *Perciformes) A small family of marine, deep-bodied fish that have a steep forehead, a produced snout, and a small mouth. The *dorsal fin is composed of strong spines as well as soft fin-rays. Pentacerotidae are usually found in deeper waters off shore and inhabit warm to temperate waters between S. Africa and the eastern Pacific coast. There are about 12 species.

pentacrinoid In some *Crinoidea, applied to a *sessile, stalked larva, whose crown eventually detaches itself to become the adult free-swimming form.

Pentacrinoid

pentadactyl Applied to a limb possessing five digits, or to one modified evolutionarily from an ancestral form which possessed five digits. The pentadactyl limb is characteristic of all living and almost all fossil tetrapods. *See also* APPENDICULAR SKELETON.

pentameral symmetry Five-sided symmetry that is strikingly exhibited by some *Echinodermata. Its evolutionary significance is unclear, but some authorities suggest that the five-sided body plan provides additional strength.

Pentapodidae (large-eye bream; subclass *Actinopterygii, order *Perciformes) A small family of marine fish, related to the *Lethrinidae, that are deep-bodied, oval-shaped, and have a continuous *dorsal fin, a forked tail fin, and large eyes. The greenish body often shows stripes, bars, or spots. Some authors have placed the fish in the family Lethrinidae, others in the *Nemipteridae. There are about nine species, occurring in the Indo-Pacific region.

Pentastomida (Linguatulida; tongue worms) An enigmatic group of 110–130 species of organisms (possibly *Crustacea) that live parasitically (*parasitism) in the lungs and nasal passages of vertebrates, mainly reptiles but also birds and mammals.

They are worm-like, 2–13 cm long, with five anterior protuberances (hence the name 'pentastomid', 'five mouths'). Four of these protuberances are leg-like, and the central one bears the mouth.

((())) **SEE WEB LINKS**
• Description of Pentastomida: tongue worms.

Pentatomidae (shieldbugs; order *He-miptera, suborder *Heteroptera) Family of bugs in which the *scutellum is large and in some forms almost completely covers the fore wings and *abdomen. A group of a few hundred species are predators, the rest feed on plants. In late summer they are often found feeding on wild berries. There are 5000 species, found in most parts of the world, but they are especially numerous in the tropics.

pentose A monosaccharide comprising five carbon atoms (e.g. deoxyribose and *ribose).

pentose-phosphate shunt See HEXOSE-MONOPHOSPHATE SHUNT.

Peodytidae (class *Amphibia, order *Anura) A family of amphibians that are related to the spadefoot toads (*Pelobatidae) but have long, leaping, hind limbs, which lack digging structures. There are two species, found in western Europe.

peppered moth See BISTON BETULARIA.

pepper-shrikes See VIREONIDAE.

pepsin An *enzyme released into the stomach that digests *proteins.

peptide A linear molecule comprising two or more *amino acids linked by *peptide bonds. The simplest peptide is $H_2N.CH_2.CO.NH.CH_2.CO_2H$ (glycylglycine).

peptide bond Chemical bond (CO.NH) by which one *amino acid molecule may be linked to the next in a chain.

peptone A partially hydrolysed *protein that is used as a nutrient in microbiological culture media.

Peracarida (phylum *Arthropoda, subphylum *Crustacea) A superorder of crustaceans that possess a brood chamber (*marsupium) beneath the thorax into which the eggs are deposited and where they are fertilized. Development of the young is direct. Most peracaridans are small. Approximately 40% of all crustaceans belong to this superorder, which includes seven orders: *Amphipoda, Cumacea, *Isopoda, Mysidacea, Spelaeogriphacea, Tanaidacea, and Thermosbaenacea.

Peramelemorpha (subclass *Theria, infraclass *Metatheria) An order (or suborder) of marsupials that contains only the *Perameloidea.

Peramelidae See PERAMELOIDEA.

Perameloidea (bandicoot; infraclass *Metatheria, order *Marsupialia) A superfamily that comprises two families, either Peramelidae (bandicoots) and Thylacomyidae (bilbies), or (as has recently been proposed) Peramelidae (the Australian bandicoots plus bilbies) and Peroryctidae (New Guinea bandicoots). They are mainly insectivorous, burrowing marsupials, in which the *marsupium opens toward the rear, the second and third digits of the hind limbs are *syndactylous and the first reduced or absent, and the dentition is *polyprotodont, the *molars being adapted for grinding. The muzzle is long, pointed, and flexible, and is used for probing in the soil for invertebrates; the hind limbs are long, and the ears are large. There are eight genera, with about 20 species, distributed throughout Australia (including Tasmania) and New Guinea.

peramorphosis Evolutionary change that results in the descendant incorporating all the ontogenetic stages of its ancestor, including the adult stage, in its *ontogeny, so that the adult descendant 'goes beyond' its ancestor. It may occur by *acceleration, *hypermorphosis, or *predisplacement.

Perca fluviatilis (perch) See PERCIDAE.

perception The appreciation of the external environment by means of the senses.

perch *See* PERCIDAE.

perching birds *See* PASSERIFORMES.

perchlet *See* CENTROPOMIDAE.

Percichthyidae (striped bass, Macquarie perch; subclass *Actinopterygii, order *Perciformes) A large family of perch-like fish found in tropical and temperate waters. Some species are exclusively marine, others inhabit freshwater systems. One of the most popular and best studied species is *Morone saxatilis* (striped bass), 1 m, of the eastern USA, which undertakes annual migrations from the sea into the rivers to spawn. There are about 40 species.

Percidae (perch; subclass *Actinopterygii, order *Perciformes) A large family of freshwater fish, including the true perch (*Perca fluviatilis*), 45 cm, that usually have a fairly deep and fully scaled body, with two separate *dorsal fins, a slightly forked tail fin, and *pelvic fins located close to the *pectoral base. Several members of this northern hemisphere family have been introduced to other continents. The family includes a number of important food fish, e.g. the perch, and *Stizostedion vitreum* (wall-eye), 90 cm, which is one of the largest species. There are at least 126 species.

Perciformes (class *Osteichthyes, subclass *Actinopterygii) An order of perch-like fish that, in terms of the number of families belonging to it (about 146), is by far the largest order of fish. All species possess a spiny-rayed first *dorsal fin and a soft-rayed second dorsal fin, with several spines preceding the soft rays of the *anal and *pelvic fins. The pelvic fins are inserted just behind or in front of the *pectoral fins. The scales are usually of the *ctenoid type. Representatives include the tiny gobies as well as the giant marlins.

Percophididae (subclass *Actinopterygii, order *Perciformes) A small family of marine fish that have a somewhat depressed head and body, large eyes, and usually two separate *dorsal fins. There are about 17 species, with world-wide distribution.

Percopsidae (troutperch; superorder *Paracanthopterygii, order *Percopsiformes) A small family of freshwater fish, of scientific interest because they are claimed to form a link between the soft-rayed and spiny-rayed fish. *Percopsis omiscomaycus* (troutperch), 20 cm, is widely distributed in N. America. There are two species.

Percopsiformes (subclass *Actinopterygii, superorder *Paracanthopterygii) An order of bony fish that includes the family Percopsidae and two other families.

Percopsis omiscomaycus **(troutperch)** *See* PERCOPSIDAE.

Perdix **(partridges)** *See* PHASIANIDAE.

Perdix perdix **(grey partridge)** *See* PHASIANIDAE.

pereiopod *See* PEREOPOD.

pereon In *Isopoda, the largest division of the body.

pereopod (pereiopod) One of the walking limbs of a crustacean.

perforate nostrils Nostrils (nares, *see* NASAL CAVITY) that are not separated by a *septum (i.e. the two nostrils are joined); the condition is found in the *Cathartidae (New World vultures). In most species of birds the nostrils are imperforate (separated by a septum).

Pergidae (sawflies; suborder *Symphyta, superfamily Tenthredinoidea) Family of symphytans which are usually less than 10 mm long. The family is large and its members have very diverse habits. In America, larvae feed extensively on oak and hickory. In Australia, species of *Perga* can cause serious defoliation of *Eucalyptus* trees.

pericardium 1. The membrane surrounding the cavity around the heart. **2.** The cavity around the heart.

pericentric inversion The inversion of a *chromosome piece (containing a block of genes) that involves the *centromere.

periderm In invertebrates, the thin sheets of material that make up the skeleton.

perignathic girdle A continuous or discontinuous ring of internal processes around the *peristome of echinoids (*Echinodermata), where the muscles which support and control the *Aristotle's lantern are attached. It is made up either of *apophyses or of *auricles.

perilymph A fluid contained in the inner ear of vertebrates, between the walls of the *bony labyrinth and the *membranous labyrinth.

perimysium *Connective tissue that surrounds *muscle fibres, binding them into bundles.

perineurium In the *peripheral nervous system, a sheath of *connective tissue surrounding a bundle of nerve fibres.

perinuclear space The region between the inner and outer *membranes of the *nuclear envelope.

peri-oral Applied to the region around the mouth.

periosteum A membrane that covers all bones.

periostracum A thin layer of dark-coloured, hardened protein on the outer surface (*ostracum) of *Brachiopoda, *Bivalvia, and *Gastropoda shells, which protects the growing edge of the shell. Being thin and composed of proteinaceous material, it is not generally preserved in fossils.

periotic bone In *Mammalia, the single bone surrounding the ear that is formed from the fused pro-otic, epiotic, and opisthotic bones.

peripatric speciation See FOUNDER EFFECT.

Peripatus **(velvet worms; phylum *Onychophora)** A genus of onychophorans that are 1.4–15 cm long, with a dry, soft, flexible, very permeable skin that is moulted, overlying a single layer of *epidermis, a thin *dermis, and three layers of muscle fibres. Its complete lack of hard parts allows *Peripatus* to squeeze into very confined spaces. The *coelom is greatly reduced and the body cavity is a *haemocoel. Each of the 14–44 segments of the trunk bears a pair of nephridia (*see* NEPHRIDIUM) and a pair of legs, each leg having an *eversible vesicle opening close to the nephridiopore and probably used to take up water. The loss of water from the body cannot be controlled and dry ground presents an insuperable barrier. Locomotion is slow and occurs by the extension of the segments in a peristaltic wave, each extension raising the legs from the surface and thrusting them forward. A pair of antennae, anterior to the mouth, tap the ground as the animal moves. There is an oral papilla on either side of the mouth at the end of which opens a gland secreting an adhesive substance used in obtaining food and in defence. The adhesive can be discharged as a stream, for up to about 50 cm, and hardens on contact with air, entrapping the prey or intruder in a tangle of sticky fibres. Most species are predatory on smaller invertebrates. All are nocturnal. *Peripatus* occurs in moist places in the tropics and the temperate regions of the southern hemisphere.

peripheral membrane protein A *protein that adheres temporarily to the *membrane with which it is associated.

peripheral nervous system (PNS) The system of nerves that are not contained within the *central nervous system. The nerves of the PNS are *segmentally arranged and connected to the spinal cord between *vertebrae. *See also* AUTONOMIC NERVOUS SYSTEM.

periphyton (aufwuchs) Organisms attached to or clinging to the stems and leaves of plants or other objects projecting above the bottom sediments of freshwater *ecosystems.

***Periplaneta americana* (American cockroach)** *See* BLATTIDAE.

periproct In *Echinodermata, the area that surrounds the anus. It is covered in tough skin in which small, calcitic plates are loosely embedded.

Periptychidae (superorder *Protoungulata, order *Condylarthra) An extinct family of primitive ungulates which are known to have lived during the *Palaeocene in N. America. The cheek teeth were *bunodont and additional cusps were present. *Periptychus* was rather smaller than a modern tapir, and had short legs and a long tail.

perisarc The dead, protein-chitinous, cylindrical sheath secreted by the epidermis which surrounds most colonial *Hydrozoa.

Perischoechinoidea (subphylum *Echinozoa, class *Echinoidea) A subclass of regular echinoids, without compound plates, in which the *perignathic girdle is absent or of *apophyses only, the teeth are grooved, and there are no gill slits or *spheridia. The number of columns of plates in *ambulacra and interambulacra is very variable except in *Cidaroida. These echinoids first appeared in the *Ordovician.

Perissodactyla (cohort *Ferungulata, superorder *Mesaxonia) An order that comprises those ungulates in which the number of functional toes is reduced to three or one, and a fourth, if present, is reduced, the weight of the animal being borne by the central digit. The order includes the three suborders *Ceratomorpha (tapirs, rhinoceroses, and their extinct relatives), *Ancylopoda (extinct forms), and *Hippomorpha (horse-like forms). They appeared in the *Eocene, derived from the *Condylarthra, and reached their zenith in the *Oligocene, when they were the most numerous of ungulates. So many fossil remains of them have been found that their evolutionary history is known in greater detail than that of any other mammalian group. Since then the order has declined dramatically, having been displaced by the *Artiodactyla, and the group as a whole is moving towards extinction.

peristalsis The rhythmic waves of muscular contraction by which matter is passed along the *alimentary canal.

peristomal Applied strictly to structures associated with the *peristome of *Echinodermata, but applied more loosely to structures just outside this region that are associated with feeding habits.

peristome In invertebrates, the mouth cavity.

peritoneum *See* MESENTERY.

Peritrichia (bell animalcules; subphylum *Ciliophora, class *Ciliatea) A subclass of ciliate *Protozoa which have a well-developed system of oral *cilia. Body cilia are mostly absent, but are present in some immature stages. Cells are variable in shape, and may be *sessile or free-swimming. Some are colonial, some have a shell-like *lorica. There is one order, with many genera. They are found in freshwater and marine habitats.

Peritrichida (class *Ciliatea, subclass *Peritrichia) An order of ciliate *Protozoa which have the characteristics of the subclass. There are two families, comprising many genera.

permeability A property of a membrane or other barrier, being the ease with which a substance will diffuse or pass across it.

permeability coefficient A quantitative estimate of the rate of passage of a solute across a membrane. In a concentration of 10 moles/litre it represents the net number of solute molecules crossing 1 cm2 of membrane per unit time.

permease One of a class of *proteins that act as carrier molecules in cells, facilitating the active or passive transport of various substances across membranes. They have many of the characteristics of enzymes with the exception that they may greatly alter the point of equilibrium of a reaction.

permeation The diffusion of a substance through a barrier (e.g. a *cell membrane).

Permian The final period of the *Palaeozoic Era, about 299–251 Ma ago. It is named after the central Russian province of Perm. The period is often noted for the widespread continental conditions that prevailed in the northern hemisphere and for the extensive nature of the southern hemisphere glaciation. Many groups of animals and plants vanished at the end of the Permian in one of the most extensive of all mass extinctions, including the Rugosa, *Trilobitomorpha, and Blastoidea.

(((•))) SEE WEB LINKS
• Detailed description of life in the Permian.

Pernis (honey buzzards) *See* ACCIPITRIDAE.

Perodictus potto (potto) *See* LORIDAE.

Peromyscus leucopus (deer mouse, white-footed deer mouse) *See* CRICETIDAE.

Peroryctidae *See* PERAMELOIDEA.

peroxidase An *enzyme that catalyses the *oxidation of certain organic compounds using hydrogen peroxide as an electron acceptor.

peroxide An inorganic or organic compound that contains linked pairs of oxygen atoms (—O—O—). Peroxides may be regarded as derivatives of hydrogen peroxide (H_2O_2) in which the hydrogen atoms are replaced by other atoms or groups.

peroxisome A round or oval, membrane-bound *organelle, 0.3–1.5μm in diameter, found in numbers in most *eukaryotic cells. Typically, it has a finely granular, relatively dense internal structure, and contains numerous *enzymes (particularly *catalase) associated with the production and degradation of peroxides. *See also* GLYOXYSOME.

persistent pulp A *pulp cavity of a tooth that remains open throughout life, permitting the continuous growth of the tooth.

pest An animal that competes with humans by consuming or damaging food, fibre, or other materials intended for human consumption or use. Many such species are harmless or ecologically beneficial (e.g. raptors, otters, and seals); others (e.g. most insect pests) are harmless until their populations increase rapidly in response to a virtually unlimited (to them) resource (e.g. a farm crop).

petaloid Resembling a petal.

Petaluridae (dragonflies; order *Odonata, suborder *Anisoptera) Archaic family of large odonates in which the eyes are well separated, and the triangles of the fore and hind wings are dissimilar. The nymphs are elongate and hairy, with short, fattened legs; and are found in bogs and swamps, where they burrow in the bottom sediments. There are 10 described extant species, found in the New World, Australasia, Oceania, and the eastern Palaearctic.

Petauridae (order *Marsupialia (or *Diprotodontia), superfamily *Phalangeroidea) A family of small marsupials that includes the gliding *Petaurus*.

Petaurista alborufa (red-and-white giant flying squirrel) *See* SCIURIDAE.

Petauroides volans (greater glider) *See* PSEUDOCHEIRIDAE.

Petaurus (sugar glider, squirrel glider) *See* PETAURIDAE.

petiole (pedicel, wasp waist) The constriction at the base of the *gaster in *Apocrita. The degree of constriction is varied, ranging from slight to extreme with elongation.

petrels 1. *See* PROCELLARIIDAE; PROCELLARIIFORMES. **2.** (diving petrels) *See* PELECANOIDIDAE. **3.** (storm petrels, *Oceanites*, *Oceanodroma*) *See* HYDROBATIDAE.

Petromyidae (rock rat; order *Rodentia, suborder *Hystricomorpha) A monospecific family (*Petromys typicus*) of squirrel-like rodents, which have ribs that can be moved in such a way as to allow the animal to compress its body dorsoventrally. The ears are short, the tail has a terminal tuft, the limbs are short, and the first digit is very short in the hind limbs and vestigial in the fore limbs. The animals shelter in crevices in rocks. They are found only in south-western Africa.

Petromys typicus (rock rat) *See* PETROMYIDAE.

Petromyzonidae (lamprey; superorder *Cephalaspidomorpha, order *Petromyzoniformes)** A small family of fish or fish-like vertebrates, which are without jaws, scales, or paired fins, but which have a funnel-shaped, sucking mouth, a protrusible tongue bearing horny teeth, seven gill openings on each side, and a single nostril between the eyes. Most species are freshwater inhabitants, and the few marine species migrate back to the rivers to spawn. The larvae (*ammocoetes) spend several years burrowed into the bottom of the river before they metamorphose into adult lampreys. There are about 31 species, found on all continents except Africa.

Petromyzoniformes (class *Cephalaspidomorphi, superorder *Cephalaspidomorpha)** An order of jawless vertebrates that includes only the family *Petromyzonidae.

Petronia (rock sparrows) *See* PLOCEIDAE.

petrosal The hard (i.e. petrous, or rock-like) part of the temporal bone that forms part of the inner ear.

pH A value on a scale 0–14 that gives a measure of the acidity or alkalinity of a medium. A neutral medium has a pH of 7; acidic media have pH values of less than 7, and alkaline media of more than 7. The lower the pH, the more acidic is the medium; the higher the pH, the more alkaline. The pH value is the reciprocal of the hydrogen ion concentration, expressed in moles per litre.

phaeomelanin A form of the pigment *melanin that produces a reddish colour in concentrated form and yellow when dilute.

Phaethon (tropicbirds) *See* PHAETHONTIDAE.

Phaethontidae (tropicbirds; class *Aves, order *Pelecaniformes)** A family of medium-sized, mainly white sea-birds which have some black on the head and wings, and long, white or red central tail feathers. They have stout, yellow or red bills, and pale legs with black, *totipalmate feet. They are highly aerial and *pelagic, feed mainly on fish caught by diving, and nest on cliffs. There are three species, including *Phaethon*, that are wide-ranging in winter and breed in the Caribbean, and in the Pacific, S. Atlantic, and Indian Oceans.

Phaethornis (hermits) *See* TROCHILIDAE.

phagocyte A cell (e.g. *macrophages and many *Protozoa) that is capable of engulfing material by *phagocytosis. Phagocytes are important in the defence mechanisms of many animals.

phagocytosis A form of *endocytosis in which a *cell membrane invaginates and encloses externally derived, solid material within a *vacuole, without disrupting the continuity of the cell surface. Subsequently this vacuole will fuse with a *lysosome and its contents will be wholly or partly digested.

phagosome A spherical *vacuole, enclosed in a *cell membrane, that is found in the cell *cytoplasm. A phagosome forms when the cell membrane folds around an invading particle such as a bacterium. As soon as the particle has been enclosed the section of membrane detaches and the cell membrane closes behind it. *Lysosomes then surround the phagosome and inject *enzymes into it, digesting the particle.

phagotroph *See* CONSUMER.

phagotrophy Mode of nutrition in which particulate food is ingested.

Phalacrocoracidae (cormorants, shags; class *Aves, order *Pelecaniformes)** A family of medium to large, mainly black and brown, aquatic birds,

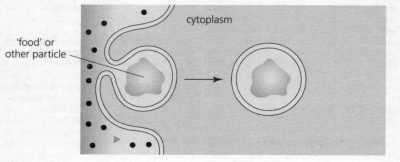

Phagocytosis

some of which have a greenish or bluish sheen. They have medium-length, slender bills with hooked tips, bare, coloured skin around the eyes, and *totipalmate feet. They have short wings, fairly long tails, and fly strongly except for *Nannopterum harrisi* (Galápagos cormorant), which is flightless. They are found on coasts, rivers, and lakes, and nest on islands, ledges, and trees. They dive for fish from the surface. There are two or three genera, and 32 species (27–31 of which comprise the genus *Phalacrocorax*), found world-wide.

Phalacrocorax (cormorants, shags) *See* PHALACROCORACIDAE.

phalange (strictly, phalanx) One of the bones of the digits.

Phalangeridae (koala, possums, cuscus; order *Marsupialia (or *Diprotodontia), superfamily *Phalangeroidea) A family of mainly arboreal, herbivorous marsupials in which the *hallux is well developed and often opposable and the tail is long and often *prehensile. The phalangers are distributed widely throughout Australia and New Guinea and north-west as far as Sulawesi. There are about six genera, and about 20–25 species.

Phalangeroidea (infraclass *Metatheria, order *Marsupialia (or *Diprotodontia)) A superfamily that comprises the families *Pseudocheiridae, *Petauridae, *Phalangeridae (koala, possums, cuscus), *Acrobatidae, and *Burramyidae. The first pair of lower *incisors is *diprotodont. In the hind foot the second and third digits are *syndactylous; typically the first hind digit is well developed and opposable, but has no claw. The *marsupium is well developed. Most Phalangeroidea are herbivores and many show modification for specialized herbivorous diets.

Phalangida (harvestmen) *See* ARACHNIDA.

phalanx *See* PHALANGE.

Phallostethidae (priapum fish; subclass *Actinopterygii, order *Atheriniformes) A family of very small, brackish-water and freshwater fish that have a relatively large, fleshy appendage (the priapum) below the head. The male uses the priapum during mating to secure internal fertilization of the eggs of the female. Almost transparent, *Phenacostethus smithi* (2–3 cm) is found in small schools in the rivers of Thailand. There are possibly four species, found in south-east Asia.

phallotreme In the males of many insect species, the long, tubular filament within the *aedeagus that is extruded during copulation and through which sperm is passed.

Phanerozoic The period of geological time that comprises the *Palaeozoic, *Mesozoic, and *Cenozoic Eras. It began 542 Ma ago at the end of the *Proterozoic and is marked by the accumulation of sediments containing the remains of animals with mineralized skeletons. The name means the period of 'visible' or 'obvious life'.

phantom larva (*Chaoborus crystallinus*) *See* CHAOBORIDAE.

pharangeal arch *See* PHARANGULA.

pharangeal groove *See* PHARANGULA.

pharangula A stage of embryogenesis in vertebrate *embryos, that follows the initial stages of *mesoderm differentiation and *organogenesis, and which is characterized by the development of pharangeal pouches. It is during this stage of embryogenesis that all vertebrate embryos display striking similarity. The sides of the embryo head become sculptured into between six and nine columns of tissue (pharangeal arches) separated by deep slits (pharangeal grooves). In nonamniotes these structures give rise to gill slits, whilst in most tetrapods these regions give rise to the *tympanic membrane, glands associated with the lymphatic system such as the *thymus gland, *parathyroid gland, carotid bodies, and tonsils. *See also* BLASTULA; GASTRULA.

Pharaoh ant (*Monomorium pharaonis*) *See* MYRMICINAE.

pharate The state of an insect that has formed a new *exoskeleton but that is still covered by the old. An insect moults at the end of each *instar, the pharate state occurring during the moulting process. There is some argument in the literature as to when each instar ends. Some say it is when the new exoskeleton is completed, others that it is when the old one has been sloughed.

Pharetronida (class *Calcarea, subclass *Calcinea) An order of sponges whose body possesses a *leucon structure. The skeleton is composed of a calcareous network, or of cemented *spicules. The order first appeared in the *Permian.

Pharomacrus mocino (resplendent quetzal) *See* TROGONIDAE.

Pharyngobdellida (phylum *Annelida, class *Hirudinea) An order of worms in which a *pharynx is present but jaws are not. All members of the order are carnivorous. Most occur in freshwater habitats, but some are terrestrial. There are five families.

pharynx 1. The tube that connects the mouth to the internal body cavity in which food is digested. In *polyps it extends well into the gastrovascular cavity. In *Enteropneusta it is extended forward to form the *stomochord. **2.** In vertebrates, the part of the gut that lies between the *buccal cavity and the *oesophagus. In primitive vertebrates (e.g. amphioxus) it occupies almost half the length of the *alimentary canal. Probably its original function was connected with the sifting of particles of food, but in lower vertebrates it is concerned with respiration, and gill slits or lungs open into it.

Phascolarctidae (order *Marsupialia (or *Diprotodontia), superfamily *Vombatoidea) A family of arboreal, leaf-eating marsupials that contains only *Phascolarctus cinereus* (koala) and its fossil precursors.

Phascolarctus cinereus (koala) *See* PHASCOLARCTIDAE.

Phascolomyidae *See* VOMBATIDAE.

phase diagram Graphical method for examining stability in biological systems, e.g. host–parasite relationships. The system variables (the numbers of host and parasite organisms) at different times are plotted against one another, and the time taken to change is shown by a line connecting the coordinate points for successive time values. The form of the resulting curve indicates the stability of the system. A line spiralling inward indicates damped oscillations favouring ultimate stability; an outward spiral suggests ultimate instability. By testing different values for the number of organisms it is possible to establish limits for stability. The technique has considerable practical relevance in agriculture.

Phasianidae (francolins, junglefowl, partridges, peafowl, pheasants, quails; class *Aves, order *Galliformes) A family of small to large gamebirds which have black and brown, cryptic or highly multicoloured plumage. Many have crests, and bare skin and *wattles on their faces. Their bills are short and stout (but francolins (40 species of *Francolinus*) have large bills), their necks short to long, their tails short to very long, and their legs short to long, often having one or more spurs. Junglefowl (four species of *Gallus* found in India, south-east Asia, and Indonesia) have bare, red, wattled faces, and the male has a large, fleshy comb on its crown; *G. gallus* (red junglefowl) is the wild ancestor of the domestic chicken. Males of the two species of *Phasianus* (pheasants) have metallic green heads, red wattles around the eyes, two ear tufts, and coppery-red or green body plumage, with a very long, black-barred, brown tail; females are dull brown. The genus is native to Asia but *P. colchicus* (common or ring-necked pheasant) has been introduced to Europe, N. America, and New Zealand, and francolins occur in Europe and Africa as well as Asia. The three species of partridges (*Perdix*), native to Europe and Asia, are brown and grey; the grey partridge (*P. perdix*) has an orange-brown mask; it has been introduced to N. America and New Zealand. The 15–18 species of *Arborophila* (tree partridges), found in Asia, lack spurs. Wood quails (14 species of *Odontophorus*) are medium-sized birds, with long crown feathers on the head, and occur only in Central and S. America. The tails of peacocks are highly elaborate for display. Members of the family are terrestrial, inhabiting woodland, grassland, and open, barren areas. They feed on seeds, berries, and insects, and nest on the ground. There are 48–50 genera, and 190 species, some migratory, found world-wide.

Phasianus (pheasants) *See* PHASIANIDAE.

Phasianus colchicus (common pheasant, ring-necked pheasant) *See* PHASIANIDAE.

Phasmatidae (stick insects, walking sticks; subclass *Pterygota, order *Phasmatodea) Family of large to very large insects, whose common name refers to their resemblance to twigs: this resemblance is often striking and is enhanced by their habit of swaying slowly from side to side. When disturbed, stick insects may become cataleptic and fall to the ground, or else rustle and expand brightly coloured hind wings. They may also produce repugnant odours and secretions. Almost all of the 2000 or so species are found in warmer regions of the world, although two species from New Zealand have established themselves in the Isles of Scilly (Great Britain).

Phasmatodea (Phasmida; class *Insecta, subclass *Pterygota) Order of large to very large, terrestrial insects, which includes the stick insects (walking sticks) and leaf insects (walking leaves). They are *exopterygote and the wing buds do not reverse their orientation in later *instars. Some species are wingless, and in most species the female has reduced wings and cannot fly. Eggs are laid by being flicked from the tip of the abdomen, and are free and seed-like, having thick, often intricately sculptured shells. *Parthenogenesis is widespread among the order. All species are foliage-feeders, and a few are of economic importance as forest defoliators. There are more than 2500 species, almost all of which are tropical or subtropical.

phasmid 1. In *Nematoda, one of a pair of sensory glands opening to either side of the tail and probably functioning as a chemoreceptor. **2.** *See* PHASMATODEA.

Phasmida *See* PHASMATODEA.

pheasants *See* PHASIANIDAE.

Phenacodontidae (superorder *Protoungulata, order *Condylarthra) An extinct family, comprising the best known of the condylarths, which was distributed throughout the *Holarctic region from the *Palaeocene to the *Eocene. Early forms were small and had claws but later forms were larger and possessed more ungulate characteristics. *Phenacodus*, an Eocene form, was about 1.2 m long, and possessed hoofs on the ends of the phalanges of all five digits, of which the third was the longest. The dentition was complete: the *canines large and the *diastema small, but the *molars square and *bunodont. In other respects *Phenacodus* was relatively unspecialized. The tail was long, and the spine curved. The skull was low and long, and the brain case small.

phenetic classification The grouping of biological organisms on the basis of observed physical similarities. *See* PHYLOGENETIC SYSTEMATICS.

phenol An aromatic compound that bears one or more hydroxyl groups.

phenotype The observable manifestation of a specific *genotype; those properties of an organism produced by the genotype in conjunction with the *environment that are observable. Organisms with the same overall genotype may have different phenotypes because of the effects of the environment and of gene interaction. Conversely, organisms may have the same phenotype but different genotypes, as a result of incomplete *dominance, *penetrance, or *expressivity.

phenotypic adaptation *See* ADAPTATION.

phenotypic variance The total variance observed in a *character.

phenylalanine An aromatic, non-polar *amino acid; it occurs in *proteins and is essential for humans.

pheromone A chemical substance, produced and released into the environment by an animal, which then elicits a physiological and/or behavioural response in another individual of the same species. For example, pheromones are released by a variety of glands on the abdomen, head, and wings of insects.

Pheucticus (grosbeak) *See* CARDINALIDAE.

Philemon (friarbird) *See* MELIPHAGIDAE.

Philepitta (asities) *See* PHILEPITTIDAE.

Philepittidae (asities, false sunbirds; class *Aves, order *Passeriformes) A family which comprises two distinct genera, *Philepitta* (asities) and *Neodrepanis* (false sunbirds). Asities are small, plump, short-tailed birds that have a yellow-green or black body and black head, and a bare *wattle around the eyes. The bill is fairly short and slightly *decurved. Asities feed on fruit and build a nest suspended from a branch. False sunbirds are small, blue and yellow, with a bare wattle around the eyes, and a long, slender, decurved bill. They feed on nectar and insects. Both genera inhabit forest bushes. There are two species in each genus, confined to Madagascar.

Philippine creepers *See* RHABDORNITHIDAE.

Philomachus pugnax (ruff) *See* SCOLOPACIDAE.

philopatry The tendency of an individual to return to or stay in its home area. Most animal species show some degree of philopatry, and one sex may show it to a greater degree than the other, e.g. male birds and female mammals tend to be more philopatric than those of the opposite sex. *Compare* DISPERSAL.

Philopteridae (order *Phthiraptera, suborder *Ischnocera) Family of chewing lice which are parasitic on birds of many orders. This family is the largest in the Ischnocera, and its members differ from those of the others in not having an entire front margin to the head, and in having two *tarsal claws to each leg. All species are believed to feed on fragments of feather and dead skin. As with other Ischnocera, many male Philopteridae use their antennae to clasp the female during copulation, and in one genus, *Harrisoniella*, the male antennae are much larger than the front legs, which have taken on a sensory function. There are 100 genera, and 1500 species.

Phlebobranchiata *See* ASCIDIACEA.

phobotaxis A random change in the direction of locomotion of a *motile microorganism or cell that is made in response to a given stimulus.

Phocidae (seals; order *Carnivora, suborder *Pinnipedia (or *Caniformia)) A family of marine carnivores in which there is no external ear, the tail is small, and the hind limbs cannot be turned forward to assist movement on land. Most feed on fish and marine invertebrates, but some feed on macroplankton, and one species feeds on penguins. Despite superficial similarities, the phocid and otariid seals are believed by some to have evolved independently and not to be closely related to one another. The phocid seals diverged from the otter branch of the *Mustelidae and can be traced back to the *Miocene. They are more completely adapted for aquatic life than the *Otariidae. Phocid seals are found in all oceans, mainly in high latitudes, and in the Caspian Sea and Lake Baikal. There are 10 genera, and 19 species.

Phocoena phocoena **(porpoise)** *See* PHOCOENIDAE.

Phocoenidae (suborder *Odontoceti, superfamily *Delphinoidea) A family of marine mammals which lack a distinct beak, the head being either rounded or conical. The dorsal fin is low or absent, the tail fin is narrow with a shallow notch, and the flippers are short. Typically, the teeth are compressed laterally and range in number from 30 to 60 in each jaw. Phocoenidae feed mainly on fish, but also on marine invertebrates. They are found throughout the northern hemisphere and in temperate regions of the southern hemisphere. There are about seven species, in three genera: *Phocoena* (*P. phocoena* is the common porpoise, distributed widely in the northern hemisphere), *Neophocoena*, and *Phocoenoides*.

Phodilus **(bay owls)** *See* TYTONIDAE.

phoebes (*Sayornis*) *See* TYRANNIDAE.

Phoenicopteridae (flamingos; class *Aves, order *Ciconiiformes) A family of large, white, pink, and vermilion birds which have black wing tips, very long necks and legs, short, webbed feet, bare faces, and a deep bill, bent down at the mid-point. The *mandibles have lamellae for sieving plankton. Flamingos inhabit shallow, freshwater and salt lakes, feed on plankton and small invertebrates, and nest colonially on mud platforms. The two or three species (sometimes regarded as a single species) of *Phoenicopterus* are typical flamingos, with rosy-pink plumage, black primaries, pink or blue legs, and a pink or yellow bill with a black tip. There are three genera, some migratory, and five or six species, found in southern Europe, Africa, southern Asia, Central and S. America, and the W. Indies.

Phoenicopterus **(flamingos)** *See* PHOENICOPTERIDAE.

Phoeniculidae (wood-hoopoes, scimitarbills; class *Aves, order *Coraciiformes) A family of medium-sized, slender, black birds, which have a green or purple gloss to their plumage. Their bills are long, slender, and *decurved, usually red or black, and they have long tails, in some species tipped white. *Phoeniculus* species (woodhoopoes) are more gregarious and noisy than *Rhinopomastus* (scimitarbills). Members of the family are arboreal and found in most African forests and woods. They feed on insects, fruit, and seeds, and nest in tree holes. The two genera contain eight species, found in Africa. They are sometimes included in the *Upupidae.

Phoeniculus **(wood-hoopoes)** *See* PHOENICULIDAE.

Phoenicurus **(redstarts)** *See* TURDIDAE.

Pholadina (class *Bivalvia, order *Myoida) A suborder of thin-shelled, unequilateral bivalves, which differ from other members of the order in possessing one or more additional plates. The foot is circular and acts as a suction disc. A small internal ligament and *chondrophore are present. Musculature is *isomyarian. The gills are elongate, with a pair of *demibranchs in most forms. Pholadina live in holes that they bore, commonly on a rock, coral, or wood substrate. They first appear in rocks of *Jurassic age.

Pholadomyoida *See* ANOMALODESMATA.

Pholidae *See* PHOLIDIDAE.

Pholididae (Pholidae; butterfish, gunnel; subclass *Actinopterygii, order *Perciformes) A small family of slender, compressed, marine fish, in which the body has very small scales, and the skin is often slimy ('buttery'). The long, low, *dorsal fin begins behind the head and terminates close to the small tail. Although the *anal fin is long-based, the *pelvic fins are minute. Commonly distributed along the shores of the cooler parts of the northern hemisphere, the fish tend to hide in crevices. There are about 13 species.

Pholidota (infraclass *Eutheria, cohort *Unguiculata) An order that comprises the single family *Manidae, the pangolins or scaly ant-eaters.

phoresy A method of dispersal in which an animal clings to the body of a much larger animal of another species and is carried some distance before releasing its grip and falling.

Phoridae (scuttle flies; order *Diptera, suborder *Cyclorrapha) Family of small or minute, grey-black or yellowish flies which run actively and have a curious, humped appearance. The antennae seem to comprise one large segment, but this conceals the other segments. The large segment bears a long apical or subdorsal *arista. Wings are often vestigial or absent. If they are present, the anterior veins are very heavily developed. Larvae live in decaying or putrefying material. Several abnormal species exist, with the possibility of some hermaphrodite adults. More than 3000 species are known to exist.

Phoronida (horseshoe worms) A phylum of marine, worm-like animals whose common name refers to their horseshoe-shaped, filter-feeding organ (a *lophophore). They are sessile suspension-feeders, most living in chitinous burrows. The gut is U-shaped and the *anus opens to the exterior on the dorsal surface above the lophophore. Horseshoe worms have a vascular system and blood containing red blood corpuscles containing *haemoglobin (a feature that distinguishes them from *Bryozoa). It is widely believed that they are ancestral to the *Brachiopoda and Bryozoa, although they first appear in rocks of *Devonian age (i.e. after the other two phyla). There are two extant genera, with 10 species.

phosphatase An *enzyme that catalyses reactions involving the *hydrolysis of esters of phosphoric acid.

phosphatide One of a large group of naturally occurring *phospholipids that are derivatives of glycerol phosphate and which normally contain a nitrogenous base.

phosphatidyl choline *See* LECITHIN.

phosphocreatine *See* CREATINE PHOSPHATE.

phosphodiester bond A bond between two molecules that is formed by phosphoric acid, which is esterified once to each molecule.

phospholipid A polar *lipid that is a lipid derivative of glycerol which contains one or more phosphate groups. Phospholipids are particularly important with regard to the structure and functioning of biological membranes because of their *amphipathic qualities.

phosphorylation The addition of a phosphate group to a compound, involving the formation of an ester bond between the reactants.

phosphotransferase An *enzyme that catalyses reactions involving the transfer of a phosphate group from one *substrate to another.

photic zone The layer of water within which organisms are exposed to daylight.

photokinesis A change in the speed of locomotion (or frequency of turning) in a *motile organism or cell that is made in response to a change in light intensity. The response is unrelated to the direction of the light source. *Compare* PHOTOTAXIS.

photoperiodism The response of an organism to periodic, often rhythmic, changes either in the intensity of light or, more usually, to the relative length of day. Many activities of animals (e.g. breeding, feeding, and migration) are seasonal and determined by photoperiodism.

photophore A luminous organ, modified from a mucous gland, that is found in the skin of fish. Rows of photophores are present in many deep-sea fish, which apparently can produce flashes of blue-green to orange light at will. The photophores secrete a compound which glows when activated, or contain colonies of phosphorescent bacteria.

photoreceptor A cell that is sensitive to light. The *neurons in the *retina of the vertebrate *eye and *ocellus are photoreceptors.

phototaxis A change in direction of locomotion in a *motile organism or cell that is made in response to a change in light intensity. The response is related to the direction of the light source. *Compare* PHOTOKINESIS.

phragmocone In *Cephalopoda, the shell that is divided by *septa into chambers (*camerae). In *Belemnitida, the septate portion of the shell which is *homologous to the external shell of other cephalopods.

phragmosis *See* SOLDIER.

Phrynomeridae (class *Amphibia, order *Anura) A small family of frogs which have finger pads and elongate fingers as do tree frogs. The pectoral girdle is vestigial. Phrynomeridae are seldom arboreal. Usually they climb and burrow in rock crevices or termite mounds. There are six species, occurring in southern Africa.

Phrynophiurida (brittle-stars; class *Stelleroidea, subclass *Ophiuroidea) An order of brittle-stars in which the articulation between the *radial shields and genital plates consists of a facet or ridge. The disc and arms are entirely covered by skin.

Phthiraptera (lice; class *Insecta, subclass *Pterygota) Order of insects which are closely related to Psocoptera, *Hemiptera, and *Thysanoptera. Lice may be distinguished by the lack of wings and *ocelli; the reduction of the antennae to three flagellomeres, *scape, and *pedicel; the loss of the metathoracic *spiracles; and the reduction of the *ovipositor. All species are obligate ectoparasites of birds and mammals, feeding on skin debris, feathers, sebaceous exudates, or blood. The eggs are attached to the fur or feathers of the host, and there are three nymphal stages. There is no *pupa, and the whole of the life cycle is spent on the body of the host. The order comprises about 3000 described species, in four suborders: *Anoplura, *Rhyncophthirina, *Ischnocera, and *Amblycera.

Phycodurus eques (leafy sea dragon) *See* SYNGNATHIDAE.

Phylactolaemata (phylum *Bryozoa, subphylum *Ectoprocta) A class of non-mineralized, freshwater ectoprocts in which the *lophophore is U-shaped. The mouth is covered by a lip. The class has a probable Late *Cretaceous representative, but otherwise is known only from the *Holocene.

phyletic evolution Evolutionary change within a lineage, as a result of gradual adjustment to environmental stimuli.

phyletic gradualism A theory holding that *macroevolution is merely the operation of *microevolution, which operates gradually and more or less continuously over relatively long periods of time. Thus gradual changes eventually will accumulate to the point at which descendants of an ancestral population diverge into separate species, genera, or higher-level taxa.

Phylliidae (leaf insects, walking leaves; subclass *Pterygota, order *Phasmatodea**) Family of large insects in which the body is flattened dorsoventrally, and the legs have broad, flattened extensions, thus giving the insect a strikingly leaf-like appearance. The family is found mainly in south-east Asia and New Guinea, although, together with the stick insects (*Phasmatidae), these insects have become popular household pets in other parts of the world. There are more than 50 species.

Phylloscopus (leaf warblers) *See* SYLVIIDAE.

Phyllostomatidae (New World leaf-nosed bats; order *Chiroptera, suborder *Microchiroptera**) A family of bats in most of which the nose leaf is simple and shaped like a spearhead, point uppermost, the nostrils being visible through it. The ears are mobile and well separated. Most are insectivorous, but some feed on vertebrates, while others are omnivorous or feed on fruit or nectar. They are distributed widely from the southern USA to northern Argentina, and in the W. Indies. There are 48 genera, and more than 120 species.

Phylloxeridae (order *Hemiptera, suborder *Homoptera**) Small family of homopterans, related to *Aphididae. Many species feed on *Quercus* species (oaks), but the best known is the vine phylloxera, which almost destroyed the European wine industry in the late 19th century. It was defeated by grafting European vines on to stocks of an American species whose roots were not affected by a subterranean stage in the insect's life cycle. Vine infestation remains a constant risk, however.

( ) SEE WEB LINKS
• Story of the Phylloxeridae infestation and its control.

phylogenetics The taxonomical classification of organisms on the basis of their degree of evolutionary relatedness.

phylogenetic species concept The view that a *species should be defined only

by its diagnosibility; i.e. that it consists of a population with a unique set of features (preferably derived). This definition was proposed by Joel Cracraft in 1982 as a more workable alternative to the *biological species concept, which implies knowledge of whether or not regular interbreeding occurs between populations.

phylogenetic systematics The study of biological organisms, and their grouping for purposes of classification, based on their evolutionary descent. *See also* CLADISTICS.

phylogenetic tree *See* CLADOGRAM.

phylogeny Evolutionary relationships within and between taxonomic levels, particularly the patterns of lines of descent, often branching, from one organism to another (i.e. the relationships of groups of organisms as reflected by their evolutionary history).

phylogerontism (racial senescence) The condition of an evolutionary lineage that is on the verge of extinction, according to an outmoded view of evolution which asserted that lineages proceed through a life cycle, from youth to senility.

phylum In animal taxonomy, one of the major groupings, coming below sub-kingdom and kingdom, and comprising superclasses, classes, and all lower taxa. Sometimes, confusingly, the term is used to mean 'lineage'. It is the root of 'phyletic', 'phylogeny', etc.

Phymosomatoida (class *Echinoidea, superorder *Echinacea**) An order of regular echinoids, which are similar to the *Hemicidaroida but have an imperforate primary *tubercle. They first appeared in the Early *Jurassic.

Phyrnosoma (horned lizards) *See* IGUANIDAE.

Physalia physalis (Portuguese man-of-war) *See* NOMEIDAE; RHIZOPHYSALIINA.

Physeter catodon (sperm whale) *See* PHYSETERIDAE.

Physeteridae (sperm whales; suborder *Odontoceti, superfamily *Physeteroidea) A family of whales in which the head is blunt and more or less rectangular in shape, containing a reservoir filled with *spermaceti. Functional teeth are present only in the lower jaw. Sperm whales feed on squid, fish, and *Crustacea, diving deeply to do so. They are distributed throughout all oceans. There are two genera, and three species, *Physeter catodon* (sperm whale), and *Kogia breviceps* and *K. simus* (pygmy sperm whales).

Physeteroidea (order *Cetacea, suborder *Odontoceti) A superfamily that comprises the families *Ziphiidae (beaked and bottle-nosed whales) and *Physeteridae (sperm whales), mammals capable of diving to great depths (2000 m in the case of sperm whales). The head contains a reservoir filled with *spermaceti.

physiognomy The form and structure of natural communities. *Compare* MORPHOLOGY.

physiological ecology The study of the functioning of organisms in relation to their environments.

physoclistous Applied to the condition in bony fish in which there is no connection or duct between the *swim-bladder and the intestinal tract. In physoclists, the gas pressure of the swim-bladder is regulated by special tissues or glands. This condition is found among the 'higher' bony fish (e.g. the perch-like fish, *Perciformes).

physogastry Extreme distension of the abdomen in female termite reproductives, caused by the growth of the ovaries, gut, and fat body.

Physophorina (class *Hydrozoa, order *Siphonophora) A suborder of *Cnidaria whose members have an apical float which lacks a pore and is attached to a stem bearing modified *zooids divided into two distinct regions.

physostomous Applied to the condition in bony fish in which the *swim-bladder is connected to the intestinal tract by a special duct (ductus pneumaticus). In physostome fish, the gas pressure can be regulated by swallowing air or by releasing it via the gut. This condition is found among the somewhat more primitive bony fish, e.g. the herrings and their relatives.

phyto- A prefix meaning 'pertaining to plants', from the Greek *phuton*, 'plant'.

phytocoenosis The primary producers (*see* PRODUCTION) that form part of the *biocoenosis in a *biogeocoenosis.

phytoflagellate Applied to *Protozoa that usually bear one or two *flagella, often contain chloroplasts, and that therefore resemble plants, so they are often classified as algae. Most are free-living. *Compare* ZOOFLAGELLATE.

Phytomastigophorea (subphylum *Sarcomastigophora, superclass *Mastigophora) In protozoan classification, a class of plant-like microscopic organisms which possess chloroplasts and are photosynthetic. These organisms are often regarded as algae. They are found in soil, shallow ponds, lakes, seas, etc.

phytophagous Feeding on plants.

Phytosauria (subclass *Archosauria, order *Thecodontia) A suborder that has only one family, the Phytosauridae, which were specialized thecodonts similar to crocodiles in general shape. They date from the latter part of the *Triassic and inhabited N. America, Europe, and India. They were displaced by the crocodiles at the end of the Triassic.

Phytosauridae *See* PHYTOSAURIA.

Phytotomidae (plantcutters; class *Aves, order *Passeriformes) A family of stocky, finch-like birds, that have grey-brown, streaked backs, some with a chestnut crown and orange under-parts. Their short wings have white wing bars and their long tails are white-tipped. Their bills are stout, with serrated edges. They inhabit woodland, open scrub, and gardens, feed on fruit, buds, and leaves, and nest in bushes. There is one genus, *Phytotomus*, with three species, found in S. America.

Phytotomus (plantcutters) *See* PHYTOTOMIDAE.

pia mater *See* MENINGES.

Picathartes *See* PICATHARTIDAE.

Picathartidae (bald crows, rockfowl; class *Aves, order *Passeriformes) A family of medium-sized birds which have grey upper-parts, white under-parts, and characteristically bare heads with yellow and black, or red, blue, and black skin. They have stout bills, fairly long tails, and rounded wings. They inhabit dense forest, feed on insects and amphibians, and build mud nests on the walls of caves. There is one genus, *Picathartes*, with two species, found in W. Africa. They are often placed in the family *Timaliidae.

Picidae (piculets, woodpeckers, wrynecks; class *Aves, order *Piciformes) A family of small to large, black, white, yellow, red, brown, or green birds, some of which are crested and many of which have yellow or red on the head. Most have long, stout bills, used for drilling holes, and long, fine tongues used for feeding. Their wings are short and rounded, and their tails are usually wedge-shaped with stiff feather shafts. Piculets (about 25 species of *Picumnus* of which one occurs in south-east Asia and the remainder in S. America) have short bills not used for drilling and tails that do not have stiffened shafts. Their legs are short, and their feet are *zygodactylous with three or four toes. *Picoides*, comprising typical woodpeckers, is often split by retaining in that genus the two three-toed species and placing the 31 four-toed species in the genus *Dendrocopos*. Woodpeckers are arboreal, climb tree trunks for insects, fruit, and sap, and nest in tree cavities, banks of earth, and termite mounds. The 15 species of *Picus* are arboreal but many feed on the ground on ants. The two species of *Jynx* (wrynecks), of Europe, Africa, and Asia, are cryptically col-

oured and rarely cling to trees (*J. torquilla*, common wryneck, is highly migratory). The 21 species of *Melanerpes*, mainly black and white, most of them with red crowns and rumps and white or black, barred, central tail feathers, occur only in America. There are 27–35 genera in the family, and about 200 species, found world-wide.

Piciformes (jacamars, puffbirds, barbets, honeyguides, toucans, woodpeckers; class *Aves) An order of small to fairly large birds which have brightly coloured plumage. The bills of each family are stout and characteristically shaped, and their toes *zygodactylous. They are mainly arboreal, feed on vegetable matter and animals, and nest in holes; one family is brood parasitic. There are six families, found world-wide.

Picoides (woodpeckers) *See* PICIDAE.

Picrodontidae (order *Primates, suborder *Plesiadapiformes) An extinct family of prosimians which lived during the *Palaeocene. They are believed to have been an evolutionary offshoot from the main line of primate evolution and to have no later descendants.

piculets (*Picumnus*) *See* PICIDAE.

Picumnus (piculets) *See* PICIDAE.

Picus (woodpeckers) *See* PICIDAE.

Pieridae (white butterflies, jezebels, orange tips, brimstones, sulphur butterflies; subclass *Pterygota, order *Lepidoptera) Family of medium-sized butterflies with predominantly white or yellow wings, in which the white colour is derived from uric-acid-based waste products. Some species are migratory. Eggs are laid singly or in groups. The larvae are not spined, but have fine *setae. The *pupa is supported by a central silk girdle.

pig *See* SUIDAE.

pigeon milk (crop milk, dove milk) A secretion from the lining of the *crop of both male and female pigeons and doves

(*Columbidae) that the birds regurgitate to feed their young. *Prolactin stimulates milk production and the milk has the appearance of cottage cheese. It consists of: 13–19% *protein; 7–13% fat; 1–2% minerals and *vitamins; and 65–81% water. *Compare* FLAMINGO MILK.

pigeons *See* COLUMBIDAE; COLUMBIFORMES.

pigfish *See* CONGIOPODIDAE.

pigment A compound (biochrome) that produces colour in the tissues of living organisms. The biological function of the compound may or may not be directly associated with this property.

pika (*Ochotona*) *See* OCHOTONIDAE.

Pikaia An early chordate (*Chordata) from the *Burgess Shale that was possibly related to the modern amphioxus (*see* BRANCHIOSTOMIDAE).

pike *See* ESOCIDAE.

pike characin (*Hepsetus odoe*) *See* HEPSETIDAE.

pike conger *See* MURAENESOCIDAE.

pikehead (*Luciocephalus pulcher*) *See* LUCIOCEPHALIDAE.

pilchard (*Sardina pilchardus*) *See* CLUPEIDAE.

pilidium larva In *Nemertini, a free-swimming larva which develops directly from the *gastrula.

pill beetle *See* BYRRHIDAE.

pill bugs *See* ARMADILLIDAE; ISOPODA.

Pilosa (order *Edentata, suborder *Xenarthra) An infra-order that comprises the sloths and ant-eaters, classified as three superfamilies: *Megalonychoidea (ground sloths, now extinct); *Myrmecophagoidea (ant-eaters); and *Bradypodoidea (tree sloths). Members of the Pilosa possess fur and are contrasted with the armadillos of the infra-order *Cingulata.

pilose Covered with fine hairs or down.

pilotage The steering of a course from one place to another by using familiar landmarks.

pilot whale (*Globicephala melaena*) *See* DELPHINIDAE.

Pimelodidae (long-whiskered catfish; superorder *Ostariophysi, order *Siluriformes) A very large family of freshwater catfish that have elongate, scaleless bodies, a small *dorsal fin, long *adipose fin, and three pairs of *barbels around the lips. Most species are nocturnal. There are 285 species, occurring in S. America.

pinacocyte In *Porifera, one of the flattened cells which comprise the outer surface and which can be expanded or contracted at their margins to allow the whole animal to alter its size slightly. At the base the pinacocytes secrete a substance which anchors the sponge to its substrate.

pincate beetle *See* TENEBRIONIDAE.

pineal body (epiphysis) In vertebrates other than humans, an outgrowth of the fore brain that secretes hormones affecting reproductive behaviour. In mammals that have a regular *oestrus cycle its secretion of melatonin inhibits development of the gonads; the pineal may have a similar function in human children.

pineal eye (epiphysial eye, median eye) In *Sphenodontidae (tuatara), some lizards, *Petromyzonidae (lamprey), and many fossil vertebrates, a structure on the top of the head that resembles an eye, with a lens and a retina, linked to the brain by a nerve. It is the larger and more developed of a pair of sacs that form by *evagination of part of the brain. Above it there is a gap in the skull covered by almost transparent skin. The function of the eye is uncertain but probably it is involved in regulating *diurnal rhythms.

pineconefish See MONOCENTRIDAE.

pine marten (*Martes martes*) See MUS-TELIDAE.

pine-tree katydids See TETTIGONIIDAE.

pinna 1. In *Mammalia, an extension of the external ear to form a trumpet-like structure, often mobile, supported by cartilage. **2.** Fin, or fin-like limb.

pinnate Borne on either side of a central stalk; like a feather in appearance.

Pinnipedia The name formerly given to a supposed suborder of *Carnivora. It comprises the families *Otariidae (sea lions), *Odobenidae (walrus), and *Phocidae (seals), together with their immediate ancestors. The families may not be related closely, the classification being based on similarities of appearance and way of life that may result from convergence (*see* CONVERGENT EVOLUTION); pinnipeds are, in any case, closely related to bears and dogs, and today are generally included in the suborder *Caniformia. All are marine carnivores in which the digits are fully webbed and, with the limbs, modified to form paddles. The hind limbs are directed to the rear while swimming, and are used for propulsion. Teeth are reduced in number.

pinocytosis A form of *endocytosis in which the material enclosed is a liquid and there is no fusion with a *lysosome. Pinocytosis may be a means by which macromolecules (e.g. hormones and proteins) are taken into a cell.

Pipa pipa (Surinam toad) See PIPIDAE.

pipefish See SYNGNATHIDAE.

pipe snakes See ANILIIDAE.

Pipidae (pipid toads, clawed toads; class *Amphibia, order *Anura) A family of tongueless toads in which the *lateral-line organs are distinct, and important in prey detection. There are five to eight *opisthocoelous vertebrae. The family includes the pipids of S. America and the clawed toads of Africa. The young of *Pipa pipa* (Surinam toad) of S. America develop encapsulated in the back of the female and hatch, already metamorphosed, in 77–136 days. *Xenopus laevis* (clawed frog or clawed toad) of tropical and southern Africa has black, horny, claw-like sheaths on the three inner toes of the large, webbed, hind feet; it feeds on fish, carrion, and invertebrates, which it pushes into its mouth with its forefingers. There are 11 species in the family, all aquatic in muddy waters.

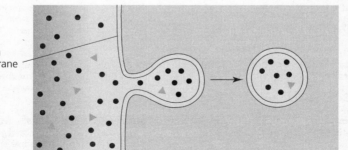

plasma membrane

Pinocytosis

pipid toads *See* PIPIDAE.

pipistrelle (*Pipistrellus pipistrellus***)** *See* VESPERTILIONIDAE.

Pipistrellus pipistrellus **(pipistrelle)** *See* VESPERTILIONIDAE.

pipits *See* MOTACILLIDAE.

Pipra **(manakins)** *See* PIPRIDAE.

Pipridae (manakins; class *Aves, order *Passeriformes) A family of small, stocky, short-tailed birds which have short, broad, slightly hooked bills, and short wings. Their short legs have the centre toe fused basally to the second or fourth toe. They are mainly black, with coloured heads and crowns. (The 16–18 species of *Pipra* are mainly glossy, black or green; the males have bright red, blue, yellow, and white heads or crowns, and brightly coloured thighs.) Manakins inhabit tropical forests, feed on fruit and insects, and nest in bushes. There are 18 or 19 genera, comprising about 60 species, found in Central and S. America.

Piranga **(tanagers)** *See* THRAUPIDAE.

piranha *See* CHARACIDAE.

Piroplasmea (phylum *Protozoa, subphylum *Sporozoa) A class of Protozoa which are parasitic in a range of vertebrates, where they infect red and white blood cells, and cells of the liver. A second (invertebrate) host is required for completion of the life cycle. There is one order, with two families. The class includes some important pathogens of domestic animals.

Pisauridae (nursery-web spiders; order *Araneae, suborder Araneomorphae) Family of medium to large spiders, with long legs. The females carry the egg sac in the *chelicerae and spin a nursery web tent for the hatchlings, which disperse after their second moult. The males often offer nuptial gifts.

Pisces In older classifications, a superclass containing the four classes of fish: the two extant classes *Chondrichthyes (cartilaginous fish, e.g. sharks) and *Osteichthyes (bony fish), the extinct class *Placodermi, and also the most primitive of the vertebrates of the class *Cephalaspidomorphi. The first protofish are known from late *Cambrian fossils, and the first true fish, an agnathan, has been recorded from earliest *Ordovician sediments. Most modern classifications omit the term 'Pisces', regarding it as artificial. *See also* GNATHOSTOMATA.

piscivorous Fish-eating.

pitfall trap A device for capturing ground-dwelling invertebrate animals, which comprises an appropriately baited vessel (e.g. a jam jar containing a small piece of rotting meat) that is buried so its mouth is at ground level. The mouth is covered by a roof, which shelters the vessel from rain and prevents larger animals from entering, raised sufficiently to allow access for target species.

pit organ *See* TEMPERATURE-SENSITIVE ORGAN.

Pitta **(pittas)** *See* PITTIDAE.

pittas (*Pitta***)** *See* PITTIDAE.

pitted-shell turtle *See* CARETTOCHELYIDAE.

Pittidae (pittas; class *Aves, order *Passeriformes) A family of small, plump, brightly coloured, short-tailed birds, which have short, rounded wings and strong, slightly curved bills. Their legs are long and their feet large. They are secretive, solitary, and terrestrial, and can hop rapidly. They inhabit jungle and forest scrub, feed mainly on insects, and build rounded nests with side entrances, placed on the ground or in low bushes. There is one genus, *Pitta*, containing 25 species, some migratory, found in Africa, Asia, and Australia.

pituitary axis The *pituitary gland and the *hypothalamus; together, this complex allows activity from the *central and *peripheral nervous systems to converge and interact with activity in the *endocrine system. *See also* DIENCEPHALON.

pituitary body *See* PITUITARY GLAND.

pituitary gland (pituitary body) A *gland at the base of the brain that secretes a range of *hormones, mostly *tropins. It is the dominant gland of, and regulates, the *endocrine system and also interacts with the *hypothalamus in the *pituitary axis. The gland comprises an anterior lobe and a posterior lobe. The anterior lobe secretes *gonadotropins, *prolactin, *thyrotropic hormone, and *adrenocorticotrophic hormone; the posterior lobe secretes *oxytocin and *vasopressin. *See also* ADENOHYPOPHYSIS; NEUROHYPOPHYSIS.

pit vipers *See* CROTALIDAE.

placenta In animals, the organ by which embryos of *viviparous species are nourished and waste products removed, formed by the fusion of embryonic and maternal tissues.

placental mammals *See* EUTHERIA.

placentotrophic A *matrotrophic mode of development in which nourishment is supplied from the mother through a vascular *placenta.

Placodermi A class of archaic, jawed (gnathostome), and heavily armoured fish, which appeared in the *Devonian Period and which were virtually extinct by its end. They were rather diverse in body form. They all possessed a head shield formed by bony plates. *Pectoral and *pelvic fins appear to have been present. Most of them were bottom-dwelling, with a depressed body terminating in a *heterocercal tail.

Placodontidae (class *Reptilia, subclass *Euryapsida) An order of *Triassic reptiles which were specialized shellfish feeders. Some were heavily armoured and strongly resembled turtles; while the more lightly armoured varieties, e.g. *Placodus*, were analogous in general form to the *Nothosauria, except for the modifications relating to their molluscan diet. *Placodus impressus* had large, flat teeth on the palate and lower jaws, webbed fingers and toes, and an elongate tail, and was adapted to feeding on shell banks. Later representatives of the placodont reptiles were more turtle-like.

Placodus impressus *See* PLACODONTIDAE.

placoid scale (dermal centicle) A type of scale that comprises the basic unit of the hard skin cover of sharks. It consists of a hard base embedded in the skin and a spiny process (cusp); these are covered by *vitrodentine. The scale projects outwards and backwards and it is the cusps that are responsible for the roughness of the shark's skin. The bulk of the scale consists of *dentine surrounding a central *pulp cavity. Unlike the scales of bony fish, placoid scales stop growing after they reach a certain size and new scales are added as the animal grows.

Placozoa A *phylum based on *Trichoplax adhaerens*, a small, multicellular organism first discovered in 1883 in a marine aquarium in Naples; since then one more species (*T. reptans*) has been discovered. *T. adhaerens* is flat, amoeboid, 2–3 mm in diameter, and has a *dorsal and *ventral layer of ciliated (*cilium) cells with a layer of loose, contractile cells between. It produces eggs and also reproduces by fission and *budding. The DNA content is less than that in any other animal.

((⊕)) SEE WEB LINKS
• Introduction to Placozoa

plains-wanderer *See* PEDIONOMIDAE.

planarians *See* TRICLADIDA.

plankton Aquatic organisms that drift with water movements, generally having no locomotive organs. The phytoplankton (plants) comprise mainly diatoms, which carry out photosynthesis and form the basis of the aquatic *food chains. The zooplankton (animals) which feed on the diatoms may sometimes show weak locomotory powers. They include protozoans, small *Crustacea, and in early summer the larval stages of many larger organisms. Plankton are sometimes divided into net plankton (more than 25μm diameter) and nannoplankton, which are too small to be caught in a plankton net. The word is derived from the Greek *plagktos*, 'wandering'.

Planktosphaeroidea (phylum *Hemichordata) An extant class of hemichordates, of which only two specimens of planktonic, ciliated larvae have been found (in the Bay of Biscay). The internal structures indicate they are hemichordates but the adults are unknown.

planogamete A motile *gamete.

plantain-eaters *See* MUSOPHAGIDAE.

plantcutters *See* PHYTOTOMIDAE.

plant hopper *See* DELPHACIDAE.

plantigrade Applied to a *gait in which the entire foot makes contact with the ground (e.g. in humans).

planula larva In coelenterates, a free-swimming, ciliated, elongated, radially symmetrical larva that has distinct anterior and posterior ends but no *gastrovascular cavity. Some authorities have hypothesized that an organism very similar to the planula larva gave rise to the *Animalia, for which reason it is also known as the planuloid ancestor.

planuloid ancestor *See* PLANULA LARVA.

plasma *See* BLOOD PLASMA.

plasmagene A *gene present in any cell structure other than the nucleus.

plasmalemma *See* CELL MEMBRANE.

plasma membrane *See* CELL MEMBRANE.

plasma thromboplastin component (PTC) *See* BLOOD CLOTTING.

plasmid A particle found in the *cytoplasm of a bacterial cell (*see* BACTERIA) that carries one or more *genes and can replicate itself autonomously. Genetic information on the plasmid is passed from each cell to its *daughter cell, and sometimes to neighbouring cells. Plasmids normally remain separate from the *chromosome, but some may become integrated into it temporarily and replicated with it incidentally. Plasmids are important in the genes they carry (e.g. they may confer resistance to particular antibiotics to their bacterial hosts).

plasmodesmata Cytoplasmic bridges, lined with a *cell membrane, that connect adjacent cells. At one time they were thought to be confined to plants, but they have now been observed in many animal cells. Almost certainly they provide major pathways of communication and transport between cells.

Plasmodiidae (order *Eucoccidia, suborder *Haemosporina) A family of parasitic *Protozoa which undergo *schizogony in the red blood cells, as well as in the cells of other tissues, in a vertebrate host. There is one genus, *Plasmodium*, with more than 50 species, which includes some very important disease-causing species: the malarial parasites. The life cycle is complex. Some stages can occur only in a second host, the mosquito *vector.

Plasmodium *See* PLASMODIIDAE.

plasmotomy A type of *asexual reproduction in which a multinucleate protozoan cell divides into two or more multinucleate daughter-cells without the occurrence of *mitosis.

plastron 1. In *Chelonia, the lower part of the shell, which is two-layered, consisting of bony plates corresponding to part of the shoulder girdle and abdominal ribs of primitive reptiles, covered by horny epidermal plates. It is joined to the *carapace by a bridge between the legs. In some families it is hinged. **2.** In some insects, a physical gill made up of *hydrofuge hairs. Insects which use a plastron when diving under water are covered with a thick felt of hairs which are bent over at the tips. The hairs trap a layer of air which is continuous with the tracheal system, and surrounds the animal. **3.** In *Echinodermata, a flattened area between the mouth and *anus, formed by the modification of the posterior *interambulacrum. It is densely covered with spines.

Platacanthomyidae (Chinese pygmy dormice, spiny dormice; order *Rodentia, suborder *Myomorpha) A family of arboreal mice which resemble dormice but have long, bushy tails. They are sometimes referred to the *Muridae as a subfamily. The first digit of the fore limb is reduced to a pad, and the hind feet are elongated. There are two species, in two genera. In *Platacanthomys* (spiny dormouse) the fur is spiny. In *Typhlomys* (Chinese pygmy dormouse) the eyes are very small, the fur is soft, and the tail is poorly haired, and scaly at its base. They are found only in southern India and southern China.

Platacanthomys (spiny dormouse) See PLATACANTHOMYIDAE.

Platalea (spoonbills) See THRESKIORNITHIDAE.

Platanista See PLATANISTOIDEA.

Platanistidae See PLATANISTOIDEA.

Platanistoidea (order *Cetacea, suborder *Odontoceti) A superfamily that comprises the one family Platinistidae (river dolphins), cetaceans which are believed to be related closely to the squalodonts (*Squalodontoidea) and to forms known from the early *Miocene, and to be more primitive than other Odontoceti. The temporal opening is unroofed. The beak is long and slender, the teeth peg-like and vary in number according to species. The cervical vertebrae are not fused, and the neck is distinct. The flippers are broad and short, the dorsal fin low, and the tail slightly notched at the mid-line. River dolphins grow to a length of 1.5–2.9 m. There are five species in four genera. *Platanista*, which is blind, is found in the Indus, Ganges, and Brahmaputra rivers of India, and there are two species; *Inia*

inhabits the Amazon and upper Orinoco rivers of S. America; *Pontoporia* is found in the estuary of the La Plata river, and along the south-east coast of S. America; *Lipotes* is found only in Lake Dongting, on the Changtze Jiang river, in China. *Lipotes vexillifer* (baiji), the Chinese river dolphin, is believed to be functionally extinct.

plated lizards See CORDYLIDAE.

platelet (blood platelet, thrombocyte) A small, red, anucleate, disc-shaped body, formed from a fragment of a larger bonemarrow cell, that is present in *blood, where there may be 250 000 per cubic millimetre. Platelets release *serotonin and are involved in *blood clotting.

Plateosaurus gracilis See PROSAUROPODA.

Platichthys flesus (flounder) See PLEURONECTIDAE.

Platyasterida (starfish; class *Stelleroidea, subclass *Asteroidea) An order of starfish that are characterized by the arrangement of arm *ossicles in transverse rows, from axial elements (*adambulacrals, *ambulacrals) laterally. Only one row of marginal ossicles is present. There are three genera, two *Palaeozoic and one *Holocene.

Platycephalidae (flathead; subclass *Actinopterygii, order *Scorpaeniformes) A family of marine fish that have a distinctly depressed or flattened body, large mouth, flat head with the eyes located on top, and *pectoral fins that project sideways. Many species are attractively coloured, and often well camouflaged when lying on the sea-bottom. There are about 55 species, occurring in the Indo-Pacific region.

Platycnemididae (damselflies; order *Odonata, suborder *Zygoptera) Ubiquitous family of damselflies whose members are characterized by having an almost rectangular discoidal cell in the wing venation, and only two postquadrilateral cells before the subnodus. The *tibia of the middle and hind legs are often dilated, especially in the male. The nymphs are generally short and broad, with thickened *tracheal gills. More than 200 extant species have been described.

Platyctenea (phylum *Ctenophora, class *Tentaculata) An order of ctenophorids which have dorsoventrally compressed bodies and two retractable *tentacles. Some can swim, but most move by creeping or are *sessile.

Platygasteridae (suborder *Apocrita, superfamily Proctotrupoidea) Family of minute (1–2.5 mm long), shiny black wasps which have a more or less flattened abdomen and almost no wing venation. Adults are very similar to *Scelionidae. Some larvae are parasites of cecidomyid gall midges and some whiteflies, others are egg parasites of *Fulgoridae and *Coleoptera. Reproduction by *polyembryony is widespread in the family.

Platyhelminthes (flatworms) A phylum of *acoelomate, *triploblastic, dorsoventrally flattened (hence their name), *bilaterally symmetrical worms in which the internal organs are well developed and *metameric segmentation is absent. The gut may be present, but there is no *anus and no blood-vascular system; protonephridia (*see* NEPHRIDIUM) are the excretory and osmoregulatory organs. Skeletal elements are rare. Morphologically they are quite diverse, and many are parasites. Most are hermaphrodites. The phylum lacks a definitive fossil record. From various evolutionary considerations it is probable that platyhelminths were in existence 1000–800 Ma ago.

platypus (*Ornithorhynchus anatinus*) *See* ORNITHORHYNCHIDAE.

platyrrhine In *Primates, applied when the nostrils face more or less to the sides and are well separated.

Platyrrhini (New World monkeys; cohort *Unguiculata, order *Primates) Although the platyrrhine (S. American) monkeys are often grouped with the catarrhine (Old World) monkeys, apes, and humans into one suborder, *Anthropoidea (or Simiiformes), the two kinds of monkeys remain distinct as far back as the *Oligocene, when they first enter the fossil record. Much of their resemblance is therefore the result of *parallel evolution. *Compare* CATARRHINI.

Platysternidae (big-headed turtles; order *Chelonia, suborder *Cryptodira) A monospecific family (*Platysternon megacephalum*) consisting of turtles found in mountain brooks in south-east Asia. They are aquatic carnivores but often travel over land. The jaws are serrated, the head is too large to withdraw into the shell, the *carapace is flattened, and the tail is long and plated.

Platysternon megacephalum (big-headed turtle) *See* PLATYSTERNIDAE.

Platystictidae (damselflies; order *Odonata, suborder *Zygoptera) Family of damselflies in which the basal cross-vein of each wing traverses the cell bounded by the second branch of the cubitus, the arculus, and the hind margin of the wing. There are also many postnodals. There are more than 130 described extant species found throughout Asia and Central and S. America.

play Voluntary, seemingly paradoxical behaviour (i.e. a goal usually associated with the behaviour is not attained because the activity is not pursued to its conclusion, or because it is misdirected), often occurring in bouts preceded by signals exchanged between participants, during which movements may be performed in apparently random succession, and in which certain sequences may be repeated many times.

pleasing fungus beetles *See* EROTYLIDAE.

pleated sheet *See* BETA PLEATED SHEET.

Plecoglossidae (ayu; subclass *Actinopterygii, order *Salmoniformes) A monospecific family (*Plecoglossus altivelis*) of fish that spend some time in the sea but return to freshwater streams to spawn. Large numbers of juveniles are caught during their annual migration, and are stocked in ponds for cultivation, making them commercially important in Japan, and to some extent China. Ayus occur in northern Asia.

Plecoglossus altivelis (ayu) *See* PLECO-GLOSSIDAE.

Plectognathi An alternative name for fish placed in the order *Tetraodontiformes.

pleiomorphism *See* PLEOMORPHISM.

pleiotropy The phenomenon in which a single *gene is responsible for a number of different phenotypic effects that are apparently unrelated.

Pleistocene The first of two epochs of the *Pleistogene, that lasted from approximately 1.81 Ma ago until the beginning of the *Holocene, about 10 000 years ago. The epoch is marked by several glacial and interglacial episodes in the northern hemisphere.

(((●))) **SEE WEB LINKS**
• Detailed description of life in the Pleistocene.

Pleistogene The present period of geologic time, which began 1.81 Ma ago at the start of the *Pleistocene. It includes the Pleistocene and *Holocene epochs.

pleomorphism (pleiomorphism) A form of *polymorphism in which distinctly different forms occur at particular stages during the life cycle of an individual (e.g. the larval, pupal, and *imago forms of many insects).

pleopod In *Crustacea, one of a number of abdominal appendages used for swimming, burrowing, obtaining food, carrying eggs, or making water currents; or in gas exchange. In males, the anterior one or two pairs are copulatory organs.

Plesiadapidae (order *Primates, suborder *Plesiadapiformes) An extinct family of primates which lived during the *Palaeocene and lower *Eocene and are known mainly from fossils found in west-

ern N. America. *Plesiadapis*, which is typical, had ears and *molar teeth reminiscent of those of lemurs, but the *incisors were chisel-like, and the digits possessed claws rather than nails. It was a squirrel-like animal with fore and hind limbs of equal length. The plesiadapids represent an evolutionary branch which died out, and they are not ancestral to any of the higher primates.

Plesiadapiformes (cohort *Unguiculata, order *Primates) An extinct suborder that comprises the families Paramomyidae, *Plesiadapidae, *Carpolestidae, and *Picrodontidae, which are the earliest known primates and flourished during the *Palaeocene. Most are known only from the most fragmentary remains, mainly of teeth and parts of jaws, though a few skulls and even partial skeletons are known, and apart from the fact that they are primates their relationship to later forms is uncertain. Probably most were too specialized to have given rise to later primates, despite their geologically ancient occurrence.

Plesiadapis *See* PLESIADAPIDAE.

Plesiadapoidea (order *Primates, suborder *Plesiadapiformes) Superfamily of middle *Palaeocene to lower *Eocene primates which resembled squirrels in their general appearance and, like squirrels, occupied a rodent-type niche. They are not in line of descent to the higher primates.

plesiomorphic Applied to a character state that is based on features shared by different groups of biological organisms and inherited from a common ancestor. The term is taken from the Greek *plesios*, 'near', and *morphe*, 'form', and means 'old-featured'. The features to which it is applied were formerly called 'primitive'. It is the opposite of *apomorphic.

plesion In taxonomy, a group of superfamilies within a suborder.

Plesiopidae (roundhead, devilfish; subclass *Actinopterygii, order *Perciformes) A small family of marine fish that have a fairly deep body, a large head with a large mouth, a single, continuous, *dorsal fin, and a rounded tail fin. The first half of the *lateral line runs close to the base of the dorsal fin. Several species are quite colourful. They tend to hide near rock ledges. The family is distributed in the Indo-Australian region.

Plesiopora (phylum *Annelida, class *Oligochaeta) An order of mainly small, aquatic worms among which ciliated forms are common. Microscopic forms are free-living, others are tube-dwellers. There are four families.

Plesiorycteropus A genus of apparently *fossorial, ant-eating mammals, known only from incomplete material found in sub-Recent deposits in Madagascar. Individuals weighed about 10 kg or less. Formerly referred to the *Tubulidentata, *Plesiorycteropus* has now been placed in a new order, *Bibymalagasia.

Plesiosauria (plesiosaurs; subclass *Euryapsida, order *Sauropterygia) A suborder of aquatic reptiles which enter the fossil record in the late *Triassic and which are common in many *Jurassic and *Cretaceous sediments. In appearance they have been likened to 'a snake strung through the body of a turtle', and some grew up to 15 m in length. There were also short-necked types, as well as the swan-necked. The former are defined as the superfamily Plesiosauroidea, the latter as the superfamily Pliosauroidea, the two making up the suborder.

Plesiosauroidea *See* PLESIOSAURIA.

plesiosaurs *See* PLESIOSAURIA.

Plethodontidae (lungless salamanders; class *Amphibia, order *Urodela) A family of amphibians in which the adults usually lack gills and lungs. They are characterized by a nasolabial groove, a glandular extension from the upper edge of the lower lip, below the nostril. They are mainly aquatic, and the family includes mountain-brook and blind, cave-dwelling forms; but some are terrestrial (e.g. burrowing worm salamanders). There are 183 species, found

in southern Europe, and N., Central, and S. America as far as 20°S.

pleura A *serous membrane that covers the *lungs. It comprises the visceral pleura, which covers the lungs, and the parietal pleura, which lines the thoracic cavity and *diaphragm.

Pleurobranchomorpha *See* NOTASPIDEA.

Pleurodira (side-necked turtles; subclass *Anapsida, order *Chelonia) A suborder of turtles which withdraw the head into the shell by bending the neck sideways. There are two families, found primarily in the southern hemisphere, in freshwater habitats.

pleurodont Applied to the condition in which the teeth are attached by one side to the inner surface of the bones of the jaw. *Compare* ACRODONT; THECODONT.

Pleuronectidae (right-eye flounder; subclass *Actinopterygii, order *Pleuronectiformes) A large family of marine flatfish, comprising the true flounders which have the eyes on the right side, the *dorsal fin starting above the upper eye, a fairly large mouth with distinct teeth, and the right *pectoral and *pelvic fins mostly larger than the left ones. The family includes many commercially important species, e.g. *Reinhardtius hippoglossoides* (Greenland halibut), 1 m, which has been heavily overfished in the N. Atlantic (catches are now restricted); *Platichthys flesus* (flounder), 50 cm; and *Pseudopleuronectes americanus* (Atlantic winter flounder), 60 cm. There are about 100 species, distributed in the Atlantic and Pacific.

Pleuronectiformes (class *Osteichthyes, subclass *Actinopterygii) An order of marine flatfish that have a highly compressed and asymmetrical body. The eyes are on the right or left side of the body. The newly hatched larvae have a normal, symmetrical fish shape, but begin to assume the peculiar flatfish features when they are about 2 cm long. The order comprises about six families, with some 520 species.

Pleuronematina (class *Ciliatea, order *Hymenostomatida) A suborder of ciliate *Protozoa in which body *cilia usually are sparse. Usually there is a single, prominent, caudal cilium. A conspicuous row of oral cilia, constituting an 'undulating membrane', extends from one side of the cell. Pleuronematina are found in marine or freshwater aquatic environments.

pleurothetic Applied to a resting position common in members of the *Bivalvia, in which individuals settle on to the substrate on their sides.

Pleurotremata An alternative, older name for the sharks, but excluding the rays (*Hypotremata).

pleurotremate Applied to the position of gill openings that are on the sides of the body anterior to the pectoral fins, as in sharks. *Compare* HYPOTREMATE.

plicate Folded or wrinkled.

Pliocene The last of the *Neogene epochs, about 5.3–1.81 Ma ago.

(((⊕))) SEE WEB LINKS
• Detailed description of life in the Pliocene.

Pliosauroidea *See* PLESIOSAURIA.

Ploceidae (weavers, sparrows, bishops, whydahs; class *Aves, order *Passeriformes) A family of small, stocky birds that have short, stout bills. The weavers (e.g. *Ploceus*, of which there are about 56 species, many kept as cage birds) are mainly black with yellow, red, or brown. Sparrows are mainly brown. Some members of the genus *Euplectes* (bishops and whydahs) have long tails. Many ploceids are gregarious, inhabit forest, grassland, desert, urban areas, and cultivated land, and feed on seeds and insects. Many build elaborate woven nests, suspended from trees, with a long entrance tunnel. Others build untidy nests in trees, grass, and buildings. Some are parasitic and most are colonial. *Passer domesticus* (house sparrow), one of 18 species in its genus, is particularly associated with humans. Rock sparrows (five species of *Petronia*) inhabit open, rocky country, bush, and forest. There are 17–19 genera, comprising 145 species, many kept as cage birds (e.g. *Euplectes*), found in Europe, Africa, Asia, Australasia, and the Pacific islands. *Quelea quelea* (red-billed quelea) is the most numerous bird in the world, numbering thousands of millions.

***Ploceus* (weavers)** *See* PLOCEIDAE.

Plotosidae (striped catfish; class *Osteichthyes, subclass *Actinopterygii) A fairly small family of marine or freshwater catfishes that have an eel-like body, tapering towards the tail, four pairs of *barbels around the mouth, a short *dorsal fin, and *pectoral fins which each bear a sharp, poisonous spine that can inflict painful wounds. There are about 30 species. The marine species are found in Indo-Pacific waters and especially around New Guinea and Australia, the freshwater species in Australian rivers.

plovers *See* CHARADRIIDAE.

plume moth *See* PTEROPHORIDAE.

plumose Branched.

plumose hair The branched type of body hair found in bees, which is adapted to pick up and retain pollen grains when a bee brushes against the anthers (male parts) of flowers.

plush-capped finch (*Catamblyrhynchus diadema*) *See* CATAMBLYRHYNCHIDAE.

***Pluvialis* (plovers)** *See* CHARADRIIDAE.

Pneumoridae (order *Orthoptera, suborder *Caelifera) Family of grasshoppers in which the abdomen of the male is inflated, and bears lateral, stridulatory ridges. The hind legs are relatively small. There are about 20 species, which occur in S. Africa.

PNS *See* PERIPHERAL NERVOUS SYSTEM.

poachers *See* AGONIDAE.

pochard (*Aythya*) *See* ANATIDAE.

pocket gopher *See* GEOMYIDAE.

pocket mice *See* HETEROMYIDAE.

Podarcis (wall lizards) *See* LACERTIDAE.

Podarcis muralis (wall lizard) *See* LACERTIDAE.

Podarcis sicula (Italian wall lizard, ruin lizard) *See* LACERTIDAE.

Podargidae (frogmouths; class *Aves, order *Caprimulgiformes) A family of brown, grey, and black, cryptically coloured birds which have broad, flat, hooked bills with a wide gape and *basal bristles. They have medium-length, rounded wings, and a long tail, are nocturnal, inhabit forest and bush, and feed on insects and small mammals caught on the ground. They nest in tree forks and the 10 species of *Batrachostomus* make their nests from a pad of down secured to a horizontal branch. There are two genera, comprising 13 species, found from south-east Asia to Australia.

Podiceps (grebes) *See* PODICIPEDIDAE.

Podicipedidae (grebes; class *Aves, order *Podicipediformes) A family of medium to large diving birds which are grey or dark brown above, and many of which are white below. Many have ear tufts and crests. The genus *Podiceps*, the largest in the family (11 species), is typical. Their legs are positioned at the rear of the body, and their feet are lobed and partially webbed. Their wings are small, some birds being practically flightless, and many have white secondaries. They have short to long, pointed bills, and most have long necks. They inhabit freshwater lakes in summer (many Central and S. American Podiceps species are confined to isolated lakes and are endangered), and some move to coastal waters in winter. They feed on fish and invertebrates, and nest on a platform of vegetation in the water. There are five genera, comprising 20 species, found in Europe, Africa, Asia, Australasia, N., Central, and S. America.

Podicipediformes (grebes; class *Aves) An order which comprises one family, *Podicipedidae.

Podopteryx mirabilis The species that is the probable ancestor of the *Pterosauria. Unlike them it was a glider, relying on the large membrane stretched between its hind legs and tail. It is known from the Early *Triassic of Soviet Kirgizstan. The name means 'wonderful foot-wing', derived from the Greek *pod-*, 'foot', and *pteron*, 'wing', and the Latin *mirabilis*, 'wonderful'.

Poeciliidae (live-bearer, guppy; subclass *Actinopterygii, order *Atheriniformes) A large family of small, freshwater and brackish-water fish that generally inhabit freshwater streams, although several species can be found in saline coastal marshes. They all have an elongate body form, a single, soft-rayed *dorsal fin on the middle of the back, with the *anal fin opposite in mid-ventral position. The tail fin is always rounded, and the *pelvic fin often small. There are about 138 species, occurring in Central and S. America.

Poecilosclerida (class *Demospongiae, subclass *Ceractinomorpha) An order of sponges which have a skeleton of *spicules, or organic fibres, or both. The dermis is often very specialized, as are the spicules and their arrangement. Many spicules are cemented together. This is a very large order which first appeared in the *Cambrian.

Pogonophora (beard worms; family *Siboglinidae) A group comprising deep-sea worms, first encountered in early *Cambrian rocks but discovered only in this century. Beard worms are *coelomate, with a superficial resemblance to *Polychaeta, and live at great depths inside chitinous tubes they secrete for themselves in soft *substrates. The body is divided into three parts. The anterior is crowned with *tentacles. Part of the body is segmented and the worms have *chaetae. Their most remarkable feature is the complete absence of a gut. This has led to difficulties with classification because it is impossible to distinguish the ventral and dorsal surfaces, and has also led to several theories concerning the method of feeding. The animals obtain nourishment by absorbing nutrients released by symbiotic bacteria. Formerly ranked as a phylum, the pogonophorans are now included in the family Siboglinidae.

poikilotherm (exotherm) An organism whose body temperature varies according to the temperature of its surroundings. Fish are poikilotherms. *Compare* ECTOTHERM.

point mutation A *mutation that can be mapped to one specific *locus. It is caused by the substitution of one *nucleotide for another.

poison beetle *See* CHRYSOMELIDAE.

Poisson distribution The basis of a method whereby the distribution of a particular attribute in a population can be calculated from its mean occurrence in a random sample of the population, provided that the population is large and the probability that the attribute will occur is less than 0.1. For a given mean the distribution can be calculated, giving the probabilities that a sample will contain 0, 1, 2, 3,...examples of the particular attribute. The distribution is named after the French mathematician S. D. Poisson (1781–1840).

(((⊕))) SEE WEB LINKS
• Explanation of the Poisson distribution.

polar body The minute cell produced during the development of an *oocyte (either primary or secondary). It contains one of the nuclei derived from the first or second meiotic divisions, but very little *cytoplasm.

polar filament *See* CNIDOSPORA.

polarity The direction of evolutionary change. The polarity of different states of a *character means whether these are *primitive or derived.

polar molecule A molecule in which, though it does not carry a net electric charge, the electrons are unequally shared between the nuclei. In the water molecule, for example, the pull of the oxygen nucleus on the shared electrons is greater than the pull of the hydrogen nuclei. As a result the oxygen end of the molecule is slightly negatively charged, and the hydrogen ends of the molecule are each slightly positively charged. The molecule is said to have a dipole moment and can attract other molecules with a dipole moment.

pole capsule In *Sporozoa, bodies located at one end of the cell and containing a long filament by means of which the spore can attach itself to the host.

Polioptilidae (gnatcatchers, gnatwrens; class *Aves, order *Passeriformes) A family of small, active birds; gnatcatchers are mainly grey and white, gnatwrens are mainly brown. Both have long, slender bills and long tails, inhabit scrub, often near water, feed on insects, and nest in trees. There are three genera, with 13 species, found in N., Central, and S. America, and the W. Indies.

***Pollachius virens* (saithe)** *See* DEMERSAL FISH.

pollen beetle *See* NITIDULIDAE.

pollex The thumb. In the *pentadactyl fore limb, it is the digit on the side of the *radius, often shorter than the other digits.

polyandry The mating of a female with more than one male at one time (usually taken to be during the course of a single breeding season). Simultaneous polyandry occurs where the female mates with more than one male to produce a clutch of eggs or brood of young bearing genes from each male. Successive polyandry occurs where the female mates with one male, lays eggs or gives birth to the young, but plays little part in their parental care (they are usually cared for by this first male), instead moving away and mating with another male. The existence of simultaneous polyandry is usually difficult to prove, and polyandry appears to be mainly of the successive type. Polyandry is generally much rarer than

p

*polygyny, presumably because the mother is usually not sexually receptive following successful mating and since she bears the young she is more involved with their care and less able to seek another mate. *See also* PROMISCUITY.

polycentromic *See* HOLOCENTRIC.

Polychaeta (bristleworms; phylum *Annelida) A class of worms which possess distinct *metameric segmentation. They have numerous *chaetae borne in sheaths on the *parapodia on each body segment. Eyes and *tentacles may be present. Most polychaetes are *dioecious but a few species are *hermaphrodites. Most are marine, although some occur in fresh water and on land. The group first appeared in the *Cambrian (polychaetes are the only annelids to have a fossil record) and it contains more than 5000 species.

polychronic species *See* POLYTOPISM.

Polycladida (class *Turbellaria, subclass *Archoophora) An order of worms which possess a *pharynx. The larvae are free-swimming.

polyembryony The production of two or more individuals from a single egg by the division of the *embryo at an early stage of development. Some parasitic wasps may produce hundreds of offspring from a single fertilized egg.

polyethism Functional specialization in different members of a colony of social insects, which leads to a division of labour within the colony. The various functions may be carried out by individuals of different morphology (caste polyethism) or of different ages (age polyethism).

polygamy In animals, a pattern of mating in which an individual has more than one sexual partner. *See also* POLYANDRY; POLYGYNY; PROMISCUITY.

polygene One of a group of *genes that together control a continuously variable character such as height or weight. Indi-

vidually each gene has very little effect on the resulting *phenotype which instead requires the interaction of many genes. Before genes had been described, a multifactorial hypothesis had been constructed to explain the absence of clear-cut segregation into readily recognizable classes that show typical Mendelian ratios, as occurs with quantitative characters. In this case a group of interacting factors corresponds to a group of interacting genes.

polygenic *See* MULTIFACTORIAL.

polygenic character A quantitatively variable character (*phenotype) which is dependent upon the interaction of a number of genes.

polygyne Applied to an ant, bee, or wasp (*Apocrita) nest that contains more than one *queen. *Compare* MONOGYNE; *see also* GYNE.

polygyny In animals, a pattern of mating in which a male has more than one female partner. *Compare* POLYANDRY; *see also* POLYGAMY; PROMISCUITY.

polyhedrosis Disease of insects caused by certain DNA-containing *viruses. The larval stages of flies, butterflies, moths, and sawflies may be affected, and the disease is frequently fatal.

polylectic Applied to bee species which collect pollen from a wide range of flowering plants.

polymer A compound that consists of many repeating structural units.

polymerase An *enzyme that catalyses the replication and repair of *nucleic acids.

polymerase chain reaction (PCR) An enzymatic technique for producing a large number of copies of a specified type of DNA (*deoxyribonucleic acid) segment. PCR may be specific to one particular DNA sequence within the *genome, or more general, as in *random amplified polymorphic DNA.

Polymixia nobilis (beardfish) *See* POLY-
MIXIIDAE.

Polymixiidae (beardfish; **superorder
Paracanthopterygii, order *Polymixiiformes)** A small family of deep-water fish
that have a deep body, a pair of long chin
*barbels, a few spines preceding the *dorsal
and *anal fins, and a forked tail fin. *Polymixia nobilis* (beardfish), 30 cm, is probably
widely distributed and may occur world-
wide in tropical seas at depths of more than
200 m. There are two species.

Polymixiiformes (**subclass *Actinopterygii, superorder *Paracanthopterygii)** An order of bony fish that includes
the family *Polymixiidae and fossil relatives.

polymorph *See* LEUCOCYTE.

polymorphic Occurring in several different forms.

polymorphism 1. In genetics, the existence of two or more forms that are genetically distinct from one another but
contained within the same interbreeding
population. The polymorphism may be
transient or it may persist over many generations, when it is said to be balanced. Classical examples of polymorphisms are the
presence or absence of banding in **Cepaea*
snails, the number of spots on the wings of
ladybirds, and eye colour in humans. All
these are visible polymorphisms that can
readily be seen in nature. Some polymorphisms, however, are cryptic and require
biochemical techniques to identify phenotypic differences. Such techniques include gel *electrophoresis of *enzymes and
other *proteins, and the fragmentation of
the *DNA molecule by restriction enzymes
(which allows the sequencing of *nucleotides), both of which operate nearer to the
level of the *genotype. **2.** In social insects,
the presence of different castes within the
same sex. *See also* POLYTYPISM.

Polynemidae (**threadfin; subclass
*Actinopterygii, order *Perciformes)** A
family of marine, coastal fish that have an
elongate body, large eyes, two separate *dor-

sal fins, a forked tail, and, typically, *pectoral
fins that are divided into two sections, the
foremost section consisting of a number of
filamentous *fin rays. The threadfins are a
commercially important group of fish. They
are distributed world-wide in tropical and
subtropical seas. There are about 35 species.

Polyodontidae (**paddlefish; subclass
*Actinopterygii, order *Acipenseri-
formes)** A small family of freshwater fish
that have many fossil relatives. They are very
elongate, with a long, paddle-shaped snout,
probably used for feeding on minute organisms sifted from the water. The skin is naked
apart from some *ganoid scales near the
tail. The American species *Polyodon spatula*
(1.8 m) is found throughout the Mississippi
system and is locally of some commercial
importance. Its only relative, *Psephurus gladius*, occurs in the Yangtze river, China.

polyoestrus In female *Mammalia, the
condition in which more than one *oestrus
cycle occurs in a breeding season or in a
year.

Polyopisthocotylea (**Polypisthcoty-
lea; phylum *Platyhelminthes, class
*Monogenea)** An order of parasitic worms
in which the mouth, *pharynx, and intestine
are present. They are bloodsuckers.

polyp The soft-bodied, usually sedentary
form of *Cnidaria, consisting of a cylindrical trunk which is fixed at one end, with
the mouth surrounded by *tentacles at the
other end. In *Siphonophora the polyp
has been modified for a *pelagic, colonial
existence.

polypeptide A linear polymer composed
of 10 or more *amino acids linked by *peptide bonds.

Polyphagidae (**subclass *Pterygota,
order *Blattodea)** Cosmopolitan family
of mostly small cockroaches in which the
females are often wingless, while the males
fly readily. Some species are burrowers in
desert regions (sand cockroaches), and one
very small species occurs in the nests of leaf-
cutter ants.

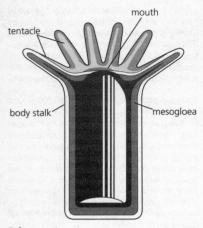

Polyp

polyphagous Applied to an organism that feeds on many types of plant or prey. *Compare* MONOPHAGOUS; OLIGOPHAGOUS.

polyphyly The occurrence in taxa of members that have descended via different ancestral lineages. True polyphyly has traditionally been distinguished from errors of classification, especially at the higher taxonomic levels, where organisms, as a result of *convergent or *parallel evolution, have been placed wrongly in the same natural group; but modern phyletic taxonomists would hold that any *taxon found to be polyphyletic is unnatural, and so an 'error', and must be disbanded.

polyphyodont Applied to a type of dentition in which the teeth are constantly replaced. Fishes, amphibians, and most reptiles have polyphyodont dentition. *Compare* DIPHYODONT, MONOPHYODONT.

Polypisthocotylea *See* POLYOPISTHOCOTYLEA.

Polyplacophora (Loricata; chitons, coat-of-mail shells; phylum *Mollusca, class *Amphineura)** A subclass of amphineurans in which seven or eight dorsal plates are generally composed of calcium carbonate and are enclosed by a spiculate *girdle. The plates articulate with one another and overlap to a varying extent. The anterior and posterior plates differ from the others, the anterior plate often being ribbed. The head is often reduced. It contains a central mouth with a *radula which most chitons use to scrape algae from rocks. The gills, which may be numerous, lie in a groove between the ventral foot and the girdle. The anatomy is typically molluscan, and the primitive nervous system has ganglionic swellings. Polyplacophora are entirely marine, often occurring interstitially beneath rocks, and first appeared in the Late *Cambrian.

polyploidy The condition in which an individual possesses one or more sets of *homologous chromosomes in excess of the normal two sets found in diploid organisms. It is caused by the replication within a *nucleus of complete *chromosome sets without subsequent nuclear division. Examples are triploidy ($3n$), tetraploidy ($4n$), hexaploidy ($6n$), and octoploidy ($8n$).

polypoid Pertaining to or resembling a *polyp.

polyprotodont In *Marsupialia, applied to the condition in which there are three or more pairs of *incisors in each jaw.

Polypteridae (bichir; subclass *Brachiopterygii, order *Polypteriformes) A small family of freshwater fish that have an elongate body, the skin covered with *ganoid scales and, typically, a *dorsal fin consisting of a series of separate *finlets each with its own stiff ray. The *pectoral fins consist of a fleshy stalk bearing a fan-like fin. Bichirs can use the modified swim-bladder as a lung. *Polypterus bichir* (Nile bichir), 70 cm, has about 15 dorsal finlets. It is a carnivorous fish often found along the weedy edges of lakes and rivers. In addition to the 11 extant species a number of fossil species are known. They occur in Africa.

Polypteriformes (class *Osteichthyes, subclass *Brachiopterygii) An order of bony fish that includes only the family *Polypteridae.

Polypterus bichir (Nile bichir) *See* POLYPTERIDAE; BRACHIOPTERYGII.

polyribosome *See* RIBOSOME.

polysaccharide A linear or branched polymer composed of 10 or more monosaccharides linked by glycosidic bonds.

polysome A group of *ribosomes that are attached to a single strand of *messenger-RNA. Apparently each ribosome is at a different stage in the synthesis of the same *polypeptide chain. Accordingly, the number of ribosomes in the polysome varies with the length of the message to be transcribed.

polytene A giant, cable-like *chromosome which consists of many identical chromosomes closely associated along their lengths. It is formed by repeated duplication without division. An example is the *salivary-gland chromosome of larval dipteran insects.

polythetic Using several or all possible criteria or attributes as the basis for each subdivision of a sample population or for agglomeration of individuals or groups. *Compare* MONOTHETIC; *see also* AGGLOMERATIVE METHOD; DIVISIVE METHOD; HIERARCHICAL AND NON-HIERARCHICAL CLASSIFICATION METHODS.

Polythoridae (damselflies; order *Odonata, suborder *Zygoptera) Family of damselflies which frequent streams. With the *Epallaginidae, they are the only odonates in which the larvae have large abdominal gills as well as the three *tracheal gills. There are more than 60 described extant species found in the New World tropics.

polytopic evolution *See* POLYTOPISM.

polytopic species *See* POLYTOPISM.

polytopism (polytopic evolution) A taxonomic anomaly in which a (usually infraspecific) *taxon occurs in more than one place. It may indicate that a more *derived *subspecies or *species has evolved in an intervening area, or that *parallel evolution has occurred to make the two or more separated populations appear alike.

polytypic Of a *species, divided into subspecies; varying geographically. *See also* POLYTYPISM. *Compare* MONOTYPIC.

polytypism The occurrence of *phenotypic variations between populations or groups of a *species that are geographically distinct. It is contrasted with *polymorphism, which is variation within a population or group. A species with systematic geographical variation (*subspecies or *clines) is said to be polytypic.

Pomacanthidae *See* CHAETODONTIDAE.

Pomacanthus imperator (emperor angelfish) *See* CHAETODONTIDAE.

Pomacentridae (damselfish, anemonefish; subclass *Actinopterygii, order *Perciformes) A very large family of mostly marine fish that have a relatively short but deep body, a small mouth, a single nostril on each side of the head, and an incomplete *lateral line. They are often very conspicuous because of their beautiful colour patterns. There are about 230 species, found mainly in the Indo-Pacific region.

Pomadasyidae (javelinfish, grunt; subclass *Actinopterygii, order *Perciformes) A large family of marine, perch-like fish that have large eyes, a truncated tail fin, a long *dorsal fin, and strong spines to the dorsal and *pelvic fins. There are about 175 species, distributed world-wide.

Pomatomidae (bluefish, tailor; subclass *Actinopterygii, order *Perciformes) A monospecific family (*Pomatomus saltator*) comprising an elongate and streamlined marine fish, 1.1 m long, that has a large mouth, short first *dorsal fin, and long-based second dorsal and *anal fins. It is a voracious predator, living in large schools in offshore waters, although it may come close to shore when following schools of prey fish. It is distributed world-wide in tropical waters.

Pomatomus saltator (bluefish, tailor) *See* POMATOMIDAE.

Pompilidae (caterpillar-hunting wasps, spider-hunting wasps; suborder *Apocrita, superfamily Pompiloidea) Family of large wasps (most 10–25 mm, but some up to 40 mm long) whose members provision their young with caught and paralysed spiders. They are long-legged, rarely fly, and run over the ground to catch their prey. Adults are usually dark-coloured, with dark wings which are not folded longitudinally at rest. They are distinguished from all other *fossorial wasps by their well-developed hind-wing venation, and prothoracic structures. Nests are cup-shaped mud structures in crevices of rocks or trees, and some may be subterranean. The host may be left in its own burrow. Some species construct the nest first, then provision it with a paralysed spider; others catch their prey before building the nest. A few pompilid species lay their own eggs on the paralysed host of another species before it is interred in the nest. *See* HUNTING WASP.

pond skater *See* GERRIDAE.

Ponerinae (suborder *Apocrita, family *Formicidae) Subfamily of ants, which have one petiolar node and a *sting. *Polymorphism is often weakly developed, the workers resembling the queen. Most are predators and scavengers. There are about 2000 species: the majority occur in the tropics, and nest in soil or rotten wood.

Pongidae The name given to the family grouping the gorilla (*Gorilla gorilla*), chimpanzee (*Pan troglodytes*), and bonobo or pygmy chimpanzee (*Pan paniscus*) of Africa and the orang-utan (*Pongo pygmaeus*) and sometimes the gibbons (*Hylobates*) of Asia. Most authorities now prefer to include the African apes, and probably the orang-utan as well, in the family *Hominidae, on the grounds that they are close to humans in phylogenetic terms, the chimpanzee being

the closest, followed by the gorilla. The former 'Pongidae' do not form a *monophyletic group.

Pongo pygmaeus (orang-utan) *See* HOMINIDAE; PONGIDAE.

pons varoli In the vertebrate brain, a transverse mass of nerve fibres connecting the two parts of the *cerebellum ventrally, located in front of the *medulla.

pony-fish *See* LEIOGNATHIDAE.

pooter A device that is used to collect and transfer small animals. It consists of a jar, sealed by a cover pierced by two tubes pointing in opposite directions. The operator sucks on the end of one tube (made flexible for convenience and the end inside the jar covered with muslin), drawing the specimen up the other (open) tube and into the jar.

population A group of organisms, all of the same *species, that occupies a particular area. The term is used of the number of individuals of a species within an *ecosystem, or (statistically) of any group of like individuals. *See also* ECOTYPE.

population dynamics The study of factors that influence the size, form, and fluctuations of individual species or genus populations. Emphasis is placed on change, energy flow, and nutrient cycling, with particular reference to homoeostatic controls. Key factors for study are those influencing natality, mortality, immigration, and emigration.

population ecology The study of the interaction of a particular *species or genus *population (or sometimes of a higher taxon) with its environment. *See also* AUTECOLOGY.

population eruption *See* POPULATION EXPLOSION.

population explosion (population eruption) A sudden rapid increase in size of the *population of a species or genus. The most violent explosions occur when a species is introduced into a new locality where it finds unexploited resources of suitable food, shelter, etc., and a lack of negative controls such as predators or parasites. *See* R-SELECTION.

population genetics The study of inherited variation in populations of organisms, and its modulation in time and space. Population genetics relates the heritable changes in populations to the underlying individual processes of inheritance and development. Such studies generally involve the estimation of *gene frequencies and the influences of *selection, *mutation, and *migration upon these frequencies in natural (and experimental) populations.

population size The number of individuals in a *population. *Compare* EFFECTIVE POPULATION SIZE.

Porcellionidae (woodlice; class *Malacostraca, order *Isopoda) Family of terrestrial isopods, including the cosmopolitan species *Porcellio scaber*, in which the *uniramous first antennae are very small and the flagella of the second antennae are two-segmented. Tracheal lungs are present on the borders of the *pleopods.

Porcellio scaber *See* PORCELLIONIDAE.

porcupine 1. (Old World) *See* HYSTRICIDAE. **2. (New World)** *See* ERETHIZONTIDAE.

porcupine fish *See* DIODONTIDAE.

porgy *See* SPARIDAE.

Porifera (Spongiaria; sponges) A phylum of aquatic *Animalia, most of which are marine. They lack definite tissues and organs, but have a filter-feeding system composed of flagellate cells, pores, and canals. Most are covered externally by either a *cortex or a dermis. The soft body may be supported by a skeleton composed of *collagen, with or without siliceous or calcareous *spicules, protein *spongin fibres, or a combination of the last two. The body shape is very variable, being influenced by the substratum and water currents (e.g. where the water movement is very strong, sponges often grow as rounded clumps, but in calm water they may assume branching shapes). Sponges have remarkable powers of regeneration, small, broken pieces being capable of growing into a whole animal. Larvae are motile and produced by sexual reproduction. The phylum ranges from the *Cambrian to the present.

porocyte In *Porifera, a cell that guards the *ostium. Each porocyte is tubular and extends from the external surface to the internal chamber of the sponge. The *lumen of the tube forms the ostium, and the outer end of the cell can be closed or opened by contraction.

porphin The parent compound of the *porphyrins: it consists of four pyrrole-like rings, linked into a ring system by four CH groups.

porphyrin A heterocyclic derivative of a *porphin, which is composed of a tetrapyrrole ring structure. As such it is capable of combining with a variety of metals and so forms part of the structure of many important biological molecules, including haemoproteins, chlorophyll, vitamin B_{12}, and cytochromes.

porpoise (*Phocoena phocoena*) *See* PHOCOENIDAE.

porrect Stretched out.

Port Jackson shark *See* HETERODONTIDAE.

Portuguese man-of-war (*Physalia physalis*) *See* NOMEIDAE; RHIZOPHYSALIINA.

Portunidae (swimming crabs; class *Malacostraca, order *Decapoda) Family of crustaceans in which the body is round, transversely oval, or square, with the anterior of the carapace not being narrowed. The fifth pair of legs is modified for swimming, being shaped as broad, flattened paddles which move in a figure-of-eight pattern. Some species use the modified legs for digging. The portunids are the most powerful and agile swimmers of all crustaceans and can swim forward, backward, and sideways with great rapidity. Males attend the females before the moulting period and carry them around beneath the body. The male releases the female to allow her to moult, and copulates with her soon after.

Porzana (crakes, rails) *See* RALLIDAE.

possum *See* PHALANGERIDAE.

postantennal organ More correctly known as the postantennal organ of Tömösvary, a structure found on the heads of *Collembola. These organs may vary from being a simple ring to a multituberculate sensory area.

postdisplacement An alteration in the *ontogeny of a descendant such that some developmental process begins later than in its ancestor, and may not have been completed by the time maturity is reached.

posterior chamber *See* AQUEOUS HUMOUR, VITREOUS HUMOUR.

postlarva 1. Stage in the development of a bony fish between the resorption (*see* RESORB) of the *yolk sac and the appearance of juvenile characteristics. **2.** The larval stage following the *nauplius stage in *Crustacea, and the one in which the full complement of trunk segments and appendages appears for the first time. The postlarvae of many crustacean groups are sufficiently distinct to have been given special names, e.g. megalops (crabs), and acanthosoma (sergestid prawns).

postnotum (postscutellum) In insects, a small *dorsal *sclerite posterior to the *notum.

postorbital bar In the skull of vertebrates, the bar of bone which forms the posterior margin of the *orbit.

post-partum (post-parturient) Applied to the period immediately following the birth of a mammal, but applied only to the mother.

postscutellum *See* POSTNOTUM.

postzygapophyses *See* VERTEBRA.

potamodromous Applied to fish that undertake regular migrations in large freshwater systems. *Compare* ANADROMOUS; CATADROMOUS. *See also* AMPHIDROMOUS; DIADROMOUS.

Potamotrygonidae (river stingray; subclass *Elasmobranchii, order *Rajiformes) A small family of freshwater stingrays in which the body has an almost circular profile ending in a fairly solid tail bearing a poisonous spine much feared by fishermen. These rays can be found partially buried in the mud of rivers. There are about 10 species, in S. America, W. Africa, and south-east Asia.

Potanichthys xingyiensis The earliest known flying fish (*potanos* is Greek for 'winged'), a fossil of which was found near Xingyi, in SW China. *P. xingyiensis* lived in the Yangtze Sea, part of the Palaeotethys Ocean, during the *Triassic period. It possessed enlarged *pectoral fins with which it could have glided and a deeply forked *caudal fin that would have generated enough power to propel the fish from the water. It was not ancestral to modern flying fishes (*Exocoetidae).

potato bug (*Stenopelmatus fuscus*) *See* STENOPELMATIDAE.

potoos *See* CAPRIMULGIFORMES.

Potoroidae (rat-kangaroos; order *Marsupialia (or *Diprotodontia), superfamily *Macropodoidea) A family of medium-sized Australian marsupials which, except for one genus (*Hypsiprymnodon*, musky rat-kangaroo), hop like small wallabies and true kangaroos (family *Macropodidae), although the fossil record shows the two families to be descended independently from quadrupedal ancestors.

potter wasps *See* EUMENIDAE.

potto (*Perodictus potto*) *See* LORIDAE.

pottoos *See* NYCTIBIIDAE.

powder-post beetle *See* LYCTIDAE.

praemorse (premorse) Having the end terminated abruptly.

praesaccus A small pouch, of uncertain function, in the gut of some colobids (*see* CERCOPITHECIDAE), at the entrance of the *saccus gastricus.

prairie dog (Cynomis) *See* SCIURIDAE.

pratincoles *See* GLAREOLIDAE.

praying mantis (Mantis religiosa) *See* MANTIDAE.

pre-adaptation Adaptation evolved in one *adaptive zone which, quite by chance, proves especially advantageous in an adjacent zone and so allows the organism to radiate into it. No selection for a future environment is implied. The concept is very similar to *exaptation, but is often thought to have teleological implications.

Precambrian A name that is now used only informally to describe the longest period of geological time, which began with the consolidation of the Earth's crust and ended with the beginning of the Cambrian Period 542 Ma ago. The Precambrian lasted approximately 4000 Ma; the rocks of this period of geological time are usually altered and few fossils with hard parts or skeletons have been found within them. Precambrian rocks outcrop extensively in shield areas such as northern Canada and the Baltic Sea. In modern usage the Precambrian is subdivided into the *Hadean, *Archaean, and *Proterozoic.

prechordal plate In a vertebrate *embryo, a thickened part of the *endoderm that is in contact with the *ectoderm and appears at the *rostral end of the *primitive streak.

precious coral (Corallicum) *See* SCLER-AXONIA.

precocial Applied to young mammals which are born with their eyes and ears open and are able to stand and walk, regulate their body temperature, and excrete without assistance. The young of grazing animals (e.g. cattle, sheep, and horses) are precocial. *Compare* ALTRICIAL.

precoracoid A bone of the *pectoral girdle, present as an element of the *coracoid in some *Amphibia and as a separate bone in many *Reptilia and some *Monotremata, but which is vestigial or absent as a distinct entity in most other *Mammalia.

predation An interaction between *species in which one organism (the predator) obtains energy (as food) by consuming, usually killing, another (the prey). Most typically, a predator is an animal that catches, kills, and eats its prey; but predation also includes feeding by insectivorous plants and grazing interactions (e.g. the complete consumption of unicellular phytoplankton by zooplankton). Predation is analogous to *parasitism in being both a means of feeding and a cause of immediate harm, but differs in that predators are often (but not always, e.g. piranha fish) larger than prey organisms and the prey is usually (but not always) wholly consumed. The distinction is not clear-cut, especially for insect populations. In detail, a continuous gradation of interactions is found from predation to parasitism.

predation analysis The mathematical modelling of predator–prey relationships (e.g. in the *Lotka–Volterra equations). Using these models, computer simulations of predator and prey behaviour can be made and tested against field results. This gives a useful insight into the development of community structure, and more particularly into the behaviour of pest organisms and other species that are important commercially or in conservation.

predation compensation The mechanisms by which a prey population may maintain a constant size despite large fluctuations in the amount of predation. Compensation occurs because intraspecific competition changes with varying predation. When predation increases, natural mortality (e.g. from lack of food) decreases and when predation decreases resources must be shared more thinly and natural mortality increases.

predator *See* PREDATION.

predisplacement An alteration in the *ontogeny of a descendant such that some developmental process begins earlier than in its ancestor, and so has progressed further by the time maturity is reached.

preen gland *See* PREENING.

preening *Grooming behaviour that is performed by birds for the maintenance of feathers. Preening cleans feathers, but also oils them with a substance secreted by a preen gland near the tail: the bird applies this substance with the tip of its bill. The oil renders the feathers supple and water-resistant, and ingested oil supplies the bird with some vitamin D.

preformation *See* EPIGENESIS.

prehensile Capable of grasping.

premaxilla The bone that forms the anterior part of the upper jaw. In *Mammalia it bears the *incisors; in *Aves it forms much of the upper bill.

premolar In *Mammalia, one of the deciduous *cheek teeth that are located between the *canines and the *molars.

premorse *See* PRAEMORSE.

pre-oral Relating to the region in front of the mouth.

Presbytis entellus (entellus langur) *See* CERCOPITHECIDAE.

pretarsus *See* ARANEAE.

prevernal In the early spring. The term is used with reference to the six-part division of the year used by ecologists, especially in relation to terrestrial and freshwater communities. *Compare* AESTIVAL; AUTUMNAL; HIBERNAL; SEROTINAL; VERNAL.

prey *See* PREDATION.

prezygapophyses *See* VERTEBRA.

Priacanthidae (big-eye, bullseye; subclass *Actinopterygii, order *Perciformes) A small family of deep-bodied marine fish that have very large eyes, an oblique mouth, and heavily spined fins. They tend to have a red colour and to become active at night. There are about 18 species, distributed world-wide.

Priapulida (priapus worms) A phylum of worm-like, marine, burrowing animals, known from *Cambrian times and ranging up to the present. Priapus worms have stout, cucumber-shaped bodies. The body cavity is a *haemocoel and there is a spined,

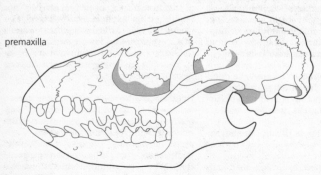

premaxilla

Premaxilla

*eversible *proboscis used to capture prey. The trunk is superficially segmented and bears peculiar posterior *tentacles that may have a respiratory function, but there are no vascular or respiratory systems.

priapum fish *See* PHALLOSTETHIDAE.

priapus worms *See* PRIAPULIDA.

prickleback *See* STICHAEIDAE.

pricklefish *See* STEPHANOBERYCIDAE.

primary One of the outer flight feathers of birds, which originate from the *manus. Most non-passerines have 10 primaries. *Ciconiidae (storks), *Phoenicopteridae (flamingos), and *Podicipedidae (grebes) have 12; *Ardeidae (herons), *Anatidae (ducks, geese, etc.), and some other groups have 11. Most *passerines have nine functional primaries, the tenth being reduced.

primary sexual character An organ that produces *gametes. Examples are the testes of male and the ovaries of female animals.

primary structure The basic structure of a *polypeptide chain. It relates solely to the type, number, and sequence of *amino acids or *nucleotides in the chain. *Compare* SECONDARY STRUCTURE; TERTIARY STRUCTURE; QUATERNARY STRUCTURE.

Primates (infraclass *Eutheria, cohort *Unguiculata) An order of mammals that have adapted to arboreal life and in some forms secondarily to life on the ground. Perhaps arising from this adaptation, eyes and ears are well developed, with binocular vision in many forms, and the sense of smell is less developed, the snout commonly being reduced. The brain case is large, the olfactory region often being reduced. The neck is very mobile, shoulders and hips permit large limb movement, and the limbs are wholly free from the body wall. The *clavicle is strong, the *scapula can be rotated, the *radius and *ulna are separate and jointed, and the *tibia and *fibula are separate (except in *Tarsiiformes). Digits are often long, with sensitive pads at their tips, and most forms have flat nails rather than claws. The *pollex is always opposable, the *hallux often so. The teeth are unspecialized; the diet usually omnivorous. In monkeys, apes, and humans the uterus is single and the foetal and maternal bloodstreams are separated only by the walls of the foetal vessels. Usually there is a single pectoral pair of mammae. The number of young produced is small and parental care continues long after birth. The Primates are divided into two suborders, *Prosimii and *Anthropoidea, or *Strepsirrhini and *Haplorrhini, but the (extinct) *Plesiadapiformes are nowadays commonly regarded as a third. The mammals diverged from shrew-like ancestors in the *Cretaceous. Both strepsirrhines and haplorrhines appeared in the *Eocene. The time at which the hominid line diverged from that of the African apes (gorilla and chimpanzee) is controversial: times of as much as 20 million and as little as 4.5 million years ago are suggested, but the latter time is now increasingly believed to be the more accurate.

primer A *substrate that is required for a polymerization reaction (e.g. the polymerization of DNA or an *enzyme) and that is structurally similar to the product of the reaction.

primer pheromones *See* PRIMING PHEROMONES.

priming pheromones (primer pheromones) *Pheromones that trigger events in the development of the animal. For example, male mice and rats release pheromones that prime the *endocrine system of females to prepare them for mating. In dairy cows a priming pheromone from the cervix of a cow in oestrus (*see* OESTRUS CYCLE) affects *ovulation in herd mates, tending to synchronize oestrus throughout the herd.

primitive Applied to a *character (as a synonym of '*plesiomorphic') or, occasionally, of a whole organism, that preserves the character states of an ancestral stage.

primitive streak In a bird or mammal *embryo, a longitudinal thickening that forms during *gastrulation.

primordium The early cells that serve as the precursors of an organ to which they later give rise by *mitosis during development.

Prinia (prinias) *See* SYLVIIDAE.

prion A proteinaceous infectious particle that is made entirely from abnormally folded *protein and that resists proteinases (*enzymes that break down proteins). Prions are believed to cause *bovine spongiform encephalopathy in cattle, scrapie in sheep and goats, transmissible mink encephalopathy in mink (*Mustelidae), chronic wasting disease in elk (*Cervus canadensis*) and mule deer (*Odocoileus hemionus*), feline spongiform encephalopathy in cats (*Felis* species), spongiform encephalopathy of the ostrich (*Struthio camelus*), and exotic ungulate encephalopathy in nyala (*Tragelaphus angasii*), oryx (*Oryx* species), and greater kudu (*T. strepsiceros*). In humans it causes Creutzfeldt–Jakob disease and kuru.

Prionace glauca (blue shark) *See* CARCHARHINIDAE.

Prionopidae *See* LANIIDAE.

prions (*Pachyptila*) *See* PROCELLARIIDAE.

Priscoan In former geologic time scales, the first period of the *Precambrian, covering the time from the formation of the Earth (about 4500 Ma ago) until the first appearance of life about 4000 Ma ago. The name has now been abandoned and this period now comprises the *Hadean (4567–3800 Ma ago).

Pristidae (sawfish; subclass *Elasmobranchii, order *Rajiformes) A small family of marine fish that, because of their elongate bodies, tend to resemble sharks, but the gill clefts are in a ventral position, as in other rays. Typically, members of the family have a long, flat snout with blunt and equal-sized teeth along the edges. They are common in all tropical seas and estuaries. There are about six species, distributed world-wide.

Pristiophoridae (sawshark; subclass *Elasmobranchii, order *Squaliformes) A small family of sharks in which the long snout (rostrum) is formed into a flat blade equipped with teeth of uneven size. Sawsharks differ from *Pristidae (sawfish) in having the gill slits on the side of the body and in possessing a pair of *barbels located half-way along the snout. Some species are quite edible. There are about four species, found in African and Indo-Pacific waters.

probable mutation effect The idea that, as most *mutations act to produce failures of development, a non-essential organ will eventually be reduced, simply by repeated mutations accumulating in a population. This idea has been used to explain the loss of hind limbs in whales, etc.

probe In genetics, a sample of radioactively labelled *nucleic acid that is used in *molecular hybridization to detect complementary sequences in the presence of a large amount of non-complementary DNA. The position of the probe may be detected by *autoradiography.

Proboscidea (cohort *Ferungulata, superorder *Paenungulata) An order that comprises the elephants and their extinct relatives, divided into four suborders: *Moeritherioidea, *Deinotherioidea, Gomphotherioidea, and *Mammutoidea. The mastodons are included in the suborder Mammutoidea. The suborder Gomphotherioidea includes two families, *Gomphotheriidae and *Elephantidae, the mammoths, modern elephants, and their immediate ancestors being included in the latter. The order was formerly highly successful and occupied the Americas, Eurasia, and Africa. Probiscideans tend toward large size. Since the late *Miocene most have possessed a long trunk: this is developed from the nose and upper lip, is operated by a complex musculature, is sensitive to chemical and tactile stimuli, and is used to obtain food and water and in making sounds (which are important in social organization). Teeth are reduced in number, adults having three *molars in each side of each jaw; these are used one at a time, old teeth being shed and replaced by those behind. The upper *incisors are enlarged to form tusks. The jaw muscles are large, and the skull short and high. The vertebrae and up to 20 ribs carry the weight of the abdomen, which is balanced on the fore limbs by the weight of the head, the hind limbs providing propulsion. The brain is well developed. Parental care of the young is prolonged, and social organization is complex.

proboscis A tubular protrusion from the anterior of an animal.

proboscis monkey (*Nasalis larvatus*) *See* CERCOPITHECIDAE.

proboscis worms *See* NEMERTINI.

Procaviidae (hyraxes) *See* HYRACOIDEA.

Procellariidae (fulmars, prions, petrels, shearwaters; class *Aves, order *Procellariiformes) A family of medium to large, white, grey, brown, or black seabirds which have long, pointed wings and short tails. Their bills are short and heavy to long and slender, and have tubular nostrils, united on the top. Their legs are short to medium, their feet webbed. They are migratory, *pelagic, and gregarious, and feed from the sea surface on fish, crustaceans, and fish offal. The 16 or 17 species of shearwaters (*Puffinus*) breed on islands in most oceans and many migrate between the southern and northern oceans. Prions (three to six species of *Pachyptila*) are oceanic and feed on zooplankton. Members of the family nest in burrows and rock crevices on cliffs and islands. Fulmars (*Fulmarus*) often follow ships. *Daption capense* (Cape petrel or Cape pigeon) nests on the ground in the Antarctic and ranges throughout the southern oceans. *Pterodroma* petrels (about 25 species) breed on the ground or in burrows on islands in the southern oceans. There are 12 genera in the family, comprising about 60 species, found in all oceans.

Procellariiformes (albatrosses, diving petrels, petrels, shearwaters, storm petrels; class *Aves) An order of small to large sea birds that have small to large bills with horny plates and tubular nostrils. They secrete oil which can be regurgitated in defence and which gives them a musty smell. They are *pelagic, marine birds, feeding on fish, plankton, crustaceans, and other invertebrates, and nesting on isolated islands and cliffs. There are four families, found world-wide.

prochoanitic Applied to a *septal neck that points forwards.

procoelous Applied to vertebrae in which there is a concavity on the anterior surface.

Procolophonia (subclass *Anapsida, order *Cotylosauria) A family of relatively advanced cotylosaurs (stem reptiles) that are distinguished from the lower types by their shorter jaws and better jaw movement. They lived in the *Triassic and *Permian.

proctodeum In a vertebrate *embryo, a depression in the *ectoderm at the tail end that develops into the *anus and anal canal.

Proctotrupidae (suborder *Apocrita, superfamily Proctotrupoidea) Small family of moderately common, shiny black insects which have characteristic wing venation. The fore wing has a costal cell with a large *stigma and a very small marginal cell. The antennae are 13-segmented, and arise from the middle of the face. Adults are 3–12 mm long, and have spindle-shaped abdomens on short *petioles. Little is known about the larval stages of some species, but some parasitize the larvae of *Coleoptera.

procumbent Projecting forward, more or less horizontally (e.g. procumbent *incisors).

procuticle *See* EXOSKELETON.

Procyonidae (raccoons; suborder *Fissipedia (or *Caniformia), superfamily *Canoidea) A family that includes the raccoon (*Procyon lotor*), coati, and perhaps the red panda. They are relatively short-legged, mainly arboreal mammals in which the gait is *plantigrade or semi-plantigrade, the claws are partly retractile or non-retractile, the *canine teeth are long and rectangular in cross-section, the *carnassials are developed only weakly, the auditory *bulla is well rounded, and the paroccipital process is prominent. The diet is omnivorous. The Procyonidae broke away from the main canid line in the early *Oligocene and were present in both Europe and N. America at that time. Their fossil record in Asia begins in the *Miocene. There is currently dispute as to whether *Ailurus fulgens* (red panda) should be referred to this family or to a separate family, *Ailuridae. If the red panda is included, there are seven genera, comprising about 19 species, found in N. and S. America and parts of Asia.

***Procyon lotor* (raccoon)** *See* PROCYONIDAE.

production The total mass of organic food that is manufactured in an *ecosystem during a certain period of time. It is the net yield of the producers and consumers and determines the amount of living matter in the ecosystem.

pro-enzyme *See* ZYMOGEN.

***Proganochelys quenstedii* (*Triassochelys quenstedii*)** The oldest known turtle, described in 1887 from the *Triassic Stubensandstein of Württemberg, Germany. It had teeth and stout ribs on the neck vertebrae.

progenesis 1. A form of *paedomorphosis in which development is cut short by the early onset of maturity and the adult descendant exactly resembles a juvenile stage of its ancestor. **2.** The early stages of *ontogeny (less common usage).

progesterone A *steroid hormone that is secreted mainly by the *corpus luteum. It is responsible for the maintenance of pregnancy.

proglottis (*pl.* proglottides) In *Cestoda, one of the segments of which the animal consists. Each proglottis is essentially a reproductive sac with both male and female organs. The posterior segments, containing fertilized eggs, break off and are passed out with the host's faeces. Proglottides are not the product of *metameric segmentation, but are self-contained, reproductive units (each metamere works in co-operation with all others making them all functionally interdependent).

prognathous Having a head that is approximately horizontal, the mouth being at the front.

progressive evolution A steady, long-term improvement of evolutionary *grade, which has allowed plants and animals to become ever more independent of the aquatic environment in which they first evolved.

progressive feeding The provision of larval food by female wasps and bees at regular intervals throughout the period of larval growth. It is practised by all the social vespid wasps and the honey-bees (*Apis* species). It is uncommon in solitary (non-social) species, though it is found in some sphecid and nyssonid wasps.

***Prohylobates* *See* CERCOPITHECOIDEA.

pro-insulin The inactive precursor of *insulin which becomes converted to the active form through the hydrolytic removal of a *peptide composed of 33 *amino acids.

prokaryote A single-celled organism in which the *cell lacks a true *nucleus and the DNA is present as a loop in the *cytoplasm rather than as *chromosomes bounded by a nuclear membrane. The prokaryotes comprise the *Bacteria, which appear in the fossil record more than 3000 Ma ago and are the oldest forms of life known, and the cyanobacteria (also Bacteria), known from slightly younger rocks.

prolactin (lactogenic hormone) A protein *hormone, secreted by the adenohypophysis (anterior *pituitary), that initiates lactation in mammals and stimulates the production of *progesterone by the *corpus luteum.

prolarva In fish, the stage of development between hatching and resorption of the *yolk sac.

prolegs (false legs) Unjointed, fleshy protuberances found on the abdomen of certain insect larvae, in contrast with true legs, which are jointed and found on the *thorax. In *Lepidoptera, prolegs, which usually bear sclerotized hooks called 'crochets', are usually found on segments 3–6 and 10, but some or all may be absent.

prolicide The killing of offspring.

proline A heterocyclic, non-polar *imino acid that is present in all *proteins studied to date.

Promeropidae (sugarbirds; class *Aves, order *Passeriformes) A family of fairly large, brown birds which have long, *decurved bills, and long, slender, elaborate tails. They inhabit mainly *Protea* bushes, feed on insects and nectar, and build an open cup nest in bushes. There is one genus, *Promeropus*, with two species, found in S. Africa.

***Promeropus* (sugarbirds)** See PROMEROP-IDAE.

Promicrops lanceolatus (Queensland grouper) See SERRANIDAE.

promiscuity A form of *polygyny or *polyandry in which a member of one sex mates with more than one member of the other sex, but each relationship is an ephemeral one and terminates after mating, whereas in the other cases a bond of some kind forms.

promoter A *nucleotide sequence within an *operon, lying between the *operator and the *structural gene or genes, that serves as a recognition site and point of attachment for the *RNA *polymerase. It is the starting-point for *transcription of the structural gene or genes in the operon, but is not itself transcribed.

pronephros In the development of an *amniote *embryo, the first part of the *kidney to appear. In *anamniotes it becomes the functional larval kidney.

pronghorn (*Antilocapra americana*) See ANTILOCAPRIDAE.

pronking Jumping high into the air several times in succession while running, a behaviour practised by springbok (*Antidorcas marsupialis*) on the African grasslands, perhaps as a warning to a predator that its presence has been noticed and it should therefore abandon its pursuit since the springbok will escape. See also STOTTING.

pronotum The dorsal, sclerotized cuticle of the first thoracic segment of an insect. See also PROTHORAX.

pro-oestrus See OESTRUS CYCLE.

propaganda pheromone A *pheromone produced by the greatly enlarged *Dufour's gland of certain *species of slave-making ants (see DULOSIS) that closely resembles secretions produced by the host species, thus facilitating entry to the host nest.

propatagium The part of the *patagium that lies to the anterior side of the fore limbs.

Prophalanopsidae (Haglidae; hump-winged crickets; order *Orthoptera, suborder *Ensifera) Family of primitive ensiferans, which includes only three living species.

prophase The first phase of *mitosis and *meiosis, divided into prophase I and prophase II, during which the *chromosomes become visible within the *nucleus, coiling up to produce a series of compact spirals. In the first division of meiosis they also undergo pairing during this phase, the phase then being divided into five successive stages: leptotene, zygotene, pachytene, diplotene, and diakinesis.

Propithecus (sifaka) See INDRIDAE.

propleurum (*pl.* **propleura**) The lateral, sclerotized cuticle of the first thoracic segment (*prothorax) of an insect. See also METATHORAX.

propodeum The structure formed by the incorporation of the first abdominal segment with the *thorax in members of the

*Apocrita. The propodeum always has a pair of large *spiracles, and there is almost always a constriction (of varying strength) between it and the rest of the abdomen.

propodium *See* APPENDICULAR SKELETON.

propolis Resins of plant origin collected by honey-bee (*Apis* species) workers and used to seal gaps in the nest or hive, or to reduce the nest entrance to a suitable size. Sometimes it is also used to entomb large trespassers into the nest (e.g. mice) which have been killed by workers acting as guard bees.

proprioceptor 1. A receptor (i.e. sense organ) that detects pressure, position, or movement. **2.** More generally, a receptor which is sensitive to bodily changes that are not caused by substances taken into the respiratory tract or *alimentary canal. *Compare* ENTEROCEPTOR; EXTEROCEPTOR.

Prosauropoda (subclass *Archosauria, order *Saurischia) A suborder of primitive, partly bipedal, saurischian dinosaurs, largely restricted to the Late *Triassic. They were heavily built herbivores (*Plateosaurus gracilis* grew to approximately 6–7 m in length and weighed two tonnes). Current theory suggests that the prosauropods were a side-branch on the sauropod (*Sauropoda) family tree, and not ancestral to later forms.

Proscopiidae (order *Orthoptera, suborder *Caelifera) A family of extremely elongate, stick-like grasshoppers, which have conical heads. About 100 species are known, all endemic to S. America.

prosector A person who prepares bodies for dissection.

prosencephalon In *Craniata, the anterior parts of the brain (forebrain), *telencephalon and *diencephalon. *Compare* RHOMBENCEPHALON.

prosepta In a *corallum, the first *septa to be formed; they are separated by *metasepta.

Prosimii (cohort *Unguiculata, order *Primates) In many classifications the primates are grouped into two suborders: *Anthropoidea, which includes the Old and New World monkeys, apes, and humans; and the Prosimii, comprising the more primitive lemurs, lorises, and tarsiers. Prosimii, therefore, is sometimes used for a suborder to include the four infra-orders: *Plesiadapiformes, of extinct primates; *Lemuriformes, of the lemurs and like forms; *Lorisiformes, of the lorises and potto; and *Tarsiiformes of the tarsiers. In modern classifications the tarsiers are included with the anthropoids in the suborder *Haplorrhini and the remaining living primates are then placed in the suborder *Strepsirrhini, the Plesiadapiformes being placed in a third suborder.

prosobranch gastropods *See* PROSOBRANCHIA.

Prosobranchia (prosobranch gastropods; phylum *Mollusca, class *Gastropoda) A subclass of gastropods which have an exterior *mantle cavity, gills, and *anus, and have therefore undergone *torsion. Commonly, they have a spiral shell and an *operculum.

prosoma In *Arachnida, the anterior portion of the body.

Prosopora (phylum *Annelida, class *Oligochaeta) An order of exclusively aquatic worms all of which have paired testes. Some members are ectoparasitic (*see* PARASITISM). There are two families.

prostaglandin One of a group of C_{20} *fatty acids, each containing a five-membered ring. Prostaglandins differ from one another in the number and position of double bonds and hydroxyl-group constituents. Their biological effects include the lowering of blood pressure and the stimulation of smooth-muscle contraction.

prosternum The ventral, sclerotized cuticle of the first thoracic segment of an insect. *See also* PROTHORAX.

prosthetic group A non-protein component of a *conjugated protein, or the *cofactor of an *enzyme to which it is bound so tightly that it cannot be removed by *dialysis.

Protacanthopterygii (class *Osteich-thyes, subclass *Actinopterygii) A superorder of bony fish that includes the *Salmoniformes.

protamine A generic term for a group of strongly basic, *globular proteins of relatively low molecular weight, which contain large quantities of the *amino acid *arginine, but no sulphur. They are found associated with *nucleic acids, particularly in sperm cells.

protandry 1. The production of sperm in males before females produce eggs (e.g. some roundworms). **2.** The arrival of males before females at breeding grounds (e.g. many territorial bird species). *Compare* PROTOGYNY. **3.** In *sequential hermaphrodites, the occurrence of the male form in younger, smaller individuals and the female in larger, older individuals.

Protanisoptera *See* ANISOZYGOPTERA.

protease An *enzyme that catalyses the *hydrolysis of the *peptide bonds of *proteins and *peptides. There are two types: endopeptidases (proteinases) primarily attack interior peptide bonds, while exopeptidases act on those adjacent to a free amino or carboxyl group.

Proteidae (mudpuppies, olm; class *Amphibia, order *Urodela) A family of aquatic, permanently larval salamanders which have an elongate body, small or rudimentary limbs, lidless eyes, and external gills retained even when lungs are present. There are two genera. *Proteus*, with one species, *P. anguinus* (olm or cave salamander) is neotenous (*see* NEOTENY), blind, unpigmented, feeds mainly on crustaceans, and occurs in caves and underground streams in south-eastern Europe; it is white with red gills, but may turn brown after long exposure to light. *Necturus* (mudpuppies or

water dogs), with seven species, are persistently neotenous, nocturnal stream and pond dwellers of southern Canada and the eastern USA. *N. maculosus* feeds on small fish and aquatic invertebrates.

protein A polymer that has a high relative molecular mass of *amino acids. Proteins occupy a central position in the architecture and functioning of living matter. Structurally they are divided into two groups, globular and fibrous proteins; and they are often found associated with a non-protein component, forming so-called conjugated proteins. Functionally they act variously as *enzymes, as structural elements (e.g. hair and *collagen), as *hormones, as respiratory pigments, as contractile elements, as *antibodies, and as hereditary factors.

proteinase *See* PROTEASE.

Proteles cristatus **(aardwolf)** *See* HY-AENIDAE.

Proteocephalidea (phylum *Platyhel-minthes, class *Cestoda) An order of parasitic worms which possess a quite complex, suckered *scolex. Two or three hosts may be inhabited during development, the final one always being a freshwater vertebrate.

proteroglyphous Applied to snakes that have fangs at the front of the maxilla (upper jaw), often with small solid teeth behind. The fangs are hollow and short, as in cobras, mambas, and coral snakes (*Elapidae). *Compare* OPISTOGLYPHOUS; SOLENOGLYPHOUS.

Proterozoic The eon of geologic time, 2500–542 Ma ago, between the *Archaean and *Phanerozoic eons, and ending in the abundantly fossiliferous *Ediacaran Period. It was the final period of the former *Precambrian.

Proteus anguinus **(olm, cave salamander)** *See* PROTEIDAE.

Proteutheria (cohort *Unguiculata, order *Insectivora) A suborder that comprises a large number of extinct forms the earliest of which lived during the late *Cretaceous. At that time the early eutherians appear to have diverged to form two distinct stocks, the Leptictidae and Deltatheridiidae, distinguished mainly by the structure of their teeth; both are included in the Proteutheria. There were many evolutionary side-branches which became extinct, but among successful proteutherians some may be ancestral to rodents and others may be ancestral to primates. Proteutheria are often excluded from the Insectivora, or even divided into two or more separate orders.

prothorax The first thoracic segment of an insect, which consists of a cuticular box made up of a *pronotum dorsally, *propleura laterally, and a *prosternum ventrally. The prothorax bears a pair of legs.

prothrombin The enzymatically inactive precursor of *thrombin, which is activated by a complex cascade of events requiring the presence of Ca^{2+} (calcium), of proteinaceous thromboplastins, and of a number of additional blood factors. *See also* BLOOD CLOTTING.

protist A member of the former kingdom Protista. *See* PROTOCTISTA.

Protista *See* PROTOCTISTA.

protobranchiate Applied to a primitive type of gill, common to certain *Bivalvia, which is simple, flat, and horizontal, so that the lower *mantle cavity is an inhalant chamber and the upper mantle cavity is an exhalant chamber.

Protoceratidae An American offshoot from the main traguloid stem of *Artiodactyla, ranging from the *Oligocene to the *Pliocene, that were deer-sized mammals with horn-like structures in the nasal region, above the orbits, and on the upper part of the brain case. The Pliocene *Synthetoceras* had two horn-pairs, the nasal pair being fused into a Y shape.

protocercal tail Possibly the most primitive type of tail found in fish. The posterior end of the vertebral column is straight, dividing the tail fin into two equal lobes supported by *fin rays.

protochordate A member of the invertebrate *Chordata; the name is applied informally to all living and fossil chordates that lack skulls and vertebrae (including *Cephalochordata and *Urochordata).

Protococcidia (class *Telosporea, subclass *Coccidia) An order of little-known *Protozoa which appear to be parasitic on marine *Annelida. Only two species have been described.

protoconch The initial (larval) shell of molluscs, often retained at the tip of the spire of the adult shell.

protocone In primitive, *triconodont, upper, *molar, mammalian teeth, the central *cusp where the three cusps form a straight line. Where they are not in line, the anterior inside cusp.

protocooperation (facultative mutualism) An interaction between organisms of different *species in which both organisms benefit, but neither is dependent on the relationship. *Compare* MUTUALISM.

Protoctista In the widely used five-*kingdom classification system for living organisms, one of the kingdoms within the superkingdom *Eukarya. In the three-domain classification system, a kingdom within the *domain Eukarya. Protoctists are aquatic eukaryotes, but they are neither animals, nor fungi, nor plants. The kingdom includes naked and shelled *amoebae, foraminiferans (*Foraminiferida), *zooflagellates, ciliates, dinoflagellates (*Dinoflagellida), diatoms, algae (including seaweeds), slime moulds, slime nets, and *Protozoa. Single-celled organisms were formerly known as 'protists' and the kingdom containing them was the Protista. This ranking was abandoned when it became evident that multicellularity evolved many times and that multicellular organisms are closely related to single-celled forms. The name Protoctista means 'first established', from the Greek *protos* (first) and *kristos* (established); 'Protista' is no longer used.

protogyny (adj. protogynous) 1. Condition in which the females develop eggs before males produce sperm, or females arrive at breeding grounds before males. *Compare* PROTANDRY. **2.** In *sequential hermaphrodites, the occurrence of the female form in younger, smaller individuals and the male form in larger, dominant individuals. Individuals that never become dominant do not become male.

protonephridium *See* NEPHRIDIUM.

Protoneuridae (damselflies; order *Odonata, suborder *Zygoptera) Family of slender, delicate damselflies, with narrow wings, which are distinguished from other damselflies by having the first anal vein fused with the wing margins. Nymphs of many species have not been described, but those which have are generally small, with short legs, and *tracheal gills of varying shape. The family occurs in the tropics, and more than 200 extant species have been described.

protoplasm The complex, translucent, colourless, colloidal substance that occurs within each cell, including the *cell membrane but excluding the large *vacuoles, masses of secretions, ingested material, etc. It is differentiated into nucleoplasm (protoplasm in the nucleus) and *cytoplasm (protoplasm in the rest of the cell).

protoplast (energid) In animals (but not plants), the entire contents of a cell.

protopodite In *Crustacea, the base to which a *biramous appendage is attached. It consists of a *coxa (coxopodite) and a basis (basopodite), the inner (endopodite) and outer (exopodite) parts of the appendage being attached to the basis.

Protopteridae (African lungfish; superorder *Cephalaspidomorpha, order *Lepidosireniformes) A small family of freshwater fish that can breathe air with the aid of a pair of air sacs or lungs. They have a somewhat rounded body and long, filamentous, *pectoral and *pelvic fins. The gills are functional, but poorly developed: the fish often have to rely wholly on their lungs as they tend to live in stagnant water low in oxygen. During the dry season they *aestivate in deep burrows in the muddy bottom of the swamp, surrounding themselves with a thin cocoon. There are four species, living in African rivers.

Protoreodon An important artiodactyl mammal recorded from the upper *Eocene of N. America. Rather pig-like in build and habits, it is considered a likely ancestor for various ruminant stocks.

Protorohippus *See* HYRACOTHERIUM.

protosoma 1. In invertebrates whose body is divided into sections (e.g. *Lophophorata and *Hemichordata), the anterior part that lies in front of the *mesosoma and the *metasoma. **2.** *See* ACARINA.

Protostomia The name sometimes given to a number of animal phyla (*Annelida, *Arthropoda, *Mollusca, and *Platyhelminthes) which are thought to share a common ancestry. The name indicates that the first embryonic opening (blastopore) develops into the mouth. *Compare* DEUTEROSTOMIA.

protostomy The condition in which the *blastopore forms the mouth of the adult animal. *Compare* DEUTEROSTOMY.

Prototheria (class *Mammalia) A subclass that comprises the extinct orders *Docodonta, *Triconodonta (although their inclusion in the subclass is controversial), and *Multituberculata, as well as the order *Monotremata, members of which (platypus and echidnas) still survive. Prototherians are distinguished from therians mainly by the skull wall, which in prototherians has a small *squamosal and *alisphenoid, and by their teeth. The two subclasses began to diverge from a common stock during the late *Triassic.

Prototroctes maraena (Australian grayling) *See* APLOCHITONIDAE, RETROPINNIDAE.

Protoungulata (infraclass *Eutheria, cohort *Ferungulata) A superorder that comprises the orders *Condylarthra, *Notoungulata, *Litopterna, *Astrapotheria, and *Tubulidentata. The superorder includes the earliest known ungulates from their appearance during the late *Cretaceous, when they departed from an insectivorous diet and began to eat plant material, until the end of the *Palaeocene by which time their radiation permits them to be grouped in distinct orders from some of which later ungulates are descended. The Tubulidentata, known to have lived in the *Miocene and possibly in the *Eocene, are animals apparently similar to condylarths and so are included in the superorder, but otherwise they are isolated zoologically and their affinities are obscure.

Protoxaea See OXAEIDAE.

Protozoa (kingdom *Protoctista) A phylum (division) of *eukaryotic, single-celled micro-organisms which show a wide variety of forms. The phylum includes some plant-like forms and animal species that are free-living, commensal (see COMMENSALISM), mutualistic (see MUTUALISM), and parasitic (see PARASITISM), and some important disease-causing organisms.

protractile mouth See PROTRUSIBLE MOUTH.

protrusible mouth (protractile mouth) In fish, a structural arrangement of the jaws that enables the animal to protrude (extend) or withdraw the mouth at will. When fully protruded, the cavity of the mouth is enlarged to form a funnel-like space facilitating the uptake of food.

Protura (subphylum *Hexapoda, class *Insecta) Subclass and order of minute (0.6–2 mm long), primitive, *apterous, white insects, which lack eyes and antennae and have reduced, *entognathous, piercing mouth-parts. Proturans are elongate, with segmented legs, the anterior pair used for sensory purposes, the middle and posterior pair for walking. The abdomen carries a pair of short styli (rudimentary limbs) on the basal three segments and the pseudocelli found on the head may represent antennal bases or postantennal organs. Species in this rare group live under bark, stones, or among rotting vegetation, and are sometimes subterranean in habit. Proturans exhibit *anamorphosis, one extra segment being added at each moult. *Metamorphosis is simple. There are 118 species, world-wide in distribution.

prowfish 1. See PATAECIDAE. **2.** (*Zaprora silenus*) See ZAPRORIDAE.

proximal Applied to that part of an organ or structure which is closest to its point of attachment to the body, or to the centre of the body. Compare DISTAL.

proximate explanation The immediate explanation, i.e. the immediate cause that produced an effect. Compare ULTIMATE EXPLANATION.

pruinose Covered with powdery granules; powdery in appearance.

Prunella modularis (dunnock) See PRUNELLIDAE.

Prunellidae (accentors; class *Aves, order *Passeriformes) A family of small, mainly brown birds which have greyish under-parts, some with chestnut flanks, and slender, pointed bills. They are found in rocky meadows, scrub, and woodland, feed on the ground on seeds and insects, and nest in bushes and on the ground. *Prunella modularis* (dunnock) is a common garden bird in Europe. There is one genus, comprising 12 species, found in Europe and Asia.

Przewalski's horse (*Equus ferus przewalskii*) See EQUIDAE.

psalterium See OMASUM.

Psaltria (pygmy tits) See AEGITHALIDAE.

Psaltriparus (bushtits) See AEGITHALIDAE.

psammon The microscopic flora and fauna of the interstitial spaces between sand grains of sea-shore and lake-shore areas.

Pselaphidae (subclass *Pterygota, order *Coleoptera) Family of tiny, red or yellowish beetles, 0.75–3.5 mm long. They have an ant-like appearance, with a narrow *thorax, and a very broad, non-flexible *abdomen. The *elytra are short, with several abdominal segments exposed. One group, with much reduced antennae and *maxillary palps, lives in ants' nests. Some feed on food regurgitated by ants and produce from the base of the abdomen a sweet substance attractive to ants. A second group has 11-segmented antennae and unusually long clubbed palps, especially in males. It is found in leaf litter, under bark, or more loosely associated with ants. Larvae are poorly known. There are 5000 species, mostly tropical.

***Psenes arafurensis** (eyebrow fish) See* NOMEIDAE.

Psettodidae (toothed flounder; subclass *Actinopterygii, order *Pleuronectiformes) A small family of marine flatfish that have spiny rays in the *dorsal fin, which begins well back on the body. The head has a mouth with strong teeth. The eyes may be on either the left or right side of the body. There are possibly five species, found in African and Indo-Pacific waters.

***Pseudaphritis urvilli** (congolli) See* BOVICHTHYIDAE.

pseudergate Member of the labouring caste in lower families of termites, which is capable of *metamorphosis into a reproductive.

Pseudidae (class *Amphibia, order *Anura) A small family of amphibians which are related to the aquatic tree frogs (*Hylidae). The vertebrae are *procoelous, they have an extra phalanx on each digit, and the thumb is opposable. They are found in S. America.

pseudobranch A small, gill-like structure that is the remnant of the primitive gill of the mandibular arch in sharks and bony fish. Despite its origin, it probably has a glandular rather than a respiratory function. In most bony fish it is found on the inner side of the base of the *gill cover.

Pseudocheiridae (ring-tailed possums; order *Marsupialia (or *Diprotodontia), superfamily *Phalangeroidea) A family of small or medium-sized, arboreal, Australian and New Guinea marsupials, of which there are three or four genera and about 20 species, including the greater glider.

Pseudochromidae (dottyback; subclass *Actinopterygii, order *Perciformes) A family of marine fish that have an elongate body, large eyes, and a large, softrayed, *dorsal fin preceded by about three spiny rays. The tail fin and *pectoral fins are rounded. Found among rocks and coral heads, they are colourful, agile fish, found in the Indo-Pacific region. There are several species. *See also* PSEUDOGRAMMIDAE.

pseudocoelom A second body cavity (the first being the gut) which occupies a space between the *mesoderm of the body wall and the *endoderm of the gut. There are no *mesenteries suspending the internal organs and no muscular layers around the gut; thus in no pseudocoelomate animal does muscular peristalsis move food through the *alimentary canal.

pseudocoelomate Applied to animals that possess a *pseudocoelom (e.g. *Nematoda).

pseudocopulation Mating behaviour that resembles copulation but in which eggs are fertilized externally. *See* AMPLEXUS.

pseudoextinction Within an *evolutionary lineage, the disappearance of one taxon caused by the appearance of the next *chronospecies in the series. The extinction is purely taxonomic.

pseudogenes *Genes that have been 'switched off' in evolution and no longer have any function. They are, therefore, entirely neutral and evolve at a constant rate. By comparing pseudogenes in related organisms a standard can be inferred against which the rate of change in other genes can be measured, enabling further inferences to be made as to the presence or absence of selection pressure in evolution.

((⊕)) SEE WEB LINKS
• Detailed information about pseudogenes.

Pseudogrammidae (dottyback; subclass *Actinopterygii, order *Perciformes) A small family of elongate marine fish, closely related to the *Pseudochromidae, that have about seven spiny rays in the *dorsal fin; the soft dorsal and the other fins have a rounded profile. They have an interrupted *lateral line and conspicuous eyes. There are at least six species occurring in Africa and the Indo-Pacific region.

pseudointerference Where consumers exhibit the *aggregative response, the decline in their rate of consumption as their density on foraging patches increases. The effect resembles *mutual interference, but in fact is a purely density-dependent consequence of high consumer density.

Pseudolestidae (Lestoideidae; damselflies; order *Odonata, suborder *Zygoptera) Family of unusual damselflies, which shows many features common to other families, but whose members can be recognized by the reduced first and second anal wing vein, and from the third intercalary, and fourth branch of the radius, arising midway between the arculus and subnodus in the wing. The larvae are short, with saccoid or lamellar *caudal gills. There are 14 described extant species, found in the eastern Palaearctic, south-east Asia, Australasia, and Central and S. America.

Pseudomyrmecinae (suborder *Apocrita, family *Formicidae) Subfamily of slender, arboreal ants, which have two petiolar nodes, a *sting, and well-developed eyes. There are about 200 species. They are restricted to the tropics, and inhabit hollow stems and twigs.

pseudopipodite In some *Crustacea, an outer lobe on a limb that gives the limb a *triramous structure.

Pseudoplesiopidae (false roundhead; subclass *Actinopterygii, order *Perciformes) A small family of marine fish that have an elongate body, large eyes, and a long, single, *dorsal fin which, like the other fins, has a rounded profile. There are several species occurring in the Indo-Pacific region.

Pseudopleuronectes americanus (Atlantic winter flounder) See PLEURONECTIDAE.

pseudopodium In amoeboid cells, a protrusion, usually temporary, of the cell body, which functions in locomotion and in the ingestion of food particles.

Pseudopotto martini (order *Primates, family *Loridae) A new genus and species of slow-climbing primates described in 1996. It is known from a skeleton and a second skull, from Cameroon; its appearance in life is unknown.

pseudorumination See REFECTION.

Pseudoryx nghetinhensis (saola, spindlehorn, Vu Quang 'ox'; superfamily *Bovoidea, family *Bovidae) A new genus and species of antelope-like animal discovered in the early 1990s in the Annamite Mountains, on the northern border between Vietnam and Laos. It is 1 m high, blackish with white markings, and has long, straight, reclined horns that point straight back. It is thought to be related to cattle, buffaloes, and nilgai.

Pseudoscorpiones (pseudoscorpions; subphylum *Chelicerata, class *Arachnida) Order of minute arachnids whose name refers to their large, scorpion-like, *chelate *pedipalps which do not, however, possess a stinging organ. Pseudoscorpions are flattened, with the oval opisthosoma joined broadly to the prosoma. Most are black, brown, or brownish-green, and have two pairs of respiratory tracheae. Few are more than 5 mm long, although the largest species reaches 12 mm. The sides of the carapace bear one or two pairs of eyes, and the small *chelicerae produce silk for the construction of moulting, overwintering, and brooding chambers. Prey items are caught by means of the large, chelate pedipalps, and are passed to the chelicerae where digestive juices are injected. Most species live in crevices and the humid conditions found under bark, plant litter, etc., and feed on *Collembola, caterpillars, and ants. Some pseudoscorpions disperse by *phoresy (clinging to the legs of passing flies and other animals). Courtship displays occur, and may involve *pheromones; and there are indications of presocial behaviour in this order of often gregarious species. 2000 species occur world-wide, of which 25 can be found in the UK. Some species inhabit houses and are cosmopolitan, hunting booklice, nymphal bedbugs, and other small insects.

pseudoscorpions See PSEUDOSCORPI-ONES.

Pseudostega (order *Cheilostomata, suborder *Anasca) A division of bryozoans in which the colony stands erect, and is cylindrical or with two layers of individuals in cross-section. The individuals are arranged in longitudinal rows. The division occurs from the *Cretaceous to the present.

Pseudostigmatidae (damselflies; **order *Odonata, suborder *Zygoptera)** Family of damselflies whose members have very slender, elongated abdomens, and are the largest extant odonates. The larvae inhabit a variety of habitats, including waterfilled holes in rotting logs and branches of trees. The *tracheal gills are often foliaceous. The family occurs in tropical rain forests, and 35 extant species have been described.

Psittacidae (lories, cockatoos, conures, macaws, parrots, pygmy parrots; class *Aves, order *Psittaciformes) A family of small to large, very colourful birds with characteristic deep bills, the upper *mandible of which is large and *decurved to form a hook, the lower mandible being upcurved and smaller. Lories (eight species of *Lorius*, found in New Guinea, Indonesia, and the

Solomon Islands) have a brush-like tip to the tongue. Parrots have large heads, short necks, thick, *prehensile tongues, and their nostrils are set in a fleshy *cere at the base of the bill. Their toes are *zygodactylous. They feed mainly on seeds and fruit, inhabit forest and more open country, and nest in holes in trees or in termite mounds. *Ara macao* (scarlet macaw) is widely kept in captivity, as are the conures of Central and S. America (18 species of *Pyrrhura*), which are mainly green with blue *primaries and grey, blue, yellow, red, and white head markings. The Australian *Melopsittacus undulatus* (budgerigar), a gregarious bird of the open plains and wooded grassland, is bred in a variety of colour forms as perhaps the most popular of all cage birds. Pygmy parrots (six species of *Micropsitta*), found only in New Guinea and nearby islands, are very small and have green upper-parts, blue and black tails, heads patterned with red, blue, or yellow, and tails with stiff, protruding feather shafts, which aid the birds in climbing trees; they inhabit mainly lowland forest, feed on lichen, fruit, seeds, and insects, and nest in termite mounds. There are about 80 genera in the family, comprising more than 340 species, found in Africa, southern Asia, Australasia, Central and S. America, and the Pacific islands.

Psittaciformes (lories, cockatoos, parrots; class *Aves) An order which comprises one family, *Psittacidae, but which is sometimes split into three: Loridae, Cacatuidae, and Psittacidae.

Psittirostra **(Hawaiian honeycreepers)** See DREPANIDIDAE.

Psocoptera (barklice, booklice; class *Insecta) An order of *hemimetabolous insects with a prominent head and a narrow neck between head and *thorax. Some barklice have two pairs of wings, the front wings larger than the hind wings. Other species are wingless, as are all booklice. Barklice live in moist environments, e.g. in litter, beneath tree bark, or beneath stones, often in small colonies, and feed on algae, fungi, and lichens; they grow up to 10 mm long and have little contact with humans. Booklice are less than 2 mm long and inhabit houses and storage buildings, feeding on grain, the glue in book bindings, wallpaper paste, and other products made from starches. Some psocopterans live in bird nests, feeding on discarded feathers and skin cells, but never parasitize the birds. There are 35 families with about 3200 species, distributed world-wide.

Psophia (trumpeter) *See* PSOPHIIDAE.

Psophiidae (trumpeters; class *Aves, order *Gruiformes) A family of large, rounded birds which have short, stout bills, long necks, rounded wings, short tails, and long legs. Their plumage is mainly black, with a purple gloss on the neck, and the head and neck feathers are very soft. Feathers on the back and inner wings have long, hair-like strands, and are white, grey, or brown. Trumpeters are gregarious and terrestrial, rarely flying, inhabit forests, feed on insects and vegetable matter, and nest in tree holes. There is one genus, *Psophia*, with three species, found in S. America.

Psychidae (bagworms; subclass *Pterygota, order *Lepidoptera) Family of moths in which the males are always winged, while in the females the wings are often reduced or absent. The larva lives within a case that is often characteristic of the species, and is often constructed of leaves, twigs, or sand grains. Most larvae feed on detritus. There are about 1350 species, distributed world-wide.

Psychodidae (moth flies, sand flies; order *Diptera, suborder *Nematocera) Family of minute, moth-like flies in which the legs, body, and wings are covered in long, coarse hairs and often scales. The larvae are mostly *amphipneustic, and usually aquatic or saprophagic. Females of some genera, e.g. *Phlebotomus*, are bloodsuckers, attacking vertebrates. These include the 'sand fly', which is known to be a vector of leishmaniasis and other related human diseases. The family has a world-wide distribution, and more than 500 species are known to exist.

Psychrolutidae (tadpole sculpin; subclass *Actinopterygii, order *Scorpaeniformes) A small family of marine fish that have a large head, stout body, but strongly tapering tail; the *pectoral fins are rather large. Some authors place these fish in the family *Cottidae. There are about five species, occurring in the N. Pacific.

Psyllidae (order *Hemiptera, suborder *Homoptera) Family of bugs in which the *rostrum arises between the bases of the front pair of legs, and the hind pair of legs is modified for jumping. Psyllids suck the sap of many different kinds of plants and some of them cause the formation of galls. The immature stages are almost immobile and cannot jump. There are about 1300 species, distributed world-wide.

PTC Plasma thromboplastin component. *See* BLOOD CLOTTING.

Pteraspidomorpha Superorder of jawless vertebrates which includes the order *Myxiniformes (hagfish).

Pteraspidomorphi (superclass *Agnatha) A class of jawless vertebrates that have an eel-like body, lack paired fins, have two *semicircular canals, and a skeleton not consisting of true *bone. The class includes the fossil order Pteraspidiformes, and the extant *Myxiniformes (hagfish). Some authors place the Myxiniformes with the class *Cephalaspidomorphi.

Pterioda (phylum *Mollusca, class *Bivalvia) An order of bivalves that are *byssate or cemented, and generally *pleurothetic. Normally they are unequilateral and unequivalve, and their variable shell composition may be crossed-lamellar, prismatic-*nacreous, or foliate. The musculature is *anisomyarian or *monomyarian. The ligament is extremely variable. The gills are *eulamellibranchiate or *filibranchiate. The order first appeared in the *Ordovician.

Pteriomorphia (mussels, scallops, oysters; phylum *Mollusca, class *Bivalvia) A subclass of bivalves, most of which are sedentary, *epifaunal, and *byssate or cemented. The most specialized genera have a reduced anterior adductor muscle and foot, and in some forms they are absent. The shells are variable in shape and composition; they may be *aragonite, *calcite, or both. They have a *duplivincular ligament, unique to certain members of the subclass, and the dentition is variable. They first appeared in the Early *Ordovician.

Pterobranchia (phylum *Hemichordata) A class of minute, fixed, colonial, deep-sea organisms which secrete an external cuticular skeleton in which they are housed. The *mesosoma is small but carries one or more pair of arms bearing *tentacles (the *lophophores). The *metasoma consists of a long *peduncle by which the individual is attached. The branchial apparatus is rudimentary. There are about 30 known species.

Pteroclidae (sandgrouse; class *Aves, order *Columbiformes) A family of medium-sized birds which have brown and grey, barred, spotted, and mottled upper-parts, yellow, chestnut, white, and black under-parts, and face markings. They have short, conical bills, short necks, long, pointed wings, and a medium to long tail. Their legs are short and feathered (in the two species of *Syrrhaptes* including the toes). They are gregarious and terrestrial, inhabit open plains, desert, and brush, feed on seeds, and nest on the ground. There are two genera, comprising 16 species, some migratory, found in Europe, Africa, and Asia. The two *Syrrhaptes* species occur only in Asia, but Pallas's sandgrouse (*S. paradoxus*) is prone to mass migration; 14 *Pterocles* species occur thoughout much of Africa, the Middle East, India, and Central Asia.

***Pterocnemia pennatus* (Darwin's rhea)** *See* RHEIDAE.

pterodactyls *See* PTEROSAURIA.

***Pterodroma* (petrels)** *See* PROCELLARIIDAE.

***Pteroglossus* (toucans)** *See* RAMPHASTIDAE.

***Pterois volitans* (lionfish)** *See* SCORPAENIDAE.

Pteromalidae (jewel wasps; suborder *Apocrita, superfamily *Chalcidoidea) Family of minute (2–4 mm long), usually black or metallic green wasps, many of which are important in the control of crop pests. This large and abundant group has a complex taxonomy and is sometimes broken into several families. Jewel wasps are parasitic on a wide range of hosts, including ants, the larvae of *Lepidoptera, *Coleoptera, and *Diptera, other chalcidoids, and the *oothecae of cockroaches and spiders, while others are *hyperparasites on the larvae of other parasitic wasps.

Pterophoridae (plume moths; subclass *Pterygota, order *Lepidoptera) Widespread family of small moths with few species. Adults are characterized by the division of the wings into plumes. Superficially they resemble the *Alucitidae, but the fore wings are cleft into two, three, or four, and the hind wings into three. Some species have undivided wings. Larvae are external or internal feeders, particularly on Asteraceae.

***Pterophyllum* (angelfish)** *See* CICHLIDAE.

Pteropoda *See* PTEROPODS.

Pteropodidae *See* MEGACHIROPTERA.

pteropods (sea butterflies) An informal, non-taxonomic name (but sometimes given the status of a superorder, Pteropoda) given to two orders of marine *Gastropoda, *Gymnosomata and *Thecosomata. *See also* OPISTHOBRANCHIA.

Pterosauria (class *Reptilia, subclass *Archosauria) A *Mesozoic order of flying reptiles (popularly known as pterodactyls) which were particularly numerous in the *Jurassic but survived until late *Cretaceous times. Their fossil skeletons suggest that they could not stand upright on land, and so it is assumed that their mode of life involved swooping over the sea to catch fish. Their remains are always associated with marine deposits.

pterostigma Pigmented spot on the anterior margin of the wing of an insect.

pterygoid One of the elements of the bar that forms the upper jaw in *Elasmobranchii and one of the bones of the upper jaw in many *Teleostei. In other vertebrates the pterygoid forms part of the roof of the mouth, but in *Squamata it continues to bear teeth.

Pterygota (subphylum *Uniramia, class *Insecta) The larger of the two subclasses of insects, whose members have wings or, if wingless, whose wings have been secondarily lost. The subclass includes all insects except for those placed in the subclass *Apterygota.

Pterygotus deepkillensis One of the eurypterids, a rather rare group of fossil *Arthropoda that includes the largest known fossil arthropods, some individuals growing to 3 m in length. *P. deepkillensis* is one of the earliest and is known from the Early *Ordovician of N. America.

pteryoglutamic acid *See* FOLIC ACID.

Ptiliidae (subclass *Pterygota, order *Coleoptera) Family of minute beetles, less than 2 mm long, which includes the smallest known beetle, *Nanosella fungi*, 0.25 mm. Usually they are convex and very shiny, but some have a conspicuous *pubescence. The wings are narrow, with much reduced venation and a fringe of long hairs. The antennae are hairy, with loose clubs. They are found in fungi, decaying vegetation, and under bark. *Ptilium myrmecophilum* lives in ant nests. There are 300 species.

Ptilinopus (fruit doves) *See* COLUMBIDAE.

ptilinum An inflatable sac above the base of the antennae in *Diptera (true flies) of the section Schizophora (suborder *Cyclorrapha). It is used by the emerging fly to break open the *puparium and to force a path through the pupation substrate.

Ptilogonatidae *See* BOMBYCILLIDAE.

Ptilonorhynchidae (bowerbirds, catbirds; class *Aves, order *Passeriformes) A family of medium-sized to large birds, which are mainly black, grey, or blue, marked with other brilliant colours. They have stout bills, short to medium, rounded wings, and short to long tails. Some species have a crest or coloured ruff of feathers. They inhabit forests, feed on insects, fruit, and seeds, and nest in trees. The males build an elaborate bower or display ground which is decorated with coloured flowers, berries, and other objects. There are eight genera, with 18 species, found in New Guinea and Australia.

Ptinidae (spider beetles; subclass *Pterygota, order *Coleoptera) Family of small beetles, 2–5 mm long, with a spiderlike appearance; the *thorax is narrow and rounded, covering the head, the *abdomen is globular, the legs long. Some are smooth and shining, but others have *elytra covered with dense *pubescence. Larvae are white, curved and fleshy, with tiny legs. They feed on dried stored products, skins, and carcasses. They are common pests in homes and warehouses. Their natural habitat is in nests of birds, mammals, and ants. There are 700 species.

ptyalin An *amylase that is present in *saliva; it catalyzes the *hydrolysis of *starch to *maltose.

Ptychodactiaria (class *Anthozoa, subclass *Zoantharia) A small order, containing two genera, of coelenterates that have primitive musculature, *nematocysts, and mesenterial structure (they lack ciliated tracts) in anemone-like *polyps. They are often regarded as a family of the order *Actiniaria.

pubescent Covered with soft hair; downy.

pubis In tetrapods, the anterior, ventral part of the *pelvis.

puff adder (*Bitis arietans*) *See* VIPERIDAE.

puffbirds *See* BUCCONIDAE; PICIFORMES.

pufferfish *See* TETRAODONTIDAE.

puffins *(Fratercula) See* ALCIDAE.

Puffinus **(shearwaters)** *See* PROCELLARI-IDAE.

pugs *See* GEOMETRIDAE.

Pulicida *See* SIPHONAPTERA.

Pulicidae *See* SIPHONAPTERA.

pulmonary Pertaining to the lungs (e.g. 'pulmonary artery' and 'pulmonary vein', the blood vessels carrying blood to and from the lungs).

Pulmonata (phylum *Mollusca, class *Gastropoda) An informal group, formerly ranked as a subclass, of mainly terrestrial gastropods, with some aquatic members, in which the anterior *mantle cavity is modified into a *lung. A few have small, secondary gills. The head usually has one or two pairs of *tentacles. The foot generally occupies the entire ventral surface, forming a flat, creeping sole. The shell, if present, can be quite variable in shape. Members of only one family have an *operculum. Individuals are hermaphroditic. The pulmonates first appeared in the *Mesozoic. Many of the species formerly included in the Pulmonata have been placed in the clade Heterobranchia.

pulp cavity The internal cavity of a vertebrate tooth (but *see also* DENTICLE), that contains *connective tissues, blood vessels, and nerves, and is connected to the tissues in which the tooth is embedded.

pulverulent Dusty; powdery; covered with powder.

pulvillus In flies (*Diptera), a small pad, of which there is one beneath each of the *unques.

pulvinate Cushion-shaped; swollen; convex.

pulvinus In insects, a spongy pad between the front of the head and the mouth.

punctate 1. Applied to any structure that is marked by pores or by very small, point-like depressions. **2.** Applied to a type of brachiopod shell structure in which fine tubes or pores extend from the inner almost to the outer surface.

punctiform Dot-like in appearance.

punctuated equilibrium The theory, first proposed in 1972 by Niles Eldredge and Stephen Jay Gould, that *evolution is characterized by geologically long periods of stability during which little speciation occurs, punctuated by short periods of rapid change, species undergoing most of their morphological changes shortly after breaking from their parent species.

pupa The stage in the life cycle of an insect during which the larval form is reorganized to form the final, adult form. Pupae are frequently inactive, forming a hard shell (chrysalis) or silken covering (cocoon) about themselves while the very large changes associated with becoming adult take place.

puparium A *pupa formed from the *exoskeleton of the final larval *instar.

pupil *See* EYE.

Purgatorius The earliest genus of the *Primates, known from the earliest *Palaeocene, and claimed also from the latest *Cretaceous.

purine A basic, nitrogenous compound that resembles a six-membered *pyrimidine ring fused to a five-membered imidazole ring. The two principal purines, adenine and guanine, are major constituents of *nucleic acids.

purse-web spiders *See* ATYPIDAE.

Pycnogonida (Pantopoda; sea spiders; phylum *Arthropoda, subphylum *Chelicerata) A class of marine chelicerates, all of which are small (total body length usually 1–10 mm but up to 6 cm in some species, with a leg span of 75 cm). The narrow body comprises a *cephalon and four to six distinct segments, the cephalon bearing a proboscis. The posterior part of the cephalon tapers to form a neck bearing four eyes mounted on a central tubercle on its dorsal surface. There is a pair of chelifores (possibly homologous to arachnid *chelicerae), a pair of *palps, and two pairs of *ovigers (reduced in some species and absent in females of some species). The four to six pairs of eight-segmented walking legs articulate with large processes, pairs of which project from each segment, giving the animal the spider-like appearance to which its common name refers. Some species have eight to ten pairs of legs, possibly as a result of *polyploidy. Most pycnogonids are carnivorous and bottom-dwellers, although some can swim. There are about 500 species found in all oceans, but more commonly in cold waters, and at all depths.

Pycnonotidae (bulbuls, greenbuls; class *Aves, order *Passeriformes) A family of medium-sized, grey, brown, and olive birds, which have areas of yellow, red, white, or black. They have short to medium-length, slender, slightly curved bills, hooked or notched in some species, with rounded wings, and fairly long tails. They have *rictal bristles, and hair-like neck feathers, and some are crested. They are mainly arboreal, inhabit forests and cultivated areas, feed on fruit and insects, and nest in trees and on the ground. *Hypsipetes* species (of which there are 20) have longer wings than other bulbuls and some are migratory. Some of the 48 species of *Pycnonotus* are popular cage birds and *P. jocosus* (red-whiskered bulbul), which inhabits more open areas and cultivated land, has become successfully established in Australia and N. America. There are 15 genera, comprising about 120 species, found in Africa, southern Asia, the Philippines, and the Moluccas.

***Pycnonotus* (bulbuls)** *See* PYCNONOTIDAE.

***Pycnonotus jocosus* (red-whiskered bulbul)** *See* PYCNONOTIDAE.

***Pygathrix nemaeus* (douc langur)** *See* CERCOPITHECIDAE.

Pygidicranidae (earwigs; subclass *Pterygota, order *Dermaptera) Family comprising the most primitive of living earwigs, which are winged, are about 1–3.5 cm long, and have long antennae. There are two subfamilies, both of which are found in Australia.

pygidium The terminal segment of an insect; the *sclerites which surround the anus. *See also* CERAPACHYINAE.

pygmy chimpanzee (bonobo; *Pan paniscus*) *See* PONGIDAE.

pygmy gliders *See* ACROBATIDAE.

pygmy hippopotamus (*Choeropsis liberiensis*) *See* HIPPOPOTAMIDAE.

pygmy locusts *See* TETRIGIDAE.

pygmy mole crickets *See* TRIDACTYLIDAE.

pygmy owls (*Glaucidium*) *See* STRIGIDAE.

pygmy parrots (*Micropsitta*) *See* PSITTACIDAE.

pygmy possums *See* ACROBATIDAE.

pygmy right whale (*Caperea*) *See* BALAENIDAE.

pygmy sperm whale (*Kogia breviceps*) *See* PHYSETERIDAE.

pygmy tits (*Psaltria*) *See* AEGITHALIDAE.

Pygopodidae (snake lizards; order *Squamata, suborder *Sauria) A family of lizards, related to the geckos (*Gekkonidae), which lack fore limbs and in which the hind limbs are reduced to scaly appendages. The eye has a vertical pupil and spectacle (*see* SERPENTES). The body is slender, with a long tail that can be autotomized (*see* AUTOTOMY). Snake lizards may mimic snake behaviour (hissing, striking, etc.). There are 13 species, occurring in Australia and New Guinea.

pygostyle (tailbone) In birds (*Aves), a plate-like bone formed by the fusion of several caudal *vertebrae that supports the *uropygium, which bears the tail feathers.

pylorus 1. In vertebrates, the region of the *stomach that lies next to the *duodenum. **2.** In insects, the first part of the hind gut, leading to the *ileum.

Pyralidae (subclass *Pterygota, order *Lepidoptera) A very large family of small to large moths, in which the structure and colour vary extensively. The hind legs are generally long. The larvae are found in terrestrial and aquatic habitats. Many are pests of crops and dried vegetable products. Distribution is wide, but the family is absent from New Zealand.

pyramidal tract One of the two masses of nerve fibres running longitudinally and connecting the *medulla with the two parts of the *cerebellum.

pyramid of biomass A diagrammatic expression of *biomass at different *trophic levels in an *ecosystem, usually plotted as *dry matter per unit area or volume. Typically this gives a gradually sloping pyramid, except where the sizes of organisms vary dramatically from one trophic level to another. In this case, the higher metabolic rate of the smaller organisms may result in a greater biomass of consumers than of producers, giving an inverted pyramid. Aquatic communities in winter typically show inverted biomass pyramids. *Compare* PYRAMID OF ENERGY; PYRAMID OF NUMBERS. *See also* ECOLOGICAL PYRAMID.

pyramid of energy A diagrammatic expression of the rates of flow of energy through the different *trophic levels of an *ecosystem. Each bar of the pyramid represents the amount of energy per unit area or volume which flows through that trophic level in a given time period. The pyramid reflects the rates of photosynthesis, respi-

ration, etc. (and not the standing crop as in the *pyramid of biomass), and can never be inverted since energy is dissipated through the ecosystem. It is the most fundamental and most useful of the three ecological pyramids. *Compare* PYRAMID OF BIOMASS; PYRAMID OF NUMBERS. *See also* ECOLOGICAL PYRAMID.

pyramid of numbers A diagrammatic expression of the numbers of individual organisms present at each *trophic level of an *ecosystem. It is the least useful of the three types of *ecological pyramid since it makes no allowance for the different sizes and metabolic rates of organisms. Typically it slopes more steeply than the other pyramids and may be inverted, e.g. when based on studies of temperate woodlands in summer. *Compare* PYRAMID OF BIOMASS; PYRAMID OF ENERGY.

Pyrgomorphidae (spear-headed grasshoppers; suborder *Ensifera, superfamily Acridoidea) Family of slender, medium-sized, mainly tropical, short-horned grasshoppers, which may be winged or wingless, and cryptically or brightly coloured; they are characterized by having a conical head and a raised process on the *prosternum.

pyridoxene (vitamin B$_6$) A *vitamin required by all mammals that acts as a *coenzyme to *enzymes involved mainly in the *metabolism of *amino acids, *glucose, and *lipids; in the synthesis of *neurotransmitters, *histamine, and *haemoglobin; and in *gene expression.

pyriform Pear-shaped.

pyrimidine A basic, six-membered, heterocyclic compound; the principal pyrimidines (uracil, thymine, and cytosine) are important constituents of *nucleic acids. *Thiamin (vitamin B$_1$) is an important pyrimidine derivative, and other derivatives play major roles in carbohydrate and lipid metabolism.

Pyrochroidae (cardinal beetles; subclass *Pterygota, order *Coleoptera)

Family of soft-bodied, *pubescent beetles, up to 17 mm long, which are often partly scarlet. The antennae are long, and *filiform to *pectinate. The beetles are found under bark or feeding at flowers. Larvae have an elongate, toughened body, the terminal segment with a pair of backward-pointing spines. The legs are well developed. There are 116 species.

Pyrophorus (firefly) *See* ELATERIDAE.

Pyrotheria (superorder *Paenungulata, order *Amblypoda) An extinct suborder of elephant-like ungulates which lived in S. America during the *Oligocene. They were the size of modern elephants, had trunks and chisel-like tusks developed from the *incisors, and their *molar teeth were similar to those of early proboscideans. The skull and skeleton were similar to that of elephants. They arose independently of the true elephants, from an early ungulate stock apparently confined to S. America, and their resemblance to elephants arises from convergence (*see* CONVERGENT EVOLUTION).

Pyrrhocoridae (order *Hemiptera, suborder *Heteroptera) Small family of plant-feeding bugs which are mainly associated with tropical plants, especially those of the family Malvaceae. *Dysdercus* (cotton stainers) damage cotton bolls in the fields. *Pyrrhocoris apterus* (firebug) lives on mallows, hollyhocks, and lime trees in Europe.

Pyrrhocoris apterus (firebug) *See* PYRRHOCORIDAE.

Pyrrhula (bullfinches) *See* FRINGILLIDAE.

Pyrrhura (conures) *See* PSITTACIDAE.

Pyrrophyta *See* DINOFLAGELLIDA.

pyruvic acid A three-carbon keto acid (i.e. containing a *ketone group) that occupies a central position in cell metabolism. It represents the final product of *glycolysis in aerobic respiration and subsequently undergoes oxidation to carbon dioxide and *acetyl coenzyme A. During anaerobic respiration it is reversibly converted to *lactic acid in animal cells. It may also be variously converted into *alanine, *malic acid, and *oxaloacetic acid.

Python reticulatus (reticulated python) *See* BOIDAE.

pythons *See* BOIDAE.

Q

quadrat A measured area, of any shape and size, that is used as a sample area in a biological survey, particularly a survey of plants or of sessile and sedentary animals.

quadrate In *Gnathostomata other than *Mammalia, the posterior cartilage-bone of the upper jaw, attached to the brain case and articulating with the lower jaw. In mammals it forms the *incus of the inner ear.

quadritubercular Applied to a *molar tooth * that has four cusps.

quadrupedal Applied to animals that walk on four feet.

quail doves (Geotrygon) See COLUMBIDAE.

quails See PHASIANIDAE.

qualitative inheritance Inheritance of a *character that differs markedly in its expression amongst individuals of a *species (i.e. variation in that species is discontinuous). Such characters are usually under the control of *major genes. A principal example is gender. Compare QUANTITATIVE INHERITANCE.

quantitative character See QUANTITATIVE INHERITANCE.

quantitative inheritance The inheritance of a *character (known as a quantitative character or trait) that depends upon the cumulative action of many *genes, each of which produces only a small effect. Examples of such quantitative characters include clutch size in birds, milk production in cattle, and weight and skin pigmentation in humans. Usually the character shows continuous variation (i.e. a gradation from one extreme to the other). Compare QUALITATIVE INHERITANCE.

quantitative trait See QUANTITATIVE INHERITANCE.

quantum evolution (quantum speciation) Traditionally, the rapid speciation that can occur in small populations isolated from the large, ancestral population, and that are therefore subject to the *founder effect and to *genetic drift. Typically it results when a population shifts into another *adaptive zone. Initially the number of individuals will be small and rapid genetic and *phenotypic change is feasible in such circumstances, particularly if the new *niche is unoccupied. The concept of quantum evolution has been given special significance in the theory of *punctuated equilibrium.

quantum speciation See QUANTUM EVOLUTION.

quasi-sympatric speciation The separation of one *species into two by the adaptation of different subpopulations into different *niches.

Quaternary A sub-era of the *Cenozoic era that began 1.81 Ma ago and continues to the present day. Because human activity has become a significant global influence during this time, the Quaternary is sometimes known as the Anthropogene. The sub-era comprises the *Pleistocene and *Holocene epochs and is noted for numerous major ice-sheet advances in the northern hemisphere. By the Pleistocene most of the faunas and floras had a modern appearance. In the most recent revision of the geologic time scale (see appendix) sub-eras have been dropped, so the term Quaternary was to have been abandoned. The name is so widely used, however, especially among scientists engaged in Quaternary research, with a scientific journal *Quaternary Research* published by the Quaternary Research Center and the International Union for Quaternary Science, that the name has been retained. 'Quaternary' is now synonymous with *'Pleistogene'.

quaternary structure The structure of a *protein that results from the interaction of two or more individual *polypeptides to give larger functional molecules. *Compare* PRIMARY STRUCTURE; SECONDARY STRUCTURE; TERTIARY STRUCTURE.

queen In a colony of social insects, the primary female reproductive.

Queensland grouper (*Promicrops lanceolatus*) *See* SERRANIDAE.

Queensland hairy-nosed wombat (*Lasiorhinus barnardi*) *See* VOMBATIDAE.

***Quelea quelea* (red-billed quelea)** *See* PLOCEIDAE.

quetzal *See* TROGONIDAE.

Quetzalcoatlus northropi A huge pterosaur (*Pterosauria), first discovered in 1975 in Texas, that had a wing-span of 10 m or more and lived as do vultures, using warm air thermals to soar high over the *Cretaceous plains.

quinone p-Dioxybenzene or a derivative thereof; many quinones act as electron carriers in *mitochondria (and chloroplasts).

quoll (*Dasyurus*) Animal formerly called 'native cat'. *See* DASYURIDAE.

quorum sensing Among *Hymenoptera, the method by which the insects determine the number of individuals a new nesting site will accommodate. If the colony comprises more insects than this, it divides, the optimum number moving into the new site, taking their queen and brood with them, while the remainder find an alternative site.

q

rabbit *See* LEPORIDAE.

rabbitfish (*Chimaera monstrosa*) *See* CHIMAERIDAE; SIGANIDAE.

raccoon (*Procyon lotor*) *See* PROCYONIDAE.

raccoon dog (*Nyctereutes procyonoides*) *See* CANIDAE.

race An interbreeding group of individuals all of whom are genetically distinct from the members of other such groups of the same *species. Usually these groups are geographically isolated from one another, so there are barriers to *gene flow. Examples are island races of birds and mammals, such as the Skomer vole and the St Kilda wren. *See* SUBSPECIES.

rachis A stem, column (e.g. vertebral column), or a cord from which a stem or column develops. In a feather, the central shaft. The word is derived from the Greek *rhakhis*, 'spine'.

Rachycentridae (cobia, black kingfish; subclass *Actinopterygii, order *Perciformes) A monospecific family (*Rachycentron canadum*) comprising a marine fish, 1.6 m long, that has an elongate body, a deeply forked tail, and long-based but low second *dorsal and *anal fins, the first dorsal fin being reduced to a series of isolated spines. The cobia usually swims alone rather than in schools. Probably it is distributed world-wide in warmer seas.

Rachycentron canadum (cobia, black kingfish) *See* RACHYCENTRIDAE.

racial senescence *See* PHYLOGERONTISM.

racquet organ (malleolus) In *Solifugae, a sensory organ on the ventral surface of the fourth pair of legs, somewhat resembling a tennis racquet.

radar tracking An observational technique, used in field studies of animal behaviour, in which subjects are tracked by radar. By watching a plan-position indicator or from time-lapse photography of the screen, information can be obtained about the movements of birds and insects. Birds fitted with metal tags can be distinguished from untagged birds because of the stronger echo returned by the tags.

radial canals 1. In the *medusae of *Cnidaria, an arrangement of the gut cavity; the mouth leads to the stomach, hanging from the centre of the umbrella, from where canals (typically there are four) extend to the outer rim of the umbrella where they are linked by a ring canal around the periphery. **2.** In *Echinodermata, ciliated canals, forming part of the water-vascular system, that extend from the *stone canal (or ring canal or water ring) into each arm, passing along the *oral side of the ambulacral groove (*see* AMBULACRUM) and ending in a small, external *tentacle.

radial shield In some brittle-stars (*see* PHRYNOPHIURIDA), pairs of large plates, positioned radially, that surround the mouth.

radial symmetry The arrangement of the body of an animal in which parts are arranged symmetrically around a central axis. Such an arrangement allows the animal to interact with its environment from all directions. It is most commonly associated with a *sessile way of life. *Compare* BILATERAL SYMMETRY.

radioimmunoassay A technique for the very precise analysis of *proteins such as polypeptide *hormones, *antigens, *antibodies, and *enzymes. It is based on the ability of unlabelled proteins to inhibit competitively the binding of labelled protein by specific antibodies (i.e. an immunological reaction). A radioactive label (e.g. radioactive iodine) may be used to follow the interaction of an antibody with an antigen. The amount of radioactivity in the antibody–antigen precipitate is a measure of the degree of inhibition. The protein concentration of the unknown sample is determined by comparing the degree of inhibition with that produced by a series of standards containing known amounts of the protein. The technique has been adapted to assay non-proteins, such as *steroids and nucleotides.

(⊕) SEE WEB LINKS
• Explanation of the radioimmunoassay technique.

Radiolaria (superclass *Sarcodina, class *Actinopodea) A subclass of *Protozoa which possess more or less elaborate skeletons of silica. The cell body is generally spherical with *axopodia radiating from the periphery. Many species are known. They are chiefly *pelagic marine organisms. Fossil radiolarians have been found in *Cambrian rocks and they are important stratigraphic fossils for *Mesozoic and *Cenozoic deep-sea sediments.

radioreceptor A *receptor that responds to electromagnetic radiation, giving rise to the heat component of the sense of touch and to the sense of sight. *See also* CHEMORECEPTOR; MECHANORECEPTOR.

radiotelemetry *See* BIOTELEMETRY.

radio tracking An observational technique used in field studies of animal behaviour, in which the animal is fitted with a battery-powered, fixed-frequency radio transmitter, usually attached to a collar, and the observer is equipped with a receiver and directional antenna.

radius 1. In tetrapods, the pre-axial bone of the fore limb. **2.** In insects, a prominent vein in the anterior of the wing. **3.** In starfish (*Asteroidea), the direction of one of the arms.

radula In most *Mollusca, a chitinous strip, constantly renewed, with rows of teeth transversely across its surface, located on the floor of the *buccal cavity and used for rasping food.

ragfish (*Icosteus aenigmaticus*) *See* ICOSTEIDAE.

rail-babbler (*Eupetes macrocercus*) *See* TIMALIIDAE.

rails *See* RALLIDAE.

rainbowfish *See* MELANOTAENIIDAE.

rainbow runner (*Elagatis bipinnulatus*) *See* CARANGIDAE.

rain frog (*Breviceps adspersus*) *See* MICROHYLIDAE.

rajah butterflies *See* NYMPHALIDAE.

Rajidae (skate, thornback ray; superorder *Batoidimorpha, order *Rajiformes) A large family of marine rays that have a relatively short tail with one or two *dorsal fins, and ending in a blunt tip without a *caudal fin. The wide body or disc is angular in shape, often showing a pointed snout (rostrum). Male skates usually show enlarged *denticles on the dorsal surface (thorny back). The fish reproduce by laying eggs encased in rectangular, horny capsules which are sometimes found among the flotsam on the beach. There are about 120 species, distributed worldwide.

Rajiformes (rays; subclass *Elasmobranchii, superorder *Batoidimorpha) An order of mostly marine fish, which differ from sharks in having a strongly depressed body, with the gill openings on the ventral (belly) side and the eyes located on top of the head. The eight families comprising the order include e.g. the *Rajidae and *Dasyatidae.

rajiform swimming A type of swimming practised by fish, such as rays and knifefish, in which long fins extending for most of the body length undulate to effect movement. *Compare* ANGUILLIFORM SWIMMING; BALLISTIFORM SWIMMING; CARANGIFORM SWIMMING; LABRIFORM SWIMMING; OSTRACIIFORM SWIMMING.

Rajiform swimming

Rallidae (rails, coots, corncrake, crakes, gallinules; class *Aves, order *Gruiformes) A family of small to medium-sized birds which have black, olive, chestnut, or buff plumage, some being barred or spotted, laterally compressed bodies, short, rounded wings, short tails, and medium to long legs with long toes. The bill is short and conical to long and *decurved, and some bills have frontal shields. Rallidae are secretive, terrestrial, and often aquatic. Most live in marshes, feed on animal and vegetable matter, and nest on the ground or low in a bush. *Fulica cornuta* builds a mound of stones as a foundation to its nest. The 7–12 species of *Gallinula* (gallinules and moorhens) and 15 species of *Porzana* (crakes and rails) are typical. Coots (nine species of *Fulica*) have a red or white shield on the forehead, replaced in *F. cornuta* (horned coot) by a muscular *wattle. Despite appearing to be a weak flier, *Crex* (corncrake) is a long-distance migrant; its population is declining due to changing agricultural practices. There are 30–40 genera in the family, with about 125 species, some migratory, found world-wide.

Ramapithecus A middle to late *Miocene ape, known from fragmentary fossils from E. Africa, south-eastern Europe, and northern India and Pakistan, dating from 14–10 Ma ago, and very similar to the E. African *Kenyapithecus*. *Ramapithecus* was regarded by many as transitional between the true Miocene apes (the Dryopithecinae) and the later *Hominidae; if this were so, then the human and ape lines must have diverged 25–15 Ma ago, prior to the late Miocene. More recent evidence, however, suggests that *Ramapithecus* and the related or identical *Sivapithecus* are nearer to the evolutionary line that led to the orang-utan.

rami *See* RAMUS.

Ramphastidae (toucans; class *Aves, order Piciformes) A family of medium to large-sized birds which have brightly coloured and contrasting black, white, red, yellow, green, and blue plumage (e.g. the nine *Pteroglossus* species have black heads, throats, and backs, with a green gloss on the back, bright red rumps, and yellow underparts with bands of red and black across the breast). They have very large, deep bills coloured black, red, and yellow, with bare skin around the eyes. Their wings are short and rounded, their tails usually long, and their legs strong, with *zygodactylous feet. They are gregarious, inhabit forests, feed on fruit, insects, and young birds, and nest in tree holes. There are six genera, with about 40 species, found in Central and S. America.

ramus (*pl.* **rami)** A branch; a projection from a bone. In some *Crustacea, one of the branches of a limb.

ramus communicans A bundle of nerve fibres that connects a spinal nerve with a *ganglion in the *sympathetic nervous system. Bundles that have a *myelin sheath are called white ramus communicans, those without a myelin sheath are grey ramus communicans.

ram ventilation The production of respiratory flow in some fish in which the mouth is opened during swimming, such that water flows through the mouth and across the gills. In fish which have a reduced or no ability to pump water buccally, such as mackerel and sharks, perpetual swimming is required to maintain ventilation.

Rana catesbiana (bullfrog) *See* RANIDAE.

Rana pipiens (leopard frog, meadow frog) *See* RANIDAE.

***Rana ridibunda* (marsh frog, laughing frog)** *See* RANIDAE.

Rana temporaria (common frog, European frog) *See* RANIDAE.

random amplified polymorphic DNA (RAPD) A technique in molecular biology for the rapid assignation of DNA-based *character states for phylogenetic analysis (*see* PHYLOGENY). The technique uses the *polymerase chain reaction to amplify any genomic region containing an arbitrary sequence. The resulting amplified DNA fragment sizes are then used as characters for phylogenetic analysis.

random assortment *See* INDEPENDENT ASSORTMENT.

random sample Sample in which each individual measured or recorded is independent of all other individuals and also independent of prominent features of the area or other unit being sampled. In ecological field surveying, random *quadrat sampling is most easily achieved by superimposing a grid over the sample area, and identifying a series of random co-ordinates, using random-number tables, at which to locate the quadrats. Alternatively, a *random-walk technique may be used.

random-walk technique A method of sampling in which the number of paces between sample points is determined by random numbers, usually drawn from random-number tables, and from each sample point a right-angle turn determines the direction of the next point, a coin being tossed to decide whether to turn left or right. Where several sets of samples are taken it is best to start each set from the same point; if this is impossible it is important that the range of numbers from which a selection is made must allow each point within the study to have an equal chance of being sampled each time.

range *See* HOME RANGE.

range management The use of extensive, open grasslands for cattle production or, where appropriate, for the exploitation for human use of semi-domesticated or wild grazing animals (e.g. *game cropping on the African savannah) by methods that maintain the *ecosystem. These require the determination of the optimum stocking density that will permit long-term cropping of the grassland without deterioration of the pasture.

***Rangifer tarandus* (reindeer)** *See* CERVIDAE.

Ranidae ('true' frogs; class *Amphibia, order *Anura) A diverse family of amphibians in which the skeleton is modified for jumping. The tongue always has a notched end. Males often develop vocal sacs. *Dendrobates* (arrow-poison frogs) of tropical S. America are notable for their extremely vivid warning colours and highly poisonous skin, which secretes *batrachotoxin. *Rana catesbiana* (bullfrog) of N. and Central America, an essentially aquatic species that feeds on other frogs, young alligators, snakes, birds, and small mammals, has paired vocal sacs (and consequent loud voice) and the *tympanic membrane is larger than the eye. *R. pipiens* (leopard frog or meadow frog) is greenish brown with rows of black, brown, or green spots, ringed with a lighter colour; it is common in southern Canada, the USA, and much of Central America. *R. temporaria* (common frog or European frog) is the most widespread European species (but it does not occur in Ireland). *R. ridibunda* (marsh frog or laughing frog), an essentially aquatic species, is the largest European frog (up to 17 cm long). There are about 300 species, occurring in all continents except Antarctica; *Rana* is the only genus with a world-wide distribution, and the only genus in N. America.

RAPD *See* RANDOM AMPLIFIED POLYMORPHIC DNA.

Raphidae (dodos, solitaires; class *Aves, order *Columbiformes) An extinct family of large, flightless birds which had rudimentary wings and short tails. They were grey or brown, and *Raphus cucullatus* (dodo) had a large head and a heavy, hooked bill, while *R. solitarius* (Rodriguez solitaire) had a smaller head and a lighter bill. They were terrestrial, probably fed on fruit, and nested on the ground. There were two genera, with three species, known only from the Mascarene Islands. They became extinct in the 17th and 18th centuries.

Raphidiidae (snakeflies; subclass *Pterygota, order *Megaloptera) Family of predatory insects, which have four membranous wings with a complex venation, similar to those of lacewings and alderflies. The characteristic prolongation of the *prothorax in the adults, to form a snake-like neck, gives the family its common name. Adults and larvae are usually found on or under bark where they feed on aphids and other small insects.

Raphidura *See* APODIDAE.

raptor From the Latin *rapere*, 'to ravish', a bird of prey of the order *Falconidae (formerly known as Raptores).

rarity The relative abundance of a species and, therefore, its vulnerability to extinction. The International Union for the Conservation of Nature and Natural Resources (IUCN), formerly known as the World Conservation Union (WCU), measures the vulnerability of a species according to five criteria: (*a*) the rate at which its numbers are observed, inferred, or projected to be declining; (*b*) in association with (*a*), whether the species occurs as a single, small population or a few, small, fragmented ones; (*c*) in association with (*a*), whether the species occupies a small geographic range or area; (*d*) the size of the population; (*e*) a mathematical estimate of the predicted risk of extinction within a specified time. From this assessment, species are allocated a position on a continuum of increasing threat with three categories: as 'critical', 'endangered', or 'vulnerable'. Species that are known to be at risk of extinction, but fail to qualify for any of the main categories, are classified as 'susceptible'.

(⊕) SEE WEB LINKS
• IUCN Red List of Threatened Species

***Rasbora maculata* (dwarf rasbora)** *See* CYPRINIDAE.

rastellum In *Mygalomorphae, a row of spines on the *chelicerae that are used for digging.

rat 1. (Old World) *See* MURIDAE. **2. (New World)** *See* CRICETIDAE.

rat chinchilla (*Abrocoma*) *See* ABROCOMIDAE.

ratel (honey badger; *Mellivora capensis*) *See* MUSTELIDAE.

ratfish *See* CHIMAERIDAE.

ratite A flightless, running bird that has no keel on its *sternum. *Aepyornithidae, *Apterygidae, *Casuaridae, *Dinornithidae, *Dromaiidae, *Rheidae, and *Struthionidae are all ratites.

rat-kangaroos *See* POTOROIDAE.

rattlesnakes *See* CROTALIDAE.

ravens *See* CORVIDAE.

ray *See* RAJIFORMES.

ray-finned fish *See* ACTINOPTERYGII.

Ray's bream (*Brama brama*) *See* BRAMIDAE.

razorfish *See* CENTRISCIDAE.

R/B ratio *See* RESPIRATION–BIOMASS RATIO.

reading-frame shift A change in the grouping of *nucleotides so they are transcribed incorrectly. In the normal *transcription of a *cistron, nucleotides are read in threes, the 'reading frame' being determined by the starting-point. Each triplet codes a specific *amino acid; the sequence of *codons therefore dictates the amino acids and their order in a given *protein. Certain *mutagens (e.g. acridine dyes), which incorporate themselves between the complementary strands of the DNA, may cause errors in replication such that the daughter DNA gains or loses a nucleotide. When this daughter DNA is transcribed, the nucleotides will be read in the correct triplets up to the point of mutation, but the extra or missing nucleotide will cause a shift in the reading frame (to left or right) thereafter, such that all subsequent nucleotides will be read in wrong triplets. The result might be a protein with the wrong amino acids (often it is terminated early) but in any case it is dysfunctional.

reading mistake In genetics, the placement of an incorrect *amino acid into a *polypeptide chain during protein synthesis.

reafference The stimulation of an animal as a result of the movements of its own body. *Compare* EXAFFERENCE.

realized niche The *niche a viable population of a species occupies in the presence of competitor species. *Compare* FUNDAMENTAL NICHE.

rear-fanged snakes *See* COLUBRIDAE.

recapitulation of phylogeny A theory, due to E. H. Haeckel, asserting that ontogeny (the development of the individual) recapitulates or reflects the phylogeny (the evolutionary history of the individual). The theory as such has been rejected as not of general applicability,

von Baer's '*biogenetic law' being sufficient explanation for the observations on which it was based. However, either *hypermorphosis, *acceleration, or *predisplacement can result in individual cases of recapitulation.

Recent *See* HOLOCENE.

receptor A cell or group of cells that are specialized to respond to particular types of stimulus. There are three main types: *chemoreceptors, *mechanoreceptors, and *radioreceptors. A fourth type, *electroreceptors, occurs in some *teleosts. When the intensity of the stimulus exceeds a threshold the receptor releases an impulse to the *neurons connected to it.

receptor-mediated endocytosis *Endocytosis in which the cell selects and ingests a particular substance from the surrounding intercellular medium. Particles of molecules destined to enter the cell bind to specific *receptors on the cell surface and are then drawn into the cell by the invagination of the *cell membrane. The process first forms a coated pit lined with *proteins called *clathrin; the invaginated pit then detaches to become a coated vesicle inside the cell.

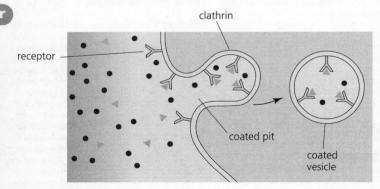

Receptor-mediated endocytosis

recessive gene A *gene whose phenotypic effect is expressed in the *homozygous state but is masked in the presence of the dominant *allele (i.e. when the organism is *heterozygous for that gene). Usually the *dominant gene produces a functional product whereas the recessive allele does not. Both one dose and two doses per nucleus of the dominant allele therefore lead to expression of its *phenotype, whereas the recessive allele is observed only in the complete absence of the dominant allele.

recipient *See* GRAFT.

reciprocal altruism (reciprocity) *Altruism in which a service performed by an individual is very likely to be reciprocated by the recipient. For example, many *Primates take turns at grooming and being groomed.

reciprocal cross In genetics, one of a pair of crosses in which the two opposite *mating types (or sexes) are each coupled with each of two different *genotypes and mated with the reciprocal combination, e.g. male of genotype A × female of genotype B (first cross), and male of genotype B × female of genotype A (the reciprocal cross). Such crosses are used to detect *sex linkage, maternal inheritance, or *cytoplasmic inheritance.

reciprocal genes Non-allelic *genes that reciprocate or complement one another. *See* COMPLEMENTARY GENES.

reciprocal predation A variety of *mutual antagonism in which two species each prey on the other.

reciprocity *See* RECIPROCAL ALTRUISM.

recognition species concept The concept that a *species is characterized by a unique fertilization system (the specific mate recognition system), which restricts gene-flow with other species. This changes the emphasis in the *biological species concept from negative (not interbreeding with other species) to positive (breeding specifically with members of the same species).

recombinant In genetics, an individual or cell with a *genotype that is produced by recombination (i.e. with combinations of genes other than those carried in the parents). Recombination results from *independent assortment or *crossing-over. *See also* GENETIC ENGINEERING.

recombination The arrangement of *genes in offspring in combinations that differ from those in either parent, and the assortment of *chromosomes into new sets.

recombination frequency The number of *recombinants divided by the total number of progeny, expressed as a percentage or fraction. Such frequencies indicate relative distances between *loci on a *genetic map.

recon The smallest unit of DNA that is capable of recombination.

recruitment pheromone A *pheromone released by a social insect, especially an ant (*see* FORMICIDAE) or a termite (*see* TERMITIDAE), that attracts other members of the nest to a food source.

rectal gill (branchial basket) Structure found only in the larvae of *Anisoptera. The gills project into the *rectum in six longitudinal rows, forming the branchial basket. Rectal gills are important for the uptake of salt, as well as for respiration.

rectum 1. In vertebrates, the part of the *alimentary canal that is adjacent to the *anus and in which *faeces are stored prior to defecation. **2.** In insects, the final section of the hind gut, leading from the *ileum.

recurved Bent backwards.

Recurvirostridae (avocets, stilts; class *Aves, order *Charadriiformes) A family of long-legged wading birds that have long bills, fly strongly, and some of which swim well. Most live near water and in many species the feet are webbed, at least partly. The plumage is usually black and white, and more or less similar in males and females. The birds feed on invertebrates, fish, frogs, and some plant material. Some authorities place the family in the order Charadriiformes, others in Scolopacidae. There are 13 species found in most parts of the world except in high northern latitudes.

red-and-white giant flying squirrel (*Petaurista alborufa*) *See* SCIURIDAE.

red-billed quelea (*Quelea quelea*) *See* PLOCEIDAE.

red coral (*Corallicum*) *See* SCLERAXONIA.

redd Gravelly spawning bed or nest in branches of rivers, made by salmon.

red deer (wapiti, American elk; *Cervus elaphus*) *See* CERVIDAE.

red fox (*Vulpes vulpes*) *See* CANIDAE.

red grouse (*Lagopus lagopus scoticus*) *See* TETRAONIDAE.

redirected behaviour Behaviour that is related to a stimulus, but is misdirected (e.g. an attack upon an inanimate object by an animal that cannot or dare not attack another animal that is the true target of its *aggression).

red junglefowl (*Gallus gallus*) *See* PHASIANIDAE.

red locust (*Nomadacris septemfasciata*) *See* LOCUST.

red mullet (*Mullus surmuletus*) *See* MULLIDAE.

redox potential A scale that indicates the *reduction (addition of electrons) and *oxidation (removal of electrons) for a given material. The position on the scale is expressed as an electric potential in millivolts, normally in the range 0–1300 or 0–1400 mV. The *pH of the sample must be known since this can alter the reading.

redox reaction A chemical reaction that involves simultaneous *reduction and *oxidation.

red panda (*Ailurus fulgens*) *See* AILURIDAE; PROCYONIDAE.

Red Queen effect An evolutionary principle, first proposed by Leigh Van Valen (1935–2010), that much of the evolution of a lineage consists simply of keeping up with environmental changes (mainly, tracking a deteriorating *environment), rather than occupying or adapting to new environments. The name is derived from the Red Queen, in Lewis Carroll's *Alice Through the Looking-Glass*, who had to run as fast as she could just to stay in the same place. *See also* ROMER'S RULE.

red squirrel (*Sciurus vulgaris*) *See* SCIURIDAE.

redstarts (*Phoenicurus*) *See* TURDIDAE.

reducing sugar A *sugar that contains an *aldehyde or potential aldehyde group and that has the ability to reduce certain inorganic ions in solution. These include the cupric ions in Fehling's and Benedict's reagents.

reduction A chemical reaction in which atoms or molecules either lose oxygen, or gain hydrogen or electrons.

reduction divisions In genetics, the two nuclear divisions in *meiosis that produce daughter nuclei each of which has one half as many *centromeres as the parental nucleus.

redundant cistron A *cistron (DNA sequence) that is present in many copies on one *chromosome, all but one of which are redundant. Examples are the cistrons in the *nucleolus organizer coding for the ribosomal (*see* RIBOSOME) RNA molecules.

Reduviidae (order *Hemiptera, suborder *Heteroptera) Family of bugs, most of which are predators of other arthropods, which they seize with the front pair of legs. They range from delicate, gnat-like forms preying on mosquitoes to large, robust insects capable of tackling heavily armoured tropical millipedes. A few suck the blood of birds and mammals. There are more than 4000 species, occurring throughout the world but especially in the tropics. *See also* ASSASSIN BUG; KISSING BUG.

Reduvius personatus *See* ASSASSIN BUG.

red-whiskered bulbul (*Pycnonotus jocosus***)** *See* PYCNONOTIDAE.

reedfish (*Calamoichthys***)** *See* BRACHIOPTERYGII.

refection (pseudorumination) In *Mammalia, a type of feeding in which food passes rapidly through the gut and is ingested for a second time as it leaves the anus. *See also* CAECOTROPHY.

reflex action An involuntary action that is made by an animal in response to a stimulus.

refuge A site, defined in space and time, within which particular organisms are sheltered from the competitive effect of other species.

Regalecidae (oarfish; subclass *Actinopterygii*, order *Lampridiformes*) A small family of oceanic fish that have a very elongate, ribbon-shaped body, the *dorsal fin originating on the head and continuing towards the tiny tail fin. The first few dorsal *fin rays are very long and form a crest or crown above the head. The *pelvic fins are reduced to a single ray trailing below the body. Reputed to reach a length of about 7 m, *Regalecus glesne* (oarfish) has a worldwide distribution. There are two genera and four species.

Regalecus glesne **(oarfish)** *See* REGALECIDAE.

regular echinoids An informal name for sea urchins (class *Echinoidea) in which the anus is enclosed within the apical system. The term includes the *Perischoechinoidea, *Diadematacea, and *Echinacea.

regular sample (systematic sample) One of a number of samples taken at regular intervals (e.g. by the use of regularly spaced *quadrats along some environmental gradient such as a valley side). Though less reliable in certain circumstances than *random sampling, regular sampling may be more practicable and more economical in the time it takes. The chief disadvantages are the possibility that the interval selected may resonate with some unsuspected environmental variable, so giving biased results, and that the form of the sample does not conform with that which is theoretically assumed for many statistical tests.

regulatory gene In the *operon theory of gene regulation, a *gene that is involved in switching on or off the *transcription of *structural genes. When transcribed, the regulatory gene produces a *repressor protein, which switches off an *operator gene and hence the operon that this controls. The regulatory gene is not part of the operon, and may even be on a different *chromosome.

Regulidae (kinglets, titwarblers; class *Aves, order *Passeriformes*) A family of tiny, loose-feathered birds which have pointed bills and short to medium-length tails. *Regulus* species (kinglets) are mainly green above, and buff below, with bright orange, yellow, or red crown patches, and inhabit coniferous woodland. *Leptopoecile* species (titwarblers) are multicoloured, the females being duller than males, and inhabit scrub and heathland. Members of both genera are insectivorous and nest in trees or bushes. There are two genera, with seven species, often placed in *Sylviidae, found in Europe, Asia, and N. America.

Regulus **(kinglets)** *See* REGULIDAE.

reindeer (caribou; *Rangifer tarandus*) *See* CERVIDAE.

reinforcement The strengthening of a response to a stimulus by reward or punishment.

Reinhardtius hippoglossoides **(Greenland halibut)** *See* PLEURONECTIDAE.

release factors In genetics, specific proteins that interact with the growing *polypeptide when a termination *codon is encountered during *transcription, and then mediate the release of the finished *polypeptide from the *ribosome.

releaser 1. A *pheromone that induces a change in the behaviour of the individual receiving it. The response is usually swift, but it soon degrades, e.g. when a mother emits a pheromone that induces her offspring to suckle, and when an individual emits a sex attractant that induces potential mates to move towards the source, and must be frequently refreshed. **2.** See SIGN STIMULUS.

relict fauna Animals belonging to groups that once had a wide distribution but now occur very locally (e.g. coelacanth).

remex (*pl.* remiges) One of the flight feathers of a bird, the remiges consisting of the *primaries and *secondaries.

remiges See REMEX.

Remizidae (penduline tits, verdin; class *Aves, order *Passeriformes) A family of small, agile, brown or yellow birds, one species of which has a black mask, and all of which have finely pointed bills. They inhabit open country with trees and bushes, or reed-beds, feed on insects and seeds, and build elaborate, globular nests with a side entrance. There are four genera, with 10 species, found in Europe, Asia, Africa, and N. and Central America.

remora See ECHENEIDAE.

renal Pertaining to the *kidney (from Latin *renes*, 'kidneys').

renal tubule See TUBULE.

reniform Kidney-shaped.

rennin A *protease enzyme, found in the abomasum (*fourth stomach) of ruminants, which causes curdling of milk through the conversion of the soluble protein caseinogen to the insoluble casein. This is then attacked by the enzyme pepsin.

repair synthesis The enzymatic excision and replacement of regions of damaged DNA. The defective single-stranded segments of DNA are cut out, correct bases are inserted using the complementary strand as a template, and these are interlocked by a DNA polymerase. A polynucleotide ligase joins the two ends of the broken strand to complete the repair. An example is the removal by a nuclease enzyme of a thymine *dimer that has been induced by ultraviolet radiation of the DNA molecule.

repetitive DNA See SATELLITE DNA.

replicase Any *enzyme that is capable of catalysing the replication of DNA or *RNA in either *prokaryotes or *eukaryotes.

replication The synthesis of new daughter molecules of nucleic acid from a parent molecule, which acts as a template.

repressor In the *operon theory of *gene regulation, a protein produced by a *regulator gene that inhibits the activity of an *operator gene, and hence switches off an *operon. The action of a repressor is determined by some other factor, an effector: either this serves as a co-repressor, and hence is required to switch off the operon; or it inactivates the repressor, and so switches on the operon.

reproductive Fertile male or female member of a colony of social insects. In ants, bees, and wasps the male (drone) dies soon after insemination, leaving the queen as the colony reproductive. In termites the male (king) lives with the queen, periodically inseminating her. Supplementary reproductives, present in termite colonies, take over colony reproduction if the founding king and queen (the primary reproductives) die.

reproductive allocation The investment of resources (e.g. of food energy and time) for the purpose of reproduction (i.e. *courtship, mating, and raising of young).

reproductive effort The proportion of its resources that an organism expends on reproduction.

reproductive isolating mechanism (RIM) The means by which different *species are kept reproductively isolated. These may be: (*a*) chromosomal (if cross-mating occurs, the incompatibility of the *karyotypes makes any hybrid inviable or sterile); (*b*) mechanical (the two species cannot mate because they are of different sizes, or because the genitalia are shaped differently); (*c*) ethological (the courtship rituals of the two species diverge at some point so that an incorrect response is given and the sequence is brought to a stop); or (*d*) ecological (the two species occupy different microhabitats and normally do not meet). Other mechanisms include the breeding seasons being out of phase, or members of one species being unattractive to members of the other. Many RIMs, especially ethological ones, amount merely to mate preference, so that in the absence of a preferred partner (of the same species) a member of a different species will be accepted. In this way hybrids between different species may be bred in captivity and may even be found to be fully fertile.

Reptilia (reptiles; phylum *Chordata, subphylum *Vertebrata) A large and diverse class of *poikilothermic vertebrates, which arose in the *Carboniferous from labyrinthodont amphibians (*Labyrinthodontia). They were the dominant animals of the *Mesozoic world and gave rise to the birds and mammals. Reptiles have a body covering of ectodermal scales, sometimes supported by bony *scutes. There is no gilled larval phase; development is by *amniote egg, but *ovovivipary is common. Reptiles are air-breathing from hatching onwards. The kidney is metanephric (*see* METANE-PHROS). The heart is incompletely divided. There are about 6000 species on all continents except Antarctica.

rescue effect A *species arriving on an island may already be represented there and so may have the effect of reducing the chance of the extinction of that species from the island (i.e. of 'rescuing' it). The rescue effect will be greater on islands which are closer to the mainland source of species than more remote islands because the immigration rate will be higher. *See also* ISLAND BIOGEOGRAPHY.

residual In statistics, data variability that is not accounted for by a particular statistical test. The residuals of individual data values, i.e. their difference from the statistically predicted value, often give ecologists insight into possible environmental influences on the individual data records.

residual body A lysosomal structure that contains only indigestible or slowly digestible material and within which enzymatic activity has become virtually exhausted.

resilifer In *Bivalvia, the extension of the valve to which the interior hinge ligament is attached.

resilium In *Bivalvia, the name sometimes given to the internal ligament.

resorb To re-absorb; i.e. to metabolize (*see* METABOLISM; METABOLITE) substances or structures that were produced metabolically by the body. For example, in some mammals *foetuses are resorbed if they are not viable. *See also* POSTLARVA.

resource partitioning *See* DIFFEREN-TIAL RESOURCE UTILIZATION.

respiration 1. Oxidative reactions in cellular metabolism that involve the sequential degradation of food substances and the generation of a high-energy compound (*ATP) in aerobic respiration with the use of molecular oxygen as a final hydrogen acceptor; ATP, carbon dioxide, and water are the products thus formed. **2.** The physicochemical processes involved in the transportation of oxygen to and carbon dioxide from the tissues. **3.** (external respiration) The act of breathing.

respiration–biomass ratio (R/B ratio) The relationship between total community *biomass and respiration. With larger

r

biomass, respiration will increase but the increase will be less if the individual biomass units or organisms are large (reflecting the inverse relationship between size and metabolic rate). Natural communities tend towards larger organisms and complex structure with low respiration rates per unit biomass.

respiration quotient (RQ) The ratio of the amount of carbon dioxide expired to the amount of oxygen consumed during the same period. (RQ = volume of carbon dioxide consumed divided by volume of oxygen utilized.)

respiratory assembly *See* CRISTA.

respiratory chain *See* ELECTRON-TRANSPORT CHAIN.

respiratory trees In *Holothuroidea, an internal respiratory apparatus, consisting of a short tube extending from the hind gut which divides into two longer tubes with numerous, complexly ramifying branches.

respirometer A device that is used to measure the *respiration rate of an animal from the exchange of oxygen and carbon dioxide. A simple version consists of a sealed chamber containing a substance that absorbs carbon dioxide (e.g. soda lime pellets or cotton soaked in potassium hydroxide) and with a *manometer to monitor the amount of air entering the chamber. The animal is placed inside the chamber. As the carbon dioxide it releases in respiration is absorbed, air is drawn in to replace it and the manometer registers the amount entering.

resplendent quetzal (*Pharomacrus mocino***)** *See* TROGONIDAE.

resting potential *See* ALL-OR-NOTHING LAW.

restriction enzyme An *enzyme that causes a molecule of foreign *DNA to break. Such enzymes occur in many bacteria, where they destroy viruses. Restriction enzymes are used widely in *genetic engineering.

rete mirabile (wonderful net, *pl.* retia mirabilia) In mammals adapted to arctic and polar climates, a network of arteries and veins that lie side by side in the lower parts of the limbs. Warm arterial blood flowing towards the extremities warms venous blood flowing towards the heart, thus allowing the temperature in the extremities to fall close to the ambient, while reducing the energy expenditure needed to maintain the core temperature of the body. For example, when the air temperature is –30°C the paws of an arctic fox (*Alopex lagopus*) are at about 0°C and its nose at 5°C, but its shoulder and chest muscles are at 37°C and 35°C, respectively.

reticular fibres (argyrophil fibres) Extremely fine fibres (cells) that form networks around and between cells in many vertebrate tissues (e.g. blood vessels, nerve cells, muscle cells, and several organs). They can be seen when stained with silver salts (hence the name 'argyrophil'—'silver-loving'). Their main constituent (reticulin) is also found in *collagen fibres, which reticular fibres resemble in some respects (e.g. they are inelastic and are not digested by *trypsin).

reticular groove A groove along the upper margin of the forestomach of ruminants and colobid monkeys (*see* CERCOPITHECIDAE), which can be closed into a channel conducting liquid foodstuffs past the fermentation chamber directly into the regions where digestion takes place.

reticular lamina *See* BASAL LAMINA.

reticulated python (*Python reticulatus***)** *See* BOIDAE.

reticulate evolution The development of a network of closely related taxa within and at the *species level, particularly by *chromosome doubling or *polyploidy. Polyploidy is more common in plants than in animals, and so reticulate evolution is more likely in the former than in the latter.

reticulate method Term sometimes used to describe a non-hierarchical clustering technique.

reticulin *See* RETICULAR FIBRES.

reticulo-endothelial system In vertebrates, a mechanism by which foreign particles are removed from *blood or *lymph by *macrophages.

reticulopodium A branching and anastomosing network of cytoplasmic threads, formed for example by foraminiferans.

reticulum In *Ruminantia, the second chamber of the stomach, having a honeycomb-like structure. This is sometimes linked with the rumen, so forming one large chamber.

retina The light-sensitive membrane in the eye on to which incoming light is focused. The retina has two layers: the outer layer is pigmented and so prevents the back-reflection of light; the inner layer contains blood vessels, nerve endings, and light-sensitive cells (*rods and *cones). *See also* TAPETUM.

retinaculum Hook or clasp that projects from the fore wing of an insect to allow wing coupling when it is engaged with a hair or spine from the posterior wing. *See* COLLEMBOLA; FRENULUM.

retinol A diterpene compound, composed of four isoprene units structurally related to carotenes, which is involved in the synthesis of the visual pigments. A deficiency of this compound results in a number of symptoms, including cessation of growth, cornification of epithelial structures, and, most notably, night blindness.

retinula cell In an *ommatidium, a long *neuron surrounded by secondary pigment cells and containing a specialized region known as a rhabdomere, from which a nerve *axon projects through the basement membrane to the optic nerve.

Retortamonadida (superclass *Mastigophora, class *Zoomastigophorea) An order of *Protozoa which live as commensals (*see* COMMENSALISM) in the gut of humans and other vertebrates. The cells have two to four *flagella and may form cysts.

retrices *See* RETRIX.

retrix (*pl.* retrices) One of the tail feathers of a bird. There are usually 12 retrices.

retrochoanitic Applied to a *septal neck that points backwards.

Retropinnidae (southern smelt; superorder *Protacanthopterygii, order *Salmoniformes) A small family of fresh- and brackish-water fish, closely related to the smelts of the northern hemisphere (*Osmeridae), that are slender, silvery, and have a single soft-rayed *dorsal fin located well back above the *anal fin, and a small *adipose fin. The eyes are rather large. The family comprises two Australian species and two New Zealand species. *Prototroctes maraena* is the Australian grayling.

retrovirus A *virus whose genetic material consists of *RNA and is able, by means of the enzyme reverse transcriptase, to manufacture its DNA equivalent and insert it into the *genome of its host species. It has been suggested that this could be a way of overriding *Weissmann's doctrine, and in effect causing Lamarckian inheritance. *See* NEO-LAMARCKISM.

reversal A form of *homoplasy; resemblance between two taxa because one of them has gained a new character, then lost it again, and the other taxon has never gained it. Reversal by character loss is common; there is much doubt about whether it ever occurs by regaining a lost character.

reverse mutation (reversion) The production by further *mutation of a pre-mutation *gene from a mutant gene. This reverse mutation restores the ability of the gene to produce a functional protein. Strictly, reversion is the correction of a mutation (i.e. it occurs at the same site); more loosely, the term is applied also to a mutation at another site that masks or suppresses the effect of the first mutation (in fact such organisms are not non-mutant, but double mutants with the same *phenotype). *Compare* SUPPRESSOR MUTATION.

reversion See REVERSE MUTATION.

rhabdites In *Turbellaria and some species of *Nemertea (indicating a link between flatworms and nemerteans), rod-shaped bodies of unknown function contained in scattered cells in or below the *ectoderm. Some workers have suggested they are defensive structures whose discharge produces a protective, sticky coat around the organism; others consider them to be excretory.

Rhabdocoela (Neorhabdocoela; class *Turbellaria, subclass *Neoophora) An order of worms which possess a simple intestine. The *pharynx is present. Most members are free-living.

rhabdome See OMMATIDIUM.

rhabdomere See RETINULA CELL.

Rhabdophorina (subclass *Holotrichia, order *Gymnostomatida) A suborder of ciliate *Protozoa in which the *cytostome occurs at the front or to one side of the cell. The organisms prey on other Protozoa which they ingest via the expansible cytostome. There are many genera, found in aquatic environments.

Rhabdopleurida (phylum *Hemichordata, class *Pterobranchia) An order of colonial animals in which *zooids are *pectocaulus. The *mesosoma bears one pair of arms. The gonads are unpaired, with a single gonadial opening on the right side. There are no branchial pores. The skeleton (coenoecium) is built of an irregularly branching system of chitinous tubes composed of regular growth bands which are clearly defined by transverse growth lines. Their first known occurrence is in the Late *Cretaceous.

Rhabdornis (Philippine creeper) See RHABDORNITHIDAE.

Rhabdornithidae (Philippine creepers; class *Aves, order *Passeriformes) A family of small birds which have brown or grey upper-parts and white under-parts, with streaked flanks. They have long, strong,

*decurved bills, brush-tipped tongues, long toes, and squared tails. They are arboreal, inhabit thick woodland, climb up and down trees in search of insects, and nest in tree holes. There is one genus, *Rhabdornis*, with two species, found only in the Philippines.

rhabdosome In *Graptolithina, a complete colony.

Rhacochilus toxotes (rubberlip seaperch) See EMBIOTOCIDAE.

Rhacophoridae (class *Amphibia, order *Anura) A family of tropical, arboreal frogs which have webbed feet with an extra cartilaginous element between the last and next to last finger and toe that aids adhesion. Vocal sacs are usually present. Eggs are generally laid in 'nests' of leaves or in foam over water. Species include the Asian 'flying' frogs, which use their large webbed feet as gliding membranes. There are many species, often brightly coloured. They are found in Africa, Madagascar, and south-east Asia.

Rhagionidae (snipe flies; order *Diptera, suborder *Brachycera) Family of flies, most of which lack bristles, and in which the third antennal segment has a terminal style. Some of the *tibiae are spurred with *pulvilli, and the *arolium is pad-like. *Squamae are virtually absent, and the wing veins are well defined. The elongate adult flies are usually predacious on other insects. The larvae are carnivorous in earth or vegetable mould. Adults of *Rhagio scolopacea* have 'swarms' of males which serve as a focal point for mating. There are more than 400 described species.

Rhamphichthyidae (knifefish; subclass *Actinopterygii, order *Cypriniformes) A small family of freshwater fish that have a very long, tapering body lacking *dorsal and *pelvic fins, but with a very long *anal fin. They have small eyes and a produced snout, with the upper jaw overlapping the lower. Knifefish are known to produce weak electrical discharges, which may assist in orientation or food location. There are several species, occurring in S. America.

Rhaphidophoridae (camel crickets, cave crickets; order *Orthoptera, suborder *Ensifera) Family of wingless, cricket-like insects, most species of which have very slender, elongate legs, and antennae which are longer than the body. They are rather hump-backed. All species live in moist environments, and they often dwell in caves. They are distributed from warmer regions to circum arctic areas.

***Rhea americana* (common rhea)** *See* RHEIDAE.

rheas *See* RHEIDAE.

Rheidae (rheas; class *Aves, order *Rheiformes) A family of large, ostrich-like birds which have long necks, wide, flat bills, short wings with soft feathers, and no tail feathers. Their body plumage is grey-brown and loose-webbed. Their legs are powerful, with three toes, enabling them to run swiftly. They are terrestrial and flightless, inhabit grassland and open brush, feed on vegetable matter and insects, and nest on the ground. *Rhea americana* (common rhea) has grey upper-parts and white under-parts, the male having a black base to its neck. *Pterocnemia pennatus* (Darwin's rhea) is similar, but smaller and with white spots. There are two *monotypic genera, found in S. America.

Rheiformes (rheas; class *Aves) An order which comprises one family, *Rheidae.

rheotaxis A change in direction of locomotion in a *motile organism or cell that is made in response to the stimulus of a current, usually a water current.

rhesus factor (Rh factor) An *antigen that occurs on the surface of the red blood cells of some individuals (who are designated rhesus positive, Rh positive, or Rh+) but not of others (who are rhesus negative, Rh negative, or Rh–). An Rh– person will develop anti-Rh *antibodies when given Rh+ blood; a subsequent transfusion of Rh+ blood will then cause *agglutination. The phenomenon was first detected in rhesus monkeys (*see* CERCOPITHECIDAE).

rhesus monkey (*Macaca*) *See* CERCOPITHECIDAE.

Rh factor *See* RHESUS FACTOR.

rhinarium In *Mammalia, the area of naked skin surrounding the nostrils.

Rhincodontidae (whale shark; subclass *Elasmobranchii, order *Lamniformes) A monospecific family (*Rhincodon typus*) comprising a shark that is thought to reach a length of more than 15 m and has the distinction of being the world's largest fish. Whale sharks are frequently seen swimming near the surface and can be recognized by the exceptionally large gill slits, which extend almost to the tip of the head. Despite their size, these sharks are not aggressive and feed on small planktonic animals. They have a world-wide distribution.

***Rhincodon typus* (whale shark)** *See* RHINCODONTIDAE.

rhinencephalon The region of the *brain concerned with the sense of smell; it is located in the *cerebrum.

Rhinobatidae (shovelnose ray, guitarfish; superorder *Batoidimorpha, order *Rajiformes) A family of marine rays that have a rounded, shark-like trunk, but a broad head with wide, ray-like, *pectoral fins, the overall body outline resembling a shovel. Often found in sandy or muddy areas, some species may reach a length of about 1 m. There are about 45 species, distributed world-wide.

rhinoceros *See* RHINOCEROTIDAE.

rhinoceros beetle *See* SCARABAEIDAE.

***Rhinoceros sondaicus* (Javan rhinoceros)** *See* RHINOCEROTIDAE.

***Rhinoceros unicornis* (Indian rhinoceros)** *See* RHINOCEROTIDAE.

rhinoceros viper (*Bitis nasicornis*) *See* VIPERIDAE.

Rhinocerotidae (rhinoceroses; suborder *Ceratomorpha, superfamily *Rhinoceratoidea) A family that includes the rhinoceroses and their fossil relatives. They were derived from the *Hyracodontidae ('running rhinoceroses') in the *Eocene and were prominent in the *Oligocene, producing a variety of different types. They are characterized by the molarization of the *premolars, the enlargement of the first upper and second lower *incisors as cutting teeth (subsequently lost in some lines) and the consequent development of a pointed lip and narrow muzzle, and a simple last upper *molar. The skeleton is similar to that of elephants, having vertebrae with long neural spines and many ribs so the spine and ribs together form a weight-bearing 'girder' resting on the fore limbs and counterbalanced by the weight of the head. All later rhinoceroses have three digits on each limb, but early forms had four. Probably early forms were hornless. While all rhinoceroses tended to large size, some aberrant forms became very large indeed. *Paraceratherium* (formerly *Baluchitherium*), a hornless form which lived in Asia during the Oligocene, was about 5.5 m tall at the shoulder, had a long neck and a skull about 1.2 m long, and must have weighed about 20 tonnes, making it the largest land mammal ever to have lived. *Elasmotherium*, which may have possessed a very large horn, was also a large form which lived on the Eurasian plains during the *Pleistocene. *Coelodonta* was the woolly rhinoceros which lived during the Pleistocene. Today there are four genera, and five species: *Rhinoceros unicornis*, the one-horned Indian form; *R. sondaicus*, the related Javan species (now very rare); *Dicerorhinus sumatrensis*, the rare two-horned Sumatran form; and the African rhinoceroses *Diceros bicornis* (black) and *Ceratotherium simum* (white).

Rhinocerotoidea (order *Perissodactyla, suborder *Ceratomorpha) A superfamily that comprises the rhinoceroses and their relatives, grouped into the families *Hyracodontidae, *Amynodontidae, and *Rhinocerotidae. They are believed to have evolved from tapir-like forms and throughout most of the *Cenozoic they were abundant over most of the northern hemisphere. Many developed horns (the name is derived from the Greek *rhino-*, 'nose', and *keras*, 'horn') made from fused hair, and many attained large size, with short legs. Early forms possessed five digits, later forms four and then three, but no member of the superfamily has possessed fewer than three. The digits have nail-like hoofs. The brain is small, eyesight is poor, the most highly developed senses being those of scent and hearing. They are mainly nocturnal, solitary, and timid. Some graze in small herds.

Rhinochimaeridae (longnose chimaera; subclass *Holocephali, order *Chimaeriformes) A small family of marine fish which differ from their relatives, the shortnose chimaeras (*Chimaeridae), in having a long and pointed snout. There are about six species, distributed in the Atlantic and Pacific.

Rhinocryptidae (tapaculos; class *Aves, order *Passeriformes) A family of small to medium-sized birds which have mainly dark brown or black plumage (the 12 species of *Scytalopus* have some barring on the underside, especially in immatures, and some have white under-parts or *supercilia). Tapaculos have short, sharp bills, short, rounded wings, and short to long tails that are frequently cocked. They have long legs with large feet, and are mainly terrestrial, walking or running. They inhabit forest undergrowth and grassland, feed on insects and seeds, and nest in holes in banks or trees, or in bushes. There are 12 genera, with about 30 species, found in Central and S. America.

***Rhinoderma darwinii* (Darwin's frog, mouth-breeding frog)** See LEPTODACTYLIDAE.

Rhinolophidae (horseshoe bats; order *Chiroptera, suborder *Microchiroptera) A family of insectivorous bats in which the nose leaf forms a disc, somewhat reminiscent of a horse-shoe in shape, with the nostrils visible within it, and above the leaf a spur projecting forward and a lance-shaped extension projecting upward above the eyeline. The ears are pointed, separate, and very mobile, with a prominent *antitragus. The tail extends to the edge of the *uropatagium. There are two genera, with about 70 species, distributed widely throughout the Old World, but most abundantly in the tropics and subtropics.

Rhinophrynidae (Mexican burrowing toad; class *Amphibia, order *Anura) A monospecific family (*Rhinophrynus dorsalis*) comprising a toad in which the tongue is free at the front and is used to lick up ants and termites. There are no teeth, no external ears, and no free ribs. There is a horny digging 'spade' on the hind feet. The toads spawn after heavy rain. The tadpoles are filter feeders with oral barbels.

Rhinophrynus dorsalis (Mexican burrowing toad) *See* RHINOPHRYNIDAE.

Rhinopithecus (snub-nosed monkey) *See* CERCOPITHECIDAE.

Rhinopoma (mouse-tailed bats) *See* RHINOPOMATIDAE.

Rhinopomastus (scimitarbills) *See* PHOENICULIDAE.

Rhinopomatidae (mouse-tailed bats; order *Chiroptera, suborder *Microchiroptera) A family of bats in which the tail is very long and thin, projecting well beyond the edge of the small *uropatagium. A fleshy, naked pad surrounds the nostrils, but is not a true nose leaf, being used only to close the nostrils between breaths. The bats are distributed throughout northern-hemisphere deserts and arid lands from Africa to Sumatra. There are three species in a single genus, *Rhinopoma*.

Rhinotermitidae (subclass *Pterygota, order *Isoptera) Family of lower termites, some species of which have *nasute*soldiers, which immobilize enemies with an adhesive suspension. There are about 150 species, with a world-wide, but mainly tropical, distribution.

Rhiphiphoridae (subclass *Pterygota, order *Coleoptera) Family of black or brown beetles, often with a yellow pattern, 3–30 mm long, in which the body is humped and tapered, frequently with reduced *elytra; some females lack wings or elytra. The head is constricted immediately behind the eyes. Males have large, fan-shaped antennae. Larvae are minute; all are insect parasites. There are 250 species.

Rhipidistia (class *Osteichthyes, subclass *Crossopterygii) A group of 'tassel-finned' fish that lived from the *Devonian to the *Permian Periods. They possessed two *dorsal fins, lobate or stalked *pectoral and *pelvic fins, a *heterocercal or *diphycercal tail fin, and internal nares. Distantly related to the living coelacanth, they are considered by some to be ancestral to the tetrapod, terrestrial vertebrates (i.e. *Amphibia).

Rhipidura (fantails) *See* MUSCICAPIDAE.

Rhipidura leucophrys (willie wagtail) *See* MUSCICAPIDAE.

Rhizocephala (parasitic barnacles; subphylum *Crustacea, class *Cirripedia) Order of naked barnacles which are entirely parasitic, mainly on *Decapoda. Rhizocephalans are recognizable as barnacles only in their larval stages. The *peduncle comprises foot-like absorption processes which penetrate the host tissues. Appendages and digestive tract are absent.

Rhizomastigida (superclass *Mastigophora, class *Zoomastigophorea) An order of *Protozoa which can form both *pseudopodia and *flagella. Most species are free-living in non-saline, stagnant water. Some species are marine, a few are parasitic.

Rhizomyidae (bamboo rats; order *Rodentia, suborder *Myomorpha) A family of burrowing rodents (often considered a subfamily of *Muridae) which feed above ground and which resemble American pocket gophers, although they lack cheek pouches. The *incisors project below the upper lip, the legs are short, *pentadactyl, and possess strong claws, and the tail is short and sparsely haired. The head is broad and flat, and the eyes and ears are small.

There are about six species, in three genera (*Rhizomys*, *Cannomys*, and *Tachyoryctes*), found in south-east Asia and tropical E. Africa.

Rhizomys See RHIZOMYIDAE.

Rhizophysaliina (class *Hydrozoa, order *Siphonophora) A suborder of *Cnidaria whose members are characterized by the possession of a large, hollow float provided with a pore. *Physalia physalis* (Portuguese man-of-war) is an example.

Rhizopoda See RHIZOPODEA.

Rhizopodea (Rhizopoda; subphylum *Sarcomastigophora, superclass *Sarcodina) A class of *Protozoa which feed and move by means of *lobopodia or *filopodia. Species are free-living in soil, freshwater, or marine environments, or are parasitic in various types of plant and animal.

Rhizostomatida (class *Scyphozoa, subclass *Scyphomedusae) An order of free *medusae lacking *tentacles and having the mouth obliterated by the growth across it of eight thick, gelatinous, oral arms, which contain numerous small mouths. These medusae are found in tropical seas, mainly in the Indo-Pacific region.

rhodopsin (visual purple) The visual pigment found in rod cells in the retina of vertebrates. It is composed of a protein, called an opsin, and retinol, and has an absorption maximum at 500 nm.

rhombencephalon In *Craniata, the posterior parts of the brain (hindbrain), *metencephalon and *myelencephalon. *Compare* PROSENCEPHALON.

Rhombozoa See DICYEMIDA.

rhopalioid See RHOPALIUM.

rhopalium (rhopalioid) In some *Scyphozoa, one of 4–16 club-shaped structures located around the margin of the bell containing concentrations of *neurons and involved in control of pulsation.

Rhopalosomatidae (suborder *Apocrita, superfamily Pompiloidea) Family of *Hymenoptera, which are sometimes referred to the Scolioidea, but whose inclusion in the Pompiloidea is based on the possession and development of a preening *calcar and *strigil on the hind leg. Adults are slender and brownish, 6–25 mm long, and have long antennae each segment of which has two apical spines. The larvae are parasitic on the abdomens of small crickets (*Gryllidae).

Rhyacichthyidae (loach goby; subclass *Actinopterygii, order *Perciformes) A possibly monospecific family (*Rhyacichthys aspro*) of freshwater fish that have a flattened head and belly, a low-slung mouth, and small eyes. The two *dorsal fins are well separated; the tail fin and *pectoral fins are rather large. The anterior part of the body, together with the *pelvic fins, form a type of suction device which enables the fish to secure its position in fast-flowing hill streams. The structure of this suction device is similar to that found in the true hillstream loaches (*Homalopteridae). There are one or two species, distributed widely in southeast Asia. The name 'loach goby' is perhaps unfortunate, as the fish is neither a loach nor a goby.

Rhyacichthys aspro (loach goby) See RHYACICHTHYIDAE.

Rhynchobdellida (phylum *Annelida, class *Hirudinea) An order of exclusively aquatic, parasitic worms in which the freshwater forms tend to be flattened in shape. They all have a proboscis. None possess jaws. There are two families.

Rhynchocephalia ('beak-heads', rhynchocephalians; class *Reptilia, subclass *Lepidosauria) An order of primitive, lizard-like reptiles, dating from the *Triassic, that contains only one family, the *Sphenodontidae, comprising one species, *Sphenodon punctatus*, the tuatara of New Zealand.

rhynchocoel In *Nemertea, a fluid-filled cavity containing the proboscis. Since it is of *mesodermal origin, it is considered a true *coelom even though its extent is limited.

Rhynchodina (subclass *Holotrichia, order *Thigmotrichida) A suborder of ciliate *Protozoa, in which body *cilia are sparse or even absent, and in which the *cytostome is replaced by a type of sucker. They are parasites or commensals in aquatic molluscs.

rhyncholite One of the beak-like structures which are considered to be the upper-jaw structures of fossil cephalopods (*Cephalopoda). They are approximately rhomboidal in shape, with a slightly concave lower surface. The anterior portion is termed the hood, the posterior portion the shaft. The first examples are found in the *Carboniferous.

Rhynchonellida (phylum *Brachiopoda, class *Articulata) An order of brachiopods that have *rostrate shells, a functional *pedicle, and a *delthyrium partly restricted by a pair of *deltidial plates. The shell is usually impunctate. They first appeared in the Middle *Ordovician. The order contains about 250 genera, most of which are extinct.

Rhyncocoela *See* NEMERTEA.

Rhyncophthirina (subclass *Pterygota, order *Phthiraptera) Suborder of lice, which are parasitic on elephants and warthogs. Members of the family are distinguished from all other lice by the elongation of the front of the head into a rigid, tubular *rostrum, at the end of which are the mouthparts. The *mandibles are toothed on the outer surface and are used to drill into the skin of the host. The insect feeds on blood, and the length of the rostrum is related to the thickness of the skin of the host: *Haematomyzus elephantis* (elephant louse) has a much longer rostrum than *H. hopkinsi* (warthog louse). The elephant louse is generally found behind the ears of its host, where the skin is thinnest. The suborder consists of only one family, Haematomyzidae, with one genus, *Haematomyzus*, containing these two species.

Rhynochetidae (kagu; class *Aves, order *Gruiformes) A monospecific family (*Rhynochetos jubatus*) which is a large, grey bird with black, white, and chestnut, barred wings. It has a long, loose crest, a long, sharp, slightly *decurved, red bill, and long, red legs with a raised hind toe. It has rounded wings, and a medium-length tail. It is nocturnal, terrestrial, and probably flightless, inhabits forests, feeds on worms, insects, and other animals, and nests on the ground. It is found in New Caledonia.

Rhynochetos jubatus (kagu) *See* RHYNOCHETIDAE.

rib Part of the *axial skeleton of vertebrates, related to the vertebral *apophyses. Ribs develop at the *myosepta *skeletagenous septa junction, articulating with *vertebral apophyses. There are two types of rib: dorsal ribs form at the myoseptum horizontal septum junction, ventral ribs form where the myoseptum intersects the *connective tissue surrounding the *coelom.

ribbonfish *See* TRACHIPTERIDAE.

ribbon worms *See* NEMERTINI.

ribitol A five-carbon sugar alcohol that forms an integral part of the structure of the flavins *riboflavin, riboflavin mononucleotide (FMN), and flavin adenine dinucleotide (FAD).

riboflavin Vitamin B_2 (*see* VITAMIN); it consists of an organic base coupled to *ribitol. It occurs widely in nature, being an integral part of the *coenzymes flavin adenine dinucleotide (FAD) and riboflavin mononucleotide (FMN).

ribonuclease An *enzyme that catalyses the *hydrolysis of *RNA, resulting in the formation of mono- and oligonucleotides.

ribonucleic acid *See* RNA.

ribose Aldopentose monosaccharide, which comprises the carbohydrate component of *RNA.

ribosomal RNA *See* RNA.

ribosome A subcellular granule, 10–20 nm in diameter, composed of RNA and protein, which is found in large numbers in all types of cells and in some subcellular *organelles. Ribosomes are the site of protein synthesis: *m-RNA attaches to them and there receives *t-RNA molecules bearing *amino acids. In *eukaryotic cells ribosomes are synthesized in the *nucleolus but are found predominantly in the *cytoplasm, singly or in chains (polysomes or polyribosomes, probably linked by the m-RNA), or attached to the *endoplasmic reticulum, which is then termed 'rough ER'. Ribosomes have a *sedimentation factor: 70S in bacteria; 80S in eukaryotic cytoplasm; and 70S in mitochondrial ribosomes. *See* MITOCHONDRION.

Ricinidae (order *Phthiraptera, suborder *Amblycera) Family of chewing lice, which are parasitic on birds of the order *Passeriformes. Members of the family are distinguished from all other Amblycera by the lack of *labial palps, by the presence of *pallettes, and by the fusion of the *mesonotum, *metanotum, and first abdominal *tergum. Many of the species feed on blood, and members of one genus, *Trochiloecetes*, have developed piercing mouth-parts analogous to those of the sucking lice. The family comprises three genera: *Ricinus*, *Trochiloecotes*, and *Trochiliphagus*, the first being worldwide in distribution and the other two confined to humming-birds in the New World. There are about 70 species.

Ricinulei (subphylum *Chelicerata, class *Arachnida) An order of small (5–10 mm long), slow-moving arachnids in which the *cuticle is thick and often sculptured and the mouth-parts and *chelicerae are covered by a hood that can be raised and lowered. Ricinuleids feed on other small arthropods. There are 45 described species, found in tropical Africa and America.

rictal bristle A stiff, modified feather, with little or no vane, found around the gape of many insectivorous birds, possibly serving to aid the capture of food in flight. *See also* BASAL BRISTLE.

right whales (*Eubalaena*) *See* BALAENIDAE.

RIM *See* REPRODUCTIVE ISOLATING MECHANISM.

ring canal *See* RADIAL CANALS.

ringed snake (*Natrix natrix*) *See* COLUBRIDAE.

Ringer's solution A physiological saline solution of variable composition, but approximately isotonic with the tissue fluids of animals. It is used extensively in experiments for the temporary maintenance of living cells.

ring-necked pheasant (*Phasianus colchicus*) *See* PHASIANIDAE.

ring species A group of subspecies that are contiguous along a *cline. Members of each *population are able to mate successfully with members of adjacent populations, but the group as a whole forms a ring, with sufficient morphological differentiation in some places to prevent interbreeding between overlapping populations. Gulls of the genus *Larus* comprise a circumpolar ring species in the northern hemisphere. Moving westwards from Britain, the herring gull (*L. argentatus*) occurs in N. America (where one variant has developed into a distinct species, *L. glaucoides*), but is somewhat different from the British race. Between central Asia and north Europe, the races increasingly resemble the black-headed gull (*L. fuscus*), and in northern Europe the two species overlap and do not naturally interbreed.

ring-tailed lemurs (*Lemur*) *See* LEMURIDAE.

ring-tailed possums *See* PSEUDOCHEIRIDAE.

Riodinidae (Nemeobiidae; subclass *Pterygota, order *Lepidoptera) Family of small butterflies in which the wings are short and broad, generally brownish, with white or blue markings, and eye spots near the margin of the hind wings. The fore legs are reduced in males but fully formed in females. Riodinidae are closely related to the *Lycaenidae. There are about 1000 species, found mainly in tropical America.

Riparia (sand martins) *See* HIRUNDINIDAE.

riparian Pertaining to the bank of a river or shore of a lake.

ritualization The modification of patterns of behaviour and often (but not always) of their *motivation and function, and their subsequent use in communication, often in stereotyped form.

river blackfish (*Gadopsis marmoratus*) *See* GADOPSIDAE.

river dolphin *See* PLATANISTIDAE.

river-jack (*Bitis nasicornis*) *See* VIPERIDAE.

river mussel *See* GLOCHIDIUM.

river stingray *See* POTAMOTRYGONIDAE.

RNA (ribonucleic acid) A *nucleic acid that is characterized by the presence of D-ribose and the *pyrimidine base uracil. It occurs in three principal forms, as *messenger-RNA, ribosomal-RNA, and *transfer-RNA, all of which participate in protein synthesis.

roach 1. (entomol.) *See* BLABERIDAE; BLATTELLIDAE; BLATTIDAE; BLATTODEA; CRYPTOCERCIDAE; DICTYOPTERA; POLYPHAGIDAE. 2. (fish, *Rutilus rutilus*) *See* CYPRINIDAE.

roadrunner, greater (*Geococcyx californianus*) *See* CUCULIDAE.

robber flies *See* ASILIDAE.

rock bristletails *See* MEINERTELLIDAE.

rock crawlers *See* GRYLLOBLATTODEA.

rock dove (*Columba livia*) *See* COLUMBIDAE.

rockfish *See* SCORPAENIDAE.

rockfowl *See* PICATHARTIDAE.

rock rat (*Petromys typicus*) *See* PETROMYIDAE.

rock slaters *See* LIGIIDAE.

rock sparrows (*Petronia*) *See* PLOCEIDAE.

rock thrushes (*Monticola*) *See* TURDIDAE.

rock whiting *See* ODACIDAE.

rod In the *retina of the vertebrate eye, the more common of the two types of light receptor (*compare* CONE), which is sensitive to very low levels of stimulation. Many rods (in humans about 150) are connected to bipolar cells to a single ganglion cell whose axon is a nerve fibre in the optic tract; stimuli to the rods are summated (i.e. added together) to produce a total stimulus that must exceed a threshold before an impulse is discharged by the ganglion cell. Because the contribution of many individual rods is summated to produce a single impulse the resulting perceived image is not clearly defined. There are estimated to be about 65×10^6 rods in a single human retina, and the eyes of some animals (e.g. opossums) have retinas containing only rods and no cones.

Rodentia (rodents; Simplicidentata; infraclass *Eutheria, cohort *Glires) An order of herbivorous or scavenging mammals in which the *incisors are reduced to one pair in each jaw (hence the name 'Simplicidentata'). The incisors have *enamel on one surface only and grow continually so that by gnawing they are simultaneously sharpened and kept at a constant length. There is a wide *diastema, the second incisors, *canines, and first *premolars being absent, and the remaining premolars and *molars are usually similar in shape. Folds of skin can be taken into the diastema to seal off the mouth into two compartments: material taken into the mouth is not necessarily swallowed. In some rodents these skin folds are developed into cheek pouches. The jaw muscles are large and divided into three parts attached so as to permit a forward and backward jaw movement. The three suborders (*Sciuromorpha, *Myomorpha, and *Hystricomorpha) are distinguished partly by the arrangement of the jaw muscles.

Apart from the modifications to the jaws and teeth, rodents are rather unspecialized. Limbs usually are *pentadactyl and clawed. Hind limbs are commonly longer than fore limbs. The digits of the fore limbs are used by many species for handling food. Gait is *plantigrade or semi-plantigrade although many species are *saltatory. Rodents emerged from eutherian stock during the late *Cretaceous or *Palaeocene and have shown little tendency to evolve to large size. Today they are distributed throughout the world. There are 1500–2000 species, so that rodents account for nearly half of all mammal species.

Rodriguez solitaire (*Raphus solitarius*) *See* RAPHIDAE.

roe The mass of eggs present in the body cavity of mature female fish.

roe deer (*Capreolus*) *See* CERVIDAE.

rollers 1. *See* CORACIIDAE; CORACII-FORMES. **2. (groundrollers, long-tailed groundroller, *Uratelornis chimaera*)** *See* BRACHYPTERACIIDAE.

Romer, Alfred Sherwood (1894–1973) An American palaeontologist and comparative anatomist, who specialized in the evolution of vertebrates. He obtained his Ph.D. in 1921 from Columbia University, then taught anatomy at the Bellevue Hospital Medical College, New York, and from 1923 to 1934 he was an Associate Professor at the University of Chicago. In 1934 he was appointed Professor of Biology at Harvard University, in 1945 the director of the biological laboratories, and in 1946 director of the Museum of Comparative Zoology, posts he held until his death.

Romer's rule The proposal, first made by A. S. Romer, that the effect of many important evolutionary changes is to enable organisms to continue in the same way of life, rather than to adapt to a new one. For example, the evolution of bony elements that strengthened the limbs of fish enabled them to crawl over land to find new ponds when the climate started to become drier.

The concept is close to that of the *Red Queen effect.

ronquil *See* BATHYMASTERIDAE.

rook (*Corvus frugilegus*) *See* CORVIDAE.

rooted tree *See* OUTGROUP.

rose beetle *See* SCARABAEIDAE.

rosefinches *See* FRINGILLIDAE.

rostellum In *Cestoda, the projection of the *scolex that bears two rows of hooks, by which the animal attaches itself to its host.

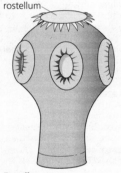

rostellum

Rostellum

rostral 1. Resembling or pertaining to a *rostrum. **2.** In the direction of the rostrum or beak (i.e. forwards).

rostrate Applied to raised or attenuated areas of shells or echinoid *tests.

Rostratulidae (painted snipe; class *Aves, family *Charadriiformes) A family of medium-sized wading birds which have olive-brown, grey, black, and white, cryptically coloured upper-parts, whitish under-parts, and a white central crown stripe. One species has chestnut sides to its head. Females are brighter than males. The bill is long and slightly *decurved, the neck and tail are short, the wings are broad, and

r

the legs are long with long toes. Painted snipe are solitary and secretive, inhabit marshes, feed on invertebrates, and nest on the ground. There are three species (*Rostratula benghalensis*, *R. australis*, and *Nyctocryhes semicollaris*), found in Africa, Asia, Australia, and S. America.

rostrum 1. A forward projection or extension of the snout (e.g. of a fish). The typical profile of an oceanic shark reveals a massive rostrum projecting well forward and above the low-slung mouth. **2.** The snout-like extension to the head that forms the specialized piercing and sucking mouth-parts of the *Hemiptera (true bugs). The *stylets are sheathed by the *labrum, which forms a jointed tube closed dorsally at the base by the labrum. In feeding, only the stylets enter the tissues of the host plant or prey. **3.** In prawns, a spine-like extension of the *cephalothorax. **4. (guard)** In *Belemnitida, a massive deposit of fibrous calcite. **5.** In *Brachiopoda, a beak-like process on the *umbo.

Rotatoria *See* ROTIFERA.

Rotifera (Rotatoria; **wheel animalcules**) A phylum of *acoelomate, unsegmented animals in which normally a complete alimentary canal is present, as is a muscular *pharynx possessing well-developed jaws. Their name is derived from the ciliated crown which in many species gives the appearance of a rotating wheel when it beats. The largest individuals reach 3 mm in length, but most are much smaller. Most are solitary and free-moving, but some are *sessile and some colonial. They occur mainly in freshwater habitats. They swim by means of their *cilia, or crawl across a substrate by muscular movements. Most are *benthic. The body is always covered by a *cuticle, which may be ornamented. There is a *pseudocoelom between the body wall and gut. Some feed on suspended matter, others are predatory on *Protozoa, rotifers, or small metazoan animals. Most are non-parasitic. All (except *Bdelloida) reproduce sexually. There are about 1800 species, grouped in two classes: Digononta, comprising the

orders *Seisonidea and Bdelloida, and containing rotifers with two ovaries; and Monogonta, with one ovary, comprising the orders Flosculariacea, Collothecacea, and Ploima.

rough ER *See* ENDOPLASMIC RETICULUM; RIBOSOME.

rough-toothed dolphin (*Steno*) *See* STENIDAE.

roughy *See* TRACHICHTHYIDAE.

round dance *See* DANCE LANGUAGE; NASANOV'S GLAND.

roundhead *See* PLESIOPIDAE.

roundworms *See* NEMATODA.

rove beetle *See* STAPHYLINIDAE.

RQ *See* RESPIRATION QUOTIENT.

***r*-selection** Selection for maximizing the *biotic potential (r) of an organism so that when favourable conditions occur (e.g. in a newly formed *habitat) the species concerned can rapidly colonize the area. Such species are opportunists. An opportunist strategy is advantageous in rapidly changing environments as in the early stages of a succession. *See also* BET-HEDGING; BIOTIC POTENTIAL; POPULATION EXPLOSION. *Compare* K-SELECTION.

rubberlip sea-perch (*Rhacochilus toxotes*) *See* EMBIOTOCIDAE.

ruby-tailed cuckoo wasps *See* CHRYSIDIDAE.

ruby-tailed wasps *See* CHRYSIDIDAE.

ruby-throated humming-bird (*Archilochus colubris*) *See* TROCHILIDAE.

rubythroats *See* TURDIDAE.

ruff (*Philomachus pugnax*) *See* SCOLOPACIDAE.

ruffe *See* CENTROLOPHIDAE.

ruffed lemur (*Varecia***)** *See* LEMURIDAE.

ruffle Part of an extended *pseudopodium by which a moving *fibroblast attaches to a surface.

rufous Reddish-brown in colour.

rugose Wrinkled; bearing many ridges.

rugulose Finely wrinkled.

ruin lizard (*Podarcis sicula***)** *See* LACERTIDAE.

rumen *See* RUMINATION.

ruminant *See* RUMINANTIA.

Ruminantia (ruminants; superorder *Paraxonia, order *Artiodactyla) A suborder that comprises those artiodactyls in which the stomach is complex and some form of *rumination occurs. The anterior teeth are much modified, with the *canines usually reduced, and the cheek teeth are *selenodont. The *mastoid region is exposed on the skull surface. Some authorities restrict the Ruminantia to the higher ruminants, excluding the modern camels and their ancestors; others allow two infraorders, *Tylopoda for the camel group and *Pecora for the remainder.

rumination In *Ruminantia, a complex digestion characterized by a stomach having three or four parts. In the rumen millions of symbiotic bacteria and other cellulase-producing micro-organisms partially digest the cellulose-rich food. The enzyme cellulase cannot be made by mammals themselves. After some time the partially digested food (cud) is assembled into pellets in the *reticulum ready for regurgitation. The regurgitated cud is chewed thoroughly before passing to the *omasum and *abomasum where normal digestion occurs. In some ruminants the rumen and reticulum form a single large chamber.

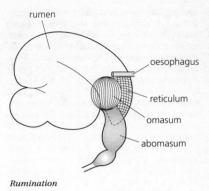

Rumination

runaway hypothesis A hypothesis proposed by R. A. Fisher (1890–1962) in 1930 to explain the consequences of female selection of a particular male trait (e.g. the length of the tail in a bird). Over successive generations such selection would favour increasingly extreme development of the trait (i.e. the tails of males would become longer) until the *fitness of the male was reduced. (This tendency has been demonstrated experimentally by shortening or lengthening the tails of male birds.) Eventually, males would be so overspecialized as to bring the *species to extinction, were it not for the restraining influence of *natural selection, which halts the development before that stage can be reached. *Compare* HANDICAP PRINCIPLE.

***Rupicola* (cock-of-the-rock)** *See* COTINGIDAE.

Ruscinian The land-mammal stage that spans the uppermost *Miocene and lower *Pliocene in Europe.

rut The annual mating season in deer (*Cervidae). In solitary species (e.g. roe deer, *Capreolus capreolus*) the male remains with one or more females during the rut. In social deer (e.g. red deer, *Cervus elaphus*) the males seek to establish a territory into which they entice a *harem of females.

Rutilus rutilus (roach) *See* CYPRINIDAE.

R value Usually, the multiple correlation coefficient, as distinct from *r*, the simple correlation coefficient.

Rynchopidae (skimmers; class *Aves, order *Charadriiformes) A family of large water birds which have black or brown upper-parts and white under-parts. The bill is long, with the lower *mandible longer than the upper. Skimmers have long, pointed wings, short, forked tails, and very short legs with small, webbed feet. They inhabit coasts and rivers, and feed on small fish and crustaceans caught by skimming over the water with the lower mandible ploughing the surface. They nest on the ground. There is one genus, *Rynchops*, with three species, found in Africa, southern Asia, and N., Central, and S. America.

Rynchops (skimmer) *See* RHYNCHOPIDAE.

s *See* SELECTION COEFFICIENT.

sablefish *See* ANOPLOPOMATIDAE.

sabretooth *See* EVERMANNELLIDAE; FELIDAE.

saccharide Alternative term for sugar.

Saccobranchidae An alternative name for the fish family *Heteropneustidae.

Saccopharyngidae (whiptail gulper; superorder *Elopomorpha, order *Anguilliformes) A small family of deep-sea fish that have an elongate body, a long, whip-like tail, a large mouth, and eyes placed close to the tip of the snout. There are possibly four species, distributed world-wide.

saccus gastricus The blind forestomach sac in Colobidae (*see* CERCOPITHECIDAE) in which bacterial fermentation of plant foods takes place; it is analogous to the rumen in ruminants (*rumination).

sacfry Recently hatched fish larvae that are still in possession of the *yolk sac.

Sacoglossa (Ascoglossa; phylum *Mollusca, class *Gastropoda) A relatively small order of *opisthobranch gastropods in which individuals are small and may be primitively shelled, bivalved, or naked. They feed with a *radula and *suctorial mouth on the cell fluids of algae. Usually they feed on green algae, but some certainly feed on red algae and there are reports of sacoglossans feeding on large diatoms and on sea grasses. Some feed suctorially on eggs of other opisthobranchs. Several genera of sacoglossans are able to retain functional chloroplasts from their food algae, and benefit nutritionally from the photosynthetic activity of the chloroplasts. Gills are usually lacking, but rarely one may be present. They are named after the sac ('ascus sac') into which worn teeth normally pass (rather than being shed as are teeth in other opisthobranchs). Sacoglossa are not known in the fossil record.

sacral vertebra *See* VERTEBRA.

sacred scarab (Scarabaeus sacer) *See* SCARABAEIDAE.

sac-winged bat *See* EMBALLONURIDAE.

sagitta *See* EAR-STONE.

Sagittariidae (secretary bird; class *Aves, order *Falconiformes) A monospecific family (*Sagittarius serpentarius*) which is a large, grey bird with black thighs and flight feathers. It has a long neck and its nape has elongated feathers. The face is bare with a short, hooked bill. The legs are very long, with short toes which are webbed basally. The wings are large and the tail long, with long central feathers. The secretary bird inhabits open plains, feeds on reptiles (particularly snakes), small mammals, and insects, and nests in a bush or tree. It is found in Africa.

***Sagittarius serpentarius* (secretary bird)** *See* SAGITTARIIDAE.

saiga (Saiga tatarica) *See* BOVIDAE.

***Saiga tatarica* (saiga)** *See* BOVIDAE.

sailfish (Istiophorus platypterus) *See* ISTIOPHORIDAE.

sailor beetle *See* CANTHARIDAE.

St Kilda wren *See* RACE.

St Mark's fly *See* BIBIONIDAE.

St Peter's fish (*Zeus faber*) *See* ZEIDAE.

saithe (*Pollachius virens*) *See* DEMERSAL FISH.

salamanders *See* SALAMANDRIDAE; PLETHODONTIDAE.

Salamandra atra (alpine salamander, black salamander) *See* SALAMANDRIDAE.

Salamandra salamandra (fire salamander) *See* SALAMANDRIDAE.

Salamandridae (newts, 'true' salamanders; class *Amphibia, order *Urodela) A family of salamanders in which the adults have long *palatal teeth, lungs, a cylindrical body, and well-developed legs and tail. The skin of some species contains powerful toxins. Some (e.g. *Salamandra atra*, alpine salamander or black salamander, of the Alps, Albania, and Herzegovina, and *S. salamandra*, fire salamander, of Europe and the Near East) are *ovoviviparous. In *S. atra* one larva develops in each *oviduct and feeds on the other eggs, so two fully developed young are produced after (in high-altitude forms) two to three years' development; in *S. salamandra* the larval phase is short. *S. salamandra* is a nocturnal carnivore, preying on terrestrial invertebrates. 'Newts' are generally placed in the genus *Triturus*. *T. cristatus* (crested newt or great warty newt) grows up to 18 cm long and has a mottled, blackish-brown back, orange-yellow belly with black spots, its skin is warty, and the male develops a high, serrated, breeding crest. *T. vulgaris* (smooth newt) is the common newt of much of Europe, its range extending into Siberia; males develop a serrated breeding crest. Males of *T. helveticus* (palmate newt) have webbed hind feet and develop a low, smooth, breeding crest that ends in a thread-like extension of the tail. Those of *T. marmoratus* (marbled newt) of south-western Europe develop black and yellow breeding crests, represented in the female by a red dorsal line;

the body colour for both sexes is green with black marbling. There are about 90 species in the family, both terrestrial and aquatic, from Europe, N. America, N. Africa, and parts of Asia.

Salangidae (icefish, glassfish; superorder *Protacanthopterygii, order *Salmoniformes) A family of slender marine fish that have cigar-shaped bodies which are scaleless and almost transparent, a pointed head, a large *anal fin, a single, soft-rayed, *dorsal fin, a forked tail fin, and a small *adipose fin. Icefish are small (about 10 cm long) but nevertheless important as food fish. There are 14 species, found in the northwestern Pacific.

Saldidae (shorebugs; order *Hemiptera, suborder *Heteroptera) Family of active, big-eyed, predatory bugs which are often found in damp places with little vegetation, e.g. shores of rivers, lakes, and estuaries. There are about 300 species, distributed world-wide.

Salenioida (class *Echinoidea, superorder *Echinacea) An order of regular echinoids, similar to the *Cidaroida but characterized by the presence of one or several large, polygonal plates developed on the *periproct, within the ring of ocular and genital plates. They first appeared in the Early *Jurassic.

Salientia *See* ANURA.

salinity A measure of the total quantity of dissolved solids in water, in parts per thousand (per mille‰) by weight, when all organic matter has been completely oxidized, all carbonate has been converted to oxide, and bromide and iodide to chloride. The salinity of ocean water is in the range 33–8‰, with an average of 35‰.

saliva A watery fluid that is secreted into the mouth by the *salivary glands when food is present in the mouth, or is anticipated. In vertebrates, saliva contains *mucus, which acts as a lubricant to aid swallowing, and in some it contains an enzyme which commences the digestion of starch. In insects, saliva is secreted to the mouth-parts and is used to lubricate food, and in some species contains digestive enzymes. In blood-sucking species, saliva usually contains anti-clotting agents and in some it contains a local anaesthetic.

salivary-gland chromosome *Polytene *chromosome found in the interphase nuclei of the salivary-gland cells in larvae of *Diptera. These chromosomes undergo complete somatic pairing to form units of identical chromosomes joined along their length (in parallel) to one another.

salivary glands In most terrestrial vertebrates, *exocrine glands situated in the lining of the mouth and throat that secrete *saliva and *amylase. Some animals, e.g. *Pelecaniformes, lack salivary glands. In blood-sucking invertebrates, e.g. some black-flies (*Simuliidae), the glands secrete a substance that prevents *blood clotting; those of leeches (*Hirudinea) secrete the anticoagulant hirudin. In caterpillars the labial salivary glands secrete silk. Cephalopods (*Cephalopoda) produce saliva containing toxins (cephalotoxins) they use to stun prey; one drop of saliva from an octopus (*Octopoda) injected into a crab will cause paralysis.

salmon *See* SALMONIDAE.

salmon-herring (*Chanos chanos***)** *See* CHANIDAE.

Salmonidae (salmon, trout; superorder *Protacanthopterygii, order *Salmoniformes) A fairly large family of freshwater fish (although a number of species spend several years in the sea before returning to the river to spawn) that have an elongate body, with the soft-rayed, single, *dorsal and *pelvic fins placed about midway along the trunk, a small *adipose fin, and a slightly forked tail fin. Most are medium to large fish, inhabiting the cold to temperate freshwater systems of the northern hemisphere. Many species are of considerable commercial importance and a number have been introduced into other countries in order to improve the local fisheries. There are about 70 species.

Salmoniformes (subclass *Actinopterygii, superorder *Protacanthopterygii) An order of bony fish that have a single *dorsal fin which, like the *pelvic fins, is placed half-way along the body or further to the back of the animal. The order incorporates some 24 families, including e.g. the *Salmonidae, the *Esocidae, and *Osmeridae.

***Salpornis spilonotus* (spotted creeper)** *See* SALPORNITHIDAE.

Salpornithidae (spotted creeper; class *Aves, order *Passeriformes) A monospecific family (*Salpornis spilonotus*), which is a small, white-spotted, brown bird, with a long, slender, *decurved bill. It is often placed in the family *Certhiidae, but differs in that its tail is rounded and lacks stiff feather shafts. It inhabits woodland and bush, feeds on insects which it locates by climbing up and down tree trunks, and nests on a branch. It is found in Africa and Asia.

salps *See* THALIACEA.

saltation The hypothesis that the derived (*apomorphic) characters of a species were all obtained simultaneously as a result of the random effects of mutation and recombination at the DNA level. Although popular in the early twentieth century, the saltation hypothesis is generally not accepted or supported by the fossil record.

saltatorial Applied to limbs that are adapted for jumping.

saltators *See* CARDINALIDAE.

saltatory Leaping.

Salticidae (jumping spiders; order *Araneae, suborder Araneomorphae) Family of spiders which have only two *tarsal claws. They are medium to small spiders in which the carapace is square-fronted. The anterior median eyes are much larger than the others and members of this family have the greatest visual acuity of any arthropod, also perhaps being sensitive to the colour and plane of polarized light. Prey is jumped on from a distance of a few centimetres by rapid elevation of the internal body pressure which extends the legs. Many species mimic ants, and all have complex visual courtship and threat displays. Most of the 4000 or so species are tropical, although there are northern temperate and even arctic species.

salt-secreting gland *See* CROCODYLIDAE; HYDROPHIIDAE.

salt-water crocodile (*Crocodylus porosus***)** *See* CROCODYLIDAE.

sand cockroaches *See* POLYPHAGIDAE.

sand crickets *See* STENOPELMATIDAE.

sand diver *See* TRICHONOTIDAE.

sand-dollars *See* ECHINOIDEA; CLYPEASTEROIDA.

sand-eel (*Ammodytes tobianus***)** *See* AMMODYTIDAE.

sanderling (*Calidris alba***)** *See* SCOLOPACIDAE.

sandfish 1. *See* GONORYNCHIDAE. 2. *See* LEPTOSCOPIDAE. 3. *See* TRICHODONTIDAE.

sand flea *See* SIPHONAPTERA.

sand flies *See* PSYCHODIDAE.

sand gobies *See* KRAEMERIIDAE.

sandgrouse *See* COLUMBIFORMES; PTEROCLIDAE.

sandhopper *See* AMPHIPODA.

sand-lance (*Ammodytes tobianus***)** *See* AMMODYTIDAE.

sand lizard (*Lacerta agilis***)** *See* LACERTIDAE.

sand martins (*Riparia***)** *See* HIRUNDINIDAE.

sand perch *See* MUGILOIDIDAE.

sandpipers *See* SCOLOPACIDAE.

sand smelts (silversides) *See* ATHERINIDAE.

sand stargazer *See* DACTYLOSCOPIDAE.

sand tiger *See* ODONTASPIDIDAE.

sand viper (*Viper ammodytes***)** *See* VIPERIDAE.

sand wasps *See* SPHECIDAE; SPHECOIDEA.

Sanger's reagent A solution of 1-fluoro-2-4-dinitrobenzene that is used for the chromatographic detection and quantification of *amino acids, *peptides, and *proteins. Its effectiveness is based on the reaction of the reagent with free alpha- and epsilon-amino groups to form yellow dinitrophenyl derivatives.

saola (spindlehorn, Vu Quang 'ox') *See* PSEUDORYX NGHETINHENSIS.

sap beetle *See* NITIDULIDAE.

saprobe *See* SAPROTROPH.

saprophage An organism that consumes other, dead, organisms. Saprophages form part of the twofold division of the *heterotrophs and consist mainly of bacteria and fungi, but also of some invertebrates, such as insect larvae (*see* CALLIPHORIDAE). They break down complex compounds obtained from dead organisms, absorbing some of the simpler products but releasing most of the products as inorganic nutrients which can then be used by other organisms.

saprophyte *See* SAPROTROPH.

saprotroph (saprobe, saprovore) Any organism that absorbs soluble organic nutrients from inanimate sources (e.g. from dead plant or animal matter, dung, etc.). If the organism is a plant or is plant-like, it is called a saprophyte; if it is an animal or is animal-like, it is called a saprozoite. *See also* CONSUMER.

saprovore *See* SAPROTROPH.

saprozoite *See* SAPROTROPH.

Sarcodina (phylum *Protozoa, sub-phylum *Sarcomastigophora) A super-class of protozoa which form *pseudopodia for feeding and locomotion. There are three classes, and many orders. The majority of species live in marine aquatic environments but some occur in fresh water (and are important members of the soil fauna) and some are parasitic in the intestinal tracts of vertebrates and invertebrates. The superclass includes the *Radiolaria, known from the *Cambrian, and the *Heliozoa, which are exclusively freshwater and have a fossil record that extends back only to the *Pleistocene.

sarcolemma The *cell membrane of a muscle cell.

Sarcomastigophora (phylum *Proto-zoa) A subphylum of protozoa which can form *flagella and/or *pseudopodia. Most species are free-living in aquatic environments; some are parasitic in plants and animals.

sarcomere That region of a *myofibril of striated muscle that lies between the *Z-lines.

Sarcophagidae (flesh flies; order *Diptera, suborder *Cyclorrapha) Family of flies that was formerly considered by some authors to be a subfamily of the *Calliphoridae. Flesh flies are associated with carrion and animal matter since these are the substrates in which their larvae feed. Adults are mainly grey, with marbled abdomens, but are rarely metallic to any extent. The *arista of the antennae are plumose on the basal half only. Many species are distinguished only by the male genitalia. The female of the common *Sarcophaga carnaria* is lar-viparous (deposits larvae rather than eggs). The larvae are well adapted to their mode of life, with rear *spiracles set in a deep pit,

the edges of which have flexible, fleshy lobes which can be used to seal off the spiracles. This is necessary when the larvae liquefy their carrion and are in danger of drowning. As the larva feeds, the fleshy lobes around the rear spiracles act as floats, allowing it to breathe. The *pupa is formed inside the cuticle of the final *instar larva (puparium). Escape from the puparium is effected by an inflatable sac (*ptilinum) on the head of the fly. This also helps in forcing a path through the pupation medium, which is often soil. More than 1600 species are known to exist.

Sarcophilus harrisi (Tasmanian devil) *See* DASYURIDAE.

sarcoplasm The *cytoplasm of muscle cells, which, although it contains the usual cellular *organelles, is largely dominated by the presence of *myofibrils.

sarcoplasmic reticulum A network of membranes that run both longitudinally and transversely across striated muscle cells. Apparently, the longitudinal channels are derived from the *endoplasmic reticulum and are not in direct communication with those running transversely (T-system), which are derived from the cell membrane. The function of the sarcoplasmic reticulum is to act as a store for calcium ions (Ca^{++}) involved in the contractile process.

Sarcopterygii The name given by some taxonomists to a group comprising the lobe-finned fish (subclasses *Crossopterygii and *Dipneusti). Others tend also to include the terrestrial vertebrates (*Tetrapoda). *See also* CHOANICHTHYES; GNATHOSTOMATA.

Sarotherodon *See* CICHLIDAE.

Sarotherodon mossambicus (cichlid) *See* BUCCAL INCUBATION; CICHLIDAE.

satellite DNA Any DNA (*deoxyribonucleic acid) which differs enough in its base composition to form a separate fraction from the majority of genomic DNA on centrifugation (*see* CENTRIFUGE). This bias in base composition is often due to highly repetitive DNA. Satellite DNA may consist of dispersed repeats or may be arranged in a *tandem array. In a tandem array it may have very short repeating units, 2–10 base pairs in the case of microsatellite DNA, or slightly longer ones, 10–100 base pairs in the case of minisatellite DNA. The number of repeating units within a single satellite tandem array changes rapidly in evolutionary terms, due to the processes of *replication slippage and *unequal crossing-over. Consequently, the length of an array can be used as a highly informative character in phylogenetic analysis (*see* PHYLOGENY) at the intraspecific level. By using many such arrays a highly specific set of *character states can be established, by which an individual organism may be identified; this is a DNA fingerprint.

satiation A process that leads to the cessation of an activity, applied most commonly to *feeding behaviour. Satiation may be associated with physiological changes and it occurs before the point at which *appetite is satisfied completely and continuation of the activity becomes physically impossible.

saturn butterflies *See* NYMPHALIDAE.

Saturniidae (atlas moths, emperor moths, moon moths; subclass *Pterygota, order *Lepidoptera) Family of large to very large moths with characteristic *eyespots on the fore wings and hind wings. The wings are very broad, the antennae are *bipectinate in both sexes, and the proboscis is absent. Larvae bear fleshy protuberances (scoli), and have hairs and spines. The cocoon is dense. Distribution is world-wide but primarily tropical.

Satyrinae (browns, satyrs, morpho butterflies, heaths, graylings, marbled whites; order *Lepidoptera, family *Nymphalidae) Subfamily of small to medium-sized butterflies, which are mostly brown, or orange and black, with *eyespots on the wings. The bases of the wings are often swollen. The butterflies are generally shade-loving. The larvae have a pair of short points at the anal end. Most larvae feed on grasses and sedges. The subfamily is distributed world-wide.

satyrs *See* SATYRINAE.

Sauria (Lacertilia; lizards; subclass *Lepidosauria, order *Squamata) A suborder of reptiles the more primitive of which are four-legged, terrestrial or arboreal forms, but limbless burrowing species have evolved in many families. The skin is generally shed in pieces. *Autotomy is common. Lizards appeared and began to diversify in the *Triassic. Their *Jurassic history is poorly known, and except for remains of the sea-lizards, e.g. the mosasaurs, this is also true of much of the *Cretaceous. However, from the late Cretaceous onwards continental deposits become more frequent again, and have yielded sufficient lizard fossils to trace the development of the group up to the present day. There are 18 extant families, found in all but the coldest parts of the world.

saurian Of or resembling a lizard. Applied loosely to lizard-like animals and also to fossils, life habits, etc. of extinct reptiles.

Saurischia (class *Reptilia, subclass *Archosauria) The order that comprises the 'lizard-hipped' dinosaurs, one of the two dinosaur orders. They included bipedal carnivores (*Theropoda) and herbivorous tetrapods (*Sauropoda). The theropods produced the largest known terrestrial carnivores, and the sauropods yielded the largest known land animals.

saurochory Dispersal of spores or seeds by snakes or lizards.

Sauropoda (subclass *Archosauria, order *Saurischia) A suborder of *Jurassic and *Cretaceous, quadripedal dinosaurs of herbivorous habit. They included *Diplodocus*, from the Late Jurassic, one skeleton of which is 26.6 m long. *Apatosaurus* was also a sauropod, and although shorter than *Diplodocus* its skeleton was more massively built.

Sauropterygia (class *Reptilia, subclass *Euryapsida) An extinct reptilian order, that comprises the suborders *Plesiosauria (plesiosaurs) and *Nothosauria (nothosaurs).

saury *See* SCOMBERESOCIDAE.

sawbills See ANATIDAE.

sawfish See PRISTIDAE.

sawflies See HYMENOPTERA; SYMPHYTA; TENTHREDINOIDEA; TENTHREDINIDAE.

sawshark See PRISTIOPHORIDAE.

Saxicola (chats) See TURDIDAE.

Sayornis (phoebes) See TYRANNIDAE.

scabrous Rough-surfaced; bearing short, stiff hairs, scales, or points.

scad See CARANGIDAE.

scala Three canals in the *cochlea. The scala vestibuli and scala tympani are connected by an opening, the helicotrema. Between the scala vestibuli and scala tympani, the scala media is filled with *endolymph.

scaled blennies See CLINIDAE.

scale insect See COCCIDAE.

scaleworms See ERRANTIA.

scallops See PTERIOMORPHIA.

scaly ant-eater See MANIDAE.

scaly dragonfish (*Stomias boa*) See STOMIATIDAE.

scaly lizards See LEPIDOSAURIA.

scaly-tailed squirrel See ANOMALURIDAE.

scampi Plural of scampo. See NEPHROPIDAE.

scampo (*Nephrops norvegicus*) See NEPHROPIDAE.

Scandentia See TUPAIIDAE.

scape The usually elongate basal segment of geniculate (elbowed) antennae in *Hymenoptera (bees, wasps, and ants).

Scaphopoda (tooth shells, tusk shells; phylum *Mollusca) A class of elongate, bilaterally symmetrical, burrowing, marine molluscs which have a tapering shell open at both ends, the larger, apertural end being anterior. Most are also curved, with the convex side ventral. The shell is secreted by the *mantle and is composed of three distinct layers of calcium carbonate. Most individuals are 2–5 cm long. The head contains the mouth with a *radula, surrounded by food-gathering appendages. There are no eyes. The mantle cavity contains paired kidneys, a liver, a simple stomach, a heart, and a solitary gonad. The nervous system is ganglionic, and there are no gills. The blood-filled foot is used for burrowing. The order contains two families. Scaphopoda are thought to have first appeared in the *Devonian, but some have been tentatively reported from the *Ordovician in Russia.

scapula 1. In tetrapods, the dorsal part of the *pectoral girdle. **2.** In *Mammalia, the shoulder blade.

scarab See SCARABAEIDAE.

Scarabaeidae (scarabs, chafers, tumblebugs; subclass *Pterygota, order *Coleoptera) Family of beetles, which are very diverse in form and habit, and are 2–150 mm long. The antennal club consists of movable, flattened plates. Chafers (e.g. *Melolontha melolontha*, the maybug or cockchafer) are nocturnal. Adults are leaf-feeders, often pests. Larvae are C-shaped, fleshy, and feed on roots, often taking up to three years to develop. Dung rollers (e.g. *Scarabaeus sacer*, the sacred scarab) have front legs without *tarsi, specialized for forming the dung ball that is rolled to an underground chamber to feed larvae. Some species remain with their young until these are mature. Cetoniines (rose beetles) are large, brightly coloured, and diurnal (e.g. *Goliathus giganteus*, Goliath beetle, which is up to 150 mm). Larvae are found in decaying plants; adults feed mostly on fruits or flowers. Dynastines, the rhinoceros beetles (e.g. the elephant beetle and Hercules beetle) are virtually all tropical. Adults are dark, shiny, and nocturnal, with large curved horns on the head and

*thorax of males. Some larvae are pests of sugar-cane and palm trees. There are 17 000 species.

Scarabaeus sacer **(sacred scarab)** *See* SCARABAEIDAE.

Scaridae (parrotfish; subclass *Actinopterygii, order *Perciformes) A fairly large family of colourful marine fish that have an oval-shaped body, continuous *dorsal fin, large scales, and with the teeth in both jaws fused into a strong cutting edge, resembling the beak of a parrot. This 'beak' is used to scrape off algae and other plant material from rocks and corals. Active during the day, parrotfish have been observed to sleep or rest during the night. Like the related wrasses (*Labridae), colour patterns can change during the day and may differ between the sexes. One of the larger species is *Scarus coeruleus* (blue parrotfish), 1.2 m, found in the western Atlantic. There are about 68 species, distributed world-wide in tropical to temperate waters.

scarious Dry and membranaceous.

scarlet macaw (*Ara macao*) *See* PSITT-ACIDAE.

Scarus coeruleus **(blue parrotfish)** *See* SCARIDAE.

scat *See* SCATOPHAGIDAE.

Scatophagidae (scat, butterfish; subclass *Actinopterygii, order *Perciformes) A small family of coastal fish, members of which frequently penetrate into fresh water. They have a deep body, two separate *dorsal fins, and, typically, four spines in the *anal fin. *Scatophagus argus* (spotted butterfish or spotted scat), 32 cm, is a fairly common species in the Indo-Pacific region. There are three species.

Scatophagus argus **(spotted butterfish, spotted scat)** *See* SCATOPHAGIDAE.

scavenger *See* DETRITIVORE.

SCE *See* SISTER CHROMATID EXCHANGE.

Scelionidae (suborder *Apocrita, superfamily Proctotrupoidea) Family of minute (usually 2 mm long or less) *Hymenoptera that are black or (rarely) brown in colour, and with *abdomens that are generally flattened, with sharp lateral margins. The antennae are elbowed or bent, arise low on the face, and are 11- or 12-segmented, with a club. The wing venation is greatly reduced. The larvae are mainly egg parasites, some particularly on grasshoppers and mantids, but the majority parasitize eggs of *Lepidoptera, *Hemiptera, and *Coleoptera. The adult female often clings to the host until eggs are laid, whereupon she dismounts and deposits her own eggs. The family is large and common.

scent brush *Eversible structure found on the *abdomen of a member of the *Lepidoptera: it is brush-like and contains a great deal of glandular twine. Scent brushes disperse *pheromones.

scent marking The use by an animal of scented secretions, faeces, urine, or saliva as a means of communication: the substance is deliberately placed on the ground, on an object, or on another animal.

Schiff's reagent A reagent, consisting of fuchsin bleached by sulphurous acid, that produces a red colour upon reaction with an *aldehyde.

Schilbeidae (superorder *Ostariophysi, order *Siluriformes) A family of freshwater catfish that have a scaleless, elongate body, strongly tapering towards the tail, a short *dorsal fin, a long-based *anal fin, and four pairs of *barbels around the mouth. There are about 40 species, in African and Indian waters. *See also* PANGASIIDAE.

Schindler fish *See* SCHINDLERIIDAE.

Schindleriidae (Schindler fish; subclass *Actinopterygii, order *Perciformes) A family comprising one genus with three species of small (about 3 cm long) fish that have a scaleless and transparent body. All the fins are very small and the *pelvic fins are lacking. Probably fairly abundant, *Schindleria praematurus* is found in surface waters of the Pacific Ocean.

schistosomiasis (bilharzia) Any one of a group of diseases of humans and some other *Mammalia caused by infestation by blood flukes (*Trematoda of the genus *Schistosoma*). The trematode requires two hosts; the first is an aquatic snail in which larvae emerging from eggs hatched in water develop through several stages, finally to fork-tailed forms (cercariae) which escape from the snail, swim freely in the water, and attach themselves to, and penetrate, the skin of their second host, a mammal. In the mammalian body they move through blood vessels, feeding upon *glycogen in the *blood plasma and growing, eventually migrating to a site that varies from species to species, where they mature, mate, and lay eggs. The eggs leave the body in urine or faeces. Symptoms vary according to the species causing the infestation, but may include swelling and tenderness of the liver, dropsy, enlargement of the spleen, watery skin eruptions, fever, cough, haemorrhage, diarrhoea, and reduced function of the affected organs. The diseases are common in low latitudes.

(SEE WEB LINKS)

• Information about schistosomiasis from the Centers for Disease Control and Prevention, US.

schizocoel A *coelom that forms by the splitting of the *mesoderm during the development of the *embryo. This distinguishes schizocoelomates from deuterostomes (*Deuterostomia) in which the coelom arises from the *coelenteron. *Mollusca, *Annelida, and *Arthropoda are schizocoelomates.

schizocoelomate An animal possessing a *schizocoel.

Schizodactylidae (order *Orthoptera, suborder *Ensifera) Small family of Old World crickets which are carnivorous, and nocturnal, burrowing into the soil during the day. There are three genera: winged forms occur in only one, *Schizodactylus*.

schizodont Applied to a type of hinge dentition, found in certain members of the *Bivalvia order Trigonioida, in which the teeth are large, and possess parallel ridges at right angles to the axis of the teeth. The left valve bears a single tooth.

schizogony A type of *asexual reproduction, occurring in some *Protozoa, in which, after a varying number of nuclear divisions, the cell divides into a number of *daughter cells. It is a form of asexual multiple fission.

Schizogregarinida (class *Telosporea, subclass *Gregarinia) An order of *Protozoa which reproduce asexually by multiple fission (*schizogony). They are parasitic on *Arthropoda and other invertebrates.

schizomids *See* Uropygi.

Schmidt layer In the arthropod *exoskeleton, the layer immediately overlying the *epidermis and below the endocuticle, which forms the membrane from which new cuticle develops at *ecdysis.

school Loosely, an aggregation of fish which are observed swimming together, possibly in response to a threat from a predator. More strictly, a grouping of fish, drawn together by social attraction, whose members are usually of the same species, size, and age, the members of the school moving in unison along parallel paths in the same direction. Sudden changes at the leading edge of the school are followed almost instantaneously by the remainder of the group, with the fish on the flank becoming the new leaders. In general, the overall size and shape of the school, as well as the cruising depth and speed, vary from one species to another.

schooling Among fish the formation of groups of individuals as a result of social attraction.

Schwann cell *See* MYELIN SHEATH.

Sciaenidae (drum, meagre; subclass *Actinopterygii, order *Perciformes) A large family of mainly marine fish that have an elongate body, two *dorsal fins, an *anal fin with two spines that is much shorter than the second dorsal fin, and a *lateral line that continues into the base of the tail fin. Many species have a *swim-bladder which can be used to produce sound. There are about 160 species, with world-wide distribution in tropical to temperate waters.

Sciaridae (European army worm; order *Diptera, suborder *Nematocera) Family of minute flies which is regarded by some authors as a subfamily of *Mycetophilidae. These may be recognized by their characteristic wing venation, with prominent anterior veins, and simple, fine, posterior veins. The *compound eyes curve to meet one another above the antennae. Some species are sexually *dimorphic, the females being normal but the males having greatly reduced wings. The larvae often move *en masse*, a phenomenon that gives them their common name. More than 1200 species have been described.

scimitarbills (*Rhinopomastus*) *See* PHOENICULIDAE.

Scincidae (skinks; order *Squamata, suborder *Sauria) A family of terrestrial and burrowing lizards in which the head is wedge-shaped, the body streamlined and elongate, and the limbs small or absent. The scales are smooth, often overlying small *osteoderms. The teeth are *pleurodont. The tongue is slightly notched. The diet is mainly insectivorous, but Australian tiliquine skinks (e.g. *Tiliqua rugosa*, stump-tailed skink) are vegetarian. *T. rugosa* is large (about 36 cm long), the shape of the tail resembles that of the head, both being short, broad, and rounded at the end, and the cobalt-blue tongue is used in threat display; stump-tailed skinks burrow in sand. There are more than 800 species of skinks, occurring in the subtropics and in desert regions throughout the world.

Sciuridae (squirrels; order *Rodentia, suborder *Sciuromorpha) A family of diurnal, mainly arboreal but also terrestrial or burrowing rodents, in which the tail is fully haired and often bushy, there are four digits

on the fore limbs and five on the hind limbs, all with sharp claws, and the eyes and ears are relatively large. The cheek teeth are low-crowned and *lophodont. Squirrels are distributed widely in Eurasia and N. America, but do not occur in Australasia or Madagascar. In addition to the familiar arboreal squirrels (e.g. *Sciurus vulgaris*, red squirrel of Eurasia, and *Petaurista alborufa*, red-and-white giant flying squirrel of southern Asia) the family includes ground squirrels (e.g. *Tamias* and *Eutamias*, N. American chipmunks; *Cynomys ludovicianus* and *C. leucurus*, prairie dogs of N. America; and *Marmota marmota*, marmot or woodchuck). Squirrels reached S. America during the *Pleistocene but are not found in the southern part of the continent. There are 47 genera, with about 250 species.

Sciuromorpha (Protrogomorpha; cohort *Glires, order *Rodentia) A suborder comprising the more primitive rodents, in which the *masseter muscle is attached mainly at the lower edge of the *zygomatic arch; and more advanced forms, placed in the family *Sciuridae, in which the middle masseter is attached to the outside of the skull in front of the orbit. The suborder includes 13 families of beavers, squirrels, marmots, and gophers. Some authorities prefer the term 'Protrogomorpha', which indicates the primitive level of development of members of the suborder, the sciurids being specialized forms rather than central to the group.

***Sciurus vulgaris* (red squirrel)** *See* SCIURIDAE.

sclera The white part of the surface of the eye.

Scleractinia *See* MADREPORARIA.

Scleraxonia (subclass *Octocorallia, order *Gorgonacea) A suborder of octocorals in which the central axial skeleton is formed from *spicules bound together more or less solidly by either horny or calcareous material. The suborder includes *Corallicum* (red or precious coral) among its 26 genera in seven families.

sclerite One of the hard components into which the external skeleton of an invertebrate may be divided (e.g. an exoskeleton plate of an arthropod (*Arthropoda) in which the sclerites are connected to one another by flexible cuticular membranes). Sclerites are made from sclerotin, which is principally *chitin and protein, and frequently contains waxes or calcium salts. The word sclerite is derived from the Greek *skleros*, 'hard'.

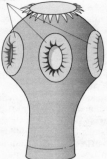

Scolex

scleroblast In *Porifera, a structure formed by the partial fusion of amoeboid cells (sclerocytes), which then separates again, its members secreting one of the spicules from which the skeleton of the sponge is composed.

sclerocyte *See* SCLEROBLAST.

Sclerodermi An older name for the filefish and triggerfish now included in the order *Tetraodontiformes.

Scleroglossa A group of reptiles that includes the *Scincidae, *Varanidae, and *Serpentes.

Scleroparei An older name for a group of fish now included in the order *Scorpaeniformes.

scleroprotein One of a group of insoluble, fibrous *proteins, which serve as structural materials within the body. The group includes the *collagens and *keratins.

sclerotin *See* SCLERITE.

sclerotization The formation and hardening of a new *exoskeleton, particularly in *Arthropoda.

sclerotome *See* SOMITE.

scolex In *Cestoda, the anterior end, which bears the hooks and suckers by which the animal is attached to its host.

scoli *See* SATURNIIDAE.

Scoliidae (suborder *Apocrita, superfamily Scolioidea) Family of *Hymenoptera in which most adults are orange-yellow with golden hairs, or occasionally dark; they can usually be found on flowers. They are large (9–36 mm long, most being 20–30 mm), robust, and densely hairy, and are parasitic on scarab larvae in the soil. The hosts are insensitized and eggs are laid externally on the abdomen. Sexual *dimorphism is often marked.

Scolopacidae (curlews, godwits, ruff, sandpipers, sanderling, snipe, stints, whimbrel; class *Aves, order *Charadriiformes) A family of small to large wading birds, most of which are cryptically patterned with brown, grey, black, and white plumage. (Snipe (15 species of *Gallinago*) have a varying number of tail feathers, the outer ones adapted to produce a vibratory sound in display flight.) Their bills are short to long and straight, *decurved, or recurved. Their necks are medium to long, their wings long, and their legs short to long. (*Calidris alba* (sanderling) lacks a hind toe.) They feed on invertebrates, and usually nest on the ground. They inhabit open areas, usually near water, flocking in coastal areas outside the breeding season. The eight species of *Numenius* (curlews and whimbrel) and 18 *Calidris* species breed in tundra and upland regions, where they are ground-nesting, and in winter they are found on shores and estuaries: they are highly migratory. Sandpipers (11 species of *Tringa*) inhabit coastal and inland waters and wet meadows. Godwits (four species of *Limosa*) breed in Arctic regions of N. America, Europe, and Asia, and migrate to S. America, Africa, southern Asia, and Australia. The ruff (*Philomachus pugnax*), in which the male is larger than the female, has a large neck ruff, variable in colour, and ear tufts in its breeding plumage; males display in a '*lek*'. Ruffs breed in wetland meadows and inhabit fields and marshes in winter. There are 23 genera in the family, with 85 species, found world-wide.

Scolopendromorpha *See* CHILOPODA.

Scolytidae (bark beetles; subclass *Pterygota, order *Coleoptera*)** Family of small, dark, cylindrical beetles, usually less than 5 mm long, in which the head is hooded by a large *pronotum. The antennae are short and clubbed. The *elytra are often incised and grooved behind, producing a device that enables them to shovel wood debris. An egg chamber is excavated under bark or deeper in wood according to species; eggs are placed in separate niches. Larvae are fleshy and legless, and tunnel away from the main chamber producing a characteristic pattern. Dutch elm disease is caused by a fungus carried by the elm-bark beetle, *Scolytus scolytus*; the fungus blocks tree vessels. Ambrosia beetles (e.g. *Xyleborus* species) penetrate wood, but larvae feed on fungi which develop on tunnel walls.

Scolytus scolytus (elm-bark beetle) *See* SCOLYTIDAE.

Scomberesocidae (saury, skipper; subclass *Actinopterygii, order *Atheriniformes*)** A small family of marine, *pelagic fish that have a very slender body, produced and beak-like jaws, and the *dorsal and *anal fins located near the tail fin and followed by a series of small *finlets. Where locally abundant, they are exploited commercially, and they may well be an important source of food for larger predatory fish. *Scomberesox saurus* (skipper or Atlantic saury) occurs in warmer seas world-wide; *Cololabis saira* (Pacific saury) is believed to occur only in the Pacific basin. There are about four species.

Scomberesox saurus (skipper, Atlantic saury) *See* SCOMBERESOCIDAE.

Scomber scombrus (Atlantic mackerel) *See* SCOMBRIDAE.

Scombridae (tuna, mackerel; subclass *Actinopterygii, order *Perciformes*)** A family of marine, *pelagic fish, found in warm to temperate waters, that often have perfectly streamlined, spindle-shaped bodies. There are two widely separated *dorsal fins; the second dorsal and the *anal fin are each followed by a series of small *finlets. The narrow *caudal peduncle bears a strongly forked or *lunate tail fin. The Scombridae are all fast-swimming, predatory fish and many species are much sought-after commercially, e.g. *Thunnus thynnus* (Atlantic tuna), 4.8 m, *Thunnus alalunga* (albacore), 1.2 m, and *Scomber scombrus* (Atlantic mackerel), 56 cm. There are about 45 species, distributed world-wide.

scopa Specialized tract of hairs between which female bees compact pollen for transport back to the nest. In solitary mining bees the principal scopa comprises *plumose hairs on the hind legs and can be massively developed. Some mining bees have additional scopae on the underside of the first abdominal (metasomal) *sternite. The families *Fideliidae and *Megachilidae have well-developed scopae on the ventral surface of the abdomen and these comprise dense fringes of stiff, straight or microscopically coiled hairs. Pollen scopae are absent in cuckoo bees, which do not forage for themselves.

Scopelarchidae (pearleye; superorder *Scopelomorpha, order *Myctophiformes*)** A small family of marine, midwater fish that have an elongate body, large mouth, large, upward-directed eyes, and the small *dorsal fin followed by an *adipose fin. Some species, e.g. *Benthalbella dentata* (23 cm), which can be found at depths of more than 1000 m, have a large, glistening spot near the eyeball. There are 18 species, distributed world-wide.

Scopelomorpha (class *Osteichthyes, subclass *Actinopterygii*)** A superorder of bony fish that includes the order *Myctophiformes.

Scopelosauridae (Notosudidae; weary-fish; superorder *Scopelomorpha, order *Myctophiformes) A small family of marine fish that have a thin, slender body, large eyes, large mouth, a single *dorsal fin, and an *adipose fin above the *anal fin. They are probably inhabitants of deep seas. There are about five species, distributed world-wide.

***Scophthalmus maximus** (European turbot) See* BOTHIDAE.

Scopidae (hammerhead, hammerkop; class *Aves, order *Ciconiiformes) A monospecific family (*Scopus umbretta*), which is a medium-sized, brown bird with a large head and a long neck crest, a long, stout bill, long wings, and a long tail. Its legs are fairly long, with partially webbed toes, the middle one being *pectinate. It inhabits marshes and streams near trees, feeds on amphibians, fish, and insects, and builds a large, roofed nest in a tree. It is found in Africa.

scops owls (*Otus*) *See* STRIGIDAE.

***Scopus umbretta** (hammerhead) See* SCOPIDAE.

Scorpaenidae (scorpionfish, rockfish; subclass *Actinopterygii, order *Scorpaeniformes) A very large family of bony fish that have moderately compressed bodies, large heads with prominent eyes, and a bony ridge or stay across the cheeks. The *gill cover is usually protected by a number of sharp spines. The *pelvic fins are placed far forward under the fan-like *pectoral fins. Scorpionfish are bottom-dwellers and poor swimmers. Although many species have intricate colour patterns, few can match the striking appearance of *Pterois volitans* (lionfish), 30 cm. Like most scorpionfish, the lionfish has venom glands at the bases of the fin spines and can inflict very painful stings. However, many species are quite edible and several are harvested commercially. There are about 330 species, distributed worldwide in tropical and temperate seas.

Scorpaeniformes (class *Osteichthyes, subclass *Actinopterygii) An order of bony fish that have a bony ridge or stay across the cheeks, rounded *pectoral fins, and spiny projections on the *gill cover. Most species are bottom-dwellers, with intricate colour patterns used to conceal the fish. The order includes 21 families, including *Cottidae (sculpins), *Scorpaenidae (scorpionfish), and *Triglidae (gurnards).

Scorpiones (scorpions; subphylum *Chelicerata, class *Arachnida) Order of what are considered to be the most primitive arachnids, which have retained the division and segmentation of the abdomen. Scorpions have changed little since the *Silurian. The prosoma is covered by a carapace and broadly joins the segmented abdomen, which is divided into a wide, anterior mesosoma (pre-abdomen), and a long, posterior metasoma (post-abdomen). The latter consists of a series of ring-shaped segments terminating in a *telson in the form of a sting with associated poison glands. The *chelicerae are very small, and the *pedipalps are armed with large pincers. Scorpions are usually yellow to brown or black, more rarely greenish or bluish, and all fluoresce in ultraviolet light. All except cavernicolous species have a pair of median eyes and two to five pairs of lateral eyes on the carapace margin. Sensory organs include *slit organs, sensory *setae, and *trichobothria, as well as *pectines responding to touch and vibration frequencies above 100 Hz. Most scorpions sting in defence or to subdue struggling prey, and although the poison may be strong and even effective against vertebrates, they themselves are immune. Members of the family Buthidae are dangerous to humans, the poisons being neurotoxic in action. Mating involves the deposition of a *spermatophore by the male, and an often lengthy and complex courtship ending with the female being led and pulled over the spermatophore, whereupon she takes the apical portion into her *gonopore. Members of the Scorpionidae are *viviparous, and as in all arachnids there is a high degree of parental care. Most of the 700 or so known species, distributed among six families, are tropical and subtropical, but the European species *Euscorpius germanus* is found in the southern Alps, *and E. flavicaudis* has been found in Kent, England. The largest known species is the African *Pandinus imperator*, which reaches 170 mm in length.

scorpionfish *See* SCORPAENIDAE.

scorpionflies *See* MECOPTERA.

scorpions *See* ARACHNIDA; SCORPIONES.

scoters (*Melanitta***)** *See* ANATIDAE.

scramble competition *Competition for a resource that is inadequate for the needs of all, but is partitioned equally among contestants, so no competitor obtains the amount it needs and in extreme cases all die. *Compare* CONTEST COMPETITION.

scrapie *See* BOVINE SPONGIFORM ENCEPHALITIS; PRION.

screamer *See* ANHIMIDAE; ANSERIFORMES.

screaming cowbird (*Molothrus rufoaxillaris***)** *See* ICTERIDAE.

screech beetle *See* HYGROBIIDAE.

screech owls (*Otus***)** *See* STRIGIDAE.

scrobicular Pertaining to *scrobicules.

scrobiculate Having many *scrobicules.

scrobicule A small pit or depression, e.g. around one of the tubercles of a sea urchin (*Echinoidea).

scrotum In male *Mammalia in which the testes descend from the abdomen, the sac in which they are contained at least during the breeding season and in which they are maintained at a temperature lower than the body temperature.

scrub-birds *See* ATRICHORNITHIDAE.

scrubfowl (*Megapodius***)** *See* MEGAPODIIDAE.

scrub wrens *See* ACANTHIZIDAE.

Scrupariina (order *Cheilostomata, suborder *Anasca) A division of bryozoans, in which the colony has an erect, upstanding habit. The individuals are tubular, and growth is uniserial. The division occurs from the *Cretaceous to the present.

sculpin *See* COTTIDAE; ICELIDAE.

scute An enlarged, bony, dermal plate or scale.

scutellum The most posterior of the three dorsal *sclerites that are found on the *mesothorax and *metathorax of many winged insects.

Scutigeromorpha *See* CHILOPODA.

scuttle flies *See* PHORIDAE.

scutum 1. In *Cirripedia, one of the two plates covering the anterior end of the sides of the *mantle. **2.** In *Insecta, the middle of the three divisions of the dorsal surface of a thoracic segment.

Scyliorhinidae (catshark; subclass *Elasmobranchii, order *Lamniformes) A family of marine sharks that have slender bodies and the first of the two *dorsal fins usually situated over or behind the *pelvic fins. There is no *nictitating membrane or oro-nasal groove. The *anal fin is present. Catsharks tend to be bottom-dwelling and slow-moving, feeding on small animals. Their eggs are enclosed in horny capsules which are often attached to seaweed. There are about 58 species, distributed world-wide in temperate to tropical seas.

scyphistoma In *Scyphozoa, the *polypoid larval stage which develops from a *planula larva. It produces more scyphistomae by budding (a process called strobolization) and, at certain times of year, produces minute *medusae (ephyrae), some of which develop into the sexual adult form.

S

Scyphomedusae (phylum *Cnidaria, class *Scyphozoa) A subclass of medusoid (*see* MEDUSA) cnidarians whose members lack a *velum and have a *coelenteron partly divided by four interradial endodermal *septa. They are first known from *Cambrian rocks.

Scyphozoa (jellyfish; phylum *Cnidaria) A class of marine, mainly *pelagic medusoids (*see* MEDUSA) which differ from the class *Hydrozoa in possessing endodermal gastric *tentacles, four-part radial symmetry, and gonads located in the gastric cavity. The *polyp stage is reduced or absent.

***Scytalopus* (tapaculo)** *See* RHINOCRYPT-IDAE.

Scytodidae (spitting spiders; order *Araneae, suborder Araneomorphae) Family of spiders whose common name refers to their unique mechanism for capturing prey. The high, domed carapace bears six eyes in three groups, and accommodates the large salivary and poison glands. Some species occupy a web and others are nocturnal, but all spit at their prey or enemies. The prosoma is lifted up and threads of sticky substance and poison are sprayed from the *chelicerae in a zigzag fashion, pinning the prey to the substrate. *Scytodes thoracica* is an inhabitant of buildings in northern latitudes, and other species are tropical.

sea anemone *See* ACTINIARIA; ANTHO-ZOA; CERIANTHARIA; CNIDARIA; ZOANTHI-NIARIA.

sea bass *See* SERRANIDAE.

sea bream *See* SPARIDAE.

sea butterflies *See* PTEROPODS.

sea carps *See* APLODACTYLIDAE.

sea chub *See* KYPHOSIDAE.

'sea cow' *See* DUGONGIDAE; SIRENIA.

sea cucumbers *See* HOLOTHUROIDEA.

sea fan *See* GORGONACEA.

sea hares *See* ANASPIDEA.

sea-horse (*Hippocampus hippocampus*) *See* SYNGNATHIDAE.

seal *See* PHOCIDAE.

sea lilies *See* ARTICULATA; BOURGUETICRI-NIDA; CRINOIDEA; ISOCRINIDA; MILLERIC-RINIDA.

sea lion *See* OTARIIDAE.

sea mice *See* ERRANTIA.

sea moth *See* PEGASIDAE.

sea peach (*Halocynthia pyriformes*) *See* ASCIDIACEA.

sea pen *See* ANTHOZOA; PENNATULACEA.

searcher *See* BATHYMASTERIDAE.

searching Behaviour by an animal that is directed toward the provision of some necessity (e.g. food, nesting material, or a mate) that has not yet been located.

searching image The mental image of an object that is apparently possessed by an animal searching for that object, and whose existence is inferred from observation of the behaviour of animals.

Searsiidae (tubeshoulder; superorder *Protacanthopterygii, order *Salmoni-formes) A small family of deep-sea fish that have a single, soft-rayed, *dorsal fin near the tail fin, above the *anal fin. The eyes and mouth are rather large. The peculiar shoulder organ is typical of the family. It is a large, black sac opening to the exterior through a tube-like process above the *pectoral fin. When the sac contents are released, a greenish glow appears beside the fish. Many species have a widespread distribution. There are about 17 species.

sea slaters *See* LIGIIDAE.

sea slugs *See* OPISTHOBRANCHIA.

sea snail *See* CYCLOPTERIDAE.

sea snakes *See* HYDROPHIIDAE.

sea spiders *See* PYCNOGONIDA.

sea squirts *See* ASCIDIACEA; UROCHORDATA.

sea urchins *See* ECHINOIDEA.

sea wasp (*Chironex fleckeri***)** *See* CUBOMEDUSAE.

seaweed flies *See* COELOPIDAE.

sebaceous gland In *Mammalia, a skin gland, projecting deep into the dermis and almost always opening into a hair follicle, which secretes a fatty substance (sebum) which helps to waterproof the hair.

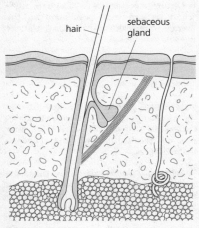

hair

sebaceous gland

Sebaceous gland

sebum *See* SEBACEOUS GLAND.

secator An enlarged, posterior, *premolar with an elongated cutting edge, as in marsupial *Phalangeridae and placental *Carpolestidae.

secondary 1. One of the inner flight *feathers of a bird, located on the trailing edge of the wing between the body and the bend of the wing. **2.** Applied to structures (e.g. the limbs of snakes, the wings of insects) that have been lost in the course of evolution, and to structures or forms (e.g. *bilateral symmetry in *Nudibranchia) that have developed.

secondary consumer A carnivore that preys upon herbivores.

secondary genitalia *See* ACCESSORY GENITALIA.

secondary sexual character A characteristic of animals that differs between the two sexes, but excluding the gonads and the ducts and associated glands that convey the *gametes. Examples are *mammary glands, external genitalia, antlers in ungulates, and certain plumage patterns (e.g. peacock's tail) in birds.

secondary structure The folding of a *polypeptide or polynucleotide chain along one axis of the molecule; it is stabilized by the formation of intramolecular hydrogen bonds along the length of the chain. *Compare* PRIMARY STRUCTURE; QUATERNARY STRUCTURE; TERTIARY STRUCTURE.

secretary bird (*Sagittarius serpentarius***)** *See* SAGITTARIIDAE.

secretion 1. The act of discharging useful materials (i.e. as opposed to the excretion of waste products) from cells. **2.** Any such material discharged from cells.

secretory vesicle (granule) A structure that forms in the *Golgi body by stages as it fills with products secreted by the Golgi apparatus. Secretory vesicles then fuse with the outer *cell membrane prior to releasing their contents to the outside of the cell.

sectorial Sharp, knife-like.

secund Arranged on one side, or curved to one side.

Sedentaria (bamboo worms, coneworms, fanworms, lugworms; phylum

*Annelida, class *Polychaeta) A subclass of entirely marine worms in which the body segments and *parapodia differ along the length of an individual. All Sedentaria are burrowers or tube-dwellers. They are first recorded from the *Ordovician. There is no acceptable ordinal classification for the Polychaeta; the class is divided into *Errantia and Sedentaria for convenience but genera belonging to each subclass may not be related.

sedentary Applied to organisms that are attached to a substrate but are capable of limited movement. *Compare* SESSILE.

sedimentation factor (sedimentation value, S factor, S value) A measure of the rate at which a molecule, *organelle, or particle settles under standard conditions of centrifugation. It is equal to the acceleration, measured in *Svedberg units, and allows particles to be separated and identified.

sedimentation value See SEDIMENTATION FACTOR.

seedeaters See FRINGILLIDAE.

seed shrimps See OSTRACODA.

seedsnipe See THINOCORIDAE.

seed weevil See BRUCHIDAE.

segmentation 1. The division of the body into a series of more or less equivalent sections along its length. *See* METAMERIC SEGMENTATION. **2.** *See* CLEAVAGE.

segregation, Mendel's law of See MENDEL'S LAWS.

Seisonidea (phylum *Aschelminthes, class *Rotifera) An order of entirely marine, free-swimming worms in which the reproductive organs are paired. Males and females are morphologically identical.

sei whale (*Balaenoptera borealis*) See BALAENOPTERIDAE.

Selachimorpha (infraclass *Holostei) A superorder of cartilaginous fish that includes all sharks.

selection The process that determines the relative share of different *genotypes which individuals possess and propagate in a population. The relative probability of survival and reproduction of a genotype is termed the *adaptive value. Selection may be natural (i.e. by nature) or artificial (i.e. by human action, e.g. plant or animal breeders).

selection coefficient (*s*) A measure of the relative excess or deficiency of *adaptive value of a *genotype compared with another genotype in the population. If $s=100$ then one out of a hundred individuals of a given genotype fails to reproduce.

selection differential The difference between the average value of a quantitative *character in a whole population and the average value of those selected to reproduce the next generation.

selection intensity A measure of the difference in *fitness values within a *population.

selection pressure Pressure exerted by the environment, through *natural selection, on evolution. Thus weak selection pressures result in little evolutionary change and vice versa.

selectively permeable membrane A *membrane that allows the passage of only certain substances, selected on the basis of the size and usefulness of molecules.

selective value See ADAPTIVE VALUE.

selenodont Applied to *molar teeth in which the grinding surfaces possess crescent-shaped ridges.

Selevinia betpakdalensis (dzhalman) See SELEVINIIDAE.

Seleviniidae (dzhalmans; order *Rodentia, suborder *Myomorpha) A

s

monospecific family (*Selevinia betpa-kdalensis*) of small, burrowing, mainly insectivorous rodents, which have somewhat rounded heads, large ears, and long, sparsely haired tails. They resemble dormice in many respects, but the hind legs are longer than the fore limbs and they move by leaping. The tympanic *bulla is very large and the *angular process of the jaw is perforated. The upper *incisors are large, the cheek teeth very small, with large central basins, and rooted. Dzhalmans are found only in the Betpakdala Desert of central Asia.

self-fertilization The fusion of male and female *gametes from the same individual (as opposed to cross-fertilization, in which the gametes come from different individuals).

selfish DNA One of a number of hypotheses advanced in an attempt to explain the presence of surplus DNA in the *genome which is not translated into protein. Three hypotheses have been put forward to account for the adaptive advantage of this apparently redundant DNA: (*a*) that extra DNA separates the genes so as to increase the cross-over frequency (*see* CROSSING-OVER); (*b*) that the possibility of varying the total amount of DNA per cell allows the control of cell volume and cell growth rate; and (*c*) (the selfish DNA hypothesis) that selection acts within the genome, favouring any method by which DNA may more rapidly replicate itself, and that this can be better achieved if *phenotypic expression can be bypassed. This is achieved, it is proposed, by DNA spreading laterally so as to be duplicated at new *loci elsewhere in the genome. In this way the DNA may be viewed as acting 'selfishly', since the apparently surplus DNA confers no advantage on the organism bearing it and therefore supplying the materials from which it is made. *See also* SELFISH GENES.

selfish genes Epithet used by some authors to reinforce their notion that organisms function as agents for replication of *genes, as opposed to genes functioning

as servants of organisms. *See also* SELFISH DNA.

selfish herd A theory proposed in 1971 by W. D. Hamilton (1936–2000) according to which the risk to an individual of predation is reduced if that individual places another individual between itself and the predator. When many individuals behave in this way an aggregation is the inevitable result and, because the risk is least near the centre and greatest at the edge, individuals of high social status will tend to occupy the centre and subordinate individuals will be pushed to the edge.

Semaeostomatida (Discomedusae; class *Scyphozoa, subclass *Scyphomedusae) An order of free *medusae that are typically flat or saucer-like in shape. The angles of the square-shaped mouth are drawn out into long, frilly lobes. They are found in great numbers in coastal waters of all oceans, in all latitudes.

semelparity (big-bang reproduction) The condition of an organism that has only one reproductive cycle during its lifetime. It is well exhibited by *Antechinus*; males die after mating (typically in autumn, in their first year of life), so the entire population overwinters as pregnant females. *Compare* ITEROPARITY.

semen The ejaculatory fluid in which *spermatazoa are suspended, produced by the *testes together with secretion from the prostate gland.

semeniferous tubules *See* TESTES.

semicircular canal In *Gnathostomata, one of the three fluid-filled cavities, semicircular in shape, arranged at right angles to one another, and located within the inner ear. They register movement of the head and are associated with balance.

semidominance *See* PARTIAL DOMINANCE.

Semionotiformes (class *Osteichthyes, subclass *Actinopterygii) An order of bony fish that has fossil and living representatives, the latter usually placed in the family *Lepisosteidae.

semipalmate In birds, applied to feet that have partially webbed toes, the webs not reaching to the ends of the toes.

semi-permeable Applied to a barrier, usually a *membrane, whose structure allows the passage of the solvent but prevents the passage of the solute. *Compare* DIFFERENTIALLY PERMEABLE MEMBRANE; SELECTIVELY PERMEABLE MEMBRANE.

semi-species A group of organisms that are taxonomically intermediate between a *race and a *species, with reduced *outbreeding and *gene flow (i.e. with incomplete reproductive isolating mechanisms). Semi-species are thought to represent advanced stages of speciation.

Semnopithecus A genus containing the Indian grey (or sacred) langurs. *See* CERCOPITHECIDAE.

senescence The complex deteriorative processes that terminate naturally the functional life of an organ or organism.

sense codon A *codon that codes for an *amino acid. *Compare* NONSENSE CODON.

sensilla ampullacea In some *Arachnida (e.g. *Solifugae), pits bearing short, sensory hairs.

sensitive period A period of time during which a young animal is most impressionable and therefore most likely to acquire learned behaviour. This is the period during which *imprinting occurs.

sensitivity *See* IRRITABILITY.

sensitivity analysis The consideration of a number of factors involved in the mathematical modelling of an *ecosystem and its components. These include feedback and control, and the stability and sensitivity of the system as a whole to changes in some parts of the system. From the analysis, predictions can be made.

sensitization 1. The increase in the likelihood that a particular and significant stimulus will produce a response in an animal repeatedly exposed to it. *Compare* EXTINCTION; HABITUATION. **2.** An increase in the specific reactivity of an organism or cell to an *antigen or *hapten.

sensory adaptation *See* ADAPTATION.

sensory neuron A *neuron that receives and transmits information from a sense organ.

Sepioidea (cuttlefish; class *Cephalopoda, subclass *Coleoidea) An order of cephalopods which have an internal *phragmocone, which may be curved or coiled. They possess eight arms and two *tentacles. They probably diverged from a belemnite (*Belemnitida) ancestor in the early part of the *Cretaceous Period.

Sepsidae (ensign flies; order *Diptera, suborder *Cyclorrapha) Small family of flies in which the adults are small and often ant-like. The head may or may not have *vibrissae, and the *palps are reduced. The most characteristic feature of adult males is their habit of collecting on vegetation, where they walk about waving their wings, which accentuates the usual black, spotted wing-tips. These aggregations may be sexual in function. The larvae are saprophagous, and have been collected from dung. Adults feed on flowers. There are about 200 species, found throughout the world.

septa 1. *See* SEPTATE JUNCTION. **2.** *See* SEPTUM.

septal neck The collar that borders the opening (foramen) in a septum.

septate Having cross-walls or *septa.

septate junction A type of adhesive cell junction, particularly characteristic of invertebrate tissues, in which the *membranes forming the junction are held very regularly 15–17 nm apart. Traversing this gap at right angles to the cell surface there are crossbars of dense material, the septa. In sections taken parallel to the cell surface, these septa are revealed as *beta pleated sheets that zigzag between the opposed membranes in parallel rows.

septomaxilla A bone at the front of the upper jaw in reptiles; monotremes are the only mammals in which this bone maintains its separate identity. *See* OBDURODON.

septum 1. (*pl.* **septa**) A cross-wall or partition. In *Cephalopoda, the shell consists of individual compartments (camerae) separated by septa. In *Annelida, the septa internally divide the individual segments of the body. **2.** *See* SEPTATE JUNCTION.

sequential hermaphrodite An organism that, during the course of its life, is capable of having both a male phase and a female phase of existence. *See also* PROTANDRY; PROTOGYNY.

sequester To bind a metal ion into a *chelate.

sera *See* SERUM.

sere A stage in a plant succession.

seriemas *See* CARIAMIDAE.

serine An aliphatic, polar, *alpha amino acid that is often associated with the *active site of an *enzyme.

Serinus (canaries, seedeaters) *See* FRINGILLIDAE.

Seriola dumerili (amberjack) *See* CARANGIDAE.

Seriolella brama (warehou) *See* CENTROLOPHIDAE.

serotinal In later summer. The term is used with reference to the six-part division of the year used by ecologists, especially in relation to terrestrial and freshwater communities. *Compare* AESTIVAL; AUTUMNAL; HIBERNAL; PREVERNAL; VERNAL.

serotonin 5-Hydroxytryptamine, a derivative of *tryptophan that is a powerful vasoconstrictor found especially in the brain, intestinal fluid, *mast cells, and blood platelets, and also in snake and toad venoms. Curiously it has also been found in nettles, bananas, and tomatoes. In humans it has been suggested as a cause of migraines.

serous membrane In vertebrates, the type of membrane that lines the surface of *coelomic cavities (i.e. body cavities that are not open to the exterior). It is composed of a single layer of plate-like cells (mesothelium) attached to the surface by a layer of *connective tissue.

Serpentes (Ophidia; snakes; subclass *Lepidosauria, order *Squamata) A suborder of elongate, limbless reptiles which evolved in the *Cretaceous from burrowing lizards. They are carnivores in which the jaws are held by ligaments and are capable of considerable extension. There are no eyelids; the eyes are covered by a spectacle (transparent scale). The skin is usually shed whole. There is no tail regeneration. The tongue is generally forked and protruded through a notch in the snout. There are 12 families.

Serranidae (sea bass, grouper; subclass *Actinopterygii, order *Perciformes) A very large family of marine fish, often with a somewhat robust, fully scaled body, and single, continuous, *dorsal fin that consists of a spiny and a soft-rayed section. The mouth is often large; the *lateral line is continuous from head to tail. Many species are *hermaphroditic, the individual fish changing sex during its lifetime. Serranidae range in size from the relatively small *Serranus cabrilla* (comber), 35 cm, of the Atlantic, to *Promicrops lanceolatus* (Queensland grouper), 3.6 m, found in the Indo-Australian region. There are at least 370 species, found in temperate to tropical seas.

Serranus cabrilla (comber) *See* SERRANIDAE.

Serrasalmus rhombeus (white piranha) *See* CHARACIDAE.

S

Serrivomeridae (superorder *Elopomorpha, order *Anguilliformes) A small family of deep-sea eels that have distinctly produced, beak-like jaws, a scaleless body, minute *pectoral fins, and the *dorsal fin originating behind the *anal fin. Many species have a world-wide distribution and may be found at depths exceeding 800 m. There are about 10 species.

Sertoli cells See TESTIS.

serum (*pl*. **sera**) The clear fluid that remains after blood is allowed to clot. It differs from *blood plasma in that it lacks *fibrinogen.

Sesiidae (Aegeriidae; clearwings; subclass *Pterygota, order *Lepidoptera) Family of fairly small moths which mimic bees and wasps both in colour and shape, and sometimes buzz in a similar way. Scales are absent from large areas of the wings. The *thorax and *abdomen usually bear wasp- or bee-like colours. Sesiidae fly rapidly and by day. Larvae often tunnel in stems of trees and shrubs. Some are pests of fruit trees. They are distributed world-wide.

sessile Attached to a substrate; nonmotile. *Compare* SEDENTARY.

seta A stiff, hair-like or bristle-like structure; 'setae' is sometimes used synonymously with *chaetae.

setation See CHAETOTAXY.

setose Bearing bristles or *setae.

sex cell See GAMETE.

sex chromatin See BARR BODY.

sex chromosome A *chromosome whose presence or absence is linked with the sex of the bearer, and that plays a role in *sex determination. It is present in all sexually reproducing, *diploid animals and plants. The sex that has a homologous pair of sex chromosomes in the *nucleus is said to be homogametic (referred to as 'XX'), whereas the sex with a dissimilar pair or with an unpaired chromosome is said to be heterogametic (referred to as 'X' or 'XY'). The homogametic sex produces *gametes that are identical in their chromosome sets, all containing one *X-chromosome. The heterogametic sex produces equal numbers of two different types of gametes, one with one X-chromosome, and one without (possessing either a *Y-chromosome or none at all). Union of gametes of the two sexes thus results in equal numbers of offspring of the two sexes. In mammals and many other animals the presence or absence of a Y-chromosome determines sex; in birds and reptiles, and in some fish, amphibians, and insects, sex is determined by the number of X-chromosomes in relation to the number of *autosomes. Occasionally a mutation will produce an extra dosage of X-chromosomes in the gamete, and so in the progeny (e.g. in Klinefelter's syndrome in humans, the individual genotype (a male) possesses an extra X-chromosome (XXY)).

sex determination The mechanism by which sex is determined. In many species, sex is determined at fertilization by the nature of the *chromosomes in the sperm that fertilizes the egg. In some species *Y-chromosome sperm produces the male (i.e. XY) *zygotes and *X-chromosome sperm female (i.e. XX) zygotes; in other species (e.g. birds) males are XX and females XY. *See also* SEX CHROMOSOME.

sex linkage The location of a *gene on a *sex chromosome. Such a gene is usually unrelated to *primary or *secondary sexual characters and is located on the *X-chromosome. The heterogametic sex (e.g. a male mammal or female bird) receives the X-chromosome and therefore all its sex-linked genes from the homogametic parent, and passes them on to offspring of the homogametic sex. Since there is no partner to the X-chromosome in the heterogametic sex, recessive genes cannot be masked by their dominant alleles and so are expressed in the *phenotype. It is for this reason that male humans manifest a larger number of recessive genes (e.g. red–green colour blindness, and haemophilia) than females.

sex ratio The relative number of males to females in a population. The measure is usually applied to a given age class. The primary sex ratio is that immediately after fertilization (normally 1 : 1); the secondary sex ratio is that at birth or hatching; and the tertiary sex ratio is that at maturity.

sex reversal Change in functioning such that a member of one sex behaves as a member of the other. Some organisms (e.g. certain molluscs) make this change as part of their normal *life cycle, with most individuals functioning as males when young and then passing through a transitional stage to a period when they function as females. Sex reversal may otherwise be experimentally induced (e.g. by hormone transplants in humans) or environmentally induced (e.g. by temperature effects on the production of male or female *zygotes in some turtles).

sexton beetle *See* SILPHIDAE.

sexual dimorphism The occurrence of morphological differences (other than *primary sexual characters) that distinguish males from females of a species of organism (e.g. male deer often have larger antlers than females, and the males of many birds have differing (often more brightly coloured) plumage).

sexual generation In *Cynipidae, the males and females that emerge from galls to mate and lay eggs that will become the *agamic generation of *parthenogenetically reproducing females. In general, sexual-stage galls follow a period of rich resources and precede a period of poor resources. Sexual individuals are generally smaller and less fecund than members of the agamic generation.

sexual reproduction Reproduction involving the fusion of *haploid nuclei, usually *gametes, which result from *meiosis.

sexual selection A theory proposed by *Darwin, that in some species males compete for mates and that characteristics enhancing their success in mating would have value in the struggle for existence. Such characteristics would be used either in male display to attract females (inter-sexual selection) or in combat between rival males (intra-sexual selection), or both at the same time, and would be perpetuated from generation to generation even if they disadvantaged the males in other ways (e.g. the cumbersome but spectacular tail of a male peacock). *See also* HANDICAP PRINCIPLE; RUNAWAY HYPOTHESIS.

(((●))) SEE WEB LINKS
• Description of sexual selection

Seychelles giant tortoise (*Testudo* (or *Geochelone*) *gigantea***)** *See* TESTUDINIDAE.

Seymouria A fossil amphibian that possessed a combination of amphibian and reptilian characters and has been referred to both groups on different occasions. One of the features separating it and its relatives from other amphibians is the presence of a large, forward-extended optic notch (the primitive site of the ear drum). A medium-sized amphibian, it existed during the Early *Permian in N. America.

S factor *See* SEDIMENTATION FACTOR.

shad *See* CLUPEIDAE.

shags *See* PHALACROCORACIDAE.

Shannon–Wiener index of diversity (information index) A measure used by ecologists when a system contains too many individuals for each to be identified and examined. A small sample is used; the index (D) is the ratio of the number of species to their importance values (e.g. *biomass or productivity) within a *trophic level or *community. $D = -\Sigma p s_i \log p_i$, where s is the total number of species in the sample, i is the total number of individuals in one species, p_i (a decimal fraction) is the number of individuals of one species in relation to the number of individuals in the population, and the log is to base-2 or base-e.

shark-toothed whales *See* SQUALODO-NTOIDEA.

sharpbill (*Oxyruncus cristatus*) *See* OXY-RUNCIDAE.

shearwaters *See* PROCELLARIIDAE; PRO-CELLARIIFORMES.

sheathbills *See* CHIONIDIDAE.

sheath of Schwann *See* NEURILEMMA.

sheep (*Ovis*) *See* BOVIDAE.

sheet-web spiders *See* AGELENIDAE.

shelducks *See* ANATIDAE.

Shelford's law of tolerance A law stating that the abundance or distribution of an organism can be controlled by certain factors (e.g. the climatic, topographic, and biological requirements of plants and animals) where levels of these exceed the maximum or minimum *limits of tolerance of that organism.

shell beak In *Bivalvia, the oldest part of the shell, located near the hinge.

shelled pteropods *See* THECOSOMATA.

shell gland 1. In *Branchiopoda, an excretory organ in which the duct is visible through the wall of the *carapace. **2.** In some *Platyhelminthes, a gland that secretes a substance that causes the hardening of egg shells.

shell plates The body covering of *Polyplacophora, consisting of eight, overlapping plates.

shield-backed grasshoppers *See* TETTIGONIIDAE.

shieldbug *See* PENTATOMIDAE.

shield-tailed snakes *See* UROPELTIDAE.

Shinisaurus crocodilurus (crocodile lizard) *See* XENOSAURIDAE.

shock disease A response to overcrowding in which physiological changes in the *endocrine system lead to physical deterioration, reduced reproductive success, and fighting, which may be followed by a catastrophic decline in numbers. Shock disease has been observed in many small mammals (e.g. field vole, *Microtus agrestis*), but is believed to be rare among invertebrates.

shoe-bill (*Balaeniceps rex*) *See* BALAENICIPITIDAE.

shorebug *See* SALDIDAE.

shore lark (*Eremophila alpestris*) *See* ALAUDIDAE.

shorthead (*Breviceps adspersus*) *See* MICROHYLIDAE.

short-horned grasshopper *See* ACRIDIDAE.

short-jawed mastodont An offshoot of the *Gomphotheriidae, the short-jawed mastodonts first appear in the *Miocene and survived in N. America until about 8000 years ago. Some of the *Pliocene representatives, belonging to the *Anancus* genus, strongly resembled the straight-tusked elephants of the *Pleistocene.

shortnosed tripodfish (*Triacanthus brevirostris*) *See* TRIACANTHIDAE.

short-winged crickets *See* GRYLLIDAE.

shortwings *See* TURDIDAE.

shovelnose ray *See* RHINOBATIDAE.

shrews *See* SORICIDAE.

shrikes *See* LANIIDAE.

shrike-vireos *See* VIREONIDAE.

shrimpfish *See* CENTRISCIDAE.

shrimps 1. *See* CRUSTACEA; DECAPODA. **2.** (mantis shrimps) *See* STOMATOPODA.

Sialidae (alderflies; subclass *Pterygota, order *Megaloptera) Family of insects which are similar to the *Corydalidae, but are usually much smaller and lack *ocelli. They are blackish-grey, resemble lacewings, and are found near ponds or streams, where their larvae develop. The larvae are similar to those of the Corydalidae, but lack hooked, anal *prolegs, and have a terminal filament.

siamang *See* HYLOBATIDAE.

sib Shortened form of '*sibling'.

sibling Offspring of the same parents; a brother and/or sister.

sibling species Species that are identical in outward appearance or very nearly so. Despite the similarity, however, they qualify as species by being reproductively isolated. *See* REPRODUCTIVE ISOLATING MECHANISM.

Siboglinidae (phylum *Annelida) A family that comprises the *Pogonophora (beard worms) and *Vestimentifera, both of which were formerly ranked as phyla.

sibship All the *siblings in a family.

sicula The skeleton of the initial *zooid of a graptolite (*Graptolithina) colony.

side-necked turtles *See* PLEURODIRA.

sidewinder (*Crotalus cerastes*) *See* CROTALIDAE.

Siegennian *See* DEVONIAN.

sifaka *See* INDRIDAE.

Siganidae (rabbitfish, spinefoot; subclass *Actinopterygii, order *Perciformes) A small family of marine fish that have an oval-shaped body, a long-based, continuous, *dorsal fin, and an *anal fin including seven spines. The *pelvic fins have two spines, separated from the three soft rays. Although these fish are considered edible, the numerous spines

can inflict painful stings (hence the name 'happy moments' reserved for these fish in northern Australia). There are about 10 species.

sigmoid growth curve *See* S-SHAPED GROWTH CURVE.

signalling pheromones *Pheromones that trigger a short-term change in behaviour, e.g. inducing *lordosis.

sign stimulus (releaser) Part of a stimulus that is sufficient to evoke a behavioural response in an animal (e.g. the patch of red colour that provokes an aggressive reaction in *Erithacus rubecula*, the European robin).

silent allele (null allele) In genetics, an *allele that has no detectable product, and so is not expressed in a *phenotype.

silent DNA *See* INTRON.

Silesian *See* CARBONIFEROUS.

silk moths *See* BOMBYCIDAE.

silky flycatchers *See* BOMBYCILLIDAE.

Sillaginidae (southern whiting; subclass *Actinopterygii, order *Perciformes) A small family of marine fish that have an elongate, almost cylindrical body, small mouth, and two separate *dorsal fins which are of about equal size. The second dorsal and the *anal fin are about equal in length. Often travelling in schools, southern whiting are found in the Indo-Pacific region. There are about seven species. They are not related to the northern whiting (*Gadidae).

Silphidae (carrion beetles, burying beetles, sexton beetles; subclass *Pterygota, order *Coleoptera) Family of fairly large beetles, up to 30 mm long, which have smooth, generally red or black *elytra. Some have exposed abdominal segments. They have prominent *mandibles; the eyes are convex; the antennae are strongly clubbed. Their names are derived from their habit of burying corpses of small animals by excavating soil beneath them, using their mandibles and spined *tibiae. Eggs are laid in connecting tunnels and larvae are fed by the female on carrion or associated fly larvae. *Silpha atrata* preys on snails, while others feed on caterpillars and root crops. There are about 200 species.

Silurian The third of six periods of the *Palaeozoic Era, approximately 443.7–416 Ma ago, whose end is marked by the climax of the Caledonian orogeny (mountain-building episode) and the filling of several Palaeozoic basins of deposition.

Siluridae (wels; superorder *Ostariophysi, order *Siluriformes) A family of freshwater catfish that have elongate, scaleless bodies, a somewhat conical head with two pairs of *barbels near the mouth, a very long *anal fin, and a forked tail fin. The single *dorsal fin is very small. The family contains many species, distributed in the northern hemisphere, from Europe to Indonesia.

Siluriformes (subclass *Actinopterygii, superorder *Ostariophysi) An order of bony fish that includes all the catfish. This very large order comprises some 2000 species of marine and freshwater fish, belonging to about 31 families. The order was known formerly as Nematognathi.

silver biddy *See* GERREIDAE.

silverbills *See* ESTRILDIDAE.

silverfish (subclass *Apterygota, order *Thysanura) Common name for members of the family *Lepismatidae; it is sometimes used to denote any member of the Thysanura.

silver hatchetfish (*Argyropelecus lychnus*) *See* STERNOPTYCHIDAE.

silversides (sand smelts) *See* ATHERINIDAE.

silver spoon effect The life-long reproductive advantage (i.e. increased *fitness) enjoyed by an individual that had access to abundant resources during the early part of its life.

Simiiformes (Anthropoidea; cohort *Unguiculata, order *Primates) A suborder (or infra-order) that comprises the monkeys (*Ceboidea and *Cercopithecoidea), apes, and humans (*Hominoidea). Old World monkeys and apes have a common ancestor and diverged in the *Oligocene or early *Miocene; New World monkeys had separated earlier (early Oligocene or late *Eocene). The so-called Dryopithecines of the succeeding *Miocene were undoubted hominoids. The traditional palaeontological view is that these Miocene apes gave rise in turn to three new lines, one leading to the gibbons, another to the great apes, the third to humans. However, on the basis of anatomical characteristics and genetic criteria, it has long been maintained that ancestors of both the gibbons and the orang-utan diverged from the ancestral line of the humans, gorillas, and chimpanzees at an early date and that only subsequently did that line split; recent palaeontological evidence now tends to support this second view. In some classifications the New World monkeys are placed in a suborder or infra-order (*Platyrrhini) separate from that used for the Old World monkeys, apes, and humans (*Catarrhini). The head is rounded, the neck mobile, the brain large with the cerebral hemispheres well developed, the olfactory region reduced, and the occipital and frontal regions enlarged. The face is hairless except for well-defined regions, the upper lip is entire, with no moist muzzle, and capable of extrusion. The face is mobile and used in expressing emotion. The orbits are forward-facing, the external ears small, their edges often rolled over, the lower *incisors vertical. There is one pair of pectoral mammae. In all families except *Callitrichidae all the digits bear flattened nails. The thumbs and big toes are opposable in some families. The tactile sense is highly developed. In many species social life is highly organized and based on sight rather than smell. Some species are arboreal, others ground-dwelling. Of some 230 species known, 180–90 survive to the present day.

similarity coefficient Any measure of the similarity of two samples. In ecological work, the similarity index devised in 1920 by

H. A. Gleason (1882–1975) has been widely used. This measures similarity as $C=2W/(a+b)$, where a and b are the quantities of all the species (or other commodity) found in the two units to be compared, and W is the sum of the lesser values for those species common to both units. Complete similarity thus scores 1, complete dissimilarity 0. *See also* AFFINITY INDEX.

simple eye Eye consisting of a single *ocellus. Such eyes are found in adult insects, and in the larvae of *hemimetabolous insects. Usually there are three, arranged triangularly on the head. Simple eyes are not thought to be involved in the formation of images, but act only as monitors of the level of light intensity.

Simplicidentata *See* RODENTIA.

simplicidentate Applied to the condition in which the single pair of upper *incisors have *persistent pulps.

Simpson, George Gaylord (1902–84) An American palaeontologist, specializing in mammalian evolution, who also studied mammal migrations. He obtained his Ph.D. from Yale University in 1926, for a thesis on *Mesozoic mammals, and then joined the staff of the American Museum of Natural History, in New York City, becoming curator in 1942. In 1945 he moved, as a professor, to Columbia University. From 1959 to 1970 he was Alexander Agassiz Professor of Vertebrate Palaeontology at the Museum of Comparative Zoology at Harvard University, and in 1967 was appointed Professor of Geosciences at the University of Arizona.

(((●))) SEE WEB LINKS
- More information about George Gaylord Simpson

Simuliidae (black-flies, buffalo gnats; order *Diptera, suborder *Nematocera) Small family of small, stout flies which have short legs and elongated mouth-parts. The wings are broad, with thickened anterior veins. The antennae are short, and 11-segmented, and *ocelli are absent. In males, the eyes are *holoptic. Females of some species are active blood-suckers, especially *Simulium damnosum* of Africa, which carries the nematode *Onchocera*, the cause of filariasis. Adults are never found far from running water, as their larvae are aquatic and prefer swift-flowing, well-oxygenated water. The larvae feed by filtering detritus and planktonic organisms from the water by means of a series of stout bristles around the mouth. The family has a worldwide distribution and in some countries is of great economic importance. More than 1300 species have been described.

sinew *See* TENDON.

single nucleotide polymorphism (SNP) A DNA (*deoxyribonucleic acid) sequence that varies by a single *nucleotide between the *genomes of individual members of a *species or between paired *chromosomes in a single individual, producing two *alleles. SNPs (pronounced 'snips') may affect the susceptibility of individuals to pathogens and how they respond to drug treatments. SNP studies are also important in genotyping (*see* GENOTYPE).

Sinornis An early *Cretaceous bird from Liaoning Province, China. It is one of the earliest to show the advanced avian features of an extensive *synsacrum and complete fusion of the distal *tarsal bones with *metatarsals.

sinuate Curved; having a wavy or indented margin.

sinus 1. A space, found e.g. in certain bones of the face of a mammal where they are filled with air and connect with the nasal cavity. **2.** In certain *Bivalvia, a recess or embayment in the *pallial line; forms with a sinus are generally burrowers.

siphon In *Bivalvia and *Gastropoda, a tube that funnels water towards and away from the gills. In bivalves siphons often occur in pairs.

Siphonaptera (fleas; class *Insecta, subclass *Pterygota) An order of wingless, parasitic insects in which the adult body is flattened laterally. The larvae (maggots) feed on organic debris, undergo complete *metamorphosis, and hatch from the *pupa when this is disturbed by a possible host. Most adults are host-specific, although some have a wide range of hosts and others will feed on any host when very hungry. There are 1400–1500 species, of world-wide distribution, grouped in two suborders: Pulicida and Apulicida. The Pulicida includes the families Pulicidae (most flea parasites of mammals, some of which are vectors for serious diseases, e.g. pneumonic plague) and Tungidae (the jigger or sand flea).

siphonate Applied to the aperture of a gastropod shell when a canal or notch for the *siphon is present.

Siphonophora (Siphonophorida, siphonophores; phylum *Cnidaria, class *Hydrozoa) An order of cnidarians whose members produce highly polymorphic colonies without skeletons and without free, sexual *medusae. They consist of floating or free-swimming assortments of *polypoid and medusoid individuals attached to a stem.

Siphonophorida See SIPHONOPHORA.

siphonozoid In some *Anthozoa, a *polyp that generates water currents.

siphuncle A long tube present in those cephalopods (*Cephalopoda) that possess external shells. It runs internally, through all the chambers, and contains the siphuncular cord of body tissue, extending from the visceral mass through a perforation (foramen) in each *septum of the shell. The siphuncle releases gas into the empty chambers of the shell, making it buoyant.

Sipuncula (Sipunculida; peanut worms) A phylum of coelomate worms which lack body segmentation. Like *Echiurida, they are cylindrical in shape, and the mouth is surrounded by small *tentacles which form part of a protrusible *introvert. They have a long, U-shaped gut with the *anus lying on the anterior dorsal surface. The nervous system includes a brain, and

sense organs are present. Males and females occur. They are exclusively marine, all are *benthic, most feed on detritus, although some are carnivores. They are first recorded in the middle *Devonian.

Sipunculida See SIPUNCULA.

Sirenia (sea cows, mermaids; cohort *Ferungulata, superorder *Paenungulata) An order of herbivorous ungulates that have adapted to a fully aquatic life. The body is streamlined, thick-skinned, and has a thick layer of blubber. The fore limbs are *pentadactyl, the digits are joined, forming large paddles, but considerable movement is possible at the wrist and elbow and in the digits. There are no hind limbs, and the pelvic girdle is vestigial. The tail is modified to form a horizontal fin. The ribs are numerous and form a barrel-shaped cage extending far back in the body, and the *diaphragm is oblique, so the lungs are large. The bones are heavy, with little marrow (pachyostosis). The skull is large, but the brain case is small. The young are born in the water and suckle from pectoral teats. Fossil sirenians are known from the *Eocene, in Egypt and Jamaica. Probably they first appeared in Africa, from the stock that also gave rise to the proboscideans; and the two surviving families, *Dugongidae and *Trichechidae, diverged possibly before the end of the Eocene and so are different in many ways from one another.

Sirenidae (sirens; class *Amphibia, order *Urodela) A family of permanently larval, aquatic amphibians which have an eel-shaped body, short fore limbs and tail, no hind limbs or pelvis, no teeth, and no eyelids. The jaws have horny plates. Gills are retained throughout life. Sirens swim and burrow in streams and ditches, and may occasionally move across land. There are three species, occurring in south-eastern USA.

sirens See SIRENIDAE.

Siricidae (horntails, woodwasps; suborder *Symphyta, superfamily Siricoidea) Family of *Hymenoptera which have very large *ovipositors used in wood boring. Most are 25–35 mm long, but some may be up to 50 mm. Adults are usually black or brownish, and sometimes metallic or with dark wings. The *abdomen has an apical, large, dorsal spine, the *ovipositor of the female being positioned below the spine. Eggs are laid under the surface, the larvae penetrating deeply into the host tree (particularly in coniferous species). The larvae have a long developmental period, sometimes emerging as adults after the timber has been used for building or furniture construction. Some species can cause serious economic damage. There are about 150 species.

siskins See FRINGILLIDAE.

Sisoridae (superorder *Ostariophysi, order *Siluriformes) The largest family of Asian freshwater catfish. They have a robust yet depressed body, small eyes, broad-based *barbels, and relatively large *pectoral fins. The flattened belly side and the pectoral fins help the fish keep its position in the fast-flowing mountain streams of the Asian highlands. There are about 100 species.

sister chromatid exchange (SCE) An event, similar to *crossing-over, that can occur between sister *chromatids at *mitosis and *meiosis. It may be detected in harlequin chromosomes (sister chromatids that stain differently so that one appears dark and the other light).

sister groups In *phylogenetics, two taxa connected through a single *internal node.

site In genetics, the position within a *cistron that is occupied by a *mutation.

site of special scientific interest (SSSI) In Britain, an area which, in the view of the Government, is of particular interest because of its fauna, flora, or geological or physiographic features. Once it has been designated, the owner of the site is required to notify the relevant authorities and obtain special permission before undertaking operations that would alter its characteristics.

Sitona lineatus (bean weevil, pea weevil) See CURCULIONIDAE; BRUCHIDAE.

Sitta (nuthatches) See SITTIDAE.

sittellas See DAPHAENOSITTIDAE.

Sittidae (nuthatches; class *Aves, order *Passeriformes) A family that comprises the one genus Sitta, although *Tichodromadidae and *Daphaenosittidae are sometimes placed in this family. Nuthatches are small birds, most of which have blue-grey upper-parts and a black crown or eye stripe. They have short tails, strong feet, and a long, pointed, tapering bill. They inhabit woodland and rocky areas, and climb up and down trees and rocks to feed on insects, spiders, seeds, and nuts. They nest in tree and rock cavities. There are 21 species found in Europe, Africa, Asia, and N. America.

Sivapithecus A genus of early *Hominoidea which probably includes the so-called *Ramapithecus. They and their relatives are known from south-eastern Europe, Turkey, Arabia, Pakistan, northern India, and southern China, from the middle and late *Miocene (16–5 Ma ago). The genus may include ancestors of great apes and humans, but certainly early relatives of the orang-utan have been identified among species of the genus. Although there are similarities with human teeth, these are probably misleading, and Sivapithecus is now regarded as an ape that was ancestral either to the orang-utan alone, or to all living great apes and humans.

Sivatherium See GIRAFFIDAE.

six-lined perch See GRAMMISTIDAE.

skate See RAJIDAE.

skeletagenous septum One of a series of segments which develop along the *notochord and are separated by *myosepta. Septa occur in two planes. Dorsal and ventral septa are plate-like processes which grow dorsally and ventrally to the notochord; horizontal septa bisect this plane perpendicularly. *Mesenchymal cells migrate to the skeletagenous septa during the *gastrula stage of embryogenesis to form sclerotomic tissue (see SOMITE). At the junction between the myoseptum and skeletagenous septum *vertebrae develop; they are consequently intersegmental structures, as are *ribs.

skeletal muscle *See* STRIATED MUSCLE.

skeleton *See* APPENDICULAR SKELETON; AXIAL SKELETON.

skimmers *See* RYNCHOPIDAE.

skin *See* INTEGUMENT.

skin beetle *See* DERMESTIDAE.

skinks *See* SCINCIDAE.

Skinner box In laboratory studies of animal behaviour, a cage in which an animal may learn that the performance of a particular activity (e.g. pressing a bar) is rewarded (e.g. with food) so that its behaviour may be conditioned. *See also* OPERANT CONDITIONING.

skip-jack *See* ELATERIDAE.

skipper 1. (*Scomberesox saurus*) *See* SCOMBERESOCIDAE. 2. *See* HESPERIIDAE.

Skomer vole *See* RACE.

skuas *See* CHARADRIIFORMES; STERCORARIIDAE.

skull *See* CRANIUM.

skunk (*Mephitis*) *See* MEPHITIDAE.

skunk bear (carcajou, glutton, wolverine; *Gulo gulo*) *See* MUSTELIDAE.

slant-faced grasshoppers *See* ACRIDIDAE.

slaters *See* CRUSTACEA; ISOPODA.

slave-making ant *See* DULOSIS.

sleeper *See* ELEOTRIDAE.

slender blind snakes *See* LEPTOTYPHLOPIDAE.

sliding-filament theory A hypothesis to explain the mechanism of contraction of striated muscle which proposes that the parallel, interdigitated, thick *myosin and thin *actin filaments slide past one another to varying degrees to shorten the muscle. This is brought about through the sequential formation of cross-bridges between the actin and myosin molecules such that the filaments move past one another in a ratchet-like manner. The energy for these processes is provided by the *hydrolysis of *ATP. The theory is applied to amoeboid and ciliary, as well as muscular, movement, and also describes chromosomal movements.

slimehead *See* TRACHICHTHYIDAE.

slipmouth *See* LEIOGNATHIDAE.

slit-faced bat (*Nycteris*) *See* NYCTERIDAE.

slit organ In *Arachnida, one of many sensory organs located in the *exoskeleton that detect changes in stress in the *cuticle and thus vibrations and movements of the animal itself.

SLOSS principle An acronym for *S*ingle *L*arge *O*ver *S*everal *S*mall, the proposal that a single large *nature reserve will contain more *species than several small reserves with the same total area. Whether this is so depends on the extent to which the species in the small reserves overlap. If there is little overlap between reserves (i.e. each small reserve contains many species that are not found in any of the other reserves), several small reserves may be preferable to a single large one. Small areas of *habitat may also afford more protection against disease or the accidental introduction of predators.

sloth *See* BRADYPODOIDEA.

slow-worm (*Anguis fragilis*) *See* ANGUIDAE.

small-eared dog (*Atelocynus microtis*) *See* CANIDAE.

smelt *See* OSMERIDAE.

Sminthopsis (marsupial mice) *See* DUNNART.

smolt The stage in the life of salmon-like fish in which the sub-adult individuals acquire a silvery colour and migrate down the river to begin their adult lives in the open sea. *See also* PARR.

smooth hound (*Mustelus mustelus*) *See* CARCHARHINIDAE.

smooth newt (*Triturus vulgaris*) *See* SALAMANDRIDAE.

smooth razorfish (*Centriscus aristatus*) *See* CENTRISCIDAE.

smooth snake (*Coronella austriaca*) *See* COLUBRIDAE.

smooth stingray (*Dasyatis brevicaudata*) *See* DASYATIDAE.

SMR *See* STANDARD METABOLIC RATE.

snaggle tooth *See* ASTRONESTHIDAE.

snake blenny *See* OPHICLINIDAE.

snake eel *See* OPHICHTHIDAE.

snakeflies *See* RAPHIDIIDAE.

snake-footed bat *See* MYZOPODIDAE.

snakeheads *See* CHANNIDAE.

snake lizards *See* PYGOPODIDAE.

snake mackerel *See* GEMPYLIDAE.

snake-necked turtles *See* CHELIDAE.

snakes *See* SERPENTES.

snap display A *courtship display in the green heron (*Butorides virescens*) in which the male points its bill downwards and snaps the mandible together. The display is derived from the action of picking up twigs for nesting and originally signified that the bird possessed a good nesting site. It now indicates that the male is ready to accept female courtship.

snapper *See* LUTJANIDAE.

snapping beetle *See* ELATERIDAE.

snapping turtles *See* CHELYDRIDAE.

sneak copulation Behaviour in which a subordinate male or in territorial species a male that has been unable to acquire a *territory of its own is nevertheless able to mate with a receptive female. In baboons (*Cercopithecidae), the subordinate male follows a dominant male and his consort, who is in oestrus (*see* OESTRUS CYCLE); if the dominant male should leave the female unguarded, the subordinate moves in quickly and mates with her. In many territorial insects, males lacking territories wait in a corner of a territory, leaving if challenged by the owner but also waiting for an opportunity to mate undetected by the territory owner. In some lizard and fish species subordinate males behave as though they were females in order to approach a genuine female without arousing suspicion. In bluehead wrasse (*Thalassoma bifasciatum*), males lacking territories wait at the edge of a territory until the owner has induced a female to spawn, then rush in to release their sperm before the territory owner can respond.

snipe 1. (*Gallinago*) *See* SCOLOPACIDAE. **2. (painted snipe)** *See* ROSTRATULIDAE.

snipe eel *See* NEMICHTHYIDAE.

snipefish *See* MACRORHAMPHOSIDAE.

snipe flies *See* RHAGIONIDAE.

snout beetle *See* CURCULIONIDAE.

snout butterflies *See* LIBYTHEIDAE.

snowfleas *See* MECOPTERA.

snowflies *See* MECOPTERA.

snow-worm *See* CANTHARIDAE.

S

SNP *See* SINGLE NUCLEOTIDE POLYMORPHISM.

snub-nosed monkey (*Rhinopithecus***)** *See* CERCOPITHECIDAE.

soapfish *See* GRAMMISTIDAE.

social behaviour The interactive behaviour of two or more individuals all of which belong to the same species.

social facilitation Intensification (i.e. *facilitation) of a behaviour (e.g. feeding) that is associated with increased population density.

social wasps (superfamily Vespoidea; family *Vespidae) The common name most usually applied to members of the subfamilies Vespinae, Rhopalidiinae, Polistinae, and Polybiinae.

society A group of individuals, all of the same species, in which there is some degree of *co-operation, *communication, and division of labour.

sociobiology The integrated study of social behaviour, based on the premise that all behaviour is adaptive. It gives special emphasis to social systems considered as ecological adaptations, and attempts a mechanistic explanation of social behaviour in terms of modern biology and, in particular, evolutionary theory.

sodium-coupled transport The entry of a *metabolite into a cell against a concentration gradient, where this entry is coupled to the movement of sodium across the membrane. Sodium readily enters cells because of the concentration of its ions outside the *cell membrane which is high compared with that inside. It is thought that carriers exist that can bind to sodium, so exploiting this feature, and also to a metabolite (such as glucose), which is thereby carried across the membrane.

sodium pump A mechanism found in higher animal cells that is responsible for the *active transport of sodium and potassium ions across membranes and against concentration gradients. Its operation is dependent upon the provision of cellular *ATP and the enzyme sodium/potassium-ATPase, and its net results are low levels of sodium ions and high levels of potassium ions inside the cell (compared with those outside). *See* COUNTER-CURRENT EXCHANGE.

soft corals *See* ALCYONACEA.

soft-shelled turtles *See* TRIONYCHIDAE.

sol A substance composed of small solid particles dispersed in a liquid to form a continuous, homogeneous phase (i.e. a type of colloid).

soldier A sterile member, specialized for defence, of an ant or termite colony. The head is usually enlarged, and it or the tip of the abdomen may be used to block entrance holes to the nest (phragmosis).

soldier beetle *See* CANTHARIDAE.

soldier flies *See* STRATIOMYIDAE.

sole *See* SOLEIDAE.

Soleidae (sole; subclass *Actinopterygii, order *Pleuronectiformes) A fairly large family of mostly marine flatfish that have an oval-shaped body outline, with the *dorsal fin originating well in front of the eyes, which are usually on the right side of the body. The small, twisted mouth is subterminal and located almost underneath the eyes. Soles are often found in shallow coastal waters. Several species are of commercial importance. There are about 117 species, distributed world-wide in temperate and tropical seas.

Solemyoida (phylum *Mollusca, class *Bivalvia) An order of bivalves whose shell has an homogeneous calcium carbonate microstructure. They are *anisomyarian, with a greatly reduced anterior adductor muscle. The hinge is *edentulous, and the gills *protobranchiate. The valves are gaping, with the posterior end shorter than the anterior. The order first appeared in the

*Devonian. There is only one extant genus, *Solemya*.

solenial *See* SOLENIUM.

Solenichthyes An alternative, older name for fish belonging to the order *Syngnathiformes.

solenium (*adj.* **solenial**) In coelenterate *polyps, a canal lined with gastrodermis that interconnects the gastric cavities.

solenocyte In some *Annelida, *Platyhelminthes, *Rotifera, and other invertebrates, a tubular terminal cell at the blind end of each tubule of a protonephridium (*see* NEPHRIDIUM). The solenocyte carries a *flagellum which wafts excretory products along the tubule. The walls of the tube are thin and consist of pillar-like rods. The excretory fluid is thought to be formed by ultra-filtration of body fluids through the wall.

Solenodon (alamiqui) *See* SOLENODONTIDAE.

Solenodontidae (alamiqui; suborder *Lipotyphla, superfamily *Soricoidea) A family of fairly large (about 30 cm long), nocturnal insectivores in which the hair is coarse and the tail is long and scaly. Each limb bears five digits with large, strong claws. The upper *molars are *zalambdodont, a feature that distinguishes them from other insectivores but which they share with the *Tenrecidae. There are two species, *Solenodon paradoxus*, occurring only in Haiti and the Dominican Republic (Hispaniola Island), and *S. cubanus*, in Cuba, the latter sometimes being placed in a separate genus, *Atopogale*.

Solenogastres *See* APLACOPHORA.

solenoglyphous Applied to snakes that have long, hollow, articulated fangs which fold against the roof of the mouth when the jaws are closed. The fangs are the only teeth in the maxilla (upper jaw) and are capable of injecting venom deep into prey. The pit vipers (*Crotalidae) and true vipers (*Viperidae) are solenoglyphous. *Compare* OPISTOGLYPHOUS; PROTEROGLYPHOUS.

Solenostomidae (ghost pipefish; subclass *Actinopterygii, order *Syngnathiformes) A small family of marine fish which resemble sea-horses, but have a tubelike snout, short body, large tail fin, and *pelvic fins which in the female are fused to form a brood pouch. There are about five species occurring in the tropical Indo-Pacific region.

Solifugae (windscorpions; subphylum *Chelicerata, class *Arachnida) Order of arachnids in which the prosoma is divided into a proterosoma and two free segments. The *abdomen is constricted, *anteriad, and the entire body is covered with long, sensory hairs. The *chelicerae are much enlarged, *chelate, and armed with teeth. The *pedipalps are leg-like, and their *tarsi, as well as the tarsi of the first pair of walking legs, contain *sensilla ampullacea. The carapace bears a large pair of median eyes and there may be vestiges of lateral eyes. The terminal segment of the last pair of walking legs bears *racquet organs which contain sensory nerve endings. The tracheal system resembles that of the *Insecta rather than Arachnida. Wind-scorpions have complex courtship behaviour, seemingly more advanced than that of scorpions but less advanced than that of spiders. Most species are crepuscular or nocturnal, and although none has specialized digging structures most of them construct burrows, and may aggregate in areas of high prey density, prey being caught by chase or ambush. Of the 800 or so species, six are found in Europe. Most, being desert-dwellers, are found in India, the W. Indies, western USA, and northern Mexico.

solitaires *See* RAPHIDAE.

solitaria *See* LOCUST.

solitary bees 1. The majority of bees, which are not social, i.e. there is no worker caste and the bees do not live in colonies. Each nest is the work of an individual female, and although nests may be aggregated, there is no co-operation among individuals. **2. (solitary mining bees)** *See* ANDRENIDAE.

solitary wasps (order *Hymenoptera, suborder *Apocrita) Name generally given to wasps that do not live in colonies, e.g. potter wasps, mason wasps, and spider-hunting wasps. Mason wasps make nests from particles of sand and soil stuck together with saliva. The nests are found in the soil, in crevices in wood, and in stone walls, and are individually provisioned with paralysed prey for the larva.

soluble RNA See s-RNA.

solute potential See OSMOTIC POTENTIAL.

soma See CYTON.

Somateria (eiders) See ANATIDAE.

somatic cell A cell that is not destined to become a *gamete, and whose genes will not be passed on to future generations; a body cell. See SOMATIC CELL HYBRID.

somatic cell hybrid A hybrid cell resulting from the fusion of two *somatic cells.

somatic crossing-over In genetics, *crossing-over during *mitosis of *somatic cells, such that parent cells *heterozygous for a given *allele, instead of giving rise to two identical heterozygous daughter cells, give rise to daughter cells one of which is *homozygous for one of these alleles, the other being homozygous for the other allele. Phenotypically differing *cell lines may result. Studies of somatic crossing-over, somatic assortment, and cell fusion make up somatic-cell genetics, a modern asexual genetic technique that allows a wide range of *in vitro* manipulation of higher cells, including human cells, as well as those of other organisms.

somatic mesoderm See MESODERM.

somatic muscles The *muscles located in the body wall.

somatic mutation A mutation that occurs in a *somatic cell. If the mutated cell continues to divide, the individual will develop a patch of tissue with a *genotype different from the cells of the rest of the body.

somatic skeleton See AXIAL SKELETON.

somatopleure In vertebrates, the part of the body derived from the *mesoderm outside the *coelom and the *ectoderm. It comprises the body wall and the outer lining of the coelom that covers the body wall, the skin (*integument), *somatic muscles, *axial skeleton, and *kidneys.

somatotropin See GROWTH HORMONE.

somite (metamere) In a vertebrate *embryo, a series of bilaterally paired blocks of *mesoderm tissue longitudinally flanking the *notochord. There are three distinct tissues within a somite: the dermatome, which will develop to form deep portions of the skin; the myotome, which will form voluntary muscle; and the sclerotome which occurs close to the notochord and gives rise to the *vertebrae. See also METAMERIC SEGMENTATION.

Somniosus microcephalus (Greenland shark) See SQUALIDAE.

song A complex pattern of sound that is produced by means of specialized organs. Such organs are used by many animals in communication.

song thrush (Turdus philomelos) See TURDIDAE.

Sordes pilosus Discovered in 1971, one of the first pterosaurs (*Pterosauria) known to have been covered in thick fur: the name means 'hairy filth'. The indication is that this reptile and its close relatives were *homoiotherms. It was found in Late *Jurassic sediments of Chimkent, Kazakhstan, and was a small, toothed pterosaur with a long tail.

Soricidae (shrews; suborder *Lipotyphla, superfamily *Soricoidea) A family of small mammals that have short, dense fur, small eyes, ears which are usually

s

visible externally, and long, mobile muzzles equipped with many sensory receptors. The upper *molar teeth have a characteristic W-shaped pattern, and the first upper *incisor is long and curved. There is no *zygomatic arch. The Soricidae are distributed throughout the world except for Australasia and most of S. America. There are 21 genera, with some 250 species.

Soricoidea (order *Insectivora, suborder *Lipotyphla) A superfamily that comprises all the members of the Lipotyphla with the exception of the hedgehogs and moon rat. The superfamily is divided into the families *Soricidae (shrews), *Talpidae (moles), *Solenodontidae (solenodon), *Tenrecidae (tenrec and otter shrew), and *Chrysochloridae (golden mole); but the last three of these families are sometimes classed as superfamilies in their own right.

Sotalia (tucuxi) *See* STENIDAE.

Sousa (**hump-backed dolphin, white dolphin**) *See* STENIDAE.

South American lungfish (*Lepidosiren paradoxa*) *See* LEPIDOSIRENIDAE.

Southern blotting A technique for transferring single-stranded (denatured) DNA fragments from a gel used for *electrophoresis to a nitrocellulose filter. A gel containing the denatured DNA fragments is placed between blotting paper and a cellulose nitrate filter is saturated with a buffer. DNA is drawn from the gel on to the cellulose nitrate, to which it binds permanently after baking at 80°C for two hours. The technique is used in *DNA fingerprinting and *genetic engineering, for testing *DNA hybridization.

southern roughy (*Trachichthys australis*) *See* TRACHICHTHYIDAE.

southern smelt *See* RETROPINNIDAE.

southern whiting *See* SILLAGINIDAE.

spadefish *See* EPHIPPIDAE.

spadefoot toad (*Pelobates fuscus*) *See* PELOBATIDAE.

Spalacidae (mole rats; order *Rodentia, suborder *Myomorpha) A family (or subfamily of *Muridae) of fairly large (up to 28 cm long) mole-like rodents which have soft, dense fur that can lie forward or backward, round heads, short fore limbs but longer, powerful, hind limbs, no externally visible tail, and vestigial eyes and ears. They live mainly below ground but often come to the surface at night. They are distributed throughout the eastern Mediterranean region and in south-eastern Europe. There are three species in a single genus, *Spalax*.

Spalax (**mole rats**) *See* SPALACIDAE.

spandrels of San Marco An analogy used, in a classic paper by Stephen Jay Gould and Richard Lewontin, to indicate how non-adaptive characters may arise in evolution. Spandrels are the spaces left between the tops of neighbouring arches in churches (in this case, St Mark's cathedral, Venice); these spaces, not related to the functional architecture, are free to be decorated in a non-functional fashion.

spangled drongo (*Dicrurus hottentottus*) *See* DICRURIDAE.

Spanish fly (*Lytta vesicatoria*) *See* MELOIDAE.

Sparassidae (order *Araneae, suborder Araneomorphae) Family of spiders in which the legs are held sideways, like those of crabs. These large and often flattened spiders run and catch prey, and are largely tropical and nocturnal.

Sparassodontia (subclass *Theria, infraclass *Metatheria) An order of S. American metatherians that became extinct in the *Pliocene or *Pleistocene. The best-known family is the Borhyaenidae.

Sparidae (sea bream, porgy; subclass *Actinopterygii, order *Perciformes) A large family of marine, perch-like, silvery fish, that have a deep, compressed and fully scaled body, continuous, single, *dorsal fin, and forked tail fin. All individuals develop testes but as they grow larger and older the fish turn into females (*see* HERMAPHRODITIC FISH). There are about 100 species, distributed world-wide in warm to temperate waters.

sparrowhawks *See* ACCIPITRIDAE.

sparrows *See* EMBERIZIDAE; PLOCEIDAE.

Spatangoida (heart urchins; subphylum *Echinozoa, class *Echinoidea) An order of irregular echinoids which have a compact *apical system and a *plastron in which the *labrum is followed by a pair of large plates. Most spatangoids have narrow bands of densely ciliated spines, which can be recognized in fossils as areas of very small *tubercles on denuded *fascioles. The Spatangoida first appeared in the lowermost *Cretaceous.

Spathebothridea (phylum *Platyhelminthes, class *Cestoda) An order of parasitic worms which possess repeated genitals. The *scolex may be present. They inhabit two hosts during their development.

spatial sorting An evolutionary process that affects rapidly invading species, e.g. *Bufo marinus* (cane toad) in Australia. As the population spreads, those individuals at the leading edge of the invasion moved faster than others in order to reach that position. Their offspring inherit their parents' ability to move rapidly and some move even faster, to become the new leading edge. Consequently, the leading edge of the invasion advances faster with each succeeding generation. The individuals able to move fastest are the first to reach unexploited resources, so evolution proceeds by the survival of the fastest.

spatial summation *See* SUMMATION.

spatulate Having an end that is broad and flattened, like a spatula.

spearfish *See* ISTIOPHORIDAE.

spear-headed grasshoppers *See* PYRGOMORPHIDAE.

special adaptation *See* GENERAL ADAPTATION.

special creation The belief that the origin of life and the diversity of life result from acts of God whereby each species was created separately. Evolution is implicitly rejected as the explanation of these phenomena. *See* CREATIONISM; CREATION 'SCIENCE'.

specialization The degree of adaptation of an organism to its environment. A high degree of specialization suggests both a narrow *habitat or *niche and significant interspecific *competition.

speciation The separation of populations of plants and animals, originally able to interbreed, into independent evolutionary units which can interbreed no longer owing to accumulated genetic differences. In *cladistics, the origin of one or more new species occurs inferentially by *cladogenesis.

species (*sing.* and *pl.*) Literally, a group of organisms that resemble one another closely. The Latin word *species* means 'appearance' or 'semblance'. In taxonomy, it is applied to one or more groups (populations) of individuals that can interbreed within the group but that do not, under natural conditions, exchange genes with other groups (populations); it is an interbreeding group of biological organisms that is isolated reproductively from all other organisms (*see* BIOSPECIES). A species can be made up of groups in which members do not actually exchange genes with members of other groups (though in principle they could do so), as e.g. at the two extremes of a continuous geographical range. However, if some *gene flow occurs along a continuum, the formation of another species is unlikely to occur. Where barriers to gene flow arise (e.g. physical barriers, such as sea, or areas of unfavourable habitat), this reproductive isolation may lead either by local selection or by random *genetic drift to the formation of morphologically distinct forms ('*races' or '*subspecies'). These could interbreed with other races of the same species if they were introduced to one another. Once this potential is lost, through some further evolutionary divergence, the races may be recognized as species, although this concept is not a rigid one. Most species cannot interbreed with others; a few can, but produce infertile offspring; a smaller number may actually produce fertile offspring. The term cannot be applied precisely to organisms whose breeding behaviour is unknown. *See also* MORPHOSPECIES; PALAEOSPECIES; REPRODUCTIVE ISOLATING MECHANISMS; RING SPECIES.

species diversity *See* DIVERSITY.

species group (superspecies) A complex of related *species that exist *allopatrically. They are grouped together because of their morphological similarities, and this grouping can often be supported by experimental crosses in which only certain pairs of species will produce *hybrids.

species longevity The persistence of *species for long periods of time (e.g. species of *Gastropoda and *Bivalvia).

species selection A postulated evolutionary process in which selection acts on an entire species population, rather than individuals. This might occur e.g. as a consequence of the geographical range of a population, which affects the population as a whole and, possibly, its longevity or development. *See also* ORTHOSELECTION.

specific immunity *See* ADAPTIVE IMMUNITY.

specific mate recognition system *See* RECOGNITION SPECIES CONCEPT.

spectacle *See* SERPENTES.

spectacled caiman (*Caiman crocodilus*) *See* ALLIGATORIDAE.

speculum In birds, a well-defined patch of coloured feathers, often iridescent, found on the secondaries or secondary coverts (e.g. of many Anatidae, particularly *Anas* species).

***Speothos venaticus* (bush dog)** *See* CANIDAE.

sperm *See* SPERMATOZOON.

spermaceti Highly vascular tissue, filled with a mixture of fatty esters (oils) which solidify at about 32°C, produced by *Physeteridae (sperm whales) and stored in the upper part of an enlarged snout. A full-grown sperm whale may carry one tonne of spermaceti. The function of spermaceti is not known with certainty, but since its density changes as it liquefies and solidifies it may permit heat loss from the body and help provide neutral buoyancy to the animal, and it may contribute to the production of the sounds by which the animal echo-locates.

spermatheca Container or sac connected to the genital tract of a female insect, which acts as a sperm store. It is of ectodermal origin and is lined with *cuticle. The spermatheca is often lined with glandular cells which are thought to supply nourishment for the sperm.

spermatid One of four *haploid cells formed during the second reduction-division of *meiosis in male animals. Without further division they undergo cytoplasmic changes and condensation of the *nucleus to give rise to spermatozoa. This process is called spermiogenesis.

spermatocyte A *diploid cell that undergoes *meiosis and forms four *spermatids. A primary spermatocyte undergoes the first meiotic division to form two secondary spermatocytes. These then divide (the second meiotic division) to produce two spermatids each (i.e. a total of four *haploid cells).

spermatogenesis The formation of sperm (spermatozoa) from a *diploid male germ cell. It includes both male *meiosis and *spermiogenesis.

spermatogonia Mitotically active cells in the gonads of male animals, which give rise to primary *spermatocytes.

S

spermatophore 1. A gelatinous cone of jelly with a sperm cap on top that is secreted by the cloacal glands of male *Urodela. It is deposited by the male during the courtship ritual and picked up by the cloacal lips of the female, thus enabling internal fertilization. **2.** Packet of sperm, produced in a wide range of insect groups, which may be deposited on the ground or transferred directly to the female. The spermatophore has a gelatinous protein capsule produced by the male *accessory genitalia.

spermatozoon (*pl.* **spermatozoa; sperm**) In animals, the male *gamete; a small, motile cell that in most species is flagellated.

sperm competition The competition to fertilize an ovum that occurs among sperm from different males when the female has mated with more than one partner. It is believed that certain sperms, different in appearance to ordinary sperms and incapable of fertilizing the ovum, but present in large numbers, are involved in repelling or disabling sperms from the rival male.

spermiogenesis The formation of *spermatozoa from the *spermatids produced during two reduction-divisions of *spermatocytes.

sperm train A phenomenon observed in wood mice (*Apodemus sylvaticus*) in which large numbers of sperm join together, using hooks on their heads or by holding on to the tail of the sperm in front, to form a train that swims at twice the speed of an individual *spermatozoon. Forming a train increases the chance that the leading sperm will fertilize the female (*see* SPERM COMPETITION). It is the only sperm able to do so, because all but the leading sperm undergo an *acrosome reaction that releases *enzymes, rendering the sperm infertile.

sperm whale (*Physeter catodon*) *See* PHYSETERIDAE.

Sphaeriidae (subclass *Pterygota, order *Coleoptera) Family of tiny beetles, less than 1 mm long, in which the body is very rounded and shining. The wings have reduced venation, and fringes of long hairs. The antennae are clubbed. *Tarsi are three-segmented, the claws unequal. Sphaeriidae are found in wet areas. Larvae are aquatic, oval, and yellowish. There are 11 species, distributed world-wide.

Sphaeritidae (subclass *Pterygota, order *Coleoptera) Family of small, convex, shiny beetles, in which the terminal abdominal segment is exposed. The antennae have three-segmented clubs. They are distinguishable from the closely related *Histeridae by the absence of elbowed antennae. Adults are associated with fungi, but little is known of their life history. There are three species.

Sphecidae (digger wasps, sand wasps; order *Hymenoptera, suborder *Apocrita) Large family of solitary hunting wasps that are very closely related to the bees and, together with them, form the superfamily *Sphecoidea. They differ from the bees in that the body hairs are unbranched and the hind *basitarsus has the same width as do the following segments (rather than being wider than them). Sphecids hunt a wide range of insect prey. A few catch spiders, others *Collembola. At the level of species, sphecids are prey-specific. There is a tendency for primitive sphecids to hunt primitive insect prey, e.g. cockroaches or crickets, while the more advanced species hunt more highly evolved insect prey, e.g. flies and beetles. There are nine subfamilies, which some recent authors have suggested should be accorded family status. There are 7600 species, with a cosmopolitan distribution. *See also* HUNTING WASP.

Sphecoidea (bees, digger wasps, sand wasps; order *Hymenoptera, suborder *Apocrita) Superfamily containing two families of hunting wasps, *Ampulicidae and *Sphecidae, and 11 families of bees, *Andrenidae, *Anthophoridae, *Apidae, *Colletidae, Ctenoplectridae, *Fideliidae, *Halictidae, *Megachilidae, *Melittidae, *Oxaeidae, and *Stenotritidae. The Sphecoidea differ from all other Hymenoptera in the structure of the *pronotum, the lateral lobes of which do not reach as far as the *tegulae, or extend to below it. The posterior pronotal lobe is rounded and limited to the spiracular cover.

Sphecotheres (Old World orioles) *See* ORIOLIDAE.

Spheniscidae (penguins; class *Aves, order *Sphenisciformes) A family of medium-sized to large, marine birds, most of which have black upper-parts and white under-parts. Their feathers are very small and dense, and some birds have crests above the eye. Their bills are short and stout to long and pointed, their legs short and set well back, with webbed toes. Their wings are modified into paddles which do not fold, and they have short tails. They are flightless and swim well under water, coming to land to breed. They walk upright on land, or slide on their bellies. They feed on fish and other marine animals, and nest in burrows or on the ground. There are six genera, with 18 species, found on the coasts of Antarctica, S. America, S. Africa, Australia, New Zealand, and sub-Antarctic islands.

Sphenisciformes (penguins; class *Aves) An order that comprises the single family *Spheniscidae.

Sphenodon punctatus (tuatara) *See* SPHENODONTIDAE.

Sphenodontidae (tuatara; subclass *Lepidosauria, order *Rhynchocephalia) A monospecific family (*Sphenodon punctatus*, tuatara) of primitive, lizard-like reptiles up to 60 cm long, which have a horny, dorsal crest. The skull is of the primitive *diapsid type, with a fixed *quadrate bone. A *pineal eye is present. The teeth are *acrodont, with a tendency to develop a beak-like structure anteriorly. There is no copulatory organ. Tuataras are carnivorous, feeding on invertebrates and small vertebrates. They survive only on a few islands in the Bay of Plenty, New Zealand, and have protected status.

spheridia Minute, club-shaped, calcareous bodies (modified spines) on short stalks, which are commonly situated in pits

in the *peristomal region of *ambulacra in echinoids. Their function is uncertain.

spheroidosis Disease of insects caused by viruses of the entomopoxvirus group. The disease may affect larval and other stages of insects of the *Coleoptera, *Diptera, *Lepidoptera, and *Orthoptera.

sphincter A circular *muscle that constricts a passage or body orifice, relaxing as required to allow material to pass through.

Sphingidae (hawk moths, sphinx moths; subclass *Pterygota, order *Lepidoptera) Family of medium-sized to very large moths that are usually fast fliers. The fore wings are elongate, narrow, and obliquely angled terminally. The proboscis is often very long and is used to suck nectar. Some species have tympanal organs ('ears') on the *palps. The larvae bear a horn at the rear. There are about 8000 species of Sphingidae, several of which have an intercontinental distribution, although the family is mainly tropical.

sphingolipid One of a group of *lipids, containing *sphingosine or its derivatives, that is found predominantly in the membranes of nerve cells.

sphingomyelin A generic term for a group of phosphorus-containing *sphingolipids that are found principally in the nervous tissue, but also in the blood. Chemically they are regarded as phosphoryl choline derivatives of a ceramide.

sphingosine An 18-carbon, unsaturated, alcohol *amine which is the parent compound of *sphingolipids.

sphinx moths *See* SPHINGIDAE.

Sphyraena barracuda (giant barracuda) *See* SPHYRAENIDAE.

Sphyraenidae (barracuda; subclass *Actinopterygii, order *Perciformes) A small family of marine fish that have a very elongate, slender body, a large mouth with strong teeth, and two widely separated *dorsal fins. Barracudas have the reputation of being aggressive and dangerous fish. Although most species rarely exceed 60 cm in length, *Sphyraena barracuda* (giant barracuda) grows to 1.7 m. There are about 18 species, distributed widely in tropical to subtropical waters.

Sphyrna zygaena (hammerhead shark) *See* SPHYRNIDAE.

Sphyrnidae (hammerhead shark; subclass *Elasmobranchii, order *Lamniformes) A small family of marine, occasionally brackish-water sharks characterized by the sideways expansion of the head with the eyes and nostrils located almost at the tip of the lateral expansions. These peculiar sharks are found principally in tropical coastal regions and tend to be large (e.g. *Sphyrna zygaena* grows to 7 m). There are about nine species.

spicate In spikes.

spicule A small needle or spine, or a small, spiky, skeletal element. In *Porifera (sponges), spicules form a skeletal framework supporting the soft, cellular mass. They exist in a variety of forms and are important in the identification and classification of species. An extensive nomenclature has developed through the use of these structures in poriferan taxonomy (e.g. they may be described as monoaxon, tetraxon, triaxon, hexaxon, polyaxon, megasclere, or microsclere). Spicules may be calcareous, siliceous, or of a fibrous material (spongin).

spider beetle *See* PTINIDAE.

spider crabs *See* MAJIDAE.

spiderhunters *See* NECTARINIIDAE.

spider-hunting wasps *See* HUNTING WASP; POMPILIDAE.

spider monkey (*Ateles*) *See* CEBIDAE.

spiders 1. *See* ARACHNIDA; ARANEAE; LIPHISTIIDAE; MIMETIDAE; SPARASSI-DAE; TRAPDOOR SPIDERS. **2. (cobweb (comb-footed) spiders)** *See* THERIDIIDAE. **3. (crab spiders)** *See* THOMISIDAE. **4. (funnel-web spiders, tarantulas)** *See* MYGALOMORPHAE; THERAPHOSIDAE. **5. (hunting (wolf) spiders)** *See* LYCOSIDAE. **6. (jumping spiders)** *See* SALTICIDAE. **7. (money spiders)** *See* LINYPHIIDAE. **8. (nursery-web spiders)** *See* PISAURIDAE. **9. (orb-web spiders)** *See* ARANEIDAE. **10. (purse-web spiders)** *See* ATYPIDAE. **11. (sheet-web spiders)** *See* AGELENIDAE. **12. (spitting spiders)** *See* SCYTODIDAE.

spikefish *See* TRIACANTHODIDAE.

spiky dory (*Neocytrus rhomboidalis*) *See* OREOSOMATIDAE.

spinal column *See* VERTEBRA.

spinal cord In vertebrates, a tube containing *neurons and bundles of nerve fibres, many of which connect to the brain, enclosed within the *spinal column.

spindle The set of microtubular fibres that appear to move the *chromosomes of *eukaryotes during cell division. The spindle is formed only at *mitosis or *meiosis, appearing at *metaphase and first arranging the chromosomes at its equator. Movement apart of *chromatids then occurs during *anaphase, probably as a result of the spindle fibres, which run from the *centromere (spindle attachment) to the spindle pole and from pole to pole, sliding past one another in a ratchet-like manner, similar to the way muscle filaments slide past each other (*see* SLIDING-FILAMENT THEORY).

spindle attachment The region of attachment on a *chromosome that links it to the *spindle at *mitosis or *meiosis. Its position determines its shape at *anaphase (a rod, or J, or V). In a few species the centromeric properties are distributed along the entire length of the chromosome; such species are said to be *holocentric (possessing a diffuse *centromere).

spindlehorn (saola, Vu Quang 'ox') *See* PSEUDORYX NGHETINHENSIS.

spinefoot *See* SIGANIDAE.

spinetails (*Synallaxis*) *See* FURNARIIDAE.

spinner *See* EPHEMEROPTERA.

spinneret One of the specialized silk-handling devices found in spiders. In general there are four pairs, two pairs on the tenth abdominal segment and two pairs on the eleventh, although few spiders retain the maximum complement in the adult stage. The anterior median pair is usually lost, but sometimes a remnant may remain in the form of a median cone (colulus) whose function is unclear. In some cases the anterior median pair fuse during embryonic development to form an oval plate covered with thousands of tiny spigots (the cribellum). Some spiders retain only two pairs of spinnerets, losing both the anterior median and posterior lateral pairs. Spinnerets are movable, sclerotized tubes composed of several segments; they vary in size, and are mostly conical in web-spinning species. The silk glands produce a protein (fibroin) as a liquid which is emitted through tiny spigots on the ends and ventral sides of the spinnerets. Polymerization of the liquid silk is not an oxidative or evaporative process, but relies on tension.

Spinosauridae (order *Saurischia, suborder *Theropoda) A *carnosaur family of the *Cretaceous of N. America and N. Africa; in *Spinosaurus* the *neural spines were elongated to give a pelycosaur-like sail, presumably an adaptation to tropical environments.

spiny ant-eater (*Tachyglossus*) *See* TACHYGLOSSIDAE.

spiny dormouse (*Platacanthomys*) *See* PLATACANTHOMYIDAE.

spiny eel *See* MASTACEMBELIDAE; NOTACANTHIDAE.

spiny fin ray The hard, pointed, unsegmented, unpaired, and unbranched *fin ray found in varying number in the *dorsal, *anal, and *pelvic fins of many bony fish (e.g. perch and scorpionfish).

spiny lobsters *See* PALINURIDAE.

spiny mouse *See* HETEROMYIDAE.

spiny rat *See* ECHIMYIDAE.

spiny-tailed lizards (*Uromastyx*) *See* AGAMIDAE.

spiracle 1. Essentially a vestigial gill slit located between the mandibular and hyoid arches in adult sharks, rays, and some primitive bony fish, and visible as an opening behind the eye. **2.** One of the pores on the body of an insect at which the *tracheae open to the outside. Spiracles are found on the thoracic and abdominal segments, and usually have some mechanism to close them.

spireme A tangle of threads, or one of those threads, that appears during the *prophase stage of cell division.

Spirotrichia (subphylum *Ciliophora, class *Ciliatea) A subclass of *Protozoa in which the system of *cilia in the oral region is well developed and conspicuous. Rows of oral cilia are arranged spirally, winding down toward the *cytostome. Body cilia usually are sparse. There are six orders, and many genera. They occupy a wide range of habitats.

spitting spiders *See* SCYTODIDAE.

spittlebug *See* CERCOPIDAE.

Spizaetus (hawk eagles) *See* ACCIPITRIDAE.

splanchic mesoderm *See* MESODERM.

splanchnocoel In vertebrate *embryos, one of a pair of temporary coelomic cavities located on either side of the body below the gut.

splanchnocranium *See* CRANIUM.

splanchnopleure In vertebrates, the part of the body derived from the *mesoderm inside the *coelom and the *endoderm. It comprises the entire digestive system and the *lungs, and the inner lining of the coelom that covers the internal organs.

spleen In vertebrates, an organ posterior to the *stomach that produces lymphocytes (see LEUCOCYTE), destroys particles of foreign matter, and acts as a reservoir for red blood cells. It breaks down old red blood cells and stores iron from them.

spondylium A curved platform for muscle attachment in the *shell-beak region of some *Brachiopoda.

sponges See PORIFERA.

Spongiaria See PORIFERA.

spongin In some *Porifera, a structure of interconnecting, coarse, *collagen fibres which forms the skeleton of the animal.

spontaneous mutation In genetics, a naturally occurring *mutation, as opposed to one artificially induced by chemicals or irradiation. Usually such mutations are due to errors in the normal functioning of cellular *enzymes.

spookfish See OPISTHOPROCTIDAE.

spoonbills See CICONIIFORMES; THRESKIORNITHIDAE.

spoon worms See ECHIURIDA.

sporosac See STYLASTERINA.

Sporozoa (phylum *Protozoa) A subphylum of Protozoa in which the life cycle includes a spore-forming or cyst-forming stage. Asexual reproduction occurs by multiple fission. All members are parasitic, parasitizing hosts throughout the animal kingdom. Some species can cause important diseases.

sporozoite In *Sporozoa, a spore produced by fission of a *zygote.

sportive lemur (*Lepilemur*) See MEGALADAPIDAE.

spotted butterfish (*Scatophagus argus*) See SCATOPHAGIDAE.

spotted creeper (*Salpornis spilonotus*) See SALPORNITHIDAE.

spotted scat (*Scatophagus argus*) See SCATOPHAGIDAE.

springtail Member of the order *Collembola, which possesses specialized leaping mechanisms (the *furcula and *retinaculum).

spur-thighed tortoise (*Testudo graeca*) See TESTUDINIDAE.

spur-throated grasshoppers See ACRIDIDAE.

squalene An open-chain *terpene, found in shark-liver oil, which acts as an intermediate in the biosynthesis of *cholesterol.

Squalidae (dogfish; subclass *Elasmobranchii, order *Squaliformes) A fairly large family of marine sharks that have two *dorsal fins, each often preceded by a solid spine. The *anal fin is lacking. Some species, e.g. *Etmopterus spinax* (Atlantic velvet belly), 90 cm, are relatively small, while others, e.g. *Somniosus microcephalus* (Greenland shark), may reach a length of about 6.4 m. All members of the family are *live-bearing species which often travel in large schools. There are at least 61 species, distributed world-wide.

Squaliformes (subclass *Elasmobranchii, superorder *Selachimorpha) An order of sharks that have two *dorsal fins of about equal size and no *anal fin. The dorsal fins may be preceded by a spine. The order includes the families *Pristiophoridae, *Squalidae, and *Squatinidae.

Squalodontoidea (shark-toothed whales; order *Cetacea, suborder *Odontoceti) An extinct superfamily of toothed whales that comprises the family

Squalodontidae and its immediate ancestors. The squalodonts were about the size of modern porpoises, which probably they resembled closely in appearance and behaviour. The teeth in the posterior part of the beak were triangular in shape and resembled those of sharks. The beak was often elongated. The skull was modified to the form characteristic of modern odontocetes and the blow-hole was positioned above and behind the eyes. The superfamily appeared in the late *Oligocene and was distributed throughout the world during the early *Miocene, but by the middle Miocene it was being replaced by more advanced porpoises, and by the *Pliocene the last squalodont was extinct.

squamae In *Diptera, flap-like appendages at the base of the wings which are often attached to the *thorax close to the wing base, and which sometimes conceal the *halteres. In some *Brachycera and *Cyclorrapha they are obvious as waxy, white flaps.

Squamata (lizards, snakes; class *Reptilia, subclass *Lepidosauria) A highly successful order which includes 95% of all living reptiles. The lizards and snakes are each given ordinal status in some classifications. The earliest lizards had appeared by the *Triassic and the snakes, which might be regarded as 'legless' lizards, diverged from the ancestral line in the *Cretaceous. The skull has lost the lower temporal arch, allowing mobility of the *quadrate and increased gape. The tongue is notched or forked. Body scales are generally small and overlapping. Limblessness is common. They are divided into two suborders (or orders), *Sauria (Lacertilia) and *Serpentes (Ophidia).

Squamiferidae (order *Isopoda, suborder Oniscidea) Family of terrestrial isopods which are covered in scales (their name being derived from the Latin *squama*, 'scale', and *fere*, 'to bear'), and are up to 6 mm long. Most species in the family lack eyes and have no tracheal system.

squamosal The bone that forms the posterior part of the side of the skull; in *Mammalia it articulates with the lower jaw.

squamous cell One of the three types of epithelial (*see* EPITHELIUM) cells (*see also* COLUMNAR CELL and CUBOIDAL CELL) that is flattened in shape.

Squamous cell

squamulose Bearing or consisting of small scales (squamules).

squaretail *See* TETRAGONURIDAE.

squashbug *See* COREIDAE.

Squatinidae (angel shark, monkfish; subclass *Elasmobranchii, order *Squaliformes) A small family of marine sharks that have a rather flattened body and large *pectoral fins. Angel sharks tend to live in fairly shallow seas, feeding on bottom-living animals. There are about 11 species, distributed in the Atlantic and Pacific.

squids *See* TEUTHOIDEA.

squirrelfish (*Holocentrus ascensionis***)** *See* HOLOCENTRIDAE.

squirrel glider (*Petaurus***)** *See* PETAURIDAE.

squirrels *See* ANOMALURIDAE; SCIURIDAE.

s-RNA Soluble RNA: transfer-RNAs that are relatively small molecules and are more soluble in acid than other *RNAs.

S-shaped growth curve (sigmoid growth curve) A pattern of growth in which, in a new environment, the population density of an organism increases slowly initially, in a positive acceleration phase; then increases rapidly approaching an exponential growth rate as in the *J-shaped curve; but then declines in a negative acceleration phase until at zero growth rate the population stabilizes. This slowing of the rate of growth reflects increasing environmental resistance which becomes proportionately more important at higher population densities. This type of population growth is termed '*density-dependent' since growth rate depends on the numbers present in the population. The point of stabilization, or zero growth rate, is termed the 'saturation value' (symbolized by K) or '*carrying capacity' of the environment for that organism. K represents the point at which the upward curve begins to level, produced when changing population numbers are plotted over time. It is usually summarized mathematically by the *logistic equation.

(⊕) SEE WEB LINKS
• More information about the S-shaped growth curve.

SSSI *See* SITE OF SPECIAL SCIENTIFIC INTEREST.

stabilimentum A conspicuous structure made of lines or spirals at the centre of the orb web spun by *Argiope* species of spiders (*see* ARANEIDAE). The purpose of the stabilimentum may be to warn birds not to fly through the web, or to hide the spider.

stabilizing selection (maintenance evolution, normalizing selection) The stabilizing influence of *natural selection in an *environment that changes little in space and time. It tends to inhibit evolutionary innovation, and accounts for the fact that many fossil groups changed very little over long periods of time. *Compare* DIRECTIONAL SELECTION; DISRUPTIVE SELECTION.

stag beetle *See* LUCANIDAE.

stalk body *See* CRISTA.

stalk-eyed flies *See* DIOPSIDAE.

standard deviation (σ) A measure of the normal variation of a population. In any given measurement, two-thirds of the population fall within one standard deviation on either side of the mean, 95% between two standard deviations, and so on; the proportion falls off sharply because of the bell-curve effect. The standard deviation is calculated as the root-mean-square deviation.

(⊕) SEE WEB LINKS
• Description of standard deviation and variance

standard metabolic rate (SMR) The minimum *metabolic rate needed to sustain life at a specified temperature. The SMR is measured in organisms that are resting in a post-absorptive state and in darkened conditions. *Compare* BASAL METABOLIC RATE.

stapes In *Mammalia, the inner auditory ossicle of the ear, stirrup-shaped because it is pierced by an artery. It is derived from the hyomandibular bone in fish, which connects the cranium and the upper jaw, and from the *columella auris of other tetrapods.

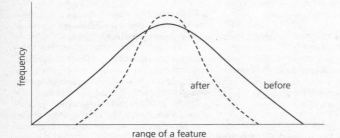

Stabilizing selection

Staphylinidae (rove beetles; subclass *Pterygota, order *Coleoptera) Family of beetles, most of which are elongate, 1–28 mm long, and generally shining, black, black and red, or metallic. The *elytra are short, leaving over half of the abdomen exposed. Some species are highly sculptured or *pubescent. The head and *mandibles are large, antennae *filiform. The abdomen is highly flexible, and is often raised in a threat gesture. Many rove beetles are strong fliers. Larvae resemble wingless adults, with the hind part of the body generally paler, without toughened plates. Adults and larvae are found in damp places, decaying matter, and dung, where they prey on other insects. Many are associated with ants' nests, some producing sweet substances highly attractive to ants. *Staphylinus olens* (devil's coach horse), about 25 mm long, is Britain's largest staphylinid. Some tropical species produce chemicals that blister the skin and cause eye irritation. There are 27 000 species, making this the second largest beetle family (after the *Curculionidae).

***Staphylinus olens* (devil's coach horse)** *See* STAPHYLINIDAE.

starch A *polysaccharide comprising many *glucose units linked by glycosidic bonds (*see* GLYCOSIDE). It consists of 20–30% *amylose and 70–80% *amylopectin, and is the principal *carbohydrate storage compound in plant tubers and the endosperm of seeds.

(⊕) SEE WEB LINKS
- 'Water structure and science: starch' by Martin Chaplin

starfish *See* ASTEROIDEA; DIPLOZONINA; FORCIPULATIDA; NOTOMYOTINA; PAXILLOSIDA; PLATYASTERIDA.

stargazer *See* URANOSCOPIDAE.

starlings *See* STURNIDAE.

startle response The reaction of an animal to sudden danger, by a *threat display (e.g. when certain butterflies and moths reveal *eyespots on the wings) or flight (e.g.

on average it takes 20 milliseconds for a puff of air from an approaching missile to set an American cockroach running).

stasigenesis The situation in which an evolutionary lineage persists through time without splitting or otherwise changing. So-called *living fossils provide examples of stasigenesis.

stasipatric speciation Rapid speciation, by chromosomal *mutation, that may occur among small breeding populations that are not completely isolated genetically or spatially. Such speciation may occur either *parapatrically or *sympatrically. M. J. D. White, who proposed the concept, pointed out that most species have their own unique *karyotype, and suggested that the karyotype changes might have been the actual promoters of reproductive isolation.

stasis A period of little or no evolutionary change; the 'equilibrium' that alternates with 'punctuations' in the theory of *punctuated equilibrium.

static allometry *See* ALLOMETRY.

static life table A *life table compiled from the age structure of a population at a particular time.

statistical method In modern usage, a method for analysing data based on probability theory. A statistical method permits the calculation of a value based on observations about some problem that may be tested for significance by comparison with the values that might be expected to arise by chance. Two main categories of statistical methods have been developed: classical or parametric tests; and the more recent non-parametric or distribution-free tests. Parametric statistical methods may be applied only to data on an interval scale, and typically they make assumptions about the background population from which the sample is taken, most often that it is normally distributed. Where data are in nominal or ordinal form, or where assumptions about the distribution of data on which a parametric test is based cannot be justified, then non-parametric (distribution-free) methods can be used. In general, parametric tests are more rigorous than non-parametric tests. Formerly, and more colloquially, 'statistical methods' embraced any form of data gathering and analysis, including *numerical method.

statoblast In *Bryozoa, a special, resistant body by which the animals reproduce asexually. It consists of a mass of cells containing stored food enclosed in a chitinous envelope, which can withstand desiccation and freezing and which may remain dormant for prolonged periods. The structure and shape of statoblasts are important taxonomic features of freshwater bryozoans.

statocyst In certain aquatic invertebrates, a vesicle containing mineral grains that stimulate sensory cells as they move in response to the movement of the animal. The statocyst is primarily a gravity receptor, acting as an organ of equilibrium that allows a swimming organism to maintain a horizontal attitude. One or more mineral grains (statoliths), often of calcium carbonate, rest in sensory hairs; any displacement relative to gravity causes the statolith(s) to stimulate the sensory hairs which send information to the nervous system.

statolith *See* STATOCYST.

Stauromedusida (Lucernarida; class *Scyphozoa, subclass *Scyphomedusae) An order of *Cnidaria in which the adults are solitary *medusae that live attached to a substrate either directly or by means of a stalk, so they appear to be *polypoid. They are inhabitants of cold seas and coastal waters.

Steatornis caripensis (oilbird) *See* STEATORNITHIDAE.

Steatornithidae (oilbird; class *Aves, order *Caprimulgiformes) A monospecific family (*Steatornis caripensis*), which is a chestnut-coloured bird, marked with white spots. It has long, pointed wings, a long tail, short legs, a short, hooked bill, a wide gape, and *rictal bristles. It is nocturnal and inhabits caves, using *echo-location to fly about in the dark. It feeds on fruit, and nests colonially on cave ledges. It is found in S. America.

Stegocephalia An informal name for a *paraphyletic group of animals comprising vertebrates other than fishes, but excluding *amniotes (*reptiles, *birds, and *mammals) and *Lissamphibia. Stegocephalians lived from the *Devonian to the *Jurassic. They resembled salamanders, but some larger forms possessed heavy dermal armour. Most are now included in the *Labyrinthodontia. Other palaeontologists class the Stegocephalia as a *clade comprising all vertebrates that possess digits rather than fins, i.e. all living terrestrial vertebrates (including those that have secondarily adopted an aquatic lifestyle) and their amphibious ancestors.

Stegodontidae (order *Proboscidea, suborder *Mammutoidea) An extinct family of mammals whose evolution paralleled that of the true elephants. The low-crowned *molar teeth had cusps arranged in transverse ridges, with traces of *cement, and the tooth succession typical of elephants was present, although the stegodonts still retained permanent *premolars. The family appeared during the *Miocene in Asia, and later invaded Africa; they died out during the *Pleistocene, having reached as far east as Timor in the late Pleistocene.

Stegosauridae (suborder *Archosauria, order *Ornithischia) A suborder of quadripedal dinosaurs, mainly *Jurassic in age, that are characterized by a double row of plates and spines along the back and tail.

stellate Star-shaped; radiating in arrangement.

Stelleroidea (phylum *Echinodermata, subphylum *Asterozoa) The only class in its subphylum, comprising echinoderms that are star-shaped and radially symmetrical.

stem cell A mitotically active cell that serves to replenish those that die during the life of a *metazoan organism (somatic stem cell) or that produces a continuing supply of *gametes (germinal stem cell).

stem flies *See* CHLOROPIDAE.

stem group In *cladistic analysis, those taxa descended from the point where an ancestral taxon split into two sister groups to the point at which a further split gave rise to an extant *crown group.

stem reptiles *See* CAPTORHINOMORPHA.

Stenidae (order *Cetacea, suborder *Odontoceti) A family of small, porpoise-like mammals in which the snout is long and merges into the forehead, the dorsal fin is hooked, and the tail is slightly notched at the mid-line. Members of the Stenidae attain lengths of about 2 m. There are four species, in three genera. *Steno* (rough-toothed dolphin) is distributed throughout the tropical and warm-temperate oceans. *Sotalia* (the tucuxi) occurs around the coast of S. America and in the rivers Amazon and Orinoco. *Sousa* (the white or hump-backed dolphin) occurs in coastal waters around Africa and southern Asia. Many authorities include the three genera in the *Delphinidae.

Steno (rough-toothed dolphin) *See* STENIDAE.

steno- From the Greek *stenos*, 'narrow', a prefix used in *ecology with adjectives that describe environmental factors, denoting a limited tolerance by an organism of those factors. *Compare* EURY-.

stenoecious Applied to an organism that can live only in a restricted range of *habitats.

stenohalic *See* STENOHALINE.

stenohaline (stenohalic) Applied to organisms that are very sensitive to changes in salinity; i.e. they are unable to tolerate a wide range of osmotic pressures. *Compare* EURYHALINE.

stenopaic Applied to an eye in which the pupil is narrow and slit-like.

Stenopelmatidae (Jerusalem crickets, king crickets, sand crickets, stone crickets; order *Orthoptera, suborder *Ensifera) Family of large, wingless insects, most species of which live in rotting logs or under stones, or burrow in sand or loose soil. They are most common in the tropics. One Indian genus is blind, and always lives below ground. The antennae are usually longer than the body. There are six genera and 38 species. *Stenopelmatus fuscus* is the Jerusalem cricket or potato bug, found only in New Mexico, Arizona, and California.

stenophagic Applied to organisms that have a highly specialized diet.

stenothermal *See* STENOTHERMOUS.

stenothermous (stenothermal) Applied to organisms that are unable to tolerate a wide temperature range.

stenotopic Tolerant of only a narrow range of environmental factors.

Stenotritidae (order *Hymenoptera, suborder *Apocrita) Family of large to very large, hairy, ground-nesting bees: they are entirely restricted to Australia. Formerly regarded as a subfamily of the *Colletidae, stenotritids are unique among bees in having the *ocelli situated low on the front of the head, nearer to the antennal sockets than to the posterior margin of the *vertex. The nest of only one species, *Stenotritus pubescens*, has been described. The cells were unlined and the provisions consisted entirely of nectar and *Eucalyptus* pollen.

Stephanoberycidae (pricklefish; subclass *Actinopterygii, order *Beryciformes) A small family of deep-sea fish which live at depths of more than 1500 m. They are small (15 cm) and have large eyes and a large mouth. Although the fins have no spines, the rough-edged scales give the fish a prickly appearance. There are three species.

stephanokont A cell that has a ring of *flagella or *cilia near one end.

Stephen Island wren (*Xenicus lyalli*) *See* XENICIDAE.

Stercorariidae (skuas; class *Aves, order *Charadriiformes) A family of medium to large, dark brown sea-birds, some colour phases of which have white underparts. Skuas have long, pointed wings, long to medium-length tails, and *Stercorarius* species have elongated central tail feathers. Their bills are medium length, strong, and hooked, their legs short, with webbed feet. They are *pelagic, occurring near coasts and breeding on islands or on inland tundra. They feed on mammals, insects, eggs, fish, and carrion, often obtained by chasing and robbing other birds. They nest on the ground. There are two genera, with six or seven species, found world-wide, breeding in northern areas of Europe, Asia, and N. America, and southern S. America and sub-Antarctic islands.

stereotaxis *See* THIGMOTAXIS.

sterile 1. Of an organism, unable to produce reproductive structures (i.e. unable to reproduce). **2.** Of land, unable to support the growth of plants, especially cultivated crops. **3.** Of an environment, object, or substance, completely free of all living organisms, including all micro-organisms of any type or form.

sterilization 1. The removal of the ability to reproduce (which may be achieved by surgery upon the gonads or their ducts, or by hormonal application). **2.** The process of killing or removing all living micro-organisms from a sample (e.g. by autoclaving).

sterlet (*Acipenser ruthenus*) *See* ACIPENSERIDAE.

Sterna (terns) *See* LARIDAE.

sternite Ventral region of a segment of the body of an insect, if the region is composed of sclerotized cuticle. (If not, the region is termed a '*sternum'.)

Sternoptychidae (hatchetfish; superorder *Protacanthopterygii, order

*Salmoniformes) A small family of marine fish that have a deep, short, and very compressed body, the outline of the belly resembling the blade of a hatchet. The mouth is large and directed upward. Many *photophores are present on the flanks of *Argyropelecus lychnus* (silvery hatchetfish), 6 cm. There are about 27 species, distributed world-wide at depths of 100–4000 m.

Sternotherus odoratus (stinkpot, common musk turtle) *See* KINOSTERNIDAE.

sternum 1. In tetrapods, the bone at the mid-ventral line of the thorax to which most of the ribs are attached at their ventral ends. The sternum is attached anteriorly to the *pectoral girdle. **2.** In *Arthropoda, a ventral cuticle on each segment that forms a thickened plate. See STERNITE.

steroid One of a group of derivatives of the fused, reduced, ring compound perhydrocyclopentanophenanthrene. As a group, steroids have a wide range of physiological functions: they include bile acids, adrenocorticoid and sex hormones, and sterols such as *cholesterol.

(((●))) **SEE WEB LINKS**
• More information about the biochemistry and physiological effects of steroids.

Steropodon (subclass *Prototheria, order *Monotremata) The earliest Australian mammal and the only known *Mesozoic representative of the monotremes. It is known only from opalized jaws and teeth found in *Cretaceous deposits at Lightning Ridge, central Australia (the name is from the Greek *sterope*, 'lightning'). The pattern of the *molar teeth suggests that monotremes, like *Eupantotheria, may be derived from *Theria and not from a separate protomammalian stock.

Stichaeidae (prickleback; subclass *Actinopterygii, order *Perciformes) A family of marine fish that have long, slender bodies, and long *dorsal fins extending to the tail end of the body and composed entirely of *spiny rays. Pricklebacks tend to live in the intertidal zone, where the females guard their eggs by curling the body around the egg mass until hatching occurs. There are about 57 species, occurring in the N. Pacific and N. Atlantic.

stick insects *See* PHASMATIDAE; PHASMATODEA.

stickleback *See* GASTEROSTEIDAE.

stifftails (*Oxyura*) *See* ANATIDAE.

stigma A darkened area on the wing of an insect.

stiletto flies *See* THEREVIDAE.

stilt *See* RECURVIROSTRIDAE.

sting The *ovipositor of *Aculeata, which has lost its egg-laying function and serves as a means of injecting venom to paralyse, but not kill, the prey of hunting wasps, and is used as a means of defence by bees. The sting of honey-bees (*Apis* species) and some social wasps is barbed and remains in the skin of the victim after the wasp or bee has become detached. In honey-bees the muscular venom sac of detached stings continues to pump venom, and it also emits an alarm *pheromone which alerts other workers, which may be recruited to join the attack.

stingless bees *See* APIDAE.

stingray 1. *See* DASYATIDAE. 2. (**river stingray**) *See* POTAMOTRYGONIDAE.

stink badger *See* MEPHITIDAE.

stink gland In most *Heteroptera, glands that produce fluids believed to be distasteful to potential predators. In the immature stages (nymphs) the stink glands are situated between the abdominal *tergites, but in adults they are in the *metathorax.

stinkpot (*Sternotherus odoratus*) *See* KINOSTERNIDAE.

stints *See* SCOLOPACIDAE.

stipe In *Graptolithina, a branch formed from overlapping *thecae.

stipitate Having a stipe or stalk.

Stipiturus (**emu wrens**) *See* MALURIDAE.

Stizostedion vitreum (**walleye**) *See* PERCIDAE.

stoat (*Mustela*) *See* MUSTELIDAE.

Stolidobranchiata *See* ASCIDIACEA.

stolon In colonial invertebrates, the stalk-like structure by which individuals are attached to the substrate. *See* CTENOSTOMATA.

Stolonifera 1. (**class *Anthozoa, subclass *Octocorallia**) An order of corals, with *polyps arising from a basal *stolon, which creeps over solid objects. Skeletal *spicules may fuse into tubes and platforms. Stolonifera are found in shallow tropical and temperate waters. They first appeared in the *Cretaceous. 2. (**class *Gymnolaemata, order *Ctenostomata**) A suborder of bryozoans in which the stolon is slender and of creeping habit. *Zooids grow by lateral budding from the stolon and are usually paired. The suborder appeared in the *Ordovician and survives to the present day.

stolotheca One of the three types of graptolite (*Graptolithina) *thecae, which encloses the main *stolon and the earliest parts of the daughter stolotheca, *autotheca, and *bitheca.

stomach The anterior region of the gut, enlarged and usually with muscular walls that churn food, and with cells in the lining that secrete digestive acids.

Stomatopoda (**mantis shrimps; class *Malacostraca, superorder Hoplocarida**) Order of crustaceans, 5–36 cm long, many of which are brightly coloured, with striped or mottled patterns. The body is flattened dorsoventrally and there is a shield-like carapace. The entire dorsal surface is armed with ridges and spines, and the abdomen is broad and segmented. The head has stalked, *compound eyes with a median *nauplius eye; large, *triramous, first antennae; and smaller third antennae. The thoracic appendages are *uniramous and subchelate, with the second pair modified for raptorial prey capture. As the common name implies, the large second thoracic appendages are equipped with a movable process armed with spines and barbs. Prey is caught or speared by a rapid extension of this process. Many species inhabit bottom burrows or crevices in coral or rock, and defend their burrows using the telson as a shield. They may leave their burrows to stalk prey (e.g. crabs) and swim using their *pleopods, steering by means of their large antennal scales and *uropods. Apart from *Isopoda, mantis shrimps are the only malacostracans that have abdominal gills. Many species exhibit parental care of their eggs. They comprise the only order of their superorder, with approximately 300 species. Most are tropical but some live in temperate waters.

Stomias boa (scaly dragonfish) *See* STOMIATIDAE.

Stomiatidae (scaly dragonfish; superorder *Protacanthopterygii, order *Salmoniformes) A small family of deep-sea fish that have long, tapering bodies covered with loosely attached scales and a very large mouth with fang-like teeth suggestive of voraciously predatory behaviour. The *dorsal and *anal fins are set close to the small tail fin. Usually dark in colour, *Stomias boa* (scaly dragonfish), 40 cm, is thought to live at depths of more than 1000 m. There are about nine species, distributed world-wide.

stomochord In *Enteropneusta, a forward extension of the *diverticulum of the dorsal part of the *pharynx.

Stomochordata *See* HEMICHORDATA.

stomodaeum (stomodeum) In many invertebrates and vertebrate *embryos, a

mouth-like modification of the anterior end of the gut formed by an indentation of the *ectoderm opening into the *enteron.

stomodeum *See* STOMODAEUM.

stone canal In *Echinodermata, a canal with walls strengthened by calcareous matter that connects the *madreporite with the water-vascular system. *See also* RADIAL CANALS.

stone crab (*Menippe mercenaria*) *See* XANTHIDAE.

stone crickets *See* STENOPELMATIDAE.

stone-curlew *See* BURHINIDAE.

stone fish *See* SYNANCEIIDAE.

stony corals *See* MADREPORARIA.

storks *See* CICONIIDAE; CICONIIFORMES.

storm petrels *See* HYDROBATIDAE; PROCELLARIIFORMES.

stotting A series of high, stiff-legged jumps made by gazelles (*Bovidae) while they are running. This behaviour alerts the herd to the presence of a predator, but the gazelle's probable reason for stotting is to warn a predator that it has been seen and its pursuit is therefore futile, because the quarry will escape. The word may be derived from the Scots word *stot* meaning 'to bounce'. *See also* PRONKING.

stouts *See* TABANIDAE.

straight-nosed pipefish (*Nerophis ophidion*) *See* SYNGNATHIDAE.

stramineous Straw-coloured.

strategy In *game theory, the course of action a contestant follows under the specified circumstances of the game.

stratified random sample (partial random sample) In statistics, a modification of the random sample that is particularly useful when obvious heterogeneity exists in the *community, area, etc. to be investigated. In such instances a simple random sample may fail to record sufficient replicates of a particular subcategory, or may do so only very inefficiently, so preventing a proper statistical monitoring of variability. In a stratified random scheme sample points are shared among the main types present, then allocated at random within those subcategories.

SEE WEB LINKS

• Explanation of the stratified random sample technique.

Stratiomyidae (soldier flies; order *Diptera, suborder *Brachycera) Family of small to large flies which are usually flattened, with bright yellow, green, white, or metallic markings. The body lacks bristles and has an *annulate third antennal segment. *Squamae are small and the *tibia is rarely spurred. The *scutellum is often strongly developed, with spines or projections. Wing veins are crowded near the *costa and are more pigmented than the posterior veins. The adults are not usually strong fliers and are often encountered sitting on flowers. The larvae are associated with aquatic environments or with dung and are carnivorous in either habitat. Most larvae have thick, leathery skin impregnated with calcareous deposits. The head is small and obvious, and there are a further 11 body segments. The family is distributed worldwide, and about 1400 species have been described.

streaming The metabolically active movement of the particulate and fluid constituents of the *protoplasm. This can occur within a cell, laterally through protoplasmic material connecting cells, and also longitudinally through files of connected cells.

Streblidae (order *Diptera, suborder *Cyclorrapha) Small family of flies in which the adults are exclusively parasitic on bats. The head is fixed to the *thorax and the eyes, if present, are small. The antennae are two-segmented and set in pits. The *palps are leaf-like, project forward, but are not

sheathed. The flies may have wings, which are shed frequently. Females deposit fully developed larvae which pupate almost immediately. In *Ascodipteron* species the female undergoes a dramatic transformation from winged insect to a bag-like sac once she is attached behind the ear of a bat of the genus *Miniopterus*. She sheds her wings and legs to achieve this transformation. The family is found mainly in the tropics and warm regions; 250 species have been described.

Strepsiptera (stylops, twisted-wing parasite; phylum *Arthropoda, class *Insecta) An order of very small (adult males are 1–1.75 mm long), beetle-like insects almost all of which are parasites whose larval stages develop inside other insects. Adult males have a single pair of wings, comparable to the hind wings of other insects, the fore wings being reduced to structures resembling the halteres of *Diptera. Females have no wings and in most species the female lives its entire life as an internal parasite of another insect, all but a small part of its body concealed inside its own last larval covering. Infestation causes changes in the body of the host (called 'stylopization') by which it can be distinguished from an uninfested individual, and causes the host to become infertile. There are nine families.

Strepsirrhini (cohort *Unguiculata, order *Primates) A suborder that includes the lemurs, bush-babies, lorises, and potto. All members of these groups possess a moist *rhinarium and a cleft upper lip bound to the gum, and these, together with other features (including the proportion of proteins they share), suggest that they form a coherent group which may be contrasted with the other suborder of Primates, the *Haplorrhini. According to this classification, Strepsirrhini and Haplorrhini replace the suborders *Prosimii and *Anthropoidea.

***Streptopelia* (turtle doves)** See COLUMBIDAE.

***Streptopelia decaocto* (collared dove)** See COLUMBIDAE.

stress A physiological condition, usually affecting behaviour, produced by excessive environmental or psychological pressures.

stretch display A display that initiates *courtship in green heron (*Butorides virescens*) in which the male turns away from the female and stretches its neck, displaying the neck with its bill pointing directly upward.

striate Marked with fine lines, ridges, or furrows.

striated muscle (striped muscle, skeletal muscle, voluntary muscle) Muscle that is composed of large, elongated cells (muscle fibres), each with many nuclei, in which the *cytoplasm is marked by fine lines (striations) at right angles to the long axis. The striations occur because the cytoplasm contains many *myofibrils, each of which is made from alternating bands of different material, and the myofibrils lie side by side with the bands aligned. Striated muscles are capable of rapid contraction and are particularly associated with voluntary movements of the skeleton.

(⊕) SEE WEB LINKS
• Detailed description and explanation of striated muscle.

stridulate To produce sound by rubbing a file across a membrane. Insects have a wide variety of mechanisms for sound production. In some of the most common examples a file on one wing rubs across a roughened surface on the other wing, or a file on the leg is drawn across the edge of the wing. The volume of sound production achieved by these mechanisms is often startling. The sound may be amplified by resonation in a wing or (e.g. in mole crickets) by causing a column of air to resonate in a chamber excavated in the ground and shaped for the purpose.

Strigidae (owls; class *Aves, order *Strigiformes) A family of small to large owls that have brown, grey, and black, cryptically marked plumage. They have facial discs, and many have 'ear' tufts (e.g. the six species of *Asio*, 12 species of *Bubo* (eagle owls), and 35 species of *Otus* (scops owls and screech owls), many of which have unfeathered legs; the 11 species of *Strix*, which have barred, spotted, or streaked under-parts, are not 'eared'). Their claws are not *pectinate. They inhabit forest, grassland, and desert, feed on fish, mammals, birds, and insects, and nest on ledges, in tree cavities, abandoned nests, or in burrows. Some pygmy owls and owlets (13–15 species of *Glaucidium*) are barely larger than a sparrow; they are partly diurnal and feed partly on birds and insects, some of which are caught on the wing. There are 22 genera in the family, with 122 species, some migratory, found world-wide.

Strigiformes (owls; class *Aves) An order of small to large, mostly nocturnal, birds of prey which have large heads with large, forward-facing eyes, short necks, and short, strong, hooked bills. Their wings are broad and rounded with soft tips for silent flight, their tails short to medium-length, and their legs short to long, with *zygodactylous toes. They inhabit forest, open country, grassland, and desert, feed on fish, mammals, birds, and insects, and nest in tree cavities or burrows. There are two families, *Strigidae and *Tytonidae, found world-wide.

strigil Brush of hairs, one of which is found on each of the fore limbs of certain insects, and which is used for grooming. The term is usually applied to structures in the *Hymenoptera and *Lepidoptera.

striped bass (*Morone saxatilis*) *See* PERCICHTHYIDAE.

striped catfish *See* PLOTOSIDAE.

striped goby (*Gobius vittatus*) *See* GOBIIDAE.

striped muscle *See* STRIATED MUSCLE.

Strix (owls) *See* STRIGIDAE.

strobila In *Cestoda, the main part of the body, behind the *scolex and neck, which consists of the linearly arranged, individual segments (proglottides, *see* PROGLOTTIS).

strobilization *See* SCYPHISTOMA.

Stromateidae (butterfish; subclass *Actinopterygii, order *Perciformes) A small family of marine fish that have a deep, oval-shaped body, long but low *dorsal and *anal fins, a small mouth, and a strongly forked tail fin. There are about 13 species, occurring in the Indo-Pacific region.

stromatolite A rock-like or firmly gelatinous structure, built up over long periods from many layers or mats of *cyanobacteria together with trapped sedimentary material. Stromatolites are found mainly in shallow marine waters in warmer regions and some are still forming (e.g. in Shark Bay, Western Australia). Fossil stromatolites dating from the *Archaean are known.

Stromatoporoidea An extinct group that has been attributed to the hydrozoans, sponges, foraminifera, bryozoans, algae, or regarded as a phylum with no modern representatives. Stromatoporoids are calcareous masses built up of horizontal laminae and vertical pillars. The upper surfaces show a pattern of polygonal markings and may have swellings and stellate grooves. They are found in limestones of *Cambrian to *Cretaceous age, often forming reefs in *Ordovician to *Devonian times.

structural gene A *gene that codes for the *amino-acid sequence of a protein. *Compare* REGULATOR GENE.

structuring method Any technique for sorting data to reveal the important patterns. *Classification and *ordination methods are examples of structuring methods.

***Struthio camelus* (ostrich)** *See* STRUTHIONIDAE.

Struthionidae (ostrich; class *Aves, order *Struthioniformes) A monospecific family (*Struthio camelus*) which is the largest living bird. Males are black with white wing and tail plumes, females are brownish-grey. Ostriches have long necks, small heads, and short, flat bills. Their legs are long with bare thighs, and their feet have only two toes. They are flightless and gregarious, running very swiftly. They feed on plants, berries, and seeds, and nest on the ground. They are farmed for their meat and decorative plume feathers. They are native to Africa.

Struthioniformes (ostrich; class *Aves) An order that comprises only the family *Struthionidae.

stump-tailed skink (*Tiliqua rugosa*) *See* SCINCIDAE.

sturgeon, common (*Acipenser sturio*) *See* ACIPENSERIDAE.

Sturnidae (starlings, glossy starlings, minas, mynas; class *Aves, order *Passeriformes) A family of medium-sized birds, most of which are black or dark in colour, many with a highly iridescent, purple, blue, green, and bronze gloss. Some have bare areas of skin and *wattles on the face, and some are crested. They have straight, fairly slender bills, short and rounded to long and pointed wings, and most have short, square tails. They are highly gregarious and are arboreal to terrestrial, running well. They are good mimics and many, especially minas (mynas), are kept as cage birds. They inhabit forests, open country, and urban areas, are omnivorous, and nest in holes in trees, banks, or buildings. *Sturnus vulgaris* (starling) has been introduced to N. America, S. Africa, Australia, and New Zealand. *Acridotheres tristis* (common mina, or myna) has become semi-domesticated and has been introduced to Australia, New Zealand, and S. Africa. *Aplonis* (19 species) is the glossy starling, widely kept as a cage bird. There are about 25 genera, with 108–110 species, some migratory, found in Europe, Africa, Asia, Australasia, and Pacific islands.

***Sturnus vulgaris* (starling)** *See* STURNIDAE.

Stylasterina (branched hydrocorals; phylum *Cnidaria, class *Hydrozoa) An order of cnidarians similar to *Milleporina but without free *medusae. *Gametes are formed on aborted medusae in special cavities (sporosacs). Stylasterina are known from the Late *Cretaceous to the present.

style 1. In some *Mollusca, a crystalline, rod-like structure which projects into the stomach where it is abraded to release enzymes. **2.** In some insects, a bristle at the end of the antenna. **3.** In some insects, a rod-like abdominal appendage.

stylet 1. In *Nemertea, a heavy barb set in the wall of the *proboscis; accessory stylets are available to replace the stylet should it be lost. **2.** In *Tardigrada, one of a pair of barbs anchored in the muscles of the *pharynx and whose points project into the anterior end of the buccal tube. **3.** In certain groups of insects (mainly *Hemiptera), one of the tubular, sucking mouth-parts formed from *mandibles and *maxillae that are modified into slender, flexible structures. The two pairs of maxillary stylets are pressed closely together and the separate salivary and food canals pass between them. The mandibular stylets slide freely and independently along the sides of the maxillary stylets. *See also* ROSTRUM.

Stylommatophora (class *Gastropoda, subclass *Pulmonata) An order of predominantly terrestrial gastropods that have two pairs of *tentacles. Forms with and without shells are common. Eyes are present at the tips of the posterior pair of tentacles. The order first appeared in the *Mesozoic.

stylopization *See* STREPSIPTERA.

stylops *See* STREPSIPTERA.

subcaudal Beneath the tail.

subcosta In insects, a thin vein behind the *costa that meets the edge of the wing along the costal margin.

subcutaneous In vertebrates, immediately below the outermost epidermal layer.

subcutaneous fat *See* PANNICULUS ADIPOSUS.

sub-imago (dun) In *Ephemeroptera (and unique among the *Insecta), a preadult, winged stage through which the insects pass. The aquatic nymphs emerge on to the surface of water or on to rocks, and take from a few seconds to several minutes to change into sub-imagos. Emergence times vary from species to species, and the sub-imago may last from minutes to a whole day before moulting to the adult stage. Some exotic species breed and die in the sub-imaginal stage.

sub-isomyarian Applied to a condition in some *Bivalvia in which the two adductor muscles are of nearly, but not quite, equal size.

sublingual glands A pair of salivary glands located beneath the tongue.

sublittoral 1. The zone of a lake or pond where the water is 6–10 m deep. **2.** The zone of the sea lying between the *littoral zone and the edge of the continental shelf or a depth of about 200 m.

submaxillary gland One of a pair of salivary glands located at the angles of the jaw.

submentum In *insects, the basal part of the *labium. *See* MENTUM.

subsessile Nearly *sessile.

subspecies Technically, a *race of a *species that is allocated a Latin name. The number of races recognized within a species and the allocation of names to them is somewhat arbitrary. Systematic and *phenotypic variations do occur within species, but there are no clear rules for identifying them as races or subspecies except that they must be: (*a*) geographically distinct; (*b*) populations, not merely *morphospecies; and (*c*) different to some degree from other geographic populations.

substitutional load In genetics, the cost in genetic deaths to the population of -replacing one *allele by another (a *mutation) in the course of evolutionary change.

substrate 1. (biochem.) The reactant acted upon by an *enzyme. **2. (substratum)** Any object or material upon which an organism grows or to which an organism is attached; an underlying layer or substance.

substratum See SUBSTRATE.

succession A progressive series of changes in the plant and animal life of a community from initial colonization to the establishment of a *climax. The term applies to animals (especially to *sessile animals in aquatic *ecosystems) as well as to plants.

succinyl coenzyme A An intermediate compound in the conversion of alpha-ketoglutaric acid to succinic acid during the *citric-acid cycle, a process linked to the formation of the energy-rich guanosine triphosphate (GTP, see GUANOSINE PHOSPHATE). Succinate is not always the final product as the succinyl coenzyme A may be employed for acylation reactions or to initiate *porphyrin synthesis.

sucker 1. An organ with which an animal attaches itself to a surface. **2.** See CATOSTOMIDAE.

sucker-footed bat See THYROPTERIDAE.

sucker loach See GYRINOCHEILIDAE.

sucking lice See ANOPLURA.

sucrase (invertase) The enzyme responsible for the catalytic *hydrolysis of *sucrose to *fructose and *glucose.

sucrose A disaccharide, composed of *fructose and *glucose, which is a common storage and transport *sugar in plants. It is known commercially as cane or beet sugar.

Suctoria (subphylum *Ciliophora, class *Ciliatea) A subclass of *Protozoa in which mature individuals lack both *cilia and a *cytostome. Mature forms typically are stalked and attached to a substrate. They

have *tentacles which are used for capturing and ingesting prey (mainly other ciliate protozoa). Reproduction occurs by budding, which produces free-swimming, ciliated, mouthless larval forms. There is one order (Suctorida), found in freshwater and marine habitats.

suctorial Adapted for sucking.

Suctorida See SUCTORIA.

sugar A member of a group of water-soluble carbohydrates that have a low molecular weight and are composed of one or more simple compounds (monosaccharides or disaccharides respectively).

sugarbirds See PROMEROPIDAE.

sugar glider (*Petaurus*) See PETAURIDAE.

Suidae (pigs; order *Artiodactyla, suborder *Suiformes) A family of omnivorous mammals that dig for roots and also eat other plant material and invertebrates. The legs are relatively short and the hair sparse and coarse. The nostrils are in a forward-facing disc at the end of the snout. The tusks, formed from the upper *canines, are curved upward, and all the canines grow persistently in the male; in the female the canines have closed roots and do not grow continuously. The skull is long, and low in front. The pig family can be traced back to the initial radiation of the artiodactyls, near the *Eocene- *Oligocene transition, and their evolution was confined to the Old World, where they are distributed widely. They never penetrated the New World except as companions to humans. There are nine or ten species, in five genera.

Suiformes (Suina; superorder *Paraxonia, order *Artiodactyla) A suborder that comprises the *Suidae (pigs), *Tayassuidae (peccaries), and *Hippopotamidae (hippopotamuses), together with ancestral forms. They are non-ruminant artiodactyls in which the upper *canine teeth form tusks, the cheek teeth are *bunodont, the stomach has three chambers (except in the Suidae where it has two), and the limbs bear four digits (except in the Tayassuidae in which the hind limbs bear three). The suborder first appeared during the *Eocene.

Suina *See* SUIFORMES.

Sula (gannets, boobies) *See* SULIDAE.

sulcal *See* SULCATE.

sulcate (sulcal) Marked with ridges, grooves, or furrows (sulci).

sulcus A furrow or groove.

Sulidae (gannets, boobies; class *Aves, order *Pelecaniformes) A family that comprises the one genus *Sula* (although three species are sometimes placed in a separate genus, *Morus*) of large sea-birds, many of which are white with black flight feathers, others brown with varying amounts of white. They have long, stout bills, lacking nostrils, fairly long necks, long, pointed wings, and long, wedge-shaped tails. Their legs are short with large, *totipalmate feet which are often brightly coloured. They inhabit islands and cliffs, feed on fish caught by diving from the air, and nest colonially on ledges, on the ground, or in trees. There are nine species, some migratory, found world-wide.

sulphur butterflies *See* PIERIDAE.

sulphydryl (thiol) The radical group -SH.

Sumatran rhinoceros (*Dicerorhinus sumatrensis*) *See* RHINOCEROTIDAE.

summation The addition of stimuli of different kinds, which may be perceived at more than one time (temporal summation) and in more than one place (spatial summation), to produce a co-ordinated response in an animal. Summation is one of the integrating properties of the central nervous system.

sun animalcule *See* HELIOZOA.

sunbathing The exposure of its body to sunshine by an animal as a means of thermoregulation (mainly in *poikilotherms) or to stimulate the production of vitamin D in the skin. Many animals prefer to rest in warm (but not hot) sunshine, thereby conserving metabolic energy that otherwise would be needed to maintain body temperature.

sunbeam snake (*Xenopeltis unicolor*) *See* XENOPELTIDAE.

sunbirds *See* NECTARINIIDAE.

sunbittern *See* EURYPYGIDAE.

Sundaland A geographical unit composed of Malaya, Sumatra, Java, and Borneo, with the intervening small islands, which are linked by the shallow-water (less than 200 m) Sunda shelf, which was exposed during periods of low sea level in the *Pleistocene. The fauna of this region is fundamentally homogeneous and differs slightly from that of areas further north-west, despite the fact that Malaya today is linked by an isthmus to the Asian mainland.

sunfish *See* CENTRARCHIDAE; MOLIDAE.

sunflies *See* SYRPHIDAE.

sungazer (*Cordylus giganteus*) *See* CORDYLIDAE.

sungrebes *See* HELIORNITHIDAE.

superb lyrebird (*Menura novaehollandia*) *See* MENURIDAE.

supercilium In a bird, a distinct mark immediately above each eye, resembling eyebrows.

superdominance *See* OVERDOMINANCE.

supergene A segment of *chromosome that is protected from *crossing-over and so is transmitted intact from generation to generation, like a *recon. The term is used to refer to several closely linked gene *loci that affect a single trait or series of interrelated traits, such as the set of gene loci controlling the colour and presence or absence of bands in the shell of the snail *Cepaea nemoralis*.

Genes involved in the same biochemical function (e.g. those controlling the synthesis of *tryptophan) are often clustered together in close linkage within the *genome to form a supergene.

supernormal stimulus An artificial stimulus that produces in an animal a response that is stronger than would be evoked by the natural stimulus it resembles. For example, in some birds incubation behaviour is stimulated by the presence of an egg, and the larger the egg the stronger the stimulus; in such birds a very large artificial egg may be incubated in preference to a much smaller real egg.

superparasite *See* HYPERPARASITE.

superspecies *See* SPECIES GROUP.

supersuppressor *See* SUPPRESSOR MUTATION.

supertramp A species that disperses efficiently, colonizes readily, and has a wide range, but is excluded by competition from habitats supporting a rich variety of species. *See also* TRAMP SPECIES.

suppression *See* SUPPRESSOR MUTATION.

suppressor mutation In genetics, a second *mutation that masks the phenotypic effects of an earlier mutation. This second mutation occurs at a different site in the *genome (i.e. it is not a strict *reverse mutation); 'intragenic suppression' results from a second mutation that corrects the functioning of the mutant gene (e.g. a mutation of a different *nucleotide in the same triplet, such that the *codon then encodes the original *amino acid); 'intergenic suppression' results from mutation of a different gene, the product of which compensates for the dysfunction in the first (e.g. a mutation that produces a mutant *transfer-RNA molecule that inserts an amino acid in response to a *nonsense codon, thus continuing a protein that would otherwise have been terminated). If a single suppressor mutation can suppress more than one existing mutation, it is said to be a supersuppressor.

supra-occipital The bone that forms the dorsal margin of the *foramen magnum.

supratragus In *Mammalia, the upper part of the *pinna.

surfperch *See* EMBIOTOCIDAE.

surf scoter (*Melanitta perspicillata***)** *See* ANATIDAE.

surgeonfish (doctorfish) *See* ACANTHURIDAE.

Suricata suricatta **(meerkat, suricate)** *See* HERPESTIDAE.

suricate (*Suricata suricatta***)** *See* HERPESTIDAE.

Surinam toad (*Pipa pipa***)** *See* PIPIDAE.

'survival of the fittest' *See* NATURAL SELECTION.

survivorship curve A graphical description of the survival of individuals in a population from birth to the maximum age attained by any one member. Usually it is plotted as the logarithm of the number of survivors as a function of age. If a population has a constant mortality rate the graph will be a straight line.

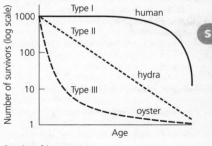

Survivorship curve

susceptible species *See* RARITY.

S value *See* SEDIMENTATION FACTOR.

Svedberg unit (S) The unit in which the *sedimentation factor is expressed. It is equal to 10–13 seconds and is written with no space between the number and the symbol (e.g. 70S). It is named after Theodor Svedberg (1884–1971), the Swedish chemist who developed the technique of centrifugation for the study of colloids and macromolecules, for which he was awarded a Nobel Prize in 1926. *See* MITOCHONDRION; RIBOSOME.

swallower (*Chiasmodon niger*) *See* BATHYPELAGIC FISH; CHIASMODONTIDAE.

swallows 1. *See* HIRUNDINIDAE. 2. (wood-swallow) *See* ARTAMIDAE.

swallowtail butterflies *See* PAPILIONIDAE.

swallow-tanager (*Tersina viridis*) *See* TERSINIDAE.

swamp eel *See* SYNBRANCHIDAE.

swan animalcule The ciliate protozoon *Lachrymaria olor*.

swan goose (*Anser cygnoides*) *See* ANATIDAE.

swans *See* ANATIDAE.

sweat bees Small bees belonging to the families *Halictidae and *Apidae (subfamily Meliponini, stingless bees) which are attracted to human sweat, especially in warm-temperate and tropical climates.

sweat flies *See* SYRPHIDAE.

sweat gland *See* ECCRINE GLAND.

sweeper *See* PEMPHERIDAE.

sweep net A finely meshed net, usually with a round mouth, used for sampling insects from vegetation. Its advantages lie in its simplicity and speed, and its ability to collect relatively dispersed insects on top of the vegetation. It cannot be used on very short vegetation.

sweepstakes dispersal route A term coined by G. G. Simpson in 1940 to describe a possible route of faunal interchange which is unlikely to be used by most animals, but which will, by chance, be used by some. It requires a major barrier that is occasionally crossed. Which groups cross and when they cross are determined virtually at random.

(()) SEE WEB LINKS

• Simpson's original 1940 paper proposing sweepstakes dispersal routes.

sweetlips emperor (*Lethrinus chrysostomus*) *See* LETHRINIDAE.

swellfish *See* TETRAODONTIDAE.

swiftlet (*Collocalia*) *See* APODIDAE.

swift moths *See* HEPIALIDAE.

swifts *See* APODIDAE; APODIFORMES.

swim-bladder (air bladder, gas bladder, hydrostatic organ) In *Actinopterygii, a thin-walled, gas-filled sac in the roof of the abdominal cavity, which allows the fish to achieve neutral buoyancy. It may (*physostomous) or may not (*physoclistous) be connected with the intestinal tract. The sac evolved from an *accessory respiratory organ and retains a respiratory function in some fish (e.g. *Lepisosteidae). It is also sometimes used for sound production (e.g. *Holocentridae), or sound reception (e.g. *Gadidae).

swimmeret In *Crustacea, one of the body appendages that are modified to increase their resistance to water, often by the possession of fringed *setae, and that are used to propel the animal through the water. In many Crustacea at least some of these appendages have become heavier and adapted for crawling or burrowing.

swimming crabs *See* PORTUNIDAE.

swordfish (*Xiphias gladius*) *See* XIPHIIDAE.

sword-tailed crickets *See* GRYLLIDAE.

S

swordtails *See* PAPILIONIDAE.

Sycettida (class *Calcarea, subclass *Calcaronea) An order of sponges whose body possesses a *sycon or *leucon structure.

sycon Applied to the body structure of a sponge if it consists of a single layer of chambers.

sycophagus Fig-eating.

***Sylvia* (Old World warblers)** *See* SYLVIIDAE.

Sylviidae (Old World warblers, cisticolas, prinias; class *Aves, order *Passeriformes) A family of small birds which have brown, grey, or greenish plumage, some having brighter colours, fine, pointed bills, medium-length wings with 10 *primaries, and fairly short to long tails. The 17 species of *Sylvia* are typical. Bush warblers (11 species of *Cettia*) and prinias (25 species of *Prinia*) have only 10 tail feathers and grasshopper warblers (nine species of *Locustella*) have graduated tails. Leaf warblers (41 species of *Phylloscopus*) have well-defined *supercilia. The six species of *Hippolais* are yellow warblers which have distinctive, peaked heads, square-ended tails, and loud, imitative songs. Warblers are mainly arboreal and insectivorous, and nest in trees, bushes, reeds, and on the ground (prinias build woven nests). There are about 50 genera, with about 320 species, many migratory, found in Europe, Africa, Asia, Australasia, and Pacific islands.

symbiont Symbiotic organism. *See also* SYMBIOSIS.

symbiosis General term describing the situation in which dissimilar organisms live together in close association. As originally defined, the term embraces all types of mutualistic and parasitic relationships. In modern use it is often restricted to mutually beneficial species interactions, i.e. *mutualism. *Compare* COMMENSALISM; PARASITISM.

Symbranchii An alternative, older name for the group of fish placed in the order *Synbranchiformes.

Symmetrodonta (subclass *Theria, infraclass *Pantotheria) An extinct order of mammals which lived during the Late *Jurassic and Early *Cretaceous and which may have appeared toward the end of the *Triassic, making them among the earliest mammals known. They had *molar teeth with three *cusps arranged in symmetrical triangles. They were very small and probably were predators. They are believed to be ancestral to the *Metatheria and *Eutheria.

symmetry *See* BILATERAL SYMMETRY; BIRADIAL SYMMETRY; PENTAMERAL SYMMETRY.

sympathetic nervous system That part of the *autonomic nervous system which generally acts to stimulate the body to cope with stress, such as increasing the rate of heart beat. Nerves of the sympathetic nervous system tend to form ganglia (*ganglion) beside the *vertebrae, preganglionic fibres leading from the spinal cord to the ganglia. The nerve endings are *adrenergic in postganglion synapses and *cholinergic in preganglion synapses. The sympathetic nervous system acts antagonistically to the *parasympathetic nervous system.

sympatric Applied to species that occupy similar *habitats or whose habitats invariably overlap. *Compare* ALLOPATRIC.

sympatric evolution (sympatric speciation) The development of new taxa from the ancestral *taxon, within the same geographic range (i.e. it is geographically possible for interbreeding to occur between the potential new taxa, but for some reason this does not happen). Because of the difficulty of envisaging what the reasons might be, few authorities until recently accepted the reality of sympatric evolution, except for certain special kinds of organisms; but recent studies have shown that chromosomal *mutation can set up a partial barrier to interbreeding sufficient to permit speciation (*stasipatric speciation); and the restricted interbreeding between ecological races of a species could promote separation. *Centrifugal speciation appears to be so common

S

in some groups of animals (e.g. mammals) that some sympatric mechanism must be occurring.

((⊕)) SEE WEB LINKS
• Information about sympatric evolution.

sympatric speciation *See* SYMPATRIC EVOLUTION.

sympatry The occurrence of *species or other taxa together in the same area. The differences between closely related species usually increase (diverge) when they occur together, in a process called character displacement, which may be morphological or ecological. *Compare* ALLOPATRY.

Symphyla (phylum *Arthropoda, subphylum *Atelocerata) A class of small (2–10 mm long), fast-moving arthropods that live in soil and superficially resemble centipedes although their mouth-parts are similar to those of insects. Most feed on decayed vegetation, but some are crop pests. There are about 120 species, found wherever temperatures never fall below −26°C.

Symphypleona *See* COLLEMBOLA.

symphysis A joint between bones, which consists of a disc of *fibrocartilage allowing limited movement.

Symphysodon (discus fish) *See* CICHLIDAE.

Symphyta (sawflies, woodwasps; subclass *Pterygota, order *Hymenoptera) The smaller suborder of the Hymenoptera, whose members are distinguished from the *Apocrita by the lack of a constriction at the base of the abdomen. The *ovipositor is large and used for piercing and sawing plant tissues. Adults have a fore wing with one to three marginal cells and a hind wing with three basal cells. In adult sawflies, the *trochanters are two-segmented. The larvae resemble the caterpillars of the *Lepidoptera, but differ in having up to seven pairs of abdominal *prolegs. The larvae are mostly herbivorous, except for members of the parasitic family

*Orussidae. Symphytans occur on flowers and in association with their host plants, some causing defoliation.

symplesiomorph Applied to a character state that is *primitive (*plesiomorphic) and shared between two or more taxa. Shared possession of a symplesiomorph character state is not evidence that the taxa in question are related.

symport *Cotransport in which two substances cross the *cell membrane both moving in the same direction.

Synallaxis (spinetails) *See* FURNARIIDAE.

Synanceiidae (stone fish; subclass *Actinopterygii, order *Scorpaeniformes) A small family of marine fish that have a relatively short, thickset body, and smooth skin covered with many wart-like projections. The large mouth is orientated obliquely and the small eyes are located near the top of the head. Using the large, paddle-shaped, *pectoral fins, the stone fish can partially bury itself and await the arrival of unsuspecting prey. The strong, *dorsal fin spines bear venom glands at their bases. (The sting of *Synanceja trachynis* (46 cm) has proved fatal to persons treading on the spines.) There are about 20 species, in the Indo-Pacific region.

synanthrope An animal that benefits from environmental modifications made by humans to such an extent that it becomes closely associated with humans. Examples include the feral pigeon (*Columba livia*), house sparrow (*Passer domesticus*), and house mouse (*Mus musculus*).

Synaphobranchidae (cut-throat eel; superorder *Elopomorpha, order *Anguilliformes) A small family of marine fish that have an eel-like body covered with scales, a large mouth, and well-developed *pectoral fins, the *dorsal and *anal fins being confluent with the tail fin. The gill openings are placed close together on the underside of the throat. They tend to be deep-sea fish. There are about 11 species, distributed world-wide.

S

synapomorphic Applied to *apomorphic features that are possessed by two or more taxa in common (i.e. the features are shared, derived). If the two groups share a character state that is not the *primitive one, it is plausible that they are related in an evolutionary sense, and only synapomorphic character states can be used as evidence that taxa are related. Phylogenetic trees are built up by discovering groups united by synapomorphies.

synapse The junction between two *neurons; the passage of an impulse along an *axon causes the secretion of a *neurotransmitter substance (e.g. *acetylcholine or *noradrenaline) into a space or cleft between the membrane of the axon and that of a *dendrite or cell body of an adjacent neuron. The neurotransmitter diffuses across the cleft where it depolarizes the membrane of the latter, thus initiating within it a new impulse; subsequently it is destroyed by an *enzyme.

synapsid A member of the subclass *Synapsida.

Synapsida (mammal-like reptiles; subphylum *Vertebrata, class *Reptilia) A subclass of reptiles which includes the *Pelycosauria and *Therapsida. The pelycosaurs appeared in the upper *Carboniferous and disappeared in Early *Permian times, displaced by the therapsids to which they had given rise. The therapsids flourished in the latter part of the Permian and in the *Triassic, but dwindled to extinction in the Early *Jurassic. The therapsids are the ancestors of the mammals, and share in common with them a synapsid skull (i.e. a skull with one temporal opening above which the postorbital and squamosal bones meet). *See also* ICHTHYOPTERYGIA.

synapsis The side-by-side pairing of *homologous chromosomes during the zygotene stage of meiotic *prophase.

synaptic accommodation *See* ACCOMMODATION.

synaptic cleft The very narrow gap that lies between two nerve cells at the *synapse and which a nerve impulse must cross (from the presynaptic to postsynaptic side) in order to stimulate the next nerve cell.

synapticulae Small rods or bars which connect opposed faces of adjacent *septa in corals. They perforate the soft *mesenteries between the septa.

Synaptosauria *See* EURYAPSIDA.

synarthrial In the skeleton of an invertebrate, applied to articulations that have a ridge on the fulcrum.

Synbranchidae (swamp eel; subclass *Actinopterygii, order *Synbranchiformes) A small family of freshwater and brackish-water fish that have a very elongate, eel-like body, small eyes, and little-developed *dorsal and *anal fins confluent with the tail fin. Because of the modifications of the gill apparatus, some species can utilize atmospheric oxygen when the swamp water is depleted of it. Swamp eels are considered valuable food fish in Asia. There are about eight species, distributed in Asia, Africa, and S. America.

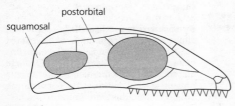

Synapsida

Synbranchiformes (class *Osteich-thyes, subclass *Actinopterygii) An order of mostly marine fish that have an eel-like body, a single gill slit, and the *dorsal and *anal fins confluent with the tail fin. The *pectoral fins are always absent.

syncitium A cell or organism that has many nuclei which are not separated by cell membranes. This condition is caused by the repeated division of the *nucleus, but not of the *cytoplasm, of the original cell.

Syndactyliformes (subclass *Theria, infraclass *Metatheria) An order (or sub-order) of marsupials that contains only the *Notoryctidae.

syndactylous In *Marsupialia, applied to the condition in which the second and third hind digits are united within a common sheath. In birds, applied to feet that have two of the front toes partly joined. The condition is found in *Alcedinidae, *Bucerotidae, *Coraciidae, and *Rupicola* species (cocks-of-the-rock, *see* COTINGIDAE).

synecology The study of whole plant and animal communities, including the study of terrestrial *ecosystems, biological aspects of oceanography, and applied problems of human management and alteration of ecosystems. *Compare* AUTECOLOGY.

Synentognathi An alternative, older name for the flying fish and its relatives, now placed in the order *Atheriniformes.

synergism The result of combined factors, each of which influences a process in the same direction but which, when combined, give a greater effect than they would acting separately.

syngamy The union of nuclei of two *gametes to produce a *zygote nucleus, following fertilization.

syngeneic *See* ISOGENEIC.

Syngnathidae (sea-horse, pipefish; subclass *Actinopterygii, order *Syngnathiformes) A large family of mainly marine fish in which the body is encased in a series of bony rings or plates. They have a tube-like snout with a minute mouth at the tip, a single *dorsal fin, and no *pelvic fins. Such species as *Nerophis ophidion* (straight-nosed pipefish), 60 cm, and *Hippocampus hippocampus* (sea-horse), 15 cm, of the eastern Atlantic, both of which are usually found among seaweeds, are known to keep their young temporarily in a brood pouch developed by the male. A few, e.g. *Phycodurus eques* (leafy sea dragon), 30 cm, of southern Australia, form elaborate leaf-like appendages, making it very hard to see the fish when they are hidden between branches of seaweed. There are about 175 species, distributed world-wide.

Syngnathiformes (class *Osteichthyes, subclass *Actinopterygii) An order of somewhat peculiar fish that have a small, toothless mouth at the end of a tube-like snout. The body is at least partly protected by a number of bony plates or rings, and the fins are generally small. The order includes about six families, e.g. *Aulostomidae (trumpetfish) and *Syngnathidae (pipefish). Most are marine but there are a few freshwater species.

Synlestidae (Chlorolestidae; damselflies; order *Odonata, suborder *Zygoptera) Family of damselflies which breed in running water. They are distinguished by the arching forward of the posterior cubitus vein as it leaves the distal end of the quadrilateral in the wing. These damselflies often have an exceptionally long *abdomen, and are metallic in colour. The family is mainly tropical, and 33 extant species have been described.

Synodidae An alternative name for the fish family *Mochokidae (upside-down catfish).

Synodontidae (lizardfish, grinner; superorder *Scopelomorpha, order *Myctophiformes) A family of marine fish that have a slightly depressed head, a large mouth with many fine teeth, and a single *dorsal and an *adipose fin opposite the *anal fin. Lizardfish are found in shallow water, sometimes burying themselves in the sand. There are about 34 species, distributed world-wide.

synonym In taxonomy, a different name for the same species or variety of organism.

synovial fluid A clear fluid, resembling egg white, that lubricates joints, *tendon sheaths, and acts as a *bursa. It consists mainly of *mucin with some *albumin and it has the property (thixotropy) of changing from a liquid to a gel under shearing stress and back to a liquid when the stress ceases.

synovial joint (diarthrosis) In vertebrates, a joint that is capable of a range of movements. *Compare* SYMPHYSIS.

synovial membrane (synovium) A layer of soft tissue that occurs in *synovial joints between the *articular capsule and the joint cavity that contains the *synovial fluid.

synovium *See* SYNOVIAL MEMBRANE.

synsacrum In birds, a structure in the pelvic *girdle formed by the fusion of certain vertebrae.

Synthemidae (Synthemistidae; dragonflies; order *Odonata, suborder *Anisoptera) Family of dragonflies which are superficially similar to the *Gomphidae. The nymphs are bottom-dwellers, and resist droughts by burying themselves very deeply. The insects frequent marshy areas, as well as fast-flowing streams. The family occurs throughout the southwestern Pacific, and 31 extant species have been described.

Synthemistidae *See* SYNTHEMIDAE.

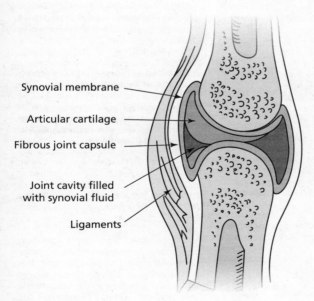

Synovial membrane —

Articular cartilage —

Fibrous joint capsule —

Joint cavity filled with synovial fluid —

Ligaments —

Synovial joint

synthetic theory The modern theory of evolution, incorporating Darwinian thinking, Mendelian genetics, and an understanding of genes and genetic change at the molecular level. *Compare* MODERN SYNTHESIS.

syntype In taxonomy, all specimens in a type-series in which no *type specimen was designated.

synzoochory *See* ZOOCHORY.

syrinx The vocal apparatus of a bird.

Syrphidae (hoverflies, sunflies, flower flies, drone flies, sweat flies; order *Diptera, suborder *Cyclorrapha) One of the largest dipteran families, whose members are quite distinctive. They are moderate to large, often brightly coloured flies, in which the outer edge of the wing has an inner margin formed by connections between the outer wing veins. The antenna has a dorsal *arista in almost all species. Adults may be striped, spotted, or banded with yellow and black, giving them a passing resemblance to wasps. Others are hairy and dark-coloured, and in many respects resemble bees. The name 'hoverfly' refers to their ability to hover in midair over a flower or mate. The larvae fall into three broad groups, and may be: (*a*) phytophagous, often feeding on roots, or inside plants; (*b*) carnivorous, perhaps best known as voracious feeders on aphids, for which they are of great economic value, but some species also eat other insect larvae; or (*c*) scavengers and detritus feeders, in mud, hymenopteran nests, or in decaying plant material. More than 5000 species have been described.

Syrrhaptes (sandgrouse) *See* PTEROCLIDAE.

Syrrhaptes paradoxus (Pallas's sandgrouse) *See* PTEROCLIDAE.

system A distinct entity that consists of a number of interacting parts such that the removal or failure of one part may incapacitate the entity as a whole.

systematic sample *See* REGULAR SAMPLE.

systemic arch In the *embryo of a tetrapod, the fourth *aortic arch which, in the adult, becomes the main source of the blood supply to all parts of the body other than the head. In amphibians and reptiles both right and left arches are present in adults. In birds only the right arch is present in adults, and in mammals only the left arch is present in adults.

systole 1. The phase of the heartbeat in which the heart contracts. **2.** The contraction of a *contractile vacuole.

syzygy In certain types of *Protozoa, an association between potential *gametocytes, prior to gamete formation, in which fusion of nuclei does not occur.

s

Tabanidae (horse-flies, gadflies, stouts, clegs, deerfly, breezes, greenhead; order *Diptera, suborder *Brachycera) Large family of flies, which are often large, stout, and lack bristles. The third antennal segment is annulated, but without a style. The eyes are very large in relation to the size of the fly, and are extended laterally. The bright, colourful, often iridescent eyes are perhaps the most obvious feature of the adult. The mouth-parts have a projecting proboscis which in the female is adapted for piercing. *Squamae are large. Adults are active on warm summer days and are rapid and often noisy fliers. They are particularly troublesome to cattle and horses, and may be responsible for the transmission of the disease surra of horses. Eggs are laid in compact masses near damp soil, mud, or other substrate suitable for larval growth. The larvae are carnivorous, feeding upon other insects. The family is distributed worldwide, and about 2000 species have been described.

Tabasco turtle *See* DERMATEMYDIDAE.

tabula In a *corallite, a horizontal plate that extends across the interior.

Tachinidae (parasitic flies; order *Diptera, suborder *Cyclorrapha) Large family of true flies whose larvae are parasites of insects and other arthropods. The adults are distinguished by a strong fan of *hypopleural bristles, and a well-developed postscutellum. The second abdominal *sternite is visible as side plates lying above the *tergite. The larvae are unlike other cyclorraphan larvae in being uniformly cylindrical, rather than tapered. There are many larval habits and methods for attacking hosts: (*a*) eggs are laid on the surface of the host; (*b*) eggs

are laid in large numbers on potential food plants for the host, and are eaten; (*c*) eggs are laid near the habitat of the host, and hatch into active larvae which seek out a host; and (*d*) larviparous females puncture the skin of the host and deposit an egg which hatches immediately. The usual hosts are larval or adult *Coleoptera, *Orthoptera, and *Hemiptera, occasionally *Hymenoptera, but most frequently *Lepidoptera. The larvae breathe either by maintaining a connection with the air outside the host, or by tapping the air supply of the host at a tracheal trunk. More than 9000 species have been described, and their distribution is worldwide.

Tachyglossidae (echidna, spiny ant-eater; subclass *Prototheria, order *Monotremata) A family of insectivorous monotremes, now recognized as being more closely related to the placental mammals (*see* EUTHERIA) than to the platypus (*see* ORNITHORHYNCHIDAE), in which the limbs are modified for digging, the body is covered with hair modified except on the belly into barbless spines, the snout is long, and the tongue is long, sticky, and has horny serrations that grind against ridges in the palate. There are no teeth, and the lower jaw is much reduced. Echidnas occur only in Australia and New Guinea. There are two species, in two genera, *Tachyglossus* and *Zaglossus*; a third genus, *Megaloglossus*, is known from the *Pleistocene.

***Tachyglossus* (echidna, spiny ant-eater)** *See* TACHYGLOSSIDAE.

Tachyoryctes *See* RHIZOMYIDAE.

Tachysuridae *See* ARIIDAE.

tachytely A rate of evolution within a group that is much faster than the average (*horotelic) rate. Such accelerated evolution typically occurs when an organism enters a new *adaptive zone and initiates an *adaptive radiation to fill the available *niches. *Compare* BRADYTELY.

tactile Pertaining to the sense of touch.

Tadorna (shelducks) *See* ANATIDAE.

tadpole The round-bodied, long-tailed larva of an anuran (*Anura). The term is also used for other larvae with this form (e.g. those of sea squirts and salps).

tadpole sculpin *See* PSYCHROLUTIDAE.

tadpole shrimp *See* NOTOSTRACA.

Taeniodontia A small mammalian order, from the *Palaeocene and *Eocene of N. America, that were derived from insectivore stock. Some forms resembled heavy dogs with rodent-like gnawing teeth.

Taenioidea *See* CYCLOPHYLLIDEA.

tagma (*pl.* tagmata) In *Arthropoda, one of the sections into which the body is divided by differences in size, shape, or function.

tagmata *See* TAGMA.

tagmosis In metamerically segmented animals (*metameric segmentation), functional specialization that leads to differentiation among the segments and the formation of tagmata (*see* TAGMA).

tailed frogs *See* ASCAPHIDAE.

tail fin *See* CAUDAL FIN.

tail-less whipscorpion *See* AMBLYPYGI.

tailor *See* POMATOMIDAE.

taipans *See* ELAPIDAE.

tallow beetle *See* DERMESTIDAE.

Talpidae (moles and desmans; suborder *Lipotyphla, superfamily *Soricoidea) A family of insectivores that are adapted for burrowing. The eyes are small and in some species covered by skin, and external ears are usually absent. The fore limbs are modified (in the moles for digging, in the desmans for swimming), and the digits form shovels or paddles and can be rotated so that they project at right angles from the body. The upper *molar teeth have a W-shaped pattern. The tail is short and the thick, short fur can lie in any direction. The Talpidae are distributed throughout temperate Europe and Asia, except Ireland and Norway, in south-east Asia, and in eastern and western N. America. There are 12 genera, with about 30 species.

talus Astragalus; ankle-bone. *See* TIBIALE.

tamarin (*Saguinus*) *See* CALLITRICHIDAE.

Tamias (chipmunk) *See* SCIURIDAE.

tanagers 1. *See* THRAUPIDAE. **2. (swallow-tanager,** *Tersina viridis*) *See* TERSINIDAE.

tandem array Repetitive DNA (*deoxyribonucleic acid), where the repeating units are contiguous. Some genes of high copy number occur in tandem arrays (e.g. ribosomal DNA in *eukaryotes). *Satellite DNA is also often arranged in tandem arrays. Such DNA regions are prone to genetic processes of *unequal crossing-over and *replication slippage.

tandemly repeated sequences A chromosomal duplication in which the duplicated segments are adjacent (i.e. in tandem) whether or not they are inverted.

Tangara (tanagers) *See* THRAUPIDAE.

tapaculos *See* RHINOCRYPTIDAE.

tapetum In many nocturnal mammals, a layer of cells either in the *retina or outside it, in the *choroid (e.g. in cats), that contain crystals of zinc and a protein (often *riboflavin, which fluoresces). The tapetum reflects light back through the retina and so increases the sensitivity of the eye to dim light. At night, when the pupil is fully dilated, reflected light from the tapetum will cause 'night shine' if the animal turns to look in the direction of a sudden bright light (e.g. car headlights).

tapeworms See CESTODA.

tapir (*Tapirus*) See TAPIRIDAE.

Tapiridae (tapirs; suborder *Ceratomorpha, superfamily *Tapiroidea) A family of solitary, nocturnal ceratomorphs which are little different from the perissodactyl stock that lived during the late *Eocene and *Oligocene. Many tapir-like forms are known from the Eocene, and sometimes are grouped as *Lophiodontidae. The limbs are short, with three digits on the hind feet and four on the fore feet. The digits bear small hoofs, but the soles of the feet also have thick pads. The *ulna and *fibula are complete and unfused. The teeth are complete in number, low-crowned, and lack *cement except for the first three *premolars. The nose has developed into a short, mobile trunk, with an associated shortening of the nasal bones. Tapirs live mainly in dense forest, feeding on fruit and vegetation, in tropical Central and S. America, and in south-east Asia. There are four species, in two genera, *Tapirus* (American) and *Acrocodia* (Asian).

Tapiroidea (order *Perissodactyla, suborder *Ceratomorpha) A superfamily that comprises the family *Tapiridae together with a number of ceratomorphs known from the *Eocene and *Oligocene and arranged in several small families. Many of these families constituted evolutionary branches from the lines leading to the modern ceratomorphs, but some were ancestral to *Protapirus*, a true tapir which lived in the Oligocene; and *Hyrachyus* was one of several Eocene forms that may be ancestral to rhinoceroses.

Tapirus (American tapir) See TAPIRIDAE.

tarantula See MYGALOMORPHAE.

Tarbosaurus bataar A carnivorous dinosaur, very similar in size and form to *Tyrannosaurus*, that was one of the largest meat-eating animals ever to have lived on Earth. Individuals from the Late *Cretaceous of Mongolia reached almost 12 m in length and weighed approximately 7 tonnes.

(((⊕))) SEE WEB LINKS
• Detailed description of *Tarbosaurus bataar*.

Tardigrada (water bears) A phylum of very small (most less than 0.5 mm long), specialized animals, most of which live in the films of water that surround the leaves of terrestrial mosses and lichens; some are found between sand grains in marine and freshwater deposits. The body is short, plump, and cylindrical, with four pairs of ventral legs that end in claws or toes. The body is *coelomate and superficially segmented, with a non-chitinous *cuticle. There are 400 species.

tarpon See MEGALOPIDAE.

tarsal bone In tetrapod vertebrates, one of the distal bones of the hind foot that articulate with the digits (metatarsals).

tarsal claw Claw carried on the final tarsal segment of an insect. There may be more than one claw on each tarsus.

tarsiers (*Tarsius*) See TARSIIDAE.

Tarsiidae (tarsiers; suborder *Haplorrhini, infra-order *Tarsiiformes) A family of arboreal, nocturnal, insectivorous or carnivorous primates, about the size of rats, in which the upper lip is whole, there is no moist *rhinarium, and the hind legs are very long and used in leaping from tree to tree. The *tibia and *fibula are fused and the *calcaneum and *astralagus elongated, but the feet retain their grasping digits. The *hallux and *pollex are opposable, all digits end in adhesive pads, and apart from the second and third hind digits, which have claws, all digits bear flattened nails. The tail is long. The snout is shortened, the face is sufficiently mobile for it to be used to express emotion, the eyes face forward and are extremely large, and the head is mounted on a neck so mobile that the animal can face directly backwards. The ears are large and mobile, and the sense of hearing is acute. The *placenta resembles that of the higher primates, but the uterus that of lemurs. Tarsiers occur only in parts of Indonesia, eastern Malaysia, and the Philippines. There are at least four species, probably more, in a single genus, *Tarsius*.

Tarsiiformes (order *Primates, suborder *Haplorrhini) An infra-order, regarded by some authorities as a suborder (Tarsii or Tarsioidea), comprising the families Omomyidae (extinct tarsier-like primates which lived during the *Eocene and *Oligocene, some surviving into the *Miocene in northern continents) and *Tarsiidae (modern tarsiers and their immediate ancestors). In some classifications the tarsioids are included as a group of a larger suborder, the Haplorrhini, which includes also the *Anthropoidea.

Tarsioidea *See* TARSIIFORMES.

Tarsius (tarsier) *See* TARSIIDAE.

tarsometatarsus A bone formed by the fusion of the *metatarsals with the *tarsal bones of the *appendicular skeleton, and found in birds (*Aves) and some dinosaurs. The beginnings of this fusion can be seen in some *Theropoda, but it is not complete until the Early *Cretaceous (*Sinornis). *See also* TIBIOTARSUS.

tarsus 1. In an insect, one of a number of small segments distal to the *tibia, with which the uppermost tarsus articulates. The tarsus is made up of two to five segments and terminates in the pretarsus. The muscles operating the tarsi are located in the tibia, and operate through the agency of tendons. **2.** In birds, the lower long bone of the leg. **3.** In *Mammalia, the collection of bones forming the ankle. **4.** In vertebrates, a plate of *connective tissue in the eyelid.

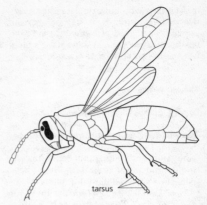

Tarsus

Tasmanian devil (*Sarcophilus harrisi*) *See* DASYURIDAE.

Tasmanian 'wolf' (*Thylacinus cynocephalus*) *See* THYLACINIDAE.

Tauraco (turacos) *See* MUSOPHAGIDAE.

Taurotragus oryx (eland) *See* BOVIDAE.

tautonym A taxonomic name in which the names for genus and species are the same.

taxa *See* TAXON.

Taxidea (badger) *See* MUSTELIDAE.

taxis The movement of an animal towards or away from a source of stimulation in response to the intensity and direction of the stimulus.

taxodont Applied to a primitive type of hinge dentition, present in certain *Bivalvia, in which teeth and sockets are small and numerous. They are arranged in a row on each side of the *shell beak, on both valves.

taxon (*pl.* taxa) A group of organisms of any taxonomic rank (e.g. family, genus, or species).

taxonomy The scientific classification of organisms.

Tayassuidae (peccaries; order *Artiodactyla, suborder *Suiformes) A family of omnivorous animals which are related to pigs but evolutionarily distinct from them for much of the *Cenozoic. The oldest fossil peccary comes from the *Oligocene of N. America. The upper canines form sharp tusks which grow downward, the *molars are simpler than those of pigs, and the *radius and *ulna are fused. There is a scent gland at the centre of the back. Probably peccaries lived in both Old and New Worlds until the *Miocene; one survived until the *Pliocene in S. Africa. During the *Pleistocene they were distributed widely in temperate N. America. They reached S. America during the Pliocene and today they occur throughout most of S. America and as far north as Texas. There are three species, in two genera: *Tayassu*, containing *T. tajacu* (the collared peccary) and *T. pecari* (the white-lipped peccary), and *Catagonus*, containing one living species, *C. wagneri* (the Chacoan peccary), known since 1930 as a fossil but in 1972 discovered to be still living in the Chaco, on the Paraguay–Argentina–Bolivia border.

Tayassu pecari (white-lipped peccary) *See* TAYASSUIDAE.

Tayassu tajacu (collared peccary) *See* TAYASSUIDAE.

T cell A type of *lymphocyte manufactured in the *thymus gland (hence the name) that is involved in the *cell-mediated response. There are several classes of T cell including: helper T cells, which aid *B cells in antibody secretion; suppressor T cells, which reduce the immunological responses of B and T cells; cytotoxic T cells, which attack virally infected or cancerous cells; and delayed hypersensitivity T cells, which participate in the delayed hypersensitivity response.

Tchagra (bush-shrikes) *See* LANIIDAE.

tectin A *chitin-like, organic substance from which the *tests of certain *Protozoa are formed.

tectrices *See* WING COVERTS.

tectrix *See* WING COVERTS.

tectum *See* MESENCEPHALON.

Tegenaria domestica (house spider) *See* AGELENIDAE.

tegmen 1. (*pl.* tegmina) The upper surface of the *theca, above and between the arms. **2.** The fore wing of an insect when this is tough and leathery.

tegmentum *See* ARTICULAMENTUM.

tegmina *See* TEGMEN.

tegu (*Tupinambis teguixin*) *See* TEIIDAE.

tegula (*pl.* tegulae) Small cuticular plate at the base of the wings of an insect, which is derived from the edge of the articular membrane.

tegulae *See* TEGULA.

Teiidae (tegus, whiptails; order *Squamata, suborder *Sauria) A family of lizards which differ from the *Lacertidae in having head plates that are separate from the skull bones. The tongue is long and forked, the tail long, and the limbs may be large and powerful, or reduced (e.g. in snake and worm teiids). *Tupinambis teguixin* (common tegu or great tegu), occurring from Guiana to Uruguay, black with bands of yellow spots, is a large (up to 1.4 m long), long-tailed, squat-bodied, ground-dwelling lizard that shelters in burrows. There are more than 200 species. They are the New World counterparts of the Lacertidae.

Telacanthura *See* APODIDAE.

telencephalon The anterior region of the *brain, part of the *prosencephalon, which includes both the *olfactory lobes and *cerebrum. In *plesiomorphic *tetrapods, the telencephalon constitutes a small portion of the total brain mass, while in more developed tetrapods the telencephalon becomes increasingly large with the development of the *neopallium within the highly corticized *cerebral hemispheres.

teleology The belief that observed phenomena may be explained in terms of a predetermined purpose or design they appear to serve.

teleonomy The hypothesis that adaptations arise without the existence of a prior purpose, but by chance may change the fitness of an organism. *Compare* TELEOLOGY.

Teleostei (teleost, teleostean) A somewhat loosely defined term (an infraclass according to some authors) that includes all the bony fish (*Osteichthyes) with the exception of a few orders of primitive fish (*Acipenseriformes, *Amiiformes, *Semionotiformes, and their fossil relatives).

Teleostomi A subgrade that comprises the 'true' fish (*Osteichthyes), i.e. those with a terminal mouth, in contrast to the shark-like fish (*Chondrichthyes) which have the mouth positioned beneath a substantial rostrum. In addition to the Osteichthyes, some authors also include the class of fossil *Acanthodii in the Teleostomi. *See also* BONY FISH.

Telestacea (class *Anthozoa, subclass *Octocorallia) An extant order of corals that have simple or branched stems arising from a creeping *stolon base, elongate *polyps produced from the body wall of the primary polyp, and numerous *sclerites that may be free or varyingly fused together. Telestaceans are found from the low-tide line to depths of 765 m.

telocentric chromosome A *chromosome the *centromere of which is at one end.

telolecithal An egg in which the yolk occurs at one end of the cytoplasm, giving rise to a vegetal pole. Mesolecithal and macrolecithal eggs are also telolecithal. *Compare* ISOLECITHAL.

telophase The fourth and final phase of *mitosis and *meiosis, during which the *spindle disappears, *nucleoli reappear, the nuclear membranes start to develop around the two groups of daughter chromosomes/chromatids, and the *chromosomes return to their extended state in which they are no longer visible. The *nuclei then enter an *interphase, the stage they were in before division occurred.

Telosporea (phylum *Protozoa, subphylum *Sporozoa) A class of parasitic protozoa, most of which live inside the cells of their hosts. They may be motile by flexing the cell body or by gliding. Some can complete their life cycle in a single host; others require more than one host species.

telo-taxis The movement of an animal in response to a stimulus, directly towards or away from the source of stimulation, guided by the use of one sense organ which, by virtue of its structure, provides information regarding the direction of the stimulus (e.g. the eye provides information regarding the direction of a light source).

telson In *Arthropoda, the final abdominal segment, present only in the embryo of insects.

Temnocephala *See* TEMNOCEPHALIDA.

Temnocephalida (phylum *Platyhelminthes) A class, or in some classifications an order, Temnocephala (class *Turbellaria, subclass *Neoophora), of platyhelminth worms which have a *pharynx and a bag-shaped intestine. The epidermis may be ciliated. They have a non-parasitic mode of life, living on the exterior of freshwater animals.

Temnopleuroida (class *Echinoidea, superorder *Echinacea) An order of regular echinoids that have a *test sculptured with ridges and/or depressions along the plate sutures, or very deep and conspicuous gill slits. They first appeared in the Early *Jurassic.

Temnospondyla (class *Amphibia) A subclass of *Carboniferous to Early *Jurassic amphibians which form a link between the primitive *Labyrinthodontia and the modern *Anura and *Urodela.

temperature-sensitive organ (pit organ) An apparatus, developed in some snakes, that can detect heat radiated from prey and air motion. Paired organs lie in front of the eye in pit vipers (*Crotalidae). The lip pits of some boids (*Boidae) perform the same function.

temporal fossa In the skull of mammals, a cavity behind the *orbit, containing muscles involved in the raising of the lower jaw.

temporal summation *See* SUMMATION.

tench (*Tinca tinca*) *See* CYPRINIDAE.

tendency The degree of behavioural *motivation that may be inferred from the observed behaviour of an animal.

tendon (sinew) A band of very strong *connective tissue that connects a *muscle to a *bone.

Tenebrio molitor (mealworm) *See* TENEBRIONIDAE.

Tenebrionidae (darkling beetles, pincate beetles, nocturnal ground beetles; subclass *Pterygota, order *Coleoptera) Family of beetles which occur as adults in a great variety of sizes and shapes. Many are flightless, with vestigial wings, and fused *elytra, which are usually brown to black. They are adapted to living in very dry conditions; some are highly specialized desert species. A number contain blistering chemicals which they spray on attackers. Larvae are long and cylindrical, the body covered with toughened plates. Several are pests of stored products (e.g. the larvae of *Tenebrio molitor* (mealworms) which feed on flour). There are 15 000 species.

tenrec *See* TENRECIDAE.

Tenrecidae (otter shrew, tenrec; suborder *Lipotyphla, superfamily *Soricoidea) A family of insectivores in which a *cloaca is present, and in which the cusps of the *molar teeth form a triangular V shape, except in the Potamogalinae, where they form a W shape. The Tenrecidae have undergone adaptive radiation to produce forms resembling shrews, moles, and hedgehogs, and *Potamogale* (otter shrew) is an aquatic form closely resembling true otters. Some are larger than most insectivores, one of the tenrecs attaining a length of 50 cm and *P. velox* (giant otter shrew) 60 cm. The Tenrecidae occur only in W. and Central Africa (Potamogalinae) and Madagascar (Tenrecinae). There are 11 genera, with 33 species.

tentacle 1. In many invertebrate animals, a long, slender, flexible structure, often bearing sense receptors, used to obtain information about the immediate environment and often to obtain food. 2. In corals and sea anemones, a movable, tubular extension of the body cavity; tentacles are arranged in a ring around the mouth. 3. In *Cephalopoda, a movable modification of the soft body; tentacles surround the mouth and are sometimes studded with sucker discs.

Tentaculata (Micropharyngea; phylum *Ctenophora) A class of ctenophorids which have *tentacles. There are four orders.

tentaculocyst In some coelenterates, a sensory organ concerned with orientation, consisting of a minute *tentacle in a pit, the pressure of the tentacle on the sides of the pit stimulating sensors.

Tenthredinidae (sawflies; suborder *Symphyta, superfamily *Tenthredinoidea) One of the largest families of symphytans, most of which are small, varying in length from 5 mm to about 20 mm. Adults have nine-segmented, thread-like antennae, and are black, brown, or brightly patterned, and can be found on flowers or other vegetation. The larvae of most species are external feeders, but some are leaf-miners or gall-makers. Eggs are inserted by the saw-like *ovipositor into leaf tissue or woody twigs, some species causing considerable damage to cultivated plants and forest trees. The well-developed abdominal *prolegs common in the symphytans are reduced or absent in leaf-mining or stem-boring species. Pupation generally takes place in the soil within a silken cocoon.

Tenthredinoidea (sawflies; order *Hymenoptera, suborder *Symphyta) Superfamily of insects, most of which belong to the family *Tenthredinidae, whose common name refers to the saw-like *ovipositors of the females. They feed on most cultivated plants of economic importance, with species specific for apple, pear, rose, pine, larch, gooseberry, turnip, etc. Some are gall-formers and stem-borers.

tER *See* TRANSITIONAL ER.

teratogenic Applied to any substance that interferes with the normal development of a foetus or embryo.

Teratornis incredibilis A condor-like vulture from the *Pleistocene Epoch of San Diego Co., California, which was the largest known bird ever to have lived on Earth. Individuals with a wing-span of more than 4 m have been recovered from the Rancho La Brea tar pits of California.

Terebrantia *See* THYSANOPTERA.

Terebratellacea (order *Terebratulida, suborder *Terebratellidina) An extant superfamily of *Brachiopoda containing six families whose members are characterized by long, complex, or recurved calcareous loops. They first appeared in the Late *Triassic.

Terebratellidina (class *Articulata, order *Terebratulida) A suborder of *Brachiopoda whose members have a calcareous loop formed by an outgrowth from both crura (*see* CRUS) and the median *septum. They first appeared in the *Devonian. The suborder comprises three superfamilies, of which only one (*Terebratellacea) is extant.

Terebratulida (phylum *Brachiopoda, class *Articulata) An order of brachiopods that have punctate shells, rounded hinge-lines, functional *pedicle, *deltidial plates, and a *lophophore support usually consisting of a pair of *crura and a calcareous loop. The three suborders comprising the order first appeared in the Early *Devonian; one became extinct at the end of the *Permian, two are extant.

tergite In *Arthropoda, the dorsal region of a segment of the body, if this region is composed of sclerotized cuticle. If it is not of sclerotized cuticle the region is termed a *tergum.

tergum In *Arthropoda, a thickened plate on the dorsal side of a segment. *Compare* TERGITE.

terminal deletion *See* DELETION.

terminating stimulus An external or internal stimulus that causes a behaviour response to end because it is completed (e.g. the visual stimulus of a completed nest will terminate nest-building).

termitaria *See* ISOPTERA.

termite *See* ISOPTERA.

Termitidae (subclass *Pterygota, order *Isoptera) Diverse family of 'higher' termites which rely on symbiotic bacteria rather than *Protozoa (as in all other termite families, the 'lower' termites) for the digestion of cellulose. Members of the African subfamily Macrotermitinae have an additional food source: a basidiomycete fungus which is cultivated on faecal pellets deposited in fungus gardens. Most Termitidae inhabit soil nests. The complex architecture of the large mounds of the Macrotermitinae ensures ventilation and regulation of the microclimate of the nest. There are five families, with about 1500 species, occurring world-wide but mainly in the tropics.

termitophile Applied to a species that must spend part of its life closely associated with termites. It may scavenge or steal food, or may prey on the termites.

ternary fission A type of cell division that results in the formation of three daughter cells from a single parent cell.

ternate Compound, and divided into three more or less equal parts.

terns *See* LARIDAE.

terpene A hydrocarbon that is composed of two or more *isoprene units. Terpenes may be linear or cyclic molecules or combinations of both, and include important biological compounds such as *vitamins A, E, and K, carotenoids, phytol, gibberellic acid, natural rubber, and some lipids.

Terrapene carolina (Carolina box turtle, eastern box turtle) *See* EMYDIDAE.

terrapins *See* EMYDIDAE.

territoriality The establishment, demarcation, and defence of an area by animals, normally during mating ritual. Once *territory has been established the animals can exist without disturbance and with sufficient food for the offspring. Evidence shows that among territorial species individuals without a territory rarely breed.

territorial pheromone A *pheromone that an animal releases to mark the boundaries of its *territory. The pheromone may be present in urine (e.g. in dogs and cats), secreted from *glands in the skin (e.g. in the visible trails left by rats), or secreted by preen glands (*see* PREENING) in some seabirds.

territory The area occupied by an animal, or by a pair or group of animals, that it or they will defend against intruders.

Tersina viridis (swallow-tanager) *See* TERSINIDAE.

Tersinidae (swallow-tanager; class *Aves, order *Passeriformes) A monospecific family (*Tersina viridis*) of birds in which the male is turquoise-blue with a black face and throat, white under-parts, and flanks barred with black, and the female is green with yellow under-parts, and flanks barred with olive. The swallow-tanager has a short bill, long wings, and short legs. It inhabits forest, feeds on fruit and insects, and nests in holes in banks and trees. It is found in Central and S. America.

Tertiary The first sub-era of the *Cenozoic Era, which began about 65.5 Ma ago and lasted approximately 63.7 Ma. The Tertiary followed the *Cretaceous and comprises five epochs: the *Palaeocene, the *Eocene,

the *Oligocene, the *Miocene, and the *Pliocene. 'Tertiary' has been abandoned as a formal name, although it continues to be used informally. The period it covered is now assigned to the *Palaeogene and *Neogene periods.

tertiary consumer A carnivore that preys upon other carnivores; a member of the topmost *trophic level of a *food web.

tertiary structure The folding of the helical coil of a *polypeptide chain.

tessellated snake (*Natrix tessellata*) *See* COLUBRIDAE.

test A protective shell that covers the cells of certain *Protozoa and the bodies of certain invertebrate animals.

Testacida *See* ARCELLINIDA.

testate Having a shell or *test.

testis The male *gonad responsible for the production of *spermatozoa. Testes are made up of semeniferous tubules in which the male *gametes develop, nourished by Sertoli cells attached to the tubule walls. Interstitial cells between the tubules produce *testosterone. *Compare* OVARY.

testosterone A *steroid hormone, secreted by the testis, that in mammals and birds is the principal male sex hormone, responsible for the development of *secondary sexual characteristics. It is also found in females as an intermediate in the synthesis of *oestrogen.

Testudinidae (land tortoises; order *Chelonia, suborder *Cryptodira) A family whose members have a high-domed *carapace, usually pillar-like legs, and blunt, heavily scaled, clawed feet. They are primarily herbivorous. *Malacochersus tornieri* (pancake tortoise), an agile climber on high, rocky slopes in E. Africa, differs from other members of the family in having a flexible, flattened shell that allows it to squeeze into rock crevices. *Testudo* (or *Geochelone*) *elephantropus* (Galápagos giant tortoise) has a carapace up to 1.5 m long and weighs up to 270 kg; there were once 14 subspecies, differing in shell shape, but many have become extinct due to human activities. *T.* (or *Geochelone*) *gigantea* (Seychelles giant tortoise) is now common only on Aldabra; it is distinguished by the median cervical plate on the anterior margin of the carapace. *T. hermanni* (Greek tortoise or Hermann's tortoise) has a horny spur on the end of the tail. *T. graeca* (spur-thighed tortoise) has a horny tubercle on the rear of each thigh, but no spur on the tail. *Gopherus polyphemus* (gopher tortoise) of the southern USA excavates long tunnels to avoid extreme heat, emerging at dusk to feed; males have elongated anterior plates on the carapace, used to lever rival males on to their backs when competing for a mate. There are about 50 species in the family, distributed in varied habitats throughout warmer parts of the world except Australia.

Testudo elephantropus (Galápagos giant tortoise) *See* TESTUDINIDAE.

Testudo gigantea (Seychelles giant tortoise) *See* TESTUDINIDAE.

Testudo graeca (spur-thighed tortoise) *See* TESTUDINIDAE.

Testudo hermanni (Greek tortoise, Hermann's tortoise) *See* TESTUDINIDAE.

Tethyan realm The *faunal region based in the region of the *Tethys Sea. Characteristically *Jurassic to *Cretaceous in age, it comprised a warm-water, tropical to subtropical fauna and flora. The name may also be applied to warm-water fauna and flora of *Mesozoic age outside the area of Tethys, especially when used in contrast to the Boreal (northern) and Austral (southern) regions.

Tethys Sea The sea that more or less separated the two great *Mesozoic supercontinents of *Laurasia, in the north, and *Gondwana, in the south. That *land bridges between the two supercontinents existed for much of the Mesozoic is attested to by the cosmopolitan character of dinosaur faunas.

Tethytheria *See* PAENUNGULATA.

tetra *See* CHARACIDAE.

Tetractinomorpha (phylum *Porifera, class *Demospongiae) A subclass of sponges many of which have a radial body structure. Most lack organic fibres in the skeleton. Most forms are oviparous.

tetrad 1. Four homologous *chromatids that occur in a bundle in the first meiotic *prophase and *metaphase. **2.** The four *haploid cells resulting from a single *meiosis.

tetrad analysis The use of *tetrads to study the behaviour of *chromosomes and *genes in *crossing-over during *meiosis. Such analyses require organisms in which the products of meiosis are held together and so can be counted as units.

(((⊕))) SEE WEB LINKS
• Detailed explanation of the tetrad analysis technique.

Tetragonuridae (squaretail; subclass *Actinopterygii, order *Perciformes) A small family of marine fish that have a very elongate, slender body, two *dorsal fins, an *anal fin with one *spiny ray, and a square or truncate tail fin. They are probably *pelagic fish, feeding on jellyfish and other planktonic organisms. There are three species, distributed widely in tropical and subtropical seas.

Tetrahymena *See* TETRAHYMENINA.

Tetrahymenina (class *Ciliatea, order *Hymenostomatida) A suborder of ciliate *Protozoa in which the *cilia in the oral region are inconspicuous. There are three families, and many genera, found in freshwater habitats. Some are parasitic. Tetrahymena species are widely used in laboratory studies of protozoa.

tetrameric *See* DIMER.

Tetraodontidae (pufferfish, swellfish; subclass *Actinopterygii, order *Tetraodontiformes) A large family of mostly marine fish that have a rotund, scaleless body, the small *dorsal fin placed well back on the body opposite the small *anal fin, and a rounded tail fin. The mouth is small, but the teeth in each jaw are fused into a solid beak, with a median division in front. They can swallow a large quantity of water, thereby inflating the body, presumably in order to discourage a predator. Some species are quite large and relished as food by inhabitants of many Pacific islands, despite the fact that the entrails and gonads are highly poisonous to man (*see* TETRODOTOXIN). Pufferfish are also appreciated in Japan, where specially trained cooks prepare them for human consumption. There are at least 130 species, distributed world-wide.

Tetraodontiformes (Plectognathi; class *Osteichthyes, subclass *Actinopterygii) An order of mainly marine bony fish that have either robust or compressed bodies, a small mouth, restricted gill openings, and scales which are often modified into plates or spines. Among the eight families included in the order are the *Balistidae (triggerfish) and the *Tetraodontidae (pufferfish).

tetra-odotoxin *See* TETRODOTOXIN.

Tetraonidae (grouse; class *Aves, order *Galliformes) A family of medium to large gamebirds which are mainly black, grey, and brown in colour, some with areas of white, others moulting into an all-white plumage in winter. (In the three *Lagopus* species brown breeding plumage is moulted into a white winter plumage in all but *L. lagopus scoticus*, red grouse.) Grouse have short, slightly *decurved bills, and many have red *wattles above the eyes. Many males have erectile head feathers and inflatable *air sacs for display purposes. They have short, rounded wings, and some have elaborate tail feathers. They inhabit forest, open country, heather moorland, and tundra, feed on buds, leaves, and other vegetable matter, and nest on the ground. *Tetrao urogallus* is the capercaillie, reintroduced into Britain after it became extinct; *Lyrurus tetrix* is the black grouse. There are six genera in the family, with 16 species, found in Europe, Asia, and N. America.

Tetrao urogallus (capercaillie) *See* TETRAONIDAE.

Tetraphyllidea (phylum *Platyhelminthes, class *Cestoda) An order of parasitic worms that have a *scolex which may be multi-suckered. Their life cycle is little understood but most inhabit three hosts during their development.

tetraploidy *See* POLYPLOIDY.

Tetrapoda An informal grouping that includes the vertebrate animals which have four limbs; i.e. the *Amphibia, *Reptilia, *Aves, and *Mammalia.

tetrapodomorph A fossil animal that is *morphologically intermediate between fishes and tetrapods (*Tetrapoda).

Tetrarhynchidea *See* TRYPANORHYNCHA.

Tetrigidae (groundhoppers, grouse locusts, pygmy locusts; order *Orthoptera, suborder *Caelifera) Family of small, dull-coloured grasshoppers whose major characteristic is an elongated *pronotum, which may cover the wings (if wings are present), and extend beyond the tip of the abdomen. Tetrigids are usually found on the ground in moist areas, and feed on algae or organically rich mud. Many species swim readily. There are about 1000 species, most of which are tropical, although 50 species are Palaearctic.

tetro-allelic Applied to a *polyploid in which four different *alleles exist at a given gene *locus.

tetrodotoxin (tetra-odotoxin) A powerful poison which can cause general paralysis in humans after ingestion. It is found in the body, particularly the intestines, liver, and gonads, of *Tetraodontidae (pufferfish) and their relatives. Only after careful filleting can these fish be eaten.

(((⊕))) SEE WEB LINKS
• Description of the properties and effects of tetrodotoxin.

Tettigoniidae (bush katydids, cone-headed grasshoppers, katydids, listroceline grasshoppers, long-horned grasshoppers, meadow grasshoppers, pine-tree katydids, shield-backed grasshoppers; order *Orthoptera, suborder *Ensifera) Family of ensiferans in which auditory communication is well developed. The *ovipositor is sword-shaped. Most species are *phytophagous, but some are predatory on other insects. It is the largest family in the suborder, with about 5000 species.

Teuthoidea (squids; class *Cephalopoda, subclass *Coleoidea) An order of cephalopods which have an internal, elongate pro-*ostracum. The *phragmocone may be present, the *rostrum is not. Eight arms and two *tentacles are present. The order probably originated in the Early *Cretaceous from a belemnite (*Belemnitida) ancestor. The squids are most closely related to the cuttlefish, both groups having ten suckered arms around the mouth.

thalamencephalon *See* DIENCEPHALON.

thalamus In a vertebrate, part of the forebrain, lying above the *hypothalamus, that is concerned with the transmission of sensory information to the *cerebrum.

Thaliacea (salps; phylum *Chordata, subphylum *Urochordata) A class of tunicates in which two *siphons occur at opposite ends of the body, so the current of water passing through the animal is used for gas exchange, feeding, and also for locomotion. Salps are adapted for a free-swimming, planktonic existence. There are six genera, most occurring in tropical or subtropical seas.

***Thamnophilus* (antshrikes)** *See* FORMICARIIDAE.

thanatosis (death feigning) Defensive behaviour in which a prey animal (e.g. an opossum or certain snakes) feigns death. It is usually employed only when escape is impossible. The technique is often effective against predators that kill only living prey. The simulated death posture is maintained for only a short time, after which the animal 'recovers'.

theca The case inside which an individual member of a colony of invertebrates lives. *See* AUTOTHECA; BITHECA; STOLOTHECA.

Thecata *See* CALYPTOBLASTINA.

thecodont Applied to the condition in which the teeth are set in sockets in the bones. *Compare* ACRODONT; PLEURODONT.

Thecodontia (class *Reptilia, subclass *Archosauria) An order of 'tooth-in-socket' (i.e. *thecodont) reptiles, and the most primitive of the archosaurs, that ranged from the Late *Permian to the Late *Triassic. *Thecodontosaurus browni* was one of the first of the group. It grew to 2–3 m in length and had a small head and neck. Essentially it was a quadruped, but it could walk on its hind legs. The Thecodontia were ancestral to the dinosaurs, pterosaurs, and crocodiles. Apart from a limited number of primitive features in the skull, *Euparkeria*, a thecodont from the Early Triassic, is a likely ancestor for most archosaurian stocks. It was a small reptile, only 60–100 cm in length.

Thecodontosaurus browni *See* THECODONTIA.

Thecosomata (shelled pteropods; phylum *Mollusca, class *Gastropoda) An order of *opisthobranch gastropods which have a planktonic mode of life, aided by fins or lobes on the foot. The *mantle cavity is large and the shell may be coiled spirally. Individuals are small in size and are ciliary feeders.

thelytoky Obligatory *parthenogenesis, such that populations consist entirely of females, with occasional functionless males. It is the only genetic system in which fertilization (the union of egg and sperm) is eliminated completely. For example, in some *Aphididae the sexual stage of the life cycle has disappeared and populations consist exclusively of females; *Eriosoma lanigerum* (woolly apple aphid), a N. American species, was introduced to Europe, where it became thelytokous on apples because it has no access to its primary host, the

American elm, which presumably supplies a chemical stimulus necessary for sexual reproduction. Some species of parasitic wasps are also thelytokous. Evolution in thelytokous species can therefore take place only by favourable mutations occurring in a single individual, and persisting in the line descending from that individual.

theory of generic cycles See GENERIC CYCLES, THEORY OF.

Theraphosidae (tarantulas; order *Araneae, suborder *Mygalomorphae) Family of very large, hairy, non-social spiders, which possess claw tufts, four *spinnerets, and eight, closely grouped eyes. They are known as 'tarantulas' in the USA. They feed on large insects and occasionally reptiles, amphibians, or nestling birds. Most species chew their prey, sometimes dipping it in water. Few, if any, of the American species are dangerously poisonous to humans. Defensive displays include the raising of *chelicerae, *pedipalps, and first walking legs, and some species *stridulate as well. Large species become mature in 3–10 years, the male living only for months but the female for up to 20 years (in captivity) after maturing. N. American species live in deep burrows and some Amazonian species are arboreal. They are basically tropical in distribution.

Theraponidae (grunter, tigerfish; subclass *Actinopterygii, order *Perciformes) A small family of marine and freshwater fish that have elongate to oblong, compressed bodies, spiny and soft-rayed parts of the *dorsal fin separated by a distinct notch, a slightly forked tail fin, and the *pelvic fins inserted behind the bases of the *pectoral fins. Possibly 14 species are found in the freshwater systems of Australia. There are about 17 species in the Indo-Pacific region.

Therapsida (class *Reptilia, subclass *Synapsida) An order of reptiles, ancestral to the mammals, that ranged from the latter part of the *Permian to the Early *Jurassic.

(((⊕))) SEE WEB LINKS
• Information about the Therapsida.

Therevidae (stiletto flies; order *Diptera, suborder *Brachycera) Small family of flies in which most adults are elongate, with dense body hair and slender legs. Their common name refers to the long, thin abdomen and broad *thorax, known as the 'blade' and 'handle' respectively. The third antennal segment has an apical *style. The *empodium is absent, or represented by a weak bristle. Adults are thought to be predacious, but observations to support this appear sparse. Larvae are predacious on other insects in leaf-mould, soil, and decaying matter. About 500 species have been described.

Theria (superclass *Gnathostomata, class *Mammalia) A subclass that comprises the three infraclasses *Pantotheria (extinct ancestral forms), *Metatheria (marsupials), and *Eutheria (placental mammals), distinguished from the *Prototheria principally by their teeth and by the structure of the sides of the skull, formed in therians by large *squamosal and *alisphenoid bones.

Theridiidae (cobweb spiders, combfooted spiders; order *Araneae, suborder Araneomorphae) Family of spiders, which have a vertical comb of serrated bristles on the *tarsi of the fourth legs which is used to throw threads of silk over the prey items caught in their webs. In general they are a small to medium (0.1–1.5 cm long), glossy species, with large, protuberant eyes, and the habit of making tangled, three-dimensional webs with sticky threads on the periphery. There are more than 1300 species, including the infamous *Latrodectus mactans* (black widow spider) whose venom is toxic to mammals and occasionally fatal to humans.

Theropithecus (geladas) See CERCOPITHECIDAE.

Theropoda (subclass *Archosauria, order *Saurischia) A suborder of dinosaurs which consists exclusively of bipedal, carnivorous forms. It includes the *Coelurosauria and the *carnosaurs and ranged from the Late *Triassic to the *Cretaceous.

thiamin Vitamin B$_1$ (*see* VITAMIN). It contributes to the formation of the important *coenzyme thiamin pyrophosphate, which is involved in the oxidative decarboxylation of alpha-keto acids and transketolase reactions. A deficiency of this compound in humans causes beriberi.

thick-headed flies *See* CONOPIDAE.

thick-knees *See* BURHINIDAE.

thigmotaxis (stereotaxis) A change in direction of locomotion in a *motile organism or cell that is made in response to a tactile stimulus (touch), and where the direction of movement is determined by the direction from which the stimulus is received. It may inhibit movement, causing the organism to come into close contact with a surface. Thigmotaxis is commonly observed in insects.

Thigmotrichida (class *Ciliatea, subclass *Holotrichia) An order of ciliate *Protozoa which have thigmotactic (*see* THIGMOTAXIS) *cilia at the anterior end of the body by means of which the organisms maintain contact with their hosts. They are parasites or commensals (*see* COMMENSALISM) in molluscs in freshwater and marine habitats. There are two suborders.

Thinocoridae (seedsnipe; class *Aves, order *Charadriiformes) A family of plump, medium-sized, brown, speckled birds, which have short, conical bills, and nostrils protected by a flap. They have long, pointed wings, a short tail, and short legs. When disturbed they have a zigzag flight, like snipe. They inhabit open country, feed on seeds, and nest on the ground. There are two genera, with four species, found in S. America.

thiol *See* SULPHYDRYL.

thixotropy *See* SYNOVIAL FLUID.

Thomisidae (crab spiders; order *Araneae, suborder Araneomorphae) Family of spiders whose common name refers to their sideways-running gait when disturbed. They have legs with only two claws, the anterior legs being held laterally, not forward. They are broad-bodied, and generally the first two pairs of legs are more robust than the posterior two pairs. The *spinnerets are small and inconspicuous. Some species live on bark, and all have a potent arthropod venom. *Misumena vatia* can change its colour to match the yellow or white flowers on which it sits in wait for prey.

Thoracica (typical barnacles; subphylum *Crustacea, class *Cirripedia) The familiar barnacles, with a mantle usually covered with calcareous plates, and with six pairs of well-developed, thoracic feeding appendages (cirri). The group includes the suborders Lepadomorpha (stalked or pedunculate barnacles), Verrucomorpha (asymmetrical sessile barnacles), and Balanomorpha (symmetrical sessile barnacles).

thoracic vertebra *See* VERTEBRA.

Thoracostei An alternative name for bony fish (including the sticklebacks) now included in the order *Gasterosteiformes.

thorax 1. The anterior portion of the body of an animal. In vertebrates it contains the heart and lungs and is separated by the *diaphragm from the abdomen. **2.** The three segments of the body of an insect that lie between the head and the abdomen. Each thoracic segment carries a pair of legs. The three thoracic segments are termed the prothorax, mesothorax, and metathorax. The mesothorax and metathorax may each carry a pair of wings.

thornback ray *See* RAJIDAE.

thornbills (*Acanthiza*) *See* ACANTHIZIDAE.

thorn bug *See* MEMBRACIDAE.

thorny catfish *See* DORADIDAE.

thorny devil (*Moloch horridus*) *See* AGAMIDAE.

thorny-headed worms *See* ACANTHOCEPHALA.

thrashers *See* MIMIDAE.

Thraupidae (tanagers; class *Aves, order *Passeriformes) A family of fairly small, brightly coloured birds, which have short to medium, rather conical bills, short to long wings, short to medium-length tails, and short legs. The nine *Piranga* species are typical, mainly red birds, some males of which have black wings, white wing bars, and yellow, grey, or green bodies; the females are yellow-green; many *Tangara* species, of which there are about 47, are brilliant blue with black face patches. Tanagers inhabit forests and bushes, feed on fruit and insects, and nest in trees and bushes. The nests of euphonias (25 *Euphonia* species) are enclosed with a side entrance and built in trees. There are 57 genera in the family, with about 250 species, many kept as cage birds, found in N., Central, and S. America.

threadfin 1. *See* POLYNEMIDAE. 2. (threadfin bream) *See* NEMIPTERIDAE. 3. (threadfin sculpin, *Icelinus filamentosus*) *See* ICELIDAE.

thread snakes *See* LEPTOTYPHLOPIDAE.

threadworms *See* NEMATODA.

threat A form of *communication by which an animal may keep rivals or potentially dangerous animals of other species at bay without fighting.

three-spined stickleback (*Gasterosteus aculeatus*) *See* GASTEROSTEIDAE.

three-toed jacamar (*Jacamaralcyon tridactyla*) *See* GALBULIDAE.

threonine An aliphatic, polar, *alpha amino acid.

Threskiornis (ibises) *See* THRESKIORNITHIDAE.

Threskiornithidae (ibises, spoonbills; class *Aves, order *Ciconiiformes) A family of medium to large birds which have white, grey-brown, or greenish-black plumage. Their bills are either long, slender, and *decurved, or (in spoonbills) long and flattened, with a spatulate tip. Their necks and wings are long and tails short. Their legs are long with long, basally webbed toes. They are gregarious, inhabit shores and marshes, feed on fish, crustaceans, reptiles, and insects, and nest in trees or on the ground. The three species of *Threskiornis* (ibises), found in Africa, Asia, Indonesia, and Australia, are all similar: large, white birds with black, *decurved bills and bare, black skin on the head and neck. Spoonbills (five species of *Platalea*) feed on fish and aquatic invertebrates caught by sweeping the bill from side to side; they are gregarious, inhabit freshwater areas, and nest colonially in bushes and trees. There are about 19 genera in the family, with 32 species, found nearly world-wide.

thrips *See* THYSANOPTERA.

thrombin A proteolytic *enzyme, involved in the clotting of blood, that is generated by the conversion of its *zymogen *prothrombin, the process being activated by the enzyme *thromboplastin, in the presence of calcium ions and other factors. Thrombin acts upon the soluble *fibrinogen in blood to produce insoluble *fibrin by catalysing the removal of the fibrinopeptides A and B from fibrinogen. *See also* BLOOD CLOTTING; BLOOD PLASMA.

thrombocyte *See* PLATELET.

thromboplastin One of a group of lipoprotein compounds apparently released by blood platelets at the site of an injury. In the presence of calcium ions and other factors, it catalyses the conversion of *prothrombin into *thrombin during the clotting of blood.

thrushes 1. *See* TURDIDAE. 2. (laughing thrushes, *Garrulax*) *See* TIMALIIDAE.

Thryonomyidae (cane rats, grasscutters; order *Rodentia, suborder *Hystricomorpha) A family of heavily built, burrowing rodents that have small eyes and ears and short, fully haired tails. The fifth digit is vestigial on all limbs. The hair is coarse and flattened, has grooves on its upper surfaces, and grows in clumps of five or six hairs. Cane rats are distributed widely throughout Africa south of the Sahara; in W. Africa they are currently undergoing domestication as they are widely relished for food. There are two species in a single genus, *Thryonomys*.

Thryonomys (cane rats, grass-cutters) *See* THRYONOMYIDAE.

thumbless bat *See* FURIPTERIDAE.

Thunnus alalunga (albacore) *See* SCOMBRIDAE.

Thunnus thynnus (Atlantic tuna) *See* SCOMBRIDAE.

Thylacinidae (order *Marsupialia, superfamily *Dasyuroidea) A monospecific family (*Thylacinus cynocephalus*, the thylacine or marsupial (Tasmanian) 'wolf' or 'tiger'), which is a highly specialized carnivore bearing many similarities to the *Borhyaenoidea of S. America, due almost certainly to convergence (*see* CONVERGENT EVOLUTION). Probably the species shares a common ancestry with the *Dasyuridae; recently discovered fossils from the *Miocene of Riversleigh, Queensland, demonstrate this common ancestry. Thylacines were present in New Guinea and Australia during the *Pleistocene, but in modern times became restricted to Tasmania and today they are believed to be extinct, the last known specimen having died at Beaumaris Zoo, Hobart, on 7 September 1936. Subsequent reports of sightings have not been confirmed.

Thylacinus cynocephalus (Tasmanian 'wolf') *See* THYLACINIDAE.

Thylacomyidae *See* PERAMELOIDEA.

thymine A *pyrimidine base that occurs in DNA.

thymus In vertebrates, an organ located in the lower neck (and formed in the *embryo from gill pouches or gill clefts) that is involved in the development of lymphoid tissue and hence of lymphocytes (*see* LEUCOCYTE). Its size decreases after puberty and is believed to function only early in life.

thyroglobulin An iodinated *protein of the gelatinous colloid of the *thyroid gland which is chiefly responsible for the storage of iodine in that gland and from which the *hormones *thyroxine and triiodothyronine are formed.

thyroid In vertebrates, an *endocrine gland, located in the neck, that secretes the *hormones *thyroxine and triiodotyrosine. The growth and activity of the thyroid are controlled by *thyrotropic hormone.

Thyroptera (disc-winged bats) *See* THYROPTERIDAE.

Thyropteridae (disc-winged bats, sucker-footed bats; order *Chiroptera, suborder *Microchiroptera) A family of bats in which there are adhesive sucker discs, borne on stalks on wrists and ankles, used when the animals roost in smooth banana leaves that have not yet opened and therefore are rolled. The first digit of each limb bears a claw. The nose is simple, with no leaf, the ears are simple and immobile, and the tail extends a little beyond the edge of the *uropatagium. Thyropteridae occur in tropical Central and S. America and in the W. Indies. There are two species in a single genus, *Thyroptera*.

thyrotropic hormone A glycoprotein *hormone, secreted by the adenohypophysis (*pituitary gland), that stimulates the thyroid gland to produce thyroid hormones and release *thyroxine.

thyroxine An iodinated *amino acid, formed in the *thyroid gland from the proteolysis of *thyroglobulin. It is the major *hormone of this gland and exerts a positive control of the *basal metabolic rate of virtually all tissues and organs in the body.

Thyrsoidea macrura (long-tailed reef-eel) *See* MURAENIDAE.

Thysanoptera (thrips; class *Insecta, subclass *Pterygota) Order of slender, minute (mostly 0.5–2.0 mm long), pale to blackish insects which have short, six- to nine-segmented antennae, short legs, and asymmetrical, sucking mouth-parts which form a conical beak on the ventral surface of the basal head area. Wings, when present, are in two pairs, with very reduced venation and long fringe hairs. The two suborders, Terebrantia and Tubulifera, differ in the shape of the abdomen and development of the *ovipositor. Most thysanopterans are herbivores and many are pests of cereals and fruit trees (e.g. onion thrips, grain thrips, pea thrips, and greenhouse thrips). A few act as vectors of plant diseases. Some thrips are predacious on other small arthropods, and many feed on fungal spores. There are about 4500 species, distributed world-wide.

Thysanura (bristletails, silverfish; class *Insecta, subclass *Apterygota) One of the two orders of the Apterygota, whose name is derived from the Greek *thysanos*, 'fringe', and *oura*, 'tail', comprising *ectognathous insects which are more or less flattened and adapted for running. The tapering body, usually less than 10 mm long, may be bare or covered with silvery scales; it is sometimes pigmented; and ends in three-segmented, bristle-like appendages. *Ocelli are absent in most species, and the *compound eyes are reduced or absent, and never contiguous as in the *Archaeognatha. The *mandibles are *dicondylar, and the *maxillary palps are five-segmented. The *thorax is not arched, and the *coxae lack *styles, although they are present on most abdominal segments. Most species are free-living and very agile, with sexual and *parthenogenetic reproduction. Post-embryonic development may be slow, or take as little as 2–3 months, and the reproductive potential is great. Thysanurans live in damp habitats, e.g. leaf-litter and rocky shores. One family, the Nicolettidae, are cavernicolous (cave-dwelling) and herbivorous. Several species are found in human habitations, where they feed on scraps, carbohydrate and dextrin compounds, and glue and sizes. Distributed throughout the world, and more diverse than the Archaeognatha, some are of economic importance. *Lepismodes inquilinus* (firebrat) inhabits warm buildings, where *Lepisma saccharina* (silverfish) also occurs, often inhabiting damp books. There are five families, with 330 species. *See also* LEPISMATOIDEA; MACHILOIDEA.

tibia 1. In a tetrapod vertebrate, the anterior long bone of the lower hind limb (the 'shin' bone). **2.** In an insect, the long and often narrow segment of the leg that articulates proximally with the *femur and distally with the *tarsus. The tibia contains the muscles that control the tarsi. The musculature operating the tibia is found in the femur, and acts through the agency of a tendon.

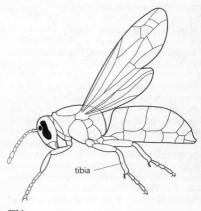

Tibia

tibiale The pre-axial proximal *tarsal bone of the ankle.

tibiotarsal joint The avian equivalent of the ankle: the joint between the *tibiotarsus and the *metatarsals.

tibiotarsus 1. The bone formed by the fusion of the *tarsal bones and the *tibia, found in birds and some dinosaurs. *See also* TARSOMETATARSUS; TIBIOTARSAL JOINT. **2.** In insects, the combined *tarsus and tibia.

Tichodromadidae (wallcreeper; class *Aves, order *Passeriformes) A monospecific family (*Tichodroma muraria*), which is a small, grey bird with brilliant crimson and black wings. It has a long, slender, *decurved bill, long, broad wings, and a short tail. Its legs are short and its feet large. It inhabits high mountains, moving lower in winter, feeds on insects, and nests in rock crevices. It is found in southern Europe and Asia.

Tichodroma muraria (wallcreeper) *See* TICHODROMADIDAE.

tiger beetle (*Cicindela*) *See* CARABIDAE.

tiger butterflies *See* DANAINAE.

tigerfish *See* THERAPONIDAE.

tiger moths *See* ARCTIIDAE.

tiger shark (*Galeocerda cuvieri*) *See* CARCHARHINIDAE.

tight junction A region of the cell surface, of variable size, and situated just below the apical border where the membranes of adjacent cells are fused to form a pentalaminar structure. In simple epithelial layers tight junctions often form a continuous layer (the zonula occludens). Their function appears to be the provision of areas of contact between cells of enhanced permeability and low electrical resistance. They occur in tissues such as the blood–brain barrier and cardiac muscle, where such features are advantageous.

tigroid body *See* NISSL BODY.

Tilapia *See* CICHLIDAE.

tilefish *See* MALACANTHIDAE.

Tiliqua rugosa (stump-tailed skink) *See* SCINCIDAE.

Tillodontia An order of mammals from the *Palaeocene and *Eocene of the northern continents, which culminated in a bear-size animal with rodent-like incisors.

Timaliidae (babblers, fulvettas, nun babblers, rail-babbler; class *Aves, order *Passeriformes) A family of fairly small birds, most of which are brown, grey, and buff, but some of which have brighter colours. They have varied bill shapes, short, rounded wings, and many have long tails. Their feathers are soft, and long on the back. They are mainly arboreal, although some are terrestrial, and inhabit forests and scrub. They feed on insects and fruit, and nest in cup-shaped or domed nests in trees, bushes, grass, or on the ground. *Eupetes macrocercus* (rail-babbler) hunts insects by running along the ground. Some (e.g. *Turdoides*) are noted for their loud and melodious song; the white-crested laughing thrush (one of the 48 species of *Garrulax*, laughing thrushes) is a popular cage bird. There are about 57 genera in the family, with 275 species, found in Europe, Africa, Asia, and Australasia.

timber beetle *See* CERAMBYCIDAE.

timber rattlesnake (*Crotalus horridus*) *See* CROTALIDAE.

Tinamidae (tinamous; class *Aves, order *Tinamiformes) A family of medium-sized, dumpy birds which have brown and grey, cryptically patterned plumage. Their bills are short to long and slightly *decurved, and they have longish necks, short, rounded wings, and very short tails. Their legs are strong and short to long. They are terrestrial, preferring to run rather than fly. They inhabit forest, brush, and grassland, feed on fruit and seeds, and nest on the ground, often in a scrape. The 20 species of *Crypturellus* are typical. There are nine genera, with 46 species, found in Central and S. America.

Tinamiformes (tinamous; class *Aves) An order of birds which comprises the single family *Tinamidae.

tinamous *See* TINAMIDAE.

Tinbergen, Nikolaas (1907–88) A Dutch-born, but later British, zoologist who shared (with Konrad Lorenz and Karl von Frisch) the 1973 Nobel Prize for Physiology or

Medicine for studies of animal behaviour under natural conditions. Tinbergen was especially noted for his studies of social organization among gulls. He became Professor of Animal Behaviour at the University of Oxford in 1966, and Professor Emeritus in 1974.

(((( ))) SEE WEB LINKS

• Biography of Nikolaas Tinbergen.

Tinca tinca (tench) *See* CYPRINIDAE.

Tineidae (clothes moths; subclass *Pterygota, order *Lepidoptera) Very large family of small, generally drab moths, the classification and definition of the family being unclear. Most larvae feed on dried organic matter, and some are case-bearers. The family includes pests of carpets, clothes, and foodstuffs, and has a world-wide distribution.

Tingidae (lacebugs; order *Hemiptera, suborder *Heteroptera) Family of plant-feeding bugs in which the *pronotum and fore wings are covered with raised reticulations, often forming fantastic and intricate shapes. Some species cause galls to form on their host plants. They are found throughout the world.

Tintinnida (class *Ciliatea, subclass *Spirotrichia) An order of ciliate *Protozoa in which the cell body is housed in a *lorica. Oral *cilia are conspicuous and extend from the lorica. Tintinnids are found in marine environments.

Tiphiidae (suborder *Apocrita, superfamily Scolioidea) Family of often shining black, brownish, or black and yellow wasps in which the adults are 10–25 mm long, and have two posterior lobes on the *mesosternum, or have rather strongly constricted *petioles. Some subfamilies contain species with wingless females, a character associated with a *fossorial habit. All are parasites of the larvae of *Coleoptera and *Hymenoptera, some specializing on tiger beetles (*see* CARABIDAE). Both sexes feed on nectar and honeydew, and some are nocturnal. More than one egg per host may be laid. The family is fairly common and widely distributed.

Tipulidae (craneflies, daddy-long-legs, leatherjackets; order *Diptera, suborder *Nematocera) Family of true flies in which the adults have long, six-segmented antennae. The legs are long and are readily discarded if trapped. The *mesonotum has a V-shaped suture on the dorsum. The distal cell is present in the wing venation. The female *ovipositor is prominent and sclerotized. The larvae live in the soil or among roots, or are aquatic, and are *phytophagous or predacious. The family includes garden pests (leatherjackets) which destroy plant roots. The family has a world-wide distribution, and at least 13 500 species are known.

tissue A group of cells of similar type that work in a co-ordinated manner towards a common function. They are normally bound together by an intercellular substance. Some fluids (e.g. blood) are also considered to be tissues.

titanothere *See* BRONTOTHERIIDAE.

titmouse *See* TITS.

tits 1. *See* PARIDAE. 2. (bushtits, long-tailed tits, pygmy tits) *See* AEGITHALIDAE. 3. (penduline tits) *See* REMIZIDAE. 4. (titwarblers, *Leptopoecile*) *See* REGULIDAE. 5. (wren-tit, *Chamaea fasciata*) *See* CHAMAEIDAE.

titwarblers (*Leptopoecile*) *See* REGULIDAE.

T$_m$ Transport maximum, i.e. the maximum ability of the *kidney to reabsorb or secrete a particular substance.

T-maze A maze shaped like a T, in which an animal introduced at the base of the shaft has to make a single choice when it reaches the crossbar. This is the simplest type of maze.

toadfish *See* BATRACHOIDIDAE.

toads 1. *See* ANURA; BUFONIDAE; DISCOGLOSSIDAE; PEODYTIDAE; PIPIDAE.

2. (Mexican burrowing toad) *See* RHINOPHRYNIDAE. **3. (spade-foot toads)** *See* PELOBATIDAE.

***Tockus* (hornbills)** *See* BUCEROTIDAE.

tocopherol Vitamin E (*see* VITAMIN); one of a group of fat-soluble *terpene compounds that function as anti-oxidants in cells. They are required for the normal growth and fertility of animals.

Todidae (todies; class *Aves, order *Coraciiformes) A family which comprises the single genus *Todus*, of small birds which have green upper-parts, red throats, and pink or yellow flanks. They have long, fairly narrow bills, with red lower *mandibles. They have large heads, short necks, and rounded bodies. Their wings are short and rounded, their tails medium-length, and their legs slender, with long, *syndactylous toes. They inhabit forests and bush, feed on insects and fruit, and nest in holes in banks. There are five species, confined to the W. Indies.

todies *See* TODIDAE.

***Todus* (todies)** *See* TODIDAE.

tokay gecko (*Gekko gecko*) *See* GEKKONIDAE.

tolerance, limits of *See* LIMITS OF TOLERANCE.

tomentose Woolly; covered with a fine mesh of hairs.

***Tomistoma schlegeli* (false gharial)** *See* CROCODYLIDAE; GAVIALIDAE.

tonguefish *See* CYNOGLOSSIDAE.

tonguesole *See* CYNOGLOSSIDAE.

tonofilament An aggregation of filaments, approximately 0.7–0.8 nm in diameter, that occurs in large numbers in certain epithelial cells (*see* EPITHELIUM). Tonofilaments are particularly conspicuous in areas immediately beneath *desmosomal plaques, which apparently they help to anchor to the cell *cytoplasm.

tool use The use by an animal of an object external to itself and not attached to any substrate, that is held, carried, or otherwise manipulated in order to achieve an objective.

toothcarp *See* CYPRINODONTIDAE.

toothed flounder *See* PSETTODIDAE.

toothed herring *See* HIODONTIDAE.

toothed whales *See* ODONTOCETI.

tooth shells *See* SCAPHOPODA.

topology In *phylogenetics, the branching pattern of a *cladogram.

topotype In taxonomy, a specimen found in the type locality of a taxon to which it is thought to belong, but that is not necessarily of that *type series.

tornaria larva The planktonic larva of *Hemichordata which resembles the *bipinnaria larva of *Asteroidea, indicating a possible evolutionary link between *Chordata and *Echinodermata.

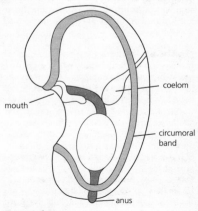

Tornaria larva

Torpedinidae (electric ray, numbfish; superorder *Batoidimorpha, order *Rajiformes) A family of marine rays that have a rounded outline, a stout trunk with a short tail fin, and a soft skin. The eyes are small or rudimentary. The spineless two *dorsal fins are well developed. All members of the family possess a pair of powerful electric organs derived from the branchial musculature and capable of delivering very unpleasant electric shocks to anyone treading on them. *Torpedo nobiliana* (Atlantic torpedo), 1.8 m, is one of the larger species of electric rays. There are about 35 species, distributed world-wide in coastal regions.

Torpedo nobiliana (Atlantic torpedo) *See* TORPEDINIDAE.

torpor A state of adaptive hypothermia used by *endotherms in order to save energy. In torpor, the body temperature of an animal may fall to within 1°C of the environmental temperature, which in some cases may itself be at about or even just below freezing. All metabolic processes slow down to as little as one-twentieth of the normal rate. A state of torpor is entered during *hibernation and when resources are insufficient to allow the maintenance of body temperature.

torrent catfish *See* AMBLYCIPITIDAE.

torsion In *Gastropoda, the twisting of the body through 180 degrees so that the digestive and nervous systems have a U shape and the *mantle cavity, *anus, *gills, and two nephridiopores (*see* NEPHRIDIUM) occupy an anterior position behind the head.

tortoise beetle *See* CHRYSOMELIDAE.

tortoises *See* CHELONIA; EMYDIDAE; TESTUDINIDAE.

tortrices *See* TORTRICIDAE.

Tortricidae (tortrices, bell moths, leafrollers, fruit moths; subclass *Pterygota, order *Lepidoptera) Large family of more than 10 350 species of small moths with broad wings, which often resemble a bell in shape when they are at rest. The *labial palps are *porrect. The larvae feed on litter, between leaves that they sew together, in fruit, or in galls. Some roll leaves. Many are pests, including the codling moth, whose larvae feed on apples. The family is distributed world-wide.

Torymidae (chalcid seed flies; suborder *Apocrita, superfamily *Chalcidoidea) Family of small to medium (2–13 mm long), elongate, usually metallic green, yellow, or black wasps whose *ovipositors may be as long as or longer than the body, and whose hind *coxae and *femora may be enlarged. Most are parasitic on dipteran and hymenopteran gall formers, although some eat seeds. The eggs are laid in galls or the seeds of trees, particularly conifers, and some are laid in caterpillars and mantid *oothecae. There are about 70 genera and more than 960 species, distributed world-wide.

totipalmate In birds, applied to feet which have all four toes joined by webs (the condition is characteristic of the *Pelecaniformes).

toucans *See* PICIFORMES; RAMPHASTIDAE.

Tournaisian *See* CARBONIFEROUS.

toxicyst In certain *Protozoa, an *organelle which resembles a *trichocyst but in which the filament carries a poison that can kill other protozoa. Toxicysts are used for capturing prey.

toxin 1. Any poisonous substance of plant or animal origin. **2.** A microbial product which is poisonous to animals or plants. The symptoms of many types of human disease are due to the production of one or more toxins by the pathogen. Toxins usually act at specific sites in the body (e.g. neurotoxins affect nerves, enterotoxins affect the gut).

Toxostoma (thrashers) *See* MIMIDAE.

Toxotes chatareus (archerfish) *See* TOXOTIDAE.

Toxotidae (archerfish; subclass *Actinopterygii, order *Perciformes) A small family of marine and freshwater fish that have a deep, oval body, the *dorsal and large *anal fins being situated well to the back of the fish. The mouth and eyes are large. *Toxotes chatareus* (archerfish), 24 cm, is capable of squirting a jet of water from its mouth and 'shooting down' insects from a height of up to 1.5 m above the water surface. There are about four species, in south-east Asia and northern Australia.

trabeculae Pillars of radiating, calcareous fibres of microscopic size, which build up the skeletal elements in corals and form cross-struts in spongy bone in *Chordata.

trace fossil *See* ICHNOFOSSIL.

trachea 1. One of the cuticular tubes that make up the respiratory system of an insect. The tracheae ramify throughout the body, terminating in fine, intracellular branches (tracheoles). *See* SPIRACLE. **2.** In air-breathing vertebrates, the 'windpipe', leading from the throat and dividing into two bronchii, which enter the lungs.

tracheal gill (caudal gill) One of the three terminal gills that are present in the larvae of most damselflies. These are richly supplied with *tracheae and are the main site of gas exchange. They are readily *autotomized, however, being replaced slowly in later *instars.

tracheal spiracle One of the two kinds of respiratory organ found in spiders (*compare* BOOK LUNGS). They are located just anterior to the *spinnerets. The fastest-moving spiders have two pairs of tracheal spiracles; these are also commonly found in small species, for which water loss is a special consideration.

tracheole *See* TRACHEA.

Trachichthodes affinis (nannygai) *See* BERYCIDAE.

Trachichthyidae (roughy, slimehead; subclass *Actinopterygii, order *Beryciformes) A small family of marine fish that have a deep and compressed body, an oblique mouth, and a slender tail with a forked tail fin. The head is large and bears slime glands in the skin, but the belly feels rough due to the presence of a series of *scutes between the *pelvic and *anal fins. Some species are offshore fish; others, e.g. *Trachichthys australis* (southern roughy), 15 cm, are near-shore reef-dwellers. There are about 14 species, distributed world-wide.

Trachichthys australis (southern roughy) *See* TRACHICHTHYIDAE.

Trachinidae (weeverfish; subclass *Actinopterygii, order *Perciformes) A small family of marine, bottom-dwelling fish that have an elongate body, a large, oblique mouth, and the eyes situated on top of the head. The short, spiny-rayed, *dorsal fin is separated from the long-based, soft-rayed, dorsal fin by a distinct notch. All weeverfish possess poison glands at the base of the spiny dorsal rays as well as on the spine of the *gill cover. These spines can inflict painful wounds. There are four species, distributed in the eastern Atlantic.

Trachipteridae (ribbonfish, dealfish; subclass *Actinopterygii, order *Lampridiformes) A small family of marine fish that have a much compressed body, gradually tapering to a very thin tail with a small tail fin. The mouth is small, but the eyes are large. Typically, they have a very long *dorsal fin, originating on the head and continuing to the tip of the tail; the *anal fin is absent. Probably living at depths of 200–1000 m. *Trachipterus altivelis* (ribbonfish, or king-of-the-salmon) of the Pacific may reach a length of about 2 m. There are seven species, distributed world-wide.

Trachipterus altivelis (ribbonfish, king-of-the-salmon) *See* TRACHIPTERIDAE.

Trachurus trachurus (eastern Atlantic scad) *See* CARANGIDAE.

Trachylina (trachyline medusae; Trachylinida; phylum *Cnidaria, class *Hydrozoa) An order of cnidarians that have no fixed hydroid stage and a mobile medusoid (*see* MEDUSA) stage that is well developed.

Marginal sense organs are formed from modified *tentacles and derived from the gastrodermis.

trachyline medusae *See* TRACHYLINA.

Trachylinida *See* TRACHYLINA.

Tragelaphus oryx (eland) *See* BOVIDAE.

Tragulidae (chevrotains; infra-order *Pecora, superfamily *Traguloidea) A family of pecorans which exhibit certain relatively primitive features. The family appears in the fossil record in the early *Miocene. They are derived from the *Hypertragulidae, the ancestral pecorans, which range from the upper *Eocene to the lower Miocene. Although *ruminants, the subdivisions of the stomach are less complex than in other pecorans. Horns are absent. The upper *incisors are absent, as in other pecorans, but the upper *canines are large, especially in the male, and form tusks. The *tibia and *fibula are fused, but the fibula is not reduced. There are four digits on each limb, which are complete although small. The cannon bone, formed by the fusion of the middle metapodials of all limbs and characteristic of higher pecorans, is complete in the hind limbs but only partial in the fore limbs of the Asian genus and absent in the fore limbs of the African genus. Chevrotains are small animals found in the forests of tropical Africa and Asia. There are four species in two genera, *Tragulus* (Asian) and *Hyemoschus* (African).

Traguloidea (suborder *Ruminantia, infra-order *Pecora) A superfamily that comprises the *Tragulidae and the extinct families *Hypertragulidae, *Protoceratidae, and Gelocidae, primitive pecorans which lived between the *Eocene and *Pliocene. The hypertragulids lived mainly in N. America and departed from the general form in the development of a lower *premolar that functioned as a *canine. The protoceratids, also from N. America, shared this feature but had longer faces and horns. The gelocids lived in the Old World during the *Oligocene and may have been transitional forms directly ancestral to the modern pecorans.

Tragulus (Asian chevrotains) *See* TRAGULIDAE.

tragus In *Mammalia, a flap developed from the lobe extending in front of the external opening of the ear.

trail pheromone A *pheromone laid down by a foraging ant that is returning to the nest with food it has found. Other members of the nest follow the trail to the food source.

trait Any detectable *phenotypic property of an organism; a *character.

tramp species Species that have been spread around the world inadvertently by human commerce. *See also* SUPERTRAMP.

transaminase (aminotransferase) An *enzyme that catalyses a *transamination reaction.

transamination The transfer of an amino group from an *amino acid to a keto acid in a reaction catalysed by a transaminase. This is the principal method in cells for the synthesis of non-essential amino acids, and requires a pyridosal or pyrido-oxamine phosphate as a *coenzyme.

transcription The polymerization of ribonucleotides into a strand of *RNA in a sequence complementary to that of a single strand of DNA. By this means the genetic information contained in the latter is faithfully matched in the former. The process is mediated by a DNA-dependent RNA *polymerase.

(((•))) **SEE WEB LINKS**
• Detailed explanation of transcription.

transect A line marked within an area that is undergoing an ecological survey to provide a means of measuring and representing graphically the distribution of organisms, especially when they are arranged in a linear sequence (e.g. up a sea-shore, or across a woodland margin). A transect is particularly useful for detecting transitions or distribution patterns. *See* BELT TRANSECT; LINE TRANSECT.

transferase An *enzyme that catalyses the transfer of a functional group from one substance to another.

transfer ribonucleic acid *See* TRANSFER-RNA.

transfer-RNA (transfer ribonucleic acid, t-RNA) A generic term for a group of small *RNA molecules, each composed of 70–80 *nucleotides arranged in a clover-leaf pattern stabilized by *hydrogen bonding. They are responsible for binding *amino acids and transferring these to the *ribosomes during the synthesis of a *polypeptide (i.e. during translation). At the ribosomes, which are attached to the *messenger-RNA (m-RNA), the 'reading frame' indicates the three m-RNA nucleotides that form the next triplet codon in the sequence: whichever t-RNA molecule carries the complementary anticodon can associate with the ribosome such that the amino acid that it bears can be joined on to the end of the growing polypeptide.

transgenic Applied to an organism which contains genetic material from another organism, usually supplied by molecular biological techniques.

transglycosylation A mechanism for glycosidic bond formation, particularly during polysaccharide synthesis; nucleoside phosphate derivatives act as 'activated' donor compounds in which the energies of their glycosidic bonds are partially conserved in the reaction products. *Glycosides cannot be synthesized spontaneously from free monosaccharides owing to the high negative free energy (–DG) of the *hydrolysis reaction.

transient polymorphism The presence in a population at a particular gene *locus of alternative *alleles in which one is progressively replaced by another. *Compare* BALANCED POLYMORPHISM.

transition In genetics, a type of *mutation (a *nucleotide-pair substitution) that involves the replacement in DNA or RNA of one *purine with another, or of one *pyrimidine with another. An example is the change of GC (guanine–cytosine) to AT (adenine–thymine).

transitional ER (tER) Adjacent to the side of the stack of *cisternae closest to the nucleus in a *Golgi body, a specialized patch of *endoplasmic reticulum that has no bound *ribosomes. Transitional ER sites often fuse with one another, but retain a constant average size, and tER is believed to give rise to cisternae that mature and move away, the tER permitting only correctly folded *proteins to pass into the Golgi apparatus.

transitional vesicle *See* TRANSPORT VESICLE.

translation The conversion of the base sequence in *m-RNA into an *amino-acid sequence in a *polypeptide chain by a process that occurs on a *ribosome and involves several small proteins, m-RNA, and *t-RNA. The sequence of amino acids in the chain is specified by that of the *nucleotides in the m-RNA (these being read as codons, i.e. in groups of three), and this in turn follows the sequence of nucleotides in the DNA.

translocation A change in the arrangement of genetic material, altering the location of a *chromosome segment. The most common forms of translocations are reciprocal, involving the exchange of chromosome segments between two nonhomologous chromosomes. A chromosomal segment may also move to a new location within the same chromosome or in a different chromosome, without reciprocal exchange; these kinds of translocations are sometimes called transpositions.

transport protein A *protein molecule to which another molecule attaches to be transported. For example, *bilirubin travels attached to *blood plasma *albumin.

transport vesicle (transitional vesicle) A structure that carries *proteins destined for export from the cell from the *endoplasmic reticulum to the side of the *cisternae closest to the nucleus in the *Golgi body, with which it then fuses.

transposable elements Chromosomal *loci that may be transposed from one spot to another within and among the *chromosomes of the complement. The process occurs through breakage on either side of these loci and their subsequent insertion into a new point either on the same or a different chromosome.

transposition See TRANSLOCATION.

transposon A DNA element that can insert at random into a *plasmid or bacterial *chromosome, independently of the host cell-recombination system. In addition to other *genes, transposons carry genes that confer new phenotypic properties on the host cell (e.g. resistance to some antibiotics).

transverse process One of the pair of lateral projections at each side of the *neural arch of a tetrapod vertebra with which the rib articulates.

transverse tubular system See T-SYSTEM.

transversion A *mutation in which a *purine is replaced by a *pyrimidine (or vice versa) in a base sequence of DNA or *RNA. This type of mutation has a lower frequency of occurrence than *transition.

trapdoor spiders (order *Araneae, suborder *Mygalomorphae) Spiders of the families Antrodiaetidae, Actinopodidae, and Ctenizidae, which have three *tarsal claws and a *rastellum. Most species are 1–3 cm long and live in a silk-lined tube dug into the ground, with a silk lid. Passing insects are attacked and pulled into the tube with great rapidity, and the tubes are enlarged as the spiders grow. Trapdoor spiders are found in the Americas, Africa, and Australia.

tree-creepers See CERTHIIDAE.

tree-crickets See GRYLLIDAE.

tree frogs See HYLIDAE.

tree hopper See MEMBRACIDAE.

tree partridge See PHASIANIDAE.

tree porcupine (*Coendou*) See ERETHIZONTIDAE.

tree shrew See TUPAIIDAE.

tree swift See APODIFORMES; HEMIPROCNIDAE.

Trematoda (flukes; phylum *Platyhelminthes) A class of flatworms, most of which are a few centimetres long but including others whose length ranges from less than 1 mm to 7 m. They have organs for adhesion, the mouth is at the anterior end and leads into a muscular *pharynx, and the *epidermis is not ciliated. All trematodes are parasitic, their epidermis protecting them against digestive enzymes secreted by their host.

Treron (green pigeons) See COLUMBIDAE.

trevalla See CENTROLOPHIDAE.

trevally See CARANGIDAE.

Triacanthidae (triplespine, tripodfish; subclass *Actinopterygii, order *Tetraodontiformes) A small family of marine, tropical fish that have a deep and compressed body, a small mouth, two *dorsal fins, and a narrow *caudal peduncle. The first dorsal fin spine and the two pectoral spines are capable of being locked firmly in an upright position when the fish is threatened. *Triacanthus brevirostris* (shortnosed tripodfish), 24 cm, is fairly common in inshore and estuarine waters. There are about seven species, distributed in the Indo-Pacific region.

Triacanthodidae (spikefish; subclass *Actinopterygii, order *Tetraodontiformes) A small family of marine tropical to subtropical fish, closely related to the *Triacanthidae (tripodfish) but with fewer *fin rays, a rounded tail fin, a pointed snout, and large eyes. There are about 19 species, distributed in the Indo-Pacific region and the western Atlantic.

Triacanthus brevirostris (shortnosed tripodfish) *See* TRIACANTHIDAE.

Triakidae *See* CARCHARHINIDAE.

trial-and-error learning *Learning in which an animal comes to associate particular behaviours with the consequences they produce. This tends to reinforce the behaviour (i.e. the behaviour is likely to be repeated if the consequences are pleasant, but not if they are unpleasant). Such learning is believed to involve a process of classical *conditioning followed by *operant conditioning.

triallelic Applied to a *polyploid in which three different *alleles exist at a given *locus.

Triassic The earliest of the three periods of the *Mesozoic Era, which lasted from 251 Ma ago to 199.6 Ma ago. As a result of the mass extinctions of the Upper *Palaeozoic, Triassic communities contained many new faunal and floral elements. Among these were the *Ammonoidea, modern corals, various *Mollusca, the *dinosaurs, and certain gymnosperms (trees).

SEE WEB LINKS
• Detailed description of life in the Triassic.

Triassochelys quenstedii *See* PROGANOCHELYS QUENSTEDII.

tricarboxylic-acid cycle *See* CITRICACID CYCLE.

Trichechidae (manatees; superorder *Paenungulata, order *Sirenia) A family of aquatic paenungulates in which the upper lip is deeply cleft and each side is capable of independent movement and is used for cropping vegetation. There are no *incisors or *canines, but the gums form horny pads. There are up to 20 cheek teeth which are peg-like and fall out when worn, probably without being replaced. The tail is unnotched. Manatees occur in coastal waters and rivers in tropical and subtropical America and W. Africa. There are three species in a single genus, *Trichechus*.

Trichechus (manatees) *See* TRICHECHIDAE.

trichiation *See* CHAETOTAXY.

Trichiuridae (cutlassfish, hairtail; subclass *Actinopterygii, order *Perciformes) A small family of marine fish that have a very elongate, slender body, the tail tapering to a point or provided with a tiny, forked tail fin. The large mouth has strong, fang-like teeth. The *dorsal fin is extremely long, but the other fins are poorly developed. Most species inhabit the upper few hundred metres of the ocean. There are about 17 species distributed world-wide in tropical and temperate seas.

trichobothrium In some *Arachnida, an upright hair, linked to nerves in the *exoskeleton and inserted in a flexible membrane, that can move in any direction. It enables the animal to detect air movements.

trichocyst A bottle-shaped *organelle, analogous to a *nematocyst to which it bears a marked resemblance, that is found orientated at right angles to the cell surface in the *pellicle of some members of the *Ciliatea. Some trichocysts are explosive, discharging a thread-like shaft surmounted by a barb through a pore to the exterior, perhaps to capture prey or to anchor the ciliate during feeding. Other types of trichocyst are fluid-filled and discharge *mucus and toxins.

Trichodectes canis (dog louse) *See* TRICHODECTIDAE.

Trichodectidae (order *Phthiraptera, suborder *Ischnocera) Family of chewing lice which are parasitic on placental mammals. Members of the family are distinguished from other Ischnocera by the presence of only one antennal *flagellomere in all males and most females, two or more hook-like *setae on the antennal *flagellum of most males, lateral abdominal tergal pits in males, and only one *tarsal claw on each leg. *Trichodectes canis* (dog louse) transmits tapeworms between dogs but otherwise members of the family are not known to be vectors of disease. Trichodectidae parasitic on sheep, goats, cattle, and dogs cause irritation by their biting, and the efforts of the host to dislodge them can lead to loss of hair. There are 20 genera, with about 350 species, distributed world-wide.

Trichodontidae (sandfish; subclass *Actinopterygii, order *Perciformes) A small family of marine fish that have a large head, a large, upward-directed mouth with fringed lips, and two *dorsal fins. The *anal fin is long and the *pectorals well developed. *Trichodon trichodon* (Pacific sandfish), 31 cm, is thought to bury its body in the sand, awaiting the arrival of potential prey. Locally important as food fish, the two species of sandfish occur only in the N. Pacific basin.

***Trichodon trichodon* (Pacific sandfish)** *See* TRICHODONTIDAE.

trichogen cell Epithelial cell beneath the *exoskeleton that secretes a cuticular hair. In many insects these hairs are sensory, and contain nerve endings.

Trichogrammatidae (suborder *Apocrita, superfamily *Chalcidoidea) Family of wasps which are unique in having three-segmented *tarsi. Apart from a few wingless females found in figs, all are egg parasites attacking a wide range of hosts, some being very valuable as biological-control agents. Adults are minute (often less than 1 mm long), fringe-winged wasps with greatly reduced venation. Most parasitize the eggs of *Lepidoptera, although some attack eggs of water beetles and water bugs, using their tiny wings to swim down to reach their hosts. There are about 80 genera with more than 840 species, distributed world-wide.

Trichomonadida (superclass *Mastigophora, class *Zoomastigophorea) An order of *Protozoa which are not bilaterally symmetrical and which have four to six *flagella. The cells swim with a jerky motion. There are several families. Most are harmless parasites which live in the intestinal or reproductive systems of vertebrates.

Trichomycteridae (parasitic catfish; superorder *Ostariophysi, order

***Siluriformes)** A large family of freshwater fish that tend to have a slender, pencil-shaped body with a scaleless skin, small fins, and two pairs of *barbels above the mouth. Many species may hide in the river bottom or in leaf-litter during the day, but attack other animals at night to feed on blood and tissues. There are about 185 species occurring in Central and S. America.

Trichonotidae (sand diver; subclass *Actinopterygii, order *Perciformes) A small family of marine fish that have an elongate, thin body, a pointed snout, and long-based second *dorsal and *anal fins. The spines of the first dorsal fin may be rather long. Sand divers are found on the sea-bottom, but tend to disappear into the mud very quickly when disturbed. There are about 19 species, occurring in tropical to subtropical seas.

Trichoptera (caddis flies, sedge flies, silver-horns, Welsh buttons; class *Insecta, subclass *Pterygota) An order of aquatic, rather moth-like insects that spend most of their lives as larvae inhabiting cases they construct from small mineral grains or plant material, the type of case often being characteristic of a species. Larvae lack abdominal legs, are free-living or spin webs to trap food particles, and pupate below the water surface in shelters attached to stones or other objects. The pupae swim to the surface, where the adult emerges. Most adults do not feed, those that do feeding on nectar. The name, Trichoptera, is derived from the Greek *trikhos*, 'hair', and *pteron*, 'wing', referring to small hairs that cover the wings. Fossil Trichoptera are known from the *Triassic. There are about 7000 species, in about 25 families. They occur in unpolluted water on all continents.

Trichostomatida (subphylum *Ciliophora, class *Ciliatea) An order of ciliate *Protozoa that are characterized by the structure of the *cytostome region, where there is a ciliated depression leading into a *buccal cavity lacking *cilia, at the base of which occurs the cytostome. Trichostomatida are found in freshwater habitats. Some are parasitic in the intestines of animals.

Tricladida (planarians; class *Turbellaria, subclass *Neoophora) An order of worms which have a divided intestine and a *pharynx. All are free-living.

triconodont Applied to a tooth, typical of primitive mammals, that has three simple, conical *cusps.

Triconodonta (class *Mammalia, subclass *Prototheria) An order that includes the earliest of all mammals, living from the *Triassic until the Early *Cretaceous and distributed over the northern continents. Typically the *molar teeth each had a row of three sharp, conical *cusps, the teeth of the upper and lower jaws forming a shearing device. *Premolars and molars were differentiated, probably with some replacement, and probably the young were fed on milk secreted by the mothers. Triconodonta may have been *homoiotherms and nocturnal, and possibly they were arboreal. They are believed to have been true carnivores rather than insectivores. *Triconodon* was the size of a modern cat. The order is believed to have evolved from therapsids (*Therapsida) independently of the main line of mammalian evolution and to have left no descendants.

(((⊕))) **SEE WEB LINKS**
• Detailed description of the Triconodonta.

tricuspid valve In the mammalian (*Mammalia) *heart, the valve that controls the flow of blood from the right *auricle into the right *ventricle. Its name refers to its structure, which usually comprises three leaflets and three muscles (although this may vary).

Tridactylidae (pygmy mole crickets; order *Orthoptera, suborder *Caelifera) Small family of minute, smooth, shiny caeliferans which are usually found on the ground or in burrows, usually near lakes or streams. The hind legs have much enlarged *femora, and are used in jumping and swimming, but not in walking. The insects probably feed on algae and other vegetable matter in the soil. About 50 species are known, four of which are Mediterranean in distribution. The genus *Tridactylus* is cosmopolitan.

triggerfish *See* BALISTIDAE.

Triglidae (gurnard; subclass *Actinopterygii, order *Scorpaeniformes) A family of marine fish that have a large, armoured head, two separate *dorsal fins, and, typically, large *pectoral fins in which the lower three rays are enlarged and free from the fin membrane. These separate rays are used as feelers for the detection of food. Several species show brilliant colour patterns on the body as well as on the pectoral fins; some species are considered commercially important. There are about 85 species, distributed world-wide in temperate and tropical seas.

Trigona (stingless bees) *See* APIDAE; HONEY-BEE.

trigonous Triangular in cross-section.

Trilobita (trilobites; phylum *Arthropoda) The most primitive arthropod class (or in some classifications a phylum, where the Arthropoda rank as a superphylum), known from more than 3900 fossil species. Inhabitants of *Palaeozoic seas, the trilobites first appeared in the Early *Cambrian, had their widest distribution and greatest diversity in the Cambrian and *Ordovician periods, and became extinct in the *Permian. The body was divided into three regions: an anterior *cephalon, comprising at least five, fused segments; a mid-body (thorax), with a varying number of segments; and a hind region (pygidium). All three regions were divided by a pair of furrows running the length of the body, giving a trilobite appearance (i.e. a median or axial lobe, flanked on either side by a lateral lobe). The mouth was situated in the middle of the central surface of the cephalon. Paired gill-bearing limbs were attached to the membranous, pleural skeleton. X-ray studies show the eyes to have resembled the *compound eyes of living arthropods. Trilobites ranged in size from 0.5 mm-long planktonic forms to those nearly 1 m in length; most species were 3–10 cm long.

trilobite A member of the fossil arthropod class *Trilobita.

Trimenoponidae (order *Phthiraptera, suborder *Amblycera) Family of chewing lice which are parasitic on *Caviidae and marsupials in S. America. Members of the family are distinguished from other Amblycera by the contraction of the first abdominal tergal plate, the reduction or absence of the first pleural plate, the loss of the *spiracles on abdominal segment 8, and the presence of two *tarsal claws on each leg (the last character being unusual in mammal lice). There are six genera, with 10 species.

trimeric *See* DIMER.

Tringa (sandpipers) *See* SCOLOPACIDAE.

Trionychidae (soft-shelled turtles; order *Chelonia, suborder *Cryptodira) A family of highly aquatic turtles that occur in fresh, brackish, and occasionally sea water. The bony shell is regressed, and covered with vascularized, respiratory skin. The head has a proboscis and fleshy lips. The limbs are oar-like. There are 25 species, occurring in N. America, Asia, and Africa.

triplespine *See* TRIACANTHIDAE.

tripletail *See* LOBOTIDAE.

triploblastic Applied to animals in which the body wall is derived from three embryonic layers: the *ectoderm; *mesoderm; and *endoderm. All *Animalia other than *Coelenterata are triploblastic.

triploid Applied to a cell that has three sets ($3n$) of *chromosomes in its *nucleus, or to an organism composed of such cells (as opposed e.g. to haploid (n) or diploid ($2n$) cells or organisms). *See also* POLYPLOIDY.

tripodfish *See* TRIACANTHIDAE.

Tripterygiidae (black-faced blenny; subclass *Actinopterygii, order *Perciformes) A family of small, blenny-like fish that have three *dorsal fins, the first two having spiny rays. The tail fin is rounded, the *anal fin is rather long, and the *pelvics are reduced to a few free rays. There are distinct differences between the colour patterns of the sexes, only the males showing a black face. These fish tend to inhabit the rocky shores and coral reefs of tropical and subtropical seas. There are about 95 species, distributed world-wide.

triramous Three-branched.

trisomic Applied to a *genome that is *diploid but which contains an extra *chromosome, *homologous with one of the existing pairs, so that one kind of chromosome is present in triplicate. The chromosome number is thus of the form $2n+1$. Trisomies give multiple doses of genes and may have deleterious effects, as in the human condition of Down's syndrome, which results from a trisomy of chromosome 21.

Triturus cristatus (crested newt, great warty newt) *See* SALAMANDRIDAE.

Triturus helveticus (palmate newt) *See* SALAMANDRIDAE.

Triturus marmoratus (marbled newt) *See* SALAMANDRIDAE.

Triturus vulgaris (smooth newt) *See* SALAMANDRIDAE.

triumph ceremony Behaviour exhibited by monogamous (*see* MONOGAMY) pairs of animals while they are establishing their bond and often following their eviction of an intruder.

triungulin larva In some parasitic insects, an active first larval *instar that hatches in the soil but that has three claws at the end of each of its legs, enabling it to climb a plant where it awaits its insect host. Some *Meloidae produce triungulin larvae.

t-RNA *See* TRANSFER RIBONUCLEIC ACID.

trochantellus An additional segment in the legs of *Hymenoptera which occurs between the *trochanter and the *femur, from which it is thought to be derived.

trochanter 1. The second joint of the leg of an insect, which articulates with the *coxa proximally and with the *femur distally. The trochanter is a small segment and can usually move in only one plane (due to the nature of its articulation with the coxa). *Odonata and *Hymenoptera have two trochanters, but the distal one appears to be derived from the femur. *See also* TRO-CHANTELLUS. **2.** A prominence on a tetrapod *femur; in mammals there are at least two (greater and lesser) and often a third (third trochanter). They are sites for muscle attachment.

Trochilidae (hermits, humming-birds, mangos; class *Aves, order *Apodiformes) A family of small, mainly green, black, or brown birds that have areas of brightly coloured, iridescent plumage on the throat, crown, head, and back. They have slender and pointed, short to long bills, which are straight or *decurved. Their wings are long and narrow, their tails rounded, square, or forked, and some tail feathers are elongated or modified. Their legs are short, with small feet. They are arboreal, inhabit forests, and feed on nectar and insects. Their flight is very rapid and they hover when feeding. They nest in trees. Hermits (22–5 species of *Phaethornis*) inhabit rain forest and make cup-shaped nests at the tips of leaves. There are about 115 genera in the family, with about 330 species, many migratory (*Archilochus colubris*, the ruby-throated humming-bird, migrates between Canada and Central America). Humming-birds are found in N., Central, and S. America, and in the W. Indies. *Mellisuga helenae* (bee humming-bird), measuring about 5 cm from the tip of its bill to the end of its tail, is the smallest bird in the world.

trochophore (trochosphere) In several invertebrates (including *Polychaeta and *Mollusca), a ciliated, usually planktonic larva. The larva is almost spherical, with a ciliated band around its equator and mouth that give it a spinning motion, like a top. The *cilia are also used for feeding. Animals which have a trochophore in their life history are believed to be related.

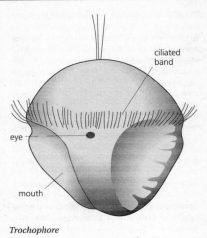

Trochophore

trochosphere *See* TROCHOPHORE.

troglodyte A cave-dweller.

***Troglodytes troglodytes* (wren, winter wren)** *See* TROGLODYTIDAE.

Troglodytidae (wrens; class *Aves, order *Passeriformes) A family of small to medium-sized birds which have brown plumage, some with white and chestnut areas. Their wings, tails, and sides are often barred, streaked, or spotted. Their bills are slender, medium to long, and slightly *decurved, their wings short and rounded, and their tails short to long, and often held cocked. Their legs and feet are strong, with long claws. They inhabit forests, brush, reed-beds, rocks, and deserts, feed on insects and spiders, and most build domed nests in trees or rock cavities, or low in grass or bushes. The four species of *Cistothorus* (marsh wrens), found in America, inhabit grassland as well as marshes. *Troglodytes troglodytes* (wren or winter wren), a small, brown bird with pale *supercilia and barred wings and tail, is the only member of the family found in Europe or Asia. There are 15 genera in the family, with 60–65 species, found in Europe, Asia, and N., Central, and S. America.

Trogon (trogons) *See* TROGONIDAE.

Trogonidae (trogons; class *Aves, order *Trogoniformes) A family of medium to large birds which have bright green or brown backs, green, blue, or violet heads and throats, and pink, red, orange, or yellow breasts. They have rounded wings, and most have long, square-ended tails. *Pharomacrus mocino* (resplendent quetzal) has very elongated upper tail coverts. The 15 species of *Trogon*, found in Central and S. America, have distinctive black and white markings on the underside of the tail. Trogons' bills are broad, with a curved *culmen. Their legs are short, with the first and second toes directed backwards. They are arboreal, inhabit forests and clearings, and feed on insects and fruit, fluttering in front of a leaf to pick off food. They nest in tree cavities and holes in termite nests. There are eight genera, with 37 species, found in Africa, Asia, and Central and S. America. The 11 species of *Harpactes*, found in India, China, south-east Asia, and Indonesia, are the only oriental representatives.

Trogoniformes (trogons; class *Aves) An order that comprises the single family *Trogonidae.

trogons *See* TROGONIDAE.

troph- A prefix, or part of a compound word (e.g. oligotrophic), derived from the Greek *trophe*, 'nourishment', and associating the word in which it occurs with food or nutrition.

trophallaxis Food sharing.

trophic level A step in the transfer of food or energy within a chain. There may be several trophic levels within a system, e.g. producers (autotrophs), primary consumers (herbivores), and secondary consumers (carnivores); further carnivores may form fourth and fifth levels. There are rarely more than five levels since usually by this stage the amount of food or energy is greatly reduced. *See also* ECOSYSTEM.

trophic level assimilation efficiency *See* ECOLOGICAL EFFICIENCY.

trophoblast *See* BLASTOCYST; CHORION.

trophozoite In *Sporozoa, the feeding stage that develops from the *sporozoite.

tropicbirds *See* PELECANIFORMES; PHAETHONTIDAE.

tropin A *hormone that induces the release of other hormones. The pituitary gland releases tropins which cause other glands of the *endocrine system to release hormones unique to those glands, often in a negative feedback relationship to the pituitary secretions.

tropomyosin A highly elongated *protein, of relative molecular mass 70 000, that is composed of two alpha-helix subunits wound about each other. It is associated *in vivo* with *actin and *troponin in the thin filaments of striated muscle and inhibits the contraction of *striated muscles when no calcium is present.

troponin A *globular protein associated *in vivo* with *actin and *tropomyosin in the thin filaments of *striated muscle.

tropo-taxis The movement of an animal, typically in a straight line, in response to a stimulus directly toward or away from the source of the stimulus. It is made possible by the possession by the animal of more than one receptor so that the strength of a stimulus to either side of its body may be detected simultaneously and compared.

troupial (*Icterus*) *See* ICTERIDAE.

trout *See* SALMONIDAE.

troutperch *See* PERCOPSIDAE.

true bugs *See* HEMIPTERA.

true flies *See* DIPTERA.

'true' frogs *See* RANIDAE.

true water beetle *See* DYTISCIDAE.

trumpeters 1. *See* LATRIDAE. **2.** *See* PSOPHIIDAE.

trumpetfish (flutemouth) *See* AULOSTOMIDAE.

Tryblidioidea (phylum *Mollusca, class *Monoplacophora) An order of mainly oval monoplacophorans, fossil forms of which have several symmetrically paired muscle scars. Shells are often thick, and sometimes ornamented with concentric growth lines. Tryblidioidea first appeared in the Early *Cambrian.

Trypanorhyncha (Tetrarhynchidea; phylum *Platyhelminthes, class *Cestoda) An order of parasitic worms in which the *scolex has suckers and *tentacles. The life cycle of Trypanorhyncha is not well studied but they are known to inhabit three hosts during their development.

Trypanosomatidae (class *Zoomasti-gophorea, order *Kinetoplastida) A family of parasitic *Protozoa in which the cells are extremely variable in form, depending on the host and the conditions. Species are parasitic in vertebrates, invertebrates, and plants.

Trypauchenidae (burrowing goby; subclass *Actinopterygii, order *Perciformes) A small family of marine and freshwater fish that have an eel-like, compressed body, a blunt head with minute eyes, and long *dorsal and *anal fins which are confluent with the tail fin. They are usually found hidden in burrows in a muddy or gravelly sea-bottom in the warmer waters of Africa and the Indo-Pacific region. There are about 10 species.

trypsin A proteinase (endopeptidase) *enzyme that acts primarily on the interior bonds of *proteins. It is produced in the pancreas as the inactive *zymogen trypsinogen, which on secretion into the intestine is converted into the active form by the intestinal enzyme enterokinase.

trypsinogen *See* TRYPSIN.

tryptophan A heterocyclic, non-polar, *alpha amino acid.

tsetse flies *See* GLOSSINIDAE.

T-system (transverse tubular system) A system of *anastomozing tubular invaginations of the *plasma membrane that run transversely across *striated muscle cells and insinuate themselves between the *myofibrils. They can undergo depolarization, and their function is to conduct impulses rapidly from the surface of the muscle cell to its interior.

t-test A test to calculate the probability that mean values for a particular measurement are significantly different in two populations.

((( ))) SEE WEB LINKS
• Detailed explanation of the t-test.

tuatara (*Sphenodon punctatus*) *See* SPHENODONTIDAE.

tube feet In *Echinodermata, hollow appendages connected to the water vascular system, used in some species for locomotion and in others for feeding.

tubercle A dome-like projection.

tubercular Knobbed or humped.

tuberous organ An *electroreceptor in weakly electrical *teleosts which detects phasic (rapidly changing) electrical discharges. *Compare* AMPULLARY ORGAN.

tubeshoulder *See* SEARSIIDAE.

tube-snout (*Aulorhynchus flavidus*) *See* AULORHYNCHIDAE.

Tubularina *See* GYMNOBLASTINA.

tubule A very small tube, e.g. a renal tubule, which secretes and collects urine that passes through it. *See* KIDNEY.

Tubulidentata (cohort *Ferungulata, superorder *Protoungulata) An order comprising the monospecific family Orycteropodidae (*Orycteropus afer*, the aardvark or ant bear), an isolated form with a skeleton similar to that of *Condylarthra, for which reason it is included in the Protoungulata. Fossil forms are known from the *Miocene of Africa and the *Pliocene of Eurasia, and some *Eocene and *Oligocene material may also belong to this group. The order derives its name from the teeth, each of which consists of many hexagonal columns of *dentine separated by tubes of pulp. There are about 10 teeth in each jaw of the adult, but there is a full complement of milk teeth in juveniles. The teeth lack *enamel. Although an ant-eater, the jaw of the aardvark is little reduced. The snout is long, the mouth round, and the tongue long. The back is highly curved. There are four digits on each fore limb and five on the hind limbs, with nails apparently intermediate between claws and hoofs, used for digging. The brain is small and primitive. The aardvark is about the size of a pig, feeds almost exclusively on ants and termites, and is distributed widely throughout Africa south of the Sahara. Supposed tubulidentates from the *Pleistocene of Madagascar have recently been placed, by Ross McPhee, in a separate order of mammals, *Bibymalagasia.

Tubulifera *See* THYSANOPTERA.

tubulin The *globular protein from which *microtubules are constructed.

Tubuliporina (class *Gymnolaemata, order *Cyclostomata) A suborder of *Bryozoa in which the colony is variable in overall form, and either encrusts hard substrates or stands erect. Members of the suborder display *polyembryony. The suborder occurs from the *Ordovician to the present.

tuco-tuco *(Ctenomys) See* CTENOMYIDAE.

tucuxi *(Sotalia) See* STENIDAE.

Tullgren funnel A device used to extract small invertebrate animals from a leaf litter or a dry soil sample. The sample is placed in a container with a base made from gauze with a mesh designed to hold soil particles but permit the animals to pass. The container is arranged over a funnel, with a light above. The heat causes the animals to move away from the top of the sample, through the gauze sheet and into the funnel from which they can be collected. Most species are collected after two hours, but complete extraction takes two to three days.

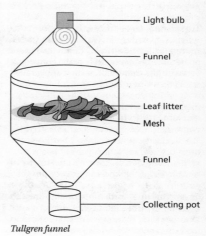

Tullgren funnel

tumblebug *See* SCARABAEIDAE.

Tumbunan division The New Guinean biota found above about 1200 m altitude, and including the Australian rain forests (*a*) between Cooktown and Mackay, and (*b*) between Gympie and Illawarra, whose fauna is closely allied to that of the New Guinea highlands. *See also* IRIAN DIVISION.

tumour An abnormal growth of new tissue that occurs anywhere in the body. Such a growth may be harmless or benign (e.g. a wart). Alternatively it may be malignant, destroying the tissue upon which it grows. It may also invade other tissues through the process of metastasis, in which cells sloughed off from the tumour are carried to other parts of the body. *See also* CANCER CELL.

tuna *See* SCOMBRIDAE.

Tungidae *See* SIPHONAPTERA.

tunic *See* UROCHORDATA.

tunicates *See* UROCHORDATA.

tunicin A substance closely resembling or identical to *cellulose that is found in the protective *test (tunic) of *Urochordata.

Tupaiidae (tree shrews; cohort *Unguiculata, order *Insectivora or (more correctly) Scandentia) A family and order of small, generally arboreal, somewhat squirrel-like mammals, which show characteristics resembling those of primates, with which some authorities still classify them although it is now generally accepted that similarities to primates are illusory or convergent (*see* CONVERGENT EVOLUTION). The brain is relatively large, with a small olfactory region, a *postorbital bar is present in some forms, the ears are small and rounded, the *hallux is slightly opposable, and the digits bear claws. There are three lower *incisors, *procumbent and pressed together to form a lemur-like 'dental comb'. The cheek teeth are primitive. Tree shrews feed on insects and fruit, and are distributed throughout India and south-east Asia. There are five genera, with about 15 species.

Tupinambis teguixin (common tegu, great tegu) *See* TEIIDAE.

turacos *See* CUCULIFORMES; MUSOPHAGIDAE.

Turbellaria (phylum *Platyhelminthes) A class of worms, most of which are not parasites. The epidermis of most is ciliated. Almost all have a cavity for the gut and a ventral mouth with a well-developed *pharynx.

turbinal bones (turbinates) In *Mammalia, delicate, scroll-like bones, covered with mucous membrane, situated in and subdividing the nasal cavity.

turbinate 1. Shaped like a top. **2.** One of the *turbinal bones.

turbot, European (Scophthalmus maximus) *See* BOTHIDAE.

Turdidae (blackbird, bluethroats, chats, nightingales, redstarts, rubythroats, shortwings, thrushes, wheatears; class *Aves, order *Passeriformes) A family of small to medium birds, most of which are brown, grey, black, olive, blue, or white, usually contrasting. Some *Turdus* species (of which there are about 63) have black-spotted, white under-parts. Some of the 29 *Zoothera* species have spotted under-parts or wing bars and many have a distinctive, white, under-wing stripe. Most wheatears (18 species of *Oenanthe*) have distinctive white rumps, and black tails. Redstarts (11–13 species of *Phoenicurus*) have orange bellies, black throats, and black, blue, or grey heads and backs; females are duller brown but both sexes have distinctive orange-red tails and rumps. Turdids have medium-length, slender to stout bills, short and rounded to long and pointed wings, and short to long tails, some forked. Their legs are of medium length, and are 'booted' (have no scales). They are arboreal and terrestrial, inhabiting forests, open country, deserts, and cultivated land, and feed on animal and vegetable matter. They nest in trees and bushes, on the ground, or in tree holes or rock cavities. The 10–12 species of *Monticola* (rock thrushes) inhabit open, rocky areas, scrub, and dry forest, and nest in rock crevices. *Luscinia megarhynchos* (nightingale) is a secretive bird noted for its song (many or all of the 7–18 *Luscinia* species (bluethroats, nightingales, and rubythroats) are often placed in the genus *Erithacus*). Shortwings (six species of *Brachypteryx*) are robin-like birds with short wings and short tails, skulking in habit, and live in dense forest undergrowth. The 10 species of chats, of which many are migratory, comprise the genus *Saxicola*. *Turdus merula* (blackbird) and *T. philomelos* (song thrush) have been introduced to New Zealand which otherwise is the only part of the world from which *Turdus* species are absent. There are about 50 genera in the family, with about 312 species, found world-wide.

Turdoides (babblers) *See* TIMALIIDAE.

Turdus (blackbird, thrushes) *See* TURDIDAE.

Turdus merula (blackbird) *See* TURDI-DAE.

Turdus philomelos (song thrush) *See* TURDIDAE.

turkeys *See* MELEAGRIDIDAE.

Turner's syndrome *See* BARR BODY.

Turnicidae (buttonquails; class *Aves, order *Gruiformes) A family of quail-like, brown birds, which have short, slender bills, short, rounded wings, and short tails. Their feet have only three toes, the hind toe being absent. They are terrestrial and secretive, inhabit grassland and open woodland, feed on seeds and insects, and nest on the ground. There are two genera, with 14 species, found in Europe, Africa, southern Asia, and Australasia.

turnover number A measure of *enzyme activity, normally expressed as the number of moles of *substrate transformed into one or more products per mole of enzyme.

Tursiops truncatus (bottle-nosed dolphin) *See* DELPHINIDAE.

turtle doves (*Streptopelia*) *See* COLUMBIDAE.

turtles 1. *See* CHELONIA. 2. (big-headed turtles, *Platysternon megacephalum*) *See* PLATYSTERNIDAE. 3. (Carolina box turtle, eastern box turtle, *Terrapene carolina*; European pond terrapin, *Emys orbicularis*; freshwater turtles, terrapins) *See* EMYDIDAE. 4. (green turtle, *Chelonia mydas*; hawksbill turtle, *Eretmochelys imbricata*; loggerhead turtle, *Caretta caretta*; marine turtles) *See* CHELONIIDAE. 5. (common musk turtle, stinkpot, *Sternotherus odoratus*; mud turtles, musk turtles) *See* KINOSTERNIDAE. 6. (hidden-necked turtles) *See* CRYPTODIRA; PELOMEDUSIDAE. 7. (leatherback, leathery turtle, *Dermochelys coriacea*) *See* DERMOCHELIDAE. 8. (matamara, *Chelus fimbriatus*; snake-necked turtles) *See* CHELIDAE. 9. (pitted-shell turtle, *Carettochelys insculpta*) *See* CARETTOCHELYIDAE. 10. (side-necked turtles) *See* PLEURODIRA. 11. (snapping turtles) *See* CHELYDRIDAE. 12. (tabasco turtles) *See* DERMATEMYDIDAE.

tusk shells *See* SCAPHOPODA.

tussock moths *See* ARCTIIDAE; LYMANTRIIDAE.

twirler moths *See* GELECHIIDAE.

twisted-wing parasite *See* STREPSIPTERA.

two-legged worm lizards (*Bipes*) *See* AMPHISBAENIDAE.

two-way table *See* CONTINGENCY TABLE.

two-winged flies *See* DIPTERA.

Tylopoda (order *Artiodactyla, suborder *Ruminantia) An infra-order that comprises the *Camelidae (camels and llamas). These are specialized remnants of a much larger group of ruminant-like artiodactyls, also included in the infra-order, that are known to have lived between the *Eocene and early *Pliocene: the Old World families Anoplotheriidae, *Cainotheriidae, *Xiphodontidae, and Amphimerycidae; and the New World families *Agriochoeridae, *Merycoidodontidae, and Dromerycidae. The anoplotheres were stoutly built animals, about 90 cm high at the shoulder; the cainotheres were rabbit-like, in size and probably in habits; the xiphodonts were possibly close to the ancestral line leading to the camels; and the amphimerycids to the line leading to the pecorans. The New World forms were rather pig-like, although the agriochoeres had long tails and redeveloped claws, leading to suggestions that they were arboreal or that they dug for roots and tubers.

tympanic bone In *Mammalia, the bone supporting the *tympanic membrane, derived from the *angular bone of the lower jaw and, in many mammals, forming the *bulla.

tympanic membrane (tympanum) The ear drum, a double layer of epidermis to either side of *connective tissue, which transmits sound waves to the *middle ear.

tympanum *See* TYMPANIC MEMBRANE.

type series In taxonomy, all the specimens on which the description of a taxon is based.

type specimen (holotype) In taxonomy, the individual animal that is chosen to serve as the basis for naming and describing a new species or variety. *Compare* LECTOTYPE; SYNTYPE.

Typhlomys (Chinese pygmy dormouse) *See* PLATACANTHOMYIDAE.

Typhlonectidae (typhlonectids; class *Amphibia, order *Apoda) A family of caecilians that are aquatic throughout life. The tail is absent. There are no dermal scales. Reproduction is *ovoviviparous. The gills are bag-like; the gill slits close before hatching. Typhlonectids feed mainly on small, aquatic vertebrates. There are 18 species, all S. American.

Typhlopidae (blind snakes; order *Squamata, suborder *Serpentes) A family of primitive burrowing snakes that have a small, rigid skull with teeth only on the upper jaw. The eyes lie beneath scales. The body is worm-like, with a short tail, often ending in a spine. A rudimentary pelvis is present. The scales are small (not enlarged ventrally). There are about 200 species, found in the Old and New Worlds, and Australia.

typhlosole A longitudinal fold formed by an *invagination in the inner wall of the *intestine found in some invertebrates, including *Annelida and bivalve molluscs (*Mollusca), and in *ammocoetes that increases the area available for absorbing nutrients.

Tyrannidae (attilas, elaenias, kingbirds, phoebes, tyrant flycatchers; class *Aves, order *Passeriformes) A diverse family of small to medium-sized birds which are usually grey, brown, or olive-green, some being more brightly coloured. The bill is usually fairly broad with a hooked tip and *rictal bristles, the wings are short and rounded to long and pointed, and the tail is usually medium-length and square, but can be greatly elongated. Kingbirds (12 species of *Tyrannus*) are typical. (Phoebes (three species of *Sayornis*) have longish tails they constantly flick downwards.) Tyrant flycatchers inhabit forests and open country, and feed on insects and fruit, some eating mice, frogs, and small birds. They nest in trees or on the ground. Some nests are open, others domed or pendant, and some are in tree holes or holes in the ground. The 17 species of *Elaenia* (elaenias) are typical. The 17 *Empidonax* species hunt from exposed perches and are recognizable by their voices. The six species of *Attila* (attilas) were formerly placed in the *Cotingidae. There are about 115 genera, with about 380 species, many migratory, found in N., Central, and S. America.

Tyrannosaurus rex (order *Saurischia, suborder *Theropoda) A giant, carnivorous dinosaur which lived during the Late *Cretaceous in N. America and Asia (Mongolia and Japan) and possibly Europe (Portugal). Individuals grew to 12 m in length, 5 m tall, and weighed about 7 tonnes. The name means 'king of the tyrant lizards', but some palaeontologists believe *T. rex* was probably a scavenger with a top speed of 16–40 km/h.

Tyrannus (kingbirds) *See* TYRANNIDAE.

tyrant flycatchers *See* TYRANNIDAE.

tyrosine An aromatic, polar, *alpha amino acid.

Tyto (barn owls) *See* TYTONIDAE.

Tytonidae (barn owls, bay owls; class *Aves, order *Strigiformes) A family of owls that have heart-shaped facial discs which meet above the bill in *Tyto* species (barn owls) but not in *Phodilus* species (bay owls). They have feathered legs, and a *pectinate middle toe. They inhabit forests and grassland, feed (mainly at night) on mammals, birds, lizards, frogs, and insects, and nest mainly in tree cavities. There are two genera, with 12 species, found in Europe, Africa, southern Asia, Indonesia, New Guinea, Australia, and N., Central, and S. America.

ubiquinone *Coenzyme Q; a generic term for a group of compounds, structurally related to *vitamin K, that function as electron carriers in the *electron-transport chain of *mitochondria.

ubiquitin A small, highly conserved regulatory *protein that occurs on the surfaces and inside of all *eukaryotic cells. Originally called ubiquitous immunopoietic polypeptide, its name refers to the fact that it is so widespread. It labels proteins that are to be degraded and controls the stability, function, and localization of a variety of proteins. It also causes certain cells to find 'roosting sites' within the body. The process of marking a protein with ubiquitin is known as ubiquitination or ubiquitylation.

ubiquitination *See* UBIQUITIN.

ubiquitylation *See* UBIQUITIN.

uintatheres *See* DINOCERATA.

ulna In tetrapods, the longer and thicker of the two bones of the fore limb; its upper end articulates with the *humerus.

ultimate explanation The explanation that lies behind the *proximate explanation. For example, many animals breed in spring. The proximate explanation for this is that changing day length and rising temperature trigger the hormonal activity that leads individuals to seek mates, build nests, and mate. But that is only a partial explanation. The ultimate explanation is that an evolutionary process has selected for animals that breed at a time of year when their young will have the best access to food and mild temperatures (that reduce energy requirements).

ultracentrifugation Centrifugation carried out at high rotor speeds (up to 75 000 rpm) and therefore under high centrifugal forces (up to 750 000 g). Analytical ultracentrifuges are employed to determine the mass and to some extent the shape of molecules; preparative ultracentrifuges are used to separate the components of mixtures on the bases of these parameters.

ultrastructure The detailed cellular structure of a biological specimen that can be studied by electron microscopy.

umbilical cord (birth cord) In mammals, the cord that connects the *embryo and *foetus to the *placenta.

umbilicus 1. (navel) In mammals, the scar marking where the *umbilical cord became detached. **2.** In shelled *Mollusca, the conical space between the coils of the shell.

umbo (*pl.* umbones) A convex protuberance or swelling in the centre. In *Brachiopoda and *Bivalvia, the first part of the shell to be formed; it occurs at the posterior of each valve in brachiopods and forms the dorsal part of the shell in bivalves.

umbonate Having or resembling an *umbo.

umbones *See* UMBO.

umbrella A flat, circular structure, such as the disc or bell of a *medusa.

umbrella mouth gulper (*Eupharynx pelecanoides*) *See* EURYPHARYNGIDAE.

umbrellar Pertaining to an *umbrella.

Umbridae (mud minnow; superorder *Protacanthopterygii, order *Salmoniformes) A small family of freshwater fish that have relatively short, stout bodies. The soft-rayed, single, *dorsal and *anal fins are placed well back on the body, near the somewhat rounded tail fin. The small *pelvics are found just anterior to the anal fin. Capable of surviving harsh conditions (e.g. low oxygen concentration and low water temperature), mud minnows have an interesting distribution: USA, Alaska, Siberia, and eastern Europe. There are about five species.

uncinate process In birds, a protrusion of bone from a rib, enabling it to join on to the rib behind it by means of ligaments. It is an avian feature that serves to strengthen the rib-cage, and it is absent only in the *Anhimidae (screamers).

unconditional stimulus See CONDITIONING.

underdominance See BALANCED POLYMORPHISM.

underwing moths See NOCTUIDAE.

unequal crossing-over In genetics, a *crossing-over after improper pairing between *homologous chromosomes that are not perfectly aligned. The result is one cross-over *chromatid with one copy of the segment and another with three copies. The phenomenon was first described at the Bar locus in *Drosophila*.

Unguiculata (subclass *Theria, infraclass *Eutheria) A cohort of mammals that includes those orders in which ancestral mammalian characteristics have been largely preserved (e.g. *Insectivora, *Chiroptera, *Dermoptera, *Edentata, *Pholidota, and *Primates).

ungulate 1. Any hoofed, grazing mammal, which usually is also adapted for running. Hoofed mammals occur in several mammalian groups, and the term 'ungulate' no longer has any formal taxonomic use. **2.** Hoof-shaped.

unguligrade Applied to a gait in which only the tips of the digits, covered with hoofs, touch the ground (e.g. in *Artiodactyla and *Perissodactyla). The limbs are moved as a whole by the action of shoulder and hip muscles.

unicuspid Applied to a tooth that has only one *cusp.

unilocular Containing one chamber.

Unionida See GLOCHIDIUM.

Unionoida (phylum *Mollusca, class *Bivalvia) An order of unequilateral, and generally equivalved, marine and non-marine bivalves whose shell shape and size is very variable and whose ornamentation is usually simple. The musculature is *isomyarian to *sub-isomyarian. The ligament is external and *opisthodetic. The dentition is quite variable, forms having one or two teeth, some pseudotaxodont and some *edentulous. They first appeared in the Late *Devonian.

uniport carrier A type of *facilitated diffusion that allows a single molecule or ion to cross a *membrane without the involvement of any other molecule or ion.

Uniramia Formerly a phylum or a subphylum of *Arthropoda, in which the appendages are unbranched (i.e. uniramous), and at one time classed as members of the *Mandibulata. The group includes the myriapods (centipedes and millipedes) and the insects, the earliest known fossils being *Devonian and resembling modern centipedes and millipedes. The Arthropoda are now held to have evolved only once (they are monophyletic) and the formal and informal use of Uniramia has been abandoned, with its members reallocated to the subphylum Atelocerata.

uniramous Unbranched.

uniserial (uniseriate) Arranged in a single row.

uniseriate See UNISERIAL.

unit-membrane model The description of all the *membranes of a *cell (i.e. the cell membrane and membranes enclosing *organelles) as having a common structure, revealed by electron microscopy as two dark bands, each about 2 nm thick, separated by a lighter band about 3.5 nm thick. Some authorities doubt that this model describes all cell membranes and that it applies to the intracellular membranes of all cell inclusions.

unit of behaviour See BEHAVIOURAL UNIT.

univalent A single *chromosome observed during *meiosis when bivalents are also present. A univalent has no pairing (synaptic) mate. An example is the *sex chromosome of an XO male.

univoltine Applied to species in which one generation reaches maturity each year.

unques In *Diptera, the 'claws' at the tip of each foot.

upright display See FORWARD DISPLAY.

upside-down catfish See MOCHOKIDAE.

Upupa epops (hoopoe) See UPUPIDAE.

Upupidae (hoopoe; class *Aves, order *Coraciiformes) A monospecific family (*Upupa epops*), which is a pinkish, chestnut bird with black and white banded back, wings, and tail. It has a long, distinctive, black-tipped crest, a long, slender, *decurved bill, broad, rounded wings, and a medium-length, square tail. Its legs are short with long toes, the third and fourth being fused basally. It inhabits fairly open country and cultivated land, feeds mainly on insects, and nests in holes in trees and banks. It is migratory, and is found in Europe, Africa, and Asia.

uracil A *pyrimidine base that occurs in *RNA.

Uraniidae (subclass *Pterygota, order *Lepidoptera) Small family of small to large moths related to the *Geometridae.

The hind wings are extended into 'tails' and resemble those of papilionid butterflies. Many species are diurnal; the nocturnal members are whitish. The larvae pupate in loosely woven cocoons. The family is predominantly tropical and occurs in both Old and New Worlds. There are 90 genera and about 700 species.

Uranoscopidae (stargazer; subclass *Actinopterygii, order *Perciformes) A family of marine, bottom-living fish that have a solid, stout body, the eyes located on the top of the large head, and a large mouth in an almost vertical position. A distinct notch separates the two parts of the *dorsal fin, with the *anal fin being larger than the second dorsal. Some species have a well-developed and poisonous spine behind the *gill cover. They often lie partially buried in the sand with only the mouth and eyes visible, wriggling a small, worm-like tentacle on the lower jaw in order to attract prey fish. There are about 25 species, distributed world-wide.

Uratelornis chimaera (long-tailed groundroller) See BRACHYPTERACIIDAE.

urceolate Flask-shaped.

urea $CO(NH_2)_2$; the compound that is formed from carbon dioxide and ammonia via the *urea cycle as a result of *amino-acid catabolism in *ureotelic animals. It is also the final product of *purine catabolism in most fish, *Amphibia, and freshwater bivalve molluscs (*Bivalvia).

urea cycle (ornithine cycle) A cyclic series of reactions by which carbon dioxide and ammonia, the latter derived from amino groups from *amino-acid catabolism, are converted to *urea prior to excretion.

((⊕)) SEE WEB LINKS
• Detailed description of the urea cycle.

urease The hydrolytic *enzyme that catalyses the reaction whereby *urea is converted to carbon dioxide and ammonia.

ureotelic Applied to organisms that excrete in the form of *urea nitrogenous waste derived from *amino-acid catabolism (e.g. *Elasmobranchii, adult *Amphibia, and *Mammalia). *See also* AMMONOTELIC; URICOTELIC.

ureter The urinary duct that drains from the kidney to the *bladder.

urethra In vertebrates, the duct through which urine flows from the bladder and out of the body. In males of most species it is also the duct through which *semen flows to the penis.

uric acid A nitrogen compound ($C_5H_4N_4O_3$) that forms the principal end-product of *amino-acid catabolism in insects, birds, and reptiles. It is also present in the urine of other organisms where it represents the final product of *purine catabolism.

uricotelic Applied to organisms that excrete in the form of *uric acid nitrogenous waste derived from *amino-acid catabolism (e.g. insects, birds, and terrestrial reptiles). *See also* AMMONOTELIC; UREOTELIC.

uridine The *nucleotide formed when *uracil is linked to *ribose sugar.

uridylic acid The *nucleotide formed from *uracil.

Urochordata (sea squirts, tunicates; phylum *Chordata) A subphylum that has a reliable fossil record from the *Permian, and possible remains in *Silurian sediments. Urochordates have tadpole-like larvae with *notochords in their tails and a dorsal, tubular nerve cord, features that link them with the Chordata. The adults have a gelatinous or leathery protective *test (the tunic) made from a chemical very similar to the cellulose found in plants. The tunic has two openings, allowing water to flow through the gill slits in the *pharynx. The water flow is maintained by ciliated cells, and food is trapped by *mucus on the pharyngeal gill slits.

Urodela (Caudata; newts, salamanders; subphylum *Vertebrata, class *Amphibia) A modern order of tailed amphibians, of which there is a Late *Jurassic representative in the fossil record. This order, and the other two living amphibian orders (collectively grouped into the subclass Lissamphibia), seem on the basis of their vertebral characteristics to be descended from the *Palaeozoic *Lepospondyli, but the teeth and other characteristics of the Lissamphibia are unlike those of any Palaeozoic amphibians. Most are four-legged and lizard-shaped, but some are elongate and eel-like, with the limbs degenerate. The tail is never lost at *metamorphosis. Fertilization is internal, *spermatophores being transferred during an elaborate courtship ritual. *Sexual dimorphism is common, with breeding colours and median-fin enlargement in the males of some species. The usual length is 7–30 cm, but the giant salamanders (*Cryptobranchidae) may reach 150 cm. Distribution is largely in the northern temperate zone, but some genera span the Equator into S. America. There are about 450 species in eight families: *Ambystomatidae; Amphiumidae; Cryptobranchidae; *Hynobiidae; *Plethodontidae; *Proteidae; *Salamandridae; and *Sirenidae.

***Uromastyx* (spiny-tailed lizards)** *See* AGAMIDAE.

uropatagium In *Chiroptera, the part of the *patagium which extends between the hind limbs.

Uropeltidae (shield-tailed snakes; order *Squamata, suborder *Serpentes) A family of primitive, burrowing snakes which resemble the blind snakes (*Typhlopidae) but possess an enlarged, modified scale near the tip of the tail (possibly an aid to burrowing or to anchor them in the ground). Reproduction is *viviparous. There are about 45 species, occurring in India and Sri Lanka.

uropod In *Malacostraca, a *ramus of the sixth abdominal appendage, comprising a large, flattened, paddle-like structure which, with the *telson, forms a tail fin used in swimming.

Uropygi (schizomids, whipscorpions; subphylum *Chelicerata, class *Arachnida) Order of small to medium-sized (up to 7.5 cm long) arachnids in which the prosoma is not divided dorsally, and the large *pedipalpi are held flexed and parallel to the ground. The last abdominal segment bears a many-segmented *flagellum. Whipscorpions possess anal spray glands used in defence, the secretions being largely acetic acid with some caprylic acid to aid cuticle penetration. Usually brownish and flat, whipscorpions live under stones, stump bark, and rubbish, and prey on small arthropods (e.g. crickets) and small toads. Schizomids are smaller, up to 7 mm long, and possess a very short, terminal, abdominal flagellum. The pedipalpi are leg-like and raptorial, not *chelate. The two small *chelicerae are tipped by small pincers. As in the whipscorpions, there are anal glands, and schizomids live under stones and leaf-litter. Uropygids possess sensory *trichobothria, *slit organs, and pits, inhabit the tropics and subtropics, and have reproduction similar to that of scorpions. There are 75 known species of whipscorpion and more than 50 known species of schizomid.

uropygium In birds, the fleshy structure attached to the *pygostyle that bears the tail feathers.

urostyle In some fish and amphibians, a long bone formed from several vertebrae fused together.

Ursidae (bears; suborder *Fissipedia (or *Caniformia), superfamily *Canoidea) A family of large, slow-moving mammals whose ancestors were the last group of canoids to diverge from the ancestral 'dog' line, some time during the *Miocene and probably soon after that line became established. They are distinctive within the order *Carnivora, for the departure involved an adaptation to an omnivorous and mainly herbivorous diet. The last *molar is absent, the remaining molars being elongated and having grinding surfaces, and the *carnassials are weakly developed. The gait is *plantigrade, the claws non-retractile. Bears are distributed throughout much of the northern hemisphere apart from Africa, and in S. America. Traditionally there have been seven species, in five genera; but nowadays the giant panda (*Ailuropoda*) is also included as a sixth genus.

uterus In female mammals (except *Monotremata), the organ in which the *embryo develops. It is paired in most species but single in *Primates, and is connected with *Fallopian tubes and to the exterior by the vagina. The uterus undergoes changes during the *oestrus (or *menstrual) cycle, and in oestrus its glandular lining provides a spongy layer in which a fertilized egg may be implanted. The outer walls of the uterus are thick and muscular, their contractions during birth forcing the *foetus through the vagina.

u

vaccine A preparation that, when introduced into the body, stimulates the production of *antibodies. A vaccine contains a substance that resembles an organism capable of causing disease and it may consist of killed or attenuated microorganisms, or products or derivatives obtained from microorganisms.

vacuity A gap between bones of the skull.

vacuole A membrane-bound sac found in many cells; vacuoles normally act as storage organs of various types.

vacuum activity Patterns of behaviour that occur in the absence of the external stimuli that normally elicit them.

vagabond butterfly fish (*Chaetodon vagabondus*) See CHAETODONTIDAE.

vagile Applied to a plant or animal that is free to move about. *Compare* SESSILE.

vagility The tolerance of an organism to a wide range of environmental conditions; vagility may be qualified as 'high' or 'low'.

valency A measure of the number of other ions in a chemical element that can be combined with a particular atom.

valine An aliphatic, non-polar, *alpha amino acid.

Valvatida (class *Stelleroidea, subclass *Asteroidea) An order of starfish most of which have suckers on their *tube feet, marginal plates, and *paxillae on the *aboral surface. Many inhabit soft substrates and these species lack suckers.

valve 1. One of the two halves of the hinged shell of *Mollusca (e.g. *Bivalvia and *Polyplacophora). **2.** A flap that can close to ensure that a fluid flows in only one direction (e.g. in a blood vessel or heart). **3.** *See* OVIPOSITOR.

vampire bat *See* DESMODONTIDAE.

Vampyromorpha (class *Cephalopoda, subclass *Coleoidea) An order of cephalopods which have eight arms and two retractable arms. An uncalcified, internal shell is also present. The order contains a single family, the Vampyroteuthidae.

Vampyroteuthidae *See* VAMPYROMORPHA.

Vanellus (plovers, lapwings) *See* CHARADRIIDAE.

vanessas *See* NYMPHALIDAE.

vangas *See* VANGIDAE.

Vangidae (vangas; class *Aves, order *Passeriformes) A family of small to medium-sized birds which have contrasting black, blue, white, brown, or grey, patterned plumage. Their bills are stout and hooked, and vary from slim and *decurved to deep and enlarged. They have short to long, rounded wings, and fairly long tails. They are arboreal, inhabit forests and brush, feed on insects and other animals, and nest in trees. There are eight genera, with 12 species, confined to Madagascar.

Van Valen's 'law' When plotted as cumulative curves on a logarithmic scale, the duration frequencies of many *species tend to show a more or less straight-line relationship with time. In effect, taxa generally become extinct regardless of their age. Duration frequencies vary greatly among different groups (e.g. on average, bivalve genera last 10 times longer than mammalian genera). The 'law' is named after its discoverer, Leigh Van Valen (1935–2010).

Varanidae (monitors; order *Squamata, suborder *Sauria) A family of large, agile, predatory lizards which typically have a pointed head, a large *tympanal membrane, a long, slender neck, a massive body, powerful, clawed limbs, and a long, thick tail. The scales are small and granular. Most monitors are good climbers and swimmers. *Varanus komodoensis* (Komodo dragon) is the largest extant lizard, males (which are larger than females) growing to 3 m long and a weight of 140 kg; adults feed mainly on deer, pigs, and monkeys (often as carrion), the young on insects and small vertebrates. Komodo dragons occur on the Indonesian islands of Komodo, Rintja, Flores, and Padar. There are about 30 species of monitors, occurring in tropical and subtropical regions of the Old World and Australia.

***Varanus komodoensis* (Komodo dragon)** *See* VARANIDAE.

***Varecia* (ruffed lemur)** *See* LEMURIDAE.

variance *See* MEAN SQUARE.

variation Differences displayed by individuals within a *species, and which may be favoured or eliminated by *natural selection. In sexual reproduction, reshuffling of genes in each generation ensures the maintenance of variation. The ultimate source of the variation is *mutation, which produces fresh genetic material.

vasomotor Applied to the nerves of the *autonomic nervous system that control the diameter of blood vessels.

vasopressin (antidiuretic hormone, ADH) A *peptide *hormone, secreted by the *neurohypophysis in mammals, that increases blood pressure and regulates the rate of water loss by the *kidneys.

vector 1. An organism that carries a disease-causing organism from an infected individual to a healthy one; the vector may transfer the pathogen passively or may itself be infected by it. **2.** In *genetic engineering, a DNA molecule, derived from a *plasmid or bacteriophage, into which fragments of DNA may be inserted. The vector carries this DNA into a host cell (e.g. a bacterium). Vector DNA contains an origin of replication so it can re-duplicate itself and the inserted DNA inside the host cell.

vegetal pole *See* ANIMAL POLE.

vegetative nucleus The macronucleus of a ciliate animal.

vein 1. A blood vessel through which blood is conveyed towards the heart. **2. (nervure)**

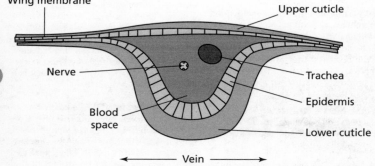

Labels: Wing membrane, Upper cuticle, Nerve, Trachea, Blood space, Epidermis, Lower cuticle, Vein

Vein

In an insect wing, one of the 'struts' that provides strength and rigidity. The wing is formed from two membranes each comprising a layer of *cuticle overlying a layer of *epidermis. Veins occur where the two epidermal layers are not in contact and are usually thickened, forming a tube between the upper and lower wing surfaces. A nerve and *trachea pass along the vein, which is also linked to the *haemocoel, so it has a blood supply. The arrangement of veins forms a pattern that is important in the classification and identification of species. See VENATION.

Veliferidae (veil-fish; subclass *Actinopterygii, order *Lampridiformes) A small family of deep-bodied, marine fish, that have long and high, multi-rayed, *dorsal and *anal fins, and a strongly forked tail fin. The mouth is small, but highly *protractile. There are three species inhabiting the deeper waters of the Indian and Pacific Oceans.

veliger In *Mollusca, a shelled larva, developed from a *trochophore, which possesses two large, semicircular folds bearing *cilia, used in swimming and feeding, lost as the veliger develops, and a *mantle, foot, shell, and other adult features. It is in the veliger stage of *Gastropoda that *torsion occurs (some scientists believe torsion developed as a larval defensive mechanism, the twisting through 180° enabling the larva to withdraw its head and *velum into the protective shell with the opening guarded by an *operculum on the end of the foot).

velum From the Latin *velum*, 'ship's sail'. **1.** In *Mollusca, the ciliated lobe possessed by the *veliger larva. **2.** In most hydroid medusae (*Cnidaria), a ridge or narrow shelf inside the border of the subumbrellar (*see* UMBRELLAR) cavity which reduces the size of the subumbrellar aperture and so increases the force of the water jet by which the *medusa swims vertically.

velvet *See* ANTLERS.

velvet ants *See* MUTILLIDAE.

velvetfish *See* APLOACTINIDAE.

velvet worms *See* ONYCHOPHORA.

vena cava In tetrapods, one of the main veins draining blood to the heart. The vena cava superior serves the fore limbs and head, usually as a pair of veins, but in many mammals only the right member of the pair persists in adults. The vena cava inferior is a single median vein, the largest in the body, serving almost all of the body behind the fore limbs. In invertebrates with a closed circulation (i.e. those in which blood circulates), a vessel that conveys blood to the heart.

venation The pattern formed by the arrangement of the *veins on the wings of insects. The veins are identified by a system of letters and numbers which allows their precise arrangement to be described. Venation is frequently used as a taxonomic character.

Veneroida (phylum *Mollusca, class *Bivalvia) An order of mainly equivalve bivalves which have true *heterodont dentition, an *opisthodetic ligament, and an *isomyarian musculature. Most are *vagile or *nestling. The shells are commonly ornamented with concentric ribbing. Veneroida first appeared in the Middle *Ordovician.

venomous lizards *See* HELODERMATIDAE.

ventral **1.** The surface of an organism or structure that is usually nearest to the *substrate. **2.** In *Chordata, the surface or structure that is furthest from the *notochord.

ventral fin *See* FIN; PELVIC FIN.

ventral tube *See* COLLEMBOLA.

ventricle **1.** A space or hollow organ. **2.** One of four interconnecting cavities in the vertebrate *brain. **3.** One of two chambers in the *heart that receive blood from the *auricles and pump it into the *arteries.

verdin *See* REMIZIDAE.

v

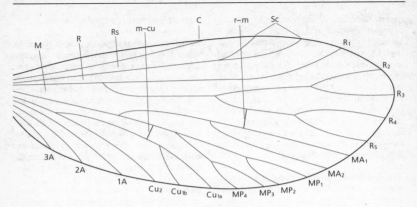

C = costa; Sc = subcosta; R = radius; Rs = radial sector; M = media; MA = anterior branch of M; MP = posterior branch of M; Cu = cubitus; A = anal; m–cu = cross vein M–Cu; r–m = cross vein R–M

Venation

Vermivora (New World warblers) *See* PARULIDAE.

vernal In the late spring. The term is used with reference to the six-part division of the year used by ecologists, especially in relation to terrestrial and freshwater communities. *Compare* AESTIVAL; AUTUMNAL; HIBERNAL; PREVERNAL; SEROTINAL.

verrucose Warty in appearance.

vertebra In the *axial skeleton of vertebrates, one of a series of bony segments formed at the *skeletagenous septum and *myoseptum junction which replace the *notochord, forming the vertebral column (or spinal column or backbone), which encases and so protects the spinal cord. Vertebrae differentiate into five types from anterior to posterior: cervical; thoracic; lumbar; sacral (in humans five vertebrae that fuse to form the sacrum); and caudal or coccygeal (in humans forming the coccyx). Cervical vertebrae facilitate the mobility of the head. The first two vertebrae of the vertebral column, the atlas and axis, are highly specialized cervical vertebrae, the former articulating with the *occipital region of the *cranium. The thoracic vertebrae articulate with *ribs that fuse with the *sternum. Lumbar vertebrae are generally larger, with abbreviated ribs fused to the centrum and supporting the posterior coelomic musculature. Sacral vertebrae fuse with the *pelvis, allowing the transfer of force to the *appendicular skeleton. Caudal vertebrae are smaller and less specialized, forming the tail of the organism. Six anatomical features are usually recognizable in vertebrae: the centrum is a solid cylinder which surrounds and often replaces the notochord, forming the central body of the vertebra; the neural arch forms a dorsal ring surrounding the spinal cord; a hemal arch grows ventrally on post-anal vertebrae, enclosing blood vessels; neural and hemal spines are anterior/posterior-oriented blades of bone that project dorsally and ventrally respectively; apophyses are bilaterally paired projections to which musculature is usually attached, including prezygapophyses and postzygapophyses, which occur on the anterior and posterior ends of a vertebra respectively and articulate with zygapophyses of adjacent vertebrae; transverse processes are bilaterally paired lateral projections at each side of the neural arch with which the rib articulates.

vertebral column *See* VERTEBRA.

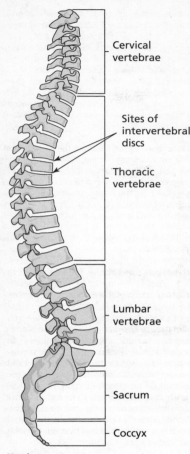

Cervical
vertebrae

Sites of
intervertebral
discs

Thoracic
vertebrae

Lumbar
vertebrae

Sacrum

Coccyx

Vertebra

Vertebrata (kingdom *Metazoa, phylum Chordata) The subphylum which includes the fish, *Amphibia, *Reptilia, *Aves, and *Mammalia. These appear successively in the fossil record, starting in the *Ordovician (although traces of fish are now known from the Late *Cambrian). The lower chordates (e.g. the living *amphioxus*) are soft-bodied and first appear in the Cambrian.

vertex The top of the head of an insect, behind and between the eyes.

verticillate Arranged in a whorl or whorls.

vesicle 1. Generally, any small bladder-like structure containing a fluid. **2.** A small, membrane-bound, fluid-filled sphere that occurs, often in large numbers, in the *cytoplasm of many *eukaryotic cells. Vesicles may be variously associated with the uptake and discharge of materials in cells and also with their transport and storage.

Vesicularina (class *Gymnolaemata, order *Ctenostomata) A suborder of *Bryozoa in which the *stolon is stout and erect. They are known only from the *Holocene.

***Vespa crabro* (hornet)** *See* VESPIDAE.

Vespertilionidae (European bats; order *Chiroptera, suborder *Microchiroptera) A family of insectivorous bats in which the nose is usually simple, the tail extends to the edge of the *uropatagium, and the ears are simple and usually immobile. Some forms have long ears and *narial pads on the nose. *Tylonycteris* has sessile adhesive pads on all limbs. *Pipistrellus pipistrellus* is the common pipistrelle bat. The family has a world-wide distribution, except for the Arctic and Antarctic regions. There are 42 genera, with about 320 species.

Vespidae (suborder *Apocrita, superfamily Vespoidea) The principal family of social wasps, containing many common species, whose members are social, predatory, and more or less *melliferous. Females can often inflict painful stings. The fore wings are usually folded longitudinally at rest. *Mandibles are usually short and broad, with overlapping apices. Most species are black and yellow with banded abdomens and yellow or white facial markings (they are often called 'yellow-jackets'; although this common name is sometimes restricted to *Vespa crabro*, the hornet). Members of some subfamilies feed their young on pre-chewed insects, and make their nests of a papery cellulose construction; other subfamilies are solitary wasps, provisioning their nests with paralysed prey.

V

vestigial organs Atrophied or non-functional organs that are well developed and fully functional in other members of the group. They result from the adoption of a way of life in which the organs are no longer required (e.g. rudimentary pelvic bones are found in some whales, but are no longer required since whales have no hind limbs as such). Such features are difficult to reconcile with the concept of *special creation (*see also* CREATIONISM; CREATION 'SCIENCE'; INTELLIGENT DESIGN).

Vestimentifera A group of animals, described in 1985 by M. L. Jones, that contains some species of large worms which had previously been assigned to the *Pogonophora. They differ from Pogonophora mainly in the detailed structure of the coelomic cavities (*see* COELOM) and the presence of metanephridia instead of protonephridia (*see* NEPHRIDIUM). They live near hydrothermal vents, mainly in the eastern Pacific. Formerly classed as a phylum, the Vestimentifera are now included in the family *Siboglinidae, together with the Pogonifera.

🌐 SEE WEB LINKS
• Detailed description of the Vestimentifera.

viability The probability that a fertilized egg will survive and develop into an adult organism.

vibrissae 'Whiskers'; in most *Mammalia, and in many other animals, stiff hairs or modified feathers projecting from the face and in some species from the limbs, vibration of which stimulates nerve sensors in the skin.

vicariad *See* VICARIANCE.

vicariance (vicariad, vicarious species) The geographical separation of a *species so that two closely related species or a species pair result, one species being the geographical counterpart of the other.

vicariance biogeography A school of biogeographical thought, derived from *panbiogeography, whose supporters maintain that the distribution of organisms depends on their normal means of dispersal (e.g. disjunctions are explicable in terms of new barriers (rivers, rises in sea level, etc.) having split formerly continuous ranges, rather than in terms of the organisms hopping over already existing barriers). Thus they reject *sweepstakes routes and similar concepts, postulating instead former *land bridges and even vanished continents where there is sufficient coincident plant and animal distribution.

vicarious distribution The distribution that results from the replacement of one member of a species pair (i.e. two closely related species derived from a common ancestor) by the other, geographically (as opposed to ecologically). There are many animals and herbaceous plants with a vicarious distribution between N. America and Europe. In zoology, subspecies are conspecific vicariants.

vicarious species *See* VICARIANCE.

Victoriapithecidae *See* CERCOPITHE-COIDEA; VICTORIAPITHECUS.

Victoriapithecus The earliest known member of the *Cercopithecoidea. Together with a related, more poorly known genus, *Prohylobates*, it is now placed in a family, Victoriapithecidae, separate from advanced Old World monkeys. So far, the family is known only from E. and N. Africa. The *bilophodonty characteristic of Old World monkeys is only partially developed in Victoriapithecidae and in other respects, too, they are very primitive.

vicuña (*Lama vicugna*) *See* CAMELIDAE.

vigilance The readiness of an animal to detect certain specified events that occur unpredictably in its environment.

Villafranchian The land-mammal stage that spans the upper *Pliocene and lower *Pleistocene (3.6–1.2 Ma ago).

villus (*pl.* villi) A small outgrowth from the surface of certain *tissues that serves to increase the surface area. In mammals, villi occur in large numbers in the *duodenum (where they are finger-like) and *ileum (where they are paddle-like); they are covered in absorptive *epithelium, each with a *lymph vessel, are supplied with blood vessels, move constantly, and absorb nutrients in solution. In mammals, villi also occur on the surface of the *chorion in the *placenta where they increase the area available for the transfer of materials between the maternal and foetal blood.

Vipera **(vipers)** *See* VIPERIDAE.

Vipera ammodytes **(horned viper, long-nosed viper, nose-horned viper, sand viper)** *See* VIPERIDAE.

Vipera aspis **(asp, aspic viper, asp viper)** *See* VIPERIDAE.

Vipera berus **(adder, common viper, northern viper, viper)** *See* VIPERIDAE.

viperfish *See* CHAULIODONTIDAE.

Viperidae (vipers; order *Squamata, suborder *Serpentes) A family of compact, sturdy snakes, most of which have a broad triangular head, small, ridged scales, a short tail, and drab coloration. The genus *Vipera*, with about 15 species, is typical: the moderately large head is broader than the neck, the eyes have vertical pupils, the tail is short and conical, body scales are strongly keeled, the anal shield is undivided, and the subcaudal plates are paired. Many vipers are *ovoviviparous (e.g. *Vipera ammodytes*, horned viper, long-nosed viper, nose-horned viper, or sand viper; *V. aspis*, asp, aspic viper, or asp viper; and *V. berus*, adder, common viper, northern viper, or viper). *V. berus* is unusual in displaying marked *sexual dimorphism, the male being more distinctively marked than the female. *V. ammodytes*, of southern Europe and the Near East, is stockily built and has a protruding horn on the snout. *Bitis arietans* (puff adder), common in sub-Saharan Africa, feeds on small vertebrates but is highly venomous and dangerous to humans; when provoked it inflates the body and hisses, while preparing to strike. The related *B. nasicornis* (rhinoceros viper

or river-jack) of Central Africa has a pair of large, prominent, horn-like scales above each nostril. *Ophiophagus hannah* (king cobra), of India, Malaya, and southern China, the longest of all venomous snakes (sometimes exceeding 5 m), was formerly included in the Viperidae; it is now placed in the *Elapidae. Viper venom is generally haemotoxic (blood-poisoning) and is injected down hollow fangs, which rotate back against the palate when the mouth is closed. There are about 100 species, occurring in Europe, Asia, and Africa.

viperine snake (*Natrix maura*) *See* COLUBRIDAE.

vipers *See* VIPERIDAE.

Vireo **(vireos)** *See* VIREONIDAE.

Vireolaniidae (shrike-vireos) *See* VIREONIDAE.

Vireonidae (vireos; class *Aves, order *Passeriformes) A family of small, brown, green, grey, and yellow birds, many of which have pale *supercilia and wing bars. (In the 25 *Vireo* species the eyes are often white or red.) Their bills are medium to long, stout, and slightly hooked. They have long and pointed to short and rounded wings. They are arboreal, inhabit forests, feed on insects and fruit, and nest in trees and bushes. There are four genera, with about 45 species, many migratory, found in N., Central, and S. America. Two groups, included here, are often regarded as separate families: Cyclarhidae (pepper-shrikes) and Vireolaniidae (shrike-vireos).

vireos *See* VIREONIDAE.

virion An individual *virus particle.

virogenes *Genes that appear to have originated as *viruses, but are now inherited by the host along with the rest of its *genome (e.g. a gene found in all species of *Felis*, which has evidently been transferred to them from Old World monkeys).

virtual blastopore *See* BLASTOPORE.

virus A type of non-cellular 'organism' which has no *metabolism of its own. It consists mainly or solely of a *nucleic acid genome (*RNA or DNA) enclosed by protein; in some cases there is also a *lipoprotein envelope. In order to replicate (multiply), a virus must infect a cell of a suitable host organism where it redirects the host-cell metabolism to manufacture more virus particles. The progeny viruses are released, with or without concomitant destruction of the host cell, and then can infect other cells. All types of organism are susceptible to infection by viruses; virus infections may be asymptomatic or may lead to more or less severe disease.

visceral arch In *Gnathostomata, one of a series of inverted, cartilaginous or bony arches between the mouth and throat, forming partitions between adjacent gill slits.

visceral pleura *See* PLEURA.

visceral skeleton That part of the *axial skeleton which includes the *gill arches and their derivatives. Developing during (and characterizing) the late *pharangula stage of embryogenesis, the visceral skeleton becomes the splanchnocranium, giving rise to gill arches in *teleosts and in higher vertebrates remaining cartilagenous as the larynx and trachea.

viscotaxis A change in direction of locomotion in a *motile organism or cell that is made in response to a change in the viscosity of the surrounding medium.

Visean *See* CARBONIFEROUS.

visible In genetics, applied to a mutant whose *phenotype may actually be observed, as opposed to lethals, whose occurrence may only be inferred from the absence of an expected class of individuals in the progeny of a cross designed to detect induced mutants.

visual purple *See* RHODOPSIN.

vital stain A stain that is capable of entering and staining a living cell without causing it injury.

vitamin An organic compound that is required in minute quantities by *heterotrophs for their normal health and development (hence the origin of the name, from Latin *vita*, life, and amine) but which they cannot themselves synthesize. Vitamins are distinct from carbohydrates, fats, proteins, and nucleic acids. They, or their derivatives, normally have a *coenzyme function and their absence from the diet or their inadequate absorption from the gut gives rise to specific deficiency diseases.

vitamin A *See* RETINOL.

vitamin B$_1$ *See* THIAMIN.

vitamin B$_2$ *See* RIBOFLAVIN.

vitamin B$_3$ *See* NIACIN.

vitamin B$_5$ *See* PANTOTHENIC ACID.

vitamin B$_6$ *See* PYRIDOXINE.

vitamin B$_7$ *See* BIOTIN.

vitamin B$_9$ *See* FOLIC ACID.

vitamin B$_{12}$ *See* COBALAMINE.

vitamin C *See* ASCORBIC ACID.

vitamin D *See* CALCIFEROL.

vitamin E *See* TOCOPHEROL.

vitamin K A generic term for 2-methyl-1,4-naphthoquinone and its derivatives; these are fat-soluble *vitamins required for the synthesis of *prothrombin and also for other clotting factors necessary for its activation. A deficiency of this vitamin causes delayed clotting, and haemorrhaging.

vitelline membrane The membrane that surrounds the ovum of animals, from which it has been secreted (e.g. in birds this is the membrane surrounding the yolk of the egg). In *Drosophila* the term is used rather specifically for the membrane that immediately surrounds the *plasma membrane of the ovum. In this case it is formed by the fusion of deposits in the intercellular space between the *oocyte and columnar follicle cells that surround it.

vitreous humour A gel that fills the interior of the eye in vertebrates and the posterior chamber of the eye in *Cephalopoda. *See also* AQUEOUS HUMOUR.

vitrodentine An enamel-like substance, containing little organic matter and harder than *dentine, found in *placoid scales.

Viverridae (civets, genets; suborder *Fissipedia (or *Feliformia), superfamily *Feloidea) A family of primitive, short-legged, long-bodied, mainly nocturnal carnivores that occupy in the Old World tropics the ecological position occupied in temperate regions by the *Mustelidae. The Viverridae resemble closely the *Miacidae of the *Oligocene, from which they are descended. Their fossil record is sparse, but extends back to the *Eocene. Some modern forms are strongly reminiscent of felids. The claws are partly retractile, and the gait is semi-*plantigrade or *digitigrade. The *canine teeth are smaller than those of most carnivores, and the *carnassials are not well developed. The skull and face are long, and the brain small. Most feed on small vertebrates and invertebrates, but some eat plant material. Viverrids are distributed widely throughout the Old World tropics and subtropics. Eight species found only in Madagascar and formerly classed in the Viverridae are now placed in the *Eupleridae. Mongooses, formerly classed in the Viverridae, are now placed in the *Herpestidae. There are 4 viverrid subfamilies containing 15 genera and 35 species.

viviparous lizard (*Lacerta vivipara*) *See* LACERTIDAE.

vivipary The method of reproduction in which young are produced at a stage of development in which they are active. The growth of the *embryo occurs within the mother's body which nourishes it. *Compare* OVIPARY; OVOVIVIPARY.

vizcacha *See* CHINCHILLIDAE.

vocalization The production of sound by the passage of air across vocal cords.

vole *See* CRICETIDAE.

voluntary behaviour Non-habitual behaviour that is associated with the somatic nervous system or (possibly in some species) with operant control of the autonomic nervous system.

voluntary muscle *See* STRIATED MUSCLE.

Vombatidae (Phascolomyidae; wombats; order *Marsupialia (or *Diprotodontia), superfamily *Vombatoidea) A family of large, burrowing, largely nocturnal marsupials in which the *incisors and cheek teeth grow continually. The body is broad and flattened, the head large and flat, and the limbs *pentadactyl, short and stout, with broad, naked soles to the feet. The tail is vestigial. Wombats occur in southern and south-eastern Australia, Flinders Island, and Tasmania. There are three species, *Vombatus ursinus* (common wombat), *Lasiorhinus latifrons* (hairy-nosed wombat), and *L. krefftii* (*L. barnardi*, Queensland hairy-nosed wombat), of which only about 90 individuals were believed to be surviving in 2007; the species is classed as critically endangered.

Vombatoidea (infraclass *Metatheria, order *Marsupialia (or *Diprotodontia) A superfamily of medium to large, heavily built, quadrupedal marsupials which have the typical enlarged *incisors in upper as well as lower jaws. There are three families: *Vombatidae; *Phascolarctidae; and the extinct *Diprotodontidae.

Vombatus ursinus (common wombat) *See* VOMBATIDAE.

vomeronasal organ (Jacobson's organ) A special 'smell-taste' organ in the hard palate of many mammals; it opens by ducts into the roof of the mouth behind the *incisors. The facial gesture called *flehmen guides olfactory molecules into the vomeronasal organ, where odours, especially sexual ones, appear to be

V

interpreted. In primates, the organ is functional in lemurs, tarsiers, and New World monkeys; in catarrhines, including humans, it begins to develop in embryos, but later degenerates. The vomeronasal organ is also present in amphibians, lizards, and snakes.

vulnerable *See* RARITY.

Vulpes (fox) *See* CANIDAE.

vultures 1. *See* FALCONIFORMES. **2. (New World vultures)** *See* CATHARTIDAE. **3. (Old World vultures, palm-nut vulture,** *Gypohierax angolensis***)** *See* ACCIPITRIDAE. **4.** *See* TERATORNIS INCREDIBILIS.

Vultur gryphus (Andean condor) *See* CATHARTIDAE.

Vu Quang 'ox' (saola, spindlehorn) *See* PSEUDORYX NGHETINHENSIS.

v

waders *See* CHARADRIIDAE; SCOLOPACIDAE.

waggle dance *See* DANCE LANGUAGE; NASANOV'S GLAND.

wagtails *See* MOTACILLIDAE.

walking catfish (*Clarias batrachus*) *See* CLARIIDAE.

walking fish (*Anabas testudineus*) *See* ANABANTIDAE.

walking leaves *See* PHASMATODEA; PHYLLIIDAE.

walking sticks *See* PHASMATIDAE; PHASMATODEA.

wallaby *See* MACROPODIDAE.

Wallace, Alfred Russel (1823–1913) An English naturalist who was a contemporary of Charles Darwin. He worked in the E. Indies and as a result of his observations there came independently to the theory of evolution by natural selection. His ideas were presented together with those of Darwin in a joint paper to the Linnean Society of London on 1 July 1858. *See also* WALLACE'S LINE.

Wallacea *See* WALLACE'S LINE.

Wallace's line The important zoogeographical division which separates the *Oriental and *Notogean zoogeographical regions. Alfred Wallace, a zoogeographer and contemporary of Charles Darwin, first demarcated the boundary between the Oriental faunal region and the Australasian region with its distinctive marsupials. The boundary is known to this day as Wallace's line. There is a zone of mixing called 'Wallacea', and strictly the line defines the extreme western limit of Australasian mammals, and the eastern limit of the main Oriental fauna.

wallcreeper (*Tichodroma muraria*) *See* TICHODROMADIDAE.

walleye (*Stizostedion vitreum*) *See* PERCIDAE.

wall lizard (*Podarcis muralis*) *See* LACERTIDAE.

walrus (*Odobenus rosmarus*) *See* ODOBENIDAE.

wandering albatross (*Diomedea exulans*) *See* DIOMEDEIDAE.

wapiti (red deer, American elk; *Cervus elaphus*) *See* CERVIDAE.

warble fly *See* GASTEROPHILIDAE; OESTRIDAE.

warblers 1. (Old World warblers) *See* SYLVIIDAE. 2. (New World warblers) *See* PARULIDAE. 3. *See* ACANTHIZIDAE. 4. (titwarblers, *Leptopoecile*) *See* REGULIDAE.

warehou (*Seriolella brama*) *See* CENTROLOPHIDAE.

warning coloration *See* APOSEMATIC COLORATION.

warthog louse (*Haematomyzus hop-kinsi*) *See* RHYNCOPHTHIRINA.

wart snakes *See* ACROCHORDIDAE.

warty angler *See* BRACHIONICHTHYIDAE.

wasps 1. *See* APOCRITA; BETHYLIDAE; CLEPTIDAE; DRYINIDAE; ENCYRTIDAE; FIGITIDAE; GASTERUPTIIDAE; HYMENOPTERA; MASARIDAE; PAPER WASPS; PLATYGASTERIDAE; PTEROMALIDAE; RHOPALOSOMATIDAE; SCOLIIDAE; SOCIAL WASPS; SOLITARY WASPS; TIPHIIDAE; VESPIDAE. 2. **(chalcid wasps)** *See* CHALCIDAE; CHALCIDOIDEA; EULOPHIDAE; LEUCOSPIDAE; TORYMIDAE; TRICHOGRAMMATIDAE. 3. **(cockroach wasp)** *See* AMPULICIDAE. 4. **(ensign wasps)** *See* EVANIIDAE. 5. **(fig wasps)** *See* AGAONIDAE. 6. **(gall wasps)** *See* CYNIPIDAE. 7. **(ichneumon wasps)** *See* BRACONIDAE; ICHNEUMONIDAE. 8. **(mason wasps)** *See* EUMENIDAE. 9. **(potter wasps)** *See* EUMENIDAE. 10. **(ruby-tailed cuckoo wasps, ruby-tailed wasps)** *See* CHRYSIDIDAE. 11. **(spider-hunting wasps)** *See* POMPILIDAE. 12. **(velvet ants)** *See* MUTILLIDAE. 13. **(woodwasps)** *See* ORUSSIDAE; SIRICIDAE; SYMPHYTA.

wasp waist *See* PETIOLE.

water bears *See* TARDIGRADA.

water beetle *See* HYDROPHILIDAE.

water boatman *See* NOTONECTIDAE.

water dog (*Necturus maculosus*) *See* PROTEIDAE.

water flea *See* CLADOCERA.

waterfowl *See* ANSERIFORMES.

water lice *See* AMPHIPODA.

water potential The difference between the energy of water in the system being considered and of pure, free water at the same temperature. The water potential of pure water is zero, so that of a solution will be negative. If there is a gradient of water potential between two *cells, water will diffuse down the gradient until equilibrium is reached.

water ring *See* RADIAL CANALS.

water scorpion *See* NEPIDAE.

water-vascular system *See* ECHINODERMATA.

Watson, James Dewey (1928–) The American geneticist who, with F. Crick and M. Wilkins, won the 1962 Nobel Prize for Physiology or Medicine for their modelling of the DNA molecule. Watson and Crick worked at the Cavendish Laboratory, Cambridge; from 1994 to 2004 Watson was president, and from 2004 until his retirement in 2007, chancellor of the Cold Spring Harbor Laboratory.

(⊕) SEE WEB LINKS
• Biography of James Dewey Watson.

Watson–Crick model The currently accepted model for the structure of DNA, as proposed by J. Watson and F. Crick in 1953. It is suggested that DNA is composed of two right-handed, antiparallel, polynucleotide chains coiled around a common axis to form a double helix. This structure is maintained by hydrogen bonds formed between the chains through the base-pairing of adenine to thymine and cytosine to guanine.

wattle In birds, a bare, fleshy area of skin, often highly coloured and pendulous, that usually hangs from the gape, around the eye, or on the throat. Wattles are used mainly in display and some are moulted annually.

wattlebirds *See* CALLAEIDAE.

wattled broadbill (*Eurylaimus steerei*) *See* EURYLAIMIDAE.

waves *See* GEOMETRIDAE.

waxbills *See* ESTRILDIDAE.

waxwings *See* BOMBYCILLIDAE.

W-chromosome The *sex chromosome that is found in female birds and *Lepidoptera.

wearyfish See SCOPELOSAURIDAE.

weasel (Mustela) See MUSTELIDAE.

weather The state of the atmosphere (e.g. temperature, pressure, and humidity) and associated phenomena (e.g. precipitation and wind) occurring at a specified time and place. Compare CLIMATE.

weatherfish (Misgurnis fossilis) See COBITIDAE.

weaver ants See FORMICINAE; OECOPHYLLA LEAKEYI.

weavers See PLOCEIDAE.

Weberian apparatus See WEBERIAN OSSICLES.

Weberian ossicles (Weberian apparatus) Structures found in bony fish belonging to the orders *Cypriniformes and *Siluriformes, and derived from the first four vertebrae. They form a link between the inner-ear region and the swimbladder, facilitating sound reception. See also OSTARIOPHYSI.

Weber's line A line of supposed 'faunal balance' between the *Oriental and the Australasian faunal regions within Wallacea. See WALLACE'S LINE.

weeverfish See TRACHINIDAE.

weevil 1. See APIONIDAE. 2. See CURCULIONIDAE.

Weichselian See DEVENSIAN.

Weismann, August Friedrich Leopold (1834–1914) A German biologist who established the improbability, if not impossibility, of the inheritance of acquired characteristics, as required by *Lamarck's theory of adaptation.

Weismann's doctrine The principle, proposed by A. Weismann, that hereditary information flows only from *genome to soma, not vice versa; the principle would be flouted were *Lamarckism true. In its refined, modern form, the principle reads that information flows always from DNA to *RNA to protein, and not back again. An enzyme, reverse transcriptase, is now known that does allow information to flow from RNA to DNA, and a category of viruses (*retroviruses) do insert their genetic component into their hosts' DNA by this means. Some genes (*virogenes) appear to be of viral origin.

wels See SILURIDAE.

western whip snake (Coluber viridiflavus) See COLUBRIDAE.

West Indian mammal subregion An *Island subregion which, according to the analysis by Charles H. Smith (1950–), is distinct from both the *Holarctic and *Latin American regions by reason of its high insular *endemism.

whale See CETACEA.

whale catfish See CETOPSIDAE.

whale lice (Cyamus) See CRUSTACEA.

whaler shark See CARCHARHINIDAE.

whale shark (Rhincodon typus) See RHINCODONTIDAE.

wheatears (Oenanthe) See TURDIDAE.

wheel animalcules See ROTIFERA.

whimbrel See SCOLOPACIDAE.

whiplash An alternative, little-used word for *flagellum.

whip-poor-will (Caprimulgus vociferus) See CAPRIMULGIDAE.

whipscorpions See AMBLYPYGI; UROPYGI.

w

whipspider *See* AMBLYPYGI.

whiptail **1.** *See* MACROURIDAE. **2.** *See* TEIIDAE.

whiptail gulper *See* SACCOPHARYNGIDAE.

whirligig beetle *See* GYRINIDAE.

whistlers (*Pachycephala*) *See* MUSCICAPIDAE.

whistling ducks (*Dendrocygna*) *See* ANATIDAE.

white ant *See* ISOPTERA.

white butterflies *See* PIERIDAE.

white-collared kingfisher (*Halcyon chloris*) *See* ALCEDINIDAE.

white-crested laughing thrush (*Garrulax eucolophus*) *See* TIMALIIDAE.

white dolphin (*Sousa*) *See* STENIDAE.

white-eyes *See* ZOSTEROPIDAE.

whitefaces *See* ACANTHIZIDAE.

whitefly *See* ALEYRODIDAE.

white-footed deer mouse (*Peromyscus leucopus*) *See* CRICETIDAE.

white-lipped peccary (*Tayassu pecari*) *See* TAYASSUIDAE.

white matter Tissue of the *central nervous system which is composed mainly of *axons of neurons surrounded by *myelin sheaths, which give rise to the white appearance. *Compare* GREY MATTER.

white piranha (*Serrasalmus rhombeus*) *See* CHARACIDAE.

white ramus communicans *See* RAMUS COMMUNICANS.

white rhinoceros (*Ceratotherium simum*) *See* RHINOCEROTIDAE.

white shark (*Carcharodon carcharias*) *See* LAMNIDAE.

white stork (*Ciconia ciconia*) *See* CICONIIDAE.

white whale (beluga; *Delphinapterus leucas*) *See* MONODONTIDAE.

white-winged guan (*Penelope albipennis*) *See* CRACIDAE.

white-winged triller (*Lalage sueurii*) *See* CAMPEPHAGIDAE.

whorl One of the coils in a molluscan shell. Whorls are most noticeable among *Gastropoda and *Cephalopoda, in which the shell is a hollow cone that grows only at the apertural end and tends to curl about its vertical axis. In most molluscs the coils form a helix in which each whorl is attached to the next, but in some the coiling is loose.

whydahs *See* PLOCEIDAE.

wide distribution (polychore distribution) The situation in which taxonomic groups of plants (and to a lesser extent animals) have a very extensive distributional range, spanning several floral kingdoms or regions (e.g. 'wides' may be cosmopolitan, subcosmopolitan, tropical, or temperate).

wildlife Any undomesticated organisms, although the term is sometimes restricted to wild animals, excluding plants.

wild-type gene The *allele most frequently observed at a given gene *locus.

willie wagtail (*Rhipidura leucophrys*) *See* MUSCICAPIDAE.

Wilson's petrel (*Oceanites oceanicus*) *See* HYDROBATIDAE.

'windpipe' *See* TRACHEA.

windscorpions *See* SOLIFUGAE.

wing coupling The joining together of the two pairs of wings of many insects, so that effectively the fore wing and hind wing operate as a single wing. *See also* FRENULUM; HAMULUS; RETINACULUM.

wing covert (tectrix, *pl.* tectrices) In birds, a feather that covers the base of a main wing feather. On the upper side of the wing each *remex is covered by a greater covert, which is covered by a median covert, and then the minor coverts. A similar arrangement occurs on the under side, although the feathers are usually less well developed.

winter wren (*Troglodytes troglodytes*) *See* TROGLODYTIDAE.

wire-worm *See* ELATERIDAE.

Wisconsinian *See* DEVENSIAN.

witch eel *See* NETTASTOMIDAE.

wobble hypothesis A theory to explain the partial degeneracy of the *genetic code due to the fact that some *t-RNA molecules can recognize more than one *codon. The theory proposes that the first two bases in the codon and *anticodon will form complementary pairs in the normal antiparallel fashion. However, a degree of steric freedom or 'wobble' is allowed in the base-pairing at the third position. Thus, for serine, six m-RNA codons may be paired with only three t-RNA anticodons.

wolf (*Canis lupus*) *See* CANIDAE.

wolf-fish *See* ANARHICHADIDAE.

wolfherring (*Chirocentrus dorab*) *See* CHIROCENTRIDAE.

wolf spiders *See* LYCOSIDAE.

wolverine (carcajou, glutton, skunk bear; *Gulo gulo*) *See* MUSTELIDAE.

wombat *See* VOMBATIDAE.

woodchuck (marmot; *Marmota marmota*) *See* SCIURIDAE.

woodcreepers *See* DENDROCOLAPTIDAE.

wood-eating cockroach (*Cryptocercus punctatus*) *See* CRYPTOCERCIDAE.

wood-hoopoes (*Phoeniculus*) *See* PHOENICULIDAE.

woodlice *See* PORCELLIONIDAE.

wood mouse (field mouse; *Apodemus sylvaticus*) *See* MURIDAE.

woodpeckers *See* PICIDAE.

wood quails (*Odontophorus*) *See* PHASIANIDAE.

wood storks (*Ibis*) *See* CICONIIDAE.

wood-swallow *See* ARTAMIDAE.

wood-warbler (*Basileuterus*) *See* PARULIDAE.

woodwasps *See* ORUSSIDAE; SIRICIDAE; SYMPHYTA.

woodworm (*Anobium punctatum*) *See* ANOBIIDAE.

woolly bear 1. (*Anthrenus*) *See* DERMESTIDAE. 2. *See* ARCTIIDAE.

woolly indri (*Avahi*) *See* INDRIDAE.

woolly rhinoceros (*Coelodonta*) *See* RHINOCEROTIDAE.

worker In social insects, the *caste that rears the brood, maintains the structure of the nest, and forages for food. Usually workers are sterile: female in ants, bees, and wasps, and either male or female in termites. In termites, a true worker caste (incapable of *metamorphosis) occurs only in *Termitidae. In the lower families colony labour is performed by nymphs and *pseudergates.

worm eel *See* MORINGUIDAE.

wormfish *See* MICRODESMIDAE.

worm lizard *See* AMPHISBAENIDAE.

worms *See* ANNELIDA; PLATYHELMINTHES.

wrasse 1. *See* LABRIDAE. **2.** (cleaner wrasse, *Labroides dimidiatus*) *See* CLEANER FISH.

wrens 1. *See* TROGLODYTIDAE. **2.** (Australian wrens) *See* MALURIDAE. **3.** (New Zealand wrens) *See* XENICIDAE.

wren-thrush (*Zeledonia coronata*) *See* ZELEDONIIDAE.

wren-tit (*Chamaea fasciata*) *See* CHAMAEIDAE.

Wright's inbreeding coefficient *See* COEFFICIENT OF INBREEDING.

wrymouth *See* CRYPTACANTHODIDAE.

wrynecks *See* PICIDAE.

Würm *See* DEVENSIAN.

w

Xanthidae (mud crabs; class *Malacostraca, order *Decapoda) A family of crabs several species of which have neurotoxins contained in their *exoskeletons. Mud crabs are oval to hexagonal, and broadened anteriorly. Many are very fecund, *Menippe mercenaria* (stone crab) producing 3–6 million eggs in less than 70 days. This species can be a serious predator of oysters, each crab taking on average up to 219 small and large oysters per year. Some species of mud crabs live on coral reefs, and three or four species of *Xantho* occur in European waters.

xanthophyll A group of compounds that are oxygenated derivatives of carotenes, which act as accessory pigments during photosynthesis in plants and are found in some protozoons (*see* DINOFLAGELLIDA).

Xantusiidae (night lizards; order *Squamata, suborder *Sauria) A family of nocturnal, insectivorous lizards in which the eye is gecko-like, with a vertical pupil and spectacle. Some species are *viviparous with a placental-type development. There are 12 species, occurring in Central America and the south-western USA.

X-chromosome The *sex chromosome that is found in double dose in the homogametic sex and in single dose in the heterogametic sex. Unlike the *Y-chromosome, it contains numerous genes (which therefore show sex-linkage).

Xenarthra (cohort *Unguiculata, order *Edentata) A suborder that comprises the infra-orders *Cingulata and *Pilosa, the S. American edentates known since the *Eocene and characterized by additional articulations in the lumbar vertebrae (the name of the suborder means 'extra-jointed'), the union of the *ischium and the caudal vertebrae, and a distinctive skull structure. The claws are well developed: in the arboreal sloths they are used for hanging below the branches of trees, in other families they are used for digging and oblige the animals to walk on the sides of their feet. The infra-orders distinguish between the forms with hair (the ant-eaters and sloths) and those with 'shells' (the armadillos).

Xenicidae (New Zealand wrens; class *Aves, order *Passeriformes) A family of small birds which are green or olive-brown above and grey or white below, with white *supercilia. They have straight, slender bills, short wings, and very short tails. Their legs and toes are long, and the third and fourth toes are joined basally. They inhabit forests and scrub, feed on insects and spiders, and nest in rock or tree cavities. There are two genera, with three species, found in New Zealand. A fourth species, *Xenicus lyalli* (Stephen Island wren) became extinct in 1894.

***Xenicus lyalli* (Stephen Island wren)** *See* XENICIDAE.

Xenocongridae (false moray; super-order *Elopomorpha, order *Anguilliformes) A small family of marine eels that have a very elongate, slender, scaleless body, small, round, gill openings, widely separated front and rear nostrils, and small *pectorals, the *pelvics being rudimentary or absent. The larvae of *Xenoconger fryeri* (45 cm) are reputed to be fairly common in the Mediterranean. There are about 15 species found in warm to tropical seas.

Xenopeltidae (sunbeam snake; order *Squamata, suborder *Serpentes) A monotypic family (*Xenopeltis unicolor*) comprising a burrowing snake that is characterized by its iridescent brown scales. It preys on small vertebrates, including other snakes. It occurs in south-east Asia.

Xenopeltis unicolor (sunbeam snake) *See* XENOPELTIDAE.

Xenopterygii An alternative name for fish belonging to the order *Gobiesociformes.

Xenopus laevis (clawed frog, clawed toad) *See* PIPIDAE.

Xenosauridae (xenosaurs, crocodile lizards; order *Squamata, suborder *Sauria) A family of robust lizards which enter water readily. They have a mixture of large and small scales. The family comprises the xenosaurs, three species of *crepuscular insect-eaters from the rain forests of Central America, and a single species from China, *Shinisaurus crocodilurus* (crocodile lizard), which feeds on fish.

Xenoturbellida A *phylum known by only a single species, *Xenoturbella bocki*, found in the N. Sea and described in 1949 by E. Westblad. It is up to 30 mm long, *hermaphrodite, and ciliated (*cilium); it has no *anus, but is *coelomate.

Xenungulata (superorder *Paenungulata, order *Amblypoda) An extinct suborder of mainly American ungulates, some of which were the size of a rhinoceros. They lived from the *Palaeocene to *Eocene.

Xestobium rufovillosum (deathwatch beetle) *See* ANOBIIDAE.

Xiphias gladius (swordfish) *See* XIPHIIDAE.

Xiphiidae (swordfish; subclass *Actinopterygii, order *Perciformes) A monospecific family (*Xiphias gladius*) of marine fish that have a long, slightly flattened snout ('sword'). These large fish can reach a length of 4.8 m. The *pectoral fins are placed low on the body, opposite the large, but short-based, *dorsal fin; the *pelvic fins are absent. Like the related marlins (family *Istiophoridae), the swordfish has a large, *lunate tail fin, and it can swim very fast. It is distributed worldwide in warmer seas.

Xiphodontidae Primitive *Eocene and *Oligocene artiodactyls, restricted to Europe, that represent a European parallel to the camels which were evolving in the New World at the same time. Probably both groups were descended from early anoplothere stock (*see* TYLOPODA). In some respects the anoplotheres represent the most primitive ruminants known.

xiphoid process *See* METASTERNUM.

Xiphorhynchus (woodcreepers) *See* DENDROCOLAPTIDAE.

xiphosternum *See* METASTERNUM.

Xyelidae (suborder *Symphyta, superfamily Xyeloidea) Uncommon, primitive, symphytan family, whose members are generally less than 5 mm long. Adults are brownish, with a very long third antennal segment, and feed on pollen from catkins of birch and other trees. The larvae are maggot-like, with short legs. Those of some species develop within pine cones, but others are external feeders on elm and hickory. They are the most ancient hymenopterans, occurring first in the *Triassic, and there are many fossil species. There are five extant genera with about 30 species, found mainly in the northern hemisphere.

Xyleborus (ambrosia beetle) *See* SCOLYTIDAE.

Xylocopa (Old World carpenter bees) *See* ACARINUM.

xylophagous Wood-eating.

yak (*Bos grunniens*) *See* BOVIDAE.

Y-chromosome The *sex chromosome that is found only in the heterogametic sex. It usually differs in size from the *X-chromosome and contains few or no major genes. Often only a short part of it pairs with the X-chromosome at *meiosis.

yellow chub (*Kyphosus incisor*) *See* KYPHOSIDAE.

yellowjacket *See* VESPIDAE.

yellowthroats (*Geothlypis*) *See* PARULIDAE.

yellow wagtail (*Motacilla flava*) *See* MOTACILLIDAE.

yellow warbler (*Dendroica petrechia*) *See* PARULIDAE.

yolk Material, consisting of a mixture of *lipids and *proteins, secreted in the form of granules, that in eggs act as a source of food for the growth and development of an *embryo.

yolk body (yolk platelet) A membrane-bound storage structure, containing *protein and *lipid, that is found in large numbers in the *cytoplasm of the eggs of all animals except mammals.

yolk platelet *See* YOLK BODY.

yolk sac In the *embryos of vertebrates, a sac arising as a diverticulum of the *alimentary canal with which, except in *Teleostei, it communicates. In *Reptilia, *Aves, and *Monotremata it contains *yolk which nourishes the embryo, and as the yolk is consumed the sac merges with the embryo.

In *Marsupialia, the sac contains no yolk, but absorbs nutrient from the uterine wall and in some cases forms a temporary *placenta. In *Eutheria the sac contains no yolk and most of it is cut off from the embryo at birth.

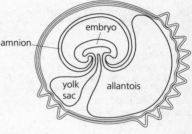

Yolk sac

Young, John Zachary (1907–97) A British zoologist who was a leading authority on cephalopods, turning later to neuroscience. His studies of the octopus brain revealed the way it stores visual and tactile memories separately and accesses them. He also discovered the giant nerve fibres of squids. He was born in Bristol and educated at Marlborough College and Oxford University. In 1945 he was appointed Professor of Anatomy at University College London, a post he held until his retirement. As well as being an excellent teacher, J. Z. Young was the author of several successful textbooks, including *The Life of Vertebrates* (1950) and *The Life of Mammals* (1957).

(((⊕))) SEE WEB LINKS

• Obituary of John Zachary Young.

Zaglossus (echidna, spiny ant-eater) *See* TACHYGLOSSIDAE.

zalambdodont Applied to *molar teeth that are V-shaped, narrow at the front and rear, with a sharp apex, and with the two *cusps partly or completely fused.

Zapodidae (jumping mice; order *Rodentia, suborder *Myomorpha) A family of mouse-like rodents in which the hind limbs are more or less elongated, permitting *saltatorial movement, and the tail is long and used for balancing. The cheek teeth are either *brachydont or semi-*hypsodont. Cheek pouches are present. They are distributed widely throughout the temperate northern hemisphere. There are four genera, with 11 species.

Zaprora silenus (prowfish) *See* ZAPRORIDAE.

Zaproridae (prowfish; subclass *Actinopterygii, order *Perciformes) A monospecific family (*Zaprora silenus*) of marine, bottom-dwelling fish that live at depths of 100–500 m, although the juveniles have been found associated with planktonic jellyfish. The prowfish grows to 88 cm long and has a stout, compressed body with a blunt head, a very long, low, *dorsal fin, rounded *pectoral and tail fins, but no *pelvic fins. It is found only in the colder waters of the N. Pacific.

Z-chromosome The *sex chromosome found in female birds and *Lepidoptera; equivalent to the *X-chromosome in mammals.

zebra *See* EQUIDAE.

Zebu cattle (*Bos indicus*) *See* BOVIDAE.

Zeidae (dories; subclass *Actinopterygii, order *Zeiformes) A family of marine fish that have a deep and compressed body, a large head with a wide and *protrusible mouth slanted upward, and large eyes. The two sections of the *dorsal fin are separated by a distinct notch. In several European countries *Zeus faber* (John Dory), 66 cm, is also known as St Peter's fish, because of the presence of a black spot ('thumb-print of St Peter') on the sides of the fish. There are several species, distributed in the Atlantic and Pacific.

Zeiformes (class *Osteichthyes, subclass *Actinopterygii) An order of bony fish that have deep, strongly compressed bodies, narrow *caudal peduncles, large, *protrusible mouths, and large eyes. They are found in inshore and offshore habitats. The largest and best-known family in the order are the *Zeidae (dories).

Zeledonia coronata (wren-thrush) *See* ZELEDONIIDAE.

Zeledoniidae (wren-thrush; class *Aves, order *Passeriformes) A monospecific family (*Zeledonia coronata*), sometimes included with the *Turdidae, which is a small, dark brown and slate bird with an orange crown. It has a short, pointed bill, short and rounded wings and tail, and long legs. It inhabits forests, feeds on insects, and builds a domed nest, with a side entrance, on a bank. It is found in Central America.

Zeugloptera *See* MICROPTERIGIDAE.

Zeus faber (John Dory, St Peter's fish) *See* ZEIDAE.

Ziphiidae (bottle-nosed and beaked whales; order *Cetacea, suborder *Odontoceti) A family of whales in which the snout is long and narrow, in some species with a high, bulging forehead, the dorsal fin and flippers are small, and the tail is only slightly notched at the mid-line. The first two cervical vertebrae are fused, and in *Hyperoodon ampullatus* (bottle-nosed whale) all the cervical vertebrae are fused. The number of teeth is variable, from one or two pairs in the lower jaw only, to 38 in the upper and 54 in the lower jaw. Ziphiids attain lengths of 4–12 m and are found in all oceans. They dive to considerable depths and feed mainly on deep-sea fish and squid. There are five genera, with 18 species.

Z-line In *striated muscle, one of the regularly spaced lines that bisect the I-bands and divide the *myofibrils into *sarcomeres.

Zoantharia (Hexacorallia; phylum *Cnidaria, class *Anthozoa) A subclass of solitary and colonial anthozoans, with or without a calcareous exoskeleton, that are characterized by paired *mesenteries being first developed in a cycle of six; and by the insertion of new pairs in two (ventrolateral), four (lateral and ventrolateral), or all six primary *exocoeles. Zoantharia first appeared in the *Ordovician.

Zoanthiniaria (class *Anthozoa, subclass *Zoantharia) An order of colonial, or more rarely solitary, sea anemones, which have one groove on the *stomodaeum. *Mesenteries are inserted only in ventrolateral *exocoeles. *Tentacles are unbranched. The body is usually encrusted with calcareous foreign bodies. Zoanthiniaria are most abundant in warm, shallow waters. There are two families, with six genera.

***Zoarces viviparus* (eelpout)** *See* ZOARCIDAE.

Zoarcidae (eelpout; subclass *Actinopterygii, order *Gadiformes) A family of marine fish that have an elongate and slender body, a relatively large head, and a slimy skin with small scales. The *dorsal and *anal fins are confluent with the tail fin; the rudimentary *pelvics are just in front of the rounded *pectorals. All are bottom-dwelling species, but while *Zoarces viviparus* (eelpout), 46 cm, is common in shallow coastal waters, *Melanostigma pammelas* (Pacific softpout), 10.2 cm, has been found at depths greater than 500 m. Although *Z. viviparus* is a *live-bearing species, others, e.g. *Macrozoarces americanus* (ocean pout), are egg-layers. There are at least 65 species, found in the temperate to cold waters of the Arctic and Antarctic.

zoea Larval stage in higher *Malacostraca, characterized by the appearance of eight pairs of trunk appendages free of the carapace.

zonal fossil *See* INDEX FOSSIL.

zona pellucida The transparent covering of an *oocyte.

zonation The spatial distribution of *species at any one time according to variations in the physical environment (e.g. the distribution of seaweeds and marine animals between the low-tide and high-tide marks on a rocky shore, and the vertical distribution of species on a mountainside).

zone In stratigraphy, a unit of rock characterized by a clearly defined fossil content. *See* INDEX FOSSIL.

zone fossil *See* INDEX FOSSIL.

zonite One of the segments into which the *cuticle is divided in *Kinorhyncha.

zonula occludens *See* TIGHT JUNCTION.

zonure, Lord Derby's (*Cordylus giganteus*) *See* CORDYLIDAE.

ZooBank The official registry of zoological names managed by the *International Commission On Zoological Nomenclature. The ZooBank is accessible online and scientists can open accounts that allow them to submit entries.

(((●))) SEE WEB LINKS
• Description of the ZooBank.

Z

zoochory (synzoochory) Dispersal of spores or seeds by animals.

zoocoenosis The secondary producers (consumers) that form part of the *biocoenosis in a *biogeocoenosis.

zooecium In *Bryozoa, the body wall.

zooflagellate Applied to *Protozoa that have one to many *flagella and contain no chloroplasts. Some are free-living but most are commensals (*see* COMMENSALISM), symbionts (*see* SYMBIOSIS), or parasites. *Compare* PHYTOFLAGELLATE.

zoogeographical region *See* FAUNAL REGION.

zoogeographical zone A geographical unit that is distinguished by a fauna which is more or less distinctive at the *species, genus, family, or order level, by virtue of its habitat, or past or present isolation.

zoogeography The study of the geographical distribution of animals at different taxonomic levels, particularly of mammals from the order down to *species level. Emphasis is given to the explanation of distinctive patterns in terms of past and/or present factors, particularly *migration routes.

zooid In colonial invertebrates, an individual member of a colony.

Zoomastigophorea (subphylum *Sarcomastigophora, superclass *Mastigophora) A class of animal-like *Protozoa which bear one or more *flagella. Some members can also form *pseudopodia. No species contains chloroplasts. There are nine orders, and many families, found in a wide range of habitats.

Zoonavena *See* APODIDAE.

zooneuston Those animals associated with the surface of water, mainly in freshwater habitats, and influenced by surface tension. The group is divided into those associated with the upper surface of the film, to which the term 'epineustic' is applied,

and those associated with the lower surface, described as 'hyponeustic'. Many organisms show adaptation to this environment (e.g. some *Hemiptera have pads at the ends of their legs that enable them to walk across the water surface). *See also* NEUSTON.

zoonosis A disease which can be transmitted from animals to humans.

zooplankton *See* PLANKTON.

Zoothera **(thrushes)** *See* TURDIDAE.

zooxanthellae Unicellular dinoflagellates (*see* DINOFLAGELLIDA) that live symbiotically (*see* SYMBIOSIS) with certain corals.

Zoraptera (class *Insecta, subclass *Pterygota) Order of small (less than 3 mm long), *exopterygote insects which superficially resemble termites and are gregarious, although they appear to lack true social organization. They live under bark, and in association with plant debris, and feed on fungal spores and small arthropods, probably dead ones. Adults may be winged and pigmented, with eyes and *ocelli; or nonpigmented, wingless, and blind. There is one genus, with 22 described species, occurring in the Ethiopian, Oriental, Nearctic, Neotropical, and Pacific regions.

zorro (culpeo, S. American dogs and foxes; *Atelocynus microtis, Cerdocyon thous, Speothos venaticus)* *See* CANIDAE.

Zosteropidae (white-eyes; class *Aves, order *Passeriformes) A family of small birds which have green upper-parts, yellow or white under-parts, and a distinctive, white eye ring. Their bills are slender, pointed, and slightly curved, and their tongues have frilled tips. Their wings are medium-length and pointed and their tails medium-length and square. They inhabit forests and brush, feed on insects, fruit, and nectar, and nest in trees and bushes. The 63 *Zosterops* species show great diversity and there are many endemic island species; they are gregarious and build semipendant nests in trees and bushes. There are 11 genera in the family, with about

84 species, found in Africa, southern Asia, the Philippines, Indonesia, Australasia, and the Pacific islands.

Zosterops (white-eyes) *See* ZOSTEROPIDAE.

Zugdisposition In birds, the preparation for migration. This includes heavy feeding and fattening. *See also* ZUGSTIMMUNG.

Zugstimmung In birds, the mood of migration (*Zug* is German for bird migration, *Stimmung* for 'mood'). During this time the bird takes to and maintains migratory flight. Prevention of migration, by caging for instance, results in migratory restlessness known as Zugunruhe (*Unruhe* means 'anxiety').

Zugunruhe *See* ZUGSTIMMUNG.

zwitterion A dipolar ion (i.e. one with both negative and positive charges and therefore no net charge). *Amino acids in solution at neutral *pH usually exist in this form, when the amino group is protonated ($-NH_3^+$) and the carbonyl group dissociated ($-COO^-$).

Zygaenidae (burnets, foresters; subclass *Pterygota, order *Lepidoptera) Family of small moths, which are generally brightly coloured, slow-flying, and diurnal. Many species are distasteful, some are toxic.

The larvae have short, dense, secondary *setae. The *cocoon is above ground, tough, and elongate. There are about 1000 species and the family is widespread.

zygapophyses *See* VERTEBRA.

zygodactylous In birds, applied to feet in which two toes point forwards, and two to the rear. The condition is found e.g. in *Cuculidae, *Picidae, *Psittacidae, and *Ramphastidae. *Compare* ANISODACTYLOUS.

zygoma *See* ZYGOMATIC ARCH.

zygomatic arch (zygoma) The bone at the side of the skull which forms an arch from beneath the *orbit to a position toward the back of the head. It is composed of the *maxilla, *jugal, and *squamosal bones, but may be incomplete. The main chewing (masseter) muscle is attached to the zygomatic arch.

zygomatic bone *See* JUGAL BONE.

zygomorphic (irregular) Bilaterally symmetrical, and therefore divisible into equal halves in only one plane.

Zygoptera (damselflies; subclass *Pterygota, order *Odonata) Cosmopolitan suborder of dragonflies, comprising insects with slender bodies and two pairs of wings. Both hind and fore wings have a

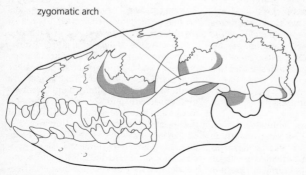

zygomatic arch

Zygomatic arch

Z

narrow base, and are folded over the abdomen when the insect is at rest. Damselflies are generally weak fliers, frequenting bankside vegetation. The aquatic nymphs have three elongate, *tracheal gills of varying form, at the end of the abdomen. There are 18 families, containing about 3000 species.

zygote The fertilized *ovum of an animal, formed from the fusion of male and female *gametes when, under normal circumstances, the *diploid chromosome number is restored, in the stage before it undergoes division.

zygotene *See* MEIOSIS; PROPHASE.

zymogen (pro-enzyme) The inactive precursor of an *enzyme, subsequently activated by specific partial proteolysis.

zymogen granule One of the dense, membrane-bound bodies containing *zymogen that is derived from the *Golgi body of cells. Zymogen granules occur in particularly large numbers in secretory cells such as the acinar cells of the pancreas.

zymogenous Applied to organisms whose presence in a given *habitat is transient; the numbers of such organisms fluctuate greatly, e.g. in response to the availability of particular nutrients.

Appendices

Endangered animals

Endangered animals are classified by the **IUCN** (International Union for Conservation Nature and Natural Resources) as: extinct (**EX**); extinct in the wild (**EW**); critically endangered (**CR**); endangered (**EN**); vulnerable (**VU**); lower risk (**LR**); data deficient (**DD**); and not evaluated (**NE**).

Within these categories, there are subdivisions in the groups CR, EN, VU, and LR.

CR

A. Population has been reduced:
  1 reduction of at least 80% has been observed, estimated, inferred, or suspected over 10 years or 3 generations, whichever is the longer, based on and specifying:
    (a) direct observation
    (b) an index of abundance appropriate for the taxon
    (c) a decline in area of occupancy, extent of occurrence, and/or quality of habitat
    (d) actual or potential levels of exploitation
    (e) the effects of introduced taxa, hybridization, pathogens, pollutants, competitors, or parasites.
  2 reduction of at least 80% projected or suspected to occur within the next 10 years or 3 generations, whichever is the longer, based on and specifying any of b, c, d, or e above.
B. Extent of occurrence estimated to be less than 100 km² or area of occupancy estimated to be less than 10 km², and estimates indicate any two of the following:
  1 severely fragmented or known to exist at only a single location
  2 continuing decline, observed, inferred, or projected, in any of the following:
    (a) extent of occurrence
    (b) area of occupancy
    (c) area, extent, and/or quality of habitat
    (d) number of locations or subpopulations
    (e) number of mature individuals
  3 extreme fluctuations in any of the following:
    (a) extent of occurrence
    (b) area of occupancy
    (c) number of locations or subpopulations
    (d) number of mature individuals.
C. Population estimated to number less than 250 mature individuals and either:
  1 an estimated continuing decline of at least 25% within 3 years or one generation, whichever is the longer, or
  2 a continuing decline, observed, projected, or inferred, in numbers of mature individuals and population structure in the form of either:
    (a) severely fragmented (i.e. no subpopulation estimated to contain more than 50 mature individuals)
    (b) all individuals are in a single subpopulation.
D. Population estimated to number less than 50 mature individuals.
E. Quantitative analysis showing the probability of extinction in the wild is at least 50% within 10 years or 3 generations, whichever is the longer.

EN

The criteria are the same as those for CR, except that:

A. 1 and 2: the population reduction is 50%
B. the extent of occurrence is less than 5000 km^2 and area of occupancy less than 500 km^2
C. the population is estimated to number less than 2500 mature individuals
C. 1: the estimated continuing decline is at least 20% within 5 years or 2 generations, whichever is the longer
D. population estimated to number less than 250 mature individuals
E. probability of extinction in the wild at least 20% within 20 years or 5 generations, whichever is the longer.

VU

The criteria are the same as those for CR, except that:

A. 1 and 2: the population reduction is 20%
B. the extent of occurrence is less than 20 000 km^2 and area of occupancy less than 2000 km^2
C. the population is estimated to number less than 10 000 mature individuals
C. 1: the estimated continuing decline is at least 10% within 10 years or 3 generations, whichever is the longer
D. population estimated to number less than 1000 mature individuals
D. 2: population occupies typically less than 100 km^2 or typically less than 5 locations
E. probability of extinction in the wild at least 10% within 100 years.

LR

1 Conservation dependent (cd). Taxa which are the focus of a continuing taxon-specific or habitat-specific conservation programme targeted toward the taxon in question, the cessation of which would result in the taxon qualifying for one of the threatened categories within a period of 5 years.
2 Near threatened (nt). Taxa which do not qualify for cd, but which are close to qualifying for VU.
3 Least concern (lc). Taxa which do not qualify for cd or nt.

Endangered animals 2012

	Number of described species	Number of species evaluated (2012)	Number of species threatened (2012)	Species threatened as % of species described	Species threatened as % of species evaluated
Vertebrates					
Mammals	5501	5501	1139	21	21
Birds	10 064	10 064	1313	13	13
Reptiles	8700	3755	807	9	21
Amphibians	6374	6374	1933	29	29
Fishes	30 000	10 590	2049	7	19
Subtotal	**60 639**	**36 284**	**7241**	**12**	**20**
Invertebrates					
Insects	950 000	4003	829	0.09	21
Molluscs	81 000	6183	1857	2.3	30
Crustaceans	40 000	2399	596	1.5	25
Others	130 000	960	288	0.2	30
Subtotal	**1 201 000**	**13 545**	**3570**	**0.3**	**26**
TOTAL	**1 261 639**	**49 829**	**10 811**	**0.9**	**22**

Threatened species are those listed as Critically Endangered (CR), Endangered (EN), or Vulnerable (VU).

The numbers and percentages of species threatened in each group do not mean that the remainder are not all threatened (i.e. are Least Concern). There are a number of species in many of the groups listed as Near Threatened or Data Deficient. These figures also need to be considered in relation to the number of species evaluated as shown in column 2.

Source: IUCN. 2012. *20127 IUCN Red List of Threatened Species.*

The universal genetic code

Amino acid	Abbreviation	Codons
alanine	Ala	GCA, GCC, GCG, GCU
arginine	Arg	AGA, AGG, CGA, CGG, CGC, CGU
asparaginine	Asn	AAC, AAU
aspartic acid	Asp	GAC, GAU
cysteine	Cys	UGC, UGU
glutamic acid	Glu	GAA, GAG
glutamine	Gln	CAA, CAG
glycine	Gly	GGA, GGC, GGG, GGU
histidine	His	CAC, CAU
isoleucine	Ile	AUA, AUC, AUU
leucine	Leu	CUA, CUC, CUG, CUU, UUA, UUG
lysine	Lys	AAA, AAG
methionine	Met	AUG
phenylalanine	Phe	UUC, UUU
proline	Pro	CCA, CCC, CCG, CCU
serine	Ser	AGC, AGU, UCA, UCC, UCG, UCU
threonine	Thr	ACA, ACC, ACG, ACU
tryptophan	Trp	UGG
tyrosine	Tyr	UAC, UAU
valine	Val	GUA, GUC, GUG, GUU
stop codon		UAA, UAG, UGA

Geologic time-scale

Eon/Eonothem	Era/Erathem	Sub-era	Period/System	Epoch/Series	Began Ma
PHANEROZOIC	Cenozoic	Quaternary	Pleistogene	Holocene	0.11
				Pleistocene	1.81
		Tertiary	Neogene	Pliocene	5.3
				Miocene	23.03
			Palaeogene	Oligocene	33.9
				Eocene	55.8
				Palaeocene	65.5
	Mesozoic		Cretaceous	Late	99.6
				Early	145.5
			Jurassic	Late	161.2
				Middle	175.6
				Early	199.6
			Triassic	Late	228
				Middle	245
				Early	251

Eon/Eonothem	Era/Erathem	Sub-era	Period/System	Epoch/Series	Began Ma
	Palaeozoic	Upper	Permian	Late	260.4
				Middle	270.6
				Early	299
			Carboniferous	Pennsylvanian	318.1
				Mississipian	359.2
			Devonian	Late	385.3
				Middle	397.5
				Early	416
		Lower	Silurian	Late	422.9
				Early	443.7
			Ordovician	Late	460.9
				Middle	471.8
				Early	488.3
			Cambrian	Late	501
				Middle	513
				Early	542
PROTEROZOIC	Neoprote-rozoic		Ediacaran		630
			Cryogenian		850
			Tonian		1000
	Mesoprote-rozoic		Stenian		1200
			Ectasian		1400
			Calymmian		1600

Eon/Eonothem	Era/Erathem	Sub-era	Period/System	Epoch/Series	Began Ma
	Palaeoproterozoic		Statherian		1800
			Orosirian		2050
			Rhyacian		2300
			Siderian		2500
ARCHAEAN	Neoarchaean				2800
	Mesoarchaean				3200
	Palaeoarchaean				3600
	Eoarchaean				3800
HADEAN	Swazian				3900
	Basin Groups				4000
	Cryptic				4567.17

(*Source*: International Union of Geological Sciences, 2004. *Note*: Hadean is an informal name. The Hadean, Archaean, and Proterozoic Eons cover the time formerly known as the Precambrian.)

SI units (Système International d'Unités)

Quantity	Name of unit	Symbol	Equivalent	Reciprocal
length	metre	m	3.281 feet	1 ft = 0.3048 m
mass	kilogram	kg	2.2 pounds	1 lb = 0.454 kg
time	second	s		
electric current	ampere	A		
thermodynamic temperature	kelvin	K	1°C = 1.8°F	1°C = 1 K
luminous intensity	candela	cd		
amount of substance	mole	mol		

Supplementary units

Quantity	Unit	Symbol
plane angle	radian	rad
solid angle	steradian	sr

Derived SI units

Quantity	Name of unit	Symbol	Equivalent	Reciprocal
frequency	hertz	Hz		
energy	joule	J	0.2388 calories	1 cal = 4.1868 J
force	newton	N	0.225 pounds force	1 lbf = 4.448 N
power	watt	W	0.00134 horse power	1 hp = 745.7 W
pressure	pascal	Pa	0.00689 pounds force/sq. inch	1 lbf/sq.in = 145 Pa
electric charge	coulomb	C		
electric potential difference	volt	V		
electric resistance	ohm	Ω		
electric conductance	siemens	S		
electric capacitance	farad	F		
magnetic flux	weber	Wb		
inductance	henry	H		
magnetic flux density	tesla	T		
luminous flux	lumen	lm		
illuminance	lux	lx		
absorbed dose	gray	Gy		
activity	becquerel	Bq		
dose equivalent	sievert	Sv		

Multiples used with SI units

Name of multiple	Symbol	Value (multiply by)
atto	a	10^{-18}
femto	f	10^{-15}
pico	p	10^{-12}
nano	n	10^{-9}
micro	μ	10^{-6}
milli	m	10^{-3}
centi	c	10^{-2}
deci	d	10^{-1}
deca	da	10
hecto	h	10^{2}
kilo	k	10^{3}
mega	M	10^{6}
giga	G	10^{9}
tera	T	10^{12}
peta	P	10^{15}
exa	E	10^{18}

Extant animal phyla

Kingdom: Animalia (Metazoa)

Group.	Phylum	Common name	Age	Approx. number of species
Protostome	Acanthocephala	Thorny-headed worms	Unknown	1191
	Annelida	Segmented worms	Cambrian to present	12 000
	Arthropoda	Arthropods	Precambrian to present	>1 106 000
	Brachiopoda	Lamp shells	Cambrian to present	234
	Bryozoa	Moss animals	Ordovician to present	5000
	Dicyemida	Lozenge animals	Recent	25
	Entoprocta	Goblet worms	Palaeogene to present	182
	Gastrotricha		Recent	744
	Gnathostomulida	Jaw worms	Recent	97
	Kinorhyncha	Mud dragons	Recent	153
	Loricifera	Brush heads	Recent	22
	Micrognathozoa		Recent	1
	Mollusca	Molluscs	Cambrian to present	150 000
	Nematoda	Eelworms, roundworms	Carboniferous to present	24 380
	Nematomorpha	Hairworms	Eocene to present	525
	Nemertea	Proboscis worms, ribbon worms	Cambrian to present	1157

Group.	Phylum	Common name	Age	Approx. number of species
	Onychophora	Velvet worms	Cambrian to present	190
	Orthonectida		Present	25
	Phoronida	Horseshoe worms	Devonian to present	14
	Platyhelminthes	Flatworms	Neoproterozoic to present	50 000
	Priapulida	Priapus worms	Cambrian to present	45
	Rotifera	Wheel animalcules	Permian to present	2000
	Sipuncula	Peanut worms	Devonian to present	147
	Tardigrada	Water bears		1015
Deuterostome	Chordata	Chordates	Cambrian to present	70 360
	Echinodermata	Echinoderms	Cambrian to present	19 000
	Hemichordata		Cambrian to present	150
	Xenoturbellida		Unknown	1
Others	Cnidaria	Cnidarians	Precambrian to present	14 500
	Ctenophora	Comb jellies		190
	Placozoa	Placozoans	Recent	2
	Porifera	Sponges	Cambrian to present	15 000
Unplaced or disputed	Acoelomorpha	Acoels	Precambrian to present	350

Group.	Phylum	Common name	Age	Approx. number of species
	Chaetognatha	Arrow worms	Carboniferous to present	200
Problematica	Gnathifera		Permian to Recent	4000
	Micrognathozoa		Permian to Recent	1
	Monoblastozoa		Recent	1

Taxonomic classification

Living organisms are classified hierarchically. The highest category is domain, in this case Eukarya, which includes all animals, plants, and fungi. Below the domain is the kingdom (Animalia, animals), and so on, down to the genus and, below that, the species, each level further narrowing the field until the final category, species, defines the organism uniquely.

Domestic cat (*Felis catus*):

Domain: Eukarya

Kingdom: Animalia

Phylum: Chordata

Class: Mammalia

Order: Carnivora

Suborder: Feliformia

Family: Felidae

Genus: *Felis*

Species: *catus*

Oxford Paperback Reference

A Dictionary of Chemistry

Over 4,700 entries covering all aspects of chemistry, including physical chemistry and biochemistry.

'It should be in every classroom and library ... the reader is drawn inevitably from one entry to the next merely to satisfy curiosity.'

School Science Review

A Dictionary of Physics

Ranging from crystal defects to the solar system, 4,000 clear and concise entries cover all commonly encountered terms and concepts of physics.

A Dictionary of Biology

The perfect guide for those studying biology — with over 5,500 entries on key terms from biology, biochemistry, medicine, and palaeontology.

'lives up to its expectations; the entries are concise, but explanatory'

Biologist

'ideally suited to students of biology, at either secondary or university level, or as a general reference source for anyone with an interest in the life sciences'

Journal of Anatomy

OXFORD

Oxford Paperback Reference

A Dictionary of Psychology
Andrew M. Colman

Over 9,000 authoritative entries make up the most wide-ranging dictionary of psychology available.

'impressive ... certainly to be recommended'
Times Higher Education Supplement

'probably the best single-volume dictionary of its kind.'
Library Journal

A Dictionary of Economics
John Black, Nigar Hashimzade, and Gareth Myles

Fully up-to-date and jargon-free coverage of economics. Over 3,400 terms on all aspects of economic theory and practice.

'strongly recommended as a handy work of reference.'
Times Higher Education Supplement

A Dictionary of Law

An ideal source of legal terminology for systems based on English law. Over 4,200 clear and concise entries.

'The entries are clearly drafted and succinctly written ... Precision for the professional is combined with a layman's enlightenment.'
Times Literary Supplement

A Dictionary of Education
Susan Wallace

In over 1,250 clear and concise entries, this authoritative dictionary covers all aspects of education, including organizations, qualifications, key figures, major legislation, theory, and curriculum and assessment terminology.

OXFORD

Oxford Paperback Reference

A Dictionary of Sociology
John Scott and Gordon Marshall

The most wide-ranging and authoritative dictionary of its kind.

'Readers and especially beginning readers of sociology can scarcely do better ... there is no better single volume compilation for an up-to-date, readable, and authoritative source of definitions, summaries and references in contemporary Sociology.'
A. H. Halsey, *Emeritus Professor, Nuffield College, University of Oxford*

The Concise Oxford Dictionary of Politics
Iain McLean and Alistair McMillan

The bestselling A-Z of politics with over 1,700 detailed entries.

'A first class work of reference ... probably the most complete as well as the best work of its type available ... Every politics student should have one'
Political Studies Association

A Dictionary of Environment and Conservation
Chris Park

An essential guide to all aspects of the environment and conservation containing over 8,500 entries.

'from *aa* to *zygote*, choices are sound and definitions are unspun'
New Scientist

OXFORD

Oxford Paperback Reference

The Kings and Queens of Britain
John Cannon and Anne Hargreaves

A detailed, fully-illustrated history ranging from mythical and pre-conquest rulers to the present House of Windsor, featuring regional maps and genealogies.

A Dictionary of World History

Over 4,000 entries on everything from prehistory to recent changes in world affairs. An excellent overview of world history.

A Dictionary of British History
Edited by John Cannon

An invaluable source of information covering the history of Britain over the past two millennia. Over 3,000 entries written by more than 100 specialist contributors.

Review of the parent volume
'the range is impressive ... truly (almost) all of human life is here'
                                                  Kenneth Morgan, *Observer*

The Oxford Companion to Irish History
Edited by S. J. Connolly

A wide-ranging and authoritative guide to all aspects of Ireland's past from prehistoric times to the present day.

'packed with small nuggets of knowledge'   *Daily Telegraph*

The Oxford Companion to Scottish History
Edited by Michael Lynch

The definitive guide to twenty centuries of life in Scotland.
'exemplary and wonderfully readable'

                                                        Financial Times

Oxford Companions

'Opening such books is like sitting down with a knowledgeable friend. Not a bore or a know-all, but a genuinely well-informed chum ... So far so splendid.'

Sunday Times [of *The Oxford Companion to Shakespeare*]

For well over 60 years Oxford University Press has been publishing Companions that are of lasting value and interest, each one not only a comprehensive source of reference, but also a stimulating guide, mentor, and friend. There are between 40 and 60 Oxford Companions available at any one time, ranging from music, art, and literature to history, warfare, religion, and wine.

Titles include:

The Oxford Companion to English Literature
Edited by Dinah Birch
'No guide could come more classic.'

Malcolm Bradbury, *The Times*

The Oxford Companion to Music
Edited by Alison Latham
'probably the best one-volume music reference book going'
Times Educational Supplement

The Oxford Companion to Theatre and Performance
Edited by Dennis Kennedy
'A work that everyone who is serious about the theatre should have at hand'

British Theatre Guide

The Oxford Companion to Food
Alan Davidson
'the best food reference work ever to appear in the English language'
New Statesman

The Oxford Companion to Wine
Edited by Jancis Robinson
'the greatest wine book ever published'

Washington Post